Encyclopedia of Engineering

Encyclopedia of Engineering

Editors

Raffaele Barretta
Ramesh Agarwal
Krzysztof Kamil Żur
Giuseppe Ruta

MDPI • Basel • Beijing • Wuhan • Barcelona • Belgrade • Manchester • Tokyo • Cluj • Tianjin

MDPI

Editors
Raffaele Barretta
University of Naples
Federico II
Italy

Ramesh Agarwal
Washington University
in St. Louis
USA

Krzysztof Kamil Żur
Bialystok University
of Technology
Poland

Giuseppe Ruta
University "La Sapienza",
& National Group for
Mathematical Physics
Italy

Editorial Office
MDPI
St. Alban-Anlage 66
4052 Basel, Switzerland

This is a reprint of articles from the Topical Collection published online in the open access journal *Encyclopedia* (ISSN 2673-8392) (available at: https://www.mdpi.com/journal/encyclopedia/topical_collections/encyclopedia_engineering).

For citation purposes, cite each article independently as indicated on the article page online and as indicated below:

LastName, A.A.; LastName, B.B.; LastName, C.C. Article Title. *Journal Name* **Year**, *Volume Number*, Page Range.

ISBN 978-3-0365-7000-6 (Hbk)
ISBN 978-3-0365-7001-3 (PDF)

Cover image courtesy of Raffaele Barretta

Contents

About the Editors

Raffaele Barretta

Professor Raffaele Barretta is a Full Professor of Solid and Structural Mechanics in the Department of Structures for Engineering and Architecture at the University of Naples Federico II. He was born in Naples (Italy) March 20, 1980. He received his Master Degree (5 years) in Civil Engineering with magna laude at the University of Naples Federico II, 2003—Title of thesis: Polar Models of Beams and Shells in Large Deformations. He received his Ph.D. in Structural Mechanics at the University of Naples Federico II, 2007—Title of thesis: Mixed Variational Methods in Elasticity. He was an Assistant Professor of Solid and Structural Mechanics at the University of Naples Federico II (2010–2015). He was an Associate Professor of Solid and Structural Mechanics at the University of Naples Federico II (2015–2021). His research interests mainly focus on Continuum Mechanics; Beams, Plates, Shells; Nano-Materials; Non-Local Elasticity; Functionally Graded Materials; MEMS/NEMS; Nanoscience and Nanotechnology. Awards: Top Italian Scientist in the Engineering Area from Single Year Career 2017 (Source paper: Ioannidis J.P.A., Baas J., Klavans R., Boyack K.W. A standardized citation metrics author database annotated for scientific field. PLoS Biology 17(8): e3000384 (2019) https://bit.ly/2LITT4P); One of the top 90 scholars of the University of Naples Federico II in the database of 100,000 top scientists published by PloS Biology: http://bit.ly/37ox6Uk. He is ranked #532 in the world and #9 in Italy among Top Scientists for 2022. The full world ranking can be found here: https://research.com/scientists-rankings/mechanical-and-aerospace-engineering. The entire ranking for Italy can be dound here: https://research.com/scientists-rankings/mechanical-and-aerospace-engineering/it.

Ramesh Agarwal

Professor Ramesh K. Agarwal is the William Palm Professor of Engineering in the Department of Mechanical Engineering and Materials Science at Washington University in St. Louis. From 1994 to 2001, he was the Sam Bloomfield Distinguished Professor and Executive Director of the National Institute for Aviation Research at Wichita State University in Kansas. From 1978 to 1994, he was the Program Director and McDonnell Douglas Fellow at McDonnell Douglas Research Laboratories in St. Louis. Dr. Agarwal received PhD in Aeronautical Sciences from Stanford University in 1975, M.S. in Aeronautical Engineering from the University of Minnesota in 1969 and B.S. in Mechanical Engineering from Indian Institute of Technology, Kharagpur, India in 1968. Over a period of 45 years, he has worked in several disciplines within mechanical and aerospace engineering, and energy and environment which include computational fluid dynamics, computational electromagnetics and acoustics, control theory, multidisciplinary design and optimization, turbomachinery and pumps, chemical looping combustion, carbon capture and sequestration, and wind energy. He is the author and co-author of over 600 publications. He has given many plenary, keynote and invited lectures at various national and international conferences in over sixty countries. He is a Fellow of 26 professional societies including the American Institute of Aeronautics and Astronautics (AIAA), American Society of Mechanical Engineers (ASME), Institute of Electrical and Electronics Engineers (IEEE), Society of Automotive Engineers (SAE), American Association for Advancement of Science (AAAS), American Physical Society (APS) and American Society for Engineering Education (ASEE). He has received many prestigious honors and national/international awards from various professional societies and organizations for his research contributions including the AIAA Reeds Aeronautics Award, SAE Medal of Honor, ASME Honorary Membership and Honorary Fellowship from Royal Aeronautical Society.

Krzysztof Kamil Żur

Professor Krzysztof Kamil Żur is a Researcher at the Faculty of Mechanical Engineering, Bialystok University of Technology. He received his Ph.D. in Theoretical and Applied Mechanics. His research is concerned with the applications of meshless methods to the dynamics of discrete–continuous composite structures with a non-linear distribution of parameters. Professor Żur is an independent European scientist who is the head of many international scientific groups working on the mechanical problems of structures and composite materials at diverse scales (from nano to macro). He is also an expert in analytical, mesh, and meshless numerical methods applied to solve different challenging linear and non-linear boundary value problems. His research is complex and strongly interdisciplinary, connecting research areas such as material science, mechanics and nanomechanics, physics (multi-field), and numerical methods. To go beyond the current state of knowledge and present new insights into the mechanics of investigated multi-scaled structures and materials, Professor Żur connects people of different abilities from around the world (US, Australia, EU, Asia, Africa) to perform high-quality interdisciplinary investigations and publish convincing and reliable results in reputable journals. He is the "Highly Cited Researcher" from the publications that rank in the top 2% by citations for field and publication year in the Web of Science™ citation index.

Professor Żur is a well-known and respected scientist within international scientific communities. He serves as one of the main editors of two reputable journals: *Engineering Analysis with Boundary Elements* (ELSEVIER) and *Applied Physics A* (Springer), and an Associate Editor, Subject Editor, and Editorial Board Member in more than 20 reputable and internationally recognized journals. He has worked for several international grants carrying out complex tasks. He works as a referee to support grant proposal assessments for a few national and international funding agencies as well as governments. He plays the role of reviewer in more than 150 journals and works as a referee for different publishers.

Giuseppe Ruta

Born in 1967, Giuseppe Ruta graduated in Mechanical Engineering from the University "La Sapienza" of Rome in 1992, receiving a PhD in Theoretical and Applied Mechanics from the same university in 1996. He received postdoc positions at the University Roma Tre and became an Assistant Professor at "La Sapienza" in 2000, then an Associate Professor in 2012, and since 2018 he is eligible for a full Professorship. He serves in the Mechanical Engineering courses and in the Ph.D. School of Theoretical and Applied Mechanics of "La Sapienza", and as a didactic expert in the Italian National Agency for University Evaluation. He is author and co-author of 150 scientific contributions, of which more than 70 are international journals. His research interests include: one-, two-, and three-dimensional continuum models; flexural-torsional instabilities for thin-walled slender elastic beams, the presence of damage and stiffeners; perturbation methods for non-linear elasticity in moderately thick cylinders; fluid–structure interactions; static and dynamic instability of beams on various elastic soils; functionally graded materials; one-dimensional continua for the dynamics of circular and parabolic arches; local damages in arches via static and dynamic measures in view of structural identification and monitoring; non-local elasticity in beams; historical–epistemological study of the works of Gabrio Piola, Enrico Betti, Luigi Federico Menabrea; and elastic models of Cauchy, Voigt, Poincaré, Born-Von Kármán, Gazis, Hrennikoff, Eringen. He is regular reviewer for several international journals (*Meccanica, Journal of Sound and Vibration, Thin-Walled Structures, Acta Mechanica, Mathematical Reviews, Journal of Vibration and Control*, a.o.); he is Academic Editor for the journal *Shock and Vibration* and an active collaborator of MDPI publishing house. He was part of the organizing committee of national and international congresses at "La Sapienza" in 2004, 2012, 2019, and 2021.

encyclopedia

MDPI

Entry

Vibration-Assisted Ball Burnishing

Ram ón Jerez-Mesa [1,*], Jordi Llumà [2] and J. Antonio Travieso-Rodríguez [1]

1 Department of Mechanical Engineering, Universitat Politècnica de Catalunya, 08034 Barcelona, Spain; antonio.travieso@upc.edu
2 Departament of Material Science and Engineering, Universitat Politècnica de Catalunya, 08034 Barcelona, Spain; jordi.lluma@upc.edu
* Correspondence: ramon.jerez@upc.edu

Definition: Vibration-Assisted Ball Burnishing is a finishing processed based on plastic deformation by means of a preloaded ball on a certain surface that rolls over it following a certain trajectory previously programmed while vibrating vertically. The dynamics of the process are based on the activation of the acoustoplastic effect on the material by means of the vibratory signal transmitted through the material lattice as a consequence of the mentioned oscillation of the ball. Materials processed by VABB show a modified surface in terms of topology distribution and scale, superior if compared to the results of the non-assisted process. Subgrain formation one of the main drivers that explain the change in hardness and residual stress resulting from the process.

Keywords: ball burnishing; acoustoplasticity; vibration-assistance; surface integrity; surface topology

Citation: Jerez-Mesa, R.; Llumà, J.; Travieso-Rodriguez, J.A. Vibration-Assisted Ball Burnishing. *Encyclopedia* **2021**, *1*, 460–471. https://doi.org/10.3390/encyclopedia1020038

Academic Editors: Krzysztof Kamil Żur, Raffaele Barretta, Ramesh Agarwal and Giuseppe Ruta

Received: 17 May 2021
Accepted: 8 June 2021
Published: 11 June 2021

1. History: From Ball Burnishing to the Vibration Assisted Version of the Process

This Encyclopedia entry deals with the main aspects and details related to the ball burnishing process assisted with a vibratory signal (namely, vibration-assisted ball burnishing or VABB henceforth). Ball burnishing is based on deforming plastically with a preloaded sphere the irregularities of a surface that has been previously machined, so that its roughness or texture features are reduced while hardness is increased due to cold deformation (Figure 1a). However, this interaction is three-dimensional and is very much influenced by the friction between the ball and the material. The main physical vector to achieve that deformation is the preload force with which the ball is preloaded and the number of passes by which the target surface is covered. By assisting the process with vibrations, a vibratory component of the force F_v is overlapped to the preload F_p, resulting in the overall vibratory burnishing force F_b, as shows Figure 1b.

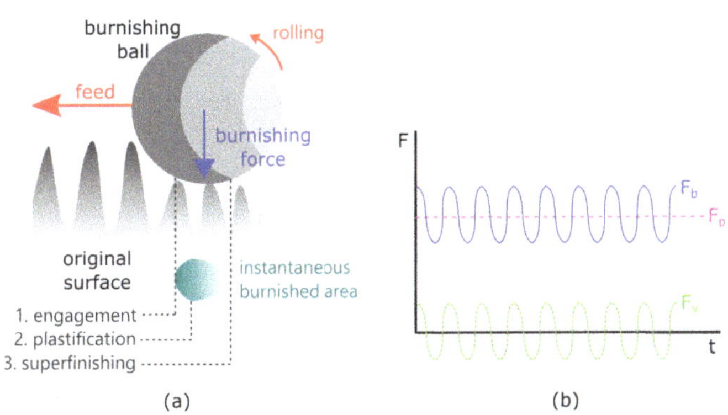

Figure 1. (**a**). General overview of a vibration-assisted ball burnishing process. (**b**). Components of the burnishing force (F_b).

The process must be understood as the upgrade of a classical operation complemented with an extra layer that introduces new dynamics and modifies the way the tool interacts with the material of the target surface. The oldest references related to ball burnishing itself refer to the processing of certain parts of the automobilistic industry in the sixties [1]. The process was described simply as a means whereby the motion of a ball or roller displaces the peaks of the surface roughness profile into the valleys. Today, we know that this apparently simple description does not account for the very complex mechanisms that are put at stake when this kind of process is deployed to provide a certain workpiece with a desired finishing state. The phenomenon whereby the material surface is modified is more likely to be compared to how the wavy surface of calm waters on the sea are smoothly moved by the effect of the wind, changing their direction, but keeping a very similar pattern all the way through.

Ball burnishing has been often cited because of its direct effects on the surface texture. The actual description of this modification can be described as a triplet:

- The surface texture features are reduced to a lower scale.
- The material that composes the surface is redistributed to a Gaussian distribution of heights.
- If enough plastic deformation is exerted, the surface features can be reoriented along the ball burnishing direction.

As the last of the described effects is only observed if the proper force and number of passes are combined to obtain the desired surface finishing, it could be said that the original explanation of ball burnishing in which material peaks were introduced in the valleys is not totally accurate.

Besides the topological effects of the process, the material also embodies other transformations that define its state after ball burnishing. Specifically, by experiencing cold deformation, the material is ultimately cold-hardened, providing the final workpiece with a reinforced outer layer with enhanced performance. Furthermore, a higher compressive residual stress profile is formed in the subsurface layers of the material. This change of mechanical state of the material is also often observed in the change of the microstructural state of the outer layers of the material itself, if a cross-section of the processed surface is observed.

The assistance of ball burnishing in the mid-twentieth century responded at the time to an extended trend in the manufacturing innovation ecosystem based on providing classical operations with extra functions that enhanced the outputs of these processes. This is how hybrid processes such as vibration-assisted machining [2] or laser-assisted ball burnishing [3] were born and are still today used in many manufacturing companies. Specifically, VABB was brought into play into the finishing operations industry, incorporating a vibratory movement to the burnishing ball simultaneous to its rolling over the surface irregularities while it runs the programmed trajectory. VABB was for the first time reported during the 1970s, designed as ultrasonic burnishing. It was assisted by 41.5 kHz vibrations and a variable amplitude from 5 to 10 μm [4]. The first detailed academic bibliography dealing with VABB dates from the 1980s [5], although some references could be found in previous years focusing on the comparison of the friction coefficient, wear rate or load bearing capacity of VABB-treated surfaces with regards to surfaces finished through other processes such as boring, grinding or even simple ball burnishing. However, these references did not focus on the phenomenology behind the results or their relation with the descriptive parameters of the surfaces themselves.

This entry is divided in three sections. The first describes the overall results observed on different materials after VABB. The second offers an insight into the physical origins of the vibratory assistance. Furthermore, finally, the hardware and physical systems reported in literature are explained.

2. Effects of VABB on Materials

In general, the affectation of surfaces after VABB can be described very similarly to the one resulting from the conventional process, namely as a comprehensive effect on the material at the surface on the topology or roughness, microhardness and residual stress with a higher affectation of the surface. The references that are included in this section show that VABB does not necessarily enhance all properties simulatenously, hence the importance and need to know the process and decide whether to use it or not and to distinguish its adequateness according to the desired surface characteristics.

The first results reported after VABB by Marakov (1973) [4] found a relevant interaction between the vibration amplitude and the obtained surface roughness. Indeed, the anticipated positive effect of higher force values on the resulting surface roughness was only observed on those mild steel specimens treated with a 2 μm amplitude. A reduction in the friction coefficient between the burnishing ball and the recipient material was also reported in those conditions. Later on, Pande and Patel (1984) [5] reported results on low-frequency (10 to 70 Hz) vibratory burnishing in contrast to the high-frequency assited process. Results provided evidence for an inverse interaction between the preload and the amplitude of the assistance, obtaining lower surface roughness values for amplitudes lower than 0.5 μm. They also found that the vibration-assistance was remarkably positive with regards to the residual hardness obtained after the tests, as an assistance with 60 Hz vibration allows the lowest preload to be successful in increasing these values compared to the unassisted process.

No other relevant research sources about ultrasonic burnishing can be found until the 2000's, when new references to the process start to be found on different materials. Bozdana et al. (2005) [6] applied the process on Ti-6Al-4V specimens assisted with 20 kHz and 6.75 μm vibration on a milling machine. It was proved that there is a critical value from which surface roughness is harmed when the VABB process is applied, and should be defined. For the ultrasonic process, that point is at a much lower preload level, probably because transient softening due to the transmission of the vibratory signal through the material favours the in situ plastic deformation. Consequently, the effects of vibration-assistance can fire back by deforming excessively the material at the surface, and should be carefully selected [7]. However, residual stress and hardness results were much more promising, as the ultrasonic process resulted in higher values with half the preload required for the non-assisted process. Therefore, it can also be stated that the process seems not to be neatly positive in affecting the surface under different perspectives, i.e., some aspects might be improved while others are harmed.

The references related to VABB since the 2010s have increased considerably, including few references applied on a lathe on different materials [8,9]. The VABB applied on milling machines clearly dominates the state of the art in this sense. A 2-kHz assistance was reported to improve the surface roughness of AISI 1038 [10] and EN AW 7078 [11] with regards to the non-assisted process (NVABB), although the results in terms of microhardness were questionable. Extensive experimental research has been performed to analyze the impact of VABB on different ball-end milled surfaces of AISI 1038 [12], Ti-6Al-4V [13], nickel-based Udimet 720 alloy [14] and AISI 306 [15]. In all cases, following the same research pattern by applying Taguchi experimental design, the authors conclude that VABB proves its effectiveness to modify effectively the surface topology of all surfaces, and redistributing the material to a Gaussian state. Other works on the AISI 316L steel has also been conducted [16]. Furthermore, the threshold value of the preload from which the surface is harmed was identified, being different for each of the tested materials. It was also noted that it seems that in applying the VABB process, the effect of the original surface is of upmost importance, as it defines the improvement potential of the surface itself. The authors conclude that the vibration-assistance should only be selected to improve surface topology if the original Sq descriptor of the surface is 5 μm or less; but on the other hand that topological improvement can be accompanied with a lower level of compressive residual stress [13].

The nanoscopical level also shows information about how the material at the surface is modified after VABB. It has been found that the process succeeds in refining the grain structure at the subsurface on many alloys such as aluminum 6061 [17] or AISI 1045 [18], and is even able to promote the phase transformation in materials such as Ti-6Al-4V [19], or austenitic metastable AISI 306 [15] (by forcing the generation of martensite by cold plastic strain) [20]. This translates into a higher residual hardening and stress. In biocompatible materials, investigators have succeeded in generating subsurfaces that favour cell adhesion, and can, therefore, increase the biocompatibility of materials [21]. Grain refinement on 17-4PH stainless steel surfaces also resulted in a wear and corrosion resistance of the material, compared to the conventional unassisted version of the process [22]. In all cases, new research seems to show that the correct direction to continue with investigations about VABB is to consider how the microstructure is changed.

Pros, Cons and Capabilities of VABB

The development of the VABB throughout the years and the very positive results obtained in research have positioned VABB as a potential process to be implemented in many kinds of industries. Although no works comparing VABB with other finishing processes have been reported, its non-assisted counterpart has proved to be superior in terms of residual stress and topological improvement if compared to it direct competitors, such as laser shock peening or shot peening [23]. If it is assumed that VABB is an upgrade of NVABB, the general superiority of VABB with regards to other competitive finishing processes can be inferred by extension.

The capabilities of the process do not only result from the effects on the material itself, but also the ease with which it can be introduced in a manufacturing routine through numerical control has to be highlighted. The authors work, for instance, with a company that is substituting their manual polishing for moulds for the automotive industry by this automatised process. The introduction of VABB in their routine not only has reduced the processing time of each part but has also allowed the owners of the company to exploit their machine tools overnight with the automatised process.

Companies from the aeronautical industry are also eligible to implement VABB in their routines. This industry is on the search of processes that can help them improve the conditions of selective surfaces that are subjected to fatigue stress. The target of VABB does not have to be a whole surface but specific sectors of the part that engineers have identified as critical, what makes the process still more interesting in comparison with other ones that cannot be so selective, such as sand blasting or laser shock peening.

The process also demonstrates disadvantages, as the equipment required to execute it is based on an external circuit that has to be branched to the VABB tool so that it can work. This wiring could be a handicap to automatise the process, or at least could be a conundrum for production engineers willing to guarantee the security of the process itself. Furthermore, it would require more space to install the external power circuit.

Theoretical models of VABB are scarce but have arisen the fact that the simultaneous improvement of texture and residual stress cannot always be possible with VABB [24]. The solution to that is to adjust very thoroughly the burnishing parameters to achieve this simultaneous effect. Therefore higher preprocessing and preparation time can also be cited as a drawback of the process and introduces a new challenge for industries willing to implement it in their routines.

The authors consider that these disadvantages cannot overshadow the evident advantages of the process itself, as it has been presented above, not only in practical terms but also in material modification and performance after being finished through VABB. It is true that the time required to define a correct selection of VABB parameters (explained in subsequent subsections) is long, but it can definitely pay off later on once the process is effectively implemented in the manufacturing routine of the adopters.

3. Physical Principles behind VABB

Originally, the introduction of vibration assistance was brought into the industry just by following the hypothesis that a vertical movement of the burnishing ball, simultaneous to its longitudinal feed movement, could have a similar effect on the material as if successive impacts were applied on the surface, i.e., due to a hammering effect on the material that composed the interface of the workpiece. The results obtained after VABB, and that shall be described in the next section, evidence that the process leads to different results compared to its original counterpart. However, it is not clear today what the phenomenological explanation is that accounts for the process results. This is caused by the fact that it is impossible to visualize how the actual engagement of the ball with the surface material is modified due to the vibratory movement, and how stress is transmitted into the material subsurface layers.

Regardless of this limitation, research during the last few decades has allowed the scientific community to obtain new insights of the technology itself. The evidence found demonstrate that there are two main causes whereby VABB offers different results with regards to it conventional counterpart, namely:

1. Due to the fact that it activates the acoustoplastic effect on the material.
2. Due to the fact that it modifies the engagement dynamics of the ball and the material during rolling.

3.1. *The Acoustoplastic Effect*

Acoustoplasticity consists of the decrease in the quasi-static stress to which a material must be subjected to be plastically deformed by means of overlapping a vibratory signal over the physical force that causes that strain. As ball burnishing is based on plastic deformation, it is, therefore, eligible to be enhanced by this effect. Acoustoplasticity was reported for the first time by Blaha and Langenecker in 1955 [25] on pure zinc crystals radiated by a 800-kHz ultrasonic wave. Hence its alternative designation as Blaha effect. It was proved later on that it can be universally observed in metals [26], although the degree of affectation varies according to the properties of the materials. For instance, the higher the acoustic impedance and the higher elastic modulus of a material, the higher its sensitivity to be affected by acosutoplasticity [27]. The acoustoplastic effect has proven to be independent of the vibration frequency [28] but its effects vary according to the vibration amplitude although the source of this influence is not clear [29,30].

The consequences of acoustoplasticity are dual because it can cause residual softening but also residual hardening [31]. Gindin et al. (1972) [32] concluded that the residual hardening is only present if an intensity threshold is surpassed. This observation is interesting to justify the assistance of ball burnishing with vibrations because it could facilitate plastic strain during the process while provoking a residual hardening of the target surface.

Despite the fact that acoustoplasticity has been experimentally observed on a high variety of materials, its actual physical source is still controversial. The conundrum is based on the fact that there is no agreement on whether acoustoplasticity has an intrinsic or extrinsic causes, i.e., whether it is provoked by a reaction inside the material's microstructure to the external source of vibrations, or by the increase of the power deployed into the system without change of the material behaviour. Ultimately, the reader shall find that acoustoplasticity has a polyhedral nature.

The intrinsic approach to the issue is based on the hypothesis that ultrasonic energy was preferentially absorbed by defects in the metal lattice (e.g., dislocations or grain boundaries). The intrinsic approach is based on the idea that these defects are actual responsible for the mechanisms of plastic deformation, hence its potential to explain why acoustoplasticity works and can be observed at a macro level. This idea was deffended by Blaha and Langenecker [25,33]. Later on, Mason (1955) [34] argued that as the lattice defects absorb the vibratory energy, dislocation mobility is enhanced and this effect allowes the metal to deform under lower loads. Based on this theory, Gindin et al. (1972) [32] justified the residual hardening observed on materials deformed through acoustoplasticity because of dislocation loops on the material lattice and the new stable vacancies accumulated in it. Pohlman and Lechfeldt (1966) [35] reinforced this intrinsic approach by observing that the force drop during ultrasonic strain was only observed during the plastic strain phase, and not in the elastic one, as plastic deformation mechanisms are related to metal lattice dynamics. Langenecker (1966) [31] proposed that the ultrasonic energy at lattice defects caused a microheating effect, facilitating the material strain. Imperfections in the metal lattice tend to look for minimal energy positions, provided that a certain threshold value of energy is exceeded. Therefore, they defended that the increase in dislocation mobility must be a thermally activated process, meaning that acoustoplasticity would only happen if a certain activation energy was surpassed. Although they contributed to the understanding of acoustoplasticity, intrinsic theories could explain why other mechanisms related to energy absorption by lattice defects such as resonance or hysteresis based on ultrasonic do not have the same effect as acoustoplasticity, and, therefore, evidenced limitations.

Chronologically simultaneous were the works undertaken by Nevill and Brotzen (1957) [36], who defended extrinsic theories. They proposed that the observed stress decrease through acoustoplasticity was independent of the temperature. Therefore, they explained the acoustoplasticity phenomenon as a result of macroscopic superposition of steady and oscillatory stresses. Kirchner et al. (1985) [37] developed an extrinsic model by a set of experiments performed with a universal testing machine on aluminium specimens, by programming different overlapped sinusoidal forces on the deforming forces at low, medium and high frequencies. However, as this model lacked total correspondence with experimental observations, the role of internal friction was introduced eventually in the system, assuming its responsibility for the equilibrium of forces that need to be satisfied during the quasi-static deformation of a material [38,39]. Still, results were not exactly consistent with experimental observations. Therefore, a purely extrinsic approach to the topic did not seem to be sufficient to explain satisfactorily the sources of acoustoplasticity.

The recent advances in microstructural analysis has allowed researchers to revive the discussion about the roots of acoustoplasticity. Vickers microindentation tests performed with a vertical 30-kHz vibrating indenter were conducted in 2011 by Siu et al. on pure aluminium [40], copper and molybdenum [41] proved that the diamond-shaped indentations for the ultrasonic-indented prints were bigger. This confirms a decrease in the hardness experienced by the material during the application of the ultrasonic plastic deformation. It was confirmed later on that this is due to dislocation annihilation [42] promoted by acoustoplasticity, as the positive vibratory cycle promotes dislocation travel to further places, and the negative cycle slows them down to favour that annihilation. Furthermore, SEM observations evidenced that subgrains are formed after indentations performed with a vibration assistance unlike the results evidenced by specimens indented quasi-statically. That explains residual hardening, as subgrains act as secondary boundaries which increase the required energy to move the dislocations because of the increase in heterogeneity in the direction of slipping planes inside the material lattice. That is associated to hardness increase. On the other hand, that strain hardening is highly unbalanced, what derives in higher residual stress [43].

This explanation that combines dislocation annihilation and subgrain formation by means of acoustoplasticity is actually a fusion of extrinsic and intrinsic theories and is so far the explanation that accounts more accurately for acoustoplasticity. Indeed, neither is acoustoplasticity just a stress addition effect, nor the preferential absorption of vibratory energy by lattice defects. It also supports the non-dependence of the softening results on the frequency [44]. In contrast to that, it seems tha the vibratory amplitude does influence the residual hardening results, as observed in 2015 by Cheng et al. (2015). In fact, this author remarked that the effect of acoustoplasticity is only conspicuous if an amplitude threshold value is surpassed [45]. This result is in line with the mid-20th century acoustoplasticity experiments explained above.

3.2. Modification of the Engagement Dynamics Ball-Material

The previous subsection has shown that the usefulness of vibration-assisted ball burnishing can be justified with the resources that material science is able to deliver. However, focusing on how the material is modified during deformation assisted with a vibration is not enough to explain why ball burnishing happens. Indeed, there is a second relevant mechanism that explains the change of the effects of ball burnishing due to vibrations related to the fact that the interaction of both solids changes as the ball moves or is moved by a dynamic mechanism. That is, the frictional behaviour of the ball and the material must be of great importance to the results, because it is that contact that enables vibratory transmission.

The described effect has been formulated guided by the extensive experimental observation and out of intuition, as it is not possible to actually see what is the interaction between the ball and the surface during the process. It is still more complicated to visualise what the impact of the vibratiory movement is. For this reason, researchers are incipiently working on finite element model that can show in detail what are these interactions and predict eventual results of the process [46]. Shen et al. (2019) have developed a 3D FE model of VABB that shoes that a forced vibration overlapped on a static force (preload), the alternative force can be understood as a dynamic hammering that derives in a higher penetration of the residual stress. They also highlight that the compressed layer will be saturated at a certain static load and that, therefore, the room for improvement after VABB is not infinite.

This line should be higher explored in the future, to better understand the dynamics of the process. Lacking the possibility of actually observing the interphase between ball and surface, numerical models are a clear alternative to understand the phenomenology of the process and know how the engagement of the ball and the material occurs during VABB.

4. Equipment to Deploy of Vibration-Assisted Ball Burnishing

To date, in this text, VABB has been explained as a single process. However, there are numerous ways whereby the vibrations can be introduced in the system and how they are technically deployed:

1. Vibrators based on electromagnets that were designed to produce a certain peak-to-peak force during their movement, and that were attached to the machine where they were executed [5]. This kind of systems are the oldest ones and their specific functioning has not been reported in the bibliography with enough detail as to understand how the system works.
2. Alternative deflection of plates subjected to variable magnetic fields, as shown in Figure 2a. The source of vibration is caused by the positive and negative deflection of the thin plate to which the ball is attached as a consequence of a variable magnetic field created by a coil excited with an external circuit. Therefore, these kind of systems have a true limitation of the frequency at which they can work because the thin plate is not able to follow an excessively high frequency for reasons of inherent stiffness. For this reason, these kinds of systems are not capable to arrive to the ultrasonic level. Although these systems exists at the experimental level, they cannot be found in the industrial level. However, their importance lies in the fact that the results that can be obtained by them can be used to establish a comparison point with the VABB process assisted with ultrasonic frequencies. For instance, Gomez-Gras et al. (2015) [47] reported asuccessful 2.1 kHz assisted system that proved to introduce the acoustoplastic effect in the system and allowed the researchers to ulteriorly report very positive effects of the process itself [10].
3. A sonotrode attached to a piezoelectric stack, which forces vibration by the expansion and contraction of a sonotrode thanks to an external power circuit that transmits an oscillatory signal [48,49]. This kind of systems is shown in Figure 2b.

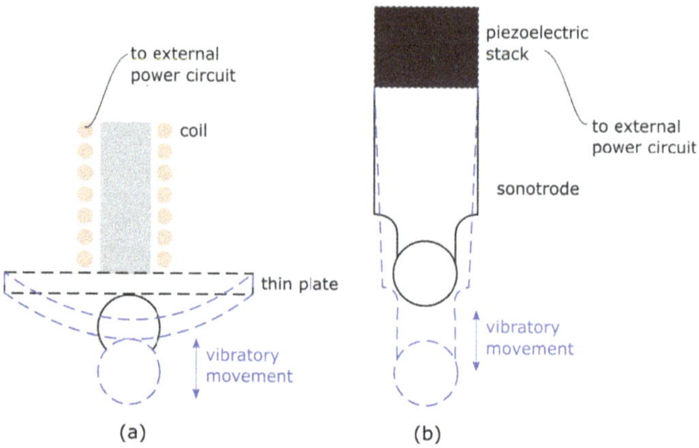

Figure 2. Schematic representations of VABB systems. (**a**). Systems based on deflective plates. (**b**). Systems based on sonotrode deformation.

The readers must take into account that all the described systems are based on the general idea of precharging the burnishing tool on the surface and then activating the vibratory system. As a consequence, the free oscillatory movement of the plate or sonotrode are restricted. That is, their normal free movement when they are excited without contact with the workpiece is dampened according to the elastic properties of the material that is being treated. As a consequence, the actual mechanical system that represents VABB is evidently complex and must be understood as a version of how the tool moves and vibrates when it is not constricted. For this reason, the correct functioning of VABB equipment should be checked to confirm that the vibratory signal originated by the vibrating tool is successfully transmitted though the material lattice. Direct dynamometric measurements or acoustic emission sensors could do the job if installed properly [48]. The challenge in this case would be to have accessibility to acquisition systems that have a high enough sampling frequency to reconstruct the signal, if the assistance is ultrasonic.

VABB Conditions

The numerous factors that can be chosen to apply the process makes it very easy to design a particular application of VABB for a certain material. Of all these parameters, some of them are directly related to the productivity of the process and the other are responsible for its technical implementation and the actual effectiveness of VABB on the target material. The reader shall find a description of all of the in the next paragraphs:

1. Preload F_p. This parameter is related to the static force that the VABB tool exerts once it makes contact with the target surface and is further pressed on it. For a correct execution of the process, it should be the mean value of the actual burnishing force F_b. Its definition is the same one for the VABB process both executed on a milling machine or a lathe.
2. Number of passes n_p. It makes reference to the number of times the process is applied on the target material. Along with the preload, it defines the degree of plastic deformation applied on the surface after the whole operation is performed.
3. Trajectories. Refer to the path that the burnishing ball follows to cover the target surface. In VABB processes programmed on a milling machine, these trajectories can either overlap, or not, and also refers to an eventual change of the feed movement along the x or the y axis. In lathe operations, it cannot be changed, as in this case VABB can be assimilated with a turning strategy that has no room for change. It has also been observed that the directionality of the passes can define the orientation of the final texture and residual stress anisotropy [12].

4. Lateral offset b. Separation between adjacent burnishing lines to cover the target surface. This value corresponds to the feed in a lathe VABB operation and the actual coordinate that the ball is laterally displaced between one pass and the next one in a milling VABB routine. It must be small enough as to gurantee that the original surface texture is covered by the process and therefore, must be defined according to the effective area of contact of the burnishing ball with the surface texture features. Therefore, it is normally defined in a preliminary assessment phase before applying the actual VABB. Furthermore, this parameter has a direct impact on productivity, as it is directly responsible for the number of adjacent passes required to cover a certain target area.
5. Feed f. It is the linear velocity by which the ball is displaced on the material. Thus far, no infulence on the actual VABB results have been reported in literature, and is therefore a mere productivity parameter.
6. Amplitude of vibration A. This parameter is defined by the vibration-assistance system and cannot usually be changed. However, at sight of the previous explanation about acoustoplasticity, it seems that it has to be high enough as to cause a change of the material by means of that effect and guarantee the transmission of the vibratory wave through the material lattice.
7. Frequency of vibration f_v. As was explained before, it seems that the effects of vibration assistance should be independent of the frequency used in the system to implement it. However, most systems do not allow the user to change this frequency, especially if it is based on resonating principles. For this reason, it is considered a parameter just for those VABB toolings where it can be adjusted, although it should be just kept constant in all cases.

From the explained parameters, the combination of preload, number of passes, trajectories and lateral offset must be defined in the NC routine implemented to apply VABB on the target surface. The former is actually the linear coordinate the ball has to be positioned at to guarantee a certain pressure on the material surface before starting the routine, whereas the three others are programmed through interpolation functions in the ISO code.

The need to define all these parameters before implementing the process is a challenge for those willing to use VABB to improve the finishing routines inside their industries. For this reason, it is necessary to follow a certain strategy to define Jerez-Mesa (2018) [50] defined after extensive work with different materials that a certain protocol has to be defined, and it depends on the alloy that has to be treated and its original surface state. Figure 3 is an extension of what can be consulted on the referenced Thesis Dissertation and it summarizes that protocol. The frequency and amplitude of the system are normally fixed by the VABB tool. Therefore, to apply the process, the user has to take into consideration what is the target material and what is its current topological state. That defines the actual preload and number of passes to be chosen to modify the topology, residual stress and hardness of the surface. On the other hand, the definition of the trajectories and the lateral offset between passes must be decided to define the desired directionality of the surface texture and preferential residual stress component. By defining these parameters, the user shall be able to master the conditions under which VABB must be executed to maximise its results. It should also be note that, in case the VABB process leads to stress relaxation, then the non-assisted process should be considered instead, although probably the adjustment of the processing conditions could lead to an eventual improvement of the effects of VABB.

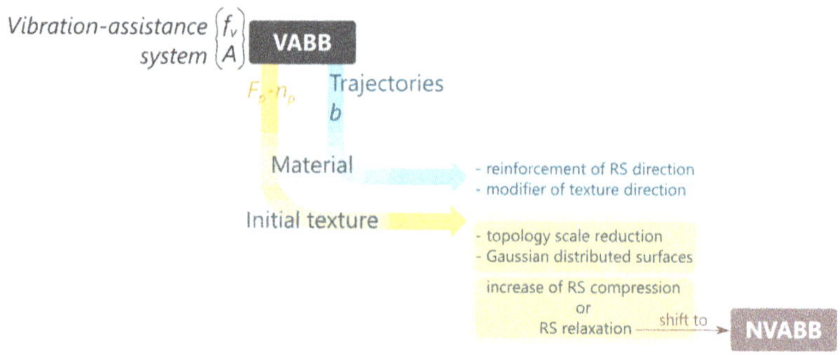

Figure 3. Recommended protocol to be followed to design the VABB processing conditions.

5. Conclusions and Prospects

The VABB process featured in this Encyclopedia entry has proved to be a procedure of interest for the industry and researchers during many decades and is starting to be more prominent now with the development of new research activities and the proliferation of practical tooling systems that are easy to manufacture. The best way to technologically implement the process in an actual environment requires a previous testing phase in which the most convenient parameters should be fixed to increase the potential of the process as much as possible and achieve simultaneous effects on texture, residual stress and hardening of the material.

VABB has all the ingredients to be the first option to be adopted as finishing technology in numerous manufacturing environments. However, the prospects of the technology are associated with certain challenges to be tackled that are related to understand the phenomenology behind the physical driver of plastic deformation under the acoustoplastic effect. Extensive experimental research has proved that the results of the process are highly dependant on the interaction of the ball and the original texture. Indeed, the interaction of the vibratory deforming body and the target material is micrometrical, and has proven to be highly influential on the actual results of the process. Increasing the theoretical knowledge of the process not only would lead to the deeper understanding of VABB—utterly important in the academic field– but could also reduce the timespan of the previous assessment phase referred to previously to plan the actual implementation of the process.

The innovation of the tooling systems to apply the process are also another innovation line to be explored in the future. The most extended systems reported in the literature to apply the actual process have been explained in this entry. However, the high frequency that these instruments vibrate at, and the dynamics of the mechanical system composed by the long tool pressed on the target surface, makes it difficult to monitor de process with conventional acquisition systems. VABB tooling must, therefore, be explored with different techniques so that the effectiveness of its systems are ratified.

All in all, VABB has proven to be an interesting process with some drawbacks in comparison with other direct competitors such as shot peening, but its capability of being applied selectively on specific areas of industrial parts and the easiness of integration in a manufacturing routine have earned it a prominent position in the present and future of finishing techniques.

Funding: Financial support for this study was provided by the Ministry of Science, Innovation and Universities of Spain, through grant RTI2018-101653-B-I00, which is greatly appreciated. Furthermore, by the regional government of Catalonia and FEDER funds for regional development through grant 2019PROD00036.

Acknowledgments: The main author Ramón Jerez-Mesa acknowledges the Serra Hunter programme of the Generalitat de Catalunya.

Conflicts of Interest: The authors declare no conflict of interest.

Entry Link on the Encyclopedia Platform: https://encyclopedia.pub/11771.

Abbreviations

The following abbreviations are used in this manuscript:

NVABB	Non-vibration-assisted ball burnishing
VABB	Vibration-assisted ball burnishing

References

1. Downes, K. Finishing of automobile components by rolling. *Prod. Eng.* **1964**, *43*, 376–382. [CrossRef]
2. Skelton, R.C. Turning with an oscillating tool. *Int. J. Mach. Tool Des. Res.* **1968**, *8*, 239–259. [CrossRef]
3. Tian, Y.; Shin, Y.C. Laser-assisted burnishing of metals. *Int. J. Mach. Tools Manuf.* **2007**, *47*, 14–22. [CrossRef]
4. Marakov, A. Ultrasonic diamond burnishing. *Russ. Eng. J.* **1973**, *53*, 58–62.
5. Pande, S.; Patel, S. Investigations on vibratory burnishing process. *Int. J. Mach. Tool Des. Res.* **1984**, *24*, 195–206. [CrossRef]
6. Bozdana, A.T.; Gindy, N.N.; Li, H. Deep cold rolling with ultrasonic vibrations—A new mechanical surface enhancement technique. *Int. J. Mach. Tools Manuf.* **2005**, *45*, 713–718. [CrossRef]
7. Bozdana, A.; Gindy, N. Comparative experimental study on effects of conventional and ultrasonic deep cold rolling processes on Ti–6Al–4V. *Mater. Sci. Technol.* **2008**, *24*, 1378–1384. [CrossRef]
8. Huuki, J.; Laakso, S.V. Integrity of surfaces finished with ultrasonic burnishing. *Proc. Inst. Mech. Eng. Part B J. Eng. Manuf.* **2013**, *227*, 45–53. [CrossRef]
9. Huuki, J.; Hornborg, M.; Juntunen, J. Influence of Ultrasonic Burnishing Technique on Surface Quality and Change in the Dimensions of Metal Shafts. *J. Eng.* **2014**, *2014*, 124247. [CrossRef]
10. Travieso-Rodriguez, J.A.; Gomez-Gras, G.; Dessein, G.; Carrillo, F.; Alexis, J.; Jorba-Peiro, J.; Aubazac, N. Effects of a ball-burnishing process assisted by vibrations in G10380 steel specimens. *Int. J. Adv. Manuf. Technol.* **2015**, *81*, 1757–1765. [CrossRef]
11. Travieso-Rodríguez, J.A.; Gras, G.G.; Peiró, J.J.; Carrillo, F.; Dessein, G.; Alexis, J.; Rojas, H.G. Experimental study on the mechanical effects of the vibration-assisted ball-burnishing process. *Mater. Manuf. Process.* **2015**, *30*, 1490–1497. [CrossRef]
12. Jerez-Mesa, R.; Landon, Y.; Travieso-Rodriguez, J.A.; Dessein, G.; Lluma-Fuentes, J.; Wagner, V. Topological surface integrity modification of AISI 1038 alloy after vibration-assisted ball burnishing. *Surf. Coat. Technol.* **2018**, *349*, 364–377. [CrossRef]
13. Jerez-Mesa, R.; Travieso-Rodríguez, J.A.; Landon, Y.; Dessein, G.; Lluma-Fuentes, J.; Wagner, V. Comprehensive analysis of surface integrity modification of ball-end milled Ti-6Al-4V surfaces through vibration-assisted ball burnishing. *J. Mater. Process. Technol.* **2019**, *267*, 230–240. [CrossRef]
14. Jerez-Mesa, R.; Plana-García, V.; Llumà, J.; Travieso-Rodriguez, J.A. Enhancing Surface Topology of Udimet® 720 Superalloy through Ultrasonic Vibration-Assisted Ball Burnishing. *Metals* **2020**, *10*, 915. [CrossRef]
15. Jerez-Mesa, R.; Fargas, G.; Roa, J.J.; Llumà, J.; Travieso-Rodriguez, J.A. Superficial Effects of Ball Burnishing on TRIP Steel AISI 301LN Sheets. *Metals* **2021**, *11*, 82. [CrossRef]
16. Salmi, M.; Huuki, J.; Ituarte, I.F. The ultrasonic burnishing of cobalt-chrome and stainless steel surface made by additive manufacturing. *Prog. Addit. Manuf.* **2017**, *2*, 31–41. [CrossRef]
17. Teimouri, R.; Amini, S.; Bami, A.B. Evaluation of optimized surface properties and residual stress in ultrasonic assisted ball burnishing of AA6061-T6. *Measurement* **2018**, *116*, 129–139. [CrossRef]
18. Amini, S.; Bagheri, A.; Teimouri, R. Ultrasonic-assisted ball burnishing of aluminum 6061 and AISI 1045 steel. *Mater. Manuf. Process.* **2018**, *33*, 1250–1259. [CrossRef]
19. Ao, N.; Liu, D.; Liu, C.; Zhang, X.; Liu, D. Face-centered titanium induced by ultrasonic surface rolling process in Ti-6Al-4V alloy and its tensile behavior. *Mater. Charact.* **2018**, *145*, 527–533. [CrossRef]
20. Zhao, J.; Liu, Z. Investigations of ultrasonic frequency effects on surface deformation in rotary ultrasonic roller burnishing Ti-6Al-4V. *Mater. Des.* **2016**, *107*, 238–249. [CrossRef]
21. Ren, K.; Yue, W.; Zhang, H. Surface modification of Ti6Al4V based on ultrasonic surface rolling processing and plasma nitriding for enhanced bone regeneration. *Surf. Coat. Technol.* **2018**, *349*, 602–610. [CrossRef]
22. Zhang, Q.; Hu, Z.; Su, W.; Zhou, H.; Liu, C.; Yang, Y.; Qi, X. Microstructure and surface properties of 17-4PH stainless steel by ultrasonic surface rolling technology. *Surf. Coat. Technol.* **2017**, *321*, 64–73. [CrossRef]
23. Shepard, M.J.; Prevey, P.; Jayaraman, N. *Effects of Surface Treatment on Fretting Fatigue Performance of Ti-6Al-4V*; Technical Report; Air Force Research Laboratory: Dayton, OH, USA, 2004.
24. Amini, C.; Jerez-Mesa, R.; Travieso-Rodriguez, J.A.; Llumà, J.; Estevez-Urra, A. Finite element analysis of ball burnishing on ball-end milled surfaces considering their original topology and residual stress. *Metals* **2020**, *10*, 638. [CrossRef]
25. Blaha, F.; Langenecker, B. Dehnung von Zink-Kristallen unter Ultraschalleinwirkung. *Naturwissenschaften* **1955**, *42*, 556. [CrossRef]
26. Izumi, O.; Oyama, K.; Suzuki, Y. Effects of superimposed ultrasonic vibration on compressive deformation of metals. *Trans. Jpn. Inst. Met.* **1966**, *7*, 162–167. [CrossRef]
27. Izumi, O.; Oyama, K.; Suzuki, Y. On the superimposing of ultrasonic vibration during compressive deformation of metals. *Trans. Jpn. Inst. Met.* **1966**, *7*, 158–161. [CrossRef]
28. Langenecker, B. Work hardening of zinc crystals by high-amplitude ultrasonic waves. *Proc. Am. Soc. Test. Mat* **1962**, *62*, 602.

29. Kozlov, A.; Selitser, S. Peculiarities in the plastic deformation of crystals subjected to the acoustoplastic effect. *Mater. Sci. Eng. A* **1988**, *102*, 143–149. [CrossRef]
30. Kozlov, A.; Selitser, S. Kinetics of the acoustoplastic effect. *Mater. Sci. Eng. A* **1991**, *131*, 17–25. [CrossRef]
31. Langenecker, B. Effects of ultrasound on deformation characteristics of metals. *Sonics Ultrason. IEEE Trans.* **1966**, *13*, 1–8. [CrossRef]
32. Gindin, I.; Malik, G.; Neklyudov, I.; Rozumnyi, O. Effect of ultrasonic vibrations on the parameters of the hardening curve for copper single crystals. *Sov. Phys. J.* **1972**, *15*, 192–196. [CrossRef]
33. Blaha, F.; Langenecker, B. Plastizitätsuntersuchungen von metallkristallen in ultraschallfeld. *Acta Metall.* **1959**, *7*, 93–100. [CrossRef]
34. Mason, W. Effect of Dislocations on Ultrasonic Wave Attenuation in Metals. *Bell Syst. Tech. J.* **1955**, *34*, 903–942. [CrossRef]
35. Pohlman, R.; Lehfeldt, E. Influence of ultrasonic vibration on metallic friction. *Ultrasonics* **1966**, *4*, 178–185. [CrossRef]
36. Nevill, G.; Brotzen, F.R. The effect of vibrations on the static yield strength of a low-carbon steel. *Proc. Am. Soc. Test. Mater.* **1957**, *57*, 751–758.
37. Kirchner, H.; Kromp, W.; Prinz, F.; Trimmel, P. Plastic deformation under simultaneous cyclic and unidirectional loading at low and ultrasonic frequencies. *Mater. Sci. Eng.* **1985**, *68*, 197–206. [CrossRef]
38. Tanibayashi, M. A theory of the Blaha effect. *Phys. Status Solidi* **1991**, *128*, 83–94. [CrossRef]
39. Malygin, G. Acoustoplastic effect and the stress superimposition mechanism. *Phys. Solid State* **2000**, *42*, 72–78. [CrossRef]
40. Siu, K.; Ngan, A.; Jones, I. New insight on acoustoplasticity—Ultrasonic irradiation enhances subgrain formation during deformation. *Int. J. Plast.* **2011**, *27*, 788–800. [CrossRef]
41. Siu, K.; Ngan, A. The continuous stiffness measurement technique in nanoindentation intrinsically modifies the strength of the sample. *Philos. Mag.* **2013**, *93*, 449–467. [CrossRef]
42. Siu, K.; Ngan, A. Understanding acoustoplasticity through dislocation dynamics simulations. *Philos. Mag.* **2011**, *91*, 4367–4387. [CrossRef]
43. Sedláček, R.; Blum, W.; Kratochvil, J.; Forest, S. Subgrain formation during deformation: Physical origin and consequences. *Metall. Mater. Trans. A* **2002**, *33*, 319–327. [CrossRef]
44. Siu, K.; Ngan, A. Oscillation-induced softening in copper and molybdenum from nano-to micro-length scales. *Mater. Sci. Eng. A* **2013**, *572*, 56–64. [CrossRef]
45. Cheng, B.; Leung, H.; Ngan, A. Strength of metals under vibrations-dislocation-density-function dynamics simulations. *Philos. Mag.* **2015**, *95*, 1845–1865. [CrossRef]
46. Liu, Y.; Wang, L.; Wang, D. Finite element modeling of ultrasonic surface rolling process. *J. Mater. Process. Technol.* **2011**, *211*, 2106–2113. [CrossRef]
47. Gómez-Gras, G.; Travieso-Rodríguez, J.A.; González-Rojas, H.A.; Nápoles-Alberro, A.; Carrillo, F.J.; Dessein, G. Study of a ball-burnishing vibration-assisted process. *Proc. Inst. Mech. Eng. Part B J. Eng. Manuf.* **2015**, *229*, 172–177. [CrossRef]
48. Jerez-Mesa, R.; Travieso-Rodriguez, J.A.; Gomez-Gras, G.; Lluma-Fuentes, J. Development, characterization and test of an ultrasonic vibration-assisted ball burnishing tool. *J. Mater. Process. Technol.* **2018**, *257*, 203–212. [CrossRef]
49. Estevez-Urra, A.; Llumà, J.; Jerez-Mesa, R.; Travieso-Rodriguez, J.A. Monitoring of Processing Conditions of an Ultrasonic Vibration-Assisted Ball-Burnishing Process. *Sensors* **2020**, *20*, 2562. [CrossRef]
50. Jerez Mesa, R. Study and Characterisation of Surface Integrity Modification after Ultrasonic Vibration-Assisted Ball Burnishing. Ph.D. Thesis, Universitat Politècnica de Catalunya, Barcelona, Spain; Université Toulouse III Paul Sabatier, Toulouse, France, 2018.

encyclopedia

MDPI

Entry

The Foundation of Classical Mechanics

Danilo Capecchi

Department of Structural and Geotechnical Engineering, Sapienza University of Rome, 00185 Roma, Italy; danilo.capecchi@uniroma1.it

Definition: Mechanics is the science of the equilibrium and motion of bodies subject to forces. The adjective classical, hence Classical Mechanics, was added in the 20th century to distinguish it from relativistic mechanics which studies motion with speed close to light speed and quantum mechanics which studies motion at a subatomic level.

Keywords: classical mechanics; fundaments; history; epistemology; analytical mechanics

Citation: Capecchi, D. The Foundation of Classical Mechanics. *Encyclopedia* **2021**, *1*, 482–495. https://doi.org/10.3390/encyclopedia1020040

Academic Editors: Ramesh Agarwal, Raffaele Barretta, Krzysztof Kamil Żur and Giuseppe Ruta

Received: 17 May 2021
Accepted: 16 June 2021
Published: 19 June 2021

1. Introduction

Classical mechanics, or more commonly *Mechanics*, is the discipline devoted to the study of the equilibrium and motion of bodies subject to forces; the adjective *classical* sets it apart from *Relativistic mechanics* which studies motion with speed close to light speed and *Quantum mechanics* which studies motion at a subatomic level. From a historical perspective *Classical mechanics* is the form of mechanics after Newton; for this reason it is often referred to, though quite improperly, as *Newtonian mechanics*. For the 16th and 17th century one speaks of *Early classical mechanics* while for previous periods the locution *Ancient mechanics* is often used.

What is now called mechanics has played a preponderant role in the development of science since ancient Greece [1]. To explain its influence it is not enough, as done just above, to define mechanics as the reasoned study, that is a science, of the phenomena of equilibrium and motion. First of all, mechanics knows how to measure the phenomena of motion: in other words, however complex their appearances may be, whatever qualitative aspects they reveal, whether it is the changing figure of a cloud, of a waterfall, deformations and resistances of an elastic solid, mechanics knows how to define entirely with the help of numbers these motions, these resistances, these deformations. It is a quantitative discipline.

As for the influence of mechanics on the development of other sciences, it is not difficult to see the reasons for this. First of all, the phenomena of motion or of equilibrium occur constantly and everywhere, whether they appear alone or whether they are accompanied by other more complex phenomena (electrical, chemical, etc.). Mechanics was therefore the necessary basis for other sciences, at least as soon as they wanted to be sufficiently precise.

There was not a historical accident which gave mechanics its preponderance. Among all the phenomena, those which are the least difficult to measure, that is to define completely with numbers, are the phenomena of motion: it is therefore mechanics which became the first of all sciences to take a quantitative form. However, by virtue of the more abstract and geometric nature of the phenomena it studies, it had to vegetate as long as it was not in a condition to assume a quantitative form in all its aspects. Let us follow, for instance, a projectile launched into the air; what will we be able to say precisely about its motion if there is no way to register it? It is therefore understandable that the development of mechanics, or better of that part which studies motion, was at the same time so late and so prodigiously rapid once it had begun, and it immediately took precedence over other sciences, even over those which, like alchemy in the Middle Ages, seemed to precede it ([2], pp. 1–4).

Classical mechanics is today generally considered a mathematical physics discipline and as such of little interest to physics researchers. It is a mathematical theory whose axioms, although they draw inspiration from the physical world, are considered as given; the only developments that can be expected are discoveries of new theorems more or less interesting for the applications of the discipline. It must be said that this point of view is not unassailable and the historical analysis of the discipline provides many elements to counter it.

> Despite the unusually distinguished and successful role Newtonian [classical] mechanics has played in the history of modern science, its foundations have been under vigorous debate since Newton first formulated his laws of motion. Moreover, although the axioms have received much more than two centuries of critical attention from outstanding physicists and philosophers, there still is wide disagreement about what they assert and what their logical status is. The axioms (or their logical equivalents) have been claimed to be either a priori truths, which can be asserted with apodictic certainty; [or] to be necessary presuppositions of experimental science though incapable either of demonstration by logic or refutation by observation; or to be empirical generalizations, "collected by induction from phenomena" ([3], p. 174).

In the 18th century, a century in which classical mechanics had reached a mature status, many important scientists like Euler and d'Alembert believed that mechanics were an *a priori* science, that needed no recourse to experience to be established. Otherwise other scientists, such as Daniel Bernoulli for instance, supported the idea of mechanics as an empiric discipline.

The possibility of this long dispute depended on the essence of mechanics. Even those who declared mechanics as an experimental science referred to the experience of everyday life and not to complicated experiments carried out in the laboratory, as was the case, even in the 18th century, for electricity and magnetism or thermology. For instance the law of lever was based on the observation that the more the heavier a body, the more it acts on an arm; the law of motion was based on the constant effect of a cause and that heavy bodies felt downward. The principle of inertia was stated based on simple thought experiments and the principle of action and reaction seemed to be self-evident. Of course no one denied the role of experience when a mechanical theory was be applied. For instance, to evaluate the law of falling of a heavy body, it is not enough to say that its speed varies proportionally to time. It is also necessary to know the value of the constant of proportionality, which is the acceleration of gravity. This can be known only by means of accurate experimentations in a laboratory. Equally the evaluation of masses, forces, times, distances may require sophisticated experiments and instruments.

Up to now no shared formulations of a completely axiomatized formulation of classical mechanics have been provided, even though some attempts have been made [4–7]. Commonly incomplete axiomatized versions circulate in which in some cases the difference between axioms and theorems is not completely clear. For example, Newton's second law of motion, $f = ma$, can be treated as a principle, a relationship between force f and acceleration a, in which both members of the relationship are considered as primitive terms of the theory, or it can be treated as a simple definition of force. This matter of fact strongly suggests that mechanics needs some more studies by physicists in the attempt to improve its comprehension.

The difference between the two kinds of axiomatic formulations, the complete and the incomplete one, is not only of a quantitative but also epistemological nature. A complete axiomatic theory is by definition a closed system, straightforwardly formalizable with the sufficiently sophisticated language of mathematical logic so that it can contain most of mathematics. It has by definition a hypothetical deductive nature; that is, no truth value is given to the axioms of the theory. What matters is that the theory is coherent, even if this coherence as argued by Gödel's theorems cannot be proved but only assumed in the absence of evident inconsistencies. Primitive terms, definitions, axioms, have no definite meaning in themselves, even if they are given names used in everyday life, such as force, mass, inertial system. These quantities take on values through some rules of correspondence, connecting the real world with the theory, which include all the conceptual difficulties of the physical theory that are avoided in the formalized theory. If the theory and the correspondence rules provide results corresponding to the empirical measurements, the theory is said to be validated (see the next sections).

In incomplete axiomatized theories, one starts from some axioms, generally treated as empirical laws and therefore considered true as such. From these axioms one derives conclusions that are true in themselves because they are deduced by true axioms with the indubitable laws of logic. If it were to be verified that the conclusions were not empirically true, the blame would not be attributed to the theory but to those who applied it, not having been able to take into account the precise manner of the different parameters involved. The incomplete axiomatic approach of contemporary treatments is from a logical point of view perfectly equivalent to the approach of Hellenistic Greece scholars, of Archimedes in particular, as will be clear from the next section. From certain points of view, therefore, it can be said that the essence of mechanics has remained unchanged over a period of more than two thousand years.

There are various formulations of classical mechanics. Perhaps the most important distinction is between a vector formulation that emphasizes the notion of force and refers to Newton and Euler and an analytic formulation that emphasizes the related notions of work and energy and refers to Leibniz and Lagrange. Up to now the equivalence of the two formulations has not been proven in a shared way, even if most scholars are convinced of it, or if they are not convinced they do not consider the fact particularly interesting. It must be said that many textbooks, even of a high level, try to prove this equivalence, unfortunately with some logical leaps. The few supporters of the non-equivalence between analytic and vector theories point out that within the analytic formulation, work and energy are concepts that can be easily derived from thermodynamics, which is not true for the vector formulation.

2. Ancient Mechanics

In ancient Greece mechanics was the name of the science dealing primarily with the study of equipments or machines (in Greek $\mu\eta\kappa\alpha\nu\eta$), to transport and lift weights. The search for equilibrium was not of practical interest and mechanics, at least at the beginning, did not care for it. From this point of view mechanics was different from modern statics which is instead seen as the science of equilibrium.

Mechanics with astronomy, music and optics was considered as a mixed mathematical discipline (Renaissance nomenclature), because it dealt with both mathematics and physics. Astronomy which studied phenomena of which only a limited direct experience could be shown, was a hypothetical deductive theory (modern terminology). That is a theory based on a principle assumed as hypothetical—for instance the diurnal rotation of the fixed stars sphere—that is not true for sure. The theory, but not the principle, was validated by its adherence to empirical observations. Different theories based on a different principle could save the same phenomenon. Mechanics, with music and optics, was instead based on indubitable axioms derived either from natural philosophy, or metaphysics or experience.

Certainly there have been very ancient reflections about mechanics; hardly any trace remains of them, however. An ancient mathematician credited with some theorizations was Archytas of Tarentum (fl. 400 BC) but the first known writing in the West is the *Mechanica problemata* attributed to Aristotle [8,9]. The attribution is doubtful, sometimes accepted, sometimes rejected by historians. Certainly a modern reader is surprised to see how a philosopher not particularly interested in mathematics, Aristotle, had taken an interest in mechanics.

The style of the *Mechanica problemata* is not the sober one of mathematicians and the fact that some scholars have seen in it the demonstration of the law of the lever while others had not, testifies to the ambiguity of the text. The first proofs of the law of the lever truly accepted by ancient mathematicians were those of Euclid and Archimedes. That of the latter is still considered by many scholars as the most convincing even if some controversial points remain, as happens in any proof of the 'axioms' of a science.

Euclid is credited as the author of a text known in the West as *The book of the balance* ([10], pp. 24–30); Archimedes was the author of the *De planorum aequlibriis* [11]. Both the authors developed a theory of a deductive kind based on some simple (empirical?) axioms. Euclid, or better the pseudo-Euclid, assumed:

1. When two equal weights are on the end of a lever of equal arms, the lever remains in equilibrium.
2. If two weights keep a lever in equilibrium, the lever remains in equilibrium when the weights are moved along vertical wires ([10], p. 24).

Archimedes on his side assumed:

1. Equal weights at equal distances are in equilibrium, and equal weights at unequal distances are not in equilibrium but incline towards the weight which is at the greater distance.
2. If, when weights at certain distances are in equilibrium, something be added to one of the weights, they are not in equilibrium but incline towards that weight to which the addition was made ([11], p. 189).

All the rest of the theory was presented by Euclid and Archimedes as a pure deduction. From the law of the lever, the Hellenistic mathematicians derived the laws of all the machines for lifting weights, or rather of their fundamental components: the pulley, wedge, winch, called simple machines, with the exception of the inclined plane. The solution of the problem posed by this last machine had to wait for the treatises of the Arabs in which a convincing demonstration of the law of the lever was resumed closer to that suggested by the *Mechanica* than by the *De planorum aequlibriis*, and lent itself naturally to the demonstration of the law of the inclined plan. Jordanus de Nemore in the 13th century was the first to exhibit a correct law for the inclined plane. Although not all historians agree, he is credited to have applied the rule: what lifts a weight np to a height h lifts a weight p to a height nh, which is a form of the principle of virtual works [12].

The ancient mechanics that in the Middle Ages was called the *Science of weights* remained substantially unchanged until the beginning of the 17th century. It was the discipline developed by mathematicians; therefore, it was of a quantitative nature that studies the equilibrium and displacement of heavy bodies.

3. Early Classical Mechanics

With the 17th century, a new perspective opened up, which at the beginning was seen as completely separate from that offered by ancient mechanics. Mathematicians began to be interested in the motion of bodies not only in that occurring with the use of lifting machines but also in the more general case such as the natural motion of falling bodies and the violent motion of an object thrown from a slingshot, whose study since then was carried out by philosophers like Plato, Aristotle in an almost exclusively descriptive way.

The novelty was made possible because mathematicians began to conceive of time as a physical magnitude that could be measured with precision; conversely, one could probably say that the need to treat time as a physical magnitude derived from the need to study the motion of bodies. The main protagonist of this development was Galileo Galilei, who in the 17th century obtained the temporal law of the free fall of a heavy body according to which the space passed downward is in the square ratio of the time.

This law seemed too complex to Galileo to be taken as a principle and he tried to deduce it from a simpler principle. He obtained it by verifying that the law of falling bodies could be derived by assuming the speed of the body increasing linearly with time. Galileo's procedure was quite elaborate—a primordial use of infinitesimal calculus indeed—and it is not completely clear, at least to me, whether it should be considered as a hypothetical deductive approach, for which the law of proportionality of speed and time is tentatively fixed and, if the passed space derived from it is in accordance with the square ratio of the time, the hypothetically assumed law is valid, or if it followed the analytical method of mathematicians, which derived one or more axioms from a theorem with purely deductive procedures. A modern mathematician would opt for this second solution because the two laws, speed proportional to time and space proportional to the square ratio of time, are obtained from each other with derivation/integration operations.

No less important and perhaps more basic was the formulation of the law of inertia—and the principle of relativity (modern terminology) strictly related to it—based no longer on metaphysical concepts as it had been in the Greek and medieval impetus theory but on thought and real experiments. Galileo's results were motivated by the defense of the heliocentric system and would not have been possible without the publication by Copernicus of the *De revolutionibus orbium coelestium* of 1543 [13].

A very clear exposition of Galileo's principle of inertia is that of the *Dialogo sopra i due massimi sistemi* of 1632, where its experimental—though idealized—nature is shown. In the following exchange, Salviati-Galileo made Simplicius to say that the motion of a body on a horizontal plane, once it has received an initial impetus, would move forever, on the condition that there are not impediments; the motion should occurs without acceleration or delay; it would thus be uniform.

> SIMP. *I cannot tell how to discover any cause of acceleration, or retardation, there being no declivity or acclivity.* [emphasis added]
> SALV. Well: but if there be no cause of retardation, much less ought there to be any cause of rest. How long therefore would you have the moveable to move?
> SIMP. *As long as that superficies, neither inclined nor declined shall last* [emphasis added].
> SALV. Therefore if such a space were interminate, the motion upon the same would likewise have no termination, that is, would be perpetual.
> SIMP. I think so, if so be the moveable be of a matter durable.
> SALV. That hath been already supposed, when it was said, that all external and accidental impediments were removed, and the brittleness of the moveable in this our case, is one of those impediments accidental ([14], p. 173. English translation in [15]).

In the *Discorsi e dimostrazioni matematiche sopra due nuove scienze* of 1638, Galileo added that, in truth, even though all the above mentioned resistances were eliminated, a further resistance would arise to contrast the motion, due to the heaviness of the body. Indeed the horizontal plane is supposed to be represented by a straight line and each point on this line equally distant from the center; this is not the case, for as one starts from the middle of the line and goes toward either end, he departs farther and farther from the center of the earth and is therefore constantly going uphill. Whence it follows that the motion cannot remain uniform through any distance whatever, but must continually diminish ([16], p. 274).

This explanation of Galileo has led many historians to affirm that Galileo did not have a clear idea of the principle of inertia and that his inertia was in fact just a circular inertia ([17,18], p. 352). It seems clear to me that in the affirmation that if resisting forces did not act on a moving body it would continue to move indefinitely, was the purest expression of the principle of inertia, even though it must be said that it is always difficult to establish a correlation between a modern concept and its older formulation. That there are in fact no bodies in nature that can move without resistance, which is another matter. Compared with Newton's formulation, Galileo's differs in one, perhaps non-trivial, way: Galileo did not admit the possibility that there were bodies on which resistant forces did not act, for example weightless bodies; Newton did. Galileo's was a terrestrial view, Newton's a cosmological one. In any case, when Galileo studied the motion of projectiles, he considered a horizontal inertia, without specifying whether he ignored the change in the direction of gravity when moving in a limited space because in this case the direction varies little, or he actually ignored the effect of gravity, thus separating the effect of weight from that of mass.

Cavalieri investigated the motion of projectiles in his *Lo specchio ustorio, overo trattato delle settioni coniche* of 1632 [19]. In this text he had paused to consider the nature of the motions to be combined.

> Moreover I say, that considering that the motion [of the body] is a straight line toward any part, if it had not other motive virtue that would push it towards another direction, [the body] should go in the place indicated by the projector along a straight line, because of the virtue impressed to it was in the straight line, from which direction it is not reasonable that the mobile deviates, as long as there is no other motive virtue that deflects it, and that when between the two terminal points there is not impediment. ([19], p. 154. My translation).

This is an expression of the principle of inertia, closer to Newton than to Galileo. Cavalieri said clearly, in a few lines, that a projectile would move in a straight line when not impeded by something (gravity for instance). The main difference compared to the sentence of Galileo is its explicitness.

Torricelli studied the motion of projectiles in Book II of the *De motu gravium*, written before 1644. For him the composition of motion became just a matter of mathematics, the composition of two motions highly idealized: a downward uniformly accelerated motion and a uniform rectilinear motion ([20], *De motu proiectorum*, p. 156).

The evolution of the principle of inertia due to Cavalieri and Torricelli is an example of the growth of scientific knowledge. A scientist (Galileo) developed a form of the principle of inertia, which he did not elaborate completely. His internal troubles transpired in his work, signaled by some uncertainties in its use. Scientists after him (Cavalieri and Torricelli) who had read his writings were not fully aware of the troubles and applied with no doubt the theory to any possible case (the angled launch of a bullet for instance). The process is similar to what Kuhn refers as gestalt switch [21].

Cavalieri and Torricelli moved ahead of Galileo because their mathematics had moved ahead too. Cavalieri and Torricelli, more inclined to mathematics, did make use of the actual infinite in mathematics; Galileo did not. For them the mathematical physical theory (broad meaning) had its reality, maintaining contact with the external world. For this reason the quasi-rectilinear but still slightly curved trajectories of bullets, moving over the earth sphere, can actually be assumed to be rectilinear and of infinite length, identifying the approximation with reality. However, Cavalieri and Torricelli were more cautious than Newton; for instance, they hardly used the quantifiers *for all* and *for ever*, as Newton did.

The concept of the principle of inertia was substantially revolutionary. It overturned completely the traditional approach to motion according to which it was caused by a force, either internal or external. From now on the force acting on a body is responsible for the change of velocity, and not displacement.

After Galilei, the attention of scientists who studied motion focused on the phenomenon of collision, with important contributions by Descartes and Huygens (with Wren and Wallis). The goal of these scholars, classifiable as mechanistic, was to provide general laws of motion that could explain all phenomena. This project of mechanical philosophy was too ambitious and was essentially aborted. However, new concepts arise from these studies especially thanks to Descartes and Huygens that will prove fundamental at the beginning of the 19th century; those of work and kinetic and potential energy.

Descartes contributed to the development of mechanics mainly as a promoter of a mechanistic philosophy (with Gassendi and Hobbes), simpler than the older philosophy which allowed many people to devote themselves to a quantitative study of nature. He contributed to spread the concept of energy/work in statics (the product of weight and vertical displacement) and in dynamics (the product $m|v|$ of mass and speed), but mainly he was the supporter of a rationalistic view of mechanics, based on evident and clear notions, with no recourse to experience. Huygens played a different but fundamental role, that for the sake of space is not commented upon here. For instance he, with Leibniz (see below), contributed to the development of the ideas of work and energy and formulated in nuce the concept of conservation of mechanical energy, developed then by Johann and Daniel Bernoulli [22,23].

John Wallis published in 1669–1671 the text *Mechanica sive de motu, tractatus geometricus* [24], where the term mechanics, until then restricted to statics, was also applied to the new science of motion. Aside from a greater wealth of content, mechanics took on a different function. It was no longer a discipline destined to solve a particular problem, of mainly technological interest, but became a new philosophy of nature with which the mathematical philosophers (that is mathematicians) and not the generalist philosophers intended to explain nature.

4. Classical Mechanics

The transformation of mechanics into a new philosophy of nature was particularly evident with Newton's *Philosophiae naturalis principia mathematica* of 1687 [25], which took up the work of Archimedes and Galileo by unifying statics and dynamics with few very simple axioms: the three famous laws of motion, considered certain and inductively derivable from experience. The approach was exactly that of the mixed mathematics of the ancient Greece, nothing substantially new after two thousand years. However, the mathematics was more refined, using modern categories it can be said that the *Principia* was a treatise of differential geometry.

Below are the first two laws of motion as reported in the 1726 edition.

Law I.
Everybody perseveres in its state of being at rest or of moving uniformly straight forward, except insofar as it is compelled to change its state by forces impressed.
Law II.
A change in motion is proportional to the motive force impressed and takes place along the straight line in which that force is impressed. ([26], p. 13. English translation in [27]).

Notice that Newton's second law is quite different from what is known today under this name. The modern relation, force equals mass by acceleration, is due mainly to Euler (see below).

Modern philosophers of science have clearly shown that Newton's laws/axioms cannot be derived from experimental data by induction, not least because many of them do not believe that induction exists. However, Newton and his followers were convinced of the empirical derivation of the laws and that all natural phenomena could be derived from them with certainty.

The comment of Ernest Nagel on the nature of Law I is worth reporting. According to him, it is clear that the first law of motion, taken in isolation, is seriously incomplete as a statement intended to have empirical content. To say that a body will persevere in its state of rest or uniform motion in a straight line, unless compelled to change its state by forces impressed on it, is to say nothing definite, if nothing further is stated as to:

1. What is the spatial frame of reference;
2. What is the way to measure time;
3. What are the marks by which the presence or absence of impressed forces is determined.

Assumed that the first point will in some way be solved, continued Nagel, what is the criterion for asserting that a body is under the action of no forces? What defines the perseverance of the body in its state of rest or uniform motion in a straight line? However, in such a case, the law is a concealed definition, a convention which specifies the conditions under which one will say that there are no impressed forces acting on a body. Moreover, in addition to assuming a definite spatial frame of reference, in its usual formulation the law takes for granted a definite system of chronometry. If, then, some method not involving the explicit or tacit use of the law were available for identifying the absence of forces, the law could be construed as an implicit definition either of uniform motion or equal time intervals ([3], pp. 175–178).

The exposition of the three laws was followed by some examples that made them plausible even though there was no reference to their empirical nature. Only in the scholium did Newton say that his laws were accepted by mathematicians and confirmed by numerous experimental results. They are attributed to Galileo

> Scholium The principles I have set forth are accepted by mathematicians and confirmed by experiments of many kinds. By means of the first two laws and the first two corollaries [the law of composition of forces] Galileo found that the descent of heavy bodies is in the squared ratio of the time and that the motion of projectiles occurs in a parabola, as experiment confirms, except insofar as these motions are somewhat retarded by the resistance of the air ([26], p. 21. English translation in [27]).

This attribution is generally considered too generous on Newton's part. In fact, Galileo had not possessed the concept of impressed force and, if he had read the *Principia*, most likely he would not have understood them. However, it should be noted that, in the 18th century, mathematicians who took up the theory, such as Varignon and Euler for instance, referred to Newton's laws as the laws of motion of Galileo.

Newton considered only the third law as original of him:

> Law III.
> *To any action there is always an opposite and equal reaction; in other words, the actions of two bodies upon each other are always equal and always opposite in direction* ([26], p. 14. English translation in [27]).

In addition to its explanations, thought experiments were reported; the simplest of them is described below.

> I demonstrate the third law of motion for attractions briefly as follows. Suppose that between any two bodies A and B that attract each other any obstacle is interposed so as to impede their coming together. If one body A is more attracted toward the other body B than that other body B is attracted toward the first body A, then the obstacle will be more strongly pressed by body A than by body B and accordingly will not remain in equilibrium. The stronger pressure will prevail and will make the system of the two bodies and the obstacle move straight forward in the direction from A toward B and, in empty space, go on indefinitely with a motion that is always accelerated, which is absurd and contrary to the first law of motion ([26], p. 25. English translation in [27]).

However, the results of a real devised experiment were also reported.

> I have tested this with a lodestone and iron. If these are placed in separate vessels that touch each other and float side by side in still water, neither one will drive the other forward, but because of the equality of the attraction in both directions they will sustain their mutual endeavors toward each other, and at last, having attained equilibrium, they will be at rest ([26], p. 25. English translation in [27]).

With his laws, Newton was able to prove old results and to find new ones. The proof of the ellipticity of the orbits of the planets about the sun, as predicted by Kepler, and the formulation of the law of universal gravitations are very interesting.

5. Post Newtonian Mechanics

Newton's mechanics proved to be incomplete as it was substantially limited to the motion of free mass points. This was particularly clear to the handful of brilliant mathematicians of the 18th century, of which the most known are Leibniz, Johann Bernoulli, Varignon, Hermann, Euler, d'Alembert, Lagrange that were able to fill the gap [28]. In the following, for the sake of space, a short hint to the contributions of Leibniz, Euler and Lagrange only is given.

Leibniz played a role similar to that of Descartes—both were great philosophers and mathematicians. His main contribution to mechanics was a critical discussion about Newton's concepts: force, time and space, the introduction of the term *dynamics* to indicate that part of mechanics devoted to the study of the motion of bodies, and, mainly, the development of a form of infinitesimal calculus more efficient in mechanics than that proposed by Newton. It is worthy signaling however that Leibniz's name, though his role is well recognized by historians, is hardly found in modern textbooks of mechanics.

Euler is usually considered as Newton's most important heir. His main merit is of having transformed mechanics from a geometric discipline into an analytical one, considered as a purely rational discipline. The incipit of his early treatise *Mechanica sive motus scientia analytice exposita* of 1636 are quite famous:

> Thus, I always have the same trouble, when I might chance to glance through Newton's *Principia* or Hermann's *Phoronomia*, that comes about in using these [synthetic methods], that whenever the solutions of problems seem to be sufficiently well understood by me, that yet by making only a small change, I might not be able to solve the new problem using this method. Thus I have endeavored a long time now, to use the old synthetic method to elicit the same propositions that are more readily handled by my own analytical method, and so by working with this latter method I have gained a perceptible increase in my understanding. Then in like manner also, everything regarding the writings about this science that I have pursued, is scattered everywhere, whereas I have set out my own method in a plain and well-ordered manner, and with everything arranged in a suitable order. Being engaged in this business, not only have I fallen upon many questions not to be found in previous tracts, to which I have been happy to provide solutions, but also I have increased our knowledge of the science by providing it with many unusual methods, by which it must be admitted that *both mechanics and analysis are evidently augmented more than just a little* ([29], Preface. English translation by Bruce I).

Notice that the title of Euler's treatise paraphrased that of Wallis, by changing *geometric* with *analytic*.

Apart from the algebraization of mechanics Euler improved much the 'primeval' Newton's development. For instance he contributed to give the second law of motion its modern form [30,31]. Moreover he applied the laws of motion to the case of extended and constrained bodies, fluid included. Euler was an extremely prolific and methodical mathematician. With him, mechanics became the discipline we know. And even today it makes sense to read some of his writings on mechanics, finding interesting topics for current research.

After Euler mechanics had become an algebraic theory; in some parts however geometry was still present, for example in the use of vectors. A concept, that of vector, today possibly introduced without any reference to classical geometry but that in the 18th century could not. Lagrange with his *Méchanique analitique* of 1788 [32] (Since the second edition of 1811 the title of Lagrange's treatise changed to *Mécanique analytique*.) undertook an important and big step toward a full algebraization of mechanics. His treatise of 1788 added little new to the development of mechanics and collected most results obtained by Lagrange himself since the 1760s. However, it was completely new because of its way of presentation and its logical–epistemological conception. Mechanics then became analytical.

The locution analytical mechanics had/has different meanings. Today its more diffuse acceptation is that of a (mathematical) theory based on algebra and Calculus, whose axioms are presented with scalar relations; this definition substantially covers *Méchanique analitique* use. According to Truesdell, analytic mechanics is limited to discrete systems (excluding thus continua, solid and fluid), and this limitation of the term is quite diffuse also [33]. Vectors could be still present, but they were introduced only through their components and mainly there could be many different local frames in the same mechanical assembly. According to the meaning assumed beforehand, that of Euler is not analytical mechanics, even though he himself referred to it as such.

In the preface of his masterpiece, Lagrange stated that his mechanical theory had become a branch of analysis. The axioms of this branch were represented by general formulas, the principle of virtual work and the principle of d'Alembert. They were enough to solve any practical problem, both in statics and dynamics. The axioms were considered to be given, and no serious interest was addressed to their derivation or proof. Accordingly the *Méchanique analitique* should be considered one of the first texts of modern mathematical physics.

In the following a large quotation of the preface of Lagrange's treatise is reported, even though well known in the literature, because of its relevance.

> I propose to condense the theory of this science and the method of solving the related problems to general formulas whose simple application produces all the necessary equations for the solution of each problem. I hope that my presentation achieves this purpose and leaves nothing lacking. In addition, this work will have another use. The various axioms presently available will be assembled and presented from a single point of view in order to facilitate the solution of the problems of mechanics. Moreover, it will also show their interdependence and mutual dependence and will permit the evaluation of their validity and scope. I have divided this work into two parts: statics or the theory of equilibrium, and dynamics or the theory of motion. In each part, I treat solid bodies and fluids separately. *No figures will be found in this work* [emphasis added]. The methods I present require neither constructions nor geometrical or mechanical arguments, but solely algebraic operations subject to a regular and uniform procedure. *Those who appreciate mathematical analysis will see with pleasure mechanics becoming a new branch of it* [emphasis added] and hence, will recognize that I have enlarged its domain ([32], Preface. English translation in [34]).

Lagrange's and Euler approaches were different in two main aspects. First, Euler's mechanics had a substantially axiomatic structure, in which concepts were explained before being used. Lagrange's mechanics did not have such an axiomatic structure. It generally used concepts and axioms already accepted by experts of mechanics. Moreover the general formulas Lagrange talked about were not true formulas but rather rules that needed interpretation. Second. Lagrange's mechanics made no reference to natural philosophy to introduce its assumptions, differently from Euler's. 'Axioms' were justified with mathematical reasoning. Where it would be necessary to exit from this ambit, Lagrange was elusory. Instead of considerations drawn from natural philosophy or metaphysics, he referred to a historical account [35]. History gave a justification of the theory and replaced the justification based on natural philosophy. Lagrange thought that mechanics were nothing but the results gained in its history, by Archimedes *in primis*. In his history there was no room for the philosophy of nature.

6. The Crisis of Classical Mechanics

In the 19th century, mechanics was a mature discipline, or at least it was considered as such by its experts. There were no substantial innovations in the understanding of the relevant phenomena. One of the main 'discoveries' was the Coriolis effect, that is the fact that in a reference system in rotation with respect to the fixed stars, apparent forces orthogonal to the motion are noted. The effort of mathematicians and physicists was concentrated on perfecting the formal aspects of the theory and in the criticism of its axioms.

At the turn of the century, after the publication of Lagrange's *Méchanique anlitique*, there was a heated debate on the logical status of the principle of virtual works, which was at the basis of Lagrange's analytical mechanics. The protagonists of this discussion were the French mathematicians and physicists, including Fourier, Ampère, Coriolis, Laplace and Poinsot ([12], pp. 317–351).The discussion showed how difficult it was to 'prove' in a rigorous way, that is according to the stringent standard of rigor of the 19th century, axioms that appeared nearly self-evident to the common people, as the composition of forces, the law of lever, the principle of virtual works. For this purpose the words by Lazare Carnot at the very end of his *Essai sur les machines en général* of 1782/86 are illuminating.

> Sciences are as a beautiful river whose course is easy to follow, when it has acquired a certain regularity; but if one wants to sail to the source one cannot find it anywhere, because it is far and near; it is diffuse somehow in the whole earth surface. The same if one wants to sail to the origin of science, one finds nothing but darkness and vague ideas, vicious circles; and one loses himself in the primitive ideas ([36], p. 107).

In the second half of the century, the analysis of the foundations concerned the whole mechanics; in particular, the Newtonian force was the subject of criticism. The protagonists were mathematicians, physicists and philosophers, such as: Saint-Venant, Kirchhoff, Hertz, Helmholtz, Reech, Mach, Poincaré, Duhem ([37], pp. 417–443).Particularly effective were the contributions of the latter three scholars. Mach in his *Die Mechanik in ihrer Entwicklung historisch-kritisch dargestellt* of 1883 scrutinized all the ideas of Newtonian mechanics, starting from the ideas of absolute space, force and mass [38]. Poincaré in the *Science et hypotèses* of 1902, commented upon the epistemology of the axioms of mechanics. According to him, mechanics is neither fully empiric nor fully a priory. For instance let us consider the principle of inertia:

> *The Principle of Inertia—A body under the action of no force can only move uniformly in a straight line.* Is this a truth imposed on the mind à priori? If this be so, how is it that the Greeks ignored it? How could they have believed that motion ceases with the cause of motion? or, again, that every body, if there is nothing to prevent it, will move in a circle, the noblest of all forms of motion. [...] Is, then, the principle of inertia, which is not an à priori truth, an experimental fact? Have there ever been experiments on bodies acted on by no forces? and, if so, how did we know that no forces were acting? ([39], pp. 112–113).

Beside the attempts to clarify the axioms, there was also the problem of the nature of mechanics in itself. Until then mechanics had been considered as the mother of the other disciplines; even electricity could be framed into a mechanical context. However, at the end of the 19th century it was clear that this was not a tenable position. Thermodynamics dealt with concepts hardly reducible to mechanics, and a similar consideration held good for electromagnetism.

Duhem championed 'energetics', holding generalized thermodynamics as foundational for physical theories; that is, thinking that all of chemistry and physics, including mechanics, electricity, and magnetism, should be derivable from first axioms of thermodynamics. His *Traité d'énergétique ou de thermodynamique générale* of 1911 [40] represented one of the first attempts to establish a physical theory on modern axiomatic basis. The author believed that the axioms of a physical theory did not require any justification apart from the assessment of their internal consistency. He considered, however, that the axioms could not be chosen by chance, but that they should benefit from formulations of past similar axioms; in this way the history of science becomes an integral part of science ([40], pp. 183–246).

The attempt by Duhem to save classical mechanics was frustrated by the emergence of serious paradoxes in the electromagnetic theory, still framed into classical mechanics for some aspects, evidenced at the end of the 19th century. They pushed to consider classical mechanics as a simple approximation of a true mechanics, the relativistic one. From then on, physicists ceased to be interested in classical mechanics, considering it scarcely interesting, or worse a dead discipline. However, this is another story.

Together with the intense work on foundation there was also an effort to clean up the theory. Hamilton and Jacobi moved in this direction, giving analytical mechanics its modern structure, based on a functional, well-known as the Hamiltonian, which allowed the treatment of statics and dynamics, rigid and continuous deformable solid bodies and fluids in a unitary way. With them the formal aspects, the elegance, the clarification of the spheres of validity of the theorems had reached the apex [41,42]. The rearrangement of mechanics also made the solutions of some problems easier and also suggested the solution of others; it is so restrictive to speak of formal perfecting only. Mechanics was now ready to be framed into the broad field of modern mathematical physics.

The expression mathematical physics had substantially two different meanings. On the one hand, it simply indicated modern physics, strongly rooted in mathematics; in this sense, the great mathematical physicists were Galileo, Kepler, Newton, Euler, etc. On the other hand, it meant that branch of science born in the 19th century which afforded some problems regulated by partial differential equations, e.g., the propagation of heat, the theory of potential, the theory of elasticity; in this sense, great mathematical physicists were Fourier, Lamé, Gauss, Piola, Beltrami, etc. Today the expression indicates an academic discipline cultivated by mathematicians which has as its basis some physical axioms whose merit was not questioned.

Examining epistemological aspects of mathematical physics gives the opportunity to stipulate an appropriate conventional meaning. As a first step, it is useful to explain the essence of physical theories. They are made up of three parts:

1. An abstract calculus which comprehends theoretical terms, definitions, axioms and inference rules.
2. A conceptual model which more or less gives a representation of the world.
3. Some correspondence rules connecting theoretical terms and theorems of the theory with the external world [22].

When the correspondence rules are missing, the physical theory turns to a mathematical physical theory as its axioms speaks about the physical world and not only of pure mathematical entities. The boundary between pure mathematics and mathematical physics is not exactly specified and various positions could be assumed. Clifford Ambrose Truesdell, for instance, does not see any difference and claims mathematical physics to be simply a branch of pure mathematics. He actually refers to his field of competence, rational mechanics, but his considerations apply to any mathematical physical theory:

> Is rational mechanics part of applied mathematics? Most certainly not. While in some cases known mathematical techniques can be used to solve new problems in rational mechanics, in other cases new mathematics must be invented. It would be misleading to claim that each achievement in rational mechanics has brought new light in mathematics as a whole as to claim the opposite, that rational mechanics is a mere reflection from known parts of pure mathematics ([43], p. 337).

Other people sustain that a theory can be called mathematical only if it is addressed to purely mathematical objects.

Quite interesting in the development of modern mathematical physics was the contribution by Carl Neumann [44]. He recognized that the theorems of mathematical physics should be confirmed by experimental data ([45], p. 130), but as a mathematician (or mathematical physicists) he found that he should focus his attention on the mathematical description of axioms and on the improvement of mathematical tools ([45], p. 127). In his time there was a substantially Aristotelian–Euclidean vision of mathematics, according to which a mathematical theory should be based on indubitable axioms so that the resulting theorems could not be disputable. According to Neumann, a mathematical physical theory was different from a pure mathematical theory because in it some 'axioms' might be hypothetical, and consequently its theorems could not necessarily be true. This possibility made mathematical physical theories very interesting for a mathematician because of their greater potential of invention (hypothetical deductive theories). These considerations by Neumann, shared by Mathieu ([46], pp. 110–111), contributed to the development of the modern concept of a mathematical theory based on arbitrary premises.

The approach by Neumann and Mathieu, not very different indeed from that of Duhem, paralleled a similar process emerging in mathematics, that will continue in the 20th century. After moving away from the Euclidean rigor for a long time, mathematics had returned to it. Therefore, demonstrations of many properties that were previously considered evident were required; indeed, this was in many cases the only way to discover the limits of their validity. The concepts of function, continuity, limit, infinity had revealed the need for a more precise determination; the negative number and the irrational number, which have long since become part of mathematics, have had to be subjected to a more precise examination of their justification. Thus, there was the tendency to give rigorous proofs everywhere ([47], p. 1).

In geometry, the attempts to prove the postulate of parallels (fifth Euclid's postulate) from the other postulates proved to the contrary its independence from them. This suggested to Lobachevsky and Bolyai the idea to replace it with a different assertion on the parallels: they could meet. This gave rise to the so called non Euclidean geometry. This form of geometry furnished results in contrast with many points with the Euclidean geometry, even though the forms of reasoning were similar and suggested the possibility of creating a mathematical theory in which the assumption of the truth of axioms was not required [48,49].

The new theories of mathematics or mathematical physics were hypothetical deductive. It is true that since the ancient Greeks there had been theories based on hypotheses and at large to be classified as hypothetical deductive, but there the hypothesis could not be fully arbitrary; they were suggested in some way by observations or deductions and were considered to assume a value of truth if the theory gave true results. Modern mathematical physics has released this requirement.

Funding: This research received no external funding.

Conflicts of Interest: The author declares no conflict of interest.

Entry Link on the Encyclopedia Platform: https://encyclopedia.pub/12056.

References

1. Allen, D.H. *How Mechanics Shaped the Modern World*; Springer: Dordrecht, The Netherlands, 2014.
2. Painlevé, P. *Les axiomes de la mécanique, examen critique; note sur la propagation de la lumière*; Gauthier-Villars: Paris, France, 1922.
3. Nagel, E. *The Structure of Science*; Harcourt: New York, NY, USA, 1961.
4. Simon, H. The axioms of Newtonian mechanics. *Philos. Mag.* **1947**, *38*, 888–905. [CrossRef]
5. Simon, H. The axiomatization of classical mechanics. *Philos. Sci.* **1954**, *21*, 340–343. [CrossRef]
6. McKinsey, J.; Sugar, A.; Suppes, P. Axiomatic foundation of classical mechanics. *J. Ration. Mech. Anal.* **1953**, *2*, 253–272.
7. Truesdell, C.A. *A First Course in Rational Continuum Mechanics*; Academic Press: New York, NY, USA, 1971.
8. Aristotle. Mechanical problems. In *Aristotle Minor Works*; Hett, W., Ed.; Harward University Press: Cambridge, UK, 1963; pp. 327–411.
9. Aristotle. *Problemi Meccanici*; Bottecchia Dehò, M.E., Ed.; Rubbettino: Catanzaro, Italy, 2000.
10. Clagett, M. *The Science of Mechanics in the Middle Ages*; The University of Wisconsin Press: Madison, WI, USA, 1959.
11. Archimedes. On the equilibrium of planes. In *The Works of Archimedes*; Heath, T., Ed.; Cambridge University Press: Cambridge, UK, 2010; pp. 189–220.
12. Capecchi, D. *History of Virtual Work Laws*; Birchäuser: Milan, Italy, 2012.
13. Copernicus, N. *De Revolutionibus Orbium Coelestium*; Petreius: Nuremberg, Germany, 1543.
14. Galilei, G. Dialogo sopra i due massimi sistemi del mondo. In *Le opere di Galileo Galilei*; Favaro, A., Ed.; Barbera: Florence, KY, USA, 1632; Volume 7.
15. Galilei, G. *Opere di Galileo Galilei Linceo*; Eredi del Dozza: Bologna, Italy, 1890; Volume 4.
16. Galilei, G. Discorsi e dimostrazioni matematiche sopra due nuove scienze. In *Le opere di Galileo Galilei*; Favaro, A., Ed.; Barbera: Florence, Italy, 1638; Volume 8.
17. Koyré, A. *Études Galiléennes*; Hermann: Paris, France, 1966.
18. Dijksterhuis, E. *The Mechanization of the World Picture (De Mechanisering van het Wereldbeeld, 1950)*; Dikshoorn, C., Ed.; Oxford University Press: New York, NY, USA, 1961.
19. Cavalieri, B. *Lo Specchio Ustorio, Overo Trattato Delle Settioni Coniche*; Ferroni: Bologna, Italy, 1632.
20. Evangelista, T. De motu proiectorum. In *Opera Geometrica*; Evangelista, T., Ed.; Masse & de Landis: Florence, Italy, 1644; pp. 154–243.
21. Kuhn, T. *The Structure of Scientific Revolution*; The University of Chicago Press: Chicago, IL, USA, 1962.
22. Capecchi, D. *The Problem of Motion of Bodies*; Springer: Dordrecht, The Netherlands, 2014.
23. Capecchi, D. *The Path to Post-Galilean Epistemology*; Springer: Dordrecht, The Netherlands, 2018.
24. Wallis, J. *Mechanica, Sive De Motu, Tractatus Geometricus*; Godbid: London, UK, 1670.
25. Newton, I. *Philosophiae Naturalis Principia Mathematica*; Jussu Societatis Regiae ac Typis Josephi Streater: London, UK, 1687.
26. Newton, I. *Philosophia Naturalis Principia Mathematica*, 3rd ed.; Innys: London, UK, 1726.
27. Newton, I. *The Principia. Mathematical Principles of Natural Philosophy*; Cohen, I.B., Withman, A., Budenz, J., Eds.; University of California Press: Oakland, CA, USA, 1999.

28. Capecchi, D. *Epistemology and Natural Philosophy in the 18th Century*; Springer: Dordrecht, The Netherlands, 2021.
29. Euler, L. *Mechanica Sive Motus Scientia Analytice Exposita (2 Vols)*; Academiae Scientiarum: Saint Petersburg, FL, USA, 1736.
30. Maltese, G. *Da "F=ma" alle Leggi Cardinali del Moto: Sviluppo della Tradizione Newtoniana nella Meccanica del 700*; Hoepli: Milan, Italy, 2001.
31. Maltese, G. On the changing fortune of the Newtonian tradition in mechanics. In *Two Cultures*; Williams, K., Ed.; Birkhäuser: Basel, Switzerland, 2006; pp. 199–222.
32. Lagrange, J.L. *Méchanique Analitique*; Desaint: Paris, France, 1788.
33. Panza, M. Mathematization of the science of motion and the birth of analytical mechanics: A historiographical note. In *The Application of Mathematics to the Sciences of Nature*; Cerrai, C., Freguglia, P., Pellegrini, P., Eds.; Springer: Dordrecht, The Netherlands, 2002; pp. 253–272.
34. Lagrange, J.L. *Analytical Mechanics*; Boissonnade, A., Vagliente, V.N., Eds.; Springer: Dordrecht, The Netherlands, 1997.
35. Capecchi, D.; Drago, A. On Lagrange's history of mechanics. *Meccanica* **2005**, *40*, 19–33. [CrossRef]
36. Carnot, L. *Essai sur les Machines en Général*; de Defay: Dijon, France, 1786.
37. Dugas, R. *Histoire de la Mécanique*; Griffon: Neuchatel, Switzerland, 1950.
38. Mach, E. *Die Mechanik in Ihrer Entwicklung Historisch-Kritisch Dargestellt*; Brockhaus: Leipzig, Germany, 1883.
39. Poincaré, H. *La Science et L'hypotèse*; Flammarion: Paris, France, 1902.
40. Duhem, P. *Traité D'énergétique ou de Thermodynamique Générale*; Gauthier-Villars: Paris, France, 1911.
41. Hamilton, W. On a general method in dynamics; by which the study of the motions of all free systems of attracting or repelling points is reduced to the search and differentiation of one central relation, or characteristic function. *Philos. Trans. R. Soc. Lond.* **1834**, *124*, 247–308.
42. Jacobi, C. *Gesammelte Werke*; Reimer: Berlin, Germany, 1882–1891; Volume 8.
43. Truesdell, C.A. *Essays in the History of Mechanics*; Springer: New York, NY, USA, 1968.
44. Disalle, R. Carl Gottfried Neumann. *Sci. Context* **1993**, *6*, 345–353. [CrossRef]
45. Schlote, K. The emergence of mathematical physics at the university of Leipzig. In *The Dialectic Relation between Physics and Mathematics in the XIXth Century*; Barbin, E., Pisano, R., Eds.; Springer: Dordrecht, The Netherlands, 2013; pp. 121–137.
46. Barbin, E.; Guitart, R. Mathematical physics in the style of Gabriel Lamé and the treatise of Emile Mathieu. In *The Dialectic Relation between Physics and Mathematics in the XIXth Century*; Barbin, E., Pisano, R., Eds.; Springer: Dordrecht, The Netherlands, 2013; pp. 97–119.
47. Frege, G. *Die Grundlagen der Arithmetik. Eine Logisch-Mathematische Untersuchung ueber den Begriff der Zahl Breslau*; Koebner: Wroclaw, Poland, 1884.
48. Kline, M. *Mathematical Thought from Ancient to Modern Times*; Oxford University Press: Oxford, UK, 1972.
49. Houzel, C. The birth of non-Euclidean geometry. In *1830–1930: A century of Geometry*; Boi, L., Flament, D., Salanskis, J.M., Eds.; Springer: Berlin/Heidelberg, Germany, 1992; pp. 1–21.

encyclopedia

Entry

Metal Nanoparticles as Free-Floating Electrodes

Johann Michael Köhler [1,2,*], Jonas Jakobus Kluitmann [1,2] and Peter Mike Günther [1,2]

1 Institute for Micro- and Nanotechnologies, Technische Universitat Ilmenau, 98693 Ilmenau, Germany; jonas.kluitmann@tu-ilmenau.de (J.J.K.); mike.guenther@tu-ilmenau.de (P.M.G.)
2 Department of Physical Chemistry and Microreaction Technology, Institute for Chemistry and Biotechnology, Technische Universitat Ilmenau, 98693 Ilmenau, Germany
* Correspondence: michael.koehler@tu-ilmenau.de

Definition: Colloidal metal nanoparticles in an electrolyte environment are not only electrically charged but also electrochemically active objects. They have the typical character of metal electrodes with ongoing charge transfer processes on the metal/liquid interface. This picture is valid for the equilibrium state and also during the formation, growth, aggregation or dissolution of nanoparticles. This behavior can be understood in analogy to macroscopic mixed-electrode systems with a free-floating potential, which is determined by the competition between anodic and cathodic partial processes. In contrast to macroscopic electrodes, the small size of nanoparticles is responsible for significant effects of low numbers of elementary charges and for self-polarization effects as they are known from molecular systems, for example. The electrical properties of nanoparticles can be estimated by basic electrochemical equations. Reconsidering these fundamentals, the assembly behavior, the formation of nonspherical assemblies of nanoparticles and the growth and the corrosion behavior of metal nanoparticles, as well as the formation of core/shell particles, branched structures and particle networks, can be understood. The consequences of electrochemical behavior, charging and self-polarization for particle growth, shape formation and particle/particle interaction are discussed.

Keywords: nanoparticles; colloidal solutions; electrical charging; self-polarization; mixed-electrode; particle growth; particle interaction

Citation: Köhler, J.M.; Kluitmann, J.J.; Günther, P.M. Metal Nanoparticles as Free-Floating Electrodes. *Encyclopedia* **2021**, *1*, 551–565. https://doi.org/10.3390/encyclopedia1030046

Academic Editors: Raffaele Barretta, Ramesh Agarwal, Krzysztof Kamil Żur and Giuseppe Ruta

Received: 10 June 2021
Accepted: 28 June 2021
Published: 12 July 2021

1. Introduction

Metal nanoparticles have attracted a lot of scientific interest in recent years. The most important practical motivations come from their interesting electronic and optical properties [1–4], their applicability for nanolabeling [5,6] and sensing [7–10] and their catalytic properties [11,12]. In addition, they are fascinating targets for basic research for understanding the nature of nano-objects and the interaction with biomolecules and living cells [13–16] and for designing new materials, as well as micro- and nanosized tools [17,18].

An important field of nanoparticle generation and handling is liquid-phase synthesis resulting in colloidal solutions of metal nanoparticles [19,20]. The existence of nanoparticles in the form of a thermodynamically stable dispersion in a liquid was firstly explained by Michael Faraday about one and a half centuries ago. Already at this time, the importance of the electrical properties of colloidal particles was recognized.

In addition to the presence of an electrical charge on metal nanoparticles, the exchange of charges and the interaction with ions are important for the generation and behavior of metal nanoparticles. Charge transfer processes can include the release of ions from the metal or the conversion of adsorbed metal cations into metal atoms. These processes, as well as oxidation and reduction reactions of other species, can be regarded as local electrochemical processes [21]. In the following, important examples of such processes will be regarded and discussed from the point of view of the electrode character of colloidal metal nanoparticles.

2. Formation of Colloidal Metal Nanoparticles by Reduction of Solution Precursors

In the past, the general model of LaMer [22] was frequently used for explaining the formation of nanoparticles in liquid media. This model postulates a two-step process, whereby in a first step, single atoms or molecules aggregate to form nuclei, and, in a second step, these nuclei grow by the attachment of further atoms or molecules, as long as the concentration of particle-forming species is above the critical saturation concentration. The concept of LaMer is based on the assumption that the critical concentration threshold for nucleation is significantly above the saturation limit. Thus, nucleation proceeds only at very high concentrations, which means in a short start phase, whereas particle growth can occur over a longer time until the exhaustion of particle-forming species above the saturation limit.

The LaMer model was not developed for the specific conditions for the reactive formation of metal nanoparticles by the reduction of a precursor. For the example of colloidal gold, it was shown by Polte et al. that the formation of spherical metal nanoparticles was caused by an aggregation and growth mechanism, which could be proved by in situ-X-ray scattering measurements [23]. The proposed mechanism of Polte et al. is very conclusive and also gives a convincing explanation for the polycrystalline character of spherical gold nanoparticles [24].

The start of reductive metal nanoparticle formation is a molecular process. The primarily formed products are clusters, and their nature is closer to small molecules than to a classical metallic solid. The simplest description is given by the following reaction equation:

$$M^{n+} + n\,e^- \rightarrow M_0 \quad (1)$$

The electrons are supplied from the oxidation of the reducing agent (RA):

$$RA \rightarrow RA^{m+} + m\,e^- \quad (2)$$

Both processes are coupled and can be written formally as a redox reaction:

$$M^{n+} + n/m\,RA \rightarrow M_0 + n/m\,RA^{m+} \quad (3)$$

Single free metal atoms are usually not stable in a liquid environment. Thus, Equation (1) should be written better as forward and backward processes or as equilibrium:

$$M^{n+} + n\,e^- \rightarrow M_0 \rightarrow M^{n+} + n\,e^- \quad (4)$$

Irreversibility is achieved if a stable cluster is formed consisting of a small number x of metal atoms; for example, z = 4:

$$z\,M^{n+} + n \times z\,e^- \rightarrow (M_z) \quad (5)$$

The above-written equations are a very formal description of the processes during the reaction of a metal precursor species in a polar solvent. It is to be assumed that the formation of clusters does not result in a neutral particle without any ligands. Instead, it has to be expected that the clusters are electrically charged and connected with ligands by analogy with the coordinatively bound or solvated metal ions of the precursor.

Metal ions are solvated or connected with ligands L in coordination compounds or complex ions. The primary formed cluster (Equation (5)) is also connected with ligands X, which can differ from the original ligands of the metal ion complex. The following equations give an example of how the formation of a stable cluster could proceed, whereby the electrons come from direct reduction by interacting molecules of the reducing agent, whereby noncharged ligands L and X and a single positive charge per metal ion are assumed, for simplification:

$$2[ML_n]^+ + e^- \rightarrow [M_2L_{(2n-a)}]^+ + a\,L \text{ (reduction in solution)} \quad (6)$$

$$[M_2L_{(2n\text{-}a)}] + b \times \rightarrow [M_2L_{(2n\text{-}a\text{-}b)}X_b] + b\ L \text{ (partial ligand exchange)} \quad (7)$$

or

$$[M_2L_{(2n\text{-}a)}] + (2n\text{-}a) \times \rightarrow [M_2X_{(2n\text{-}a)}]^+ + (2n-a)\ L \text{ (complete ligand exchange)} \quad (8)$$

The following steps are written for the example without ligand (L by X) exchange:

$$[M_2L_{(2n\text{-}a)}]^+ + [ML_n]^+ \rightarrow [M_3L_{(3n\text{-}a\text{-}b)}]^{2+} + b\ L \text{ (metal ion adsorption)} \quad (9)$$

$$[M_3L_{(3n\text{-}a\text{-}b)}]^{2+} + e^- \rightarrow [M_3L_{(3n\text{-}a\text{-}b)}]^+ \text{ (reduction step)} \quad (10)$$

$$[M_3L_{(2n\text{-}a\text{-}b)}]^+ + [ML_n]^+ \rightarrow [M_4L_{(4n\text{-}a\text{-}b\text{-}c)}]^{2+} + c\ L \text{ (metal ion adsorption)} \quad (11)$$

$$[M_4L_{(4n\text{-}a\text{-}b\text{-}c)}]^{2+} + e^- \rightarrow [M_4L_{(4n\text{-}a\text{-}b\text{-}c)}]^+ \text{ (reduction step)} \quad (12)$$

and so on.

The equations above illustrate the alternation of steps of metal ion or complex, respectively, adsorption and reduction. It is not necessary to assume the formation of metal(0) atoms that have to aggregate spontaneously to form a nucleus that can subsequently grow. It is much more probable that there is a direct way from the formation of a coordinated cluster of very few atoms to a continuous growth process. This growth process is always composed of the reduction of adsorbed metal ions into metal(0) and the electron supplying oxidation of the reducing agent. From this point of view, there is no classical nucleation process, as postulated by the LaMer model. Instead of this, the reductive formation of metal nanoparticles can be described by the coupling of reduction and oxidation from the very beginning of the first metal–metal interaction.

After completion of the particle growth, the particles in a stable colloidal solution are in a thermodynamic equilibrium state. This means that cathodic and anodic elementary processes are further running but with equal rates in both directions. As a result, the surfaces of metal nanoparticles experience a fluctuation in the adsorption and release of metal ions, as well as in the oxidation and reduction of residual active redox species in solution. Each elongation of the thermodynamic equilibrium state—for example by changing temperature, pH, metal precursor, ligand or reducing agent concentrations—induces shifts in the intensity of the electrochemical processes on the particle surfaces and, therefore, changes in the size or shape of particles.

3. Forming Colloidal Metal Nanoparticles as Mixed-Electrode Objects

The reaction system of reductive metal nanoparticle formation consists of two different object types: the ion-containing solution with the character of an electrolyte and the metallic clusters and nanoparticles. Bulk metals and metal films and metal clusters are marked by high internal charge carrier mobility. This is the essential feature of the metallic state. Together with the electrolyte, an electrode is formed.

Each electrode is marked by its electrode potential. This potential can be measured, in principle, by the voltage between the electrode and a reference electrode. However, such a direct measurement is impossible in the case of nanoparticles.

The electrochemical potential of a metal particle during its formation results at least from the metal deposition and the oxidation of the reducing agent (RA). The metal reduction represents the cathodic partial process as formulated by Equation (1). The anodic partial process can be written as:

$$RA \rightarrow RA^+ + e^- \quad (13)$$

Due to the, at least, two electrode processes, the particle is a mixed-electrode object (Figure 1). The absence of any outside current source means that this potential is only controlled by the ongoing electrode processes. Metal nanoparticles have a free-floating potential.

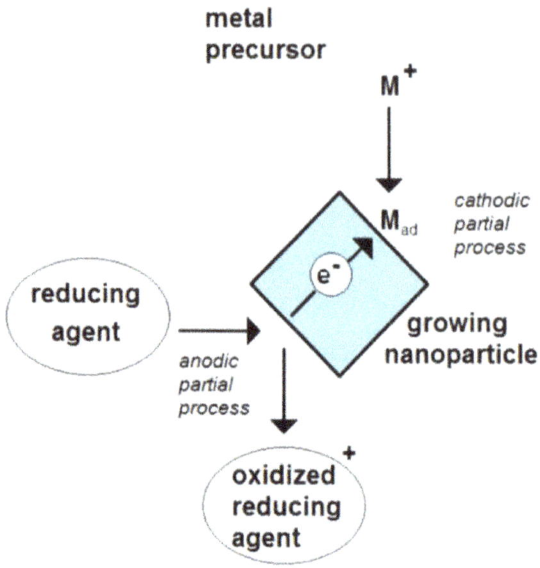

Figure 1. Scheme of the formation of an electrochemical mixed potential of a growing nanoparticle by the coupling of a cathodic and an anodic partial process.

The potential of clusters and nanoparticles of z atoms M_z is modified, in addition, by the adsorption of ions, ligands and other reaction partners, for example by following ligand exchange reaction:

$$(M_zL_n) + X^- \rightarrow (M_zL_n\text{-}1X)^- + L \tag{14}$$

The small size of nanoparticles causes a high sensitivity of the particle potential against the exchange of single charges. The clusters and the growing nanoparticles behave as electron confinements with a fluctuating potential. Each adsorption or desorption of an ion and each uptake or release of an electron changes the electrochemical potential. An increase in potential enhances the probability of oxidation of an adsorbed or colliding molecule of the reducing agent. A decrease in potential enhances the probability of metal deposition.

The order of magnitude of potential change by the transfer of one elementary charge qe can be estimated by regarding the particle surface as an electrical capacitor with the capacitance C, an electrode distance in the order of magnitude of the electrochemical double-layer thickness d and the surface area A:

$$C = \varepsilon_0 \times \varepsilon_r \times A/d \tag{15}$$

The stored charge on such a capacitor Q depends on the capacitance and the voltage:

$$Q = C \times U \tag{16}$$

From these equations, the potential U can be estimated:

$$U = Q \times d/(A \times \varepsilon_0 \times \varepsilon_r) \tag{17}$$

For the nearly atomic distances, ε_r can be approximated for the vacuum (=1), and the expected double-layer thickness is in molecular dimensions (about 0.5 nm). Then, the change of nanoparticle potential U_{NP} by one elementary charge q_e can be approximated by:

$$U_{NP} \approx q_e \times 0.5\ \text{nm}/(A \times \varepsilon_0) \tag{18}$$

Significant shifts in electrochemical potential should be expected for small nanoparticles, which means in the early stage of growth. In this phase, the reacting sites of the particle surface and their stochastic interactions with species from the solution have a large effect on the electrochemical state of a particle (Figure 2). It has to be kept in mind that changes of some tens of millivolts typically have a significant effect on the intensity of electrochemical processes, and changes of several hundred millivolts are related to strong changes in the electrochemical process intensities, often connected with drastic changes in the general chemical behavior. The estimation shows that such strong effects caused by fluctuations of single elementary charges must be expected for small particles with sizes below about 2 to 4 nm.

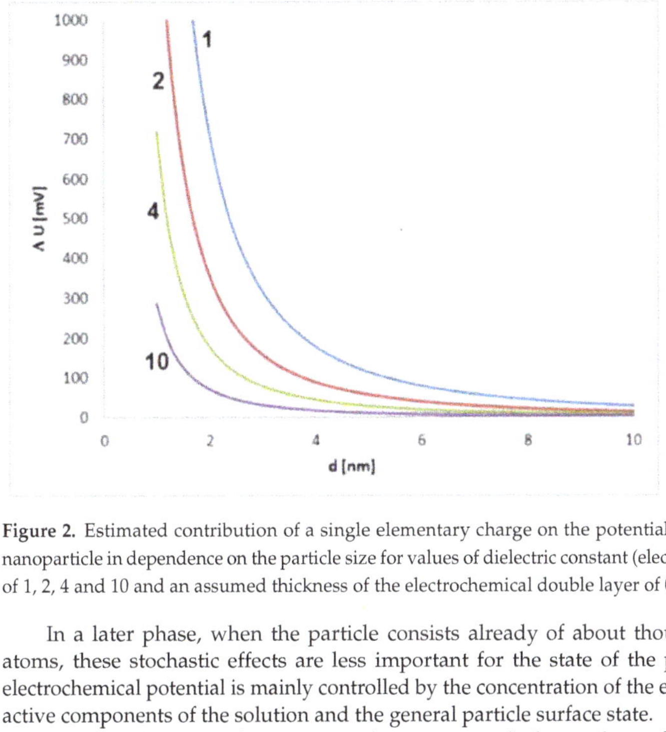

Figure 2. Estimated contribution of a single elementary charge on the potential of a single metal nanoparticle in dependence on the particle size for values of dielectric constant (electrical permittivity) of 1, 2, 4 and 10 and an assumed thickness of the electrochemical double layer of 0.5 nm.

In a later phase, when the particle consists already of about thousands or more atoms, these stochastic effects are less important for the state of the particle, and its electrochemical potential is mainly controlled by the concentration of the electrochemically active components of the solution and the general particle surface state.

The particle potential remains nearly constant in the latest phase of particle growth if the particle is larger and the changes of concentrations of reacting components in the solution become negligible. In the case of constant electrochemical potential, the sum of all partial currents of electrochemical processes is zero. That means that the currents of all cathodic partial processes I are completely compensated by the electrochemical currents of anodic partial processes I^+:

$$|I^+| = |I^-| \quad (19)$$

The electrochemical potential of a growing metal nanoparticle increases when the intensity of cathodic partial process (metal deposition) exceeds the intensity of the anodic partial process:

$$|I^+| < |I^-| \rightarrow \text{increasing particle potential} \quad (20)$$

The electrochemical potential of a growing metal nanoparticle decreases when the intensity of the anodic partial process (oxidation of reducing agent) exceeds the intensity of the cathodic partial process:

$$|I^+| > |I^-| \rightarrow \text{decreasing particle potential} \quad (21)$$

4. Self-Polarization Effects of Nonspherical Metal Nanoparticles

Metal nanoparticles are marked by a high mobility of charges. This matches the electron gas model of metallic solids and is valid for charged nanoparticles (or ion-like clusters), too. The electrical excess charges of metal nanoparticles are fluctuating, but the fluctuation of single elementary charges is coupled with the others due to electrostatic repulsion. Therefore, the charges are concentrated at the surface of the particle. Spherical particles are marked by a regular distribution of the excess charges at their surface due to the tendency to maximize the distances between all single charges.

The symmetry in the charge distribution gets lost if the shape symmetry of a sphere is broken. The distribution of charges then depends on the particle shape and two or more charge centers of gravity appear. Some examples of such geometry-dependent charge distributions are shown in Figure 3.

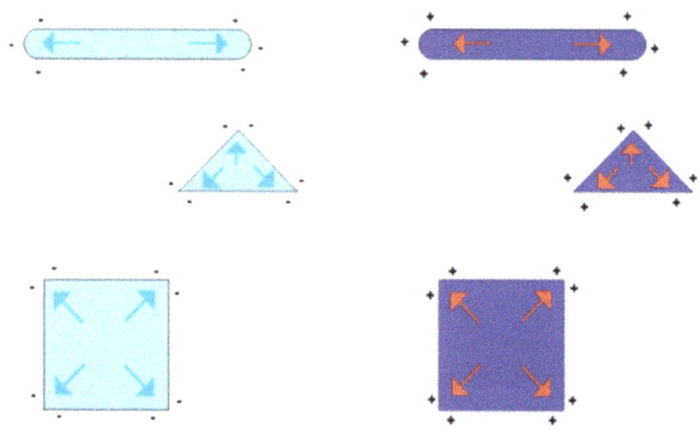

Figure 3. Formation of charge centers of gravity in the case of electrically charged nonspherical metal nanoparticles (blue symbolizing negative excess charge, red symbolizing positive excess charge).

This effect of the self-polarization of charged nanoparticles is in strong contrast to the charge distribution of macroscopic metallic objects. Due to the smaller dimensions of the nanoparticles, its charge distribution is closer to the conditions in molecules with delocalized electrons than to the electrical conditions in a larger metal body, in which the charge distribution is dominated by thermal fluctuations for temperatures near room temperature. This effect can easily be illustrated by the comparison between the change of thermal energy ΔT and electrostatic energy for single elementary charges in dependence on charge distance r:

$$[1/(4\varepsilon\pi)] \times q_e^2/r = k_B \times \Delta T \quad (22)$$

$$\Delta T = [1/(4\varepsilon\pi)] \times q_e^2/(r \times k_B) \quad (23)$$

The strength of the effect depends on the relative permittivity ε_r. This is a dimensionless material-specific parameter that can be assumed to be in the range between about 5 and 100. For this range, significant self-polarization effects that are not leveled by thermal fluctuations should be expected for nonspherical particles between about a few tens and few hundreds of nanometers size depending on the material (Figure 4).

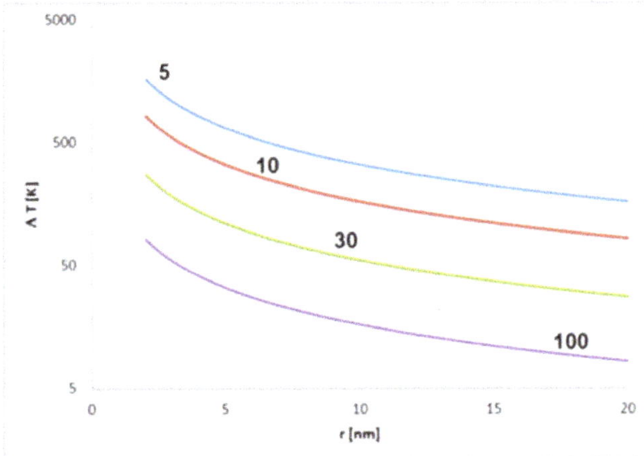

Figure 4. Estimated temperature differences T corresponding to the thermal energy equal to the electrostatic energy of a pair of elementary charges in distance r (for permittivity of 5, 10, 30 and 100).

The self-polarization effect has important consequences for the growth and for the corrosion behavior of metal nanoparticles. The simplest case is the self-polarization of a nanorod, and this case can serve as an example to explain the principle effects. Dictated by the repulsion of equal charges, the charges are concentrated on the poles of the nanorod.

In the case of a positive excess charge, the potential is higher on the poles and lower in the center of the particle (Figure 5a). This means that in the case of corrosion, the poles are corroded faster than the center due to the higher intensity of the potential-dependent anodic partial on the pole. In the case of deposition (particle growth), a positive excess charge lowers the deposition on the poles and enhances the deposition in the center region. Thus, it is expected that at a positive excess charge, corrosion and deposition result in the lowering of the aspect ratio.

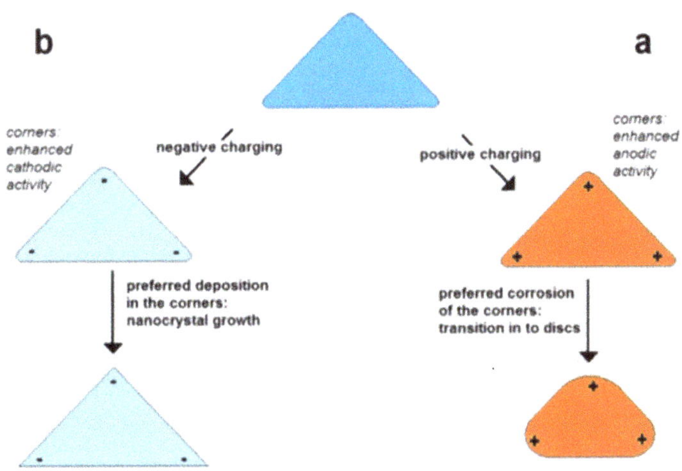

Figure 5. Self-polarization effect: effect of excess charge on the change of particle shape in dependence on charge sign shown for the example of a flat triangular metal nanoprism (schematically); (**a**) enhanced positive charge in the corners of triangular particle, (**b**) enhanced negative charge in the corners of particles.

In the case of a negative excess charge, the potential at the poles is decreased stronger than in the center of a nanorod (Figure 5b). This means that the deposition rate is enhanced on the poles and lowered in the center. This promotes the growth in the axial direction or the forming of "dog bone"-like geometries. In the case of corrosion, the material loss is lowered on the poles but enhanced in the center leading to a thinning of the central region. Both processes result in the tendency of an increase in the aspect ratio or the bone-shaping of nanorods.

5. Factors in Shape Control of Growing Metal Nanoparticles

Some general mechanisms can be formulated for the evolution of metal nanoparticle shape during particle growth [25]. It has to be kept in mind that surface energy plays a crucial role in metal deposition and etching in addition to the electrochemical potential [26]. Therefore, the lattice structure of the growing or corroding nanorods and the crystallographic planes at the particle surface are further important factors for the resulting geometries. However, considering the above-mentioned size-dependent effects, it can be concluded that a stronger influence of the lattice structure on the geometry evolution in the case of larger particles can be expected, whereas small particles are probably strongly dependent on the electrochemical potentials and self-polarization effects. It is well known that halogen ions determine the character of metal nanoparticle growth by the formation of adsorbates [27,28]. In addition, surfactants and their specific interaction with crystal planes are important for the shape development of metal nanoparticles [29,30]. In the case of the specific adsorption of shape-steering additives on preferred lattice planes and their inhibition of local metal deposition during the nanoparticle formation, it can be assumed that electrochemical effects such as self-polarization modulate the anisotropic particle growth and shaping.

The formation of regular crystals can be expected if the adsorbing metal ions or the formed metal adatoms have a high surface mobility. In this case, they tend to move to a thermodynamically preferred place. The formation of regular crystal lattices is then controlled by the minimization of lattice energy.

In the case of lower surface mobility, there is a mixture between a random adsorption of metal ions and adatoms and energy-controlled translocation taking place. This can result in more or less regular crystal structures. A typical result is the formation of branched crystals and dendritic nanocrystals with local crystallinity and a combination of different lattice orientations in different domains of the nanoparticle. In addition to the random adsorption of metal ions, an accidental local binding of ligands or other molecular effectors can influence the growth behavior of nanoparticles and also the formation of branched or polycrystalline structures.

Random effects in the attachment of adsorbing metal ions on the nanoparticle surface and the tendency of binding on thermodynamically excellent sites are superposed by the electrostatic self-polarization effect. On the one hand, local electrical fields direct approaching ionic species into preferred regions of the growing particle. On the other hand, the distribution of electrical charges on the particle surface modulates the local surface energy. The formation of different aspect ratios or the appearance of dog-bone-like nanorod structures is a typical consequence of this effect.

In addition to the shape control, the size control in the formation of metal nanoparticles is also very important because the electronic and optical properties of metal nanoparticles are size dependent. For the growth of larger nanoparticles from seed particles, the ratio of seed concentration to the concentration of precursor is generally the key parameter for steering the particle size. In the case of the complete consumption of the precursor during a homogeneous nanoparticle growth, this ratio determines the particle size directly. High seed concentrations result in small final product particles, and low seed concentrations cause the formation of large particles.

6. Electrically Controlled Assembly of Metal Nanoparticles

The use of attractive forces of oppositely charged nanoparticles is a simple strategy for their assembly. However, the mixing of particles of opposite charges involves the possibility of fast coagulation. A particle of one type can act as "particle glue" for connecting the particles of another type. Therefore, it is important to work with suitable particle concentrations (Figure 6).

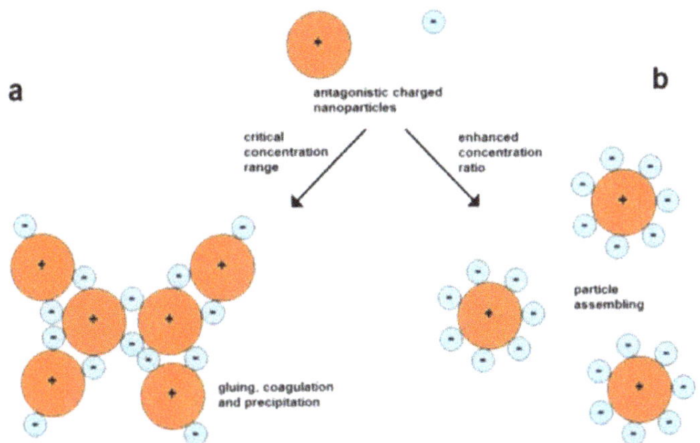

Figure 6. Coagulation and assembly of oppositely charged nanoparticles; (**a**) high probability of coagulation at lower concentration ratios, (**b**) forming of dispersed assemblies at higher concentration ratios between small negative and larger positive particles.

Assembly can also take place in the case of equally charged particles. Therefore, forces are needed for overcoming the barrier of electrostatic repulsion. The electrostatic repulsion of charged nanoparticles is well known for the stabilization of the colloidal state. The domination of kinetic energy and binding over repulsion leads to the aggregation of particles, which can finally result in complete coagulation and precipitation. In other cases, moderate lowering of the particle velocity while maintaining the particle charges can cause the limited assembly and restabilization of colloids. Such a mechanism can cause the formation of polycrystalline metal nanoparticles by an in situ assembly process during Au nanoparticle formation, as discovered by Polte et al. [24], and is also a probable reason for the formation of star-like Au/Ag nanoparticles [31].

Discharging of colloidal particles can be induced by:

- Raising the general ion concentrations, which means the ionic strength ("salt-out effect");
- Targeted addition of antagonistic ions, which means metal cations with high surface affinity in the case of negatively charged particles;
- Targeted addition of surface-affine anions or electron-donor ligands in the case of positively charged particles;
- Lowering the electrochemical potential of particles in the case of positively charged particles by enhancing the concentration of reducing agents;
- Raising the electrochemical potential in the case of negatively charged particles by the addition of oxidizing agents.

In all cases, a restabilization mechanism has to take place in order to avoid a continuation of aggregation, the uncontrolled formation of large aggregates and precipitation. In principle, this restabilization can be induced by controlling the electrochemical particle potential, too. Therefore, a "backward strategy" or a "forward strategy" can be applied. In the "backward strategy", the addition of complementary effectors invert the primary-induced potential shift for the stabilization of colloids after the temporary destabilization. The idea behind the "forward strategy" is to continue the primary-induced potential shift in the same direction to overcome the zero-potential point quickly and to achieve the colloidal restabilization by inversion of the charge sign. Thus, the addition of adsorbing anions or the stepwise enhancement of the reducing agent concentration can lower the positive potential of colloidal metal nanoparticles in a first step but can lead to restabilized colloids by negatively charged particles. On the other hand, a colloid of negatively charged metal nanoparticles can be destabilized by the addition of oxidizing agents or metal cations but restabilized by their further addition in the positive potential range. It has to be mentioned that the addition of oxidizing agents has to be performed carefully in order to avoid the corrosion or complete dissolution of the colloidal particles. It seems that reductants can also cause a disruption of particles, especially strong ones such as $NaBH_4$, by inducing a Rayleigh instability [32].

Self-polarization or the induced polarization of nonspherical particles causes different conditions for particle–particle interactions dependent on position. This can decide the formation of compact, linear or branched assemblies. Such different behavior was observed on polarizable polymer nanoparticles carrying polyionic macromolecules [33], and it can be, obviously, the reason for the formation of linear branched and network structures in the in situ assembly of metal nanoparticles.

7. Secondary Metal Deposition

An electrostatic model was proposed for the formation of networks of Au/Ag nanoparticles during the deposition of silver on preformed spherical gold nanoparticles [34]. It describes the preferred bonding of spherical particles on the poles of nanorods due to the induced polarization by the electrostatic interaction of the charged metal nanoparticles. This preferred bonding on the poles leads to the successive prolongation of primary formed nonspherical aggregates. The intrinsic reason for this linear aggregation is the symmetry break by binding of two spherical nanoparticles, which means a transition from a spherical geometry into a body with an enhanced aspect ratio. This initial process is responsible for the symmetry break in the distribution of surface charges. High repulsion forces result in a strict linear growth. Lower charging means a certain probability of occasional binding events on intermediate positions, working as branching points.

In general, higher electrical charges lead to more stable colloids and a separate growth of single nanoparticles, whereas lower electrical charges enhance the tendency of aggregation. The aggregation can end with restabilization after a few binding events after the formation of larger aggregates or networks or lead to complete coagulation and precipitation [35]. These effects can well be observed in the deposition of silver on preformed gold nanoseeds. In the case of a strong dominance of positively charged silver cations, the colloid remains stable at a higher potential. In the case of a reduced charge, particles can join each other. Due to the mobility of electrical charges, self-polarization takes place, and a preferred axial growth results in rod- and astragal-like structures (Figure 7). Therefore, not only are compact aggregates observed but also single particles. Further in situ aggregation during silver deposition can lead to branched astragal-like structures and to the formation of nanoparticle networks (examples in Figure 8).

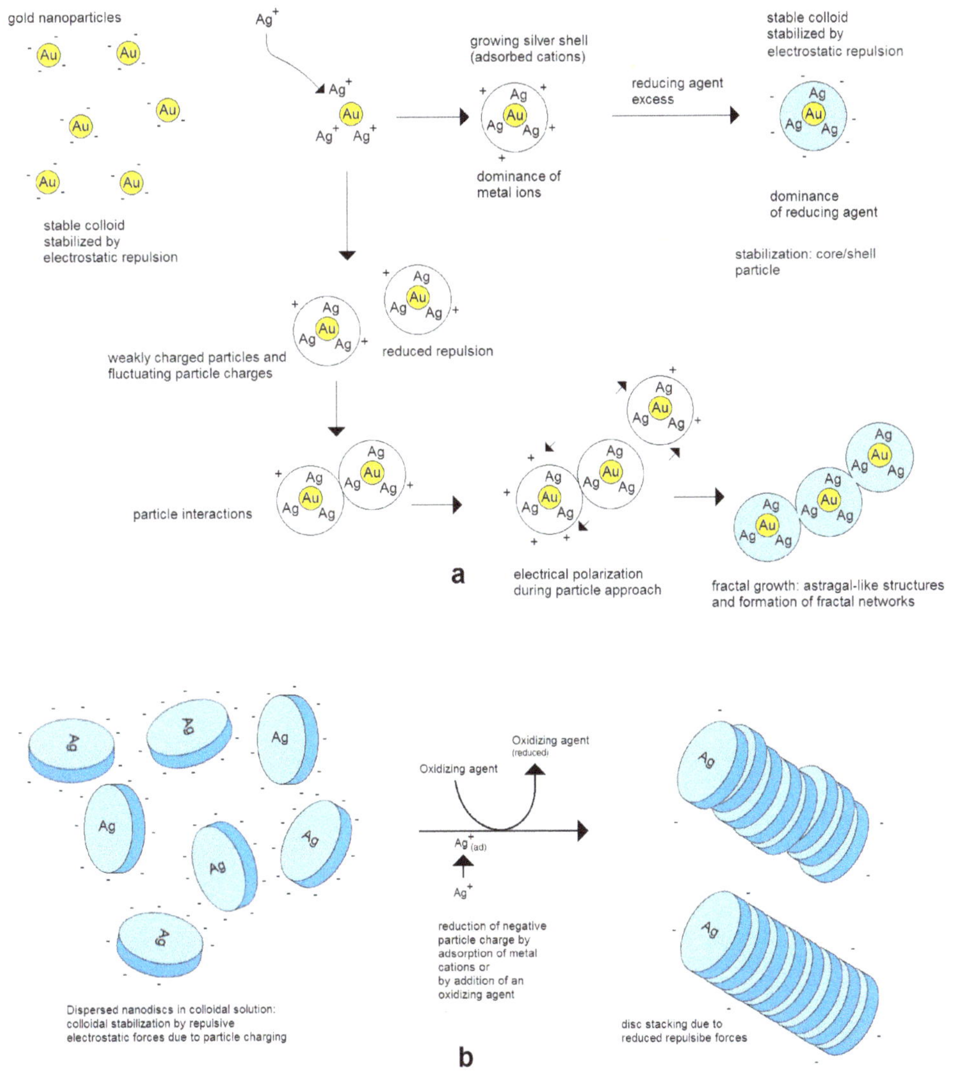

Figure 7. Assembly behavior of charged metal nanoparticles: (**a**) spherical nanoparticles; (**b**) nonspherical metal nanoparticles, for example nanodiscs.

The effect of electrical control and self-polarization is well reflected by the formation of nanoframe structures by galvanic replacement reactions [36,37]. The deposition of gold on flat silver nanotriangles starts on the edges, at the preferred cathodically active sites, whereas the corrosion of silver starts in the center of the particles, at the preferred anodically active area. As a result, a central hole is formed, and the original prismatic particle shape is transformed into a frame structure (Figure 9).

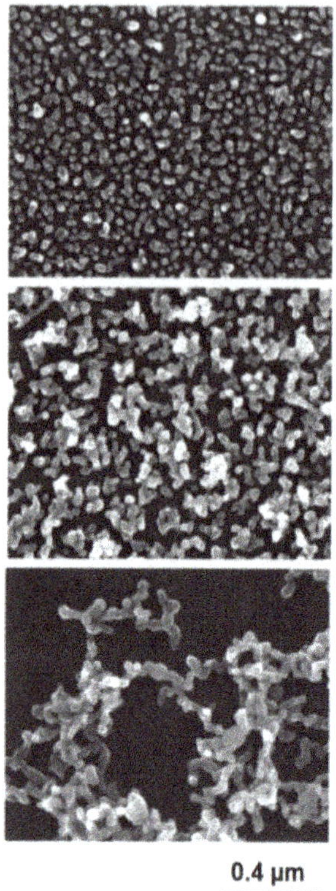

Figure 8. In situ formation of aggregate particles during the chemical deposition of silver nanoparticles on gold seeds: (**a**) small aggregate particles; (**b**) enhanced aggregate size; (**c**) network formation.

The interplay between lattice-directed metal deposition and accidental deposition can be nicely observed in the case of the formation of Au/Pt nanorods [38]. The seed-like gold nanorods are positively charged due to adsorbed cetyltrimethylammonium cations (CTAB). This charge and, therefore, the electrochemical potential of the nanoparticles, are lowered by the addition of the reducing agent, which can be oxidized on the particle surface. If a slight reduction of potential only occurs, the nucleation of platinum can only proceed on crystallographic exposed points on the gold surface. As a result, the structure of the deposited platinum nanocrystallites follows the borderlines between the crystallographic surface planes of gold nanorods. At low deposition rates, the nucleation of platinum starts on these borderlines between the crystallographic planes of the single-crystalline gold nanorods. The decoration of these crystallographic lines by the platinum crystallites is clearly visible in the SEM images (Figure 10a). At moderately enhanced platinum deposition activities (enhanced platinum precursor concentration), the nucleation of platinum also takes place on the crystallographic planes; the decoration effect of the borderlines between the planes, however, remains visible (Figure 10b). In this case, a more stochastic distribution of platinum crystallites in the shell is formed. At a higher precursor concentration, the nucleation probability generally increases, the dominance of the borderline-directed nucleation disappears, and a comparatively dense shell of platinum crystallites is formed around the gold core (Figure 10c).

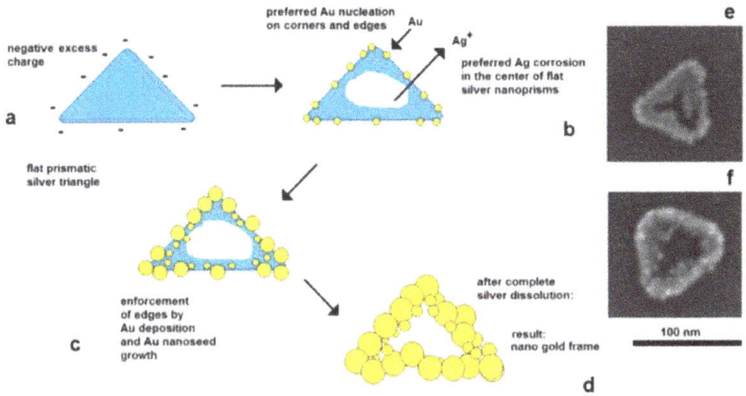

Figure 9. Electrically controlled formation of metal frame particles by the substitution of a less-noble metal by a noble metal. Deposition of gold on flat silver nanotriangles: (**a**) negatively charged silver nanotriangles in the colloidal state; (**b**) starting gold deposition on the particle edges (lowest potential) and starting silver corrosion in the center of the flat prism (highest potential); (**c**) enforcement of gold in the edge regions of particles; (**d**) formation of a gold frame by the complete dissolution of silver (schematically); (**e**) silver triangle with a hole formation in the particle center; (**f**) advanced silver dissolution and edge-directed gold deposition (SEM images).

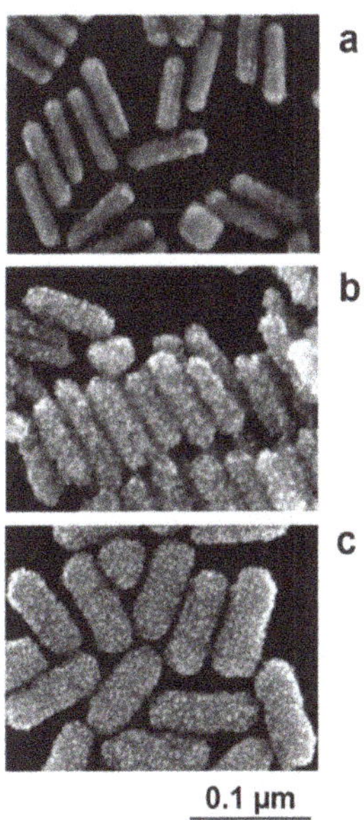

Figure 10. Interplay between the lattice-controlled and accidental nucleation of platinum on gold nanorods: (**a**) lowest platinum precursor concentration: decoration of borderlines between crystallographic planes; (**b**) mediate precursor concentration: additional nucleation on the crystallographic planes between the excellent borderlines; (**c**) high platinum precursor concentration: densified platinum deposition as a shell around the gold core.

8. Conclusions

Metal nanoparticles in colloidal solutions are electrically charged and electrochemically active objects. The electrochemical point of view allows for an understanding of many phenomena related to the growth, formation and interaction of metal nanoparticles in colloidal solution. It illustrates the analogy between macroscopic electrochemical open-circuit and mixed-electrode systems and metal nanoparticles in the colloidal state. However, this point of view also reflects the specificities of electrochemical processes that are related to the small, meaning nearly molecular, size of the electrode-like nano-objects, the role of small numbers of elementary charges, self-polarization effects of nonspherical nanoparticles and the interplay between lattice-dominated, potential-dominated and accidental elementary events. The electrochemical point of view is well suited for understanding the stabilization, destabilization and restabilization of spherical nanoparticles in colloidal solution, as well as the growth characteristics of nonspherical metal nanoparticles, the formation of binary nanoparticles and particle/particle interaction. The combination of electrochemical activity and self-polarization of charged metallic nano-objects also gives a conclusive explanation of the aggregation behavior of nanoparticles in colloidal solution and for the formation of linear, branched and network-like structures.

Funding: The financial support of the experimental work on binary metal nanoparticles by the Deutsche Forschungsgemeinschaft (DFG) (Kz. Ko1403/45-1) is gratefully acknowledged.

Conflicts of Interest: The authors declare no conflict of interests.

Entry Link on the Encyclopedia Platform: https://encyclopedia.pub/13014.

References

1. Homberger, M.; Simon, U. On the application potential of gold nanoparticles in nanoelectronics and biomedicine. *Philos. Trans. R. Soc. A-Math. Phys. Eng. Sci.* **2010**, *368*, 1405–1453. [CrossRef]
2. Link, S.; El-Sayed, M.A. Spectral properties and relaxation dynamics of surface plasmon electronic oscillations in gold and silver nanodots and nanorods. *J. Phys. Chem. B* **1999**, *103*, 8410–8426. [CrossRef]
3. Senthil Kumar, P.; Pastoriza-Santos, I.; Rodriguez-Gonzalez, B.; Garcia de Abajo, F.J.; Liz-Marzan, L.M. High-yield synthesis and optical response of gold nanostars. *Nanotechnology* **2008**, *19*, 015606. [CrossRef]
4. Amendola, V.; Pilot, R.; Frasconi, M.; Maragò, O.M.; Iatì, M.A. Surface plasmon resonance in gold nanoparticles: A review. *J. Phys.* **2017**, *29*, 203002. [CrossRef] [PubMed]
5. Reichert, J.; Csaki, A.; Kohler, J.M.; Fritzsche, W. Chip-based optical detection of DNA hybridization by means of nanobead labeling. *Anal. Chem.* **2000**, *72*, 6025–6029. [CrossRef]
6. Csaki, A.; Kaplanek, P.; Moller, R.; Fritzsche, W. The optical detection of individual DNA-conjugated gold nanoparticle labels after metal enhancement. *Nanotechnology* **2003**, *14*, 1262–1268. [CrossRef] [PubMed]
7. Penn, S.G.; He, L.; Natan, M.J. Nanoparticles for bioanalysis. *Curr. Opin. Chem. Biol.* **2003**, *7*, 609–615. [CrossRef] [PubMed]
8. Khoury, C.G.; Vo-Dinh, T. Gold Nanostars For Surface-Enhanced Raman Scattering: Synthesis, Characterization and Optimization. *J. Phys. Chem. C* **2008**, *112*, 18849–18859. [CrossRef] [PubMed]
9. Köhler, J.M.; März, A.; Popp, J.; Knauer, A.; Kraus, I.; Faerber, J.; Serra, C. Polyacrylamide/silver composite particles produced via microfluidic photopolymerization for single particle-based SERS microsensorics. *Anal. Chem.* **2013**, *85*, 313–318. [CrossRef]
10. Gao, Z.; Ye, H.; Tang, D.; Tao, J.; Habibi, S.; Minerick, A.; Tang, D.; Xia, X. Platinum-decorated gold nanoparticles with dual functionalities for ultrsensitive colorimetric in vitro diagnostics. *Nano Lett.* **2017**, *17*, 5572–5579. [CrossRef]
11. Liu, L.; Corma, A. Metal Catalysts for Heterogeneous Catalysis: From Single Atoms to Nanoclusters and Nanoparticles. *Chem. Rev.* **2018**, *118*, 4981–5079. [CrossRef]
12. Kawahara, K.; Inoue-Kahino, N.; Namie, K.; Kato, Y.; Tomo, T.; Shibata, Y.; Kashino, Y.; Noguchi, T. A gold nanoparticle conjugate with photosystem I and photosystem II for development of a biohybrid water-splitting photocatalyst. *Biomed. Spectrosc. Imaging* **2020**, *9*, 73–81. [CrossRef]
13. Haes, A.J.; Zou, S.L.; Schatz, G.C.; van Duyne, R.P. Nanoscale optical biosensor: Short range distance dependence of the localized surface plasmon resonance of noble metal nanoparticles. *J. Phys. Chem. B* **2004**, *108*, 6961–6968. [CrossRef]
14. Kong, F.-Y.; Zhang, J.-W.; Li, R.-F. Unique role of gold nanoparticles in drug delivery, targeting and imaging applications. *Molecules* **2017**, *22*, 1445. [CrossRef]
15. Alim, S.; Vejavan, J.; Yusoff, M.; Kafi, A.K.M. Recent use of carbon nanotubes & gold nanoparticles in electrochemistry with applications in biosensing: A review. *Biosens. Bioelectron.* **2018**, *121*, 125–136. [PubMed]
16. Liu, N.; Liedl, T. DNA-assembled advanced plasmonic architectures. *Chem. Rev.* **2018**, *118*, 3032–3053. [CrossRef]
17. Singh, R.; Belgamwar, R.; Dhiman, M.; Polshettiwar, V. Dendritic fibrous nano-silica-supported gold nanoparticles as an artificial enzyme. *J. Mater. Chem. B* **2018**, *6*, 1600–1604. [CrossRef] [PubMed]
18. Li, G.; Zhao, S.; Zhang, Y.; Tang, Z. Metal–Organic Frameworks Encapsulating Active Nanoparticles as Emerging Compo-sites for Catalysis: Recent Progress and Perspectives. *Adv. Mater.* **2018**. [CrossRef]
19. Capek, I. Noble metal nanoparticles: Preparation, composite nanostructures, biodecoration and collective properties. *Nanostructure Sci. Technol.* **2017**, *30*, 211–316.
20. Lee, S.H.; Jun, B.-H. Silver nanoparticles: Synthesis and application for nanomedicine. *Int. J. Mol. Sci.* **2019**, *20*, 865. [CrossRef]

21. Koehler, J.M.; Visaveliya, N.; Knauer, A. Controlling formation and assembling of nanoparticles by control of electrical charging, polarization, and electrochemical potential. *Nanotechnol. Rev.* **2014**, *3*, 553–568. [CrossRef]
22. Whitehead, C.B.; Saim, O.; Finke, R.G. LaMer's 1950 model of particle formation: A review and critical analysis of its classical nucleation and fluctuation theory basis, of competing models and mechanisms for phase-changes and particle formation, and then of its application to silver halide, semiconductor, metal, and metal-oxide nanoparticles. *Mater. Adv.* **2021**, *2*, 186–235.
23. Polte, J.; Ahner, T.T.; Delissen, F.; Sokolov, S.; Emmerling, F.; Thuenemann, A.F.; Kraehnert, R. Mechanism of Gold Nanoparticle Formation in the Classical Citrate Synthesis Method Derived from Coupled In Situ XANES and SAXS Evaluation. *J. Am. Chem. Soc.* **2010**, *132*, 1296–1301. [CrossRef]
24. Polte, J.; Erler, R.; Thuenemann, A.F.; Sokolov, S.; Ahner, T.T.; Rademann, K.; Emmerling, F.; Kraehnert, R. Nucleation and growth of gold nanoparticles studied via in situ Small Angle X-ray Scattering at millisecond time resolution. *ACS Nano* **2010**, *4*, 1076–1082. [CrossRef]
25. Sohn, K.; Kim, F.; Pradel, K.C.; Peng, Y.; Zhou, F.; Huang, J. Construction of evolutionary tree for morphological engineering of nanoparticles. *ACS Nano* **2009**, *3*, 2191–2198. [CrossRef]
26. Xia, Y.; Xia, X.; Peng, H.-C. Shape-controlled synthesis of colloidal metal nanocrystals: Thermodynamic versus kinetic products. *JACS* **2015**, *137*, 7947–7966. [CrossRef]
27. Chen, Z.; Balankura, T.; Fichthorn, K.A.; Rioux, R.M. Revisiting the polyol synthesis of silver nanostructures. Role of chloride in nanocube formation. *ACS Nano* **2019**, *13*, 1849–1860. [CrossRef] [PubMed]
28. Khanal, B.P.; Zubarev, E.R. Chemical transformation of nanorods to nanowires: Reversible growth and dissolution of anisotropic gold nanostructures. *ACS Nano* **2019**, *13*, 2370–2378. [CrossRef] [PubMed]
29. Khlebtsov, B.N.; Khanadeev, V.A.; Ye, J.; Sukhourukov, G.B.; Khlebtsov, N.G. Overgrowth of gold nanorods by using a binary surfactant mixture. *Langmuir* **2014**, *30*, 1696–1703. [CrossRef] [PubMed]
30. Sau, T.K.; Murphy, C.J. Seeded high yield synthesis of short Au nanorods in aqueous solution. *Langmuir* **2004**, *20*, 6414–6420. [CrossRef] [PubMed]
31. Köhler, J.M.; Romanus, H.; Hübner, U.; Wagner, J. Formation of star-like and core-shell AuAg nanoparticles during two- and three-step preparation in batch and in microfluidic systems. *J. Nanomater.* **2007**, 98134. [CrossRef]
32. Novo, C.; Mulvaney, P. Charge-induced Rayleigh instabilities in small gold nanorods. *Nano Lett.* **2007**, *7*, 520–524. [CrossRef] [PubMed]
33. Visaveliya, N.; Knauer, A.; Köhler, J.M. Application of polyionic macromolecules in micro flow syntheses of nanoparticles. *Macromol. Chem. Phys.* **2017**, *218*, 1600371. [CrossRef]
34. Koehler, J.M.; Kluitmann, J.; Knauer, A. Metal Nano Networks by Potential-Controlled In Situ Assembling of Gold/Silver Nanoparticles. *ChemistryOpen* **2019**, *8*, 1369–1374. [CrossRef] [PubMed]
35. Kohler, J.M.; Wagner, J.; Albert, J. Formation of isolated and clustered Au nanoparticles in the presence of polyelectrolyte molecules using a flow-through Si chip reactor. *J. Mater. Chem.* **2005**, *15*, 1924–1930. [CrossRef]
36. Mahmoud, M.A.; El-Sayed, M.A. Gold nanoframes: Very high surface plasmon fields and excellent near-infrared sensors. *JACS* **2010**, *132*, 12704–12710. [CrossRef]
37. Cheng, S.; Su, H.; Wang, Y.; Wu, W.; Zeng, J. Size-controlled synthesis of platinum-copper hierarchical trigonal bipyramid nanoframes. *Angew. Chem. Int. Ed. Engl.* **2015**, *54*, 108–113. [CrossRef] [PubMed]
38. Kluitmann, J.J.; Köhler, J.M. Tuning the morphology of bimetallic gold-platinum nanorods in a microflow synthesis. *Colloids Surf. A Physicochem. Eng. Asp.* **2021**, *626*, 127085. [CrossRef]

encyclopedia MDPI

Entry

Machine Learning for Additive Manufacturing

Dean Grierson *, Allan E. W. Rennie and Stephen D. Quayle

Engineering Department, Lancaster University, Lancaster LA1 4YW, UK; a.rennie@lancaster.ac.uk (A.E.W.R.); s.quayle@lancaster.ac.uk (S.D.Q.)
* Correspondence: d.grierson@lancaster.ac.uk

Definition: Additive manufacturing (AM) is the name given to a family of manufacturing processes where materials are joined to make parts from 3D modelling data, generally in a layer-upon-layer manner. AM is rapidly increasing in industrial adoption for the manufacture of end-use parts, which is therefore pushing for the maturation of design, process, and production techniques. Machine learning (ML) is a branch of artificial intelligence concerned with training programs to self-improve and has applications in a wide range of areas, such as computer vision, prediction, and information retrieval. Many of the problems facing AM can be categorised into one or more of these application areas. Studies have shown ML techniques to be effective in improving AM design, process, and production but there are limited industrial case studies to support further development of these techniques.

Keywords: machine learning; supervised learning; unsupervised learning; reinforcement learning; additive manufacturing; design for additive manufacturing; additive manufacturing process; additive manufacturing monitoring

Citation: Grierson, D.; Rennie, A.E.W.; Quayle, S.D. Machine Learning for Additive Manufacturing. *Encyclopedia* **2021**, *1*, 576–588. https://doi.org/10.3390/encyclopedia1030048

Academic Editors: Raffaele Barretta, Ramesh Agarwal, Krzysztof Kamil Żur and Giuseppe Ruta

Received: 16 June 2021
Accepted: 14 July 2021
Published: 19 July 2021

Publisher's Note: MDPI stays neutral with regard to jurisdictional claims in published maps and institutional affiliations.

1. Introduction

Additive manufacturing (AM) is the name given to a family of manufacturing processes where materials are directly joined to make parts from 3D modelling data [1]. This is generally done in discrete planar layers, but non-planar processes also exist [2]. AM enables various advantages, particularly when compared with traditional manufacturing techniques, the enablement of mass part customisation and greater part complexity on the macro-, meso-, and micro-scales [3]. Other advantages include not requiring any hard tooling and enablement of on-demand manufacturing [4]. Despite these benefits, drawbacks include a lack of inherent repeatability [5] which has led to difficulty in gaining certification in some sectors [3]. Another drawback is the lack of widespread design knowledge and tools tailored specifically for AM and the above-mentioned benefits that are enabled [3].

While, to some extent, applicable to all major industries, AM development has been largely driven by the aerospace, automotive and medical sectors [4]. The major driver in the aerospace and automotive sectors is to reduce component mass whilst not hindering performance [4]. A wider range of motivations is seen for medical applications of AM, although patient customisation, improved biocompatibility and performance is a common theme [4]. AM is also often used in consumer products, with mass customisation and light-weighting both being common motivations [4]. Machine learning (ML) is a branch of artificial intelligence concerned with training programs to automatically improve their performance. With this broad definition in mind, there are different types of ML which may be classified as supervised, unsupervised, semi-supervised, or reinforcement learning [6]. Shinde and Shah identified five key application domains for ML [7]:

1. Computer vision.
2. Prediction.
3. Semantic analysis.
4. Natural language processing.

5. Information retrieval.

Of these five domains, computer vision, prediction, and information retrieval have applications in AM. More varied exploration into these areas has been enabled by recent advances in graphics hardware which have allowed for faster optimisation of ML algorithms on large training sets [7]. These advances have allowed for the implementation of ML solutions within AM environments.

Across design, production, and process, improvements to current practices in AM require significant expertise in operators and designers [8]. To leverage the benefits of AM, the design, process, and production become significantly more complex [8]. In design, mass customisation requires deep knowledge of the links between the variables being changed as well as the requirements of the part. Similarly, increasing part complexity, either to lighten the weight or deliver improved performance, greatly increases the difficulty in designing suitable part topologies. As a result, these goals often come with large time and/or computation trade-offs.

There is little orthogonalisation in AM parameters: for example, in material extrusion, increasing extrusion temperature may improve layer adhesion but may also increase stringing. As a result, optimising process parameters for specific parts or new materials can be a time consuming and costly procedure [9]. Furthermore, part consistency is essential in sectors where AM adoption is most likely, such as aerospace, but variation in part quality both between and within machines and builds presents a barrier to more widespread adoption. Variations can include inconsistent part geometries, porosity, and functional performance. These issues encompass the management and interpretation of large amounts of data and knowledge. Such problems may be eased through the proper application of ML methods by reducing the amount of human or computational effort required to deliver satisfactory results.

2. Additive Manufacturing

This section details the various AM processes and use cases to provide an understanding of the current position of AM and the barriers limiting adoption. This understanding will be further discussed in Section 4. There are a number of different AM processes currently available—ASTM categorises these into seven types [1]:

1. Binder jetting: a liquid bonding agent or adhesive is selectively deposited to join powdered materials together.
2. Directed energy deposition (DED): focused thermal energy (e.g., laser, electron beam, plasma arc) is used to fuse materials by melting as they are deposited.
3. Material extrusion: material is selectively dispensed through a nozzle or orifice onto a substrate.
4. Material jetting: droplets of build material are selectively deposited.
5. Powder bed fusion (PBF): thermal energy selectively fuses regions of a powder bed.
6. Sheet lamination: sheets of material are bonded layer-upon-layer to form a part.
7. Vat photopolymerisation: a vat of liquid photopolymer is selectively cured by light-activated polymerisation.

There are three primary reasons for the adoption of AM techniques, including production of more complex geometries (than are achievable with conventional manufacturing approaches) without increasing cost, mass customisation of components, and supply chain disintermediation.

2.1. Complex Geometries

AM's layer-by-layer approach to manufacturing means that it can produce geometries that would be impossible for traditionally recognised manufacturing techniques. This has three main applications: light-weighting, performance optimisation and part consolidation.

Light-weighting may be achieved through topology optimisation (TO) or latticisation. In TO, a part is analysed based on an objective function, such as stiffness, and material is removed from the design that contributes least towards this function. This produces a more topologically complex part that can more precisely, and with less mass, meet its specification. Latticisation involves permeating a unit cell throughout the internal volume of a part. This can be done to remove mass but may also be leveraged to produce custom material properties or improve biofunctionality, such as with the design of auxetic structures for implants [10].

Fundamentally altering a part with a design for the additive manufacturing (DfAM) approach can enable further performance optimisation that would be impossible if AM was not used as the manufacturing technology of choice. Using hydraulic manifolds as an example, traditional designs may use drilled out through holes and plugs to create the required internal channels. This creates undesirable energy and pressure losses whilst resulting in parts with unnecessarily large masses [11]. By redesigning the manifolds for AM, without the constraints of straight, orthogonally intersecting channels, pressure losses have been found to reduce by up to 29.6% [11] and part mass has been reduced by as much as 91% [12]. Despite this, recognising which features require redesign as well as how to go about said redesign is very often dependent on the expertise of the designer.

Many large assemblies are designed in that way due to limitations in the ability of the manufacturing process to produce complex geometries (e.g., tool paths and undercuts). Reducing the number of parts in an assembly can reduce maintenance requirements, lead-time, weight, and production and non-recurring costs. A case study for AM part-consolidation is presented in Table 1. While the benefits of part consolidation are tangible, it may be difficult for designers, especially those inexperienced in AM, to select appropriate candidate parts.

Table 1. Profitability analysis of the effects of AM-enabled part consolidation of a high-bandwidth, direction tracking antenna array carried out by Optisys LLC [13].

Item	Improvement
Part count	100 pieces down to a 1-piece integrated assembly
Lead time	11 months down to 2 months
Weight	95% savings
Production costs	20–25% reduction
Non-recurring costs	75% reduction

2.2. Mass Customisation

AM's lack of reliance on tooling combined with its expensive materials and low production speeds (relative to feedstocks for more conventional manufacturing processes) make it best suited for high-value parts with low production volumes. This feature of AM is most beneficial when the part under consideration is bespoke, thus enabling cost effective production of parts customised for an individual consumer's needs or wants. The three sectors where this is most commonly used are medical/dental, packaging, and consumer products [3]. A selection of examples from each industry is summarised in Table 2.

Table 2. Prominent case studies of mass customization across its three primary application areas.

Sector	Case Study	Source(s)
Medical	Bespoke hearing aids	[14]
	Hard and soft tissue biomedical implants	[3]
	Custom casts, splints, orthotics, and prostheses	[3,15]
Packaging	Custom-fit 'direct digital packaging'	[3]
Consumer products	Custom-fit running shoes and insoles	[16,17]

While scanning allows for conformal shapes to be easily designed into a part [14], sophisticated design of the part's mesostructure is dependent on designer experience and computationally expensive tools, providing a hurdle for mass customised parts due to the lack of tools for this task.

2.3. Supply Chain Disintermediation

Due to its ability to manufacture parts on-site, AM allows for lean and agile manufacturing. This is particularly of note for requirements such as spare parts, where the demand is highly variable. Liu et al. [18] concluded that AM has the ability to increase supply chain efficiency for spare parts in aerospace [18] while Hernandez et al. [19] concluded that AM has the ability to significantly reduce supply chain disruption for the United States Navy. These benefits may be further leveraged should the production of AM parts be made more repeatable through greater part consistency and geometrical accuracy.

3. Machine Learning

This section details the different types of ML and outlines common use cases relevant to AM to highlight the potential of the field when applied to AM limitations.

Supervised learning algorithms fit hypotheses to labelled training datasets, those where there is a known output. The trained algorithm may then be applied to unlabelled cases to predict the corresponding label. Supervised learning is itself split into two categories: classification and regression. Classification problems have qualitative labels, i.e., classes such as whether an image is that of a cat or not, whereas regression problems have quantitative labels, such as estimating the age of a cat based on an image [20].

Neural networks (NNs) are popular tools for supervised learning, especially with large datasets. These algorithms seek to emulate a brain by implementing layers of connected neurons, as shown in Figure 1. NNs map an input space onto an output space, which is usually of different dimensions. The implementation of NNs allows for non-linear decision boundaries to be inferred in a computationally efficient manner. More specialised NNs also exist for specific application areas. For example, convolutional neural networks (CNNs) utilise convolutional layers to identify features that may be present throughout the input space. CNNs are most often used in computer vision tasks, where similar features, for example vertical lines, may occur anywhere in the input space.

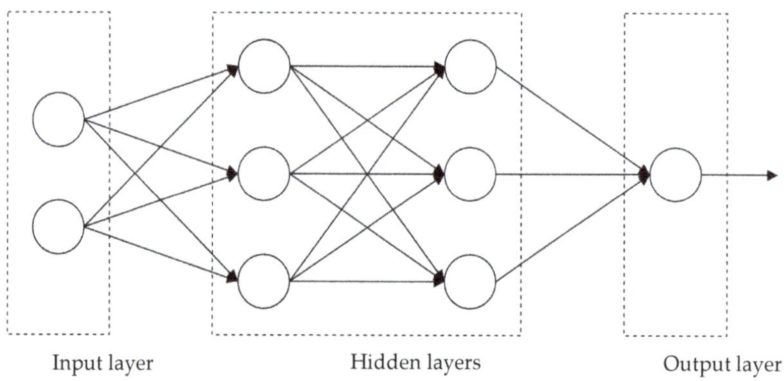

Figure 1. A fully connected NN with one input layer, two hidden layers, and one output layer. Circles represent neurons, which apply activation functions to the total sum of the products of the previous layer's activation and weight pairs.

Support vector machines (SVMs) are used in supervised learning tasks: traditionally classification but they are capable of regression. They calculate relationships between data in a higher dimensional space using a kernel function and placing a hyperplane decision boundary between classes to maximise the margins [20]. In regression tasks, the hyperplane is selected to best fit the data [20]. SVMs perform well for high-dimensional data but when there are far more features than training examples, they are prone to overfitting the training data. However, this can often be overcome through careful selection of an appropriate kernel function or through regularisation.

Unsupervised learning problems attempt to infer patterns from unlabelled data. Since these problems do not have a label to compare against, they are often more difficult to evaluate. There are various types of unsupervised learning algorithms, with the two most common types being clustering and association rules. Clustering, also called data segmentation, algorithms group data into clusters. The data within each cluster are such that they are more closely related to each other than any data in a different cluster [20]. Association rule analysis, often called market basket analysis, seeks to identify prototype values for a feature set such that the probability density at those values is relatively large [20].

Reinforcement learning considers the optimisation of an agent's interaction with its environment through a reward signal [6]. While supervised learning extrapolates its output from a set of known scenarios, reinforcement learning operates without known training scenarios for the reward signal [6]. Despite this, reinforcement learning is disparate from unsupervised learning in that it is not attempting to discern hidden structure within the data but optimises said reward signal [6].

4. Machine Learning for Additive Manufacturing

This section presents prominent research and findings of ML applications in AM. Applications have been grouped by their area of application with AM, in accordance with Wang et al. [8].

4.1. Machine Learning for Design for Additive Manufacturing

Multiple studies, notably by Sosnovik and Oseledets [21], have utilised ML tools to accelerate the TO process. This was enabled through the implementation of CNNs whereby:

1. The input volume was the result of the topology optimisation after just a few iterations of the Solid Isotropic Material with Penalisation (SIMP) algorithm.
2. The output volume was a prediction of the part topology after 100 iterations of the SIMP algorithm.

Whilst not removing the need for the SIMP algorithm, the method presented by Sosnovik and Oseledets [21], demonstrated that ML can be effective in drastically reducing the computational workload required to run TO. Results ranged from 92% intersection over union accuracy for input volumes of just five SIMP iterations, to 99.2% after 80 [21]. This work has been extended in further studies by Banga et al. [22] and later by Harish et al. [23], which looked at 3D implementations for cantilevered beams, with Banga et al. [22] obtaining a binary accuracy of 96%. The implications of these studies show that effective implementation of ML into the TO workflow may allow for:

1. A reduction in the time taken to carry out TO with a similar quality of output.
2. Higher resolution TO being efficiently used in practice thus enabling superior quality outputs.
3. Greater complexity in constraints applied, such as those specific to AM (e.g., supportless structures).

Biomimetic designs are enabled by AM but it is a major challenge to generate metamaterial structures with desired properties. Gu et al. effectively utilised CNNs to be able to generate structures with predictable strength and toughness [24]. This study built up three unit-cell "building blocks" from two jetted materials: one hard and one soft. These three building blocks were then randomly assigned to each of the 64 cells in an 8×8 array and then simulated to create a training example. In total, 80,000 training examples, or 10^{-8}% of the possible combinations, were created to train the CNN which was able to predict material properties. The model had a normalised root mean square deviation from the simulated results of 0.4926 on the test set. AM specimens were then produced to validate these findings. Here, the use of ML allowed the researcher to accurately screen the whole design space in hours compared to years for conventional simulation-only approaches [24].

Recent works have also leveraged NNs to generate lattice structures based on required mechanical properties. Jiang et al. developed a methodology for designing an ML implementation to determine optimal design geometries to produce desired properties, for example, stress–strain curves [25]. The methodology was implemented through the design of an ankle brace with three zones, each with its own distinct mechanical requirements: maximum torque, range of motion, as well as stress and strain in two orientations. An NN was developed to define the design parameters for a pre-determined unit cell that would meet these requirements. To train said network, simulations of random sets of the design parameters within acceptable ranges were carried out to determine their stress–strain response under the given maximum torque. This was then used as training data for their model. The resulting regression model, with the same amount of data available, was capable of generating structures with comparable errors to conventional methods, but with improved computational efficiency [25]. Due to the nature of the NN architecture used, however, performance could likely be further improved through gathering additional data.

An additional problem faced in DfAM is in knowledge dissemination: helping designers learn how to best design their parts for fabrication using AM technologies [3]. With the number of AM-specific design features growing, tools for designers are increasingly required [26]: ML has successfully been implemented in creating such tools. Yao et al. developed a flexible hybrid ML tool to recommend design features to inexperienced designers based on three designer-coded categories: loadings, objectives, and properties [26]. The tool produced a dendrogram for the designer's part and compared it against a hierarchically clustered database of existing design features [26]. An SVM was then used to target features based on their similarity to the designer's coding for the part [26]. Despite only considering functionality-specific features, after testing it with inexperienced designers, the tool was found to effectively enable the exploration of AM design freedoms [26].

4.2. Machine Learning for Additive Manufacturing Process

Within the domain of AM processes, ML is applied in two main areas: parameter optimisation and process monitoring [8]. This section will outline the current states of these areas.

4.2.1. Parameter Optimisation

Process parameter optimisation is often a manual and time-consuming process, making it costly. Similarly, manual process monitoring/control also creates additional costs. Since manual parameter optimisation requires the production of large numbers of samples, there is readily available data for the production of ML tools. Said tools, which make up a plurality of the research on ML for AM [27], largely take the route of optimising key parameters for a particular quality indicator or set of indicators [8]. PBF and material extrusion are the most common AM technologies to have had this type of tool developed.

Many studies have been carried out to optimise parameters for powder bed fusion processes. A summary of the parameters studied as well as respective quality indicators are shown in Figure 2, with porosity and fatigue life identified as having the largest variety of process parameters linked to them. In addition, porosity is the leading quality indicator in the research, with at least five studies implementing machine learning algorithms to optimise process parameters [28–32]. Liu et al. [32] built on previous efforts to develop a "physics-informed" model rather than a conventional "setting" model. The two model types, along with a third combined model, were found to provide similar performance to one another (within error margin). In addition, the novel "physics-informed" model was identified as being more easily generalised to other machines, although this was not tested.

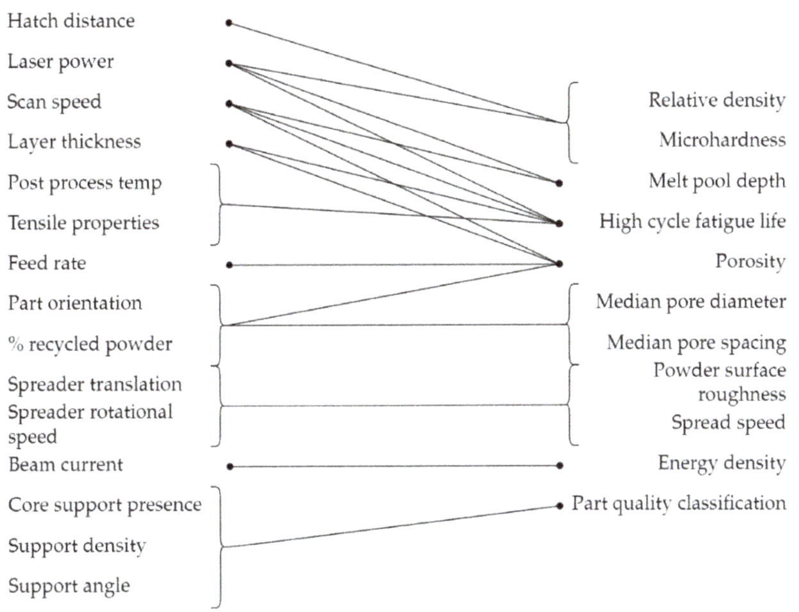

Figure 2. Matching diagram relating PBF process parameters to quality indicators from successful implementations. Adapted from Wang et al. [8].

Many studies have also been carried out to predict and/or optimise quality indicators for the material extrusion process. Figure 3 shows a summary of which process parameters have been linked to which quality indicators. Comparing Figure 2 to Figure 3, it can be seen that a much greater number of parameters are considered for each quality indicator in material extrusion over PBF. This may be attributed to lower part costs in material extrusion over PBF, thus allowing for more training samples and potentially more complex ML implementations. NNs can capture non-linearities more easily than simpler regression models, making them the prevailing algorithm for material extrusion process parameter studies.

A key area for optimisation in material extrusion is component surface roughness [4]. Li et al. [33] built a predictive model for surface roughness based on build plate and extruder temperature, build plate and extruder vibration, and melt pool temperature. The model used AdaBoost, a learning algorithm that is an ensemble of various weaker algorithms, and was able to predict surface roughness with a root mean square error of 0.7 μm [33].

DED processes have also had parameters optimised as discussed in the literature. The bulk of these works focus on controlling aspects of the melt pool and resulting tracks, including: track width and height [34–36], melt pool geometry [37], and thermal history of the melt pool [38]. More recent works have studied the properties of the final part: Narayana et al. [39] built a NN to predict built part height and density from laser power, scan speed, powder feed rate, and layer thickness. It was found that these parameters were all of significant importance for density whereas scan speed and feed rate had the largest effect on build height. These findings were reinforced by the model's prediction accuracy of 99%. Similar to material extrusion, DED produces parts with poor surface roughness [40]. Xia et al. [40], used an NN to model and predict surface roughness based on overlap ratio, welding speed, and wire feed speed with a root mean square error of 6.94%. A small training set was identified as a major limiter on the model's accuracy [40].

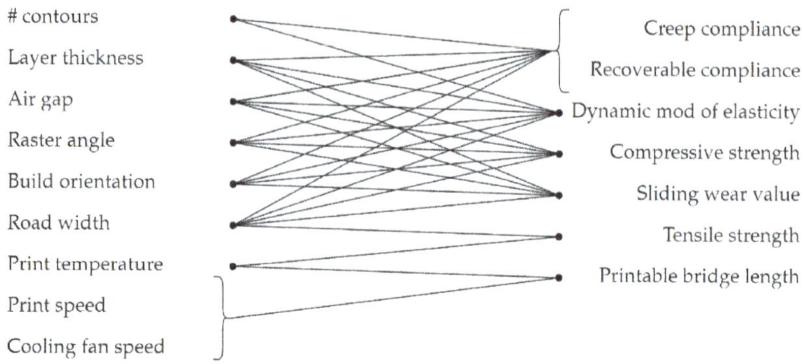

Figure 3. Matching diagram relating material extrusion process parameters to quality indicators from successful implementations. Adapted from Wang et al. [8].

ML for process parameter optimisation in binder jetting has a lesser volume of literature associated with these aspects: Chen and Zhao [41] developed an NN powered software tool to recommend layer thickness, printing saturation, heater power ratio and drying time based on user defined preference for surface roughness as well as dimensional accuracy along the Y and Z axes. Their software also predicted surface roughness as well as Y and Z shrinkage. The error in these predictions was found to be 1.98%, 5.38%, and 16.58%, respectively. It was suggested that improvements could be made through use of a larger training set or by reducing the model's independence by implementing an element of physical modelling [41].

4.2.2. Process Monitoring

While parameter optimization may help to improve process predictability, it cannot eliminate failures entirely [42]. With print failures contributing significantly to the cost of AM parts [43], process monitoring techniques able to detect build failures and defects are necessary. Various ML implementations have sought to solve this problem and fall into two categories depending on their input data type: optical and acoustic [8].

Optical monitoring solutions are the most widely used, with the data often coming from digital, high speed, or infrared cameras [8]. In PBF processes, where the bulk of monitoring research is currently concentrated, the most common target of these computer vision tasks is the melt pool. From thermal data of the melt pool, Kwon et al. [42] trained a CNN-based program to differentiate between high, medium, and low quality builds with a failure rate of under 1.1% [42], allowing for potential time and cost savings. Other works have used optical data from laser melting plumes [44–46] for similar quality classification tasks, with Zhang et al. [45,46] finding that the best results are achieved when melt pool, plume, and spatter data are used together to classify part quality. The most recent work found a type of NN called a long–short term memory network, to be most effective in prediction, with a root mean squared error of 13.9% [46].

Optical monitoring has also been implemented for other AM processes, including binder jetting and material extrusion. Gunther et al. [47] used optical monitoring to detect faults in binder jet parts based on the frequency and density of brightly coloured pixels. The algorithms employed here were not provided, nor was there discussion of the accuracy of the employed model. In material extrusion, optical monitoring has been implemented for in situ defect detection [48]. Wu et al. [48] used a classification algorithm to identify the presence of infill print defects in material extrusion, allowing for greater confidence in final part's quality. The study achieved an accuracy of over 95% but did not consider other vital quality indicators such as precision and recall. Li et al. [49] used in situ optical monitoring of a material extrusion process to determine dimensional deviation with zero mean error and a standard deviation of 0.02mm.

Acoustic monitoring is a newer and less widely adopted method of monitoring a build mid-print. These techniques rely on characteristic acoustic signals that relate to part porosity [50] and melt states [51] in PBF processes as well as process failures in material extrusion [52]. Advantages of acoustic monitoring solutions include the lower cost sensors compared to optical monitoring techniques [50]. ML algorithms applied also vary from supervised CNNs to clustering solutions. Researchers in these fields have achieved confidence of up to 89% for porosity classification [50] and 94% for melt-pool related defects [51], showing acoustic monitoring to be an effective method for flagging problem builds with less need for post-print examination and testing [51]. Acoustic monitoring has also been applied to material extrusion processes. Wu et al. [53] used acoustic monitoring and an SVM classifier to determine if the extruder was successfully pushing out material with an accuracy of 100%. The SVM was also able to identify extruder blockages (normal, semi-blocked, or blocked) with a 92% accuracy.

4.3. Machine Learning for Additive Manufacturing Production

In AM production, as well as implementing strategies discussed in Sections 4.1 and 4.2, additional tools have been developed to aid in general production planning using NNs, and manufacturability through a variety of methods. In addition, research has been conducted on ML implementations to replicate CAD geometry from acoustic signals produced during manufacturing, creating concerns around data security [54,55].

Printability and Dimensional Deviation Management

Models have been developed to aid in the identification of printability of components in material extrusion [56] and PBF processes [57] utilising CNNs and SVMs, respectively. The use of NNs has also been shown to aid in reducing print time estimation error for PBF processes from 20–35% to 2–15% [58], enabling improved management of these machines.

There are three sources of dimensional deviation in AM parts [59]: the material (shrinkage and warpage), the machine, and the file preparation (e.g., resolution reduction due to conversion from CAD model to STL file format). ML has been used to correct these in PBF [60], where NNs were implemented to first optimise part orientation, reducing deviation due to the machine, then modify CAD geometry to account for thermal effects on the material.

Khanzedah et al. [61] implemented an unsupervised learning algorithm, a self-organising map, to analyse and assess point-cloud data for the dimensional deviation of parts made with material extrusion processes. Their implementation was able to sort the part's deviations into discrete clusters based on the severity of the deviations present, which allowed sub-optimal process conditions to be identified. Noriega et al. [62] used NNs to compensate for dimensional deviation by modifying the part's scale for material extrusion. Their two NN implementations were used to reduce deviation by 50% for external dimensions and 30% for internal dimensions. In a more robust study using the AM of a T4 spinal vertebrae as a case study, Charalampous et al. [63] were able to achieve a 25% reduction in dimensional deviation at a 1:1 scale and 33% at a 3:1 scale.

Other studies have looked to correct dimensional deviation in binder jetting and DED processes. Shen et al. [64] presented a study using CNNs to predict dimensional deviation and compensate for it via translation, scaling, and rotation of dental crown CAD geometry to be manufactured by binder jetting processes. The CNN used a voxel-based approach where, in the analysis of the implementation, each voxel was deemed to be correct or incorrect. This was then used to generate F1 scores: a single-value metric used to evaluate the model's recall and precision. The predictive model and compensatory model both averaged 94% for their F1 scores. Despite this, no physical samples were manufactured so these findings are unverified.

In DED, the dominant method of dimensional deviation correction is to optimise the geometry of individual tracks through the optimisation of process parameters [33,35,36,64]. This partially corrects material and machine errors but not those associated with the file preparation process. Caiazzo and Caggiano [37] developed one such model to tune laser power, scanning speed, and powder feed rate to achieve a specified track geometry with mean absolute errors of 2.0%, 5.8%, and 5.5%, respectively. An alternative approach was presented by Choi who built a predictive NN for multi-track height based on laser power, powder feed rate, and coaxial gas flow rate with an accuracy of 96.63%. Choi [65] queried their model for optimal results: 300 W, 3.7 g/min, and 6 l/min, respectively.

The vulnerability of material extrusion machines to intellectual property theft was identified by Al Faruque et al. [54] and Hojjati et al. [55]. Al Faruque et al. [54] showed that the noise emitted by stepper motors during printing can be recorded and processed to infer features of the print process:

1. Axis of motion.
2. Direction of motion.
3. Speed of hot-end motion
4. Distance travelled in each axis.
5. Extrusion amount.

This information was used in a physical-to-cyber attack to reconstruct the CAD geometry being fabricated. Both studies utilised non-descript supervised classification and regression models to determine the features and feature values, respectively. Geometry was successfully reconstructed with an average axis prediction accuracy of 78% and an average length prediction error of 18% [54]. Unlike most applications discussed, IP theft is an aspect of AM that is harmed by ML. This must be resolved if AM, particularly material extrusion, is to be made viable for security-sensitive applications.

5. Conclusions and Prospects

In research, ML has been shown to be effective in furthering AM design, process, and production. In AM design, ML has been used to accelerate tools, explore new materials, enable the identification of property–structure relationships, and aid novice designers. TO acceleration and material exploration have limited scope in their current states and need further development to work with larger design spaces or with finer spatial resolutions. Property–structure relationships have the potential to be useful in functional lattice design, but current implementations have inadequate transitional regions and accuracies that may be too low for some industries. There are also insufficient case studies to support further adoption of these techniques. Design feature recommenders in their current forms are quite robust but require curation to remain relevant and, like functional lattice design systems, need further case studies to be developed to support their adoption. In AM process, most work focuses on process parameter optimisation. These are effective in optimising process parameters for one or multiple quality indicators. Despite this, these optimisers are machine specific, and no studies have been identified that attempt to produce more general models. Without such developments, many samples will have to be produced whether the final process used for optimisation is manual or through ML. ML has been successfully used to improve the predictability of AM production techniques but has also been shown to be able to exploit AM processes and enable IP theft. While there have been no real-world reports of this sort of theft taking place, it remains a hurdle to be solved to make AM, particularly material extrusion, a secure process.

Overall, ML has had a positive impact on the prospects of furthering AM adoption and improving its value proposition. That said, most ML applications for AM are not robust or trusted enough to be adopted in industry. As a result, research efforts should focus on further developing these tools for real-world use and reporting industry case studies to build confidence in their efficacy.

Author Contributions: Conceptualization, D.G. and A.E.W.R.; formal analysis, D.G.; investigation, D.G.; resources, D.G.; writing—original draft preparation, D.G.; writing—review and editing, A.E.W.R. and S.D.Q.; visualization, D.G.; supervision, A.E.W.R. and S.D.Q.; project administration, D.G. and A.E.W.R. All authors have read and agreed to the published version of the manuscript.

Funding: This research was funded by the European Regional Development Fund, through the Greater Innovation for Smart Materials Optimisation (GISMO) Project (grant reference number 03R18P02671).

Data Availability Statement: Dara sharing not applicable.

Conflicts of Interest: The authors declare no conflict of interest.

Entry Link on the Encyclopedia Platform: https://encyclopedia.pub/13191.

References

1. ISO/ASTM. ISO/ASTM52900-15, Standard Terminology for Additive Manufacturing—General Principles—Terminology. *ASTM Int.* **2015**, *1*, 1–9. [CrossRef]
2. Ahlers, D.; Wasserfall, F.; Hendrich, N.; Zhang, J. 3D printing of nonplanar layers for smooth surface generation. In Proceedings of the IEEE International Conference on Automation Science and Engineering, Vancouver, BC, Canada, 22–26 August 2019; pp. 1737–1743. [CrossRef]
3. Thompson, M.K.; Moroni, G.; Vaneker, T.; Fadel, G.; Campbell, R.I.; Gibson, I.; Bernard, A.; Schulz, J.; Graf, P.; Ahuja, B.; et al. Design for Additive Manufacturing: Trends, opportunities, considerations, and constraints. *CIRP Ann. Manuf. Technol.* **2016**, *65*, 737–760. [CrossRef]
4. Gibson, I.; Rosen, D.; Stucker, B.; Khorasani, M. *Additive Manufacturing Technologies*, 3rd ed.; Springer: Berlin/Heidelberg, Germany, 2021; ISBN 978-3-030-56127-7. [CrossRef]
5. Dowling, L.; Kennedy, J.; O'Shaughnessy, S.; Trimble, D. A review of critical repeatability and reproducibility issues in powder bed fusion. *Mater. Des.* **2020**, *186*, 108346. [CrossRef]
6. Sutton, R.S.; Barto, A.G. *Reinforcement Learning: An Introduction*, 2nd ed.; The MIT Press: Cambridge, MA, USA, 2018; ISBN 9780262039246.
7. Shinde, P.P.; Shah, S. A Review of Machine Learning and Deep Learning Applications. In Proceedings of the 2018 4th International Conference on Computing, Communication Control and Automation (ICCUBEA), Pune, India, 16–18 August 2018. [CrossRef]
8. Wang, C.; Tan, X.P.; Tor, S.B.; Lim, C.S. Machine learning in additive manufacturing: State-of-the-art and perspectives. *Addit. Manuf.* **2020**, *36*, 101538. [CrossRef]
9. Qi, X.; Chen, G.; Li, Y.; Cheng, X.; Li, C. Applying Neural-Network-Based Machine Learning to Additive Manufacturing: Current Applications, Challenges, and Future Perspectives. *Engineering* **2019**, *5*, 721–729 101016/jeng201904012. [CrossRef]
10. Darwish, S.M.H.; Aslam, M.U. Auxetic cellular structures for custom made orthopedic implants using additive manufacturing. *Int. J. Eng. Adv. Technol.* **2014**, *4*, 2249–8958.
11. Xie, G.; Dong, Y.; Zhou, J.; Sheng, Z. Topology optimization design of hydraulic valve blocks for additive manufacturing. *Proc. Inst. Mech. Eng. Part C J. Mech. Eng. Sci.* **2020**, *234*, 1899–1912. [CrossRef]
12. Diegel, O.; Schutte, J.; Ferreira, A.; Chan, Y.L. Design for additive manufacturing process for a lightweight hydraulic manifold. *Addit. Manuf.* **2020**, *36*, 101446. [CrossRef]
13. GE Additive. For the Ready. Launch Your Additive Manufacturing Program with Help from GE Additive. 2020. Available online: https://www.ge.com/additive/sites/default/files/2020-11/GE_Space_Brochure_8.5x11_R5_0.pdf (accessed on 16 June 2021).
14. Redwood, B.; Schoffer, F.; Garret, B. *The 3D Printing Handbook: Technologies, Design and Applications*, 1st ed.; 3D Hubs: Amsterdam, The Netherlands, 2018; ISBN 9789082748505.
15. Chen, R.K.; Jin, Y.-A.; Wensman, J.; Shih, A. Additive manufacturing of custom orthoses and prostheses-A review. *Addit. Manuf.* **2016**, *12*, 77–89. [CrossRef]
16. EOS Shoe Soles from the 3D Printer | EOS GmbH. Available online: https://www.eos.info/en/3d-printing-examples-applications/people-health/sports-lifestyle-consumer-goods/shoes-soles (accessed on 28 May 2021).
17. Dong, G.; Tessier, D.; Zhao, Y.F. Design of Shoe Soles Using Lattice Structures Fabricated by Additive Manufacturing. *Proc. Des. Soc. Int. Conf. Eng. Des.* **2019**, 5–8. [CrossRef]
18. Liu, P.; Huang, S.H.; Mokasdar, A.; Zhou, H.; Hou, L. Production Planning & Control The Management of Operations The impact of additive manufacturing in the aircraft spare parts supply chain: Supply chain operation reference (scor) model based analysis. *Prod. Plan. Control.* **2014**, *25*, 1169–1181. [CrossRef]
19. Hernandez, B.R.; Housel, T.; Ford, D. An Investigation into the Use of £D Scanning and PRinting Technologies in the Navy Collaborative Product Lifecycle Management. Master's Thesis, Naval Postgraduate SCHOOL, Monterey, CA, USA, 2013. Available online: https://calhoun.nps.edu/bitstream/handle/10945/38945/13Dec_Hernandez_Benjamin.pdf?sequence=1&isAllowed=y (accessed on 16 June 2021).
20. Hastie, T.; Tibshirani, R.; Friedman, J. *Springer Series in Statistics The Elements of Statistical Learning Data Mining, Inference, and Prediction*, 2nd ed.; Springer: Berlin/Heidelberg, Germany, 2008; ISBN 978-0-387-84857-0. [CrossRef]

21. Sosnovik, I.; Oseledets, I. Neural networks for topology optimization. *Russ. J. Numer. Anal. Math. Model.* **2019**, *34*, 215–223. [CrossRef]
22. Banga, S.; Gehani, H.; Bhilare, S.; Patel, J.; Kara, B. 3D Topology Optimzation Using Convolutional Neural Networks. *arXiv* **2018**, arXiv:1808.07440.
23. Harish, B.; Eswara Sai Kumar, K.; Srinivasan, B. Topology optimization using convolutional neural network. In *Proceedings of the Lecture Notes in Mechanical Engineering*; Springer: Berlin/Heidelberg, Germany, 2020; pp. 301–307. [CrossRef]
24. Gu, G.X.; Chun-Teh Chen, A.; Richmond, D.J.; Buehler, M.J. Bioinspired hierarchical composite design using machine learning: Simulation, additive manufacturing, and experiment. *Mater. Horizons* **2018**, *5*, 939. [CrossRef]
25. Jiang, J.; Xiong, Y.; Zhang, Z.; Rosen, D.W. Machine learning integrated design for additive manufacturing. *J. Intell. Manuf.* **2020**, 1–14. [CrossRef]
26. Yao, X.; Moon, S.K.; Bi, G. A hybrid machine learning approach for additive manufacturing design feature recommendation. *Rapid Prototyp. J.* **2017**, *23*. [CrossRef]
27. Mahmood, M.A.; Visan, A.I.; Ristoscu, C.; Mihailescu, I.N. Artificial Neural Network Algorithms for 3D Printing. *Materials* **2021**, *14*, 163. [CrossRef] [PubMed]
28. Silbernagel, C.; Aremu, A.; Ashcroft, I. Using machine learning to aid in the parameter optimisation process for metal-based additive manufacturing. *Rapid Prototyp. J.* **2019**, *26*, 625–637. [CrossRef]
29. Kappes, B.; Moorthy, S.; Drake, D.; Geerlings, H.; Stebner, A. Machine learning to optimize additive manufacturing parameters for laser powder bed fusion of Inconel 718. In *Proceedings of the Minerals, Metals and Materials Series*; Springer International Publishing: Berlin/Heidelberg, Germany, 2018; Volume 2018-June, pp. 595–627. [CrossRef]
30. Garg, A.; Lam, J.S.L. Measurement of environmental aspect of 3-D printing process using soft computing methods. *Meas. J. Int. Meas. Confed.* **2015**, *75*, 210–217. [CrossRef]
31. Garg, A.; Lam, J.S.L.; Savalani, M.M. A new computational intelligence approach in formulation of functional relationship of open porosity of the additive manufacturing process. *Int. J. Adv. Manuf. Technol.* **2015**, *80*, 555–565. [CrossRef]
32. Liu, R.; Liu, S.; Zhang, X. A physics-informed machine learning model for porosity analysis in laser powder bed fusion additive manufacturing. *Int. J. Adv. Manuf. Technol.* **2021**, *113*, 1943–1958. [CrossRef]
33. Li, Z.; Zhang, Z.; Shi, J.; Wu, D. Prediction of surface roughness in extrusion-based additive manufacturing with machine learning. *Robot. Comput. Integr. Manuf.* **2019**, *57*, 488–495. [CrossRef]
34. Xiong, J.; Zhang, G.; Hu, J.; Wu, L. Bead geometry prediction for robotic GMAW-based rapid manufacturing through a neural network and a second-order regression analysis. *J. Intell. Manuf.* **2014**, *25*, 157–163. [CrossRef]
35. Li, Y.; Sun, Y.; Han, Q.; Zhang, G.; Horváth, I. Enhanced beads overlapping model for wire and arc additive manufacturing of multi-layer multi-bead metallic parts. *J. Mater. Process. Technol.* **2018**, *252*, 838–848. [CrossRef]
36. Lu, Z.; Li, Ã.; Lu, B.; Zhang, A.; Zhu, G.; Pi, G. The prediction of the building precision in the Laser Engineered Net Shaping process using advanced networks ARTICLE IN PRESS. *Opt. Lasers Eng.* **2010**, *48*, 519–525. [CrossRef]
37. Caiazzo, F.; Caggiano, A. Laser Direct Metal Deposition of 2024 Al Alloy: Trace Geometry Prediction via Machine Learning. *Materials* **2018**, *11*, 444. [CrossRef]
38. Mozaffar, M.; Paul, A.; Al-Bahrani, R.; Wolff, S.; Choudhary, A.; Agrawal, A.; Ehmann, K.; Cao, J. Data-driven prediction of the high-dimensional thermal history in directed energy deposition processes via recurrent neural networks. *Manuf. Lett.* **2018**, *18*, 35–39. [CrossRef]
39. Narayana, P.L.; Kim, J.H.; Lee, J.; Choi, S.-W.; Lee, S.; Park, C.H.; Yeom, J.-T.; Reddy, N.G.S.; Hong, J.-K. Optimization of process parameters for direct energy deposited Ti-6Al-4V alloy using neural networks. *Int. J. Adv. Manuf. Technol.* **2021**, *114*, 3269–3283. [CrossRef]
40. Xia, C.; Pan, Z.; Polden, J.; Li, H.; Xu, Y.; Chen, S. Modelling and prediction of surface roughness in wire arc additive manufacturing using machine learning. *J. Intell. Manuf.* **2021**, 1–16. [CrossRef]
41. Chen, H.; Zhao, Y.F. Learning Algorithm Based Modeling and Process Parameters Recommendation System for Binder Jetting Additive Manufacturing Process. *Proc. ASME Des. Eng. Tech. Conf.* **2016**, *1*. [CrossRef]
42. Kwon, O.; Kim, G.H.; Ham, M.J.; Kim, W.; Kim, G.-H.; Cho, J.-H.; Kim, N.I.; Kim, K. Kangil A deep neural network for classification of melt-pool images in metal additive manufacturing. *J. Intell. Manuf.* **2020**, *31*, 375–386. [CrossRef]
43. Baumers, M.; Dickens, P.; Tuck, C.; Hague, R. The cost of additive manufacturing: Machine productivity, economies of scale and technology-push. *Technol. Forecast. Soc. Chang.* **2016**, *102*, 193–201. [CrossRef]
44. Ye, D.; Hsi Fuh, J.Y.; Zhang, Y.; Hong, G.S.; Zhu, K. In situ monitoring of selective laser melting using plume and spatter signatures by deep belief networks. *ISA Trans.* **2018**, *81*, 96–104. [CrossRef] [PubMed]
45. Zhang, Y.; Hong, G.S.; Ye, D.; Zhu, K.; Fuh, J.Y.H. Extraction and evaluation of melt pool, plume and spatter information for powder-bed fusion AM process monitoring. *Mater. Des.* **2018**, *156*, 458–469. [CrossRef]
46. Zhang, Z.; Liu, Z.; Wu, D. Prediction of melt pool temperature in directed energy deposition using machine learning. *Addit. Manuf.* **2021**, *37*, 101692. [CrossRef]
47. Gunther, D.; Pirehgalin, M.F.; Weis, I.; Vogel-Heuser, B. Condition monitoring for the Binder Jetting AM-process with machine learning approaches. In Proceedings of the 2020 IEEE Conference on Industrial Cyberphysical Systems (ICPS), Tampere, Finland, 10–12 June 2020; pp. 417–420. [CrossRef]

48. Wu, M.; Phoha, V.V.; Moon, Y.B.; Belman, A.K. Detecting malicious defects in 3d printing process using machine learning and image classification, ASME 2016 International Mechanical Engineering Congress and Exposition. *Am. Soc. Mech. Eng. Digit. Collect.* **2016**. [CrossRef]
49. Li, L.; McGuan, R.; Isaac, R.; Kavehpour, P.; Candler, R. Improving precision of material extrusion 3D printing by in-situ monitoring & predicting 3D geometric deviation using conditional adversarial networks. *Addit. Manuf.* **2021**, *38*, 101695. [CrossRef]
50. Shevchik, S.A.; Kenel, C.; Leinenbach, C.; Wasmer, K. Acoustic emission for in situ quality monitoring in additive manufacturing using spectral convolutional neural networks. *Addit. Manuf.* **2018**, *21*, 598–604. [CrossRef]
51. Ye, D.; Hong, G.S.; Zhang, Y.; Zhu, K.; Fuh, J.Y.H. Defect detection in selective laser melting technology by acoustic signals with deep belief networks. *Int. J. Adv. Manuf. Technol.* **2018**, *96*, 2791–2801. [CrossRef]
52. Wu, H.; Yu, Z.; Wang, Y. A New Approach for Online Monitoring of Additive Manufacturing Based on Acoustic Emission. *ASME Int.* **2016**. [CrossRef]
53. Wu, H.; Wang, Y.; Yu, Z. In situ monitoring of FDM machine condition via acoustic emission. *Int. J. Adv. Manuf. Technol.* **2016**, *84*, 1483–1495. [CrossRef]
54. Abdullah Al Faruqye, M.; Chhetri, S.R.; Canedo, A.; Wan, J. Acoustic side-channel attacks on additive manufacturing systems. In Proceedings of the 7th International Conference on Cyber-Physical Systems (ICCPS '16), Vienna, Austria, 11–14 April 2016; pp. 1–10.
55. Hojjati, A.; Adhikari, A.; Struckmann, K.; Chou, E.J.; Nguyen, T.N.T.; Madan, K.; Winslett, M.S.; Gunter, C.A.; King, W.P. Leave your phone at the door: Side channels that reveal factory floor secrets. *Proc. ACM Conf. Comput. Commun. Secur.* **2016**, 883–894. [CrossRef]
56. Tang, Y.; Dong, G.; Zhou, Q.; Zhao, Y.F. Lattice Structure Design and Optimization with Additive Manufacturing Constraints. *IEEE Trans. Autom. Sci. Eng.* **2018**, *15*, 1546–1562. [CrossRef]
57. Zhang, Y.; Dong, G.; Yang, S.; Zhao, Y.F. Machine learning assisted prediction of the manufacturability of laser-based powder bed fusion process. In Proceedings of the ASME Design Engineering Technical Conference; American Society of Mechanical Engineers (ASME), Anaheim, CA, USA, 18–21 August 2019; Volume 1. [CrossRef]
58. Munguía, J.; Ciurana, J.; Riba, C. Neural-network-based model for build-time estimation in selective laser sintering. *Proc. Inst. Mech. Eng. Part B J. Eng. Manuf.* **2009**, *223*, 995–1003. [CrossRef]
59. Omairi, A.; Ismail, Z.H. Towards machine learning for error compensation in additive manufacturing. *Appl. Sci.* **2021**, *11*, 2375. [CrossRef]
60. Chowdhury, S.; Mhapsekar, K.; Anand, S. Part Build Orientation Optimization and Neural Network-Based Geometry Compensation for Additive Manufacturing Process. *J. Manuf. Sci. Eng. Trans. ASME* **2018**, *140*. [CrossRef]
61. Khanzadeh, M.; Rao, P.; Jafari-Marandi, R.; Smith, B.K.; Tschopp, M.A.; Bian, L. Quantifying Geometric Accuracy with Unsupervised Machine Learning: Using Self-Organizing Map on Fused Filament Fabrication Additive Manufacturing Parts. *J. Manuf. Sci. Eng. Trans. ASME* **2018**, *140*. [CrossRef]
62. Noriega, A.; Blanco, D.; Alvarez, B.J.; Garcia, A. Dimensional accuracy improvement of FDM square cross-section parts using artificial neural networks and an optimization algorithm. *Int. J. Adv. Manuf. Technol.* **2013**, *69*, 2301–2313. [CrossRef]
63. Charalampous, P.; Kostavelis, I.; Kontodina, T.; Tzovaras, D. Learning-based error modeling in FDM 3D printing process. *Rapid Prototyp. J.* **2021**, *27*, 507–517. [CrossRef]
64. Shen, Z.; Shang, X.; Zhao, M.; Dong, X.; Xiong, G.; Wang, F.Y. A learning-based framework for error compensation in 3D printing. *IEEE Trans. Cybern.* **2019**, *49*, 4042–4050. [CrossRef]
65. Choi, T.-Y. *Machine Learning Based Predictive Modelling of Dimensional Quality in Depostion with SUS316L*; Graduate School of UNIST: Ulsan, Korea, 2020.

encyclopedia

MDPI

Entry

Challenges for Nanotechnology

Johann Michael Köhler [1,2]

[1] Institute for Micro- and Nanotechnologies, Technische Universitat Ilmenau, 98693 Ilmenau, Germany; michael.koehler@tu-ilmenau.de; Tel.: +49-0-3677-69-3629; Fax: +49-0-3677-69-3179
[2] Department of Physical Chemistry and Microreaction Technology, Institute for Chemistry and Biotechnology, Technische Universitat Ilmenau, 98693 Ilmenau, Germany

Definition: The term "Nanotechnology" describes a large field of scientific and technical activities dealing with objects and technical components with small dimensions. Typically, bodies that are in–at least–two dimensions smaller than 0.1 μm are regarded as "nanobjects". By this definition, a lot of advanced materials, as well as the advanced electronic devices, are objects of nanotechnology. In addition, many aspects of molecular biotechnology as well as macromolecular and supermolecular chemistry and nanoparticle techniques are summarized under "nanotechnology". Despite this size-oriented definition, nanotechnology is dealing with physics and chemistry as well as with the realization of technical functions in the area between very small bodies and single particles and molecules. This includes the shift from classical physics into the quantum world of small molecules and low numbers or single elementary particles. Besides the already established fields of nanotechnology, there is a big expectation about technical progress and solution to essential economic, medical, and ecological problems by means of nanotechnology. Nanotechnology can only meet these expectations if fundamental progress behind the recent state of the art can be achieved. Therefore, very important challenges for nanotechnology are discussed here.

Keywords: limits of nanotechnology; nanofacility shrinking; modularity; sustainability; hierarchical organization; entropy export; time scales; life cycles

Citation: Köhler, J.M. Challenges for Nanotechnology. *Encyclopedia* **2021**, *1*, 618–631. https://doi.org/10.3390/encyclopedia1030051

Academic Editors: Raffaele Barretta, Ramesh Agarwal, Krzysztof Kamil Żur and Giuseppe Ruta

Received: 25 June 2021
Accepted: 21 July 2021
Published: 25 July 2021

1. Introduction

About half a century ago, nanotechnology was not much more than a vision [1]. However, during the last decades, it has developed quickly, and there are many branches of science and technology which are related to nanotechnology [2]. From the point of application, two fields are particularly far developed: on the one hand, the creation and production of nanomaterials [3] and, on the other hand, the production of electronic chip elements, which play a crucial role in nearly every recent field of advanced technology due to their key role for computer and communication technology, for machine control, sensing, and for many other technical devices.

Highly integrated electronic solid-state devices are built up by billions of single nanostructures, recently. This ultimate degree of integration is based on a very high level of circuit design, micro, and nanolithography, and a lot of special preparation and measurement technologies and sophisticated materials are needed [4]. The enormous power of recent computers is a direct consequence of stepwise downscaling of the minimal structures in microlithography and the continuous improvement of all related technological steps and the equipment for manufacturing over the past five decades. These result in critical structure sizes below about 20 nm in production and below 10 nm in advanced development, which is not very far from the dimension of small molecules [5], and are in the order of magnitude of 1 nm. The basis for this successful development is the general convention of planar technology and a consequent down-scaling of functional structures in the frame of this proved concept.

The opposite development was inspired by the vision of realizing a bottom-up approach instead of a downscaling of microtechniques into the nanometer range. This approach is motivated by the insight into the ability of nature to form very complex functional structures by arranging small molecular units in living cells. The understanding of the chemical structure of complex biomolecules and the causing nature of chemical bonds behind–as it was explained by L. Pauling, F. Crick, J. Watson, and others–gave hope to constructing complex molecular machinery and to build complex functional systems from ultimate small modules artificially [6]. Supermolecular chemistry and strategies of controlled molecular self-assembling [7] are the main approaches for realizing molecular machines [8]. Indeed, these ideas led to a lot of interesting investigations and inventions during the last decades. However, it must be said that the dream of bottom-up nanotechnical manufacturing was not fulfilled, up to now. Thus, very important challenges remain, and some of the important ones will be discussed in the following.

2. Overcoming the Limitations of Planar Technology

Planar technology is a technical convention that allows the efficient production of high and extremely high integrated chip devices. The convention ensures to achieve the required exact positioning of large numbers of smallest functional structures on macroscopic carriers, for example, silicon wafers. By the conventions of planar technology, it becomes possible to develop extremely complex architectures of integrated circuits containing billions of single semiconductor elements and ensure their reliable function over trillions of electronics operations. This concept was very successful for more than a half-century. Its fundament is the connection of the macroscopic and microscopic scale in two dimensions, but a strict restriction of the third dimension to the microscopic scale. The complete industry of integrated semiconductors–that means nearly all communication and computer technology–is based on this concept.

The restriction on two lateral dimensions is important for device fabrication and thin-film and lithographic techniques. Technological tools as projection photolithography, lithography with focused or shaped electron and ion beams, and, in particular, the lithographic alignments between subsequent lithographic layers in functional multilayer systems are reliant on the strict application of the planar technological conventions. Planarization steps in the multilayer technology support a high homogeneity and reproducibility in film thicknesses and in the related electronic and other physical properties of micro and nano-patterned structures.

Besides the manufacturing, the planar architectures of semiconductor chip devices are also crucial for operating them. The high surface-to-volume ratio ensures, for example, a sufficiently high heat exchange. In addition, the planar structure supports the integration of sensing components, optical arrays, and other interface components.

The restriction on two dimensions means a restriction in the topology of interconnections, too. On the one hand, a very large and hierarchically structured network of connections is realized in two dimensions. In the third dimension, it is impossible to realize a high number of connections or complex structures because only a few layers can be used for designing them. The restriction to two dimensions means a low degree of connectivity. This is, probably, the most important difference between the architectures of computer chips and brains. The wirings of chips are mainly marked by series connections, which means strongly limited connectivity of logical elements. In contrast, the three-dimensional network of synapses in the brain represents high connectivity and allows huge numbers of parallelized operations.

It has to be remarked, that the electronic switches in integrated circuits are operated in the sub-nanosecond range. They are very fast in comparison to the electrical processes in synapses which are marked by the release and transport of ions and result in a time scale in the order of magnitude of milliseconds. Thus, semiconductor switches are operated about a million times faster than nervous connections. It is a fascinating vision to combine the advantages of both systems: the fast electron transport in technical nanodevices and the enormous parallelization in the three-dimensional brain-like networks.

The extension of micro and nanofabrication from planar technology into the third dimension demands an extension of production methods into the third dimension, on the one hand. On the other hand, the third dimension of devices needs architectures allowing a fast transfer of power, heat, signals, and–probably–masses, too. The thermodynamic power density has to be lowered drastically in comparison with recent electronic standard devices. Architectures with a strongly enhanced degree of connectivity via three-dimensional networks have to be developed. The thin-film technology and the plane-related lithography have to be substituted by three-dimensional patterning and assembling strategies.

There is no convincing concept of how these challenging developments could be initiated, up to now.

3. Shrinking of Production Facilities

A second important problem is a blatant disproportion between the sizes of production facilities and functional nanodevices. Integrated solid-state devices are produced, recently, in large cleanroom facilities. In general, the size of facilities and their investment volumes increased with the decrease of the size of lithographic structures during the last decades. This trend must be inverted.

New strategies are needed which allow for creating production facilities for nanodevices which can be downscaled to small dimensions. Future nanotechnology should not only produce nano-scaled objects but also have to use nano-scaled production tools. There are required concepts on how the size of manufacturing systems can be limited to nearly the same order of magnitude as the operated objects and the generated products.

At the moment, it seems to be a crazy illusion to shrink nanofabrication facilities into the nanometer range. However, an important step in this direction is the size reduction from large industrial buildings to table-top machines. The next step leads into the matchbox scale, hopefully, followed by steps into milli and micro-manufacturing systems. It is clear that such a development demands a revolution in production strategies and in the designing of production facilities.

It is very probable that these steps cannot be continued with the keeping of all traditional architectures for devices and facilities and for using identical materials. However, the need for miniaturization of production tools is not the only reason for changing materials and technologies. At least from the point of view of sustainability (see below) a strong re-thinking about the character of the recent industry is required, too. The challenges from the point of view of shrinking of nanotechnical production systems and the requirement of environment-adapted production procedures point in the same direction.

4. Completed Sustainability

During the last decades, a lot of discussions and demands concerned the conversion of traditional industrial energy production into sustainable energy management. In the beginning, this discussion was mainly devoted to saving fossil resources. Meanwhile, the need for atmosphere protection became so urgent that the limitation of fossil resources such as coal, gas, and oil stepped into the background. The arguments for closing all power plants which use fossil energy resources as a substitution for fuel-driven machines and cars by electrically driven are coming mainly from the insight into the non-reversible changes in earth climate due to continuous burning of fossil resources. The insight into rising danger for human life on earth, climate change-induced desertification, human miseries, and misery-driven migration are strong arguments for re-enforcing nuclear power despite its non-sustainable character and all unsolved safety and waste-deposition problems.

Sustainable energy production is related to the choice of exploiting spot-concentrated resources or using large surface areas. The classical energy production using coal, oil, gas, and uranium–and to a certain extent, the use of waterpower, too–are spot related. The concentration of energy carrying matter on small spots made energy production comparatively cheap, convenient, and profitable. However, it is connected with deep artificial impacts into the local natural situations and into the global natural material cycles.

Sustainable energy production by solar-thermal, photovoltaic, and wind energy is related to the earth's surface. Larger areas have to be involved in energy production. This aspect moves these technologies close to the basis of photosynthesis by plants and photosynthetic microorganisms. Here, nanotechnology is obviously required to contribute to sustainable energy production by the use of the sunlight illuminating the earth's surface. The question is if using giant windmills and large semiconductor arrays for photovoltaics is the right way for the future.

With a closer look at the arising global problems, the problem of sustainable production of energy is recognized as only the tip of the iceberg. For the future, the rearrangement of all production and consumption processes for achieving complete sustainability is on the agenda. In the future, we have to close the mines and quarries, we have to substitute many plastics and inorganic semiconductors and metals as far as possible, and we have to construct closed-loop strategies for all needs in industry and in everyday life [9]. This demand concerns all devices and production facilities and also concerns the materials used for the construction of windmills and photovoltaic cells. Finally, all human activities have to become adapted to the natural material cycles and their intrinsic time scales [10].

For solving the connected problems and to find new solutions, important hopes are directed to smart technologies, among them nanotechnology [11]. In principle, the problems should be solvable by nanodevices. Living nature shows that biomolecular nanomachines can be based on completely recyclable components. They are able to synthesize a large spectrum of substances, food for many different organisms, and technically usable materials. This type of machinery is able to collect energy and convert it into the energetically charged matter and molecular building blocks by photosynthesis. The sustainability of these processes is kept stable as long as all human activities are part of the natural material cycles, known from traditional agriculture and use of forests, but it gets lost in the case of the most recent types of industrial use.

Recently, there is no clear picture of how nanotechnology can offer the right perspectives for the development of sustainability for industrial production. Nanotechnical devices as well as the majority of applied nanomaterials are made by production concepts that are very far from sustainability including the use of non-renewable resources and the distribution of toxic side products and wastes. For a fundamental change, the most inorganic components have to be substituted by organic–or better–biological materials which can easily be recycled by natural environmental processes. A simple picture might illustrate how far we are from this goal: Imagine a car, an airplane, or even a computer that can be composted in forest soil or in a garden's compost heap!

This strange picture also gives a spotlight on the state of the bottom-up strategy of nanotechnology. The basic idea includes the possibility to substitute classical machines with molecule-scale devices, inorganic macroscopic bodies by organic by filigree nanoscaled molecular architectures. Macroscopic shaping and assembly by macroscopic tools should be substituted by highly specific molecular interaction and self-assembling. The vision of a transition from top-down strategies with its inorganic basis and harsh operations to the soft materials and self-controlled careful processes in the frame of the bottom-up concept is more than three decades old. However, a real change in system designs and industrial production concepts is not recognizable, up to now.

An honest look into the recent state of nanotechnology discloses a disillusioning picture: On the one side stands the top-down strategy which is technically and economically successful but far away from sustainability. On the other side stands the bottom-up strategy which has created a lot of fascinating research projects for more than three decades but brought no breakthrough for sustainable production in any important direction.

5. Learning from Nature

5.1. The Bottom-Up Approach

The idea of constructing complex nanomachines by chemical techniques, macromolecular, supermolecular, and biomolecular chemistry, was pushed forward in the frame of the nanotechnical "bottom-up approach" [12]. The main stimulation for the bottom-up concept has deviated from the fascinating model of functional nanostructures in living nature. Biomacromolecules, microorganisms, and highly specialized cells of different tissues are motivated to search for the realization of analog or similar nanosystems by technical means. The fascinating world of natural nanosystems was opened by a deep insight into the chemical structure of molecules and a detailed understanding of molecular biological and biochemical mechanisms. The most important message from living nature is that we can learn a lot about nanotechnology [7].

This learning starts with insight into the mechanisms of natural sustainability. All technologies with the claim of sustainability have to respect the natural cycles of matter and their time scales. Life on earth can be protected, and the future of mankind can be ensured if we work with and not against the natural material and life cycles. Future material management has to be integrated completely into the natural cycles of matter. Future technologies including nanotechnology have to be based on this integration into this natural network of matter flows.

The most amazing aspect of natural nanoworld is the realization of reliable and reproducible processes of molecular self-organization under conditions of strong thermal fluctuations resulting in bridging the level of molecular building blocks with the micrometer-scaled size of cells containing billions of these smallest units. Meanwhile, many details of the molecular processes are well understood and can be modified in the frame of advanced biochemistry, molecular and cell biology, and molecular biotechnology [13]. However, despite this impressive scientific state of the art [14], there is no new creation of comparable systems by technical means. Therefore, the question remains, what makes natural nanosystems so unique and efficient (Figure 1).

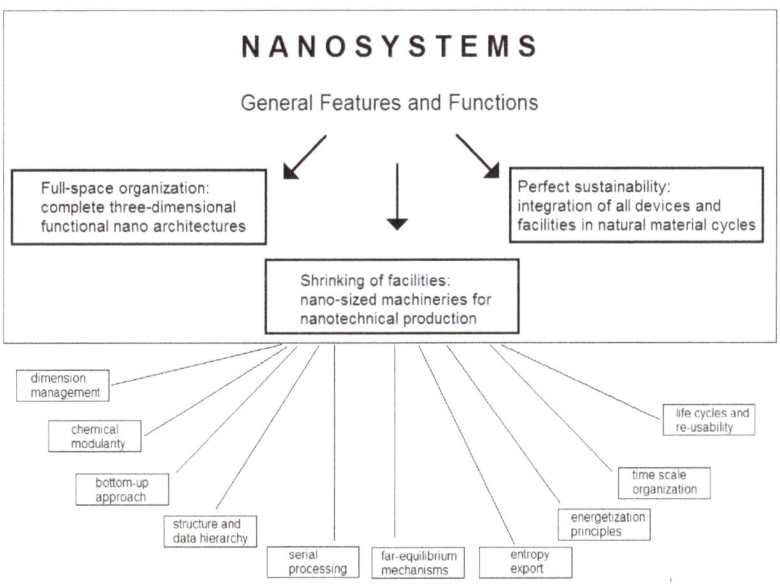

Figure 1. Fundamental requirements for future nanosystems.

5.2. Chemical Modularity–Syntheses Using Molecular Standard Building Blocks

From a chemical point of view, biological cells are extremely complex. Despite this enormous complexity it is fascinating to see, that nature is able to organize cellular activities, metabolic networks, responsivity against changes in environmental conditions, and steering the cell cycle by highly parallelized biochemical reactions with high efficiency and reliability. The fundament is the fact that key molecules and their processing–synthesis, application, and decomposition–follow very clear rules and mechanisms. The restriction on certain classes of key molecules and reactions is still more important than the high number of different chemical species.

The molecular processes are strictly based on standardization and modularity. The conventions for this molecular standardization are not only fixed for one organism or for one species but for the complete system of life–from the simplest bacteria up to highly developed plants and animals. This standardization is perfectly represented by nucleic acids and proteins. All DNA molecules are constructed by four different standard modules, only–the four different nucleotides. Single molecular strands of DNA present a linear arrangement of these few standard building units. The simple construction principle, the availability of enzymes as natural tools for processing of DNA fragments, and the principle of molecular recognition by base-pairing allowing for manipulation of DNA in the artificial molecular construction is an impressive example for the realization of a bottom-up approach [15,16].

Besides nucleic acids, proteins are built by a modular chemical principle, too. All proteins are primarily synthesized from a set of 20 alpha-amino acids. The huge variability in structures and functions of proteins is made from this standardized set with only slight post-translational modifications. It is very astonishing what an incredibly large spectrum of biochemical activities and cellular functions is implemented by the arrangement of these 20 amino acids in the right linear chain, only. Meanwhile, the structure and properties of natural proteins can not only be enlightened but modified in the frame of molecular engineering [17].

The limitations of these few standard modules have less to do with the unique and expedient properties of the amino acids as substance class but must be specially explained by the ultimate need of limitation. Probably, other of the millions and billions of different chemical substances could be suited for the creation of a powerful modular system. The deciding trick is the restriction to one special molecular system which was optimized in the early biomolecular evolution.

5.3. *Management of Geometric Dimensions*

Molecular tools are demanded special structures, recognition, and binding sites, for special chemical functionalities and combinations of stiff and mobile molecular components. It is clear, that these requirements can only be met by three-dimensional molecular architectures. Natural chain molecules as nuclei acids and peptides are primarily synthesized as linear objects. These constitutional elements form a chain that can be described by a sequence of characters. However, despite functioning as linear information-carrying molecules, both substance classes are able to form complex three-dimensional structures, spontaneously [18]. The folded proteins–and also catalytic nucleic acid molecules folded by internal base-pairing–represent, indeed well-defined and functional three-dimensional geometries.

Although the three-dimensional character is very important for the biomolecular function of most proteins, they are not synthesized by the three-dimensional molecular mounting of the basic modules. In opposite, they are formed as linear objects, primary, and only folded secondary in secondary and tertiary structures. The mastery of three-dimensional space is based on a primary restriction to one dimension. Later only, this linear structure is self-organizing into a three-dimensional architecture by a program encoded in the order of building units inside the original linear chain.

Recently, technical systems are available to build such molecular chains by automated syntheses. They are well suited for the generation of DNA libraries which can be used for many purposes, for example for DNA nanotechnology and DNA origami arrangements [16]. DNA can also be used for molecular labeling and for the creation of DNA-encoded chemical libraries [19,20]. In principle, oligopeptides and–to a certain length- polypeptides can be generated by automated solid-phase syntheses in a similar way. However, besides these biologically created substance classes of sequence molecules like nucleic acids and proteins, no other substance systems for molecular construction by folding (one-dimensional) molecular sequences into a large spectrum of there-dimensional molecular architectures are available. The usual synthetic polymers, copolymers, block copolymers, and related systems are far away from intelligent dimension management and self-optimized three-dimensional folding.

It is an urgent challenge to evaluate which artificial modular chemical systems could be of interest for technical purposes and fulfill the following requirements:

- simple chemical building units
- automatable free synthesis of sequence molecules (sequence generation)
- high variability of chemical, mechanical and electronic properties with a small set of different building units
- self-organized folding into well-determined three-dimensional functional structures
- accessibility of characterization by automated sequencing (sequence read-out)
- transferability of sequence information into other classes of sequence molecules
- replicability of sequence molecules

5.4. Serial Processing

The crucial advantage of sequence molecules is their one-dimensional character. It ensures the possibility of direct transfer of chains of commands or characters into linear spatial arrangements and vice versa. This linear principle is of central importance in technical processes and found, for example, in the linear character of production in assembly lines, in the sequence of characters, words, and sentences in texts as in the linear structure of computer programs. It is also essential for the biological synthesis and replication of key molecules like DNA, RNA, polysaccharides, and proteins. All these biomolecules are generated or copied in a linear process, by stepwise addition of building units.

In future molecular nanotechnology, systems for serial processing are required, too. There have to be developed micro or nanomachines for molecular manufacturing and conversion (Figure 2). This includes the following functions:

(A) copying of sequence molecules: There is a nanotechnical system required which is able to recognize a certain type of sequence molecules formed by–at least–two classes of molecular building units, to use it as a template, and to catalyze the synthesis of an identical molecule, which means, with the identical order of molecular building units inside the molecular chain. (In analogy to replication of nucleic acids, for example [21,22]).

(B) readout of molecular sequence information: There is required a nanotechnical system that is able to recognize a sequence of molecules, distinguish different molecular building units inside the molecular chain of this sequence molecule, and transfer this sequence information into a physical signal readable by technical standard systems like an optical or an electronic sensor. (Miniaturized sequencing machine).

(C) synthesis of sequence molecules by program-controlled stepwise addition of building units: There is a nanotechnical system required that is able to receive a sequential data set from a technical standard data system like optical or electronic devices, and to translate, this information into a synthesis program for chemical coupling of molecular building units into a molecular chain with a sequence of building units exactly corresponding to the input data set [23]. (Miniaturized synthesis machine).

(D) translation of the information content of one sequence molecule to another type of sequence molecule: There is a nanotechnical system required that is able to recognize a certain type of sequence molecules formed by—at least—two classes of molecular building units and to use it as a template for the catalytic synthesis of sequence molecules of another type of sequence molecules. (In analogy to transcription or translation process in molecular biology).

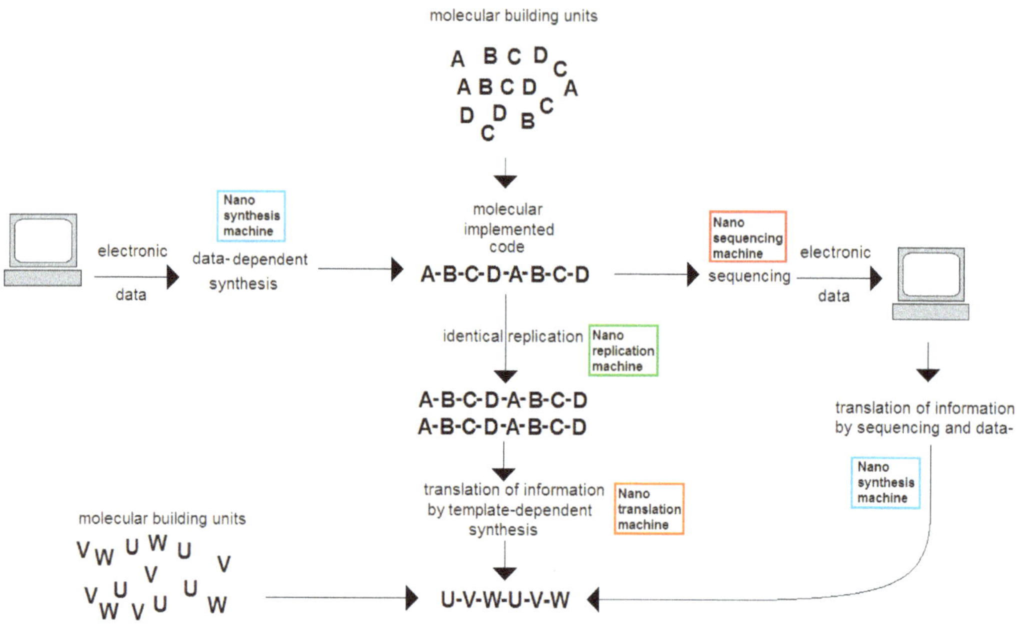

Figure 2. Basic processes for conversion of data and molecular-encoded information and realization and modification of molecular functionality.

5.5. Hierarchy of Molecular Structures and Information Units

It is a trivial fact that the required complex nanotechnical machinery consists of large numbers of atoms and building units. It is impossible to control a large number of components without having an ordering mechanism. Therefore, a hierarchical structure is required. Such a hierarchy has to involve two simple main aspects: (1) Lower levels of objects (or sub-systems) in the hierarchy are used as components for the construction of higher levels. (2) The knot strength in the hierarchy–which means the number of elements belonging to one common parent unit–should not be too large and in the same order of magnitude for all organization levels, which means within the complete hierarchy.

The English language is a nice example of a hierarchical organization of information-carrying units: The first level is formed by the letters (characters). The second level is formed by words, the third level by sentences, the fourth level by paragraphs, the fifth level by chapters, and the sixth level by books, typically. The whole system is forming a one-dimensional data set, meaning a structured line of characters that can be sequentially written and read. Synthetic molecular systems should be one-dimensionally constructed in the same way: Nature shows that nucleic acids are structured similarly to language using the nucleotides as letters forming triplets (like words), genes (like sentences or paragraphs), gen clusters (like chapters), chromosomes (like books) and complete genomes (like a library).

Artificial systems for molecular information storage and processing should be structured in a hierarchical order, too: At the first level, building units have to be defined, for example by limited sets of different monomers which can be used for creating sequence molecules by them. At the second level, modules consisting of several monomers should get a certain meaning, and finally, macromolecules representing large data sets have to be synthesized.

The natural organization of proteins impressively shows how such a hierarchy can work in the nano cosmos [24]: each proteinogenic amino acid consists of four groups of a low number of atoms: three constant groups (carboxyl group COOH, amino group NH_2, and central methin group CH) and one variable group. Twenty different amino acids are used for coding all information and chemical functions which are required for the three-dimensional self-organized construction of the folded protein, the so-called tertiary structure. In dependence on the type of protein, one or two further organization levels exist between the level of single amino acids and the complete protein. The first is the formation of secondary structure elements like betasheets and alpha-helices. The second is the formation of domains in the case of proteins which are organized in domains–typically between 2 and about a dozen of these subunits. Finally, two or several proteins are assembled into supermolecules forming the quarternary structure of proteins.

The combination of modularity and hierarchical organization is also an important precondition for consequent sustainability. It supports the organization of life cycles, the de-assembling of systems to different levels of integration, and the re-use of elementary building units as well as of larger modules and sub-systems (Figure 3).

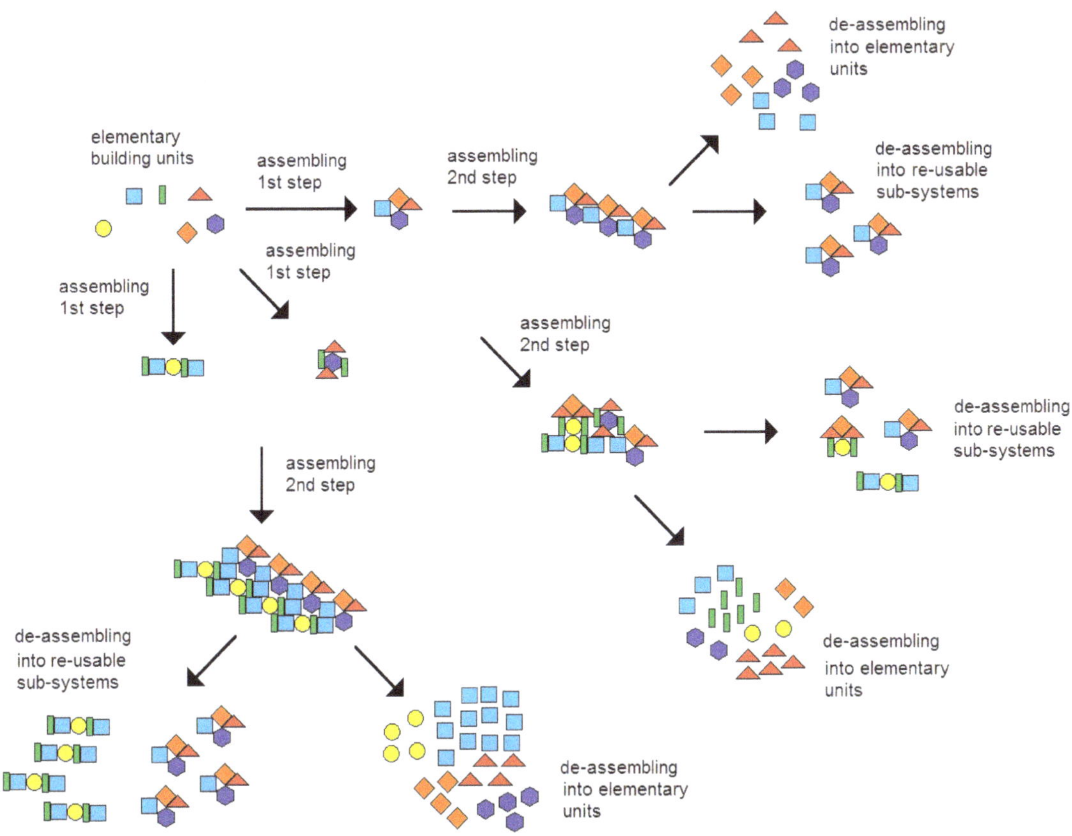

Figure 3. Nested assembling and de-assembling of modular and hierarchical constructed systems for a sustainable circular economy (schematically).

5.6. Hierarchy of Bond Strengths and Coupling of Near- and Far-Equilibrium Processes

Despite the fact that the processes of creation, read-out, and processing of information-carrying molecules are under the control of supermolecular nanomachinery, molecular self-organization is crucial for well-determined molecular processes [25]. All elementary processes proceed with the background of Brownian motion and thermal fluctuation of chemical reactions. Surprisingly, complex directed development processes are based on chemical reactions and can be controlled in spite of the unpredictable single motions in the noise of thermal fluctuations.

How can such a system work? The solution is found by the combination of reactions with more and less distance to the thermal equilibrium. Highly reversible elementary processes are combined with strictly non-reversible procedures. On the one hand, there are reactions marked by low activation thresholds, which permanently run for the adaption of chemical equilibria. On the other hand, there are thermodynamically more demanding reactions marked by higher activation thresholds and are running into a preferential direction as far as the chemical system is kept at a certain distance to the related chemical equilibrium.

These thermodynamic differences present a fundamental condition for controlling complex biochemical reaction networks in living cells. The thermodynamic boundary conditions determine the kinetics of the ongoing chemical processes. They are responsible for the realization of typical time constants in biochemical reactions and in response time to perturbations from outside.

The molecular basis for these different time constants is given by the differences in the bond strengths. Biomolecular processes are using systems of stepwise-structured bond strength. Living nature has realized a complementary system of a hierarchy of structures, a hierarchy of (bio)-chemical time constants, and a hierarchy of molecular bond strengths. The example of proteins impressively shows the connection between the hierarchical molecular structure and the hierarchy of bond stability. Non-polar covalent bonds dominate the lowest structural level, the single amino acids. The internal bonds of the single amino acids are very stable and cannot be split by hydrolyzation. At the next level, connections between single amino acids are formed by peptide bonds, which present a polar covalent bond. They are forming the primary structure of oligo and polypeptides, which can be split by hydrolyzation again into amino acids. At the third level of bond strength hierarchy, dense regular polyvalent structures of hydrogen bridges ("H-bridges") are responsible for the formation of comparative stable secondary structures as helices and leaflets. Finally, a combination of second-order interactions as dipole interactions, H-bridges, and non-polar interactions contribute to the formation of tertiary and quarternary protein structures.

Despite the fact that, meanwhile, the principal molecular structures and mechanisms of proteins and protein biochemistry are well known, there is no comparable new artificial molecular system that is comparable with peptides and proteins. The translation of the recognized principles of protein structure, protein synthesis, and protein chemistry to other types of molecules is a still-unsolved problem.

5.7. Time-Scale Management

The hierarchical organization of bond strength and bond sensitivity is strongly connected to time-scale management. The chemical strategy of controlling time scales is the control of reaction rates. In principle, the rate of a chemical reaction can be controlled by temperature. However, it has to be reconsidered, that the degree of freedom for varying temperatures is not very compatible with living systems should be achieved, on the one hand. On the other hand, a temperature shift would accelerate or delay all chemical reactions in a complex reaction network. However, a possibility for independent control of individual reaction rates are needed.

Living nature has developed a very subtle instrument for the individual adaptation of reaction rates. The trick consists of the fine-tuning of activation energy. This fine-tuning is achieved by small differences in the efficiency of the biocatalysts. Small variations in the sequence and structure of enzymes result in more or less strong changes in the catalyzed reactions.

For molecular nanotechnology, analogous instruments are required. Efficient catalysts must be constructed in such a way that small variations in their structure can be used for tuning catalytic activity. It is hard to imagine that this challenge can be met simply by inorganic solid-state catalysts as they are recently mainly used in technical heterogeneous catalysis. Instead, enzyme-analog technical catalysts have to be developed. They could help to realize nanotechnical time-scale management in analogy to biochemical reaction networks.

5.8. Active Drive by a Universal Energy Conversion System

Future nanosystems cannot only work by passive chemical or bio-analogous catalytic processes alone. In addition, driven partial processes are required, which allows to elongate reaction systems in controlled directions from the thermodynamic equilibrium.

All living beings are such driven far-equilibrium systems. Working machines as well as working computers, for example, are far-equilibrium systems, as long as they are running. Machines as well as living beings convert input energy in order to build and to maintain a far-equilibrium state.

The introduction of well-controlled driven far-equilibrium processes demands a reliable energy supply. In technical environments, electrical power is used, therefore, mainly for a standardized energy supply. The energy flow of technical systems is, normally, adapted, to the supply and consumption of electrical energy. In contrast, organisms and cells are using chemical energy for maintaining far-equilibrium states. However, these systems need a standardized power supply too. The central role of adenosine triphosphate (ATP) in living cells represents such an important standardized chemical power supply. A high number of driven enzymatic processes are based on the activation of ATP. Cells are producing ATP as a universal energy source enabling them to drive a lot of essential processes against the general thermodynamic time arrow.

Artificial nanosystems need an energy supply too. An electrical power supply by a permanent wire connection is not suited if the nanosystems should be mobile. The advantage of the chemical energizing as it is realized by the ATP system is that the power-carrying molecule itself is not permanently integrated into the nanotechnical systems, but is picked up from the environment. Cells are "charging" their internal system by the "power molecules" and the bionanomachines make a self-service of them. Similar approaches are needed for artificial nanosystems, too. The feeding can be realized by chemical energizing, but also by physical energizing from outside, for example by light or by electricity. In the last-mentioned cases, the outside energy source could be combined with a small nanosystem-internal storage for temporal accumulation of small amounts of energy like a "nano battery system".

5.9. Entropy Export Management

Energizing, mechanisms with driven processes, and energy conversion are key features of a living or a working system. Each process of energy conversion is connected with the production of entropy. From a thermodynamic point of view, each living being and each working machine produces entropy [26]. These systems have to be open in a thermodynamical sense. They have to be supplied by convertible energy and they have to have the ability to export the produced entropy. The transfer of entropy from the living system to the environment is absolutely required for keeping the system alive and to realize a further development, for accumulation of information and increasing complexity. Active machines, as well as living beings, are "Entropy-exporting" systems.

Without entropy export, running systems would lose their driving forces and could be destroying themselves. Therefore, active nanosystems–like all working machines–have to be equipped with an entropy export mechanism. Typically, this entropy export is marked by the input of "high-value" energy like electrical or chemical energy and output of low-value" energy as heat.

Besides energy conversion, the entropy export can also be realized by chemical reactions or by the distribution of substances. Light-driven systems are producing exporting entropy by uptake of short-wavelength photons and a release of heat or long-wavelength photons.

A key issue is the coupling of all driven nanosystem activities with the entropy export mechanism. Most desirable is a standard entropy management and export strategy which can be connected with different forms of nanosystem activities including information processing, chemical conversion, and synthesis, as well as active mobility and directed motion.

5.10. Local Information Processing, Communication, and Controlled Functional Autonomy

An efficient entropy export management is also the precondition for constructing nanosystems with individual mobility and functional autonomy. Besides the mechanical drive, energy is needed for an internal signal and information processing, too.

Functional autonomy of nanosystems [27] demands internal data storage [28] and information processing. Such nanosystems must be able to receive signals from the environment to convert the primary signals, evaluate them, make a decision about the response to a signal from the outside, and initiate the response activity. The most simply "nano-brain" activity would consist of the realization of case-sensitive activity programs. A higher level of "technical intelligence" could be achieved if autonomous nanosystems have implemented learning mechanisms, for example by algorithms of artificial intelligence.

Technical nanosystems should not operate completely independent of the control from outside. Technical systems are built and released with a certain purpose. Therefore, activity control from outside is necessary. Thus, these functional nanosystems must include communication competence. The data exchange between the nanosystem and a master system would allow for the use of all advantages of operational autonomy of the nanosystem with safe control of its activities. This ability for communication can also be used for decentralized cooperation of several or many partial autonomous nanosystems in the form of swarm activities.

These functions likely cannot be realized by the typical tools of classical construction of technical systems or by the usual strategies of solid-state electronics. It could be rather realized by functional bio-analogous super-molecular assemblies. Obviously, we are at the very beginning of the development of such molecular-based systems.

5.11. Establishing Life Cycles by Controlled Self-Assembling, Dissembling, and Re-Use

One of the most challenging points for future nanosystem development is the absolute need for sustainability. A short look into what the components of recent computer, optical communication, and sensor systems teach us is that the recent material use, as well as the structures of our nano-sized devices, is far away from sustainability. We need special mineral resources for their production, and their recycling is energy-demanding and incomplete. Lost or released components are often a danger for animals, plants, and soils.

It has to be expected, that we cannot completely dispense with solid-state devices, metals, semiconductors, and lithographic technologies. However, future constructions of nanosystems must solve their difficult recycling problems.

The best solution would consist of a complete substitution of inorganic metals and semiconductors, compound semiconductors, as well as doped silicon–by functional organic materials. Therefore, new types of synthetic macromolecules, supermolecules, and functional self-assembling systems are required. A very promising field is the development of new devices based on functional derivatives of graphene and carbon nanotubes. Probably, metals cannot be completely excluded because they are needed for special electronic, optical, and catalytic properties. However, their content in devices should be reduced drastically down to a level comparable with the absolutely required metal ion content in metalloenzymes, for example, enabling them as highly efficient biocatalytic tools.

In addition, and meanwhile, we have to think about nanosystem designs including strategies for easy dissembling in a recycling operation. Still, better would be nanosystems with self-dissembling mechanisms. Such mechanisms could directly supply the raw material for new production processes. Dissembling and production would be realized in one integrated factory. The automated splitting of complex devices in their components and the separation of all different materials would include the possibility to re-use them also in case of completely new designed products. Therefore, conceptual modularity in all devices is required, which includes the construction of state-of-the-art systems and then disassembly of these systems after use to the elementary modules and pure materials for the next generation of use.

The recent "life cycles" of technical devices are determined by their reliable operability and by their level in the process of technological innovation. These criteria have to be completed by the criterion of "recycling lifetime". This means, that future nanosystems have to be constructed for becoming systematically recycled after a reasonable time of use.

6. Outlook: Integrability of Technical Matter in Natural Material Cycles

Recently, the term "recycling" stands for an improvement of material use and protection of natural resources. However, for many technical and everyday applications, only a small number of cycles can be realized. The disaster of overusing the earth's natural resources can be delayed by the recent strategies of "recycling", but it cannot be avoided. Therefore, a completely new approach is demanded.

Above everything stands the problem of keeping the earth's biosphere viable. This includes the global cycles of organic and inorganic materials by the non-biological transport and conversion processes and by the entirety of metabolic activities in the biosphere. These natural cycling processes must become the ultimate scale for a sustainable economy.

A real sustainable solution means the complete integration of all production and consumption processes into natural life cycles. Therefore, the type of objects, the applied materials, the combination of materials, and the placing and distribution of all materials, devices, and all residual technical or personal waste must fulfill the criterion of integrability into natural material cycles. This integration is related to the qualities as well as to the quantities of matters and has to further reconsider the very different process rates and environmental conditions in different parts of the earth.

We have to adapt our technical products and product application to the conversion processes in living nature. We have to reconsider the converted materials, their conversion paths, transport mechanisms, and their feedback on the development of living nature in the concerned regions and places. Resource management, material use, and recycling are directly coupled with all aspects of species protection and maintenance of biocenoses. Therefore, we need new thinking about interfaces between the technical and natural world.

Nanotechnology is particularly asked for the development of these new interfaces. It could supply keys for the adaptation of advanced technical solutions to ecological requirements. The future progress in nanotechnology should lead this technical field closer to the conditions and relations of living and ecological systems.

This future convergence between technology and nature will probably become a process in which most technologies have to respect and support the original natural mechanisms. However, in a further future, we will learn how much we can modulate the natural cycles of life and matter without risking the natural global viability. It is to be assumed, that the fusion of nanotechnology with biotechnology and supermolecular chemistry will be a deciding step in this direction.

Funding: This research received no external funding.

Conflicts of Interest: The author declares no conflict of interest.

Entry Link on the Encyclopedia Platform: https://encyclopedia.pub/13441.

References

1. Feynman, R.P. There's plenty of room at the bottom. *Eng. Sci.* **1960**, *23*, 22–36.
2. Köhler, M.; Fritzsche, W. *Nanotechnology. An Introduction to Nanostructuring Techniques*; Wiley VCH: Weinheim, Germany, 2007; pp. 1–11.

3. Vollath, D. Nanomaterials. In *An Introduction to Synthesis, Characterization and Processing*; Wiley VCH: Weinheim, Germany, 2008.
4. Madou, M. *Fundamentals of Microfabrication*; CRC: Boca Raton, FL, USA, 1997.
5. Li, L.; Liu, X.; Pal, S.; Wang, S.L.; Ober, C.K.; Giannelis, E.P. Extreme ultraviolet resist material for sub-7-nm patterning. *Chem. Soc. Rev.* **2017**, *46*, 4855–4866. [CrossRef] [PubMed]
6. Drexler, K.E. *Nanosystems*; John Wiley & Sons: New York, NY, USA, 1992.
7. Lindsey, S. Self-assembly in synthetic routes to molecular devices. Biological principles and chemical perspectives: A review. *New J. Chem.* **1991**, *15*, 153–180. [CrossRef]
8. Balzani, V.; Credi, A.; Venturi, M. *Molecular Devices and Machines. A Journey into the Nanoworld*; Wiley VCH: Weinheim, Germany, 2003.
9. Genovese, A.; Acquaye, A.A.; Figueroa, A.; Koh, S.C.L. Sustainable supply chain management ant the transition towards a circular economy; evidence and some applications. *Omega Int. J. Manag. Sci.* **2017**, *66*, 344–357. [CrossRef]
10. Köhler, J.M. The ecological time scale violation by industrial society and the chemical challenes for transition to a sustainable global entropy export management. *Green Process. Synth.* **2014**, *3*, 33–45.
11. Nizetic, S.; Djilali, N.; Papadopoulos, A.; Rodrigues, J.J.P.C. Smart technologies for promotion of energy efficiency, utilization of sustainable resources and waste management. *J. Clean. Prod.* **2019**, *231*, 565–591. [CrossRef]
12. Lieber, C.M. Nanoscale science and technology: Building a big future from small things. *MRS Bull.* **2003**, *28*, 486–491. [CrossRef]
13. Gopfrich, K.; Platzmann, I.; Spatz, J.P. Mastering complexity: Towards bottom-up construction of multifunctional eukaryotic synthetic cells. *Trends Biotechnol.* **2018**, *36*, 938–951. [CrossRef]
14. Tasbas, M.N.; Sahin, E.; Erbas-Cakmak, S. Bio-inspired molecular machines and their biological applications. *Coord. Chem. Rev.* **2021**, *443*, 214039. [CrossRef]
15. Seeman, N.C.; Belcher, A.M. Building nanostructures from the bottom up. *Proc. Natl. Acad. Sci. USA* **2002**, *99*, 6451–6455. [CrossRef]
16. Han, D.; Qi, X.; Myhrvold, C.; Wang, B.; Dai, M.; Jiang, S.; Bates, M.; Liu, Y.; An, B.; Zhang, F.; et al. Single-stranded DNA and RNA origami. *Science* **2017**, *358*, eaa02648. [CrossRef] [PubMed]
17. Glasscock, C.J.; Lucks, J.B.; DeLisa, M.P. Engineered protein machines: Emergent tools for synthetic biology. *Cell Chem. Biol.* **2016**, *23*, 45–56. [CrossRef] [PubMed]
18. Ellis, T.; Adie, T.; Baldwin, G.S. DNA assembly for synthetic biology: From parts to pathways and beyond. *Integr. Biol.* **2011**, *3*, 109–118. [CrossRef] [PubMed]
19. Dumelin, C.E.; Scheuermann, J.; Melkko, S.; Neri, D. DNA-encoded chemical libraries. *J. Biotechnol.* **2006**, *126*, 568–581. [CrossRef]
20. Neri, D.; Lerner, R.A. DNA-encoded chemical libraries: A selection system based on endowing organic compounds with amplifiable information. *Ann. Rev. Biochem.* **2018**, *87*, 479–502. [CrossRef]
21. Quan, J.Y.; Tian, J.D. Circular polymerase extension cloning for high-throughput cloning of complex and combinatorial DNA libraries. *Nat. Protoc.* **2011**, *6*, 242–251. [CrossRef]
22. Stratmann, S.A. DNA replication at the single-molecule level. *Chem. Soc. Rev.* **2014**, *43*, 1201–1220. [CrossRef] [PubMed]
23. Badi, N.; Lutz, J.F. Sequence control in polymer synthesis. *Chem. Soc. Rev.* **2009**, *12*, 3383–3390. [CrossRef]
24. Köhler, J.M. What proteins teaching us on fundamental strategies for molecular nanotechnology? *Nanotechnol. Rev.* **2015**, *4*, 145–160. [CrossRef]
25. Schmidt, H.W.; Wurther, F. A periodic system of supramolecular elements. *Angew. Chem. Int. Ed.* **2020**, *59*, 8766–8775. [CrossRef]
26. Yildiz, C.; Semerciöz, A.S.; Yalçınkaya, B.H.; Ipek, T.D.; Ozturk-Isik, E.; Özilgen, M. Entropy generation and accumulation in biological systems. *Int. J. Exergy* **2020**, *33*, 444–468. [CrossRef]
27. Lagzi, I. Chemical robots—Chemotactic drug carriers. *Centr. Eur. J. Med.* **2013**, *8*, 377–382.
28. Martens, S.; Landuyt, A.; Espeel, P.; Devreese, B.; DuPrez, F. Multifunctional sequence-defined macromolecules for chemical data storage. *Nat. Commun.* **2018**, *9*, 4451. [CrossRef] [PubMed]

encyclopedia

MDPI

Entry

Novel Bioactive Extraction and Nano-Encapsulation

Shaba Noore [1,2], Navin Kumar Rastogi [3], Colm O'Donnell [2] and Brijesh Tiwari [1,2,*]

1 Department of Food Chemistry & Technology, Teagasc Food Research Centre, Ashtown, D15 DY05 Dublin, Ireland; shaba.noore@ucdconnect.ie
2 School of Biosystems and Food Engineering, University College Dublin, Belfield, D04 V1W8 Dublin, Ireland; colm.odonnell@ucd.ie
3 Department of Food Engineering, CSIR-Central Food Technological Research Institute, Mysuru 570020, India; nkrastogi@cftri.res.in
* Correspondence: Brijesh.Tiwari@teagasc.ie

Definition: An extraction technology works on the principle of two consecutive steps that involves mixture of solute with solvent and the movement of soluble compounds from the cell into the solvent and its consequent diffusion and extraction. The conventional extraction techniques are mostly based on the use of mild/high temperatures (50–90 °C) that can cause thermal degradation, are dependent on the mass transfer rate, being reflected on long extraction times, high costs, low extraction efficiency, with consequent low extraction yields. Due to these disadvantages, it is of interest to develop non-thermal extraction methods, such as microwave, ultrasounds, supercritical fluids (mostly using carbon dioxide, SC-CO_2), and high hydrostatic pressure-assisted extractions which works on the phenomena of minimum heat exposure with reduced processing time, thereby minimizing the loss of bioactive compounds during extraction. Further, to improve the stability of these extracted compounds, nano-encapsulation is required. Nano-encapsulation is a process which forms a thin layer of protection against environmental degradation and retains the nutritional and functional qualities of bioactive compounds in nano-scale level capsules by employing fats, starches, dextrins, alginates, protein and lipid materials as encapsulation materials.

Keywords: non-thermal extraction; bioactive compounds; nanoencapsulation; ultrasound; cold plasma; high-pressure processing; supercritical extraction; pulse electric field

Citation: Noore, S.; Rastogi, N.K.; O'Donnell, C.; Tiwari, B. Novel Bioactive Extraction and Nano-Encapsulation. *Encyclopedia* **2021**, *1*, 632–664. https://doi.org/10.3390/encyclopedia1030052

Academic Editors: Raffaele Barretta, Ramesh Agarwal, Krzysztof Kamil Żur and Giuseppe Ruta

Received: 11 June 2021
Accepted: 20 July 2021
Published: 26 July 2021

1. Introduction

Bioactive compounds also known as secondary metabolites are widely present in plant matrix and over the past few decades, several in vitro and in vivo reports including epidemiological, and cohort studies provide evidence that consumption of plant-based food provides protection against several diseases. These bioactive extracts are also capable of treating chronic diseases including cancer, cardiovascular and diabetes mellitus (DM). Nutraceutical and pharmaceutical sectors use these extracts to develop functional food- and plant-based medicines, which have a potential to cure and deliver health benefits. According to the World Health Organization (WHO), about 80% of the global population depends on natural medicines. The initial steps followed to use these active compounds from plant matrix are extraction followed by pharmacological testing, isolation, characterization and clinical evaluation. Figure 1 represents a detailed flow chart of bioactive compound extraction from plant matrix.

The quality and yield of the bioactive compounds depend on two important factors: (a) the method opted for its extraction, (b) its extraction parameters including plant matrix type, solvent used, time and temperature. The most conventional method employed for bioactive extraction is Soxhlet extraction, maceration and hydro-distillation. Although, these techniques are commercially employed, but excessive use of solvents and longer processing times are the downsides of these technologies. Presently, demand for sustainable, chemical-free, advanced extraction processes with enhanced overall yield of bioactive compounds, also known as "green techniques", which include ultrasound-assisted, enzyme-assisted, microwave-assisted, pulsed electric field-assisted, high-pressure processing, supercritical fluid and pressurized liquid extraction processes are gaining attention. Treating the plant matrix with these green technologies helps in breaking the cell structure, which allows the bioactive compound to leach or rinse out from the cell wall through solvents; as a result, it enhanced extraction yield. Further, purification of the extracted bioactive poses another technological challenge as each of these compounds has a unique molecular structure depending on their type, source and biological activity. The extracted compounds can be further purified, employing super critical CO_2 isolation, by addition of a co-solvent including ethanol, water to isolate the respective bioactive compound efficiently at an optimized temperature and pressure [1]. In addition, it is essential to protect the extracted bioactive compound post extraction and purification, as these compounds are highly sensitive to environment exposure including moisture and high temperature (sensitive under heat, light, oxygen). Therefore, protection techniques such as nanoencapsulation are used to ensure that biological activity of these compounds is preserved until they reach and perform their function at the targeted location in the human body.

Encapsulation plays a vital role in protecting the bioactive compounds from getting degraded. At present, there are two kinds of encapsulation including microencapsulation and nanoencapsulation. The reason why nanoencapsulation is preferred over microencapsulation is due to its nano-scale size, as the smaller the size of the capsules, the higher their bioavailability and their release can be modified and controlled in a better way comparatively. Nanoencapsulation provides a protective shield around bioactive compounds. It is a system where a suitable nano-carrier, resistant to enzymatic degradation especially in gastrointestinal tract including chitosan, zein, and alginate, are widely used to encapsulate bioactive compounds employing several delivery methods including association colloids, nano-particles, nano-emulsions, nano-fibers/nano-tubes, nano-laminates. The selection encapsulation method is based on two main factors: (a) nature of the core material; (b) nature of wall material including wall material size, thickness, solubility, permeability and its rate of delivery. Basically, these techniques are classified into three main genres including chemical (emulsion and interfacial polymerization), physical–chemical (emulsification and coacervation) and physical–mechanical methods (spray-drying/spray-cooling/spray-congealing/prilling, freeze-drying, electrodynamic methods and extrusion) [2]. However, in certain cases, combinations of these techniques are practiced as in the case of emulsification; first using homogenization, the emulsions are prepared and later converted to dry powder state using spray-drying and/or freeze-drying techniques [3]. Reports indicate that about 80–90% of flavor encapsulation is done using spray-drying, while 5–10% by spray-chilling, 2–3% by melt extrusion and ~2% by melt injection [4]. Castro et al. [5] reported electro-spinning encapsulation as a heat-free technique to encapsulate fragrance and flavor, which is extremely promising for heat-sensitive compounds.

To recapitulate, this chapter provides a comprehensive summary on several aspects of bioactive extraction using non-thermal technologies and its nanoencapsulation. A brief description on nano-carriers employed for encapsulation is also discussed along with the detailed description of their application in food systems. Various opportunities and future challenges are also outlined.

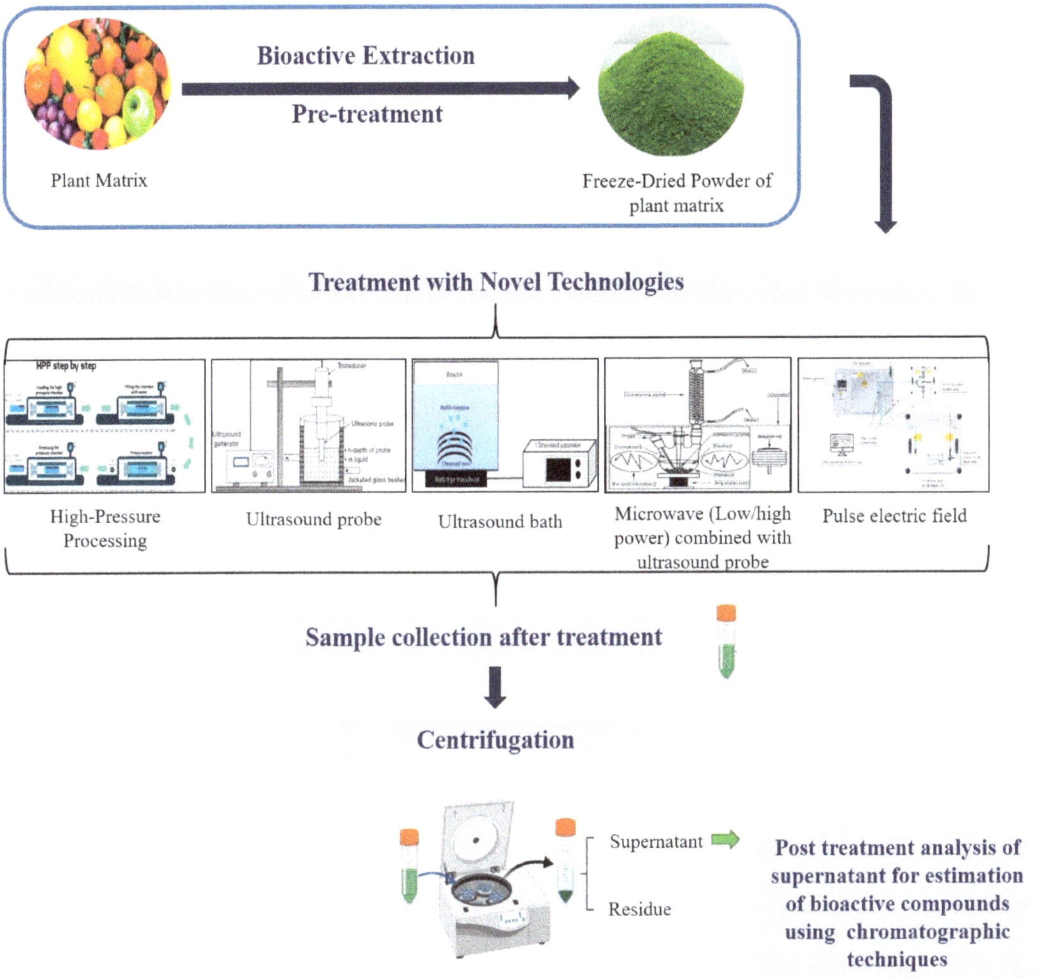

Figure 1. Illustration for the extraction of bioactive compound using novel strategies.

2. Bioactive Compounds from Plant Materials

Ever since the beginning of human existence, plants have always been a boon for living a healthy life, as they not only provide a healthy environment to live, but most importantly, they provide food and bioactive compounds for medicinal use. In the beginning, plants and plant-related foods were used as a source of food and nutrition; later, their medicinal properties were discovered, which were able to cure diseases. Vinatoru et al. [6] reported that Egyptian papyruses extracted oil from coriander and caster and used it in several applications including medicine, cosmetics and as preservatives. Further, Paulsen et al. [7] reported that, during the Roman and Greek era, herbal plants were used by several therapeutics. According to literature [8], bioactive compounds comprise three different categories including terpenes/terpenoids, alkaloids and phenolics. Basically, the chemical structure of these three categories differs, as shown in Figure 2, and maximum bioactive compounds extracted from plant matrix belong to the terpenoids family.

Additionally, these compounds are classified based on their clinical and toxicological attributes as follows.

2.1. Glycosides

Glycosides are generally bonded by a mono/oligosaccharide or uronic acids. The part that is bonded with saccharide is called glycone and the other part is termed as aglycone, which consists of pentacyclic triterpenoids/tetracyclic steroids. The major subgroups of glycosides include cardiac glycosides, saponins, anthraquinone, glucosinolates and cyanogenics. Moreover, flavonoids commonly exist as glycosides. These glycosides are broken down in colon post ingestion; however, hydrophobic glycosides tend to get absorbed by the muscle cells. Cardiac glycosides are generally found in plants such as the *Scrophulariaceae* family, specifically in *Digitalis purpura* and in *Convallaria majalis* from *Convallariaceae* family. Additionally, cyanogenic glycosides can be found in the *Prunus* spp. of *Rosaceae* family as well as saponin, a bitter-tasting compound is found in glycosides. These saponin glycosides, found widely in the *Liliaceas* family (*Narthesium ossifragum*), is comprised of bigger molecules attached to hydrophilic glycone as well as hydrophobic aglycone, which creates forming quality, and thus it is used in the production of soap/detergent. Saponins also play an important role in modulating immune system and reducing blood sugar level. Besides, anthraquinone glycosids found in the *Rumex crispus* and *Rheum* spp. of polygonaceae family help in electrolyte secretion as well as induction of water and peristalisis in colon. Moreover, flavonoids are comprised of tri-ring at the center of the structure and proanthocyanidin is an oligomer in flavonoids. These two groups of compounds can also exist as glycosides. These are responsible for liberating antioxidant properties, inflammation and anti-carcinogenic activities. They are also responsible for the pigments in a wide range of plants. In addition, isoflavones are considered as a nutritional supplement type of bioactive molecules produced almost exclusively by the Fabaceae (Leguminosae or bean) family. They are basically a precise group of molecules, popularly known as phytochemicals, natively found in legumes and spices such as red clove. They are also considered as antioxidant molecules, as they help in the reduction of damage caused by oxygen in the body. Additionally, it plays a vital role in fighting against cancer cells.

2.2. Tannins

Tannins are widely found in plants, especially in the *Fagaceae* and *Polygonaseae* family. They are basically divided into two types, including condensed and hydrolysable tannins. Condensed groups of tannins are comprised of bigger polymers of flavonoids, while hydrolysable groups of tannins are clusters of monosaccharide (glucose) bonded with various derivatives of catechin. Tannin molecules tend to indiscriminately bind with protein molecules. Larger groups of tannins are used as medicine for treating skin bleeding, diarrhea and transudates.

2.3. Mono/Sesqui-Terpenoids and Phenylpropanoids

Synthesis of terpenoids takes place by a penta-carbon isoprene. In case of monoterpenoids, two units of isoprene are found, while in sespui-terpenoids there are three units of isoprene. They are popularly known for their low molecular weight and wide range of categories (more than 25,000). However, phenylpropanoids comprises group of molecules where the basic carbon skeleton starts from nine and more, with strong odor, flavors and are volatile in nature. Generally, these compounds are commonly called volatile oils, widely found in the *Lamiaceae* family. It is used as an herbal medication including antineoplastic, antiviral and antibacterial effects. Besides, it also helps in gastrointestinal stimulation. In addition, diterpenoids are lipophilic non-volatile (odorless) compound and is a cluster of 4 units of isoprene with a strong flavor. It is widely found in various plants including *Coffee arabica* and popularly known for its antioxidant qualities.

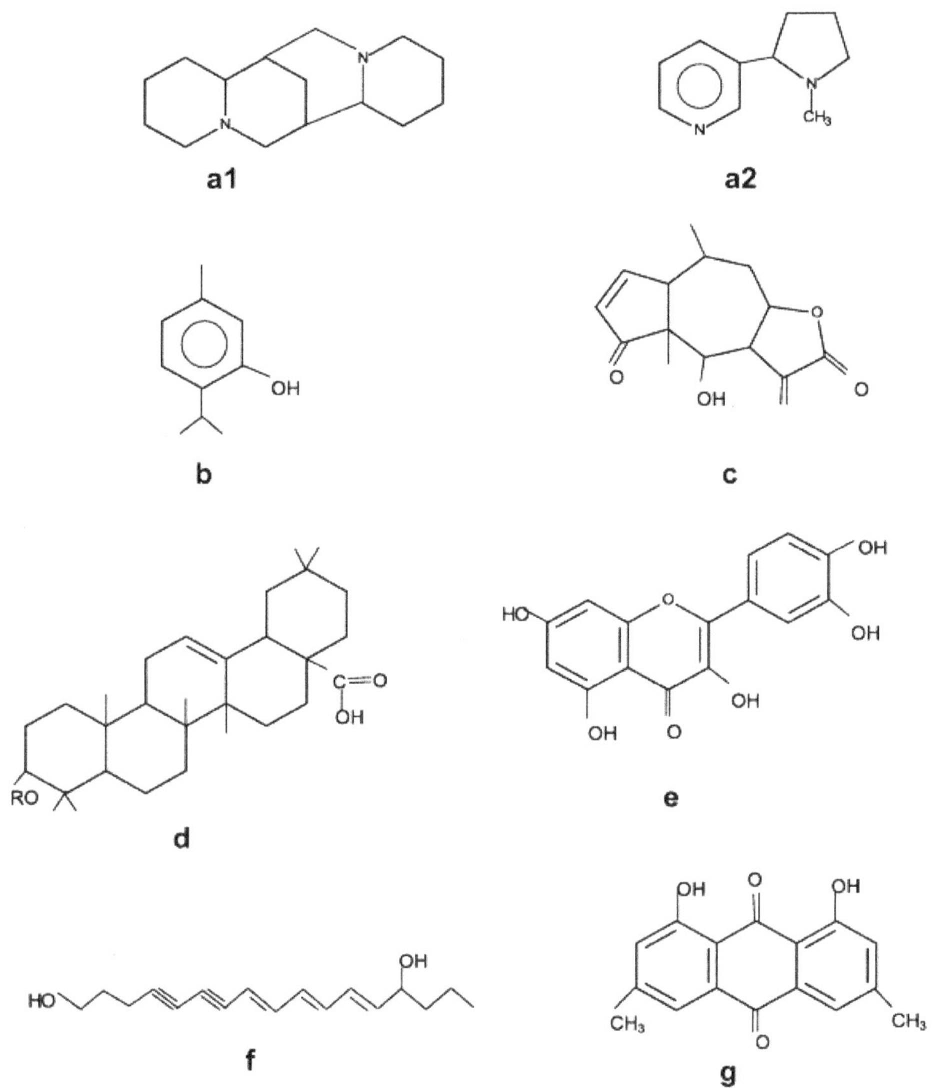

Figure 2. Basic structures of plant bioactive compounds alkaloids (**a1**,**a2**), monoterpenes (**b**), sesqueterpenes (**c**), triterpenes, saponins, steroids (**d**), flavonoids (**e**), polyacetylenes (**f**), polyketides (**g**) (Adopted with permission from Wink et al. [9].

2.4. Resins

Resins are composite mixtures which comprise both volatile as well as non-volatile attribute compounds; as well, they are comprised of a lipid soluble group of compounds. Non-volatile resins consist of diterpenoid and triterpenoid compounds, while volatile resins are equipped with mono/sesquiterpenoids. These resins are broadly found in herbaceous plants and are popularly known for their wound healing and antimicrobial properties.

2.5. Alkaloids

Alkaloids, a bitter-tasting and nitrogen-holding compound is a heterocyclic with limited spread in the plant kingdom. The *Solanaceae* family, including *Atropa belladonna*,

Datura spp as well as *Hyoscyamus niger*, consists of tropane alkaloids with anticholinergic properties. It is widely used for reducing muscle pain. Besides, pyrrolizidine alkaloids belong to the *Asteraceae* and *Boraginaceae* family, especially in *Senecio* spp. It comprises of a wide range of application including treating cancer cells, stimulating bone marrow leucocytes and myocardial contractility. In addition, methylxanthine alkaloids are distributed in *Coffee arabia* as well as *Theobroma cacao*.

2.6. Proteins

Plant proteins have gained significant popularity in the field of food and medicinal sectors as they are a major source of nutrient for humans and animals. The *Euphorbiaceae* family as well as *Fabaceae* and lentils are known to contain a high content of protein.

3. Novel Strategies for the Extraction of Bioactive Compounds from Plant Matrix

3.1. Individual Strategies

3.1.1. Ultrasound-Assisted Extraction (UAE)

Bioactive extraction from plant matrix using ultrasound has been widely employed over the past few decades [10]. It works on the principal of mechanical wave with a frequency ranging from 20 kHz to 100 MHz, which passes through a medium at a cycle of expansion and compression. In the case of liquid medium, cavitation bubbles are formed, at high acoustic pressure [11]. This phenomenon is known as "acoustic cavitation" as it enhances the extraction yield as the high shear force is induced by the cavitation, which leads to mass transfer of bioactive compounds by turbulent mixing and acoustic flow [12,13]. Ultrasound (US) extraction works on four basic parameters including ultrasound power, ultrasonic intensity, mode of working (e.g., non-pulsed/pulsed) and acoustic energy density [14]. In addition, it is divided into two different set-ups such as the US-bath and US-probe systems. In the case of an ultrasound bath system, the ultrasonic transducer array is placed below at the bottom of the extraction bath, which can also be attached at the side walls of the US-bath or inside the bath as a transducer array box. The transducer array box can be placed at any direction as per the requirement based on sample matrix. Whereas, in the case of the US-probe system, the probe is bonded with the transducer, which is submersed in liquid medium, enabling direct distribution of US waves, hence resulting in minimum loss of US energy. In addition, US intensity is an important factor affecting the yield of bioactive extraction, hence, it is important to consider the type of US employed, especially the probe diameter and the design of transducer employed as per the requirement [15].

Over the past few years, US has evolved, as from the fixed US power system, now it is possible to adjust the acoustic power. Most probe-type US devices usually control the amplitude of the probe vibration, and some of them can apply busters to increase its maximum amplitude. Alexandru et al. [16] developed a continuous US system for scale-up extraction to industrial level. In this system, a huge capacity of samples can be extracted continuously by feeding into a relatively small tank with multi-horn ultrasonic reactor. Elevated amplitude/intensity can enhance the sonochemistry but it leads to degradation of the transducer, which leads to increased agitation and reduced cavitation level. Hence, high amplitude/intensity is not essential to improve the extraction efficiency of cavitation level. However, the sample with high viscosity needs high amplitude as the high viscous samples tend to decrease the effect of sonication or cavitation [17]. Therefore, to achieve the required level of cavitation, it is necessary to enhance the level of amplitude [18]. Further, enhanced US frequency results in reduction in cavitation level. US develops cavitation bubbles, which take time to be initiated after the compression–rarefaction cycles. At high frequencies, it is challenging to produce acoustic cavitation, as the cycles of compression–rarefaction are too short to allow the growth of the cavitation bubbles. Therefore, higher amplitude and intensity of ultrasonic devices are needed to produce acoustic cavitation at high frequency [19]. Apart from physical properties, chemical properties including solubility and stability of the target compound in selected solvent also play an essential role in the extraction yield and efficiency of bioactive compounds from plant matric using US. Parameters like time and temperature are also considered to influence the level of extraction [20,21].

Several bioactive compounds have been effectively extracted using US from a plant-based matrix (fruits, vegetables/their by-products) [22–24]. Pan et al. [25] extracted bioactive compounds from pomegranate peel using conventional and US extraction techniques. The results indicate that a continuous US-pulsed system enhanced the level of antioxidant extraction to 22–24% and reduced the extraction time to 90%. Due to the enhanced extraction yield and reduced time as well as energy consumption, US is considered as an alternate and green technology for the extraction of bioactive compounds. Apart from fruits and vegetables, US is also employed to extracted bioactive compounds from medicinal herbs, spices and oleaginous seeds [26–29].

3.1.2. Microwave-Assisted Extraction (MAE)

Over the past few years, microwave (MW) has gained popularity in the extraction of bioactive compounds from plant-based matrix [30–33]. It works on the principal of electromagnetic waves and frequencies, mostly 915 MHz and 2450 MHz frequencies employed for industrial and domestic applications. The mechanism behind this technique is that MW works on heating effect, which results in higher extraction temperature, causing faster mass transfer [34]. MW has the tendency to penetrate inside the sample matrix causing interaction in polar components, thus causing direct/bulk heating effect to the solvent and the sample matrix [12]. Further, this direct/bulk heating by MW also helps in the reduction of time and solvent used especially in industrial level extraction. Moreover, in MW extraction, the direct heating enters inside the matrix and increases local temperature and pressure, leaching out the target bioactive compounds from the sample matrix to the solvent solution. Two different setups are available for MW extraction: (a) open system, (b) closed system, where the pressure can be adjusted, e.g., increase or set at atmospheric level, respectively. The closed system MW extraction is carried out in a sealed vessel with constant MW heating, at controlled pressure and temperature. This closed system helps in reaching higher temperatures than the open system, as the amplified pressure in the closed vessel increases the boiling point of the extraction solvent [35]. Although high temperature and pressure lead to efficient, high and fast extraction yield with less solvent consumption, they also escalate the safety risks (Figure 3). In addition, this particular system is limited to certain bioactive compounds as maximum compounds are heat-sensitive and tends to degrade at elevated temperature, thus, the open system is promoted widely [36].

MW extraction depends on several ranges of factors including microwave power, frequency, exposure time, moisture content, particle size of sample matrix, type and composition of solvent, dilution ratio, extraction temperature, extraction pressure and the number of extraction cycles. The detailed description of these factors has been reported by some reviews. The most critical factor is selection of extraction solvents as its solubility with sample matrix, its dielectric constant and dissipation plays a major role in extraction process. Solvents with higher dielectric constant like water and polar solvents can store more microwave energy than nonpolar solvents, thus water and polar solvents are reported to be better for MW extraction [37]. Besides, the dissipation factor, which converts electromagnetic energy into heat, is considered a significant factor for MW extraction. Ajila et al. [38] reported that solvents including ethanol and methanol, which contain a higher dissipation factor, are better solvents than water in the extraction of phenolics. Although, water possesses a high dielectric constant as compared to ethanol and methanol, but due to its low dissipation factor, it fails to heat up the sample matrix in depth. Therefore, for this reason, combination of solvents (water with ethanol or methanol), which contains high dielectric constant as well as a high dissipation factor, can be used to enhance the extraction efficiency.

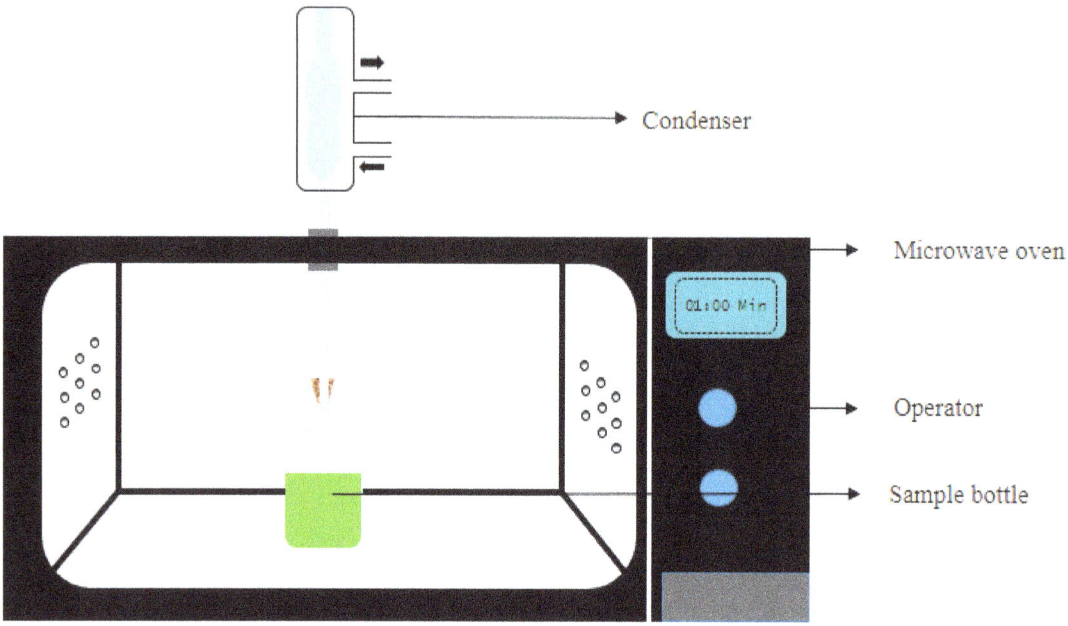

Figure 3. Schematic representation of the microwave-assisted extraction apparatus.

MW extraction comprises a wide range of advantages over conventional extraction techniques including reduction in extraction time, solvent used and extraction cost with a significant level of enhancement in extracted compounds. Shu et al. [39] extracted ginsenosides from ginseng root using MW extraction techniques for 15 min and resulted in a higher yield compared to conventional extraction, which took 10 h to complete. Dhobi et al. [40] extracted flavolignin and silybinin from Silybum marianum employing MW extraction and results revealed that the extraction efficiency was enhanced to 60% compared with the conventional solvent extraction methods. Similarly, Asghari et al. [41] extracted cinnamaldehyde and tannin from some medicinal Asian plants and their results reflected in a quicker and easier technique compared with conventional extraction techniques.

3.1.3. Enzyme-Assisted Extraction (EAE)

EAE is considered as one of the most efficient, eco-friendly, and non-thermal extraction strategies over conventional extraction techniques. It has been employed by several food industries for the extraction of various bioactive compounds saponin, carotenoid, anthocyanin and many more [41]. Incorporation of certain enzymes including pectinases, cellulases and hemicellulases during extraction can significantly enhance extraction efficiency/yield of bioactive compounds by the principal of degradation in cell wall and membrane interiority [42,43]. In this technique, adequate knowledge about the catalytic specificity and its mode of action is necessary to acquire as well as to know its optimum conditions suitable for the enzymes to act on the plant matrixes. In order to use enzymes efficiently for EAE, it is essential to understand their catalytic specificity and mode of action, as well as investigate optimal conditions and which enzyme or enzyme combination is more suitable for the raw materials [44]. Several significant factors including enzyme composition and concentration, type of extraction solvent, solid-to-liquid ratio, enzyme/substrate ratio, pH, extraction temperature and time play a vital role in activation of enzyme reaction and extraction of bioactive compounds [12,45].

Temperature is one of the important factors influencing the rate of extraction, but excessive increase of temperature may also inactivate enzymes. Moreover, a wide range of compounds are heat-sensitive, and therefore require mild temperature throughout the extraction process [46]. Furthermore, pH is one of the major reaction conditions where the enzyme gets activated and stated to degrade the cells of the sample matrix. The optimum pH for each enzyme has already been reported; however, it may vary depending upon the matrix used and reaction conditions applied [19,47]. Enhancement in ratio between enzyme/substrate tends to improve catalytic reaction rate, but due to this amount of enzyme used, increases, which results in increased extraction cost. Further, the solvent used for the extraction may not be suitable for the enzymes used in the extraction process, for instance, several enzymes used for extraction of bioactive compounds from plant matrix are active in water, which may be not be active at higher concentrations of solvents including methanol and ethanol. Many bioactive compounds are highly soluble in concentrated methanol and ethanol; therefore, in such cases, major concern needs to be given to the selection of enzyme and its reaction conditions to achieve the desired extraction yield.

EAE is broadly employed for the extraction of bioactive components from a wide range of plant matrix including flavonoid from the peel of citrus, fructans from Dasylirion wheeleri, curcumin from turmeric, anthocyanin from beetroot, total phenolics from pomegranate peels and Cassia fistula pods, polysaccharide from seaweed and Cedrela sinensis, lycopene from tomato tissues and fatty acids from microalgae [48–53].

3.1.4. Pulse Electric Field-Assisted Extraction (PEF)

Over the years, PEF has become one of the most promising non-thermal and cost-effective bioactive extraction techniques to extract compounds used by the nutraceutical and pharmaceutical sector. The concept of PEF began in 1999 by a researcher named Ganeva and co-workers. They treated beer yeast with 2.75 kV/cm pulse electric power and kept it for macerating for 5 h for the extraction of protein. Further, they found that, after treating the sample with PEF, the dissolution enhanced, which was reflected in a significant increment in protein extraction. This gave a light of hope to the other researchers, as PEF has the ability to improve the mass transfer rate through cell membrane. Based on this belief, many researchers carried out several experiments for the extraction of bioactive compounds employing PEF and found that this non-thermal technology, when compared with several other extraction techniques, indicated shorter treatment time (some microseconds) with higher extraction yield [54]. In addition, this technique can easily be employed at the industrial level for continuous flow of extraction. The system is equipped with a high-voltage pulse generator, a sample holding chamber and a controller. The treatment chamber contains two electrodes with a gap. One of the electrodes is connected to the pulse generator and the other is earthed (Figure 4). Before treating the sample, it is subjected to fine powder, followed by agitation in the selected solvent. After agitation, the sample is pooled in the treatment chamber, led by the treatment parameter settings such as pulse number or pulse width (μs), electrode voltage (24 kV), energy input (kJ), frequency (Hz). Usually, at high electric voltage, the extraction is higher, but in the case of some compounds such as polysaccharides, it tends to decompose, which results in low extraction yield [54]. Hence, it is important to decide the parameters of the extraction based on the targeted compounds. In addition to this factor, including solvent conductivity, polarity, solubility with the target compounds also plays a vital role in the extraction process. Generally, conductivity, solubility of the solvent and dilution ratio of solvent to solute as well as pulse duration/width is directly propositioning to the extraction yield. However, if we further increase these conditions beyond the requirement, then it tends to reflect negative results as when a high amount of electric pulse is applied, the targeted compound may also start to degrade, which can reduce the extraction yield and thus selection of the correct solvent, and operation factors are extremely significant [55]. Bioactive compounds including protein, saccharides, calcium and others are easily extracted using PEF.

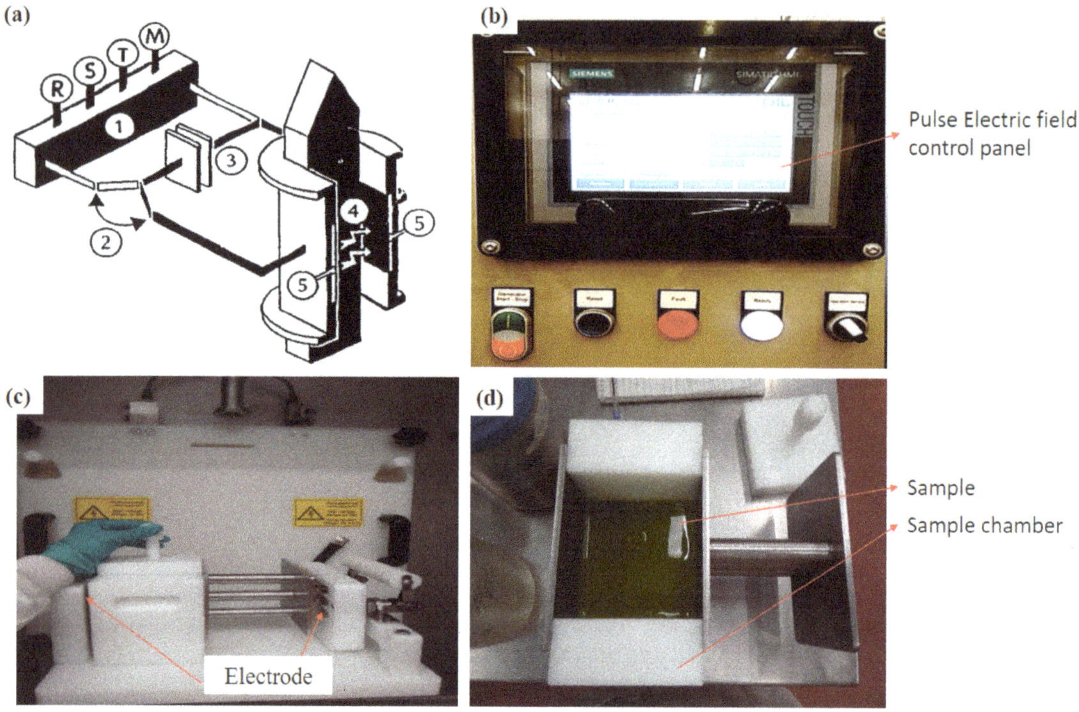

Figure 4. Simplified schematic diagram of PEF (**a**) 1. High-voltage generator; 2. Switch; 3. Capacitor; 4. Medium; 5. Electrodes: and R, S, T, M, connector points for the main supply; (**b**) control panel of PEF; (**c**) Pulse electric field electrodes and treatment chamber; (**d**) sample chamber with sample. (Modified and adopted with permission from Knorr et al. [56].

3.1.5. Moderate Pressure Application

Supercritical-Assisted Extraction (SFE)

Supercritical fluid extraction is employed over the conventional extraction method, as the solvents used in this technique differ in physicochemical properties including density, diffusivity, viscosity and dielectric constant. In addition, these properties play a critical role in the extraction process, as these supercritical fluids are low viscous and high in diffusivity, thus movement of solvent through the plant matrix becomes easy and results in a faster rate of exchange. There are a wide range of compounds/solvents including carbon dioxide, ethane, ethane, methanol, nitrous oxide, n-butene, n-pentane, sulphur hexafluoride and water used as supercritical fluid. However, carbon dioxide is considered the most promising fluid for the extraction of bioactive compounds due to few reasons, such as, it is harmless to the environment and human health, its favorable critical temperature (31.2 °C) which helps to extract heat-sensitive compounds easily and the extracted compounds are preserved from oxidation when exposed in air [57]. As carbon dioxide is at room temperature in its gaseous state of matter, it is easily eliminated after the extraction is completed, and the achieved compounds are left as solvent-free extraction.

Basically, the SFE process of extraction is divided into two main categories: (a) solubilization of bioactive compounds in the extraction fluid; (b) its separation into the fluid of supercritical. Throughout the extraction, the fluid passed through the cells of plant matrix, solubilizing the compounds present in cell membrane, thereby resulting in the extraction chamber with the solubilized target compound. Further, by release of pressure and change in temperature, the fluid becomes separated from the compound and, as a result, the pure form of the compound is extracted. Brunner et al. [58] reported that the initial step of extraction (solubilization) takes place in several stages. In the beginning, the plant matrix absorbs the supercritical fluid, which leads to swelling of its cellular membrane followed by expansion in intracellular passages, which results in mass transfer, and thus the solubilized compounds moved from inner cell membrane to outer surface, and lastly, it is separated from the fluid. In order to optimize the treatment conditions of the SCFE technique, intense knowledge in thermodynamic (solubility and selectivity) and kinetic data (mass transfer coefficients) is essential. The kinetic illustration of SFE is achieved by extraction curve graph, illustrated in Figure 5, which educates about the extraction yield subjected to extraction time (t). Overall extraction curve (OEC) is divided into three different time phases of mass transfer, (a) constant extraction rate (CER) and the phase is called tCER, wherein the compound is packed inside the solute, therefore leading to convection mass transfer; (b) falling extraction rate (FER) and the phase is called tFER, wherein convection is combined with diffusion mechanism as the external lipid layer of the cell membrane fails to remain intact; (c) diffusion control (DC) and the phase is called tDC. In tDC, the lipid layer is completely corroded and the diffusion starts inside the plant matrix; hence, maximum extraction is achieved [59–61]. Moreover, the evaluation of the extraction curve was carried out using spline model, where the extraction takes place from constant phase of extraction to falling extraction rate and then finally to diffusion rate [62]. In addition, tCER tFER and tDC (in min) indicate the time span of CER, FER and DC respectively. As the sample passes through these phases, the bioactive compounds are extracted with the increase in time at a specific pressure and temperature. At higher temperature with low pressure (approximately 20 MPa), extraction yield is amplified. However, the rise in temperature reduced the characteristics of compounds extracted.

Due to the promising outcome of SFE techniques, it provides a wide range of application in the food, cosmetics and pharmaceutical sectors, as it helps in the extraction of flavors, analgesics and anti-inflammatory drugs. It is also helpful for the development of drugs for the treatment of chronic diseases such as stroke, cancer and Alzheimer [63,64].

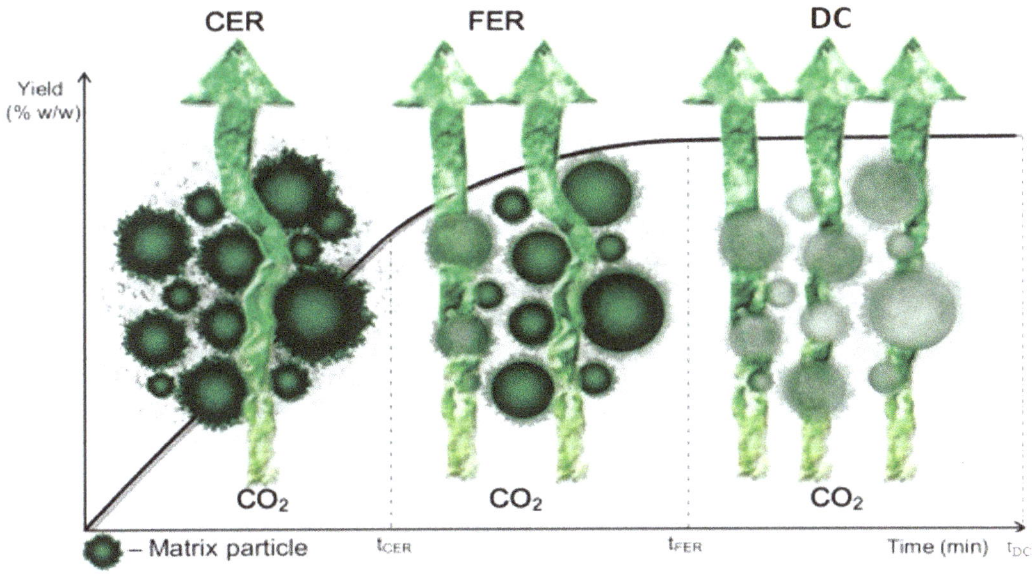

Figure 5. Graphical representation of mass transfer of bioactive compounds from cell membrane to solvent (modified and adopted with permission from Rui et al. [62]).

3.1.6. High-Pressure Application

High Hydrostatic Pressure-Assisted Extraction (HHPAE)

This novel extraction technique works on the principle of combination of pressure (100 to 500 MPa) and temperature (20–50 °C) which results in enhancements to mass transfer rate. According to the US Food and Drug Administration, this technique possesses environmentally friendly attributes, and hence has gained popularity in a wide range of food and nutraceutical sectors [65]. The mild temperature employed in HHPAE has positively reflected promising results for the extraction of heat-sensitive bioactive compounds [65–67]. HHPAE improved the rate of extraction by enhancing mass transfer rate by rupturing the cell membrane and organelles at minimum consumption of solvent and time [68–71]. HHPAE was first reported by a German researcher, Knorr et al., 1999 [72], for the extraction of caffeine from coffee seeds. Further, in 2004, Sanchez-Moreno et al. [73] stabilized a protocol for the extraction of carotenoid from tomato at 100–400 MPa. Over the last few years, HHPAE has evolved rapidly for the extraction of these bioactive compounds. Firstly, the plant matrix is dried and milled followed by sieving (40–60 mesh) to secure an even particle size of sample for the extraction. Secondly, an appropriate solvent is selected based on the solubility of the target bioactive compound in that solvent. Lastly, fine plant powder is incorporated in solvent in a sterile polyethylene bag. The bag is further vacuum-sealed and kept inside the pressure vessels equipped with a pressure and temperature regulator (thermocouple) attached at the top and bottom of the vessel to maintain the desired temperature (Figure 6). Further, the pressure vessel is filled with water to create a pressure by a pressure pump attached to the vessel. Post extraction, the mixture is filtered and the solid particles are removed. The filtered extract is further centrifuged at 4000 to 8000 rpm for 10–15 min. Post centrifugation, the supernatant is collected and passed through a 0.45 µm membrane for characterization and quantification analysis of the targeted compound [74,75]. Liu et al. [76] studied the effect of high-pressure treatment on cell membrane of ginseng roots using scanning electron microscopy (SEM), and the results reveal that the damaged/ruptured cell membrane was clearly identified in the high-pressure-treated sample compared to untreated samples, and therefore it is concluded that HHPAE enhances the extraction efficiency of bioactive compounds in plant matrix. The efficiency of the extractions is based on a few important parameters including pressure applied, time/temperature combination, dilution ratio of solvent and solute, particle size of solute and polarity of the solvent used [77]. Solvents with similar polarity to the targeted compounds significantly give better extraction yield.

HHPAE has been proven to be promising for enhancing the diffusion capacity of the solvent inside the plant matrix by rupturing the cell membrane, which results in improved permeability, hence increased extraction yield [78]. It also breaks the hydrophobic bonds and denatures the protein molecules, thereby making the extraction better [79,80]. Moreover, based on dissolution mass transfer theory (mass transfer rate = pressure/resistance of mass transfer), the dissolution is higher in HHPAE [81]. The more the pressure/temperature (30 °C, 50 °C, 70 °C) is applied, the more is the solvent dissolution to cell, which enables the compounds to leach out the membrane. In addition, pressure-holding time helps in maintaining the equilibrium of solvent between inside and outside of the cell membrane. However, long pressure-holding time may damage the biological activities of the plant matrix; therefore, it is necessary to maintain a suitable time according to the target compound. If the dry sample has to be given high-pressure treatment, it will facilitate cell enlargement, leading to swelling and opening of pores in the cell membrane [82–85].

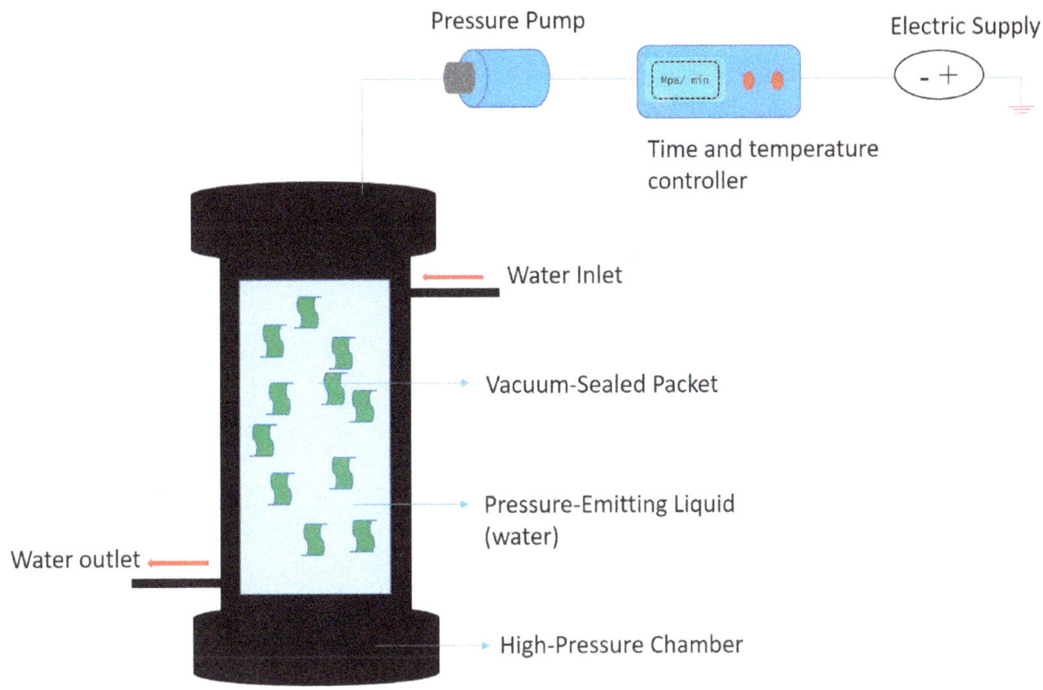

Figure 6. Schematic representation of high-pressure processing.

3.2. Combination of Novel Stategies

In the past few decades, researchers have been trying to enhance the extraction yield of bioactive compounds using novel technologies, but in order to further increase and purify the yield, combination strategies have been implemented, which mainly includes ultrasound-enzyme-assisted extraction, ultrasound-microwave-assisted extraction, microwave-enzyme-assisted extraction, ultrasound/microwave-enzyme-assisted extraction, pulse electric field enzyme-assisted extraction, supercritical fluid enzyme-assisted extraction and high-pressure enzyme-assisted extraction [86–89]. Each individual technique, such as ultrasound when applied on plant matrix, tends to enhance its mass transfer rate by rupturing the cell membrane; however, some portion of the cell membrane is still hindering the path, which can be removed by hydrolysis of the sample using enzyme treatment. Therefore, the combination treatment gained popularity and gave promising results. Table 1 illustrates extraction of bioactive compound saponin from several novel treatments individually, as well as in various combinations.

Table 1. Novel extraction strategies for the extraction of bioactive compounds.

Strategies	Bioactive Compounds	Matrix	Solvent	Dilution Ratio	Extraction Parameters	Time/Temp	Yield	Ref.
Individual Strategies								
Ultrasound-assisted Extraction	Saponins	*Zizyphus jujuba*	Ethanol 50%	1:36	38 W	50 min	15.54%	[90]
		Quinoa	Ethanol	1:10	60% A	15 min	5.51 g/100 g	[91]
		Lentil					10.63 g/100 g	
		Fenugreek					12.90 g/100 g	
		Soyaben					4.08 g/100 g	
		Lupin					4.55 g/100 g	
		Alfalfa (*Medicago sativa*)	Ethanol 78.2%	1:11.4	112.0 W	2.84 h/76.8 °C	1.61%	[92]
		Eclipta prostrasta L.	Ethanol 50%	1:14	100 W	3 h/70 °C	2.096%	[93]
	Total phenolic compounds	pomegranate (*Punica granatum* L.) peel	Distilled water	1:4	400 W/60% A	10 min	69 mg/g	[94]
	Total phenolic compounds	Tamarillo fruit (*Solanum betaceum*)	Acetone 80 %	1:50	50% A	12 min	23.96 mg/g	[95]
	Flavonoids	*Lycium barbarum* L. fruits	Mixture of choline chloride and p-toluene sulfonic acid	1:2 M	60% A	1.5 h	Myricetin (57.2 mg/g)Morin (12.7 mg/g)Rutin (9.1 mg/g)	[96]
	Flavonoids and phenolic compounds	*Ocimum tenuiflorum* leaves	Ethanol 55. 34%	1:40	50 W/0.26 W/cm^3	11.71 min	Gallic acid (6.81 µg/mL) Vanillic acid (8.52 µg/mL) Rutin (14.21 µg/mL) Sinapic acid (11.37 µg/mL) Quercetin (3.49 µg/mL) Luteolin (1.76 µg/mL) Apigenin (8.66 µg/mL)	[97]
	Total flavonoids (TF) and total phenolic (TP) compounds	wild garlic (*Allium ursinum* L.)	Ethanol 50%	1:5	28.8 W/L	80 min/80 °C	TP 1.61 g GAE/100 DW; TF 0.41 g CE/100 g DW	[98]

Table 1. *Cont.*

Strategies	Bioactive Compounds	Matrix	Solvent	Dilution Ratio	Extraction Parameters	Time/Temp	Yield	Ref.
	Total phenolic compounds	Peaches and Pumpkins	Ethanol 100%	1:10	44.60% A	27.86 min/ 41.45 °C	Gallic acid 55 mg/100 g	[99]
	Phenolic compounds	Sparganii rhizoma	Ethanol 80%	1:22.74	300 W	33.54 min	ρ-Hydroxybenzaldehyde (2.12 μg/g) Vanillic acid (21.66 μg/g) ρ-Coumaric acid (4.32 μg/g) Ferulic acid (2.08 μg/g) Rutin (10.06 μg/g) Kaempferol (106.35 μg/g)	[100]
	Flavonoids	Olive (Olea europaea) Leaves	Distilled water	1:41	270 W	50 °C/50 min	74.95 mg RE/g dm	[101]
	Flavonoids	Grapefruit (*Citrus paradisi* L.)	Ethanol	1:8	100 W	25 °C/3 min	75 mg gallic acid equivalents/g dw	[102]
Microwave-assisted Extraction	*Saponins*	*Furcraeaselloa var. marginata*	Water	23.54 mL/g	2.45 GHz/1000 W	9 min/90 °C	5.77%	[103]
		Physalis alkekengi L. *var. franchetii (Mast.) Makino*	Ethanol 80 %	1:32	300 W	29 min	6.41 mg/g	[104]
		starfish *Echinaster sepositus*	Ethanol 50 %	1:40	200 W	3 min	60.3 mg/g	[105]
		Phyllanthus amarus	Methanol 100%	1:50	360 W/4 s/Min Irradiation time	50 min	227.9 mg/g	[106]
		Aralia elata (Miq.)	Ethanol 100%	1:20	560 W/3 cycles	50 min/40 °C	1.22 mg/g	[107]
		Gac (Momordica cochinchinensis Spreng.) Seeds	Ethanol 100%	1:30	360 W/Irradiation cycle 10 s on 15 s off/cycle/3 cycles	75 S	26 mg/100 g	[108]
		Ganoderma atrum	Ethanol 95%	1:25	33 kHz	5 min/90 °C	0.968%	[109]
		P. ginseng	Ethanol 45%	1:20	88 W	180 s/83.7 °C	1.31%	[110]
		Cacao pod husk	Methanol 85%	1:50	600 W 6 s/min irradiation time	40 min	69.9 mg EE/g (Escin Equivalant)	[111]
		Paramignya trimera root	Methanol 100%	1:100	360 W/5 s/2 min Irradiation time	40 min	520.5 mg escin equivalents (EE)/g	[112]

Table 1. *Cont.*

Strategies	Bioactive Compounds	Matrix	Solvent	Dilution Ratio	Extraction Parameters	Time/Temp	Yield	Ref.
	Anthocyanin	Blueberry powder	Ethanol 55.5%	1:34	7 min Irradiation time	47 °C	73.73 %	[113]
	Anthocyanins	Purple corn (*Zea mays* L.) cob	1.5 M HCl–95% Ethanol	1:20	555 W/19 min	55 °C	185.1 mg/100 g	[114]
	Rutin and Quercetin	Stalks of *Euonymus alatus* (Thunb.) Sieb	Ethanol 50%	1:40	170 W/6 min	Room Temperature	Rutin 0.225 mg/g; Quercetin 0.012 mg/g	[115]
Enzyme-assisted Extraction	Saponins	*Gomphrena celosioides* Mart	Water	1:25	Cellulose 0.7% (*v/w*)	4 h/60 °C/5	1.550%	[116]
		Pseuderanthemum palatiferum (Nees) Radlk. Dry Leaf Powder	Water	1:7.5	Viscozyme 7.5%	75 min/55 °C	2.267 mg/g	[117]
	Flavonoids	Grape skins	Water	1:10	Lallzyme EX-V (10.52 mg/g) pH 2.0	3 h/45 °C	0.012 mg/g	[118]
	Total flavonoids and total phenolic compounds	Bay leaves (*Laurus nobilis* L.)	Methanol 100%	1:5	30 mg of the ternary enzyme mixture (cellulase: hemicellulase: xylanase; 1:1:1	1 h at 40 °C	TPC 5.87 mg GAE/g; TFC 5.18 mg QE/g	[119]
	Flavonoid compounds	*Stevia rebaudiana*	n-hexane	6:1	pectinases and β-glucanases	100 W/140 °C	Luteolin 0.657 mg/g	[120]
Pulse electric field-assisted extraction	*Saponins*	Acanthophyllum Roots	Water	1: 6	7 kV/60	-	9.1%	[121]
High-pressure processing extraction	*Saponins*	Bitter Melon	Ethanol 70%	1:45.3	423.1 MPa	7 min/30 °C	3.270 g Rg1 equivalents/100 g (Ginsenoside standard)	[122]

Table 1. *Cont.*

Strategies	Bioactive Compounds	Matrix	Solvent	Dilution Ratio	Extraction Parameters	Time/Temp	Yield	Ref.
Combine Techniques								
Ultrasound-assisted surfactant Extraction	*Saponins*	*Panax notoginseng*	ethanol	1:6.25	1% (*w*/*v*) of DTAB/SDS vesicle	20 min		[123]
Microwave-assisted surfactant extraction	*Saponins*	*Momordica charantia* L.	Water	1:12	400 W/SDS 0.05 concentration	5 min	4.69%	[124]
Pulse electric field-assisted enzyme extraction	*Saponins*	Panax ginseng	Water/β-glucosidase 2% (4.7 pH/50 °C)	1:50	15 kV/2 µs/10		38.15 mg/g	[125]
High-pressure-processing-assisted enzyme extraction	*Saponins*	Ginseng Root	Ethanol 70%/Cellulose 60 µL/30 mL	1:5	100 MPa	12 h/50 °C	40.02 mg/g	[126]
Conventional Extraction								
Reflux extraction	*Saponins*	Bitter Melon	Ethanol 70%	1:35	-	3 h	2.478 g/100 g	[127]
Solvent extraction		starfish *Echinaster sepositus*	Methanol 70%	1:50	Magnetic agitator	24 h/Room Temperature	7.7 mg/g	[105]
Maceration	Total Phenolics	Blackberry (*Rubus fruticosus* L.) residues	Ethanol 50%	1:40	-	80 °C	3.66 mg GAE/g FR	[102]

4. Nanoencapsulation of Bioactive Compounds

Over the past few years, nanoencapsulation has gained popularity in the field of food science. It is a process of encapsulation whereby a bioactive compound as a core matrix is captured inside a wall matrix which can withstand the environmental and enzymatic degradation. Employing a significant wall matrix/nano-carrier provides protection against bioactive compounds, which are extremely sensitive to heat and digestive enzymes present in the stomach and gastrointestinal tract of the human body. Besides, these wall materials help in maintaining nutritional activity of the compound as well as helping in masking the undesirable taste of some of compounds [128]. Liposomes or lipid bilayer shells are considered best for encapsulation and delivery of bioactive compounds as they protect the compounds for a longer duration of time compared to other types of nano-carriers. In addition to this, casein micelles are considered promising to encapsulate minerals such as calcium and phosphate [129]. Nano-capsules generally range from 1 to 100 nm in particle size. Apart from the food industry, nanoencapsulation is also practiced by the packaging industry for packaging of meat and fruits to extend their shelf life with retained nutritional qualities [130].

Several nano-carriers including protein, casein, chitosan, gelatin, zein, polyethylene glycol, arabinogalactan, poly-D, L-lactide-co glycolide, poly L-lysine and polyaniline are used for encapsulation. According to literature, chitosan-based glycolipid nano-carriers have the tendency to enhance the anticancer activity of fucoxanthin by 25.8-fold, as they help to keep the compound active for a longer period of time [131]. In different studies, rutin was encapsulated using poly (lactic-co-glycolic acid) and zein as a wall matrix, which resulted in a slow delivery rate (25% after 60 h) of rutin at the targeted location in the human body [132]. Presently, one of the researchers used chitosan and alginate to encapsulate bioactive compounds from the waste of grapes, in which, interestingly, the bioactivity of the targeted compound was enhanced and the encapsulation protected the compound from degradation in the gastrointestinal tract [133]. Besides, compounds including peptides are encapsulated using lipid-based nano-carriers or liposomes, which are comprised of double-layer protection of surfactant molecules and aqueous fluid [134]. Furthermore, nano-carriers of hybrid structure (combination of liposome and chitosan) are also in practice to encapsulate compounds like caffeine to enhance the encapsulation efficiency [135].

The techniques employed to encapsulate these compounds are divided into categories based on power consumed for encapsulating, such as top-down and bottom-down techniques as well as their combined treatment [136]. In the case of the top-down method of encapsulation, high power consumption takes place as it is equipped with instruments like spray-drying, ultra-sonication, homogenizer and many more, while the bottom-down method is practiced in minimum consumption of energy; for example, precipitation, micro-emulsification, conjugation, interchange of atoms, etc. Major factors which play a vital role in selecting a particular technique for nanoencapsulation are delivery motive, delivery rate of release, solubility and stability of the nano-carrier, as well as cost of production [131].

4.1. Encapsulating Carriers for Bioactive Compounds

4.1.1. Polymeric Nano-Carriers

Polymeric nano-carriers are considered to be extremely suitable material for encapsulation and delivery of bioactive compounds. Presently, natural-based nano-carriers including casein, starch, chitosan, whey protein and albumin are maximum used. In 2018, Ravi et al. [132] used chitosan as a wall material to encapsulate marine carotenoid fucoxanthin. It resulted in enhanced anticancer activity of the bioactive compound and increased the caspase-3 activity 25.8-fold. Further, Gagliardi et al. [133] carried out a comparative study using synthetic and natural-based nanoparticles, namely, poly (lactic-co-glycolic acid) and zein for the encapsulation of rutin. The results indicated that zein loaded with 0.8% rutin concentration reflected slower release (25%) after 60 h as compared to poly (lactic-co-glycolic acid) (100%). Further, in 2021, Portugal researcher Costa et al. [134] encapsulated bioactive grape pomace extract using chitosan and alginate nanoparticles, which protected the bioactive compounds from hydrolysis in the gastrointestinal tract as well as enhanced its bioactivity.

4.1.2. Lipid-Based Nano-Carriers

Lipid-based nano-carriers, known as vesicular carriers, include nano-liposomes, niosomes and particulate carriers (solid lipid nanoparticles and nano-structured lipid carriers). It is a spherical bilayer developed by the reaction between surfactant molecule and aqueous fluid. It is used to encapsulate various bioactive compounds including peptide. Solid lipid nanoparticles are fabricated by mixing solid lipid in internal phase, whereas nano-lipid particles are developed by mixing liquid and solid lipid together [137]. Chaudhari et al. [138] encapsulated piperine and quercetin using Compritol as a solid lipid, while squalene as liquid lipid and span 80 as well as tween 80 as emulsifiers and co-emulsifier. These encapsulated bioactive compounds reflected slower release due to slower erosion of lipid wall matrix (12 h). Another study conducted by Abd-Elhakeem et al. [139] illustrated about improving the bioavailability and oral target delivery of eplerenone by using lipid-based nanoencapsulation. Eplerenone-loaded nano-lipid capsules reflected in improved permeability up to two folds higher compared to conventional aqueous drug in rabbit intestine after the period of 24 h.

4.1.3. Hybrid Nano-Carriers

Hybrid nano-carriers consist of two main networks including internal (metallic materials and polymers) and external (single/multi-lipid layer) networks. The outer layer of this nano-particle acts as a protection against deterioration and water diffusion. These organic–inorganic and lipid-polymer carriers are basically developed for the treatment of cancer cells with controlled release of bioactive compounds. Seyedabadi et al. [136] developed a slow-release encapsulated caffeine using chitosan coated in nano-liposomes as compared to nano-liposome without chitosan covering; hence, combination of chitosome proved better for the encapsulation of caffeine.

4.2. Nanoencapsulation Techniques for Encapsulation of Bioactive Compounds

Nanoencapsulation of bioactive compounds is significantly more complex as compared to the micro-encapsulation process. It is divided into three main categories, which consist of top-down, bottom-down and a blend of both [131]. Top-down techniques require high-energy-efficiency instruments including spray-drying, ultra-sonication, high-pressure homogenization, etc., while the bottom-down technique is limited to low-energy-consumption techniques, such as precipitation, micro-emulsification, conjugation, layer-wise accumulation, interchange of atoms and molecular into nano-size level, etc. [137]. Several factors influence the choice of a particular technique for tailoring nano-capsules, which include delivery motive, delivery rate of release, solubility and stability of the nano-carrier, as well as cost of production.

In the present review, several encapsulation techniques including ultra-sonication, high-pressure homogenization, microfluidization, nano-fluidics, nano-spray-drying, electrospinning, electro-spraying, milling and vortex fluidic have been discussed.

4.2.1. Electrospinning

In electrospinning, fluids with electric charges are processed by passing high-voltage electricity to polymeric fluid, which results in the development of dry micro- and nano-structures. Basically, the instrument comprises three major parts including a syringe pump, stainless steel electrified needle, and the collector plate inside a chamber. The feed fluid is pushed with a definite flow-rate inside the nozzle/needle, which is subjected to high-voltage electricity. The experiments are carried out at ambient temperature and after the nano-fibers are collected on the electric collector plate, they are kept in a desiccator prior to packaging [140]. The principal element responsible for the characteristic of nano-fiber includes operating conditions such as spinning fluid properties, polymer attributes and mechanical parameters of the instrument with its several nozzle setups [141]. The diameter of nano-fibers ranges from 1 μm to numerous nano-meters. These nano-fibers are in demand as a reinforced material in food packaging, drug delivery, and biosensing due to its substantial surface area, flexibility to develop into several structures and magnified porosity [142]. There are five different types of electrospinning strategies including blend electrospinning, coaxial electrospinning, emulsion electrospinning, high-throughput electrospinning and polymer-free electrospinning. In the blend and emulsion electrospinning method, the core (bioactive compound) and the wall (polymeric) solutions are blended together for electrospinning using single-nozzle, which works efficiently in controlling the release of bioactive compounds. Hydrophilic and hydrophobic molecules can be easily encapsulated using this technique [143]. On the contrary, in the case of a coaxial electrospinning setup, the component consists of a syringe with a twin-compartment, where two different nozzles are attached to one syringe outlet pump for electrospinning the core and wall solution together, which results in achieving encapsulated fibers [144]. However, high-throughput electrospinning is a needless technique applicable for the fabrication of ultrathin fibers via emulsions subjected to centrifugal pressure of polymeric fluid. In 2018, Kutzli produced glycoconjugates using high-throughput electrospinning. Apart from all these techniques, polymer-free electrospinning is an enhanced version of electrospinning, wherein the polymeric fluid with high molar mass is injected through a pump on the surface to achieve higher yield than the conventional techniques [145]. Xiao et al. [146] indicated that this technique was able to fabricate nano-fibers of a diameter ranging from 87 to 57 nm, whereas other reports by Moreira et al. [147] indicated that polymer-free electrospinning is useful for the food industry to improve the uniformity of nano-fibers derived from spirulina. Further, Poornima et al. [148] used electrospinning to encapsulate resveratrol with poly(ε-caprolactone) and poly (lactic) acid and the results proved to be effective in controlled drug release delivery, while another recent study by Leena and Anandharamakrishnan et al. [149] achieved the highest encapsulation efficiency (96.9%) of resveratrol using Zein. Researchers also indicated that these nano-fibers are useful as edible nano-films for oral delivery.

4.2.2. Electrospraying

Elcetrospraying, popularly known as electro-hydro-dynamic atomization (EHDA), is an alternate solution for the drying-encapsulation technique. It runs on high-voltage electric current at ambient temperature. The primarily principle for both the techniques (electrospinning and the electrospraying) is similar, the unique difference between these technologies lies in intermolecular cohesion of polymeric fluid which is remarkably low in the case of electrospraying, and thus results in breakage of jet into fine droplets. The jet particles, when exposed in air, gain a spherical shape in view of the surface tension. A report by Bhushani and Anandharamakrishnan [140] revealed that electrospraying helps to enhance the permeability and bioactive releasing attributes in catechins from green tea with the help of zein as a wall material. Later in 2019, Jayan and Anandharamakrishnan also revealed that resveratrol, when nanoencapsulated using the same wall material, results in 68% encapsulation efficiency [150].

4.2.3. Nano-Spray Dryer

The nano-spray-drying technique is quite similar to that of conventional spray-drying, where the fluid is subjected to droplet formation and further dried by the heated drying gas to form dry particles. In the case of nano-spray, the little modification has been made which includes a particular nozzle, which fabricates nano-droplets using a constant flow of drying gas used from laminar. Further, the nano-droplets are subjected to vibrating mesh of 4.0,

5.5, and 7.0 μm size holes. The fragment size of the spray-dried sample totally depends on the concentration of the fluid, temperature of drying gas, spray velocity and the size of the droplets [151]. Adel et al. [152] encapsulated curcumin in hydroxypropyl beta-cyclodextrin using the nano-spray-drying technique for the pulmonary delivery of curcumin in lung tissues and the results indicated significant reduction of proinflammatory cytokines compared to the pure drug. Further, Mozaffar et al. [153] dried nano-structured lipid carriers of 3% palm oil and 3% tween 80 (60 g/L) mixed with sodium chloride. The results revealed that the presence of sodium chloride protected nano-structured lipid carriers from aggregation during the process of spray-drying and about 50% of lipid molecules were encapsulated in salt particles.

4.2.4. Micro-/Nano-Fluidics

The primary concept of micro-/nano-fluidics is based on interfacial interaction between the core and wall fluids. It helps in the formation of spherical drops and slows down the release of the bioactive compounds. In addition to this, it helps in the production of similar-size accurate nano-droplets [154]. It consists of components including a molded set of channels fabricated on a base of polydimethylsiloxane (PDMS) of glass. The fluid is passed through this channel as these are interconnects from all directions. Gas and liquid are injected from a syringe using hydrostatic pressure. Generally, this technique incorporated four types of emulsion devices including single-, double-, multi- and flow-focusing nano-/microfluidic devices. It is employed for the fabrication of nano-emulsions, nano-liposomes and nanoencapsules. Jafari et al. [155] developed an oil-in-water nano-emulsion using microfludization at an optimized condition of 42–63 MPa microfluidization pressures with 1–2 cycles. The results of the study indicated smaller droplets of fish oil in developed nano-emulsion. After a couple of years, Wang et al. [156] modified citrus pectin using microfluidation, and the results revealed that properties of nano-emulsion were enhanced, thereby protecting cholecalciferol from UV degradation as compared to the original pectin. In addition, the molecular weight and hydrodynamic diameter of modified pectin was also reduced to 237.69 kDa and 418 nm, respectively.

4.2.5. High-Pressure Homogenization

High-pressure homogenization refers to the production of nano-fragments of a homogenous size in a fluid under a specific high pressure. It comprises 10–15-fold higher pressure (100–400 MPa) than conventional homogenizer. According to the research, this technique is excessively utilized by the milk and milk products industries to upgrade its texture, taste, flavor, and improved shelf life characteristics with enhanced anti-microbial quality (in activation of Salmonella spp., Listeria monocytogenes, Staphylococcus aureus, and Escherichia coli). About 3 and 4 log cycles reduction was reported at 200 and 300 MPa homogenization pressure at 30 °C and 40 °C inlet temperature [157]. It also acts as a substitute against thermal processing, as it is significantly effective for the inactivation of enzymes and microbial activity [158]. Additionally, it also helps in the development of nano-emulsion, which is stable at ambient condition. Fernandez-Avila et al. [159] developed a soy protein isolate-stabilized emulsion employing high-pressure homogenization, and the results revealed that emulsion treated with 100 to 200 MPa with 20% soybean oil was most stable with its improved physical stability in terms of particle size and rheology.

4.2.6. Ultrasonication

"Ultrasound" is a cluster of sound waves beyond human hearing frequency (>16 kHz). Basically, it is divided into two streams, low- and high-intensity waved; low sound waves are generally used for the detection purpose (sonography), whereas high sound waves are used for the modification in molecules including its size reduction, it also helps in the development of emulsification and extensively utilized by the food industries [160]. In nanotechnology, it is basically utilized for the development of several genres of nano-structure. The components of ultrasound comprise an electric generator, piezoelectric transducer for transforming electrical energy into sound energy and a sound emitter of titanium horn shape for conveying the ultrasonic waves into the sample or medium [161]. Various ranges of nano-delivery techniques have been developed, incorporating lipid and surfactant molecules as a wall material for the fabrication of nano-emulsions, nano-liposomes, niosomes, etc., using ultrasound [162–164]. Nano-carriers such as biopolymeric and polymeric have also been developed for the encapsulation of various food bioactive compounds by the food industry.

4.2.7. Supercritical-Based Technologies

It is an alternate green method for the development of nano-particles [165]. Principally, the temperature of critical solvent is above its critical limit or at single phase irrespective of pressure. Carbon dioxide is commonly used as a supercritical fluid as it is nontoxic, low cost and non-flammable. In this process, a liquid solvent is employed, which could absolutely mix with the super critical fluid such as carbon dioxide so as to liquefy the solute subjected to micro-ionization. Now, due to the insolubility of solute in the super critical fluid, instant precipitation takes place, which further brings out the outcome of developed nano-particles [166]. Presently, methods including micronization via rapid expansion of supercritical (RESS) solution, supercritical antisolvent (SAS), supercritical melt micronization (ScMM), spray coation, supercritical CO_2 coating, etc. are employed for encapsulation of bioactive compounds using supercritical system. Several researchers indicate that this technology can be useful in various ways including development of encapsulated products depending upon properties of wall matric and the active ingredients. Based on the behavior of the core and wall material to be encapsulated, the processing technique using supercritical CO_2 is decided. For example, the interaction of CO_2 with the active material, wall material and the solvent used. In case of a biopolymer drug delivery system, the interface between the supercritical CO_2 and the polymer used plays a vital role in encapsulation process. Whereas polymer including polylactide (PLA) is good for SAS treatment, in the case of RESS, it is difficult to solubilize in supercritical CO_2 [167,168].

4.2.8. Polymerization

In this technique of nano- encapsulation, firstly, to form nano-particles, monomers of an aqueous fluid are polymerized then the bioactive compound is mixed in it. Further, when the encapsulation takes place, the nano-particles are purified by removing extra stabilizer and surfactant settled on the surface of the nano-particles. This technique is generally incorporated for developing poly butylcyanoacrylate nano-particles [169,170].

4.2.9. Coacervation or Ionic Gelation

A technique with wide range of hydrophilic compounds (gelatin, sodium alginate and chitosan) was incorporated for the development of nano-particles. Two kinds of aqueous phase fluids are prepared using chitosan polymer (propylene oxide) and polyanion sodium tripolyphosphate. A nano-size coacervation is formed when the tripolyphosphate (negative charged) reacts with chitosan (positively charged) and, due to the reaction of ions, liquid is converted into gel [171].

5. Physicochemical Properties of Encapsulated Bioactive Compounds

5.1. Particle Size

The foremost properties of nano-particles are the size of its particle and its distribution. These properties are responsible for overall quality including its delivery ability, stability, viscosity, etc. [172]. Due to the relative mobility and significantly tiny size, the intracellular holding capacity is relatively higher in nano-particles than micro-particles. Reports reveal that nano-particles of 100 nm reflected 2.5 times higher holding capacity compared to 1 μm micro particles [173].

Several types of microscopes are used to detect the size and structure of the nano-particles. Optical properties of nano-particles including single/double/multi-emulsions and micro/nano-capsules capsule are measured by scanning electron microscope/laser diffraction based on its properties (wet/dry). To determine the number of pores, surface study is required which needs extremely strong analyzers such as transmission electron microscopy. Several compound locations can be detected using confocal or fluorescence microscopy by mixing fluorophores to dye the bioactive compound. Dynamin light scattering, also known as photon correlation spectroscopy, is extensively utilized to detect size of the nano-particles, which are in the range of 1000 nm. It helps in finding the range of particle size along with its concentration in the given matrix [174]. It is operated to identify the charge attributes of the nano-particles. It indicates the electrical capacity of the nano-particles, which can further be modified by changing the composition of compounds mixed in the aqueous fluid. Zeta potential more than (+/−) 30 mV of a nano-particle is known to be stable. Interestingly, with the help of the Zeta potential test, it can be identified whether the wall material has been encapsulated inside the nano-capsule or covering its outside structure [175]. Desai et al. [176] reported that nano-particles can diffuse through the submucosal layers in rats while a micro-particle is only limited till the epithelial lining.

5.2. Stability of Encapsulated Bioactive Compound

"Stability of nano-particles" refers to the strength and balance needed by nano-particles to remain intact inside the wall matrix until the desired time and place of release. Nano-emulsions have better stability due to their morphological structure of tiny droplets. Additionally, the strength of bioactive compound to remain stable can be examined by placing them in a modified environment including high/low temperature, different ionic charge fluids and different pH range [177].

5.3. Encapsulation Efficiency and Loading Capacity

It is defined as the amount of bioactive compound encapsulated inside the wall matrix. The amount of compound encapsulated can be quantified by using techniques including high-performance liquid chromatography, UV-Vis spectroscopy and UV-Vis spectroscopy [178]. A perfect nano-particle is the one which has maximum compound loading capacity with minimum quantity of wall material. Loading of bioactive compounds can be carried out using two methods including incorporation and absorption methods. The capacity of entrapment is basically dependent on solubility of the compound encapsulated in the wall material, specifically the interaction in the molecules of bioactive compound-polymer, molecular weight and the availability of functional groups [179]. Proteins and macromolecules at isoelectric degree are reported to possess maximum holding capacity. Moreover, in the case of smaller molecules, the ionic reactions between compound and polymer can help in increasing the holding capacity of the matrix.

5.4. Control Release

The release of the bioactive compounds encapsulated in a particular matrix depends on several aspects including compound solubility, surface bound/adsorption, and diffusion from the matrix, matrix degradation and combination of both diffusion and degradation. Nano-spheres with even distribution of bioactive compounds tend to release by erosion of wall material. The process of release is totally controlled by diffusion if degradation of wall material takes place at a slow rate. The quick release of the compound results in poor wall material or low bounding capacity of the compound [180]. It was reported that the mixing method practiced plays a vital role on release profile of nano-capsules as it will slow the release of compounds [181]. In contrast to this, if the compound is protected by the polymer coating, then the release takes place by the diffusion method from inside to matrix outside. Further, there are several techniques including ultra-filtration, reverse dialysis bag, dialysis bag and diffusion of cell with synthetic of artificial membrane that is practiced for the compound release in vitro.

6. Application of Nanoencapsulation in Food Industries

Application of nanoencapsulation in several industries including nutraceutical, pharmaceutical, food, packaging and preservation has exponentially expanded in the last few years. It is extremely promising for the fragrance and flavor industries as the volatile compounds tend to evaporate during processing time. Besides, these compounds undergo chemical changes due to oxidation at atmospheric condition, which results in degradation of the compounds, thus encapsulation helps to stabilize as well as retain the natural color, flavor and fragrance while enhancing the shelf life of the compound. An excellent example is retaining the fresh aroma of brewed coffee through microencapsulation of flavor compounds such as ketones, pyrazines, furans, pyridines, etc., using food starch derived from waxy maize as encapsulating material. Nanoencapsulation of tea compounds (caffeine, theanine and catechins) using various proteins, lipids and carbohydrates have been reported to increase the effectiveness of their health imparting properties such as anticancer, antidiabetic and anti-inflammatory [182–184].

It has been researched and reported that during gastrointestinal digestion, encapsulated catechin showed improved retention of its biological properties as compared to free catechin [185]. Rojas-Graü et al. [186] revealed that the nanoencapsulation technique can be utilized as an anti-browning technique for food industries. Nanoencapsulated plant-based compounds are tagged as anti-browning compounds. Tyrosinase enzyme, also known as catalase B, accelerates the unwanted chemical reaction during food processing including enzymatic browning of fruits, vegetables and beverages that causes adverse effects on its organoleptic characteristics (Polyphenol Oxidase + $O_2 \rightarrow$ Melanin). Zheng et al. [187] isolated tyrosinase inhibitors from Artocarpus heterophyllus to retard browning reacting in fresh-cut-apple slices. The results revealed that apple slices treated with Artocarpus heterophyllus extract along with 0.5% ascorbic acid reflected no browning reaction after 24 h.

Further, nanoencapsulation is popularly appreciated for improving the water stability, solubility and bio-accessibility of hydrophobic bioactive compounds including curcumin by binding it with naturally existing proteins (legume oligomeric globulins, ferritin and casein micelles) by a hydrophobic interactions bond which acts as a nanocarrier for hydrophobic nutraceuticals in drug delivery [188]. Human Serum Albumin-curcumin nano-particles can be used for cancer treatment as they reflect effective antioxidant activity with enhanced antitumor properties [189]. Luo and co-workers [190] carried out ex vivo and in vivo adhesion experiments employing Tannic acid/IR780-nano-particles, which were tailored with anti-ulcerative colitis properties (encapsulated Curcumin). Reports reveal that tannic acid-loaded curcumin nano-particles can be utilized for drug delivery to treat ulcerative colitis as tannic acid possesses degradable adhesive properties which can accumulate on the surface of inflamed mucosa. In Figure 4, ulcerative colitis mice were orally ingested with three different solutions including free IR780, IR780 nano-particles and tannic acid/IR780- nano-particles. A gradual decrease in adherence level was noticed in all the three samples at different time intervals (3 h, 6 h, 12 h, and 24 h). Mice with tannic acid/IR780-nano-particles reflected the maximum adherence towards inflamed mucosa. Additionally, nanoencapsulation is also considered as an effective technique for encapsulating antidiabetic synthetic compound insulin. In 2021, Hadiya et al. [191] encapsulated insulin using chitosan to 170–800 nm size of spherical shape and about 15–52% of delivery efficiency was achieved.

Nanoencapsulation is also useful in active packaging, which helps to prolong shelf life and preserve quality of the food products. It is employed to retain and enhance the nutritional and organoleptic attributes of the product with extended shelf life [192]. Encapsulated bioactive compounds rich in antioxidants and antimicrobials including vitamin C, Vitamin E, carotenoids, etc., are incorporated to edible films as active compounds during the fabrication as it helps in maintaining the level of freshness of the products. The organoleptic attributes of the coated product can be maintained by incorporating additional flavor, colors and sweeteners. Nano-fibers of cinnamon EO encapsulated in polyvinyl alcohol/β-cyclodextrin are obtained to use as a film for coating the packaging box. The film coating (1.5 cinnamon EO-β-cyclodextrin) acts as an antimicrobial barrier to suppress bacterial and fungal spoilage of the food sample while extending its shelf life by 5 d at 10 ± 0.5 °C [193]. Adel et al. [194] prepared a bio-composite using β-cyclodextrin citrate (50%), and oxidized nano-cellulose (7%) in chitosan solution. The prepared film reflected in reduced water vapor permeability 2.09 ± 0.08 (10–11 g m^{-1} s^{-1} Pa^{-1}). In addition, the film was fortified with clove EO nano-particles, which resulted in higher activity of Gram-negative bacteria than Gram-positive. Further, Xiao et al. [195] developed a nano-composite film by encapsulating pesticide/insecticide (iprodione) using poly (ethylene glycol)-poly (ε-caprolactone and chitosan. The encapsulation of iprodione reflected in its improved efficacy by 2-fold along with its reduced dosage.

7. Conclusions and Prospects

At present, due to the tremendous growth of deadly diseases, it has become necessary for humankind to build up a strong immune system and this could be only possible though ingestion of bioactive compounds extracted from plant sources. This scenario has encouraged researchers to search for the conventional and sustainable extraction method for collection of bioactive compounds from plant matrixes. Nevertheless, application of these novel extraction techniques still remains a challenge due to its working principal as it works on several parameters; therefore, in-depth knowledge is required to follow this technique on the larger scale at the industrial level to achieve a promising output. However, due to the evolution of these novel techniques, the extraction has become easy and the extraction yield has been significantly enhanced in maximum bioactive compounds. Novel technologies have helped in enhancing the extraction of bioactive compounds without significant degradation. Moreover, encapsulation techniques offered to deliver the bioactive compounds at the site where it is needed. The combination of these two techniques offers a synergistic effect for extraction as well as targeted delivery.

Author Contributions: S.N., writing—original draft preparation; N.K.R., writing—review and editing; C.O., writing—review and editing; and B.T., writing—review and editing. All authors have read and agreed to the published version of the manuscript.

Funding: The authors and their work were supported by the Irish Department of Agriculture, Food and the Marine, under the Food Institutional Research Measure to the U-Protein project under grant No. 2019PROG702.

Conflicts of Interest: The authors declare no conflict of interest.

Entry Link on the Encyclopaedia Platform: https://encyclopedia.pub/13442 the entry published on the encyclopedia platform.

References

1. Talmaciu, A.I.; Ravber, M.; Volf, I.; Knez, Z.; Popa, V.I. Isolation of bioactive compounds from spruce bark waste using sub- and supercritical fluids. *J. Supercrit. Fluids* **2016**, *117*, 243–251. [CrossRef]
2. Perinelli, D.R.; Palmieri, G.F.; Cespi, M.; Bonacucina, G. Encapsulation of flavours and fragrances into polymeric capsules and cyclodextrins inclusion complexes: An update. *Molecules* **2020**, *25*, 5878. [CrossRef]

3. Barbosa, R.F.S.; de Souza, A.G.; Rangari, V.; dos Santos Rosa, D. The influence of PBAT content in the nano-capsules preparation and its effect in essential oils release. *Food Chem.* **2021**, *344*, 128611. [CrossRef]
4. Gupta, S.; Khan, S.; Muzafar, M.; Kushwaha, M.; Yadav, A.K.; Gupta, A.P. 6-Encapsulation: Entrapping essential oil/flavors/aromas in food. *Encapsulations* **2016**, *2*, 229–268. [CrossRef]
5. Coelho, S.C.; Estevinho, B.N.; Rocha, F. Encapsulation in food industry with emerging electrohydrodynamic techniques: Electrospinning and Electrospraying—A review. *Food Chem.* **2020**, *339*, 127850. [CrossRef]
6. Vinatoru, M. An overview of the ultrasonically assisted extraction of bioactiveprinciples from herbs. *Ultrason. Sonochem.* **2001**, *8*, 303–313. [CrossRef]
7. Paulsen, B.S. Highlights through the history of plant medicine. In Proceedings of the Symposium Held at the Norwegian Academy of Science and Letters, Oslo, Norway, 13–14 November 2010. Available online: C:/Users/shaba.noore/Downloads/31110_Bioactive_compounds_in_plants%20(1).pdf (accessed on 11 June 2021).
8. Croteau, R.; Kutchan, T.M.; Lewis, N.G. Natural products (secondary metabolites). In *Biochemistry and Molecular Biology of Plants*; Buchanan, B., Gruissem, W., Jones, R., Eds.; American Society of Plant Physiologists: Rockville, MD, USA, 2000; pp. 1250–1318.
9. Wink, M. Evolution of secondary metabolites from an ecological and molecular phylogenetic perspective. *Phytochemistry* **2003**, *64*, 3–19. [CrossRef]
10. Chemat, F.; Rombaut, N.; Sicaire, A.-G.; Meullemiestre, A.; Fabiano-Tixier, A.-S.; Abert-Vian, M. Ultrasound assisted extraction of food and natural products. Mechanisms, techniques, combinations, protocols and applications. A review. *Ultrason. Sonochem.* **2017**, *34*, 540–560. [CrossRef]
11. Kovačević, D.B.; Maras, M.; Barba, F.J.; Granato, D.; Roohinejad, S.; Mallikarjunan, K.; Montesano, D.; Lorenzo, J.M.; Putnik, P. Innovative technologies for the recovery of phytochemicals from stevia rebaudiana bertoni leaves: A review. *Food Chem.* **2018**, *268*, 513–521. [CrossRef]
12. Azmir, J.; Zaidul, I.S.M.; Rahman, M.M.; Sharif, K.M.; Mohamed, A.; Sahena, F.; Jahurul, M.H.A.; Ghafoor, K.; Norulaini, N.A.N.; Omar, A.K.M. Techniques for extraction of bioactive compounds from plant materials: A review. *J. Food Enginr.* **2013**, *117*, 426–436. [CrossRef]
13. Chemat, F.; Huma, Z.-e.; Khan, M.K. Applications of ultrasound in food technology: Processing, preservation and extraction. *Ultrason. Sonochem.* **2011**, *18*, 813–835. [CrossRef]
14. Tiwari, B.K. Ultrasound: A clean, green extraction technology. *TrAC Trends Anal. Chem.* **2015**, *71*, 100–109. [CrossRef]
15. Rostagno, M.A.; Prado, J.M. *Natural Product Extraction: Principles and Applications*; The Royal Society of Chemistry's: London, UK, 2013.
16. Alexandru, L.; Cravotto, G.; Giordana, L.; Binello, A.; Chemat, F. Ultrasound-assisted extraction of clove buds using batch-and flow-reactors: A comparative study on a pilot scale. *Innov. Food Sci. Emerg. Technol.* **2013**, *20*, 167–172. [CrossRef]
17. Wen, C.; Zhang, J.; Zhang, H.; Dzah, C.; Zandile, M.; Duan, Y.; Ma, H.; Luo, X. Advances in ultrasound assisted extraction of bioactive compounds from cash crops—A review. *Ultrason. Sonochem.* **2018**, *48*, 538–549. [CrossRef] [PubMed]
18. Capelo-Martínez, J. (Ed.) *Ultrasound in Chemistry: Analytical Applications*; John Wiley & Sons: Hoboken, NJ, USA, 2009.
19. Marić, M.; Grassino, A.N.; Zhu, Z.; Barba, F.J.; Brnčić, M.; Brnčić, S.R. An overview of the traditional and innovative approaches for pectin extraction from plant food wastes and by-products: Ultrasound-, microwaves-, and enzyme-assisted extraction. *Trends Food Sci. Technol.* **2018**, *76*, 28–37. [CrossRef]
20. Shirsath, S.; Sonawane, S.; Gogate, P. Intensification of extraction of natural products using ultrasonic irradiations—A review of current status. *Chem. Eng. Process. Process. Intensif.* **2012**, *53*, 10–23. [CrossRef]
21. Rastogi, N.K. Opportunities and challenges in application of ultrasound in food processing. *Crit. Rev. Food Sci. Nutr.* **2012**, *51*, 705–722. [CrossRef] [PubMed]
22. Sharayei, P.; Azarpazhooh, E.; Zomorodi, S.; Ramaswamy, H.S. Ultrasound assisted extraction of bioactive compounds from pomegranate (*Punica granatum* L.) peel. *LWT* **2019**, *101*, 342–350. [CrossRef]
23. Setyaningsih, W.; Saputro, I.E.; Carrera, C.A.; Palma, M. Optimisation of an ultrasound-assisted extraction method for the simultaneous determination of phenolics in rice grains. *Food Chem.* **2019**, *288*, 221–227. [CrossRef]
24. Song, J.; Yang, Q.; Huang, W.; Xiao, Y.; Li, D.; Liu, C. Optimization of translutein from pumpkin (cucurbita moschata) peel by ultrasound-assisted extraction. *Food Bioprod. Process.* **2018**, *107*, 104–112. [CrossRef]
25. Pan, Z.; Qu, W.; Ma, H.; Atungulu, G.G.; McHugh, T.H. Continuous and pulsed ultrasound-assisted extractions of antioxidants from pomegranate peel. *Ultrason. Sonochem.* **2011**, *18*, 1249–1257. [CrossRef] [PubMed]
26. Assami, K.; Pingret, D.; Chemat, S.; Meklati, B.Y.; Chemat, F. Ultrasound induced intensification and selective extraction of essential oil from *Carum carvi* L. Seeds. *Chem. Enginr Process. Process. Intensif.* **2012**, *62*, 99–105. [CrossRef]
27. Jacotet-navarro, M.; Rombaut, N.; Fabiano-Tixier, A.-S.; Danguien, M.; Bily, A.; Chemat, F. Ultrasound versus microwave as green processes for extraction of rosmarinic, carnosic and ursolic acids from rosemary. *Ultrason. Sonochem.* **2015**, *27*, 102–109. [CrossRef]
28. Sereshti, H.; Heidari, R.; Samadi, S. Determination of volatile components of saffron by optimised ultrasound-assisted extraction in tandem with dispersive liquid–liquid microextraction followed by gas chromatography–mass spectrometry. *Food Chem.* **2014**, *143*, 499–505. [CrossRef] [PubMed]

29. Hossain, M.B.; Brunton, N.P.; Patras, A.; Tiwari, B.; O'Donnell, C.P.; Martin-Diana, A.B.; Barry-Ryan, C. Optimization of ultrasound assisted extraction of antioxidant compounds from marjoram (*Origanum majorana* L.) Using response surface methodology. *Ultrason. Sonochem.* **2012**, *19*, 582–590. [CrossRef]
30. Kaderides, K.; Papaoikonomou, L.; Serafim, M.; Goula, A.M. Microwave-assisted extraction of phenolics from pomegranate peels: Optimization, kinetics, and comparison with ultrasounds extraction. *Chem. Eng. Process. Process. Intensif.* **2019**, *137*, 1–11. [CrossRef]
31. Pettinato, M.; Casazza, A.A.; Perego, P. The role of heating step in microwave-assisted extraction of polyphenols from spent coffee grounds. *Food Bioprod. Process.* **2019**, *114*, 227–234. [CrossRef]
32. Pimentel-Moral, S.; Borrás-Linares, I.; Lozano-Sánchez, J.; Arráez-Román, D.; Martínez-Férez, A.; Segura-Carretero, A. Microwave-assisted extraction for hibiscus sabdariffa bioactive compounds. *J. Pharm. Biomed. Anal.* **2018**, *156*, 313–322. [CrossRef]
33. Rodsamran, P.; Sothornvit, R. Extraction of phenolic compounds from lime peel waste using ultrasonic-assisted and microwave-assisted extractions. *Food Biosci.* **2019**, *28*, 66–73. [CrossRef]
34. Vinatoru, M.; Mason, T.J.; Calinescu, I. Ultrasonically assisted extraction (uae) and microwave assisted extraction (mae) of functional compounds from plant materials. *TrAC Trends Anal. Chem.* **2017**, *97*, 159–178. [CrossRef]
35. Mandal, V.; Mohan, Y.; Hemalatha, S. Microwave assisted extraction—An innovative and promising extraction tool for medicinal plant research. *Pharmacogn. Rev.* **2007**, *1*, 7–18.
36. Chan, C.-H.; Yusoff, R.; Ngoh, G.-C.; Kung, F.W.-L. Microwave-assisted extractions of active ingredients from plants. *J. Chromatogr.* **2011**, *1218*, 6213–6225. [CrossRef] [PubMed]
37. Wang, L.; Weller, C.L. Recent advances in extraction of nutraceuticals from plants. *Trends Food Sci. Technol.* **2006**, *17*, 300–312. [CrossRef]
38. Ajila, C.M.; Brar, S.K.; Verna, M.; Tyagi, R.D.; Valero, J.R. Solid-state fermentation of apple pomace using phanerocheate chrysosporium—Liberation and extraction of phenolic antioxidants. *Food Chem.* **2011**, *126*, 1071–1080. [CrossRef]
39. Shu, Y.Y.; Ko, M.Y.; Chang, Y.S. Microwave-assisted extraction of ginsenosides from ginseng root. *Microchem. J.* **2003**, *74*, 131–139. [CrossRef]
40. Dhobi, M.; Mandal, V.; Hemalatha, S. Optimization of microwave assisted extraction of bioactive flavonolignan-silybinin. *J. Chem. Metrol.* **2009**, *3*, 13.
41. Asghari, J.; Ondruschka, B.; Mazaheritehrani, M. Extraction of bioactive chemical compounds from the medicinal asian plants by microwave irradiation. *J. Med. Plants Res.* **2011**, *5*, 495–506.
42. Puri, M.; Sharma, D.; Barrow, C.J. Enzyme-assisted extraction of bioactives from plants. *Trends Biotechnol.* **2012**, *30*, 37–44. [CrossRef]
43. Laroze, L.; Soto, C.; Zúñiga, M.E. Phenolic antioxidants extraction from raspberry wastes assisted by-enzymes. *Electron. J. Biotechnol.* **2010**, *13*, 11–12. [CrossRef]
44. Gómez-García, R.; Martínez-Avila, G.C.G.; Aguilar, C.N. Enzyme-assisted extraction of antioxidative phenolics from grape (*Vitis vinifera* L.) Residues. *Biotechnology* **2012**, *2*, 297–300. [CrossRef]
45. Zhu, Z.; Li, S.; He, J.; Thirumdas, R.; Montesano, D.; Barba, F.J. Enzyme-assisted extraction of polyphenol from edible lotus (nelumbo nucifera) rhizome knot: Ultra-filtration performance and hplc-ms2 profile. *Food Res. Int.* **2018**, *111*, 291–298. [CrossRef]
46. Nadar, S.S.; Rao, P.; Rathod, V.K. Enzyme assisted extraction of biomolecules as an approach to novel extraction technology: A review. *Food Res. Inter.* **2018**, *108*, 309–330. [CrossRef] [PubMed]
47. de Moura, J.M.L.N.; Campbell, K.; Mahfuz, A.; Jung, S.; Glatz, C.E.; Johnson, L. Enzyme-assisted aqueous extraction of oil and protein from soybeans and cream de-emulsification. *J. Am. Oil Chem. Soc.* **2008**, *85*, 985–995. [CrossRef]
48. Dominguez, H.; Nunez, M.; Lema, J.M. Enzyme-assisted hexane extraction of soya bean oil. *Food Chem.* **1995**, *54*, 223–231. [CrossRef]
49. Marathe, S.J.; Jadhav, S.B.; Bankar, S.B.; Dubey, K.K.; Singhal, R.S. Improvements in the extraction of bioactive compounds by enzymes. *Curr. Opin. Food Sci.* **2019**, *25*, 62–72. [CrossRef]
50. Choudhari, S.M.; Ananthanarayan, L. Enzyme aided extraction of lycopene from tomato tissues. *Food Chem.* **2007**, *102*, 77–81. [CrossRef]
51. Passos, C.P.; Yilmaz, S.; Silva, C.M.; Coimbra, M.A. Enhancement of grape seed oil extraction using a cell wall degrading enzyme cocktail. *Food Chem.* **2009**, *115*, 48–53. [CrossRef]
52. Haider, W.; Sultana, B.; Mushtaq, M.; Bhatti, I.A. Multi-response optimization of enzyme-assisted maceration to enhance the yield and antioxidant activity of cassia fistula pods extracts. *J. Food Meas. Charact.* **2018**, *12*, 2685–2694. [CrossRef]
53. Oh, M.J.; Lee, J.; Seo, S.; Yoon, S.; Seo, M.; Park, S.; Kim, H.-S.; Ha, D.-W.; Lee, S.; Jo, Y.J. Vortex pinning in artificially layered ba(fe,co)2as2 film. *Cryogenics* **2018**, *92*, 1–4. [CrossRef]
54. Eduardo, P.; Iñigo, M.d.M. Olive oil pilot-production assisted by pulsed electric field: Impact on extraction yield, chemical parameters and sensory properties. *Food Chem.* **2015**, *167*, 497–502.
55. Rastogi, N.K. Application of high-intensity pulsed electrical fields in food processing. *Food Rev. Int.* **2003**, *19*, 229–251. [CrossRef]
56. Knoor, D.; Geulen, M.; Grahl, T.; Sitzmann, W. Food application of high electric field pulses. *Trends Food Sci. Technol.* **1994**, *5*, 71–75. [CrossRef]

57. Barbosa, H.M.A.; de Melo, M.M.R.; Coimbra, M.A.; Passos, C.P.; Silva, C.M. Optimization of the supercritical fluid coextraction of oil and diterpenes from spent coffee grounds using experimental design and response surface methodology. *J. Supercrit. Fluids* **2014**, *85*, 165–172. [CrossRef]
58. Brunner, G. Supercritical gases as solvents: Phase equilibria. In *Gas Extraction*; Springer: Berlin/Heidelberg, Germany, 1994; pp. 59–146.
59. Jesus, S.P.; Calheiros, M.N.; Hense, H.; Meireles, M.A.A. A simplified model to describe the kinetic behavior of supercritical fluid extraction from a rice bran oil byproduct. *Food Public Health* **2013**, *3*, 215–222. [CrossRef]
60. Mezzomo, N.; Martínez, J.; Ferreira, S.R. Supercritical fluid extraction of peach (prunus persica) almond oil: Kinetics, mathematical modeling and scale-up. *J. Supercrit. Fluids* **2009**, *51*, 10–16. [CrossRef]
61. Minozzo, M.; Popiolski, A.; Prá, V.D.; Treichel, H.; Cansian, R.L.; Oliveira, J.V.; Mossi, A.; Mazutti, M.A. Modeling of the overal kinetic extraction from maytenus aquifolia using compressed CO_2, braz. *J. Chem. Eng.* **2012**, *29*, 835–843.
62. Rui da Silva, R.P.; Rocha-Santos, T.A.P.; Duarte, A.C. Supercritical fluid extraction of bioactive compounds. *TrAC Trends Anal. Chem.* **2016**, *76*, 40–51. [CrossRef]
63. Mesomo, M.C.; Corazza, M.L.; Ndiaye, P.; Santa, O.R.D.; Cardozo, L.; Scheer, A.D.P. Supercritical CO_2 extracts and essential oil of ginger (zingiber officinale r.): Chemical composition and antibacterial activity. *J. Supercrit. Fluids* **2013**, *80*, 44–49. [CrossRef]
64. Lindy, J. *Supercritical Fluid Extraction (Technology, Applications and Limitations)*; Nova Science: New York, NY, USA, 2014.
65. Sovilj, M.N.; Nikolovski, B.G.; Spasojević, M.Đ. Critical review of supercritical fluid extraction of selected spice plant materials. *Maced. J. Chem. Chem. Eng.* **2011**, *30*, 197–220. [CrossRef]
66. Joo, C.G.; Lee, K.H.; Park, C.; Joo, I.W.; Choe, T.B.; Lee, B.C. Correlation of increased antioxidation with the phenolic compound and amino acids contents of camellia sinensis leaf extracts following ultra-high pressure extraction. *J. Ind. Eng. Chem* **2012**, *18*, 623–628. [CrossRef]
67. Xi, J.; Wang, B.S. Optimization of ultrahigh-pressure extraction of polyphenolic antioxidants from green tea by response surface methodology. *Food Bioprocess Technol.* **2013**, *6*, 2538–2546. [CrossRef]
68. Lee, H.S.; Lee, H.J.; Yu, H.J.; Ju, D.W.; Kim, Y.; Kim, C.-T.; Kim, C.-J.; Cho, Y.-J.; Kim, N.; Choi, S.-Y.; et al. A comparison between high hydrostatic pressure extraction and heat extraction of ginsenosides from ginseng (panax ginseng ca meyer). *J. Sci. Food Agric.* **2011**, *8*, 1466–1473. [CrossRef] [PubMed]
69. Zhang, S.T.; Wang, Z.Y.; Cheng, C.H.; Wang, T.S.; Ni, Z.; Lin, J.M. Optimization for ultra-high pressure extraction of scutellaria barbata by central composite designresponse surface methodology. *Afr. J. Biotechnol.* **2011**, *10*, 14637–14643.
70. Chen, R.; Meng, F.; Zhang, S.; Liu, Z. Effects of ultrahigh pressure extraction conditions on yields and antioxidant activity of ginsenoside from ginseng. *Sep. Purif. Technol.* **2009**, *66*, 340–346. [CrossRef]
71. Xi, J.; Shen, D.J.; Ye, L.; Rui, Z. Micromechanism of ultrahigh pressure extraction of active ingredients from green tea leaves. *Food Control.* **2011**, *22*, 1473–1476.
72. Knorr, D. Process assessment of high pressure processing of foods: An overview. In *Processing Foods: Quality Optimisation and Process Assessment*; Oliveira, F.A.R., Oliveira Jorge, C., Eds.; CRC Press: Boca Raton, FL, USA, 1999; pp. 249–267.
73. Sanchez-Moreno, C.; Plaza, L. Effect of combined treatments of high-pressure and natural additives on carotenoid extractability and antioxidant activity of tomato puree (*Lycopersicum esculentum* mill.). *Eur. Food Res. Technol.* **2004**, *219*, 151–160. [CrossRef]
74. Prasad, K.N.; Yang, B.; Zhao, M.; Wei, X.; Jiang, Y.; Chen, F. High pressure extraction of corilagin from longan (*Dimocarpus longan* Lour.) Fruit Pericarp. *Sep. Purif Technol.* **2009**, *70*, 41–45. [CrossRef]
75. Zhang, S.T.; Wang, Z.Y.; Wang, T.S.; Zheng, N. Optimization of central composite design-response surface methodology in ultra-high pressure extraction of scutellaria baicalensis. *J. Medicina. Plant. Res.* **2012**, *6*, 373–378.
76. Liu, F.; Wang, D.; Liu, W.; Wang, X.; Bai, A.; Huang, L. Ionic liquid-based ultrahigh pressure extraction of five tanshinones from salvia miltiorrhiza bunge. *Sep. Purif. Technol.* **2013**, *110*, 86–92. [CrossRef]
77. Xi, J.; Zhao, S.; Lu, B.; Zhang, R.; Li, Y.; Shen, D.; Zhou, G. Separation of major catechins from green tea by ultrahigh pressure extraction. *Int. J. Pharmaceut.* **2010**, *386*, 229–231.
78. Corrales, M.; García, A.F.; Butz, P.; Tauscher, B. Extraction of anthocyanins from grape skins assisted by high hydrostatic pressure. *J. Food Eng.* **2009**, *90*, 415–421. [CrossRef]
79. Rastogi, N.K.; Raghavarao, K.S.M.S.; Balasubramaniam, V.; Niranjan, K.; Knorr, D. Opportunities and challenges in high pressure processing of foods. *Crit. Rev. Food Sci. Nutri.* **2007**, *47*, 69–112. [CrossRef]
80. Khan, S.A.; Aslam, R.; Makroo, H.A. High pressure extraction and its application in the extraction of bioactive coumounds: A review. *J. Food Process. Engr.* **2018**, *42*, e12896. [CrossRef]
81. Zhang, S.; Xi, J.; Wang, C. High hydrostatic pressure extraction of flavonoids from propolis. *J. Chem. Technol. Biot.* **2005**, *80*, 50–54.
82. Le Noble, W.J. *Organic High Pressure Chemistry*; Elsevier: Amsterdam, The Netherlands, 1988; Volume 37, 489p.
83. Richard, J.S. *High Pressure Phase Behaviour of Multicomponent Fluid Mixtures*; Elsevier: Amsterdam, The Netherlands, 1992.
84. Ganeva, V.; Galutzov, B. Electropulsation as an alternative method for protein extraction from yeast. *Fems. Microbiol. Lett.* **1999**, *174*, 279–284. [CrossRef] [PubMed]

85. Yin, Y.; Cui, Y.; Ding, H. Optimization of betulin extraction process from inonotus obliquus with pulsed electric fields. *Innov. Food Sci. Emerg.* **2008**, *9*, 306–310. [CrossRef]
86. Luo, X.; Bai, R.; Zhen, D.; Yang, Z.; Huang, D.; Mao, H.; Li, X.; Zou, H.; Xiang, Y.; Liu, K.; et al. Response surface optimization of the enzyme-based ultrasound-assisted extraction of acorn tannins and their corrosion inhibition properties. *Ind. Crop. Prod.* **2019**, *129*, 405–413. [CrossRef]
87. Chen, F.; Zhang, Q.; Gu, H.; Yang, L. An approach for extraction of kernel oil from pinus pumila using homogenate-circulating ultrasound in combination with an aqueous enzymatic process and evaluation of its antioxidant activity. *J. Chromatogr. A* **2016**, *1471*, 68–79. [CrossRef]
88. Amiri-rigi, A.; Abbasi, S.; Scanlon, M.G. Enhanced lycopene extraction from tomato industrial waste using microemulsion technique: Optimization of enzymatic and ultrasound pre-treatments. *Innov. Food Sci. Emerg. Technol.* **2016**, *35*, 160–167. [CrossRef]
89. Ladole, M.R.; Nair, R.R.; Bhutada, Y.D.; Amritkar, V.D.; Pandit, A.B. Synergistic effect of ultrasonication and co-immobilized enzymes on tomato peels for lycopene extraction. *Ultrason. Sonochem.* **2018**, *48*, 453–462. [CrossRef]
90. Sun, Y.; Zhang, Y.; Qi, W.; Xie, J.; Cui, X. Saponins extracted by ultrasound from zizyphus jujuba mil var. Spinosa leaves exert resistance to oxidative damage in caenorhabditis elegans. *Food Meas.* **2021**, *15*, 541–554. [CrossRef]
91. Navarro Del Hierro, J.; Herrera, T.; García-Risco, M.R.; Fornari, T.; Reglero, G.; Martin, D. Ultrasound-assisted extraction and bioaccessibility of saponins from edible seeds: Quinoa, lentil, fenugreek, soybean and lupin. *Food Res. Interna.* **2018**, *109*, 440–447. [CrossRef] [PubMed]
92. Hadidi, M.; Ibarz, A.; Pagan, J. Optimisation and kinetic study of the ultrasonic-assisted extraction of total saponins from alfalfa (medicago sativa) and its bioaccessibility using the response surface methodology. *Food Chem.* **2020**, *309*, 125786. [CrossRef]
93. Hu, T.; Guo, Y.-Y.; Zhou, Q.-F.; Zhong, X.-K.; Zhu, L.; Piao, J.-H.; Chen, J.; Jiang, J.-G. Optimization of ultrasonic-assisted extraction of total saponins from *Eclipta prostrasta* L. using response surface methodology. *J. Food Sci.* **2012**, *77*, 975–982. [CrossRef] [PubMed]
94. Nafiunisa, A.; Aryanti, N.; Wardhani, D.H. Kinetic study of saponin extraction from sapindus rarak dc by ultrasound-assisted extraction methods. *Bull. Chem. React. Eng. Catal.* **2019**, *14*, 468–477. [CrossRef]
95. Rohilla, S.; Mahanta, C.L. Optimization of extraction conditions for ultrasound-assisted extraction of phenolic compounds from tamarillo fruit (Solanum betaceum) using response surface methodology. *Food Meas.* **2021**, *15*, 1763–1773. [CrossRef]
96. Ali, M.C.; Chen, J.; Zhang, H.; Li, Z.; Zhao, L.; Qiu, H. Effective extraction of flavonoids from *Lycium barbarum* L. fruits by deep eutectic solvents-based ultrasound-assisted extraction. *Talanta* **2019**, *203*, 16–22. [CrossRef] [PubMed]
97. Upadhyay, R.; Nachiappan, G.; Mishra, H.N. Ultrasound-assisted extraction of flavonoids and phenolic compounds from Ocimum tenuiflorum leaves. *Food Sci. Biotechnol.* **2015**, *24*, 1951–1958. [CrossRef]
98. Tomšik, A.; Pavlić, B.; Vladić, J.; Ramić, M.; Brindza, J.; Vidović, S. Optimization of ultrasound-assisted extraction of bioactive compounds from wild garlic (*Allium ursinum* L.). *Ultrason. Sonochem.* **2016**, *29*, 502–511. [CrossRef]
99. Altemimi, A.; Watson, D.G.; Choudhary, R.; Dasari, M.R.; Lightfoot, D. Ultrasound assisted extraction of phenolic compounds from peaches and pumpkins. *PLoS ONE* **2016**, *11*, e0148758. [CrossRef]
100. Wang, X.; Wu, Y.; Chen, G.; Yue, W.; Liang, Q.; Wu, Q. Optimisation of ultrasound assisted extraction of phenolic compounds from Sparganii rhizoma with response surface methodology. *Ultrason. Sonochem.* **2013**, *20*, 846–854. [CrossRef] [PubMed]
101. Wang, B.; Qu, J.; Luo, S.; Feng, S.; Li, T.; Yuan, M.; Huang, Y.; Liao, J.; Yang, R.; Ding, C. Optimization of ultrasound-assisted extraction of flavonoids from olive (Olea europaea) leaves, and evaluation of their antioxidant and anticancer activities. *Molecules* **2018**, *23*, 2513. [CrossRef] [PubMed]
102. Garcia-Castello, E.M.; Rodriguez-Lopez, A.D.; Mayor, L.; Ballesteros, R.; Conidi, C.; Cassano, A. Optimization of conventional and ultrasound assisted extraction of flavonoids from grapefruit (*Citrus paradisi* L.) solid wastes. *Food Sci. Technol.* **2015**, *64*, 1114–1122. [CrossRef]
103. Ramli, N.H.; Yusup, S.; Quitain, A.T.; Johari, K.; Bin Kueh, B.W. Optimization of saponin extracts using microwave-assisted extraction as a sustainable biopesticide to reduce pomacea canaliculata population in paddy cultivation. *Sustain. Chem. Pharm.* **2019**, *11*, 23–35. [CrossRef]
104. Gao, W.; Chen, F.; Li, H.; Wang, X.; Meng, Q. Microwave-assisted extraction of total saponins from *Physalis alkekengi* L. var. Franchetii (mast.) Makino and their in vitro anti-inflammatory activity. *Food Meas.* **2019**, *13*, 2921–2934. [CrossRef]
105. Dahmoune, B.; Houma-Bachari, F. Microwave assisted extraction of bioactive saponins from the starfish echinaster sepositus: Optimization by response surface methodology and comparison with ultrasound and conventional solvent extraction. *Chem. Eng. Process. Process. Intensif.* **2021**, *163*, 108359. [CrossRef]
106. Nguyen, V.T.; Bowyer, M.C.; Van Altena, I.A.; Scarlett, C.J. Microwave-assisted extraction as an advanced technique for optimization of saponin yield and antioxidant potential from phyllanthus amarus. *Sep. Sci. Technol.* **2017**, *52*, 1–11. [CrossRef]
107. Tian, Y.; Zhao, H. Comparison of different extraction techniques and optimization of the microwave-assisted extraction of saponins from aralia elata (miq.) Seem fruits and rachises. *Chem. Pap.* **2020**, *74*, 3077–3087. [CrossRef]

108. Le, A.V.; Parks, S.E.; Nguyen, M.H.; Roach, P.D. Optimisation of the microwave-assisted ethanol extraction of saponins from gac (momordica cochinchinensis spreng.) Seeds. *Medicines* **2018**, *5*, 70. [CrossRef]
109. Chen, Y.; Xie, M.-Y.; Gong, X.-F. Microwave-assisted extraction used for the isolation of total triterpenoid saponins from ganoderma atrum. *J. Food Engr.* **2007**, *81*, 162–170. [CrossRef]
110. Kwon, J.-H.; Lee, G.-D.; Bélanger, J.M.R.; Paré, J.R.J. Effect of ethanol concentration on the efficiency of extraction of ginseng saponins when using a microwave-assisted process (maptm). *Int. J. Food Sci. Technol.* **2003**, *38*, 615–622. [CrossRef]
111. Nguyen, V.T.; Le, M.D.; Nguyen, T.T.T.; Khong, T.T.; Nguyen, H.N.; Huynh, B.N.D.; Tran, H.T.M.; Trang, T.S. Microwave-assisted extraction for optimizing saponin yield and antioxidant capacity from cacao pod husk (*Theobroma cacao* L.). *J. Food Process. Preserv.* **2021**, *45*, e15134. [CrossRef]
112. Nguyen, V.T.; Vuong, Q.V.; Bowyer, M.C.; Van Altena, I.A.; Scarlett, C.J. Microwave-assisted extraction for saponins and antioxidant capacity from xao tam phan (paramignya trimera) root. *J. Food Process. Preserv.* **2016**, *41*, e12851. [CrossRef]
113. Zheng, X.; Xu, X.; Liu, C.; Sun, Y.; Lin, Z.; Liu, H. Extraction characteristics and optimal parameters of anthocyanin from blueberry powder under microwave-assisted extraction conditions. *Sep. Purif. Technol.* **2013**, *104*, 17–25. [CrossRef]
114. Yang, Z.; Zhai, W. Optimization of microwave-assisted extraction of anthocyanins from purple corn (*Zea mays* L.) cob and identification with HPLC–MS. *Innov. Food Sci. Emerg. Technol.* **2010**, *11*, 470–476. [CrossRef]
115. Zhang, F.; Yang, Y.; Su, P.; Guo, Z. Microwave-assisted extraction of rutin and quercetin from the stalks of Euonymus alatus (Thunb.) sieb. *Phytochem. Anal.* **2008**, *20*, 33–37. [CrossRef] [PubMed]
116. Yen, N.T.; Quoc, L.P. Chemical composition and cellulase-assisted extraction of total saponins from gomphrena celosioides mart. *Biotechnologia* **2020**, *101*, 361–367. [CrossRef]
117. Huy, T.B.; Phuong, N.T.L.; Nga, B.K.; Oanh, H.N.; Hieu, N.H. Enzyme-assisted extraction of triterpenoid saponins from pseuderanthemum palatiferum (nees) radlk. Dry leaf powder and bioactivities examination of extracts. *Chem. Sel.* **2019**, *4*, 8129–8134. [CrossRef]
118. Tomaz, I.; Maslov, L.; Stupić, D.; Preiner, D.; Ašperger, D.; Kontić, J.K. Recovery of flavonoids from grape skins by enzyme-assisted extraction. *Sep. Sci. Technol.* **2016**, *51*, 255–268. [CrossRef]
119. Boulila, A.; Hassen, I.; Haouari, L.; Mejri, F.; Ben Amor, I.; Casabianca, H.; Hosni, K. Enzyme-assisted extraction of bioactive compounds from bay leaves (*Laurus nobilis* L.). *Ind. Crop. Prod.* **2015**, *74*, 485–493. [CrossRef]
120. Abhari, H.; Rad, A.H.; Abhari, A.; Karajian, H. Extracting of saponin from acanthophyllum roots using pulsed electric field (pef) pretreatment and comparing it with thermal method and studying of its antioxidant and foaming properties. *J. Biochem. Technol.* **2018**, *2*, 166–176.
121. Ji, H.; Zhang, L.; Li, J.; Yang, M.; Liu, X. Optimization of ultrahigh pressure extraction of momordicosides from bitter melon. *Int. J. Food Eng.* **2010**, *6*. [CrossRef]
122. Wang, Q.-Y.; Dong, X. Vesicle based ultrasonic-assisted extraction of saponins in panax notoginseng. *Food Chem.* **2020**, *303*, 125394. [CrossRef]
123. Kai, T.; Jie, Y.; Xin, H. A technological research on synergetic extraction of *Momordica charantia* L. Saponin using microwave-surfactant. *Appl. Mech. Mater.* **2014**, *563*, 384–390. [CrossRef]
124. Lu, C.-W.; Yin, Y.-G.; Yu, Q. Optimized extraction of ginsenosides from ginseng root (panax ginsengc.a. meyer) by pulsed electric field combined with commercial enzyme. *J. Food Process. Preserva.* **2016**, *41*, e12766. [CrossRef]
125. Tzima, K.; Brunton, N.P.; Lyng, J.G.; Frontuto, D.; Rai, D.K. The effect of pulsed electric field as a pre-treatment step in ultrasound assisted extraction of phenolic compounds from fresh rosemary and thyme by-products. *Innov. Food Sci. Emerg. Technol.* **2021**, *69*, 102644. [CrossRef]
126. Sunwoo, H.H.; Gujral, N.; Huebl, A.C.; Kim, C.-T. Application of high hydrostatic pressure and enzymatic hydrolysis for the extraction of ginsenosides from fresh ginseng root (panax ginseng c.a. myer). *Food Bioprocess. Techn.* **2013**, *7*, 1246–1254. [CrossRef]
127. Machado, A.P.D.F.; Reátegui, J.L.P.; Barbero, G.F.; Martínez, J. Pressurized liquid extraction of bioactive compounds from blackberry (*Rubus fruticosus* L.) residues: A comparison with methods. *Food Res. Int.* **2015**, *77*, 675–683. [CrossRef]
128. Kiss, E. Nano-technology in food systems: A review. *Acta Aliment.* **2020**, *49*, 460–474. [CrossRef]
129. Shimoni, E. Nano-technology for foods: Delivery systems. *Glob. Issues Food Sci. Technol.* **2009**, 411–424. [CrossRef]
130. Liu, T.; Liu, L.; Gong, X.; Chi, F.; Ma, Z. Fabrication and comparison of active films from chitosan incorporating different spice extracts for shelf life extension of refrigerated pork. *LWT* **2021**, *135*, 110181. [CrossRef]
131. Katouzian, I.; Jafari, S.M. Nano-encapsulation as a promising approach for targeted delivery and controlled release of vitamin. *Trends Food Sci. Technol.* **2016**, *53*, 34–48. [CrossRef]
132. Ravi, H.; Kurrey, N.; Manabe, Y.; Sugawara, T.; Baskaran, V. Polymeric chitosan-glycolipid nano-carriers for an effective delivery of marine carotenoid fucoxanthin for induction of apoptosis in human colon cancer cells (caco-2 cells). *Mater. Sci. Engr.* **2018**, *91*, 785–795. [CrossRef]
133. Gagliardi, A.; Paolino, D.; Costa, N.; Fresta, M.; Cosco, D. Zein- vs plga-based nano-particles containing rutin: A comparative investigation. *Mater. Sci. Eng. C* **2020**, *118*, 111538. [CrossRef] [PubMed]
134. Costa, J.R.; Xavier, M.; Amado, I.R.; Gonçalves, C.; Castro, P.M.; Tonon, R.V.; Cabral, L.M.; Pastrana, L.; Pintado, M.E. Polymeric nano-particles as oral delivery systems for a grape pomace extract towards the improvement of biological activities. *Mater. Sci. Eng. C* **2021**, *119*, 111551. [CrossRef] [PubMed]

135. Zare, M.; Norouzi Roshan, Z.; Assadpour, E.; Jafari, S.M. Improving the cancer prevention/treatment role of carotenoids through various nano-delivery systems. *Crit. Rev. Food Sci. Nutr.* **2020**, *61*, 1–13. [CrossRef] [PubMed]
136. Seyedabadi, M.M.; Rostami, H.; Jafari, S.M.; Fathi, M. Development and characterization of chitosan-coated nano-liposomes for encapsulation of caffeine. *Food Biosci.* **2021**, *40*, 100857. [CrossRef]
137. Assadpour, E.; Jafari, S.M. Nano-encapsulation: Techniques and developments for food applications. *Nano Mater. Food Appl.* **2019**, 35–61. [CrossRef]
138. Chaudhari, A.K.; Singh, V.K.; Das, S.; Dubey, N.K. Nano-encapsulation of essential oils and their bioactive constituents: A novel strategy to control mycotoxin contamination in food system. *Food Chem. Toxicol.* **2021**, *149*, 112019. [CrossRef] [PubMed]
139. Abd-Elhakeem, E.; El-Nabarawi, M.; Shamma, R. Lipidbased nano-formulation platform for eplerenone oral delivery as a potential treatment of chronic central serous chorioretinopathy: In-vitro optimization and ex-vivo assessment. *Drug Deliv.* **2021**, *28*, 642–654. [CrossRef]
140. Bhushani, J.A.; Kurrey, N.K.; Anandharamakrishnan, C. Nano-encapsulation of green tea catechins by electrospraying technique and its effect on controlled release and in-vitro permeability. *J. Food Engr.* **2017**, *199*, 82–92. [CrossRef]
141. Ghorani, B.; Tucker, N. Fundamentals of electrospinning as a novel delivery vehicle for bioactive compounds in food nanotechnology. *Food Hydrocoll.* **2015**, *51*, 227–240. [CrossRef]
142. Torres-Giner, S.; Perez-Masia, R.; Lagaron, J.M. A review on electrospun polymer nano-structures as advanced bioactive platforms. *Polym. Engr. Sci.* **2016**, *56*, 500–527. [CrossRef]
143. Zhang, C.; Zhang, H. Formation and stability of core-shell nano-fibers by electrospinning of gel-like corn oil-in-water emulsions stabilized by gelatin. *J. Agri. Food Chem.* **2018**, *66*, 11681–11690. [CrossRef]
144. Vyslouzilova, L.; Buzgo, M.; Pokorný, P.; Chvojka, J.; Míčková, A.; Rampichová, M.; Kula, J.; Pejchar, K.; Bilek, M.; Lukáš, D.; et al. Needleless coaxial electrospinning: A novel approach to mass production of coaxial nano-fibers. *Int. J. Pharm.* **2017**, *516*, 293–300. [CrossRef]
145. Kutzli, I.; Gibis, M. Formation of whey protein isolate (WPI)-maltodextrin conjugates in fibers produced by needleless electrospinning. *J. Agric. Food Chem.* **2018**, *66*, 10283–10291. [CrossRef] [PubMed]
146. Xiao, Q.; Lim, L.-T. Pullulan-alginate fibers produced using free surface electrospinning. *Int. J. Biol. Macromol.* **2018**, *112*, 809–817. [CrossRef]
147. Moreira, J.B.; Lim, L.-T. Microalgae protein heating in acid/basic solution for nano-fibers production by free surface electrospinning. *J. Food Eng.* **2018**, *230*, 49–54. [CrossRef]
148. Poornima, B.; Korrapati, P.S. Fabrication of chitosan-polycaprolactone composite nano-fibrous scaffold for simultaneous delivery of ferulic acid and resveratrol. *Carbohydr. Polym.* **2017**, *157*, 1741–1749. [CrossRef] [PubMed]
149. Leena, M.; Yoha, K.S. Edible coating with resveratrol loaded electrospun zein nano-fibers with enhanced bioaccessibility. *Food Biosci.* **2020**, *36*, 100669. [CrossRef]
150. Jayan, H.; Maria Leena, M.; Sundari, S.S.; Moses, J.; Anandharamakrishnan, C. Improvement of bioavailability for resveratrol through encapsulation in zein using electrospraying technique. *J. Funct. Foods* **2019**, *57*, 417–424. [CrossRef]
151. Maa, Y.-F.; Nguyen, P.-A.; Sit, K.; Hsu, C.C. Spray-drying performance of a bench-top spray dryer for protein aerosol powder preparation. *Biotech. Bioengr.* **2000**, *60*, 301–309. [CrossRef]
152. Adel, I.M.; ElMeligy, M.F.; Abdelrahim, M.E.; Maged, A.; Abdelkhalek, A.A.; Abdelmoteleb, A.M.; Elkasabgy, N.A. Design and characterization of spray-dried proliposomes for the pulmonary delivery of curcumin. *Int. J. Nano Med.* **2021**, *16*, 2667–2687. [CrossRef] [PubMed]
153. Mozaffar, S.; Radi, M. A new approach for drying of nano-structured lipid carriers (NLC) by spray-drying and using sodium chloride as the excipient. *J. Drug Deliv. Sci. Technol.* **2021**, *61*, 102212. [CrossRef]
154. Datta, S.S.; Abbaspourrad, A.; Amstad, E.; Fan, J.; Kim, S.-H.; Romanowsky, M.; Shum, H.C.; Sun, B.; Utada, A.S.; Windbergs, M.; et al. 25th anniversary article: Double emulsion templated solid microcapsules: Mechanics and controlled release. *Adv. Mater.* **2014**, *26*, 2205–2218. [CrossRef]
155. Jafari, S.M.; He, Y.; Bhandari, B. Optimization of nano-emulsions production by microfluidization. *Eur. Food Res. Technol.* **2007**, *225*, 733–741. [CrossRef]
156. Wang, S.; Ye, X.; Sun, Y.; Liang, J.; Yue, P.; Gao, X. Nano-complexes derived from chitosan and whey protein isolate enhance the thermal stability and slow the release of anthocyanins in simulated digestion and prepared instant coffee. *Food Chem.* **2021**, *336*, 127707. [CrossRef] [PubMed]
157. Donsi, F.; Ferrari, G.; Maresca, P. High-pressure homogenization for food sanitization. *Glob. Issues Food Sci. Technol.* **2009**, 309–352. [CrossRef]
158. Georget, E.; Miller, B.; Aganovic, K.; Callanan, M.; Heinz, V.; Mathys, A. Bacterial spore inactivation by ultra-high pressure homogenization. *Innov. Food Sci. Emerg. Technol.* **2014**, *26*, 116–123. [CrossRef]
159. Fernandez-Avila, C.; Trujillo, A. Ultra-high pressure homogenization improves oxidative stability and interfacial properties of soy protein isolate stabilized emulsions. *Food Chem.* **2016**, *209*, 104–113. [CrossRef]
160. Lorimer, J.P.; Mason, T.J. Sonochemistry. Part 1—The physical aspects. *Chem. Soc. Rev.* **1987**, *16*, 239–274. [CrossRef]
161. Bermudez-Aguirre, D.; Mobbs, T.; Barbosa-Cánovas, G.V. *Ultrasound Applications in Food Processing*; Springer: New York, NY, USA, 2011; pp. 65–105. [CrossRef]

162. Rafiee, Z.; Nejatian, M.; Daeihamed, M.; Jafari, S.M. Application of different nano-carriers for encapsulation of curcumin. *Crit. Rev. Food Sci. Nutr.* **2019**, *59*, 1–77. [CrossRef] [PubMed]
163. Rezaei, A.; Fathi, M.; Jafari, S.M. Nano-encapsulation of hydrophobic and low- soluble food bioactive compounds within different nano-carriers. *Food Hydrocoll.* **2019**, *88*, 146–162. [CrossRef]
164. Rostamabadi, H.; Falsafi, S.R.; Jafari, S.M. Nano-encapsulation of carotenoids within lipid-based nano-carriers. *J. Control. Release* **2019**, *298*, 38–67. [CrossRef]
165. Thote, A.J.; Gupta, R.B. Formation of nano-particles of a hydrophilic drug using supercritical carbon dioxide and microencapsulation for sustained release. *Nanomed. Nanotechnol. Biol. Med.* **2005**, *1*, 85–90. [CrossRef] [PubMed]
166. Reverchon, E.; Adami, R. Nano-materials and supercritical fluids. *J. Supercrit. Fluids* **2006**, *37*, 1–22. [CrossRef]
167. Lee, L.Y.; Wang, C.H.; Smith, K.A. Supercritical antisolvent production of biodegradable micro- and nanoparticles for controlled delivery of paclitaxel. *J. Control. Release* **2008**, *125*, 96–106. [CrossRef]
168. Lee, L.Y.; Smith, K.A.; Wang, C.-H. Fabrication of controlled release devices for anticancer agents using supercritical antisolvent method. In Proceedings of the AIChE Annual Meeting, Cincinnati, OH, USA, 30 October–4 November 2005.
169. Zhang, Q.; Shen, Z.; Nagai, T. Prolonged hypoglycemic effect of insulin-loaded polybutylcyanoacrylate nano-particles after pulmonary administration to normal rats. *Int. J. Pharm.* **2001**, *218*, 75–80. [CrossRef]
170. Boudad, H.; Legrand, P.; Lebas, G.; Cheron, M.; Duchêne, D.; Ponchel, G. Combined hydroxypropyl-[beta]- cyclodextrin and poly(alkylcyanoacrylate) nano-particles intended for oral administration of saquinavir. *Int. J. Pharm.* **2001**, *218*, 113–124. [CrossRef]
171. Calvo, P.; Remunan-Lopez, C.; Vila-Jato, J.L.; Alonso, M.J. Chitosan and chitosan/ethylene oxide-propylene oxide block copolymer nano-particles as novel carriers for proteins and vaccines. *Pharm. Res.* **1997**, *14*, 1431–1436. [CrossRef]
172. Panyam, J.; Labhasetwar, V. Biodegradable nano-particles for drug and gene delivery to cells and tissue. *Adv. Drug Deliv Rev.* **2003**, *55*, 329–347. [CrossRef]
173. Desai, M.P.; Labhasetwar, V.; Walter, E.; Levy, R.J.; Amidon, G.L. The mechanism of uptake of biodegradable microparticles in Caco-2 cells is size dependent. *Pharm. Res.* **1997**, *14*, 1568–1573. [CrossRef]
174. Alexander, M.; Dalgleish, D.G. Dynamic light scattering techniques and their applications in food science. *Food Biophys.* **2006**, *1*, 2–13. [CrossRef]
175. Couvreur, P.; Barratt, G.; Fattal, E.; Legrand, P.; Vauthier, C. Nano-capsule technology: A review. *Crit. Rev. Ther. Drug Carrier Syst.* **2002**, *19*, 99–134. [CrossRef] [PubMed]
176. Desai, M.P.; Labhasetwar, V.; Amidon, G.L.; Levy, R.J. Gastrointestinal uptake of biodegradable microparticles: Effect of particle size. *Pharm. Res.* **1996**, *13*, 1838–1845. [CrossRef] [PubMed]
177. Comunian, T.A.; Jafari, S.M. Production of food bioactive-loaded nano-structures by micro-/nano-fluidics. *Nano Encap. Food Ingr. Speci. Equip.* **2019**, *3*, 213–250. [CrossRef]
178. Jafari, S.M.; Assadpoor, E.; He, Y.; Bhandari, B. Encapsulation efficiency of food flavours and oils during spray drying. *Drying Technol.* **2008**, *26*, 816–835. [CrossRef]
179. Panyam, J.; Williams, D.; Dash, A.; Leslie-Pelecky, D.; Labhasetwar, V. Solid-state solubility influences encapsulation and release of hydrophobic drugs from PLGA/PLA nano-particles. *J. Pharm. Sci.* **2004**, *93*, 1804–1814. [CrossRef] [PubMed]
180. Magenheim, B.; Levy, M.Y.; Benita, S. A new in vitro technique for the evaluation of drug release profile from colloidal carriers—Ultrafiltration technique at low pressure. *Int. J. Pharm.* **1993**, *94*, 115–123. [CrossRef]
181. Puglisi, G.; Fresta, M.; Giammona, G.; Ventura, C.A. Influence of the preparation conditions on poly(ethylcyanoacrylate) nano-capsule formation. *Int. J. Pharm.* **1995**, *125*, 283–287. [CrossRef]
182. Cai, Z.-Y.; Li, X.-M.; Liang, J.-P.; Xiang, L.-P.; Wang, K.-R.; Shi, Y.-L.; Yang, R.; Shi, M.; Ye, J.-H.; Lu, J.-L.; et al. Bioavailability of tea catechins and its improvement. *Molecules* **2018**, *23*, 2346. [CrossRef] [PubMed]
183. Zanin, R.C.; Smrke, S.; Kurozawa, L.E.; Yamashita, F.; Yeretzian, C. Modulation of aroma release of instant coffees through microparticles of roasted coffee oil. *Food Chem.* **2021**, *341*, 128193. [CrossRef] [PubMed]
184. Rana, A.; Rana, S.; Kumar, S. Phytotherapy with active tea constituents: A review. *Environ. Chem. Lett.* **2021**, *19*, 2031–2041. [CrossRef]
185. Ahmad, M.; Mudgil, M.; Gani, A.; Hamed, F.; Masoodi, F.A.; Maqsood, S. Nano-encapsulation of catechin in starch nano-particles: Characterization; release behavior and bioactivity retention during in-vitro digestion. *Food Chem.* **2019**, *270*, 95–104. [CrossRef] [PubMed]
186. Rojas-Graü, M.A.; Soliva-Fortuny, R.; Martín-Belloso, O. Effect of natural antibrowning agents on color and related enzymes in fresh-cut fuji apples as an alternative to the use of ascorbic acid. *J. Food Sci.* **2008**, *73*, 267–672. [CrossRef]
187. Zheng, Z.-P.; Cheng, K.-W.; To, J.T.-K.; Li, H.; Wang, M. Isolation of tyrosinase inhibitors from Artocarpus heterophyllus and use of its extract as antibrowning agent. *Mol. Nutr. Food Res.* **2008**, *52*, 1530–1538. [CrossRef]
188. Tang, C.-H. Strategies to utilize naturally occurring protein architectures as nano-vehicles for hydrophobic nutraceuticals. *Food Hydrocoll.* **2021**, *112*, 106344. [CrossRef]
189. Hao, T.; Wang, K.; Zhang, S.; Yang, S.; Wang, P.; Gao, F.; Zhao, Y.; Guo, N.; Yu, P. Preparation, characterization, antioxidant evaluation of new curcumin derivatives and effects of forming HSA-bound nano-particles on the stability and activity. *Eur. J. Med. Chem.* **2020**, *207*, 112798. [CrossRef] [PubMed]

190. Luo, R.; Lin, M.; Zhang, C.; Shi, J.; Zhang, S.; Chen, Q.; Hu, Y.; Zhang, M.; Zhang, J.; Gao, F. Genipin-crosslinked human serum albumin coating using a tannic acid layer for enhanced oral administration of curcumin in the treatment of ulcerative colitis. *Food Chem.* **2020**, *330*, 127241. [CrossRef]
191. Hadiya, S.; Radwan, R.; Zakaria, M.; El-Sherif, T.; Hamad, M.A.; Elsabahy, M. Nano-particles integrating natural and synthetic polymers for in vivo insulin delivery. *Pharm. Dev. Technol.* **2021**, *26*, 30–40. [CrossRef]
192. Cruz-Romero, M.; Murphy, T.V.; Morris, M.; Cummins, E.; Kerry, J. Antimicrobial activity of chitosan, organic acids and nano-sized solubilisates for potential use in smart antimicrobially-active packaging for potential food applications. *Food Control.* **2013**, *34*, 393–397. [CrossRef]
193. Pan, J.; Ai, F.; Shao, P.; Chen, H.; Gao, H. Development of polyvinyl alcohol/β-cyclodextrin antimicrobial nano-fibers for fresh mushroom packaging. *Food Chem.* **2019**, *300*, 125249. [CrossRef] [PubMed]
194. Adel, A.M.; Ibrahim, A.A.; El-Shafei, A.M.; Al-Shemy, M.T. Inclusion complex of clove oil with chitosan/β-cyclodextrin citrate/oxidized nano-cellulose biocomposite for active food packaging. *Food Packag. Shelf Life* **2019**, *20*, 100307. [CrossRef]
195. Xiao, N.-Y.; Zhang, X.-Q.; Ma, X.-Y.; Luo, W.-H.; Li, H.-Q.; Zeng, Q.-Y.; Zhong, L.; Zhao, W.-H. Construction of EVA/chitosan based PEG-PCL micelles nano-composite films with controlled release of iprodione and its application in pre-harvest treatment of grapes. *Food Chem.* **2020**, *331*, 127277. [CrossRef] [PubMed]

encyclopedia

MDPI

Entry

High-Speed Railway

Inara Watson

School of Engineering, London South Bank University, London SE1 0AA, UK; watsoni2@lsbu.ac.uk

Definition: Union Internationale des Chemins (UIC) defines the high-speed railway (HSR) as a high-speed railway system that contains the infrastructure and the rolling stock. The infrastructure can be newly built dedicated lines enabled for trains to travel with speed above 250 km/h or upgraded conventional lines with a speed up to 200 or even 220 km/h. HSR requires specially built trains with increased power to weight ratio and must have an in-cab signalling system as traditional signalling systems are incapable of above 200 km/h.

Keywords: speed; infrastructure; rolling stock; in-cab signaling system; absence of level crossing

Citation: Watson, I. High-Speed Railway. *Encyclopedia* **2021**, *1*, 665–688. https://doi.org/10.3390/encyclopedia1030053

Academic Editors: Raffaele Barretta, Ramesh Agarwal, Krzysztof Kamil Żur and Giuseppe Ruta

Received: 14 June 2021
Accepted: 19 July 2021
Published: 27 July 2021

1. HSR Technologies

HSR systems were divided into four groups depending on their relationship with conventional rail [1]; dedicated line, mixed high-speed line, conventional mixed line, and fully mixed [2]. Each of these types of HSR has some advantages and disadvantages.

Dedicated HSR represents a line that is fully separated from a conventional line, has a high capacity, high safety, and no level crossings. The line has fences all along the line, often built on viaducts or in long tunnels, and has a high construction cost, such as the case in Japan and Taiwan.

Mixed HSR lines have a wider area to serve, increased accessibility as high-speed trains run on dedicated and conventional lines, high capacity of HS lines stretch over larger areas, reduced building costs. HSR trains can use conventional rails in city centres in areas where land is more expensive to build dedicated lines. However, stretches of conventional lines have less capacity and can be a bottleneck for increased traffic, reduces safety, increases maintenance costs, whilst the rolling stock must be equipped with two signalling systems for HSR and conventional rails, such as the case in France and China.

Mixed conventional rail represents lines that are used by HSR trains and by conventional trains. Mixed traffic reduces the capacity of the line because of big differences in the speed of trains, and it also reduces safety. It can be a suitable solution if a country has a different gauge from the standard gauge size to be part of the European railway network and supports interoperability of international services, such as the case in Spain. This type is more difficult and expensive to maintain, needs special rolling stock, which is also more expensive to purchase and maintain.

Fully mixed lines represent lines used by all types of trains, including freight, have maximum flexibility to be used to full capacity, reduces safety, reliability, and punctuality, and increases maintenance costs. An example of such lines as those used in Germany.

There are two ways to develop the HSR system: build new systems or upgrade conventional railways. Building lines, operating, and maintaining them is an expensive business, but it gives an opportunity to develop a system that can operate at a higher speed and with bigger time savings [3].

Each new project includes planning, land purchasing, infrastructure building, and rolling stock costs. Upgrading existing lines will exclude the need for land purchasing, which may bring big savings. Upgrading conventional rail creates lots of disruption for traffic, and it does not allow reaching the required speed on the new lines. However, it is less expensive as it costs approximately US$4.37 million/km in 2007 prices [4]. Table 1 shows the HSR technologies in selected countries. Despite this difference, there is a lot in common in all HSR systems. All are powered by electricity, have a continuous welded rail, which reduces the noise level and the track vibration, in addition to being built on ballast track or on concrete slab tracks.

Table 1. High-speed rail technologies in selected countries Adapted from ref. [5].

Country	Japan	France	Germany	Italy	Spain	Korea	UK
Line	Tokyo-Osaka	Paris-Lyon	Hannover-Wurzburg	Roma-Paris	Madrid-Barcelona	Seoul-Pusan	London-Birmingham
Length of line in km	515	427	327	260	522	412	225
Max. speed (km/h)	260/300	300	250	250	300	250	360
Travel time	2 h 30 min	1 h 50 min	2h	1 h 35 min	2 h 30 min	1 h 55 min	52 min
Radius of curvature R_{min} (m)	2500	4000	7000	3000	4000	7000	7200
Max. longitudinal gradient (%)	20	35	12.5	8	30	35	N/A
Distance of axes of two tracks (m)	4.2	4.2	4.5	4.2	N/A	5.0	5.0
Superelevation (mm)	200	180	150	160	N/A	N/A	N/A

All HSRs must have advanced signalling systems and automated train control systems. The Automated Train Control (ATC) systems were first developed in Japan and introduced for Shinkansen trains. The system was named Digital Communication and Automatic Train Control (DS-ATC) system. In Europe, it is the European Train Control System (ETCS). The next step in the development of the control systems was the introduction of ERTMS. The ERTMS system was first introduced in Italy on a 204 km line between Rome and Naples. The ERTMS combines GSM-R (communication) and ETCS (signalling) systems. With rapid progress from 2G to 5G network UIC is working on developing the successor of GSM-R, the Future Rail Mobile Communication System (FRMCS) [4]. The system can be introduced to railways as early as 2025. Another common thing for all HSRs is that they are very expensive to build [5] and only two of them recovered construction costs, Shinkansen in Japan, and Paris-Lyon line in France [6].

Rolling Stock is another part of the HSR systems. High-speed trains have a large variety in axle loading ranging from 11.4 t for the Hitachi train to 23 t axle loading for Bombardier and Acela Express [7]. This large difference can be explained by the type of railway that uses this rolling stock. The Shinkansen line that uses Shinkansen-Series 700 is fenced throughout to secure the entire length of the track. In contrast to these, the Acela Express operates on an upgraded line with level crossings. Amtrak trains are equipped with an anti-collision structure to meet USA crash standards. Zefiro, a high-speed train manufactured by Bombardier, is one of the most efficient and advanced trains in the world.

2. Differences between HSR and Conventional Rail

The fundamental principle is the same, but the biggest difference lies within speed and capacity. To increase the capacity of the line, it is very important that trains that operate on the line must not have a big difference in operational speeds.

HSR and conventional rail have the following technological differences:

In track quality: HSR requires a specific design, higher standards of surface, welded rails, and a different and more advanced signalling system. Most HSRs are built on slab tracks. HSR has a larger curve radius than conventional rail. For a speed of 300 km/h, the minimum radius is 4000 m, grades of 3.5% rather than 1–1.5% for existing lines with mixed traffic [8]. HSR track is protected by fences from wild and farm animals. HSR lines do not have any level crossings and have fewer stops than conventional lines.

In traction power: For higher speeds, there is a need to have a more powerful rolling stock.

In signalling system: The signalling system depends on the speed of trains. Railway lines with speeds under 160 km/h use trackside signals to control the safe movements of trains, but if speeds exceed 160 km/h, the driver cannot reliably read signals placed on the trackside. So, for speed above 160 km/h, use onboard signalling [9]. When the speed of the HSR is above 200 km/h, the traditional signalling systems become inefficient [1]. HSR trains require in-cab signalling systems.

In power supply: HSR needs at least 25 kV, but conventional rail voltage can be lower. There are three main electrification systems that are in use for HSR, and they are: 3 kV DC, 15 kV AC, and 25 kV AC. The lines electrified at a higher voltage have lower losses during the transformation and transmission of the energy from the power station to the train. It was found that for trains on lines electrified by 3 kV DC it is necessary to produce 22.6% more than the energy received by the pantograph. For lines electrified by 25 kV AC, it is necessary to produce only 8% more energy than received by a pantograph. Another advantage of using 25 kV AC is the possibility to supply power to high-speed trains with a greater distance between substations, which means a reduction in construction and maintenance costs [10]. HSR can be powered by lower voltage, but it will increase the amount of energy that it needs to produce to power high-speed trains, increase the carbon dioxide emissions, and increase the cost of construction and maintenance of HSR lines.

Technological innovations can promote a faster and less costly transition to the decarbonisation of railways. To speed up the decarbonisation of HSR, the technical innovation will play a significant role. One of these innovations relates to using the Flexible Medium Voltage DC Electric Railway System (MVDC-ERS). It will increase the efficiency of the power distribution and support and consolidate the renewable energy sources [11].

3. Selected HSRs from Europe, Asia, and USA

According to the UIC on 2020, there were 20 countries worldwide that have HSR in operation: 13 countries in Europe (Austria, Belgium, Denmark, Finland, France, Germany, Italy, Poland, Czech Republic, Spain, Switzerland, the Netherlands, and the U.K.), 3 countries in Asia (China, Taiwan-China, Japan, and South Korea), 4 other countries (Saudi Arabia, Morocco, Turkey, and the USA), and another 24 countries worldwide planning to build HSR. The total length of HSR in operation, under construction, and planned to build is 104,413 km worldwide [7]. Table 2 shows all countries with HSR systems in operation worldwide in 2020.

3.1. HSR in Japan

The first high-speed line was opened in Japan on 1 October 1964. It took five years to construct a 515 km line between Tokyo and Osaka. It provided a tremendous boost to the economy in Japan and encouraged countries around the world to develop HSR in their own countries. The core of the success of the Shinkansen lies in the decision to build dedicated lines for high-speed trains. A new double-track high-speed standard-gauge line for speeds up to 250 km/h. It was innovative to operate commercial services at this speed. Because of geological and geographical conditions in Japan, HSR requires many long tunnels and bridges. The disadvantage of this decision was the very expensive civil engineering work. Approximately 13% (86 km) of the line is in tunnels, and 33% (174 km) on bridges and viaducts [12] (p. 14). HSR infrastructure in Japan must be resistant to earthquakes, floods, and deep snow.

From the beginning of building Shinkansen, it was decided not to use outside signals to provide information for the driver but to provide indications inside the cabin. It was achieved by transmitting coded signals to the train through track circuits and receivers installed at intervals along the track. In-cab signalling is mandatory today for all high-speed systems around the world. The track of Shinkansen was divided into sections, and if only one section ahead is clear, the train can move forward. This means that the train ahead was only one block away. In this case, the train can move only with reduced speed, and if the train ahead is two or more blocks ahead, then the train behind can move on at full speed [12] (p. 18). This system increases the capacity of a track.

Table 2. High-speed lines in operation worldwide Adapted from ref. [7].

Year of Introducing HSR System	Country	Length of HSR Line in Operation in 2020 (km)
1964	Japan	3041
1977	Italy	921
1981	France	2734
1991	Germany	1571
1992	Spain	3330
1995	Finland	1120
1997	Belgium	209
2000	USA	735
2003	United Kingdom	113
2003	China	35,388
2004	Switzerland	144
2004	South Korea	893
2006	The Netherlands	90
2007	Taiwan	354
2009	Turkey	594
2012	Austria	254
2015	Poland	224
2018	Saudi Arabia	449
2018	Morocco	200
2019	Denmark	56
N/A	Czech Republic	64

Along with the new infrastructure, there is also a new generation of a train that has been built. One innovation of the Shinkansen rolling stock was that instead of putting traction equipment in a locomotive, it was spread along the length of the train. This reduces the maximum axle load to 16 tonnes [12] (p. 20). This increases the speed of the train, minimises infrastructure maintenance, and reduces energy consumption. However, with distributed power, the noise in passenger saloons is higher than with concentrated power (locomotive) trains [13]. The maximum length of trains in Japan is the same as in Europe, which is 400 m. In Japan, high-speed rolling stock (HSRS) was designed for a 15–20-year life cycle, but in Europe, it is around 30 years [13], and because of this, HSRSs in Japan do not need major renovation or technological innovations within their life cycle. This is an example and a suitable balance between economic benefits, shorter asset life, and low maintenance costs.

In 1964, it took four hours to travel from Tokyo to Osaka. By 2021, it reduced the time to 2 h 30 min after speeds were increased to 285 km/h [14]. In 2015, the number of departures of the Shinkansen bullet from Tokyo had increased to 14 times per hour (358 departures per day), and daily ridership increased to 445,000 with an average delay per train of 0.2 min. Totally 355 million passengers use a high-speed railway every year [15]. E5 series Shinkansen train has 10 cars with a capacity of 731 seats, that is more seats per train than in Germany or France [16]. This is one reason for the profitability of the Shinkansen. Figure 1 shows the plan of the Shinkansen HSR lines in Japan.

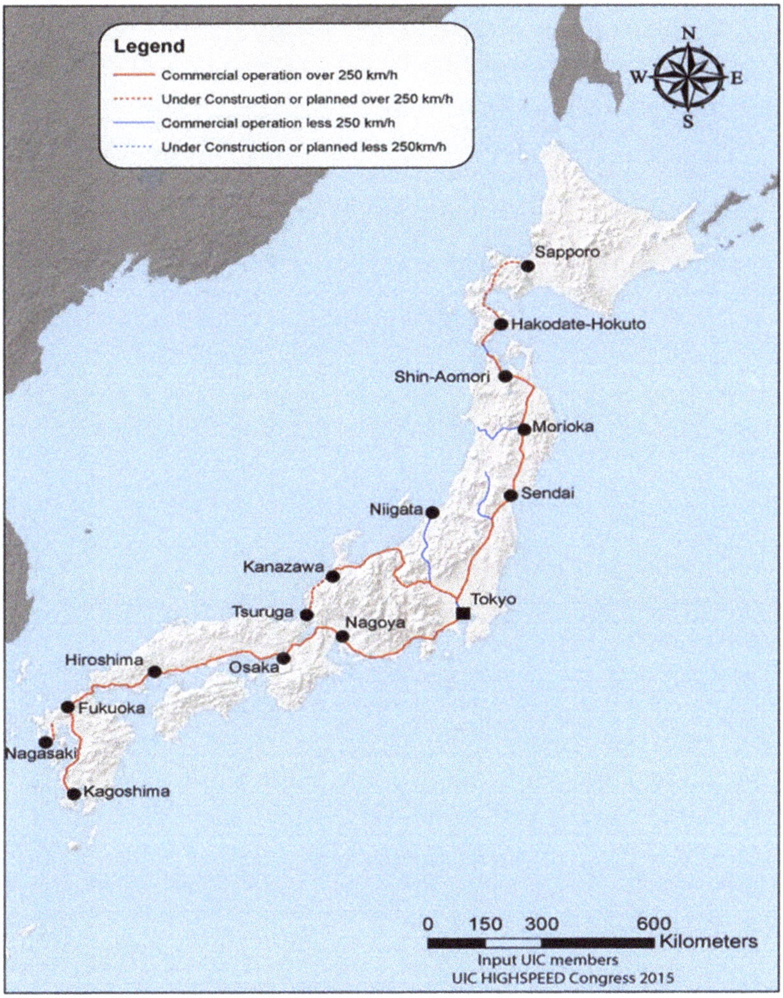

Figure 1. HSR network in Japan, in 2015 Reprinted from ref. [7].

The safety records and punctuality on Shinkansen lines are outstanding. The HSR carried out thousands of scheduled journeys per day, and the average delay across all high-speed lines is 0.2 min per train, including weather-related delays. For over 50 years of operation and carrying over 10 billion passengers without an accident in which passengers could be killed or injured, Shinkansen trains are the most safe and reliable in the world. Shinkansen is the great economic success of Japan, and today the Shinkansen has an operational revenue of around US$19 billion a year [17]. This success has many reasons, and one of them is that the railway company owns the entire infrastructure, stations, the rolling stock, the track, and land around the railways. There is less bureaucracy, less management, and decisions are taken more quickly. With the future decline in Japan's population, there will be a slowdown in new developments of HSR, as there will not be enough demand for high-speed trains.

3.2. HSR in France

The first high-speed trains were introduced in Europe, in France over three decades ago in 1981, and they have carried over 2 billion passengers during these years [18]. It was an immediate successful implementation of TGV. It caused a decrease in air and road traffic, especially for flights. TGV offered a shorter trip time, higher comfort, frequent services, and competitive prices. Figure 2 shows the HSR network in France in 2015.

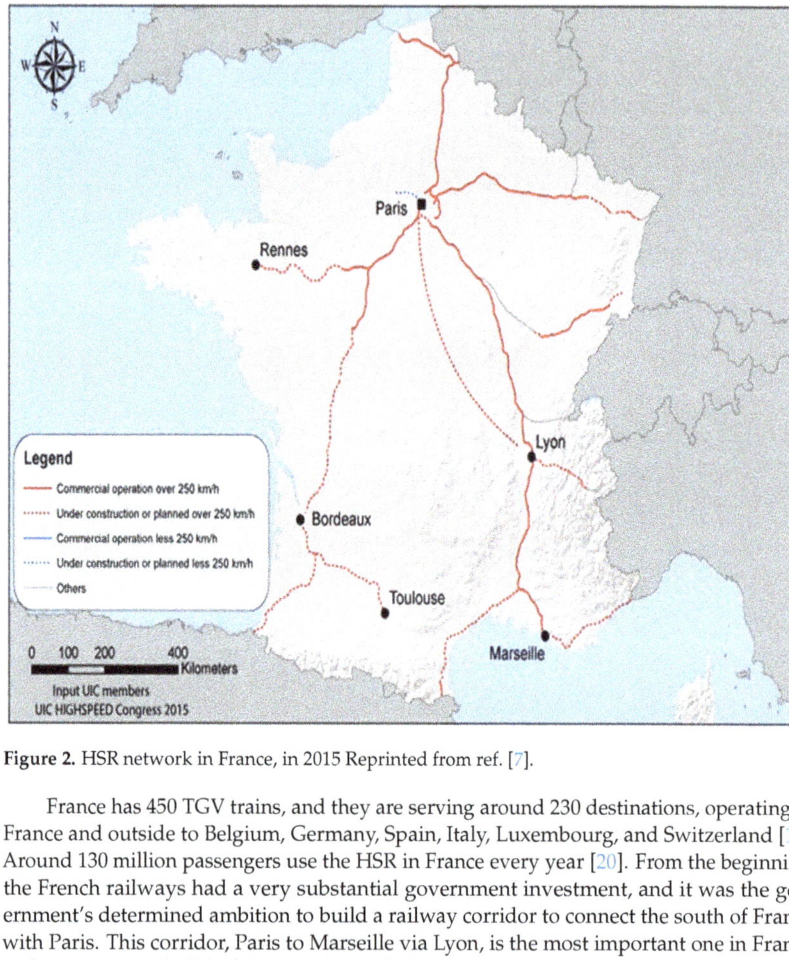

Figure 2. HSR network in France, in 2015 Reprinted from ref. [7].

France has 450 TGV trains, and they are serving around 230 destinations, operating in France and outside to Belgium, Germany, Spain, Italy, Luxembourg, and Switzerland [19]. Around 130 million passengers use the HSR in France every year [20]. From the beginning, the French railways had a very substantial government investment, and it was the government's determined ambition to build a railway corridor to connect the south of France with Paris. This corridor, Paris to Marseille via Lyon, is the most important one in France and serves around 40% of the population [21].

After more than 30 years of operating high-speed trains in France, almost 40 percent of these trains travel on conventional track [19]. Around 60 percent of TGV trains are travelling on new lines designated only for TGV, and all other traffic was prohibited. The new lines have higher gradients unsuitable for freight traffic. In France, the new HSRs were designed to avoid tunnelling, and this gives the benefit of the possibility to implement the double-deck trains. The TGV Duplex was introduced in 1996. This train can travel at speeds up to 300 km/h [22]. The big achievement of the HSR system in France is that the TGV system is compatible with existing conventional railways.

National Society of French Railways (SNCF) has one of the fastest train services in the world. In April 2007, TGV test train reached 574.8 km/h [23]. The latest TGV Duplex Oceane trains have a maximum operational speed of 320 km/h, and they comprise two powered cars, one at each end and eight carriages with a capacity of 556 seats, with the same number of staffs on the train as the TGV. The length of the TGV Duplex is 200 m. These two trains can be coupled together to increase the capacity of the train on busy lines [7].

There are two different ways to power high-speed trains: it can be as in Japan, with distributed traction, or as in France, with TGV, centralised traction. With increasing awareness about the damaging effect caused by transport on the environment and looking for ways to increase the efficiency of transport, it looks more economically appropriate to use distributed traction to power high-speed trains.

There are other ways to increase the capacity of trains, and one of them can be double-deckers (France, Germany, and Japan) or the widening of carriages (Sweden, Japan). Increasing the capacity of trains gives the possibility to reduce operational costs, and this can reduce the railways' fares. To increase the capacity of railway lines, it is needed to electrify the line, implement more advanced signalling systems, increase the speed of trains, and have dedicated lines for high-speed trains only.

One of the most important developments in the construction of TGV-PSE was the introduction of articulated suspension between passenger vehicles. Using this innovation can reduce the weight of the train, reduce the aerodynamic drag, decrease the level of noise from the train, and improve passenger comfort [24]. Figure 3 shows non-articulated and articulated bogies that are used in high-speed rolling stock.

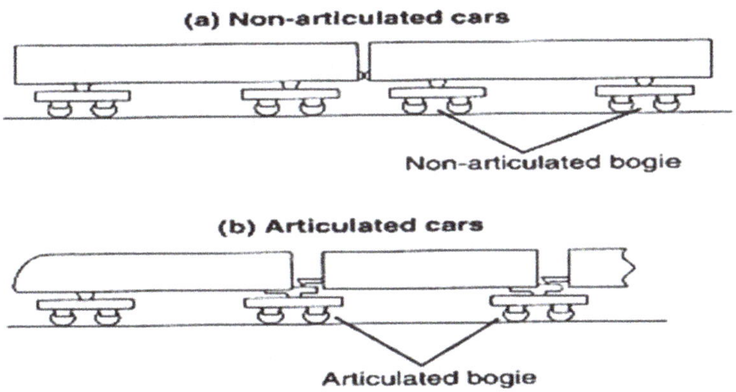

Figure 3. Non-articulated and articulated bogies Reprinted from ref. [12]. (**a**) Non-articulated cars. (**b**) Articulated cars.

A traditional coach has two bogies, and each bogie has two axles; on TGV-PSE coaches, each bogie supports two coaches. Advantages of articulated trains are a more comfortable ride, passengers on the train are less affected by running noise, but the downsides of articulated trains are increased axle load, and that maintenance of these bogies is more difficult. TGV uses the 25 kV AC electrification system, but it can also work on 1500 DC. All TGV trains have at least two electrification systems. To extend track formation life and increase the speed of trains, the weight on one axle was restricted to 17 tonnes [7]. Minimising the axle load will reduce infrastructure and other structural maintenance, reduce construction costs, and reduce the noise level.

The new generation of TGV trains are lighter, have a 15% lower energy consumption, and are designed to be 98% recyclable with a maximum speed of 360 km/h [3]. Reducing the weight has been achieved by using aluminium instead of steel and by using articulated bogies. The train has a regenerative braking system that recovers 8% to 17% of electricity and puts it back in the network [25]. It sufficiently reduces CO_2 emissions and increases the efficiency of high-speed trains.

TGV trains have been installed with Automatic Protection System (APS) and been fitted with in-cab signalling system TVM430. The tracks have been divided into 1500-metre sections, and the in-cab signalling system TVM430 informs the driver of the maximum speed possible on any section. If the train's current speed is higher than the speed limit for that section, then ATP applies the brakes automatically. The TVM430 signalling system allows three minutes of headway, and this increases the capacity of the track to 22,000 passengers in one hour in one direction [26]. This capacity of rail track is equal to a six-lane motorway.

The safety of passengers for any railway network is a crucial requirement. To improve the safety of high-speed trains in France, the lines were redesigned without level crossings. In addition, lines were fully fenced, and advanced equipment was fitted to detect obstructions that occur on a railway line.

From 2008, the profitability of TGV steadily declined, and it was pronounced that there was a need to reduce several stations served by TGV to make the HSR network profitable. HSR services carry only 7% of passengers but account for 61% of the total French rail network traffic [27]. Since the economic crisis in 2007, the number of passengers using high-speed trains continues to decline.

Most HSRs around the world are not profitable but need to look at what benefit they can bring in areas where they pass through, and one example of this can be the city of Lille. It is a suitable example of how a city can flourish from the bypassing of high-speed trains. The development of the TGV HSR and TGV station brought prosperity to the city, blooming commercial activities, and tourism [3].

High-speed trains are the most efficient mode of transport. This is one reason that society continues to fund HSR services. HSR saves time and energy, improves accessibility, increases economic activity, and generates employment [20]. However, with a low-density population in France and only a few larges populated urban areas, it looks unlikely that in the current condition, the HSR will be profitable.

3.3. HSR in Spain

Construction of the HSR in Spain began in 1989 in the corridor between Madrid and Seville, and high-speed trains (AVE) started to run in 1992. Spain's HSR network is one of the widest in the world. In 2020, the length of the HSR network was 3330 km, 1293 km under construction, and 676 planned to build [7], with an average cost of €14 m per kilometre. HSR in Spain has a fleet of 229 trains with an average age of 11 years and in 2015 carried over 35 million passengers [28].

The Spanish HSR has a standard gauge of 1435 mm and is electrified with 25 kV AC and represents 16% of the total Spanish network [29]. Figure 4 shows the HSR network in Spain in 2015.

One of the advantages of 25 kV AC is the possibility to supply power to high-speed trains with a greater distance between substations (approximately 50 km apart), which means reduced construction and maintenance costs. In Spain, they have 59 AC substations and 384 DC substations [30].

The Spanish railway network has two different sizes of rail line gauges: standard gauge (1435 mm) and Iberian gauge (1668 mm). Talgo trains have automatic gauge changing equipment, which allows for the change from one type of gauge to another without stopping at speeds up to 15 km/h and have operational speeds of 220 km/h [31]. Spanish National Railway Network (RENFE) has a punctuality level of 98%. If a train arrives with a delay of over five minutes, all passengers will receive a 100% refund of the ticket price [32].

The Talgo 350, class S102 has the nickname "The Duck", which has served since 2005 and has an operating speed of up to 300 km/h. It is one of the fastest trains in Europe. These trains have 12 coaches and two locomotives with a capacity of 314 and a maximum axle load of 17 tonnes [7]. The new generation of Talgo trains is Talgo Avril. This train is designed for speeds up to 380 km/h, with a low floor to improve the accessibility for vulnerable passengers [33].

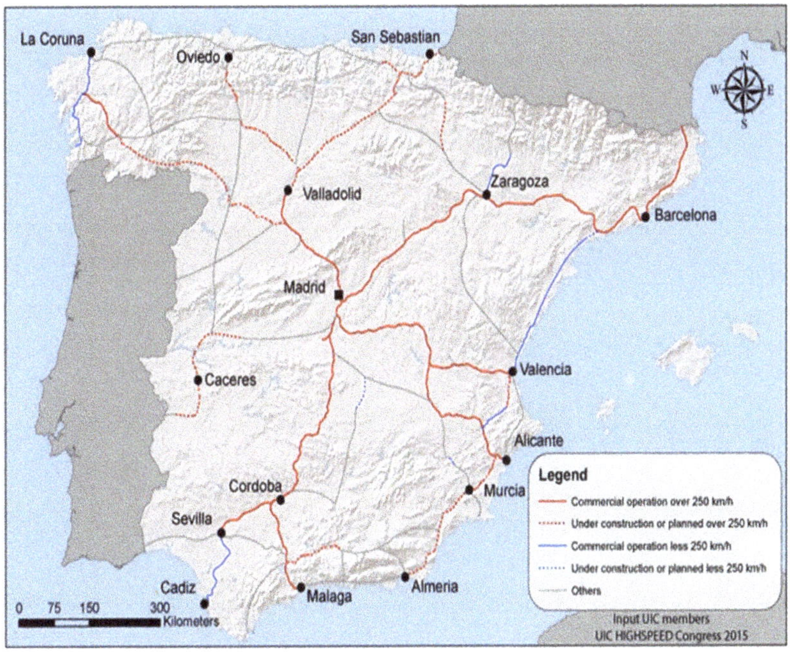

Figure 4. HSR network in Spain, in 2015 Reprinted from ref. [7].

This train will consume less energy because of its lightweight construction, will produce less noise and vibration, and generate less carbon dioxide emissions. The Talgo Avril train comprises two powered cars, one in the front and one at the back, and 12 carriages with a capacity of up to 600 passengers. The train can automatically switch between different track gauges (1435 or 1668 or 1520 mm) and can be run on diesel or electric or both [34]. In addition, this train can use AC or DC electrification systems to run the train.

HSR in Spain on many routes is not profitable because of the low occupancy of trains. There are a few reasons for low occupancy: high unemployment, high ticket prices, many towns with HSR stations are small and can only generate a few passengers. Spanish Government subsidises HSR heavily, and the cost is around US$3 billion per year [35].

3.4. HSR in Germany

Germany has 1571 km HSR in operation, 147 km under construction, and 291 km in the planning stage [7]. The development of HSR in Germany relieved the increasing demand for air and auto travel. The HSR now connects all the largest cities of Germany, and it is in the centre of the country's transport system. Germany has twice the density of the population of France and taking into consideration that the territory of Germany is smaller than that of France, this can be a suitable foundation for the success of the HSR system.

The Intercity Express (ICE) trains were designed and built by a Siemens-led consortium and are operated by the German Federal Railways (DB). ICE1 was introduced into service in 1991, and it composes of 2 power cars, one in the front and one at the back, and 12 carriages between them. It has a maximum axle load of 19.5 tonnes, 358 m long, and has a capacity of 703 seats. Germany has 59 sets of ICE1. The train has a 280 km/h maximum operational speed, powered by electricity from the overhead catenary 15 kV. ICE1 has three signalling systems, LZB, PZB, and ZUB, which are suitable for traffic to Switzerland [7].

The latest ICE3 trains were introduced in 2000 with a maximum operational speed of up to 300 km/h. The power system of ICE3 has been moved from the two ends to the underside of the cars, which is the same system as used in Japan. The train comprises eight coaches, four of which are powered. As trains do not have a power car, it increases the capability of trains as more seating is available for passengers; it has 429 seats per train [7]. The distributed power system has other advantages, and one of them is the low axle load, which is 16 tonnes per axle [7], and this will reduce maintenance costs.

ICE3 has one of the best braking systems. Braking equipment on ICE3 comprises three systems, and one of them is the regenerative system. In addition, ICE3 has a smaller loading gauge than previous ICE1 and ICE2. These changes have been made to allow operation on the European network [36]. For example, France did not allow ICE trains on their network before as ICE was too wide and too heavy. French railways have a restriction of 17 tonnes on one axle for HSR lines.

All railway lines in Germany have a standard gauge of 1435 mm and are electrified at 15 kV AC 16.7 Hz. The new lines are designed for speeds up to 300 km/h [7]. Figure 5 shows the HSR network in Germany in 2015.

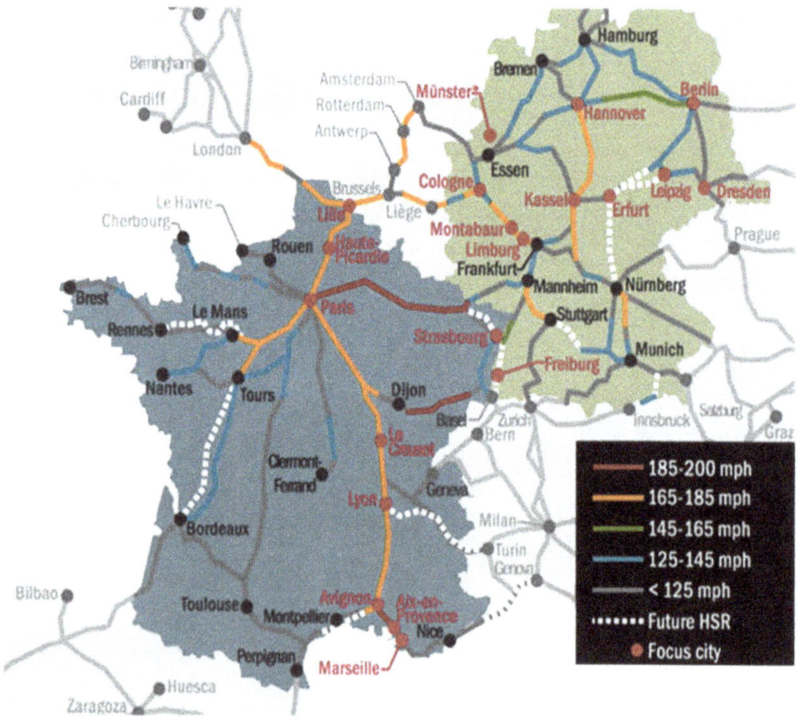

Figure 5. HSR network in Germany and France Adapted from ref. [37].

From the beginning, DB allowed all categories of trains, including high-speed trains and freight trains to use the conventional lines and HSR lines also, but only some trains must run at a lower speed. This decision to allow freight trains to use the HSR was due to the amount of income that freight transportation brought into Germany. This decision differed from Japan and France, where HSR lines are dedicated only to passenger traffic. The mix of traffic brings some disadvantages. For example, if trains with different speeds use the same line, it will decrease the line capacity and can increase safety problems.

It is problematic to produce a timetable that satisfies both passenger traffic and freight traffic because of the significant speed differences. Most of the daylight time was allocated to passenger trains, but during the night, the line needed to be maintained. This caused some serious delays for freight transport. Increasing the number of real-time sensors to monitor the condition of infrastructure and rolling stock and implementing proactive maintenance instead of reactive will decrease the cost and time that needs maintenance.

Many railways' tracks have been upgraded in Germany, but with the mix of traffic, German HSR lines cannot compare with the French network. To travel fast, there is a need to have not only advanced rolling stock but modern infrastructure too. Travelling time in France will be half of that in Germany for the same length of a journey on railways. For example, the travelling time from Paris to Marseille (661 km) is 3 h and 17 min [38], but a similar distance between Hamburg and Munich (791 km) will take 6 h [39]. Another reason that affects travelling time on DB is that there are too many stops for high-speed trains in rural areas with not enough demand for passengers travelling on trains.

The biggest environmental impact from HSR is noise pollution. In Germany, noise legislation for railways has been in force since 1974 [40]. The maximum noise levels for the new build or upgraded transport infrastructure in Germany are as follows in Table 3.

Table 3. German maximum noise level in dB(A) for newly built or upgraded transport infrastructures Reprinted from ref. [41].

Location	L Day	L Night
Near schools, hospitals	57	47
Residential areas	59	49
Central or mix areas	64	54
Industrial areas	69	59

Germany spends annually €150 million to mitigate noise pollution from railways, and by 2020, they reduced the noise level by 10 dB [42]. Because of this legislation, some parts of newly built high-speed lines have been built in cut-and-cover tunnels to reduce the noise level and visual impact for areas with a high density of population.

The safety of passengers travelling on the railways is paramount for every railway. At the present time, only ICE trains and Eurostar have been fitted with a warning system that can detect damages in bogies and wheels early. ICE has been equipped with a train control system (LZB) [14]. This signalling system provides the driver with information for several kilometres ahead. The LZB system improves passenger safety and allows an increase in track capacity. Similar signalling systems for high-speed trains have been developed in Japan and France. Apart from the advanced train control systems, passenger safety in Germany was secured by not having level crossings on HSR lines.

3.5. HSR in Italy

Italy was the second country after Japan that introduced high-speed trains, and the first train went into operation in 1977, but the line was only completed in 1992. Italy has 921 km of HSR in operation and 327 km under construction [7]. Italy was the only country in the world that opened the HSR network for competition. From 2012, NTV (Nuovo Transporto Viaggiaton), a private HSR company, began to operate [43].

ETR460 Pendolino, a tilting train that went into service in Italy in 1988 [7]. For a country with many mountains, it was convenient to use tilting technology on conventional lines. By introducing tilting technology, the train can travel around 30% faster. This active tilting technology soon spread around the world, and now around 70% of all high-speed trains are using it [44]. Pendolino is electric powered with 3 kV DC with a designed maximum operational speed of up to 250 km/h [7].

The train is formed from nine cars. The maximum axle load of an unloaded train is 13.5 tonnes, a train length of 237 m with 480 seats [7]. Different variations of this train are now used in Germany, Spain, and the USA. The ETR500 high-speed train went into service in 1995. It was the first high-speed train that was designed in Italy and only to be used inside

Italy. It has concentrated power, two locomotives, one at each end and 12 trailers, and the total length of the train is 354 m. There were 59 sets of trains manufactured [7]. There was another version of ETR500 designed and built for the Italian and French railway systems. The train was designed for maximum operational speeds up to 300 km/h with improved aerodynamic performances, and the maximum capacity of the train is 671 passengers. The train can run on 3 kV DC and 25 kV AC [7]. There was some back down in the integrated Italian HSR network in the European system, as Italy has some HSR lines using 3 kV DC electrification instead of the standard European system of 25 kV AC [7].

The latest high-speed train in Italy, the ETR1000, went into operation in 2015. Trenitalia is investing €1.5 billion in these trains and will build 50 sets of them. At that moment, they produced 13 sets of these trains. Trains can travel up to 400 km/h with a maximum operational speed of 300 km/h, and it is the fastest train in Europe [45]. ETR1000 is 202 m long with distributed traction along the carriages, four motored coaches, and four trailer coaches [7].

The train has been designed to be compatible with different signalling systems and different electrification systems within the European HSR network. The train has compatibility with the European Traffic Control System (ETCS). The ETR1000 can carry 457 passengers. The cost of this train is around US$40 million [46].

With expanding the HSR network and upgrading conventional lines, HSR in Italy is getting more attractive to customers. HSR lines in Italy run from Turin to Salerno, and Italy has more HSR lines in development: from Milan to Venice, from Milan to Genoa, and from Naples to Bari. Figure 6 shows the HSR network in Italy in 2015.

Figure 6. HSR network in Italy in 2015 Reprinted from ref. [7].

On the HSR line from Rome to Florence operated by the "Frecciarossa", ETR500 trains have four departures every hour from Rome with a maximum operational speed of 300 km/h. The Italian railway has a standard track gauge of 1435 mm and is electrified by 3 kV DC. It will take only 1 h and 30 min on HSR to travel from Rome to Florence compared to a conventional train that takes 3 h and 20 min [47]. The majority of the lines were built close to existing corridors to reduce the environmental impact of the projects. The Italian government had lots of investments put into developing the HSR system in Italy. Table 4 shows the construction costs of the selected HSR lines in Italy.

Table 4. Capital costs of HSR in Italy Reprinted from ref. [1].

HSR Line	Construction Costs ($ Billions)	Miles	Cost per Mile ($Millions)
Toronto-Milano	11.68	78	130.0
Milano-Bologna	10.73	115	77.0
Bologna-Firenze	8.82	49	163.0
Roma-Napoli	8.48	129	58.0

The Rome-Florence line is mostly straight and with no level crossings, with one line in each direction. It is planned to upgrade this line by changing the electrification system from 3 kV DC to the European standard of 25 kV AC [1]. The same trains also operate on the Rome to Naples line, which is 205 km with two departures every hour and a top speed up to 300 km/h [7]. This line has 39 km of tunnels and 39 km of viaducts and bridges. Rome-Florence line was the first to introduce the ERTMS [1].

ERTMS is the most advanced signalling technology in the world. This system uses wireless technology to replace the signals along the railway track. A computer inside the train cabin controls the speed limit of the train and braking distance. The ERTMS system can automatically reduce the speed of the train if it exceeds the maximum allowed speed on the line [48].

One of the significant features of the Italian HSR network is the wide introduction of ERTMS that it integrated into interconnection with the European railway network. It opened the possibility to drive the same rolling stock with the same team around Europe with no need to change on the border and carry on at speeds up to 300 km/h. With building the HSR lines in Italy, they delegated conventional railways to freight transport and to serve the regional passengers. This system increases the safety of passengers travelling by train as it prevents human error. It reduces operational and maintenance costs, as there is no need to install and maintain signals along the railway tracks, and it increases the capacity of tracks. This system is known around the world as the most advanced and safe signalling system for high-speed trains.

3.6. HSR in USA

USA has only one HSR, the Northeast Corridor (NEC) from Boston to Washington D.C., 735 km long, and the same track is shared by freight and passenger trains with much lower speeds. In the future, HSR networks in the USA will expand, with 763 km under construction, and is planning to build another 2108 km [7]. There are many reasons the USA is behind other countries by implementing the HSR system. One of those reasons is that in the USA, the land that the tracks are on is regulated by the individual states, but transportation decisions are regulated through federal policy. The development of land and infrastructure is not run by one department, and often the local interest opposes national interest. Another reason is that policies in the USA encourage car ownership: larger subsidies in highway construction, low density of suburban housing, and cheap fuel.

It was December of 2000 when Amtrak introduced a new train, Acela Express, the first high-speed train in the USA that can travel at a maximum speed up to 240 km/h. Amtrak is the private company that owns the line and provided the railway service from 1971 with very limited federal subsidies. The Acela Express runs between Washington D.C. and Boston. This area has a very high density of population. At the end of 2012, Acela Express brought around 25% of the total Amtrak's service revenue [49]. On the northeast corridor from Washington, D.C. to New York, high-speed trains carried over 3.5 million rail passengers every year and have 76% percent of the market share between Washington, D.C., and New York [1].

The track Washington D.C.-Boston stretches through areas with a very high density of population and does not have fences to protect trains from frequently encountered debris. In addition, the line between Washington D.C. and Boston has many level crossings. The trains in these circumstances must be built to ensure the safety of drivers and passengers. As the Amtrak needs to have an anti-collision structure, trains are 45% heavier than the similar French TGV trains [7]. Figure 7 shows the service map of Acela Express.

Figure 7. Acela Express service map Adapted from ref. [50].

The Acela Express train comprises two powered cars, one at each end, and six passenger carriages, and in use are 20 sets of trains. The maximum axle load is 23 tonnes, a length of 203 m with a seating capacity of 304 seats. A total of 44 of those are first class, and 260 are second class [7]. The Acela Express uses the conventional line that has been upgraded, and this limits the maximum speed to 240 km/h, but by using the tilting-train technology, it provided the possibility to cut journey time [51]. Avelia Liberty, the new trains, will start operating in 2021, and there have been orders to manufacture 28 new trains. Avelia Liberty will comprise 2 locomotives and 10 passenger cars with a total seat capacity of 512 [7].

Avelia Liberty will be a tilted train with concentrated power and a maximum operational speed of 257 km/h [7]. The new trains will cut travel time by approximately 30 min. Tilting-track technology can be an alternative option for building new tracks or straightening existing ones. The tilting technology prevents passengers from having some discomfort from the lateral acceleration, also much cheaper than having to build new tracks [52]. This technology allows the train to have higher speeds in curves, which can reduce the journey time. Figure 8 shows the tilting bogie technology called passive tilt.

When the train runs on a curve, the tilting system tilts the body of the carriage up to a 5° gradient [12]. In addition, another tilting system has been developed, the controlled tilt system, called active tilt (Figure 9). Active tilt technology was first implemented in the United Kingdom. The onboard train computer stores all information about the curve radius, alignment, elevation, and the railway line where the train will run. Tilting the carriages start approximately from 30 to 40 m before the carriages enter the curve as it was found that it will reduce the passengers feeling the sense of motion sickness [12]. The Acela Express has active tilt technology [7].

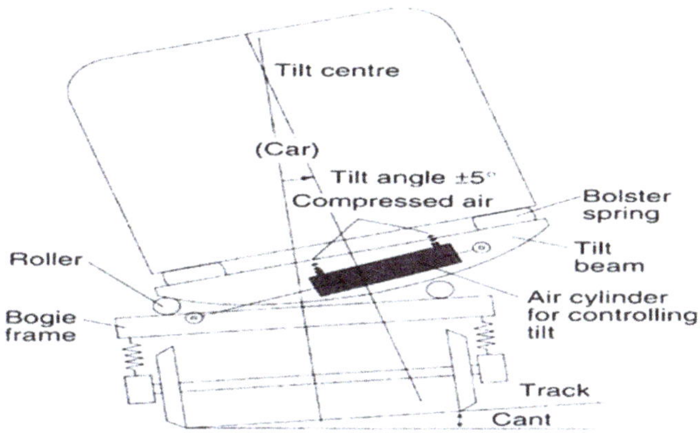

Figure 8. Roller-type tilt system Adapted from ref. [12].

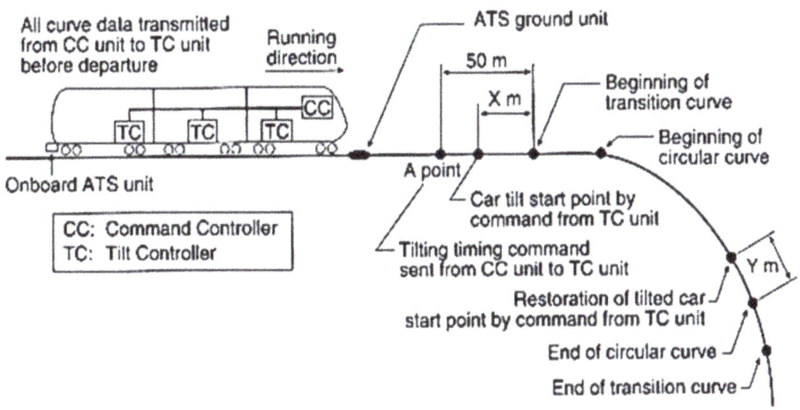

Figure 9. Principles of controlled tilt systems Reprinted from ref. [12].

3.7. HSR in China

The first HSR line, 405 km between Qinhuangdao and Shenyang North, went into operation in 2003, with a maximum speed of 250 km [7]. By 2020, China has the largest HSR system in the world, approximately 60% of the length of the world's HSR network. In 2020, China had 35,388 km of newly built passenger dedicated HSR lines, 5250 km under construction, and is planning to build another 1328 km [7]. HSR network in China will consist of eight horizontal and eight vertical HSR lines, which will connect the largest cities of China that have a population of 500,000 or more [53]. It will connect the south and north of China and west and east. The next step will be to connect China with Taiwan by an underwater tunnel [54].

China has some unproportioned population density, with more people living in the east part of China than in the west. The development of a new HSR will help to satisfy increasing demand in travel to the east of China. The most famous line in China is a line between Beijing and Shanghai. The line's length is 1318 km, it cost around €28 billion to build, and it is the third longest HSR in the world [55]. It took less than two years to build this line. HSR started operating in 2011. Trains can run at speeds up to 350 km/h. The line has 23 stops, and in the first year after opening, over 52 million passengers travelled on it [53].

Around 90 trains depart every day between Beijing and Shanghai. There are two types: super-fast with speed up to 300 km/h and just fast with speed up to 250 km/h. The journey time by the fastest train from Beijing to Shanghai is 4 h 48 min [55]. HSR lines can transport freight during night-time. Figure 10 shows the HSR network in China in 2015.

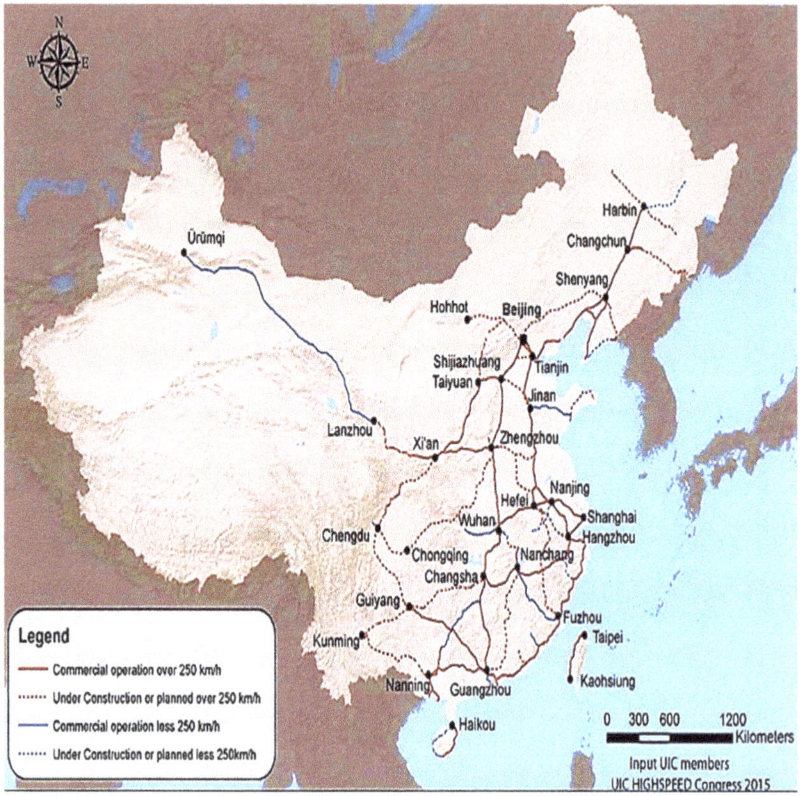

Figure 10. HSR network in China, in 2015 Reprinted from ref. [7].

China has around 2500 high-speed train sets. The train CRH380-AL has been designed and manufactured in China, and it first went into service in 2011. It has a maximum operational speed of up to 300 km/h. CRH380-AL can be formed from 8 to 16 carriages with a capacity of up to 1061 passengers. The 16-carriage train has distributed power with 14 motored coaches and 2 trailer coaches, and it is 403 m long [7].

This train, equipped with a 25 kV AC 50 Hz electric system, has two braking systems, and one of them is the regenerative system, which generates electricity when the braking system slows the train down [7]. The train can accelerate up to 380 km/h [56].

3.8. HSR in Turkey

The first HSR line, 221 km between (Ankara-)Sincan and Eskisehir, went into operation in 2009 with a maximum speed up to 250 km [7]. With the opening of HSR, the travel time between Ankara and Istanbul has been reduced by more than half to just 3 h from 6 h and 30 min. The HSR lines are equipped with an ETCS Level 1 signalling system. In 2016, there were 38 high-speed trains per day, but by 2023, it will increase to 300 services per day, which will carry 120,000 passengers, and passenger numbers can increase to 945 million per year [57]. The development of HSR is funded mostly by the government and by credits from foreign banks [58]. Turkey has the fifth largest HSR network in Europe and the ninth largest in the world. Figure 11 shows the HSR network in Turkey in 2015.

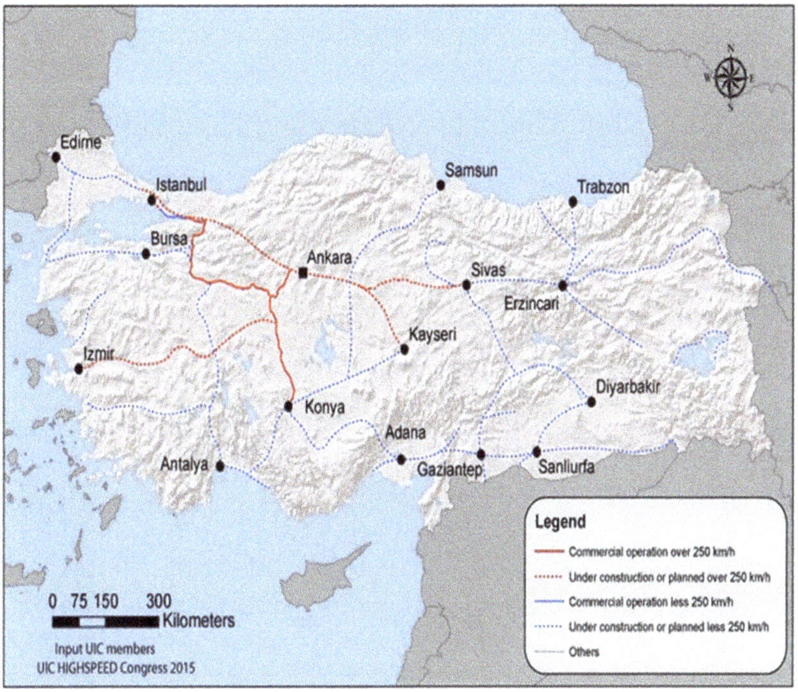

Figure 11. HSR network in Turkey, in 2015 Reprinted from ref. [7].

Ankara-Eskisehir CAF (Spanish railway vehicle constructor) manufactured trains Class HT65000. The trains have a maximum operational speed of 250 km/h, have distributed power, comprising four motor coaches and two trailer coaches, and have a capacity of 419 seats [7]. In 2013, the State Railways of the Republic of Turkey (TCDD) ordered Siemens to manufacture 17 EMU (Electric Multiple Unit) Velaro-D Class HT80000, with distributed power, with four motored coaches and four trailer coaches, and with a maximum operational speed of 300 km/h [59]. In addition, Turkey intends to purchase another 106 high-speed trains [60]. However, one of Turkey's targets is to manufacture all HSR stock in-house and convert existing RS to be compatible with HSR.

In 2020, Turkey had 594 km of HSR in operation, 1652 under construction, and is planning to build another 5173 km [7]. HSR will connect 16 of the largest cities, and the government is planning to connect 55% of the population within an HSR network by 2023 [61]. Turkey wants to integrate into the EU and be a part of the European Single Market and part of the European Transport Network.

3.9. HSR in Taiwan

Figure 12 shows the route of HSR in Taiwan. The HSR in Taiwan comprises one double-track line 354 km long [7]. The line runs along Taiwan's western corridor from the capital of Taiwan (Taipei) to the main industrial city of Kaohsiung on the south of the island. These are the two largest cities in Taiwan. The first 345 km of HSR in Taiwan went into operation on 5 January 2007 with a maximum operational speed of 300 km/h from Taipei to Kohsiung [7]. In 2016, an additional 9 km was built from Taipei to Nangang [7]. In September 2018, the government announced that by 2029 they will build a new HSR line from Kaohsiung to Pingtung 17.5 km long. The estimated cost of a new line is US$1.78 billion [62].

Figure 12. HSR network in Taiwan Adapted from ref. [63].

The line has a standard gauge of 1435 mm and is built on the slab track. To avoid crossing roads on the same level, 90.7% of the line was built in tunnels and on viaducts. The total cost of HSR was estimated to be US$15 billion [64]. Figure 13 shows the ratio of HSR structure type by length.

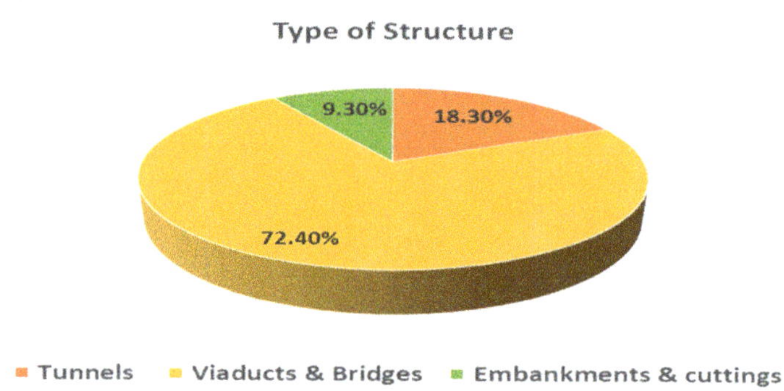

Figure 13. Ratio of HSR structure type by length Adapted from ref. [65].

This line links towns and cities with 94% of the total population in Taiwan. The development of this line reduced the number of car journeys and the number of flights. As a result, it reduced the dependency of Taiwan on the export of petroleum and reduced CO_2 emissions from the transport sector.

In 2013, ridership in a single day reached 200 thousand, and in 2018 it increased to 285 thousand. In 2017, the number of passengers using HSR increased to 60.57 million with 198 daily services. The total number of trains reached 51,751 with a 65.16% occupancy rate [63]. THSRC has an excellent punctuality rate of 99.72% [64].

The HSR in Taiwan is based on the technology of the Shinkansen. This is a new line, and as often to reduce the cost of the project, the stations are located outside of cities. Passengers can reach a station by a free shuttle bus from the centres of towns. In 2017, it provided 393,819 free shuttle journeys [64].

According to UIC HS Rolling Stock tables, Taiwan has in operation 34 trains, Class T700. The design of trains is based on the Series 700 Shinkansen and modified for THSR. Trains went into operation in 2007 [7]. The number of trains will increase to 54 by 2033. Each train comprises nine motor coaches and three trailer coaches. The maximum operational speed is 300 km/h. Each EMU has a length of 304 metres and has 989 seats, one coach first class with 66 seats and 11 coaches second class with 923 seats [7]. The train is powered by an overhead electric line with a voltage of 25 kV 50 Hz, equipped with a signalling system that allows operation on both tracks in both directions. The HSR line is equipped with an Automatic Train Control (ATC) system. The trains are equipped with an earthquake early warning system. This system allows the train to stop or slow down 10 s before an earthquake hits [64].

To cover 354 km without stops, it takes only 96 min, and in 2007 there were 61 trains every 24 h in each direction, from 7 am to 9:06 pm. In 2015, the number of stations reached 11, and the running time of trains stopped at each station was reduced to 138 min. In 2017, the number of stations increased to 12 [66]. Table 5 shows the increased ridership on HSR in Taiwan from 2013 to 2017.

Table 5. Operational statistics Reprinted from ref. [64].

Year	Total Ridership (Millions)	Train Services per Year	Occupancy (Passenger-km/Seat-km)	Punctuality Rate (% of Trains Arriving within Five Minutes of Scheduled Time)	Passenger-km (Millions)
2013	47.49	48,859	57.59%	99.38%	9.118
2014	48.02	50,467	57.12%	99.61%	9.235
2015	50.56	50,532	59.65%	99.66%	9.655
2016	56.59	51,106	63.52%	99.43%	10.488
2017	60.57	51,751	65.16%	99.72%	11.103

3.10. HSR in South Korea

Figure 14 shows the HSR network in South Korea. In 2020, the total length of HSR lines in operation was 893 km, with another 49 km planned to build. The first HSR line from Seoul to Dongdaegu went into operation on 1 April 2004 [7]. The HSR network comprises two corridors, one from Seoul to Busan, an area where over 70% of the population lives, and another one from Seoul to Gwangju. The lines have a standard gauge of 1435 mm and are built on ballasted track with concrete sleepers. The estimated cost of 411 km HSR from Seoul to Busan was US$17 billion [67]. It was an expensive project, as 190 km is in tunnels and 120 km on bridges and viaducts. The HSR network gives an opportunity for commuters to travel to most parts of the country in 1 h 30 min. Figure 15 shows the ratio of structure type by the length of HSR in the corridor Seoul-Busan.

With the opening in 2004 of HSR, the travel time on the Seoul-Busan route was reduced from 4 h and 10 min to 2 h and 40 min and after developing a new line between Daegu and Busan times was cut to 1 h 46 min [69]. Table 6 shows the HSR lines in South Korea.

In the first 10 years, from 2004 to 2014, KTX carried around 414 million passengers with an average of 150,000 per day [70].

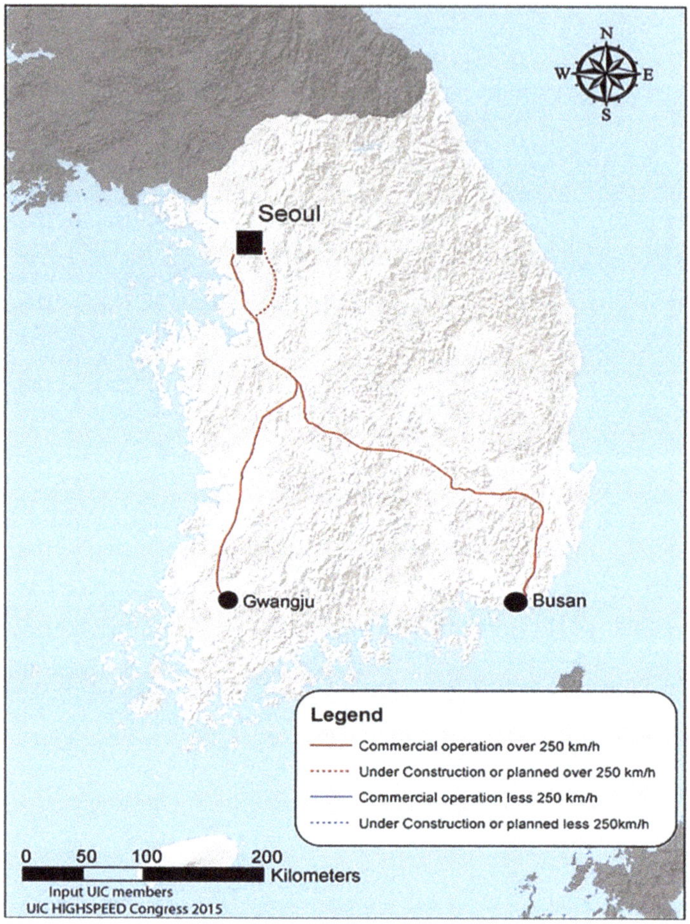

Figure 14. HSR network in South Korea, in 2015 Reprinted from ref. [7].

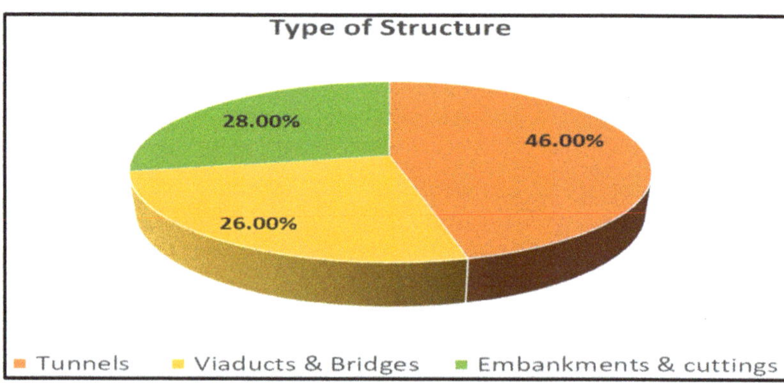

Figure 15. Ratio of structure type by length of HSR Adapted from ref. [68].

Table 6. High-speed lines in South Korea on 1 October 2019 Adapted from ref. [1].

Section	Status	Date	Length (km)	Max. Operational Speed (km/h)
Seoul-Dongdaegy	completed	2004	288	305
Dongdaegy-Busan	completed	2010	130	305
Osong-Gwangju	completed	2015	184	305
Suseo-Pyoengtaek	completed	2016	61	305
Seoul-Gangneung	completed	2017	230	250
Gwangju-Mokpo	planned	2025	49	305

According to UIC HS Rolling Stock tables, Korea has in operation 117 KTX (Korea Train Express) trains, with a maximum operational speed of 300 km/h [7]. The train design was based on the French TGV Reseau. The first 46 trains went into operation in 2004. A total of 12 of them were supplied by Alstom and 34 manufactured in Korea by Hyundai Rotem. Korea was the fourth country in the world after Japan, France, and Germany, which uses its own technology to build trains with a maximum speed of 330 km. The first 46 trains have a set formula of 2 locomotives, 2 motorised trailers, and 16 trailer coaches. The trains are 388 m long with a maximum number of seats of 935 and with maximum traction power of 13.200 kW. The latest 71 trains manufactured by Hyundai Rotem comprise two locomotives and eight trailer coaches. They are 201 m long with 363 seats, 30 seats in first class and 333 seats in second class, and a maximum traction power of 8800 kW [7].

These trains a lighter as the car body is made from aluminium alloy, and the powered car body is made from mild steel, compared to the first 46 trains, in which the bodies were made from steel. The weight of 10 cars is 434 tonnes. The train is powered by an overhead electric line with a voltage of 25 kV 60 Hz, equipped with an Automatic Train Control (ATC) system, which is continuously checking the speed of the train [7].

4. Conclusions and Prospects

In the present chapter, different HSR systems worldwide have been discussed and what they have in common, and how they differ from each other. HSRs worldwide have some technical and organisational differences, such as a difference in operating voltages (Spain), different gauges (Spain, Japan), signalling systems (France, Germany), and languages (Eurostar). The most challenging of them is the Eurostar, which crosses three different railway networks. It has three different power collection systems and signalling systems for Belgium, France, and the U.K. The first HSR was built in 1964, but now 44 countries worldwide have constructed or are planning to construct HSR systems. Amongst them are countries such as India, Indonesia, South Africa, Iran, Brazil, and Russia, and the numbers are growing.

High-speed trains can be very competitive for trips between 150 and 700 km for links between urban centres with a high density of population. It can break the dependency of the transport system on fossil fuels. HSR is powered by electricity and can be almost zero carbon, but it must be powered by renewable energy, as the energy type can significantly influence the level of environmental impact. To support the decarbonisation of railways, there is a need to improve the operational strategies, optimise scheduling, and maximise the use of rolling stock and infrastructure.

The role of the HSR for moving people and goods will steadily increase. HSR must be cheap to build, cheap to maintain, and with high occupancy of seats. Sometimes, it would be more beneficial to upgrade the existing railway network than build new HSR lines. There is a need to ask, are the conventional lines profitable, and if not, why would the new HSR perform any different?

The HSR network in Europe was developed in different countries at different times and standards and to link all the railway networks into one system. There is a need to have compatible technical standards. The standardisation and harmonisation of track gauge, maximum load per axle, systems of electric traction, signalling systems, and line profiles is crucial to operating effectively across Europe. The differences in railway technical standards in Europe create an additional cost for all railway systems. By reducing the number of different HSR technologies and increasing the standardisation, it will improve safety, reduce the capital cost, and increase the compatibility of railways and, in the long term, will reduce the journey costs.

Funding: This research received no external funding.

Conflicts of Interest: The authors declare no conflict of interest.

Entry Link on the Encyclopedia Platform: https://encyclopedia.pub/13532.

References

1. Feigenbaum, B. *High Speed Rail in Europe and Asia: Lessons for the United States*; Reason Foundation: Los Angeles, CA, USA, 2013. Available online: https://reason.org/wp-content/uploads/files/high_speed_rail_lessons.pdf (accessed on 22 July 2021).
2. Campos, J.; de Rus, G. Some stylized facts about high-speed rail: A review of HSR experiences around the world. *Transp. Policy* **2009**, *16*, 19–28. Available online: https://www.sciencedirect.com/science/article/abs/pii/S0967070X09000109 (accessed on 13 June 2021). [CrossRef]
3. UIC-High Speed Department. High Speed Rail & Sustainability. Available online: https://www.shop-etf.com/en/high-speed-rail-sustainability (accessed on 22 July 2021).
4. UIC-International Union of Railways. FRMCS, UIC-International Union of Railways Website. 2021. Available online: https://uic.org/rail-system/frmcs/ (accessed on 21 July 2021).
5. Hasegawa, D. Reducing Land take and Energy use of High-Speed Railways through the Robust Design of Operations. Ph.D. Thesis, University of Birmingham, Birmingham, UK, April 2016. Available online: https://core.ac.uk/download/pdf/83926257.pdf (accessed on 4 August 2019).
6. Wiggins, J. Bullet Trains 'not Suited for Australia', Australian Financial Review Website. 2020. Available online: https://www.afr.com/companies/infrastructure/bullet-trains-not-suited-for-australia-20200522-p54vmn (accessed on 12 June 2021).
7. UIC-International Union of Railways. High-Speed Database & Maps, UIC-International Union of Railways Website. 2021. Available online: https://uic.org/passenger/highspeed/article/high-speed-database-maps (accessed on 22 July 2021).
8. Track, T. The Railway Track, Tuxdoc Website. 2017. Available online: https://tuxdoc.com/download/the-railway-track_pdf (accessed on 22 July 2021).
9. TGVweb. The TGV Signaling System, TGV Website. 2021. Available online: http://www.railfaneurope.net/tgv/signals.html (accessed on 6 July 2021).
10. UIC-High Speed Department. *High Speed, Energy Consumption and Emissions*; International Union of Railways: Paris, France, 2010. Available online: http://www.shop-etf.com/en/high-speed-energy-consumption-and-emissions.html (accessed on 12 July 2017).
11. Brenna, M.; Foiadelli, F.; Kaleybar, H. The Evolution of Railway Power Supply Systems Toward Smart Microgrids: The concept of the energy hub and integration of distributed energy resources. *IEEE Electrif. Mag.* **2020**, *8*, 12–23. Available online: https://ieeexplore.ieee.org/document/9024071/authors#authors (accessed on 5 July 2021). [CrossRef]
12. Railway Technical Research Institute. *Japanese Railway Technology Today*; Railway Technical Research Institute: Tokyo, Japan, 2001; p. 241.
13. De Vos, P. *Railway Noise in Europe, State of the Art Report*; International Union of Railways: Paris, France, 2016. Available online: https://uic.org/IMG/pdf/railway_noise_in_europe_2016_final.pdf (accessed on 22 July 2021).
14. Shinkansen Trains. Osaka to Tokyo Shinkansen Schedule & Tickets Cost, Shinkansen Trains Website. 2021. Available online: https://www.shinkansentrains.com/osaka-to-tokyo.html (accessed on 22 July 2021).
15. Nippon. The Tōkaidō Shinkansen's World-Class Safety, Reliability, and Frequency, Nippon Website. 2018. Available online: https://www.nippon.com/en/features/h00201/ (accessed on 12 June 2021).
16. Railway Technology. E5 Series Shinkansen Bullet Train, Railway Technology. 2021. Available online: https://www.railway-technology.com/projects/shinkansen-bullet-train-high-speed-japan/ (accessed on 22 July 2021).
17. ISO. The Reason for Rail. 2021. Available online: https://www.iso.org/news/2013/11/Ref1793.html (accessed on 22 July 2021).
18. Bunn, M. The History of the French High Speed Rail Network and TGV, Electric Railway Society Website. 2015. Available online: https://electric-rly-society.org.uk/the-history-of-the-french-high-speed-rail-network-and-tgv/ (accessed on 22 July 2021).
19. Barrow, K. France Faces Tough Choices over Future of TGV, International Railway Journal Website. 2015. Available online: https://www.railjournal.com/in_depth/france-faces-tough-choices-over-future-of-tgv/ (accessed on 22 July 2021).
20. Briginshaw, D. SNCF Welcomes 2 Billionth TGV Passenger, International Railway Journal Website. 2013. Available online: http://www.railjournal.com/index.php/high-speed/sncf-welcomes-its-2-billionth-tgv-passenger.html (accessed on 22 July 2021).

21. Scordamaglia, D. *High-Speed Rail in the EU*; European Union: Brussels, Belgium, 2015. Available online: https://www.europarl.europa.eu/RegData/etudes/BRIE/2015/568350/EPRS_BRI(2015)568350_EN.pdf (accessed on 12 June 2021).
22. Railway Technology. Talgo Avril Very High Speed Train, Railway Technology. 2019. Available online: http://www.railway-technology.com/projects/talgo-avril-very-high-speed-train/ (accessed on 7 August 2019).
23. Bernard, A. French TGV Train Breaks Record for Rail Speed, The New York Times Website. 2007. Available online: https://www.nytimes.com/2007/04/03/world/europe/03iht-train.4.5130569.html (accessed on 12 June 2021).
24. Railway Technology. Resistance Is Futile: How Aerodynamics Inform Train Design, Railway Technology Website. 2020. Available online: https://www.railway-technology.com/features/feature-aerodynamic-train-design-arrowedge-union-pacific-bombardier-zefiro/ (accessed on 12 June 2021).
25. Climate Technology Centre & Network. Regenerative Braking in Trains, Climate Technology Centre & Network Website. 2016. Available online: https://www.ctc-n.org/technologies/regenerative-braking-trains (accessed on 22 July 2021).
26. Railway Technology. TGV France High Speed Railways Operated by SNCF, Railway Technology Website. 2019. Available online: https://www.railway-technology.com/projects/frenchtgv/ (accessed on 8 August 2019).
27. Watson, I.; Ali, A.; Bayyati, A. Social Sustainability of High Speed Railways-Comparative Study. In Proceedings of the Stephenson Conference 2017, London, UK, 25–27 April 2017; ImechE: London, UK, 2017.
28. Puente, F. Spain Pushes Ahead with High-Speed Plans, International Railway Journal Website. 2016. Available online: https://www.railjournal.com/in_depth/spain-pushes-ahead-with-high-speed-plans (accessed on 22 July 2021).
29. Alvarez, A.G. *Automatic Track Gauge Changeover for Trains in Spain*; Fundación de los Ferrocarriles Españoles: Madrid, Spain, 2010. Available online: https://www.vialibre-ffe.com/pdf/track_gauge_changeover.pdf (accessed on 12 June 2021).
30. International Union of Railways. Reversible Substations in Spanish Conventional-DC lines, International Union of Railways Website. 2017. Available online: https://events.uic.org/IMG/pdf/07_renfe_reversible_substations_conventional_lines_ed2_delafuente.pdf (accessed on 22 July 2021).
31. Shael, E. Talgo 250. Slideshare Website. Available online: https://www.slideshare.net/ErShael/talgo-250 (accessed on 8 August 2019).
32. Watson, I.; Ali, A.; Bayyati, A. Investigation of the operational reliability of high-speed railways and possible measures of improvement. In Proceedings of the Railway Engineering Conference 2017, Edinburgh, UK, 21–22 June 2017; University of Edinburgh: Edinburgh, UK, 2017.
33. Talgo. Talgo Begins Rail Tests for its Very High-Speed Train Talgo Avril, Talgo Website. 2021. Available online: https://www.talgo.com/-/talgo-comienza-las-pruebas-en-via-del-tren-de-muy-alta-velocidad-talgo-avril/ (accessed on 22 July 2021).
34. Talgo Home Page. Available online: https://www.talgo.com/en/rolling-stock/very-high-speed/avril/ (accessed on 22 July 2021).
35. Bodman, A. Debts and Subsidies for High Speed Rail, STOP HS2 Website. 2021. Available online: http://stophs2.org/news/5712-debts-subsidies-high-speed-rail (accessed on 22 July 2021).
36. Railfaneurope. ICE 3: The Third Generation, Railfaneurope Website. 2021. Available online: http://www.railfaneurope.net/ice/ice3.html (accessed on 12 June 2021).
37. Eidlin, E. *Making the Most of High-Speed Rail in California: Lessons from France and Germany*; The German Marshall Fund of the United States: Washington, DC, USA, 2015. Available online: https://www.gmfus.org/publications/making-most-high-speed-rail-california (accessed on 22 July 2021).
38. Eurorailways. Paris to Marseille Train Tickets by Euro Railways, Eurorailways Website. 2019. Available online: http://www.eurorailways.com/train-tickets/D-FR-FR-MRS-PAR-R-paris-to-marseille.html (accessed on 8 August 2019).
39. Eurail. TGV High-Speed Train, Eurail Website. 2019. Available online: http://www.eurail.com/europe-by-train/high-speed-trains/tgv (accessed on 7 August 2019).
40. Umwelt Bundesamt. Traffic Noise, Umwelt Bundesamt Website. 2012. Available online: https://www.umweltbundesamt.de/en/topics/transport-noise/traffic-noise#noise-pollution (accessed on 12 June 2021).
41. European Parliament. Directorate—General for Internal Policies. 2012. Available online: http://www.europarl.europa.eu/RegData/etudes/etudes/join/2012/474533/IPOL-TRAN_ET(2012)474533_EN.pdf (accessed on 8 August 2019).
42. Dovetail Games. Class 605 ICE TD. 2014. Available online: https://steamcdn-a.akamaihd.net/steam/apps/277755/manuals/DB_BR_605_ICE_TD_Manual.pdf?t=1460545078 (accessed on 12 June 2021).
43. Cascetta, E.; Coppola, P.; Velardi, V. High-Speed Rail Demand: Before-and-After Evidence from the Italian Market. *disP Plan. Rev.* **2013**, *49*, 51–59. Available online: https://www.researchgate.net/publication/258450620_High-Speed_Rail_Demand_Before-and-After_Evidence_from_the_Italian_Market (accessed on 12 June 2021). [CrossRef]
44. Zhou, L.; Shen, Z. Progress in high-speed train technology around the world. *J. Mod. Transp.* **2011**, *19*, 1–6. Available online: https://link.springer.com/article/10.1007/BF03325733 (accessed on 9 August 2019). [CrossRef]
45. Railway Gazette. Trenitalia unveils Frecciarossa 1000, Railway Gazette Website. 2019. Available online: http://www.railwaygazette.com/news/passenger/single-view/view/trenitalia-unveils-frecciarossa-1000.html (accessed on 9 August 2019).
46. Railway Technology. Frecciarossa 1000 Very High-Speed Train, Railway Technology Website. 2019. Available online: https://www.railway-technology.com/projects/frecciarossa-1000-very-high-speed-train/ (accessed on 6 September 2019).
47. Eurail. How to Get from Rome to Florence by Train, Eurail Website. 2019. Available online: https://www.eurail.com/en/get-inspired/popular-routes/how-get-rome-florence-train#train-options (accessed on 9 August 2019).

48. Palumbo, M.; Ruscigno, M.; Scalise, J. The ERTMS/ETCS Signaling System, Enterprise Europe Scotland Website. 2016. Available online: http://www.enterprise-euope-scotland.com/ProfileImages/TRNL20160511001/ERTMS-ETCS%20Signalling%20System.pdf (accessed on 12 June 2021).
49. Amtrak. Happy 15th Anniversary, Acela Express, Amtrak Website. 2015. Available online: https://history.amtrak.com/blogs/blog/happy-15th-anniversary-acela-express (accessed on 22 July 2021).
50. Edelmann, U. Acela Express, TMC Reisen Website. 2021. Available online: https://www.tmc-reisen.de/usa/zugrundreisen/acela-express.html (accessed on 22 July 2021).
51. Persson, R.; Goodall, R.; Sasaki, K. Carbody tilting—Technologies and benefits. *Veh. Syst. Dyn.* **2010**, *47*, 949–981. [CrossRef]
52. Railway Technology. The Avelia Liberty: Transforming the Northeast Corridor, Railway Technology Website. 2016. Available online: https://www.railway-technology.com/features/featurethe-avelia-liberty-transforming-the-northeast-corridor-5660922/ (accessed on 12 June 2021).
53. International Union of Railways; Community of European Railway and Infrastructure Companies. *Rail Transport and Environment: Facts & Figures*; UIC/CER: Paris, France, 2015. Available online: http://www.cer.be/sites/default/files/publication/Facts%20and%20figures%202014.pdf (accessed on 22 December 2016).
54. Asia Times. China Wants to Build a Massive Underwater Tunnel to Taiwan. What Could Go Wrong? The National Interest Website. 2018. Available online: https://nationalinterest.org/blog/the-buzz/china-wants-build-massive-underwater-tunnel-taiwan-what-25314 (accessed on 13 June 2021).
55. Seat61. Beijing to Shanghai by Train or Flight, Seat61 Website. 2021. Available online: https://www.seat61.com/beijing-to-shanghai-by-train.htm (accessed on 22 July 2021).
56. Expert Magazine. 10 Fastest Trains in the World, Expert Magazine Website. 2021. Available online: https://jrexpertmag.com/10-fastest-trains-in-the-world (accessed on 13 June 2021).
57. Smith, K. Expansion and Reform Reshape Turkey's Railways, International Railway Journal Website. 2016. Available online: http://www.railjournal.com/index.php/high-speed/expansion-and-reform-reshape-turkeys-railways.html (accessed on 9 August 2019).
58. ITE Transport & Logistics. Turkey and Eurasia, ITE Transport & Logistics Website. 2017. Available online: http://www.transport-exhibitions.com/Market-Insights/Turkey-and-Eurasia (accessed on 9 August 2019).
59. Siemens. Velaro Turkey High-Speed Train, Siemens Website. 2021. Available online: https://press.siemens.com/global/en/feature/velaro-turkey-high-speed-train (accessed on 22 July 2021).
60. ITE Transport & Logistics. Turkish High Speed Rail Construction Update ITE Transport & Logistics, ITE Transport & Logistics Website. 2017. Available online: https://www.transport-exhibitions.com/Market-Insights/Turkey-and-Eurasia/Turkey-high-speed-railway-network (accessed on 13 June 2021).
61. Ilie, E. Turkey, Enormous Transport Potential at the Heart of Eurasian Platform, Railway PRO Communication Platform Website. 2017. Available online: https://www.railwaypro.com/wp/turkey-enormous-transport-potential-heart-eurasian-platform/ (accessed on 9 August 2019).
62. Bradsher, K. Taiwan's Bullet Trains Can't Outrun Controversy, New York Times Website. 2007. Available online: https://www.nytimes.com/2007/01/04/world/asia/04taipei.html (accessed on 22 July 2021).
63. Globthailand. ไต้หวัน. Globthailand Website. 2017. Available online: https://globthailand.com/market/%E0%B9%84%E0%B8%95%E0%B9%89%E0%B8%AB%E0%B8%A7%E0%B8%B1%E0%B8%99/ (accessed on 4 July 2021).
64. Shima, T. High-Speed Railways in Asia, Taiwan High Speed Railway. *Jpn. Railw. Transp. Rev.* **2007**, *48*. Available online: https://web.archive.org/web/20110613163935/http://www.jrtr.net/jrtr48/pdf/f40_Shi.pdf (accessed on 22 July 2021).
65. Taiwan High Speed Rail. Taiwan High Speed Rail, 2017 Annual Report, Taiwan High Speed Rail Website. 2018. Available online: http://www.thsrc.com.tw/event/tmp/pdf/2017_annual_EN.pdf (accessed on 23 October 2019).
66. Taipei Times. THSRC to Increase Runs, Buy more New Cars: Ou Chin-der, Taipei Times Website. 2008. Available online: http://www.taipeitimes.com/News/taiwan/archives/2008/11/22/2003429230> (accessed on 22 July 2021).
67. Hye-mi, K. Bullet Trains Coming to a Town Near You by 2020, Korea JoongAng Daily Website. 2020. Available online: https://koreajoongangdaily.joins.com/news/article/article.aspx?aid=2925474 (accessed on 22 July 2021).
68. Thenicee, T. Highly Speed Korea Train Express Is in South Korea's (ktx), Thenicee Website. 2019. Available online: https://www.thenicee.com/Article/highly-speed-korea-train-express-is-in-south-korea (accessed on 22 July 2021).
69. Railway Gazette International. South Korea's Growing Network, Railway Gazette International Website. 2008. Available online: https://www.railwaygazette.com/news/single-view/view/10/south-koreas-growing-network.html (accessed on 22 July 2021).
70. Chosunilbo & Chosun. KTX Clocks up over 414 Million Passengers in a Decade, Chosunilbo & Chosun Website. 2014. Available online: http://english.chosun.com/site/data/html_dir/2014/03/31/2014033101291.html (accessed on 22 July 2021).

encyclopedia

MDPI

Entry

Geometric Design of Suburban Roundabouts

Saša Ahac * and Vesna Dragčević

Faculty of Civil Engineering, University of Zagreb, 10000 Zagreb, Croatia; vesna.dragcevic@grad.unizg.hr
* Correspondence: sasa.ahac@grad.unizg.hr

Definition: A modern roundabout is an intersection with a circulatory roadway at which the vehicle speed is low, and the traffic is continuous and circulating in one direction around the central island towards the exits at the approach legs. Modern roundabout design is an iterative process that is composed of the following steps: (1) the identification of the roundabout as the optimal traffic solution; (2) the definition of the number of lanes at the intersection based on the required capacity and the level of service; (3) the initial design of the roundabout geometry; (4) design vehicle swept path, the fastest path analysis, and visibility performance checks; and (5) detailed roundabout design if the results of the performance checks are in line with the design recommendations. Initial roundabout geometry design elements are not independent of each other; therefore, care must be taken to provide compatibility between them. An overview and a comparative analysis of the initial geometric design elements for suburban single-lane roundabouts defined in roundabout design guidelines and norms used in Croatia, Austria, France, the Netherlands, Germany, Serbia, and Switzerland is given in this entry.

Keywords: approach alignment; outer radius; circulatory roadway; apron; splitter island; roundabout entry; roundabout exit; longitudinal slope

Citation: Ahac, S.; Dragčević, V. Geometric Design of Suburban Roundabouts. *Encyclopedia* **2021**, *1*, 720–743. https://doi.org/10.3390/encyclopedia1030056

Academic Editors: Raffaele Barretta, Ramesh Agarwal, Krzysztof Kamil Żur and Giuseppe Ruta

Received: 17 June 2021
Accepted: 4 August 2021
Published: 5 August 2021

1. Introduction

The development of modern roundabouts began in the 1960s in the United Kingdom with the adoption of the yield-at-entry rule, which gave the circulating traffic priority over entering traffic [1]. Modern roundabouts spread to other parts of Europe in the 1980s [2]. Intensive construction of roundabouts has begun in Europe in the last 30 years. European countries that stand out in the total number of roundabouts are France (63,212), Spain (36,762), and Italy (30,917) [3], and states like the Netherlands, Sweden, Switzerland, Denmark, Finland, Germany, and Austria are also pursuing policies of mass roundabout construction.

Modern roundabout design is an iterative process, and it begins with the identification of the roundabout as the optimal traffic solution in the given conditions. The initial roundabout design refers to (1) defining the size of the intersection by selecting the outer radius, (2) laying the approach leg axes, and (3) defining the geometry of the design elements on roundabout entry and exit lanes, circulatory roadway, and central island. The initial design of the roundabout is usually followed by three performance checks: the examination of the design vehicle swept path, the fastest path analysis, and the visibility checks. If the results of these checks are not in line with the design recommendations, the geometry of the elements applied in the initial design phase is modified.

In this entry, guidelines for the initial design of individual geometric elements at a suburban single-lane roundabout, given in design guidelines and norms that are used in Croatia [4], Austria [5], France [6], the Netherlands [7,8], Germany [9], Serbia [10,11], and Switzerland [12–15], are presented. These documents are selected for the following reasons. Firstly, they all define the following geometric elements of suburban roundabouts: the outer radius, the circulatory roadway, the apron, the splitter islands, entry and exit design, and the longitudinal slopes of the approaches and/or the intersection plane. Secondly, the geography and the terrain in the countries these documents originate from are different, which affects the intersection design elements' dimensions and shapes. For instance, predominantly flat terrain can be found in the Netherlands, whereas predominantly mountainous terrain is found in Switzerland and Austria. At the same time, all terrain types are represented in France, Germany, Croatia, and Serbia. The third reason for the selection of the documents is the year in which they were issued, ranging from 1991, when Swiss guidelines were published, to the year 2014, when the newest edition of Croatian guidelines for suburban roundabouts was issued. Namely, this 23-year range was marked by the mass construction of roundabouts in all the above-mentioned countries, so we believe it would be interesting to observe whether and how roundabout design approaches changed in that period. The Federal Highway Administration (FHWA) guidelines are not included in this entry because the American design vehicles used in the swept path analysis (and, consequently, the dimensions of roundabout design elements) are larger than the vehicles that can be found on European roads. The reason the United Kingdom and Australian guidelines are omitted from this entry is the fact that the UK and Australian traffic drives on the left.

2. Suburban Roundabout Design Elements

Modern roundabout design is a process of determining the optimal balance between safety provisions, operational performance, and accommodation of the design vehicle. The criteria for the acceptability of roundabouts are usually defined in national guidelines and are adapted to the local circumstances [16]. These criteria include a functional criterion, capacity criterion, spatial criterion, design and technical criteria, traffic safety criterion, and economic criterion [4,7,10]. According to [4], to determine whether a roundabout is the best solution at a particular location, it is necessary to examine what the primary function of the planned intersection is (source-destination or transit traffic), what its role in the traffic network is, and its position concerning settlements (urban or suburban), as well as in the wider traffic network. The capacity criterion should examine whether the roundabout is an acceptable solution for the existing and expected circumstances, given the traffic flows and the distribution of traffic. The spatial criterion examines the availability of space for the geometric elements of the intersection. Design and technical criteria examine the geometry of the intersection, the position of the approaches, the number of the approaches, and the angle between the axes of the consecutive approaches. Traffic safety criteria should examine whether the roundabout in the existing conditions is a solution that guarantees safety to all road users. To achieve the required level of traffic safety, roundabouts must be recognizable in the traffic network and their geometry must force traffic to enter and circulate at a slow speed. Therefore, we should strive for the application of (1) uniform design solutions and (2) standardized design elements that provide sufficient deflection around the central island, with the minimal adaptation of these elements to the limitations arising from the specific location of the planned intersection.

According to the above-mentioned criteria, relevant parameters of roundabout design are spatial requirements and limitations, traffic load, design speed, and traffic flow structure. Based on these parameters, the geometric elements of roundabouts, which are described below, are defined. As these geometric elements are not independent of each other, care must be taken to provide compatibility between them to meet the design vehicle swept path requirements, overall safety, and capacity objectives. Geometric elements of modern suburban single-lane roundabouts are presented in Figure 1 and described below.

The central island is a raised physical barrier with a (usually) circular ground plan, placed in the center of the roundabout around which traffic circulates.

The circulatory roadway is a lane at which vehicles circle the central island in a counterclockwise direction. Vehicles at a circulatory roadway have priority over vehicles entering a roundabout.

The outer radius is the radius of the outer edge of the circulatory roadway. It is the sum of the central island radius and the circulatory roadway width.

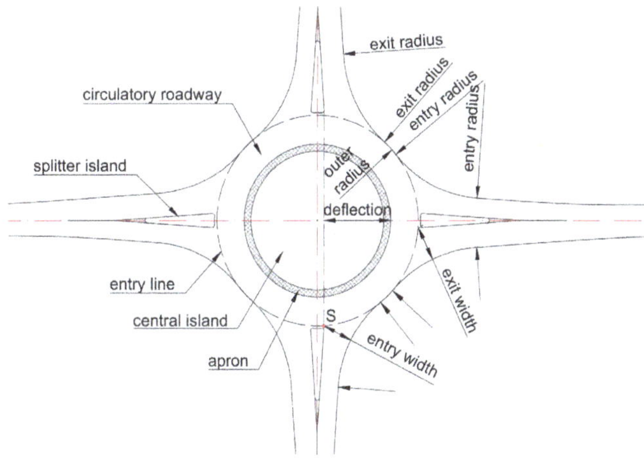

Figure 1. The geometric elements of single-lane roundabouts.

The apron is the traversable part of the central island that may be needed to allow long vehicles to negotiate a roundabout. It differs from the circulatory roadway by the cross slope, material used for the final layer, and/or color.

The entry line marks the point of entry into the circulatory roadway. This line is an extension of the circulatory roadway edge line, and it functions as a yield or give-way line: Entering vehicles must yield to any circulating traffic coming from the left before crossing this line into the circulatory roadway.

The splitter island is a raised or drawn element on the roundabout approach intended to channel traffic flows at the entrance and exit of the intersection.

The roundabout entry is bounded by a curb or edge of pavement consisting of a curve leading into the circulatory roadway. This curve is tangential to the outside edge of the circulatory roadway.

The roundabout exit is bounded by a curb or edge of pavement consisting of a curve leading away from the circulatory roadway. This curve is tangential to the outside edge of the circulatory roadway. The exit radii are usually larger than the entry radii to minimize the likelihood of congestion and crashes at the exits.

The entry width is the shortest distance between the intersection of the line connecting the vertices of the opposite dividing islands and the entrance line (point S) and the right curb or edge of the pavement at the entrance to the roundabout.

The deflection is the distance between a straight line joining the vertices of opposite dividing islands and its parallel that is tangent to a central island.

2.1. The Approach Alignment

Roundabout approach alignment affects the curvature of the vehicle's trajectory when passing through the intersection, the accommodation of the design vehicles, and the viewing angles on the adjacent approaches. The optimal design of a roundabout depends on the size and position of the roundabout concerning the approach alignment.

The standard approach alignment on suburban roundabouts is radial, where the approach axes intersect in the center of the outer radius of the intersection (Figure 2a). This alignment allows the appropriate design of the geometric elements of single-lane roundabouts, which ensures low vehicle velocity. The radial alignment of the approaches ensures the visibility of intersections in the traffic network and minimizes the need to modify the geometric elements on the approaches. An alternative to radial alignment is a displacement of the intersection approach axes, which can be carried out either by the displacement of this intersection to the left of the center of the outer radius (Figure 2b), or by laying the approach axes so they intersect to the right of the center of the outer radius of the roundabout (Figure 2c). Displacement to the left increases the curvature of the entrance path of the vehicle, i.e., provides better speed control at the intersection. Displacement to the right (Figure 2c) should be avoided, as it does not provide sufficient curvature of the vehicle entry path, i.e., this alignment allows vehicles to enter the intersection too fast, which increases the risk of collision.

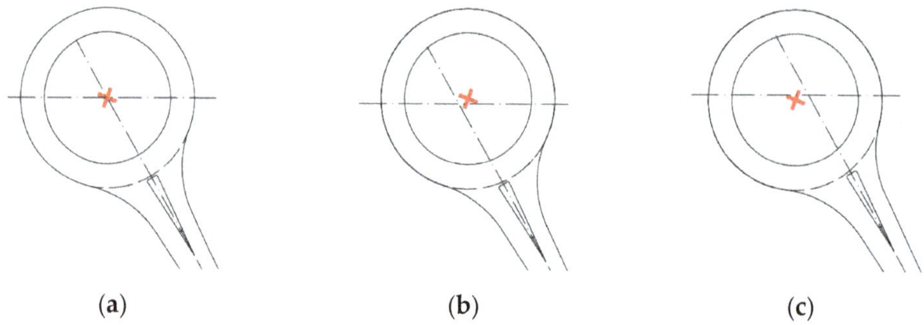

Figure 2. The approach alignment variants according to [8]: (**a**) radial alignment geometric elements of single-lane roundabouts, (**b**) displacement to the left, and (**c**) displacement to the right. The red "×" marks the center of the roundabout outer radius.

The optimal angle between the approach axes is 90°. If the axes intersect at an angle significantly greater than 90°, the permitted driving speed may be exceeded for right turns. If the approach axes intersect at an angle significantly less than 90°, the required accommodation of long vehicles is disabled. Increasing the radius of curve at the right edge of the pavement at the entrance to such intersections (which must be done to accommodate long vehicles) increases the entrance width, which reduces the level of intersection safety, as it can result in higher vehicle velocity. Reducing the angle between the approach axis results in the need to increase the outer radius of the roundabout to meet the conditions of the design vehicle swept path and speed requirements at an intersection.

2.2. The Outer Radius

At single-lane roundabouts, the size of the outer radius depends on the spatial requirements and limitations, the number and alignment of the approaches, the traffic requirements of the design vehicle swept path (the outer radius must allow the passage of the design vehicle), and the design speed (the curvature of the vehicle's trajectory must limit driving speeds to ensure low vehicle speeds when passing through the intersection).

When examining the swept path of the design vehicles, it must be demonstrated that the design vehicle can pass through the intersection using the available lane width while ensuring minimum lateral clearance (usually between 25 and 50 cm wide) and respecting road markings. The design vehicle is a vehicle of a certain type and dimensions that characterize a certain group of vehicles and fully complies with the legal regulations on vehicle dimensions, i.e., international recommendations. Design vehicles are used to examine the possibility of vehicles of a certain category passing through an intersection. They are most often selected according to the position of the intersection in the road network, depending on the structure and category of vehicles that appear on the observed section of the road.

The recommended and limit values of the outer radii at suburban single-lane roundabouts listed in the analyzed documents are shown in Figure 3.

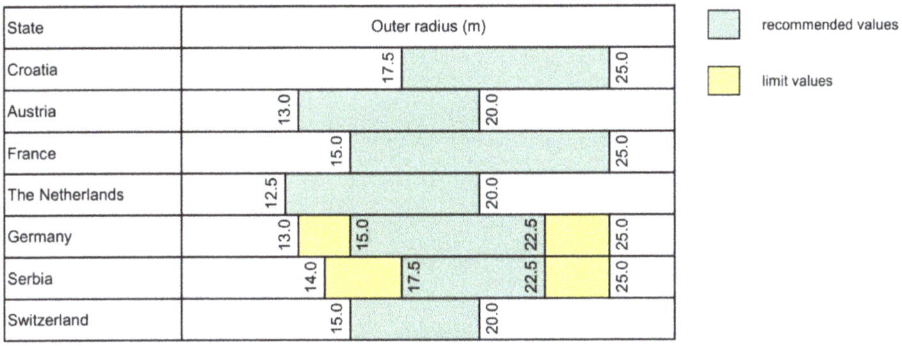

Figure 3. Recommended outer radii for single-lane roundabouts according to [4–12].

2.3. The Circulatory Roadway

At single-lane roundabouts, the width of the circulatory roadway depends on the swept path requirements of the design vehicle, whereas the cross slope of the circulatory roadway depends on surface drainage and topographic conditions. The cross slope of the circulatory roadway is usually directed towards the outside of the intersection. Such a direction is suitable for the following reasons: It increases the safety of intersections by raising and improving the visibility of the central island, it reduces vehicle speed on the path around the central island, it minimizes cross slope changes at the entry and exit, and it helps surface drainage.

According to [4], the minimal width of a circulatory roadway is determined based on the swept path of a selected two-axle design vehicle for driving in a full circle. Recommended values for the width of the circulatory roadway range from 4.5 to 6.0 m, and the recommended value of the cross slope of the circulatory roadway according to [5] is smaller than or equal to 2.5%.

According to [5], the width of the circulatory roadway ranges from 6.5 to 9.0 m. When determining the width of the circulatory roadway, it is necessary to ensure the deflection of the vehicle driving around the central island. If sufficient deflection is not achieved, it is necessary to plan the construction of a larger central island with an apron. The recommended value of the cross slope of the circulatory roadway according to [5] is 2.5%, where the slope is directed towards the outer edge of the intersection, as shown in Figure 4a. If, due to topographic conditions, it is necessary to construct an intersection with a longitudinal slope greater than 2.5%, the circulatory roadway is placed in one plane, with a slope less than or equal to 4% (Figure 4b). The minimum value of the cross slope is 1.5%, which ensures sufficient surface drainage of the intersection.

According to [6], a circulatory roadway should not look like a one-way multi-lane road, but a single lane of sufficient width to ensure the accommodation of the design vehicle. On single-lane roundabouts, the width of the circulatory roadway depends on the outer radius and the width of the widest entrance to the intersection. The width of the circulatory roadway is fixed and 20% larger than the widest entrance, with the minimum width of the circulatory roadway being 6 m and the usual width 7 m. The use of an 8-m-wide circulatory roadway is justified at roundabouts where the design vehicle is a tractor with a semi-trailer. The cross slope of the circulatory roadway is also constant along its entire length and its value can range from 1.5 to 2%. The cross slope is directed towards the outer edge of the intersection. These values do not apply to roundabouts on steep slopes, which should in any case be avoided. Additionally, the maximum cross slope of the roundabout is 3%.

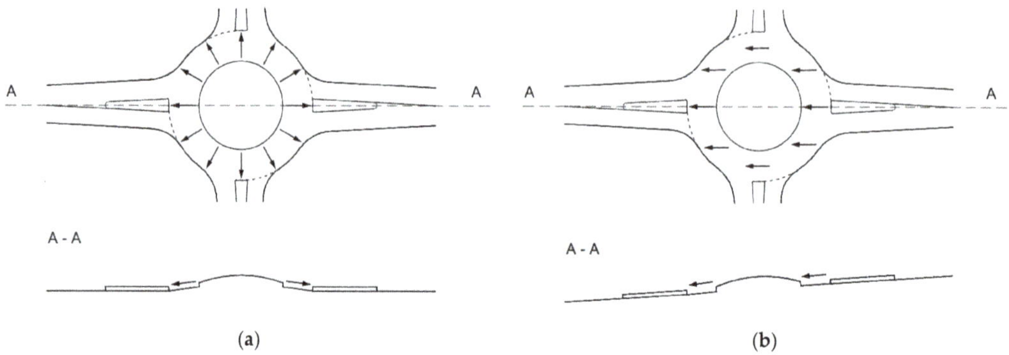

Figure 4. Circulatory roadway cross slope according to [5]: (**a**) recommended cross slope direction and (**b**) circulatory roadway in one plane. Slope direction is represented by the arrows.

According to [7,8], the standard width of the circulatory roadway at single-lane suburban roundabouts is 5.25 m, regardless of the dimensions of the design vehicle. It is noted that these standard values must not be adopted as absolute: The width of the circulatory roadway must meet the requirements of the design vehicle, but also the condition of limiting the speed of the vehicle passing through the intersection. Circulatory roadway width must be fixed along the entire length of the lane. Figure 5 shows the values of the width of the circulatory roadway ("B") as a function of the outer and inner radius of the roundabout ("R_{bu}" and "R_{bi}") for single-lane roundabouts. The cross slope of the circulatory roadway according to [7] is 2.0 to 2.5% and is directed towards the outer edge of the roundabout.

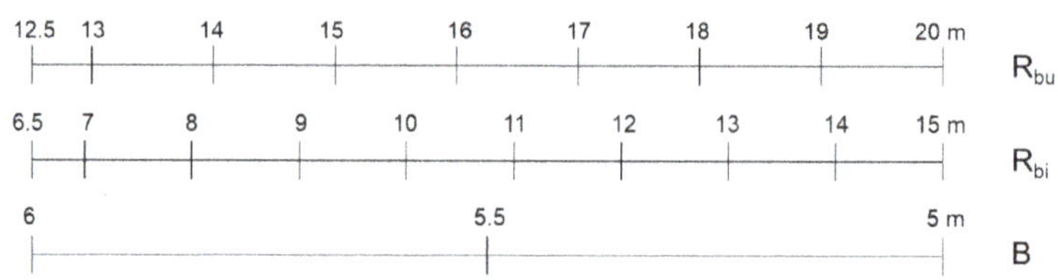

Figure 5. Circulatory roadway width ("B") depending on the outer and inner radius ("R_{bu}" and "R_{bi}") [8].

According to [9], the circulatory roadway, together with the apron, which is not included in the width of the circulatory roadway, forms a circular ring of intersection (width "B_k"). The circulatory roadway is of constant width and a constant cross slope. The circulatory roadway width depends on the size of the outer radius—the widths of the circular ring of intersection ("B_k") for single-lane roundabouts depending on the size of the outer radius are given in Table 1. For suburban roundabouts, the widths of the circular ring of intersection ("B_k") shown in the table correspond to the widths of the circulatory roadway, since the aprons are not constructed at these intersections. At suburban intersections with increased truck traffic, it is possible to apply larger widths of the circulatory roadway than those given in the table. The cross slope of the circulatory roadway is directed towards the outer edge of the intersection and must be at least 2.5% to ensure proper surface drainage of the pavement.

Table 1. Circular ring width ("B_k") depending on the outer radius [9].

Outer Radius (m)	Circular Ring Width "B_k" (m)
13	9.0
15	8.0
17.5	7.0
≥20	6.5

According to [10], the width of the circulatory roadway ("b_k") arises from the swept path requirements of the design vehicle and the driving conditions. It can be standardized to some extent for typical suburban road network conditions for single-lane and two-lane roundabouts. The standard widths of the roundabout ("b_k") for single-lane roundabouts (Figure 6) are defined depending on the diameter of the inscribed circle of a roundabout ("D") and include marginal strips that are 0.20 m wide.

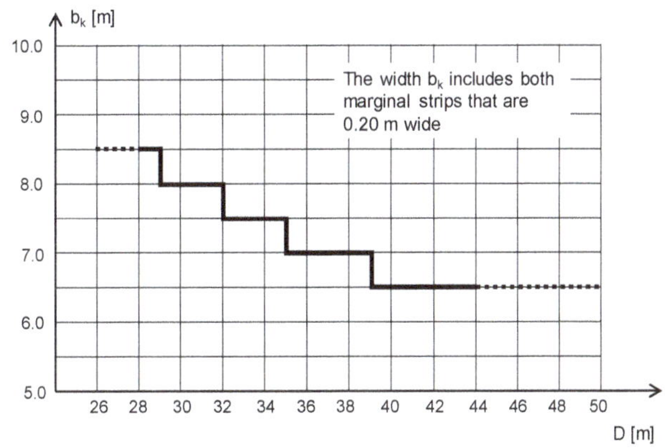

Figure 6. Circulatory roadway width ("b_k") for single-lane roundabouts [10].

According to [10], the standard cross slope of the circulatory roadway is 2.5%. The cross slope is directed towards the outer edge of the intersection. The maximum cross slopes of the circulatory roadway ("i_p") as a function of the slope of the intersection plane ("i_{Nkr}") is shown in Table 2. At the same time, the largest total slope is 4% [10].

Table 2. Circulatory roadway slope [10].

Longitudinal Slope "i_{Nkr}" (%)	Maximum Cross Slope "i_p" (%)
$0 < i_{Nkr} \leq 2.5$	−2.5
$2.5 < i_{Nkr} \leq 4.0$	2.5
$4.0 < i_{Nkr} \leq 5.7$	4.0

According to [12], the total width of the circulatory roadway and apron depends on the outer diameter of the roundabout and the swept path conditions of the design vehicle. Figure 7 shows a diagram used to determine the minimum widths depending on the outer diameter of the intersection. These widths are determined by swept path analysis for the design vehicle defined according to [13]. These widths do not include lateral clearances. Larger circulatory roadway widths should be avoided for safety reasons. It is recommended that for a width of the circulatory roadway greater than 5.5 m, the apron be used, intended only for tractors with semi-trailers and trucks with trailers. The maximum cross slope of the circulatory roadway is 5% (exceptionally 7%, in demanding topographic conditions). The minimum cross slope of the circulatory roadway is limited to 3% due to the pavement surface drainage conditions.

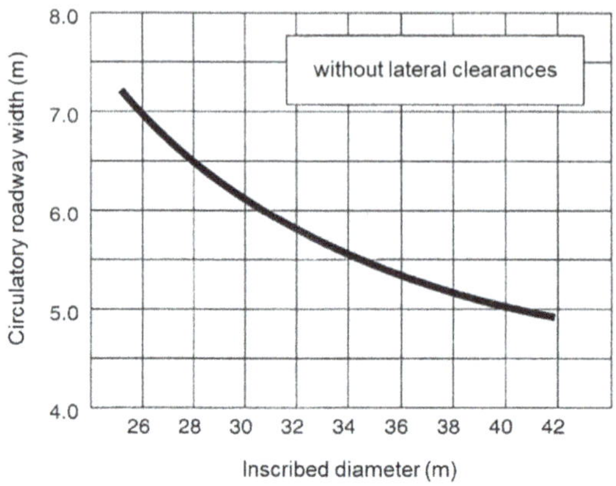

Figure 7. Minimum circulatory roadway width (together with apron width) depending on the roundabout inscribed diameter [12].

Standard circulatory roadway widths for suburban single-lane roundabouts listed in the analyzed documents are shown in Figures 8 and 9. The values in Figure 8 refer to single-lane roundabouts with and without the apron, the application of which depends on the choice of design vehicle: The apron is intended exclusively to accommodate trucks (tractors with semi-trailers, trucks with trailers).

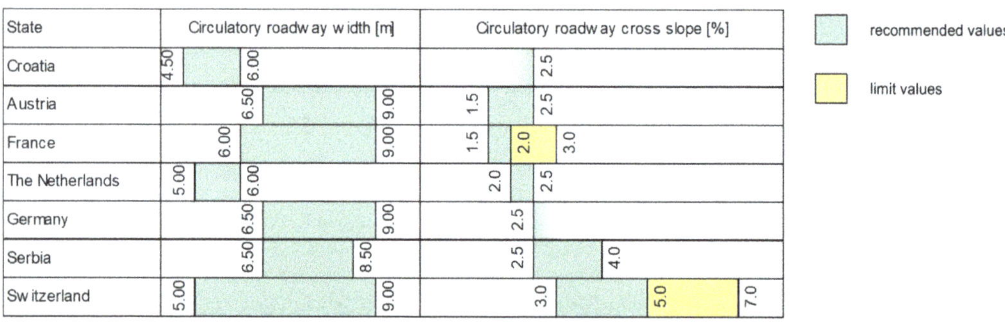

Figure 8. Recommended and limit circulatory roadway widths and cross slope according to [4–12].

According to the data shown in Figure 9, the highest values of the width of the circular lane are defined by the German guidelines and the lowest values by the Dutch guidelines. The width of the circular lane determined by the diagrams shown in Figure 9 is usually rounded up to a larger 0.25 m. According to [9–12], the width of a circular lane includes an additional element—an apron, the transit edge of the central island, described below.

2.4. The Apron

At roundabouts, to ensure the low driving speed of passenger cars while meeting the traffic conditions of long vehicles, it is sometimes necessary to provide an apron on the edge of the central island.

According to [4], the apron width is determined based on the swept path of a design vehicle for driving in a full circle. For suburban roundabouts, this design vehicle is a 16.5 m-long tractor with a semi-trailer.

According to [5], the apron is constructed only when the required deflection cannot be achieved due to the conditions of the swept path for the design vehicle. The widths and other design features of the apron are not described in the document.

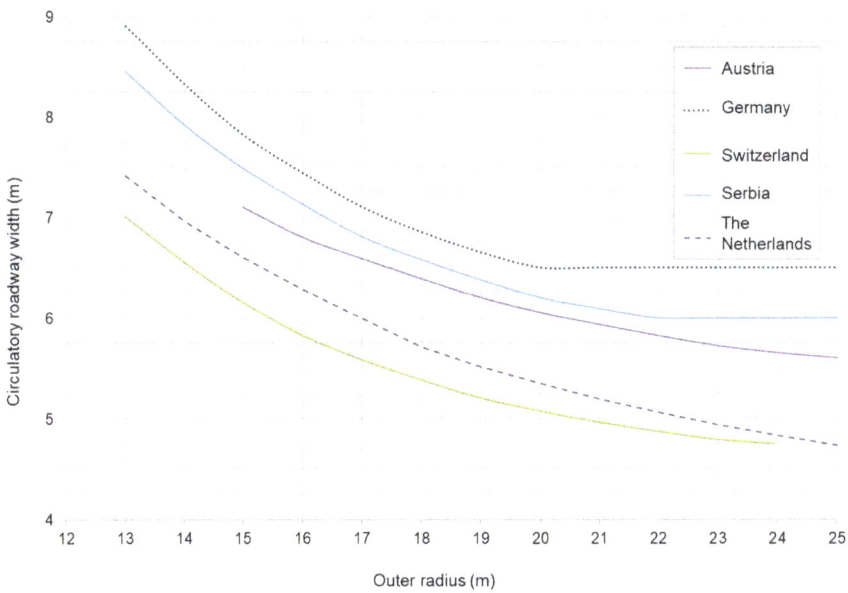

Figure 9. Standard circulatory roadway widths for suburban single-lane roundabouts [8,11].

According to [6], the apron is constructed at roundabouts with an outer radius in the range of 12 to 15 m. The apron width is in the range of 1.5 to 2.0 m (Figure 10). The apron may also be constructed at roundabouts with an outer radius larger than 15 m if convoys of special cargo pass through the intersection. The cross slope of the apron is directed towards the outer edge of the intersection and ranges from 4 to 6%. It is recommended to raise the apron above the circulatory roadway by applying a low curb (3 cm high) and to make a final layer of stone cubes so that the edge is visible in both day and night driving conditions.

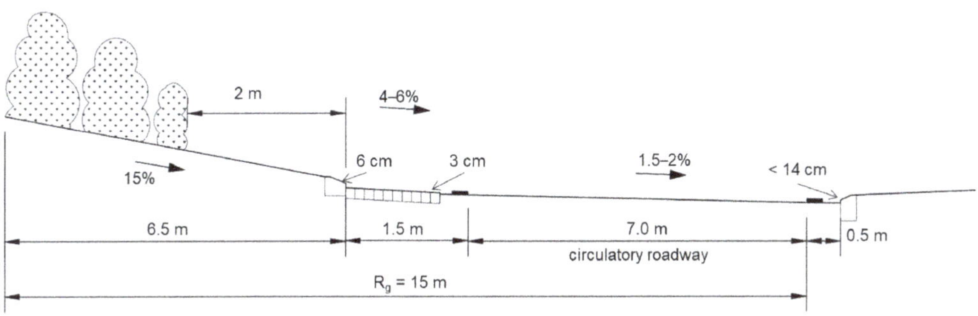

Figure 10. Standard circulatory roadway cross-section for roundabouts with an outer radius of 15 m, according to [6].

According to [7,8], the apron width depends on the design vehicle and the combination of dimensions of the remaining geometric elements of the roundabout. The standard cross slope of the apron is 1%. The apron is constructed with a low curb with a slight slope (the height difference compared to the circulatory roadway pavement is 5 cm at most) and with a paved surface (cubes) [8]. The standard apron width at suburban single-lane roundabouts is 1.5 m. Depending on the dimensions of the design vehicle, the apron width is 3 m (for a vehicle 22 m long) or 4 m (for a vehicle 27 m long) [7].

According to [9], aprons are not constructed on suburban roundabouts.

According to [10], the inner edges of the circulatory roadway are formed by applying elements of a different structure (e.g., paving elements, a small cube). The central island is shaped with one hog curve at the center of the island and two sag curves at the outer edge of the island (Figure 11), and the slope of the tangents between these elements is 4%. The document in question, which refers to suburban roundabouts, does not envisage the construction of the apron. According to [11], the apron width is determined based on swept path analysis for the design vehicle, with the minimum apron width being 1 m. This element is raised above the circulatory roadway by 3 cm, and its cross slope is 4% (3% in exceptional cases).

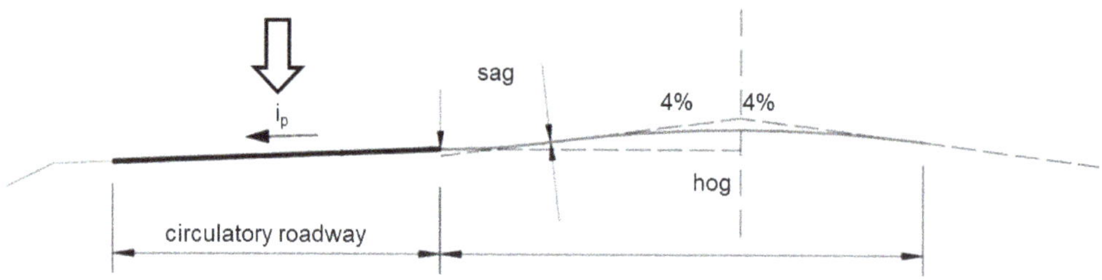

Figure 11. Central island design [10].

According to the standard [12], for circulatory roadway widths greater than 5.5 m, it is recommended to use the apron, intended only for trucks. The design of the apron is not defined in the specified standards. According to [15], the usual width of the apron (paved with stone cubes) at single-lane roundabouts is 1.5–2.0 m.

The standard widths and cross slopes of the apron at suburban single-lane roundabouts listed in the considered documents are shown in Table 3.

Table 3. Apron width and cross slope according to [4–12].

State	Apron Width (m)	Apron Cross Slope (%)
Croatia	defined based on the design vehicle swept path	not defined
Austria	defined based on the design vehicle swept path	not defined
France	1.5–2.0 (for outer radius between 12 and 15 m)	4.0–6.0
The Netherlands	defined based on the design vehicle swept path; 1.5–4.0	1.0
Germany	aprons are not constructed	–
Serbia	aprons are not constructed	–
Switzerland	1.5–2.0	not defined

2.5. The Splitter Island

Splitter islands are mandatory elements of modern suburban roundabouts. They provide better control of the vehicle speed by channeling traffic flows and provide space for vertical signalization. When designing roundabouts, it is necessary to determine the space for the splitter island before defining the width of the entrance and exit lanes.

The basic functions of the splitter islands are as follows:

- Timely warning of drivers about a roundabout in the traffic network.
- Separation of traffic flows.
- Providing space for the installation of traffic signs, lighting, and other traffic equipment.
- Preventing a left turn into a roundabout.

According to [4], the use of the triangular or funnel-shaped splitter island is mandatory on suburban roundabouts. These splitter islands can be raised above the pavement by 15 cm, or drawn, which is a more flexible solution at intersections with heavy truck traffic.

According to [5], the splitter islands must be provided on all approaches to the intersection. In addition, at roundabouts located in unfavorable topographic conditions (resulting in convex curves in vertical alignment or large turning angles on the approaches), it is possible to ensure the required recognizability of intersections by applying long splitter islands. The minimum width of the splitter island is 2 m, but this guideline recommends the use of a wider island to separate the entry and exit lanes. The rounding of the top of the splitter islands is formed by a radius of at least 0.75 m, and the distance of the island from the outer edge of the roundabout is at least 0.25 m [5].

According to [6], raised splitter islands are used for the physical separation of entry and exit lanes on the roundabout approaches. The splitter island on the suburban roundabout is usually funnel shaped. The standard length of the island ("H") is equal to the value of the outer radius of the roundabout ("R_g"), as shown in Table 4. The splitter island on suburban roundabouts (where the outer radius is equal to or larger than 15 m) is shifted to the left to ensure that the approach axis passes through the center point at the top of the island (point C, Figure 12). The minimum width of the splitter island is 2 m, and the usual width at the entrance to the intersection ("B") is equal to a quarter of the outer radius of the intersection.

Table 4. Splitter island elements according to [6].

Outer Radius "R_g" (m)	Splitter Island Length "H" (m)	Splitter Island Width "B" (m)	Shift "d" (m)	Radius at the Top of the Splitter Island "r" (m)
15	15	3.75	0.40	0.30
20	20	5.00	0.45	0.40
25	25	6.25	0.50	0.50

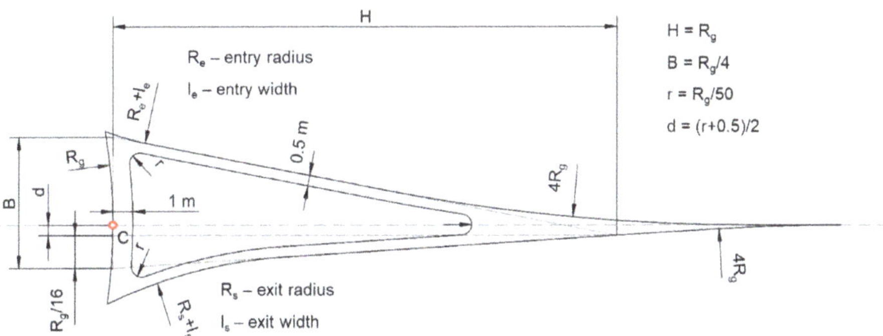

Figure 12. Splitter island elements according to [6]. The center point at the top of the island (point C) is represented by the red circle.

According to [7,8], splitter islands are placed on each approach to separate incoming and outgoing traffic flows. They can be raised or drawn, which is a more flexible solution at intersections with heavy truck traffic. The distance between the edge of the raised splitter island and the outer edge of the roundabout is approximately 1 m, to improve the traffic conditions at the intersection. In terms of their layout, the splitter islands used in the Netherlands are exclusive of the radial type (Figure 13), i.e., the edges of the islands are parallel to the axis of the approach, which passes through the center of the roundabout. The advantages of such a layout are as follows:

- Entry and exit speeds are low due to provided deflection.
- The shape of the splitter island forces the driver at the entrance to the intersection to respect the right of way of the vehicle at the intersection.
- The angle between the entry lane and the circulatory roadway remains 90° (which is advantageous in terms of visibility).

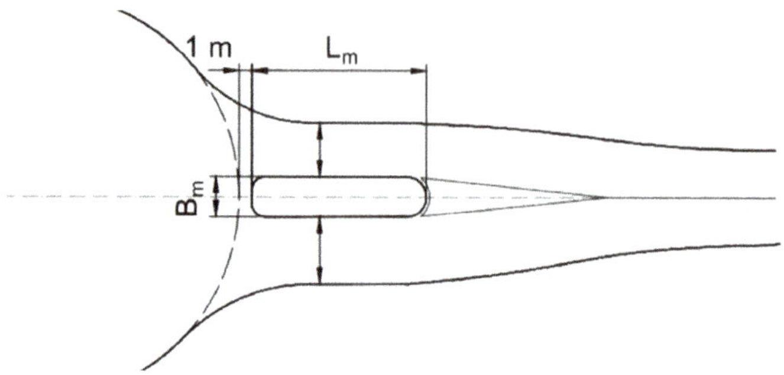

Figure 13. Radial splitter island elements for suburban roundabouts according to [7,8].

The recommended length of the raised part of the splitter island at single-lane suburban roundabouts ("L_m") is 10 to 15 m, and the standard width of the island ("B_m") is 3 m (Figure 13).

According to [9], the axis of the splitter island is perpendicular to the outer edge of the roundabout, with the smallest width of the island being 1.6 m. At suburban roundabouts, funnel-shaped splitter islands are used to achieve the required lane width at the entrance to the intersection and to ensure that the curvature of the splitter island drawn edge follows the curvature of the vehicle path (Figure 14).

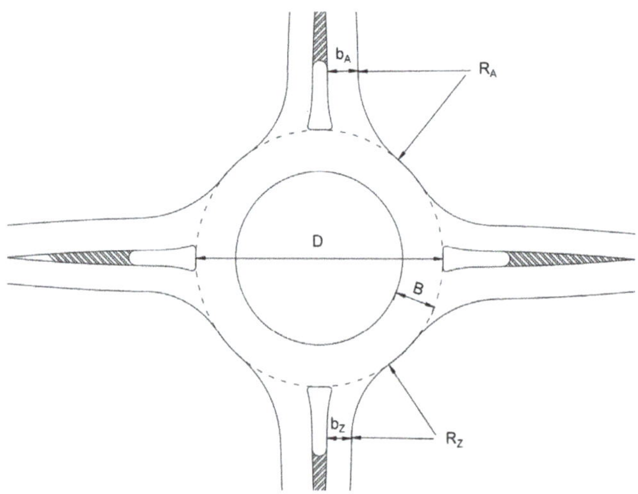

Figure 14. Funnel-shaped splitter islands on suburban roundabouts according to [9].

According to [10], the shape of the splitter island is conditioned by the desired level of traffic-flow channeling. If at single-lane roundabouts the outer radius of the intersection is greater than 20 m and/or the maximum design speed is equal to or greater than 60 km/h, the highest level of channeling is applied (Figure 15). A medium level of channeling (Figure 16) can be applied at a roundabout with an outer radius between 14 and 20 m with a maximum design speed between 50 and 60 km/h. The lowest level of channeling (Figure 17) is applied at the intersections of access roads and intersections of collector and access roads when an outer radius is smaller than 15 m and the highest design speed is below 50 km/h.

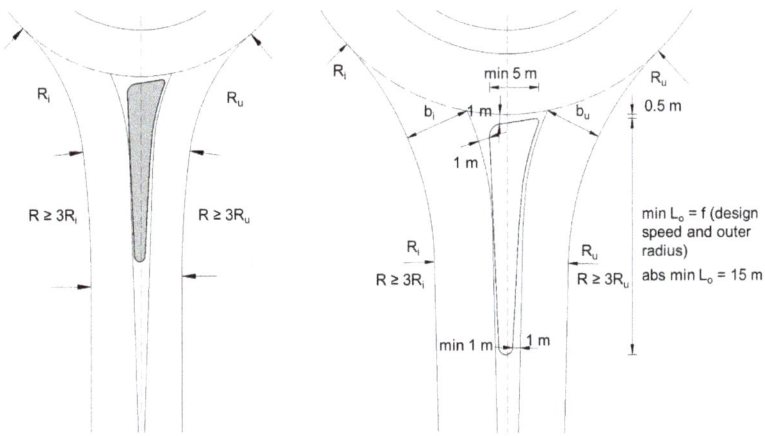

Figure 15. Splitter islands for a high level of traffic-flow channeling [10].

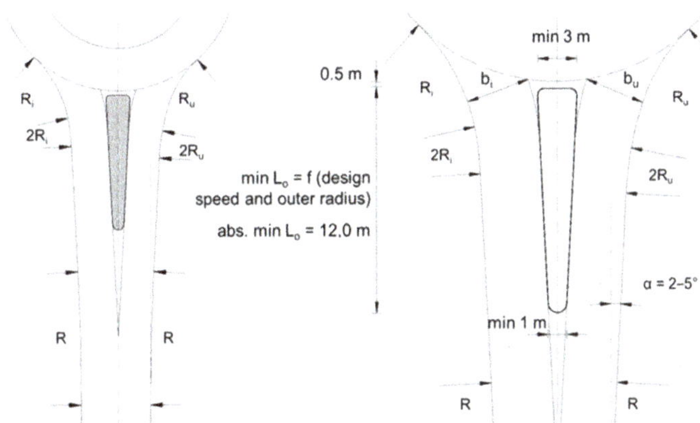

Figure 16. Splitter islands for a medium level of traffic-flow channeling [10].

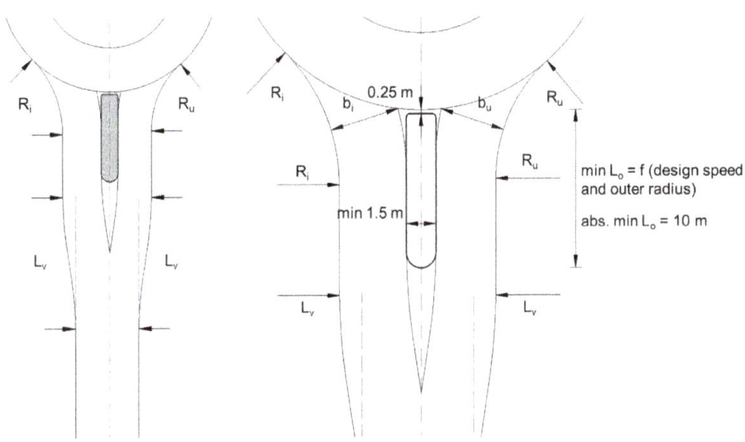

Figure 17. Splitter islands for a low level of traffic-flow channeling [10].

According to [12], the shape and dimensions of splitter islands depend on the widths of the entry and exit lanes, and the smallest dimensions are defined by the norm [14]. The minimum allowable width of splitter islands is 1.2 m, and the length of the island ranges from 30 to 50 m. A splitter island at a suburban roundabout defined according to the guidelines [15] is shown in Figure 18—the use of funnel-shaped islands with a minimum width of 5 m is recommended.

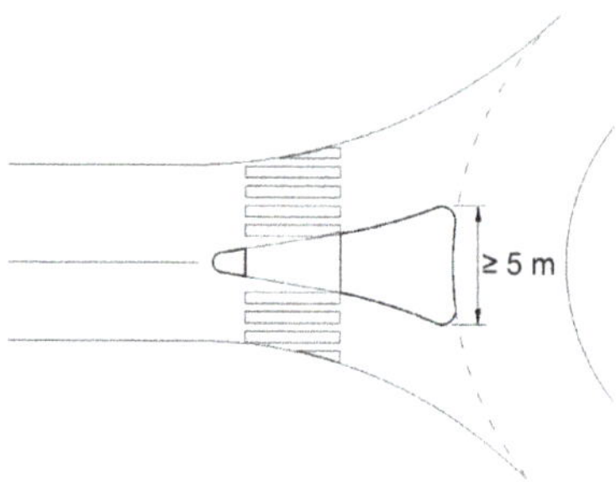

Figure 18. Splitter island according to [15].

An overview of the recommended dimensions of splitter islands listed in the considered documents is given in Table 5.

Table 5. Recommended design shape and dimensions of splitter islands according to [4–12].

State	Splitter Island Shape	Splitter Island Length (m)	Splitter Island Width (m)	Splitter Island Distance from the Outer Edge of the Roundabout (m)
Croatia	triangle	–	≥2.0	≤0.50
	funnel	–	≥2.0	≤0.50
Austria	triangle	–	≥2.0	≥0.25
France	funnel	15–25	≥2.0 or (outer radius/4)	1.0
The Netherlands	radial	10–15	≥3.0	1.0
Germany	funnel	–	≥1.6	0.0
	radial	≥10	≥1.5	0.25
Serbia	triangle	≥12	≥3.0	0.5
	funnel	≥15	≥5.0	0.5–1.0
Switzerland	funnel	30–50	≥1.2 (5.0)	0.0

2.6. The Entry and Exit Design

The design of the roundabout entry and exit refers to the definition of the widths of lanes and the radius of curvature of the right edge of the pavement at the roundabout entry and exit. The widths of the entry and exit lanes are determined based on the swept path requirements of the design vehicle. Additional widening of the entry and/or exit lane, necessary to accommodate the design vehicle, is performed by rounding the right edge of the pavement or curb using one or two radii of the appropriate size.

According to [4], a prerequisite for unobstructed vehicle movement on roundabout entry and exit is a proper design of the right pavement edge and selection of the following design elements: entry and exit radii ("R_{ul}" and "R_{iz}"), entry and exit widths ("e" and "e'"), and circulatory roadway width ("u"). The right pavement edge can be designed in two different ways:

1. With a shorter effective widening length ("l'")—the pavement edge is composed of a circular arc and a straight line that is parallel to the side of the triangular splitter island (Figure 19a), and the selected widening ("Δš") is equal to zero;
2. With a longer effective widening length ("l'")—the pavement edge is composed of a circular arc and a straight line that is not parallel to the side of the triangular splitter island (Figure 19b), and the selected value of widening ("Δš") is larger than zero.

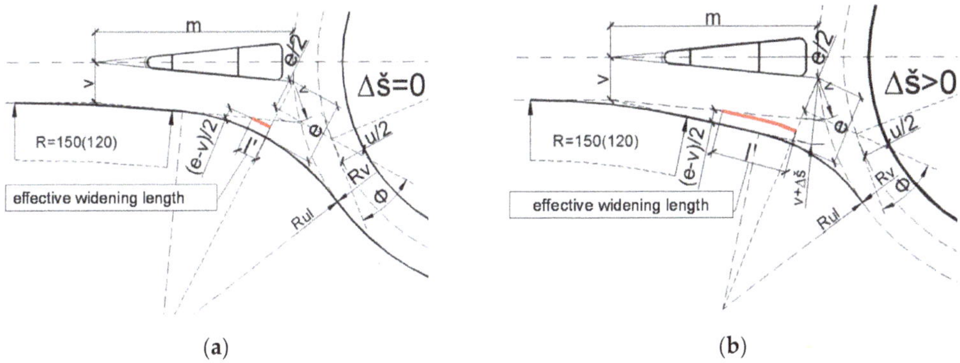

Figure 19. Entry design according to [4]: (**a**) shorter effective widening length and (**b**) longer effective widening length; "v" is the approach lane width, "m" is the splitter island length, "R_v" is the roundabout outer radius, and "R_{ul}" is the entry radius.

The designer must choose the way that will ensure unobstructed vehicle movement, and the decision must be based on the design vehicle swept path analysis. Additionally, entry design should ensure that the entrance angle ("Φ"), which is the tangent angle between the vehicle paths at the roundabout entry, is around 30°. In terms of the ratio between entry and exit radius, Croatian guidelines [4] recommend that the exit radius should be greater than or equal to the entry radius. The roundabout exit width ("e′") depends on the swept path width made by the design vehicle (Figure 20).

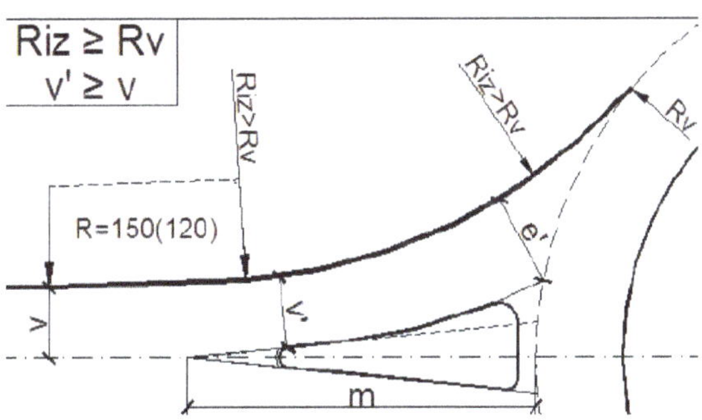

Figure 20. Exit design according to [4]: "v′" is the approach lane width, "m" is the splitter island length, "R_v" is the roundabout outer radius, and "R_{ul}" is the entry radius.

According to [5], on single-lane entrances to roundabouts, the width of the entry lane along the splitter island is at least 3.75 m. At the same time, the width of the exit lane along the splitter island is at least 4 m. The right edge of the pavement at suburban roundabouts is rounded with a radius varying from 12 to 16 m at the entrance (entry radius "R_E"), whereas a radius varying from 15 to 25 m is applied at the exit (exit radius "R_A"), as shown in Figure 21. The radii at the entry are smaller than the radii at the exit to reduce the speed of vehicles at the entrance while facilitating the exit of long vehicles and buses from the intersection. The radii applied must meet the design vehicle swept path requirement.

According to [6], the recommended entry width ("l_e") at single-lane entrances, measured between the boundary lines, is 4 m, and the entrance radius ("R_e") must always be less than or equal to the outer radius of the roundabout ("R_g"). The standard values of the entry radius range from 10 to 15 m, depending on the position of the approaches. The entry lane is bounded by boundary lines. The standard configuration of the single-lane entrance is shown in Figure 22 (for an outer radius of 20 m). The width of the exit lane ("l_s") ranges from 4 to 5 m for single-lane approaches and depends on the size of the outer radius of the roundabout ("R_g"). The exit radius ("R_s") must be larger than the inner radius of the intersection ("R_i"), with a minimum recommended value of 15 m and a maximum of 30 m. The values of these parameters depend on the size of the outer radius of the intersection ("R_g") and are shown in Table 6.

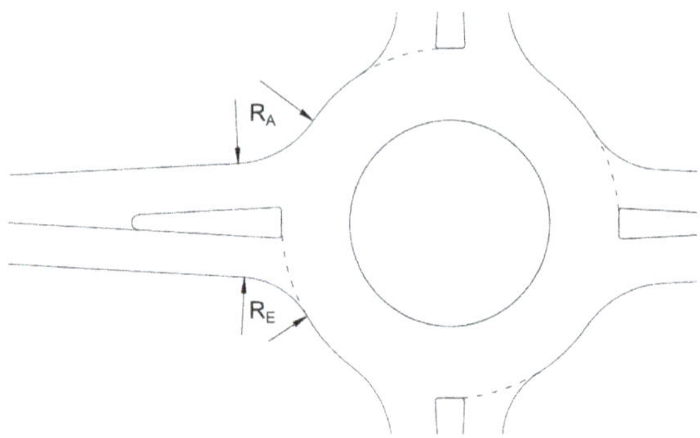

Figure 21. Right edge of the pavement at the roundabout entry and exit according to [5].

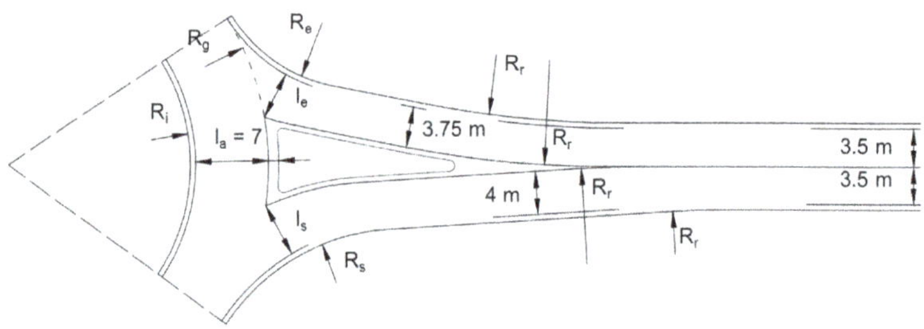

Figure 22. Single-lane approach for an outer radius of 20 m according to [6].

Table 6. Design elements for roundabout entry and exit according to [6].

Outer Radius "R_g" (m)	Entry Width "l_e" (m)	Entry Radius "R_e" (m)	Exit Width "l_s" (m)	Exit Radius "R_s" (m)	Transitional Radius "R_r" (m)
12	4	12	4.0	15	48
15	4	15	4.0	20	60
20	4	15	4.5	20	80
25	4	15	5.0	20	100

According to [8], the widths of entry and exit lanes do not have a large impact on the speed at the roundabout but affect the accommodation of design vehicles and visibility at the roundabout. The width of the entry lanes ("B_t") should be in the range of 3 to 4 m—wider lanes encourage drivers to drive around the roundabout at higher speeds, which reduces safety. The width of the exit lanes ("B_a") depends on the design vehicle swept path requirements, and ranges from 3.75 to 4.50 m. The standard values of the entry radii ("R_t") at single-lane suburban roundabouts range from 8 to 12 m, and the exit radii ("R_a") from 12 to 15 m. Design elements of entry and exit at suburban single-lane roundabouts are given in Figure 23.

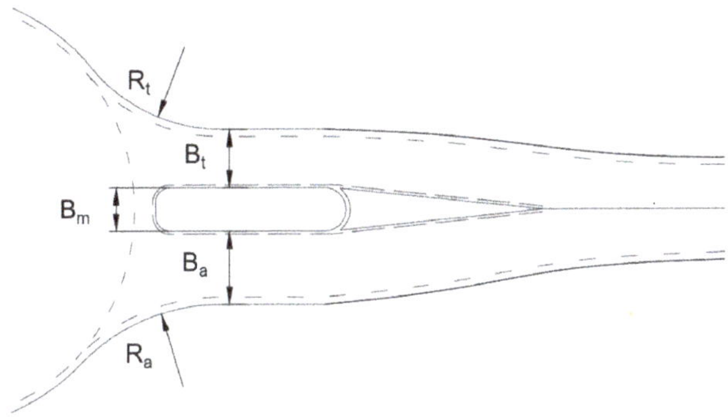

Figure 23. Design elements of entry and exit at suburban single-lane roundabouts according to [8].

According to [9], the approaches should be perpendicular to the roundabout, which is achieved by radially laying the approach axis to the outer radius of the intersection. The center of the roundabout must be as close as possible to the intersection of the approach axis. At suburban roundabouts, the recommended entry lane width in the vicinity of the splitter island ("B_Z") ranges from 3.50 to 4.00 m, whereas the recommended exit lane width ("B_A") is in the range of 3.75 to 4.50 m (Figure 23). The rounding of the right edge of the pavement at the roundabout entry and exit lane is formed by applying a circular arc, the size of which depends on the desired speed limit and the design vehicle swept path requirements (Figure 24). The standard radius of the circular arc at the entry lane ("R_Z") ranges from 14 to 16 m, whereas the radius of the circular arc at the exit lane ("R_A") is in the range of 16 to 18 m. The stated values of the exit radii can be increased by 30% at suburban roundabouts (i.e., "R_A" is in the range of 21 to 23 m).

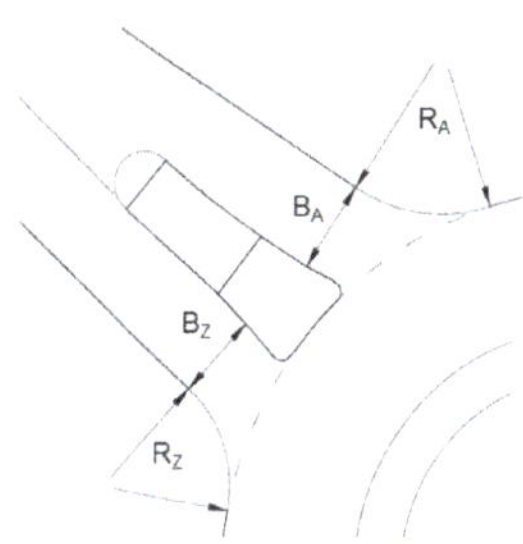

Figure 24. Design elements of roundabout entry and exit according to [9].

According to [10], the geometric elements of entry and exit are the starting point for the design vehicle swept path analysis and the definition of the roundabout design speed, thus directly affecting the capacity and safety of the roundabout. The standard width of the entry lane ("b_u") at single-lane roundabouts is 3.5 to 4.0 m, whereas the standard width of the exit lane ("b_i") is 3.75 to 4.50 m. Regardless of the required level of traffic-flow channeling, the initial condition for the entry radius ("R_u") is as follows: The radius ("$R_u + b_u$") can ultimately tangent (but not intersect) the edge of the roundabout, as shown in Figure 25. The standard values of the entry radius ("R_u") are in the range of 12 to 16 m. The condition of the ratio of entry and exit speeds requires that the exit radius ("R_i") be larger than the entry radius by 2 m. The standard values of the exit radius are in the range of 14 to 18 m.

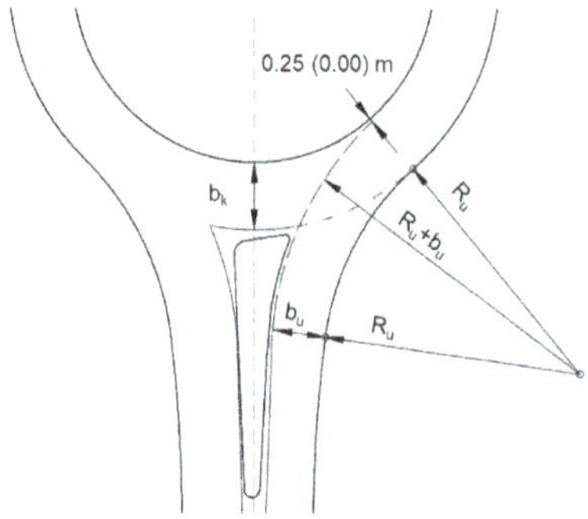

Figure 25. Initial design parameters for single-lane roundabout entry according to [10].

The number and size of the radius of curvature at the right edge of the pavement at the roundabout entry and exit and the design of the transition from the width of the approach lane to the width of the entry and exit lane depend on the desired level of traffic-flow channeling traffic.

- For the highest level of channeling, with a funnel-shaped splitter island, the rounding off the right edge of the pavement at the entry and exit of the intersection is performed using the radii "R_u" and "R_i," whereas the transition from lane width at the approaches is performed using the radius "R," which is at least three times larger than "R_u" for the entry and at least three times larger than "R_i" for the exit.
- For a medium level of channeling, with a triangular splitter island, rounding off the right edge of the pavement at the entry is performed using radii "R_u" and "$2R_u$," and rounding of the right edge of the pavement at the exit is performed using radii "R_i" and "$2R_i$." The transition from the lane width at the approach to the entry lane width is performed by applying the radius "R," whose tangents are a straight line that defines the edge of the pavement at the entry and a straight line parallel to the edge of the triangular splitter island and at a distance from that edge "b_u."
- For the lowest level of channeling, with a radial splitter island, the rounding of the right edge of the pavement at the entry and exit of the intersection is performed using the radii "R_u" and "R_i," whereas the transition from the approach lane width to the entry lane width is achieved by applying a curve of length "L_v."

According to [12], a value of 3.0 to 3.5 m has been defined as a suitable entry lane width ("b_e") for single-lane roundabouts in terms of safety, whereas a value of 3.5 to 4.5 m is defined as a suitable width of the lane at the exit ("b_a"). According to [12], the right edge of the pavement at the roundabout entry and exit is formed by applying two radii, as shown in Figure 26. At suburban roundabouts, the recommended size of the inner entry radius ("R_{e2}") is 12 m, whereas the outer entry radius ("R_{e1}") is five times larger. The recommended size of the inner exit radius ("R_{a2}") is 14 m, whereas the outer exit radius ("R_{a1}") is four times larger.

2.7. The Longitudinal Slopes at Roundabouts

Longitudinal slopes at a roundabout (longitudinal slopes of the approaches and the circulatory roadway) reflect the topography of the area.

According to [5], the longitudinal slopes of the circulatory roadway and the approaches at 20 m from the outer edge of the roundabout must not exceed 4% (2.5% if a large proportion of heavy vehicles at the intersection is foreseen). The difference between the longitudinal slope of the approaches and the cross slope of the circulatory roadway must not exceed 5%.

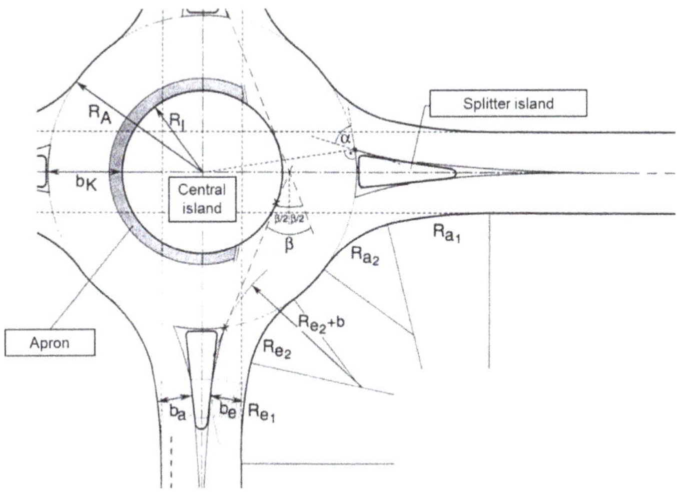

Figure 26. Design elements for single-lane roundabout entry and exit according to [12].

The recommended values for the design elements on entry and exit at suburban single-lane roundabouts listed in the analyzed documents are shown in Table 7.

Table 7. Design elements for roundabout entry and exit according to [4–12].

State	Entry Width (m)	Exit Width (m)	Entry Radius (m)	Exit Radius (m)
Croatia	4.00–7.00	4.00–7.00	8–20	10–25
Austria	≥3.75	≥4.00	12–16	15–25
France	3.75–4.00	4.00–5.00	10–15	15–30
The Netherlands	3.50–4.00	4.00–4.50	8–12	12–15
Germany	3.50–4.00	3.75–4.50	14–16	16–18
Serbia	3.50–4.00	3.75–4.50	$3R_u$ or $2R_u$ $R_u = 8–12$	$3R_i$ or $2R_i$ $R_i = R_u + 2$
Switzerland	3.00–3.50	3.50–4.50	$R_{e1} = 60$ $R_{e2} = 12$	$R_{a1} = 56$ $R_{a2} = 14$

According to [6], the construction of roundabouts on sections of roads with longitudinal slopes of up to 3% is not considered an issue. At longitudinal slopes between 3% and 6%, instability of trucks passing through an intersection can occur. For longitudinal slopes greater than 6%, it is necessary to relocate the intersection or modify the longitudinal profile of the road. Therefore, the longitudinal slopes of the roundabout approaches must not exceed 3%.

According to [9], the longitudinal slope of the roundabout approach should not exceed 6%. If the terrain slopes are higher than 6%, it is necessary to place the entire surface of the intersection in one plane.

According to [10], roundabouts should be in locations where terrain slopes are less than or equal to 2.5%. Longitudinal slope changes (sag or hog curves) of the approaches should be located at approximately half of the distance "L_o" from the outer edge of the roundabout (Figure 27, "V_{ras}" is the design speed). If the longitudinal slopes of the approaches are greater than 2.5%, it is necessary to mitigate them, as shown in Figure 28.

The minimum values of the vertical hog and sag radii are determined according to the relevant speed of entry or exit. The tangent of the vertical curve should end at the edge of the circular pavement along the splitter island. Mitigation of longitudinal slopes to a value equal to or below 2.5% (Figure 28) is mandatory at all one-lane roundabouts with an outer radius equal to or greater than 20 m and/or when the maximum design speed is equal to or larger than 60 km/h [10]. Mitigation of longitudinal slopes to a value between 2.5% and 4.0% (Figure 28) can be applied in conditions of spatial constraints at single-lane roundabouts with a design speed below 50 km/h [10]. The highest longitudinal slopes in the roundabout zone (Figure 28) are applied only under strict spatial restrictions at intersections with an intersection design speed below 50 km/h and low probability of occurrence of trucks and buses in the roundabout (the share of trucks and buses at the roundabout is equal to or lower than 2%) [10].

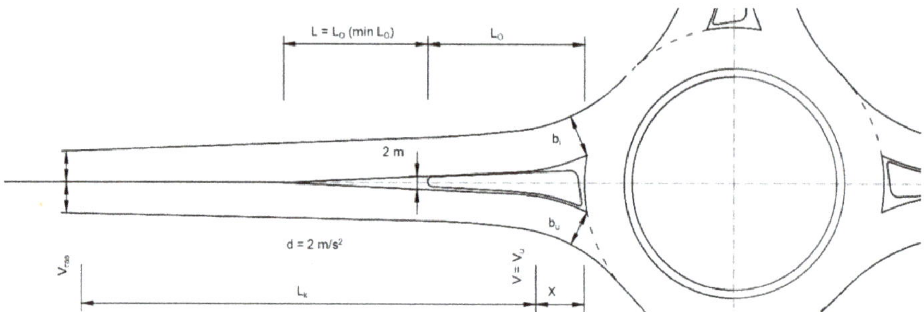

V_{ras} [km/h]	L_k [m]	X [m]	V_u [km/h]	L_O [m]	min L_O [m]
50	30	7	30	15	10
60	45	10	35	20	15
70	65	12	40	25	20*
80	85	15	45	35	25*
90	105	20	50	45	30*

* min 15 m

Figure 27. Roundabout approach design elements according to [10].

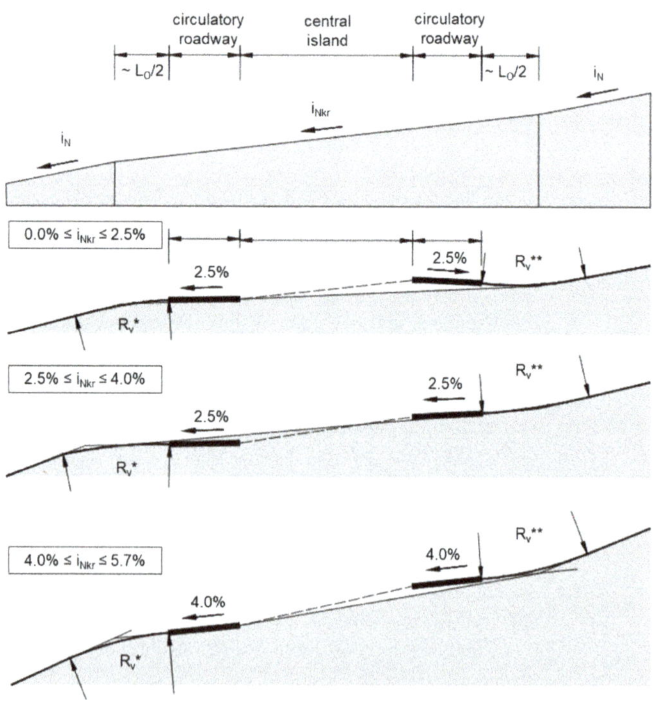

Figure 28. Mitigation of longitudinal slopes on roundabout approaches (Rv*—hog vertical curve radius; Rv**—sag vertical curve radius) [10].

According to [12], the intersection must be laid in one plane, the slope of which corresponds to the longitudinal and transverse slopes of the approaches. The maximum slope of the intersection surface is 5%, but under demanding topographic conditions, the slope can be as high as 7%. In this case, it is necessary to provide low traffic speeds at the roundabout by deflection.

The recommended values for the longitudinal slopes at roundabouts listed in the analyzed documents are shown in Table 8. According to the considered documents, in demanding topographic conditions (in hilly terrain) the entire surface of the intersection should be placed on one plane with a maximum slope of 2.5 to 7%. The maximum permitted longitudinal slopes of the approaches are in the range of 3 to 7%.

Table 8. Longitudinal slopes at roundabouts according to [4–12].

State	Maximum Slope of the Intersection Plane (%)	Maximum Longitudinal Slope of the Approaches (%)	Longitudinal Slope Changes–Distance to the Roundabout Outer Edge (m)
Croatia	≤4	≤4	≥6.0
Austria	≤4	≤4	20.0
France	≤6	≤3	not defined
The Netherlands	not defined	not defined	not defined
Germany	≤6	≤6	not defined
Serbia	≤2.5–5.7	not defined	7.5
Switzerland	≤5 (7)	≤5 (7)	not defined

3. Comparative Analysis

According to the analyzed documents, the design of modern roundabout geometric elements is composed of the following steps: (1) selection of the roundabout size, (2) selection of the circulatory roadway (and apron) width, (3) selection of the splitter island shape, (4) selection of the shape and elements of the right edge of the pavement at the entry and exit, and (5) final control of the roundabout geometry.

Roundabout size is defined by the value of the outer radius. These values depend on the spatial requirements and limitations, the number and the alignment of the approaches, the traffic requirements of the design vehicle swept path, and the design speed [17–19]. The recommended values of the outer radius given in the analyzed documents range from 12.5 to 25 m. Smaller values of outer radii are detected in Austria, the Netherlands, and Switzerland. This dissipation primarily reflects the spatial requirements and limitations, as well as the dimensions and types of the design vehicles used in the swept path analysis.

In terms of the circulatory roadway width, the main difference between the analyzed documents is the approach applied in the definition of this width, the selection of the design vehicle, and the application of the apron. Namely, according to the French guidelines, the width of the circulatory roadway is fixed and 20% larger than the widest entrance, whereas all other analyzed documents define this width according to the design vehicle swept path analysis. When defining the circulatory roadway width, Croatian guidelines recommend the use of a two-axle design vehicle, whereas types of design vehicles given in other analyzed documents that should be used in this analysis are not specifically defined.

The design of an apron on suburban roundabouts is described in Austrian, Croatian, French, Dutch, and Swiss guidelines and norms. German and Serbian guidelines do not envisage these elements in suburban locations. According to these documents, the apron width is defined based on the (long) design vehicle swept path analysis. French, Dutch, and Swiss documents recommend an apron width of at least 1.5 m. Apron cross slopes given in the analyzed documents range from 1 to 6%. Concerning the apron cross slope, one should be aware that the apron shape should deter car (and bus) drivers from crossing it, so the apron must be raised relative to the circulatory roadway and constructed with a cross slope greater than the circulatory roadway cross slope [20]. The apron also differs from the circulatory roadway in the finish layer material and/or color used.

According to the analyzed documents, splitter islands (raised or drawn) are mandatory elements of modern suburban roundabouts. In terms of their layout, the splitter islands used in the Netherlands are exclusively of the radial type. Radial splitter islands are also given in the Serbian document—they are used for the lowest level of channeling. Other analyzed documents define the use of triangular (Croatian, Austrian, Serbian) and/or funnel-shaped splitter islands (Croatian, French, German, Serbian, Swiss) in suburban roundabouts.

Among the analyzed documents, the most elaborate approach to the design of the roundabout entry and exit is given in the Croatian guidelines. This approach is based on the design vehicle swept path analysis and therefore should result in a design that provides unobstructed vehicle movement even in the initial roundabout design. The only drawback of this approach is the fact that it can be very time-consuming [17]. The approach given in Swiss guidelines is more straightforward—additional widening of the entry and exit lane, necessary to accommodate the design vehicle, is performed by rounding the right edge of the pavement or curb using two radii of the appropriate size. According to the other analyzed documents, entry and exit lane widening is performed by rounding the right edge of the pavement using only one radius of the appropriate size.

4. Concluding Remarks

The documents analyzed in this entry were issued in a period of over 20 years, which was marked by the mass construction of roundabouts in their countries of origin. Even though the analysis has shown that all analyzed documents have similar design approaches for suburban roundabouts, the dissipation of the recommended and limit values of roundabout geometric elements is evident, and in some cases, even significant. This dissipation is the result of the following: (1) the differences in the geography and terrain in the countries these documents originate from, which affects the intersection design elements' dimensions and shapes, and (2) the differences in types and dimensions of design vehicles used in the swept path analysis, which is the basis for the definition of roundabout lane widths.

The following recommendations for suburban single-lane roundabouts can be given as concluding remarks for this entry:

- The roundabout outer radius must be selected by considering the spatial limitations and the requirements of the design vehicle swept path.
- The swept path analysis for the design vehicle should be the basis for the determination of the minimum required width of the circulatory roadway.
- When defining the width of a circulatory roadway at single-lane roundabouts, widths above 5.5 m should be avoided so that drivers do not use a wide roadway as two lanes, and so that sufficient deflection is achieved.
- When the swept path conditions for the design vehicle require a large width for the circulatory roadway, it is necessary to introduce an apron.
- The apron is intended for the passage of trucks only—buses use only the width of the circulatory roadway to ensure driving comfort.
- The design of the splitter island, i.e., the channeling of traffic flows, should depend on the traffic conditions (the main traffic flow should be undisturbed), the type of road (the road cross-section outside of the intersection zone), horizontal alignment (good visibility must be provided), and the position of intersection in the network (for suburban roundabouts, the channelization must be stretched and well signaled, otherwise raised islands can be a source of danger due to higher driving speeds).
- The entry and exit radii, along with the width of the circulatory roadway and the approach lanes, should enable the fulfillment of the design vehicle swept path requirements.
- It is recommended to use the smallest possible entry and exit radii since larger radii can result in higher speeds at the intersection.
- The exit radius should be larger than the entry radius to increase the exit speed.

- Particular attention should be paid to wide entrances: An entry width greater than 5.5 m or greater than the width of a circulatory roadway can lead drivers to interpret a wide single-lane entrance as two lanes, which increases the risk of collision when entering a single-lane intersection.

To complete a modern roundabout design, a fastest path analysis and visibility checks must be conducted. If the results of these checks are not in line with the design recommendations, the geometry of the elements applied in the initial design phase must be modified. After fulfilling the conditions of the performance checks, a detailed roundabout design is conducted: the definition of signalization, lighting, and other equipment of the intersection; the final examination of the conditions of visibility; and final modification of the design elements, if necessary. This iterative process could be simplified by the introduction of the design approach that is described, to some extent, in the Croatian guidelines. Namely, these guidelines recommend that the design of the following geometric elements on suburban roundabouts be based primarily on the design vehicles swept path analyses: the outer radius, the circulatory roadway width, the apron width, and the right pavement edge at the roundabout entry and exit. This approach, which incorporates the optimization technique rather than the iterative (trial-and-error) process, could be an effective way to determine the optimal design parameters in geometric design—as similar research has shown [18,19,21]. With today's development of software for vehicle movement simulation that allows easier and faster construction and modification of design vehicles' movement trajectories, this could be achieved easily [17,21].

Author Contributions: Conceptualization, S.A.; formal analysis, S.A.; resources, V.D.; writing—original draft preparation, S.A.; writing—review and editing, V.D.; supervision, V.D. Both authors have read and agreed to the published version of the manuscript.

Funding: This research received no external funding.

Conflicts of Interest: The authors declare no conflict of interest.

Entry Link on the Encyclopedia Platform: https://encyclopedia.pub/13862.

References

1. Waddell, E. Evolution of Roundabout Technology—A History-Based Literature Review. In Proceedings of the Institute of Transportation Engineers 67th annual Meeting, Boston, MA, USA, 3–7 August 1997.
2. Brilon, W. Studies on Roundabouts in Germany: Lessons Learned. In Proceedings of the 3rd International Roundabout Conference, Carmel, IN, USA, 17–20 May 2011.
3. Beyondthemaps. Available online: https://beyondthemaps.wordpress.com/2017/08/26/roundabouts-in-europe-2017-edition/ (accessed on 16 June 2021).
4. Smjernice za Projektiranje Kružnih Raskrižja na Državnim Cestama, Hrvatske Ceste. 2014. Available online: https://hrvatske-ceste.hr/hr/stranice/tehnicka-dokumentacija/dokumenti/45-smjernice-za-projektiranje-kruznih-raskrizja (accessed on 16 June 2021).
5. *Plangleiche Knoten–Kreisverkehre*; (RVS 03.05.14); Österreichische Forschungsgesellschaft Straße-Schiene-Verkehr (FSV): Wien, Austria, 2010.
6. *Aménagement des Carrefours Interurbains sur les Routes Principales, Carrefours Plans—Guide Technique*; Service d'Etudes Techniques des Routes et Autoroutes, Centre de la Sécurité et des Techniques Routières: Bagneux, France, 1998.
7. *CROW: Eenheid in Rotondes*; CROW publication no.126; CROW: Ede, The Netherlands, 1998.
8. *Roundabouts—Application and Design: A Practical Manual*; (Dutch) Ministry of Transport, Public Works and Water Management, Partners for Roads: Amsterdam, The Netherlands, 2009.
9. *Merkblatt für die Anlage von Kreisverkehren, K 10000*; FGSV (Forschungsgesellschaft für Strassen-und Verkehrswesen): Köln, Germany, 2006.
10. Pravilnik o Uslovima Koje sa Aspekta Bezbednosti Saobraćaja Moraju da Ispunjavaju Putni Objekti i Drugi Elementi Javnog Puta, Službeni Glasnik RS, br. 50/2011. Available online: https://www.mgsi.gov.rs/sites/default/files/Pravilnik%20o%20uslovima%20koje%20sa%20aspekta%20bezbednosti%20saobracaja%20moraju%20da%20ispunjavaju%20putni%20objekti%20i%20drugi%20elementi%20javnog%20puta.pdf (accessed on 16 June 2021).
11. Maletin, M.; Andjus, V.; Katanić, J. *Tehnička Uputstva za Projektovanje Površinskih Raskrsnica (PGS-PR/07)*; Građevinski Centar: Beograd, Serbia, 2010.

12. *Knoten, Knoten mit Kreisverkehr*; Schweizer Norm (SN 640 263); Vereinigung Schwiezerischer Strassenfachleute (VSS): Zürich, Switzerland, 2000.
13. *Kontrolle der Befahrbarkeit*; Schweizer Norm (SN 640 271a); Vereinigung Schwiezerischer Strassenfachleute (VSS): Zürich, Switzerland, 1991.
14. *Knoten, Knoten in Einer Ebene (ohne Kreisverkehr)*; Schweizer Norm (SN 640 262); Vereinigung Schwiezerischer Strassenfachleute (VSS): Zürich, Switzerland, 1999.
15. Bovy, P.; de Aragao, P.; Blanc, P.H.; Veuve, L. *Guide Suisse des Giratoires*; Fonds de Securite Routiere, Institut des Transports et de Planification, Ecole Polytechnique Federale de Lausanne: Lausanne, Switzerland, 1991.
16. Tolazzi, T. *Alternative Types of Roundabouts*; Springer: Berlin/Heidelberg, Germany, 2015.
17. Stančerić, I.; Ahac, S.; Bezina, Š.; Vlaović, F. Design limits for intersection angles between approach legs of suburban roundabouts. *Građevinar* **2019**, *71*, 389–399. [CrossRef]
18. Ahmed, H.; Easa, S.M. Multi-Objective Evaluation Model of Single-Lane Roundabouts. *Transp. Res. Rec.* **2021**. [CrossRef]
19. Diachuk, M.; Easa, S.M. Guidelines for roundabout circulatory and entry widths based on vehicle dynamics. *J. Traffic Transp. Eng.* **2018**, *5*, 361–371. [CrossRef]
20. Montella, A.; Turner, S.; Chiaradonna, S.; Aldridge, D. International overview of roundabout design practices and insights for improvement of the Italian standard. *Can. J. Civ. Eng.* **2013**, *40*, 1215–1226. [CrossRef]
21. Ahac, S. Design of Suburban Roundabouts Based on Rules of Vehicle Movement Geometry. Ph.D. Thesis, University of Zagreb, Zagreb, Croatia, April 2014.

encyclopedia

Entry

Knowledge Integration in Smart Factories

Johannes Zenkert *, Christian Weber, Mareike Dornhöfer, Hasan Abu-Rasheed and Madjid Fathi

Department of Electrical Engineering and Computer Science, Institute of Knowledge Based Systems and Knowledge Management, University of Siegen, 57076 Siegen, Germany; christian.weber@uni-siegen.de (C.W.); m.dornhoefer@uni-siegen.de (M.D.); hasan.abu.rasheed@uni-siegen.de (H.A.-R.); fathi@informatik.uni-siegen.de (M.F.)

* Correspondence: johannes.zenkert@uni-siegen.de

Definition: Knowledge integration is well explained by the human–organization–technology (HOT) approach known from knowledge management. This approach contains the horizontal and vertical interaction and communication between employees, human-to-machine, but also machine-to-machine. Different organizational structures and processes are supported with the help of appropriate technologies and suitable data processing and integration techniques. In a Smart Factory, manufacturing systems act largely autonomously on the basis of continuously collected data. The technical design concerns the networking of machines, their connectivity and the interaction between human and machine as well as machine-to-machine. Within a Smart Factory, machines can be considered as intelligent manufacturing systems. Such manufacturing systems can autonomously adapt to events through the ability to intelligently analyze data and act as adaptive manufacturing systems that consider changes in production, the supply chain and customer requirements. Inter-connected physical devices, sensors, actuators, and controllers form the building block of the Smart Factory, which is called the Internet of Things (IoT). IoT uses different data processing solutions, such as cloud computing, fog computing, or edge computing, to fuse and process data. This is accomplished in an integrated and cross-device manner.

Keywords: smart factory; cloud computing; fog computing; edge computing; knowledge integration; knowledge management; data analytics; text analytics; knowledge graph

Citation: Zenkert, J.; Weber, C.; Dornhöfer, M.; Abu-Rasheed, H.; Fathi, M. Knowledge Integration in Smart Factories. *Encyclopedia* **2021**, *1*, 792–811. https://doi.org/10.3390/encyclopedia1030061

Academic Editors: Raffaele Barretta, Ramesh Agarwal, Krzysztof Kamil Żur and Giuseppe Ruta

Received: 15 July 2021
Accepted: 12 August 2021
Published: 16 August 2021

1. Introduction

In the wake of the Industry 4.0 development, the concept of Smart Factories and related technologies such as Cyber–Physical Systems (CPS) or the application of Internet of Things (IoT) in an industrial context emerged in the span of just ten years. Cyber–Physical Systems combine the analogue or physical production world with the digital world in a newfound complexity. Consequently, data and knowledge are playing an increasingly bigger role, supporting and leading to data-driven manufacturing (e.g., [1,2]).

Industry 4.0 has first been published on a larger scale as a (marketing) concept in 2011 at the Hannover fair in Germany. What followed was the backwards view on how to define the previous epochs of Industry 1.0 to 3.0 and their respective historical focus (e.g., [3]). Industry 4.0 presents a forward view of how the concept may be used to transform the current production environment and integrate digital solutions to improve aspects such as performance, maintenance, manufacturing of individualized products or to generate transparency over the whole production process or value chain of a company. Zhong et al. (2017) conclude that "*Industry 4.0 combines embedded production system technologies with intelligent production processes to pave the way for a new technological age that will fundamentally transform industry value chains, production value chains, and business models*" [4]. The technological advance also requires interaction with and integration of skilled workforces, even though this is often not addressed [5]. In this light, Industry 4.0 can be further defined as a network of humans and machines, covering the whole value chain, while supporting digitization and fostering real-time analysis of data to make the manufacturing processes more transparent and simultaneously more efficient to tailor intelligent products and services to the customer [6]. Depending on the type of realization and number of data sources, there might be a requirement for Big Data analysis [1,2].

Intensive research has been conducted on how to make the existing factories "smarter". In this context, the term "smart" refers to making manufacturing processes more autonomous, self-configured and data-driven. Such capabilities enable, for example, gathering and utilizing of knowledge about machine failures, to enable predictive maintenance actions or ad hoc process adaptations. In addition, the products and services which are manufactured often are aimed to be "smart" too, meaning they contain the means to gather data which may be used to improve functionalities or services through continuous data feedback to the manufacturer.

A generic definition of the term Smart Factory is still difficult, as many authors provide definitions based on their specific research area [7]. It can be concluded from this that the Smart Factory concept is targeting a multi-dimensional transformation of the manufacturing sector that is still continuing. Based on their analysis, Shi et al. (2020) conclude on four main features of a Smart Factory: (1) sensors for data gathering and environment detection with the goal of analysis and self-organization, (2) interconnectivity, interoperability and real-time control leading to flexibility, (3) application of artificial intelligence (AI) technologies such as robots, analysis algorithms as well as (4) virtual reality to enhance *"human–machine integration"* [7]. To target the diversity of topics related to the term Smart Factory, Strozzi et al. (2017) conducted a literature survey of publications between 2007 and 2016 and concluded from more than 400 publications direct relations between smart factories and the topics of real-time processing, wireless communication, (multi)-agent solutions, RFID, intelligent, smart, flexible and real-time manufacturing, ontologies, cloud computing, sustainability and optimization [8], identifying main areas but also enablers for a Smart Factory.

The overall question this work tries to answer is "How do Industry 4.0 environments or Smart Factory plants of the future look like and what role does data and knowledge play in this development?" Tao et al., (2019), referencing Zhong et al. (2017) [4], summarize that *"Manufacturing is shifting from knowledge-based intelligent manufacturing to data-driven and knowledge-enabled smart manufacturing, in which the term "smart" refers to the creation and use of data"* [9]. This shift has to be considered with the help of concepts known from the disciplines of data analytics, knowledge management (KM) and knowledge integration, machine learning and artificial intelligence. It presents a change from "knowledge-based", explicitly represented, qualitative data to the consideration of quantitative data in which meaningful patterns trigger manufacturing decisions, while being informed by supporting knowledge representations, such as ontologies. Especially the pronounced roles of data and knowledge are key aspects of future manufacturing environments and products.

Before talking more about this change the terms data, information and knowledge as well as knowledge management will be introduced briefly, giving a better background of understanding [10]: From a knowledge management perspective, the three terms are closely related, whereas data are the basis being formed out of a specific alphabet and grammar/syntax and may be structured, semi-structured or unstructured. Information builds on top of data which are used and interpreted in a certain (semantical) context, while knowledge is interconnected, applied or integrated information and oftentimes relates to a specific application area or an individual. That is why the terms of individual and collective knowledge are important factors for knowledge management, a discipline supporting, e.g., the acquisition, development, distribution, application or storage of knowledge within an organization. Different KM models or processes may be established and manage targeting human, organizational and technological aspects. Frey-Luxemberger gives an overview of the KM field [10].

The changes and role of data in manufacturing detailed above, is motivated or required by the rising customer demands of customized or tailored orders [11]. From an outside perspective the change of market demands requires hybrid solutions which not only focus on the manufacturing of physical devices or products but an accompanying (smart) service [12], which is only possible if the product generates data to be analyzed and used for offering said service. At the same time, the interconnected technologies require a change in knowledge management. Bettiol et al. (2020) conclude: *"On the one hand advanced, interconnected technologies generate new knowledge autonomously, but on the other hand, in order to really deploy the value connected to data produced by such technologies, firm should also rely on the social dimension of knowledge management dynamics"* [13]. The social dimension will be discussed later when reflecting on the changing role of employees in Smart Factories.

To meet a lot size of one, while offering extensive automated configuration abilities throughout the production process, the Smart Factory has to offer configuration and adaptation possibilities in a scalable way. At the same time these have to be manageable by the human workers, as well as being aligned to the underlying business processes. The realization is only possible by collecting and using data and knowledge throughout the manufacturing and documentation process, as well as by deploying automated data analytics and visualization tools to enable real time management and reconfiguration. It is expected that in the future workers inside Smart Factories will have to fulfill different roles or tasks in different processes or together with (intelligent) machines (e.g., [14,15]). Furthermore, instead of only administering one isolated machine, they will be supporting overarching tasks as the surveying and monitoring of interconnected production machines or plants, as well as flexible automation solutions. This again requires knowledge about inter-dependencies in the production process as well as about consequences for multiple production queues, e.g., in case of a failure of an intermediate machine. In this context the topic of predictive maintenance (e.g., [16]) is another major issue, as the gathered data inside the Smart Factory can and has to be used to minimize the times of failures in the more complex manufacturing environment, deploying analytics strategies (e.g., [16]) or machine learning algorithms to detect potential failures or maintenance measures. The concept of a digital twin (DT) (e.g., [9]) might be used here to fuse data and simulation models to create a real-time digital simulation and forecast the real environment, supporting the early detection of potential problems and real-time reconfiguration.

In the following, the different aspects of a Smart Factory including computing, analytics and knowledge integration perspectives will be discussed in more detail.

2. Smart Factory

2.1. Smart Factory Environment

Before the different levels of integration and computing in *Smart Factories* are explicitly addressed, a conceptual overview of a Smart Factory environment is given and explained. Figure 1 summarizes the integration levels and technologies related to a Smart Factory. The icons indicate the interconnectivity and communication between them and are detailed in the following sub-sections. In this setting, the horizontal perspective is focused on the shop floor level of the manufacturing facilities where products are manufactured, and related data are collected for later integration and analysis. The vertical perspective is a perspective of knowledge integration where gathered data from the manufacturing environment are fused, aggregated and integrated with the knowledge about the underlying business processes. The goal is to derive at the lowest level a real-time perspective on the state of manufacturing, while enabling at the top level a predictive business perspective. These alignments are reflected both horizontally and vertically in the RAMI 4.0 architecture of the i40-platform [17] and vertically in the 5C model [18].

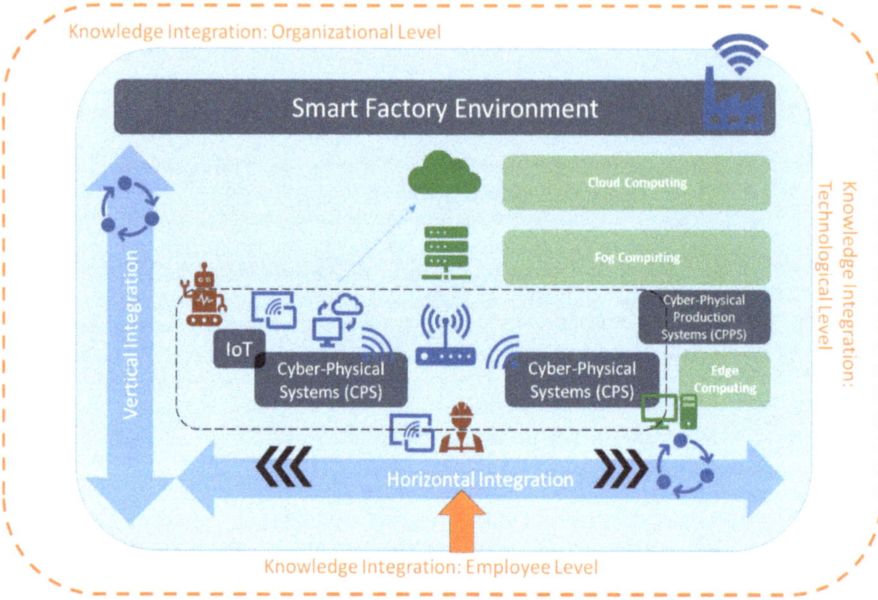

Figure 1. Smart Factory Environment and Knowledge Integration.

Depending on the author, either one or both terms of Cyber–Physical Systems (CPS) or Cyber–Physical Production Systems (CPPS) are used as the main building blocks of a Smart Factory. While the term CPS is sometimes used synonymously with CPPS, the term CPPS is also used to refer to a higher-level system that consists of multiple singular CPS [2,6]. In this paper, we follow the second interpretation. We use the term CPPS to refer to a wider scope, in which interconnected CPSs build the CPPS. In a practical scenario, these may be different production machines connected to one production process, where each machine resembles a unit as an integration of the "cyber", meaning the computation and networking component, as well as the "physical" electrical and mechanical components. Karnouskos et al. (2019) named autonomy, integrability and convertibility as the main CPPS pillars [19].

Next to CPS or CPPS, other main building blocks are IoT devices, which may be any devices that are able to gather data through embedded sensors as an interconnected entity within a wider network, and subsequently integrating the collected data into the Smart Factory network. This way, previously closed manufacturing environments or passive objects may be transformed into an active role inside the network and, as such, the manufacturing or monitoring processes.

Different communication protocols, technologies and standards may be used for the identification of objects, realization of connectivity or transmission of data, e.g., RFID, WLAN, GPS, OPC UA or specific industrial communication protocols (e.g., [6]). The identification of individual objects inside the production process is essential for individualization of products and automation of processes [20]. Soic et al. (2020) reflect in their work about context-awareness in smart environments that it requires the interconnection of *"physical and virtual nodes"*, whereas the nodes relate to *"an arbitrary number of sensors and actuators or external systems"* [21]. Fei et al. (2019) conclude that data gathered from *"interconnected twin cybernetics digital system[s]"* supports prediction and decision making [22]. Gorodetsky et al. (2019) view *"digital twin-based CPPS"* as a basis for *"real-time data-driven operation control"* [23].

IoT architecture models such as the Industrial Internet of Things (IIoT) architecture published by the Industrial Internet Consortium (IIC) [24] or the RAMI 4.0 architecture of the i40-platform [17] are approaches to standardize the implementation of a Smart Factory from the physical layer up to the application layer similar to the OSI 7 layer model.

Each CPS (e.g., through embedded sensors), or IoT device, produces or gathers data which need to be processed and analyzed to generate immediate actions or further derive knowledge about the status of the devices or the manufacturing process in focus. This may happen directly inside the device or with the help of a local computing device attached to the machine, called edge component and *Edge Computing*, respectively, on a local network level, called *Fog Computing*, or with the help of a central *Cloud Computing* platform. The Cloud Computing solution is beneficial when there are different production plants and the data need to be gathered at one common, yet transparently distributed, place to be analyzed. Edge or Fog Computing are better in cases of immediate processing or processing inside a local plant. In this context, it is important to define if data need to be analyzed immediately, e.g., for monitoring purposes or for cases where an analysis of a specific time frame is necessary. In the first case, *Stream Processing* or Stream Analytics can be used where the data are gathered and immediately analyzed "in place", meaning while being streamed, while in the second case *Batch Processing* or Batch analysis takes place, where data are collected and processed together with the option of aggregating over time and features. In Section 3, the different aspects are discussed in more detail.

2.2. Multi-Dimensional Knowledge Integration in Smart Factories

Theorem 1. *From the perspective of Knowledge Management and considering the three associated main pillars of (1) human workforce, (2) organizational structures or processes as well as (3) technology, the establishment of a Smart Factory requires a Multi-dimensional Knowledge Integration Perspective targeting the aforementioned three pillars, while considering the knowledge associated with, as well as the interconnected manufacturing processes and automation requirements to support a "Smart" solution.*

Based on Theorem 1, the following sub-sections detail the different aspects of how to create such a multi-dimensional knowledge integration perspective. We consider the human–organization–technology (HOT) (e.g., [10], or [25]) perspectives the target for this multi-dimensional knowledge integration [26]. Examples of the human (employee) perspective are roles like involved knowledge experts, lifelong learning, human–machine interaction, etc. Examples on the organizational perspective are the transformation of a hierarchical into a network-based structure, organizational learning and a data-driven business process integration; and finally, from the technological perspective we have the variance of knowledge assets (e.g., textual content, sensor data) as well as analysis processes, which will be detailed in Section 3.3. Figure 1 summarizes the perspectives around the Smart Factory environment.

3. Knowledge Integration

3.1. Knowledge Integration on Organizational Level: Horizontal and Vertical Integration

The Smart Factory integration or transformation from a regular factory into a Smart Factory bases on a horizontal and a vertical integration perspective. The horizontal perspective targets the production and transportation processes along the value chain, while the vertical perspective has a look at how the "new" or "transformed" production environment fits and interrelates with the other organizational areas such as production planning, logistics, sales or marketing, etc. The platform i4.0 visualized this change between Industry 3.0 [27] towards Industry 4.0 [28] with the change from a production pyramid towards a production network. Hierarchies are no longer as important in Industry 4.0 as the concept transforms the organization into an interconnected network, where hierarchical or departmental boundaries are resolved or less significant. The important aspect is the value chain and all departments working towards customer-oriented goals. This of course requires a change in the way the employees work and interact with each other. While before employees in the same hierarchy level or working on the same subject were mostly communicating with each other, this organizational change also leads to a social change inside the company. In addition, job profiles are changing towards a "knowledge worker". This will be discussed in Section 3.3.2 as it requires knowledge management aspects to support this change process.

Going back to the horizontal and vertical integration, both aspects are not only related to the organizational changes necessary to establish a Smart Factory. They also target the data perspective. Data play a central role for the Smart Factory, as it is needed to automate processes or exchange information between different manufacturing machines or between divisions as production and logistics, etc. These data are gathered mostly with the help of sensors, embedded into IoT devices which are part of the CPS. A sensor might be attached directly at a specific manufacturing machine or at different gates where the products or pre-products come along during their way through the manufacturing environment. RFID tags may be used to automatically scan a product, check its identity and update the status or next manufacturing step. If the data are gathered and analyzed along the manufacturing process, then they are part of the horizontal integration, as the results of analysis might directly influence the next production steps and focus on a real-time intervention. Data analytics [16] from a vertical perspective have a broader focus and gather and integrate data from and at different hierarchical levels and different IT systems, as, e.g., the enterprise resources planning (ERP) system, as well as over a longer time period to, e.g., generate reports about the Smart Factory or a specific production development for a specified time.

3.2. Knowledge Integration: Employee Level

If we consider the organizational changes leading to a Smart Factory, it is important to consider the role of the current and future employees in this environment. This applies especially to those employees whose production environments have been analogue or not interconnected before and who now need to be part of the digitized production process. As such, an essential factor in this transformation process is lifelong learning and needs to be considered in all knowledge management activities of the Smart Factory [29].

From the perspective of knowledge integration, it is recommended to involve knowledge engineers or managers to support the transformation process but to also consider the concerns or problems the employees have and to acquire and provide the training they need. Another aspect is to establish the knowledge exchange between different involved engineering disciplines, as well as computer science for the design and understanding of complex systems such as CPS.

The employees in Smart Factories may be concerned about their future role and tasks as there will be continuous shifts in human–machine interaction (HMI), where the mechanized counterparts of a human worker in the future may be different CPS, a robot or an AI application. The roles that workers execute may be the controller of a machine, peer or teammate up to their replacement by an intelligent machine [30]. Ansari, Erol and Shin (2018) differentiate HMI into human–machine cooperation and human–machine collaboration in the Smart Factory environment: "*A human labor on the one side is assisted by smart devices and machines (human–machine cooperation) and on the other should interact and exchange information with intelligent machines (human–machine collaboration)*" [14]. This differentiation indicates that the human worker has benefits which will make their work process easier due to assisted technical and smart solutions, but also pose challenges as the worker needs to learn to work together with this new technical and digitized work environment. Vogel-Heuser et al. (2017) highlight that the human worker is now interconnected with CPS with the help of multi-modal human–machine interfaces [29]. The automation and application of AI are factors which question (a) the role of the human worker in the work process as well as (b) their abilities in decision making as it might be influenced by or contrary to the recommendations or actions of the AI application. An AI "*would always base its decision-making on optimizing its objectives, rather than incorporating social or emotional factors*" [15], posing a challenge of who is the main decision maker in the Smart

Factory and if the explicit and implicit knowledge and experience of the human worker are more valuable than the programmed AI logic, e.g., based on data and the execution of machine learning algorithms. Seeber et al. (2019) resume that *"The optimal conditions for humans and truly intelligent AI to coexist and work together have not yet been adequately analyzed"*, leading to future challenges and requirements to create recommendations or standards for said coexistence and collaboration between human and machine teammates [15]. One approach is a "mutual learning" between human and machine, supported by *human acquisition, machine acquisition, human participation* and *machine participation* leading to the execution of shared tasks between human and machine [14]. North, Maier and Haas (2018) envision that *"in the future expertise will be defined as human (expert) plus intelligent machine"*, with the challenge being how they learn and work together [30]. A possible system, showcasing this synergetic collaboration is the implementation of cobots, passive robots that are tailored for collaboratively inhabiting a shared space for the purpose of operating processes together with humans [31].

3.3. Knowledge Integration: Technological Level

In the introduction, the central role of data and knowledge in modern manufacturing has been motivated. Tao et al. are even speaking of "data-driven smart manufacturing" [1]. This means a continuous generation of data streams, leading to big data which require processing and analysis [2]. The following Table 1 summarizes how these data may be used or processed to generate knowledge inside the Smart Factory.

Table 1. Overview of Knowledge Integration on a Technological Level.

Knowledge Processing	Realization in Smart Factory	Benefits	Exemplary Technologies [1]
Data Computing/ Processing	Cloud Computing	Software as a Service (SaaS), central software applications without local installation	Amazon AWS, Microsoft Azure
	Fog Computing	Enhance low-latency, mobility, network bandwidth, security and privacy	Cisco IOx
	Edge Computing	Offload network bandwidth, shorter delay time (latency)	Cisco IOx, Intel IOT solutions, Nvidia EGX
Data Analytics	Descriptive Analytics	Status and usage monitoring, reporting, anomaly detection and diagnosis, modeling, or training	RapidMiner, RStudio Server, Tableau
		Stream Analytics: Real-Time Analysis	Apache Kafka/Flink, Elasticsearch and Kibana
		Batch Analytics: Monitoring/Reporting	Apache Spark/Zeppelin, Cassandra, Tensorflow, Keras
	Predictive Analytics	Predicting capacity needs and utilization, material and energy consumption, predicting component and system wear and failures	RapidMiner, RStudio Server, Tensorflow, Keras, AutoKeras, Google AutoML
	Prescriptive Analytics	Guidance to recommend operational changes to optimize processes, avoid failures and associated downtime	RapidMiner, RStudio Server, Tensorflow, Keras, AutoKeras, Google AutoML
Simulation	Digital Twin Concept	Real-time analysis, simulation of scalable products and product changes, wear and tear projection	MATLAB Simulink, Azure Digital Twins, Ansys Twin Builder

Table 1. *Cont.*

Knowledge Processing	Realization in Smart Factory	Benefits	Exemplary Technologies [1]
Semantic Knowledge Representation	Knowledge Graphs	Contextualization of multi-source data, Semantic relational learning	Neo4j, Grakn, Ontotext GraphDB, Eccenca corporate memory, Protégé
Text Mining	Intelligent Documents	Integration of lessons learned from reporting and failure logs in new design or production cycles	tm (R), nltk (Python), RapidMiner, Text Analytics Toolbox (MATLAB), Apache OpenNLP, Stanford CoreNLP

[1] Technologies and products are listed here exemplary and do not represent an exhaustive market overview.

Based on the table above, the mentioned technologies may be brought into context from a data to a knowledge integration perspective. The interaction and integration of these technologies are shown in Figure 2. It is assumed that the processing from data to information to knowledge, an integral concept of knowledge management, is applied here as well. Data might be in the form of structured streaming data, mostly generated from sensors or in logging units, or unstructured in form, such as texts, process documentation or reports created by the experts inside the Smart Factory environment. The structured data are mostly processed as incoming data streams which are stored in a database (locally or inside a cloud) and may either be analyzed in real-time or at a later stage in a batch. Machine learning methods might be applied for the task of learning and reasoning, especially if the application requires tasks such as predictive maintenance, decision support, or providing recommendations. Unstructured data can be utilized by pre-processing with the help of text mining methods such as entity and relationship recognition or semantic enrichments before being composed in a knowledge graph. A knowledge graph allows an enrichment with results from the analyzed structured data stream as well. All results can contribute to different applications in the context of smart manufacturing, such as monitoring, digital twin simulations or decision support.

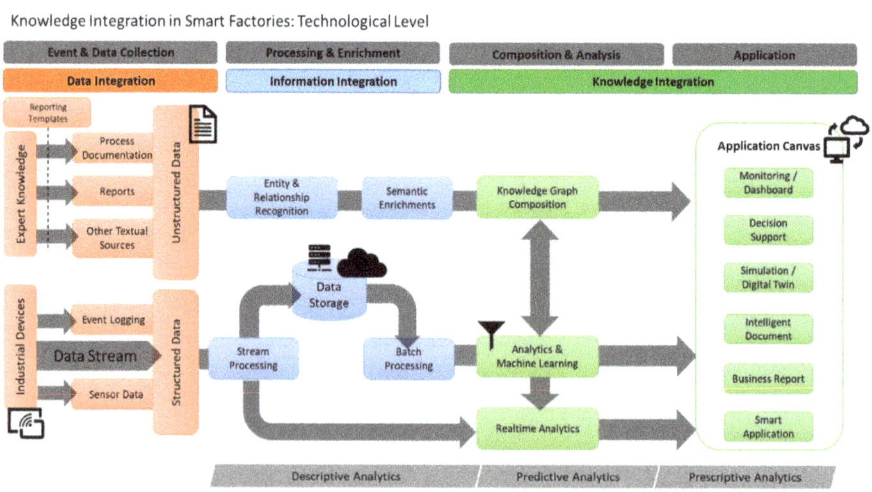

Figure 2. Smart Factory Environment—Technological Level.

In the following sub-sections, the technological level will be discussed in more detail.

3.3.1. Data Computing/Processing in Smart Factories

Smart Factories apply different kinds of computing levels to meet requirements regarding, e.g., real-time processing or big data analysis. The deciding factor for how to gather and process data streams inside the Smart Factory is oftentimes the available time or the amount of data. As a rule of thumb, the faster an analysis should be executed, the closer the necessary computing unit needs to be to the production machine. On the other hand, large amounts of data oftentimes need a central storage to be fused and aggregated before they are processed. Therefore, Smart Factories consider three computing tiers or levels: Cloud, Fog and Edge Computing, e.g., [32]. In the case of a Smart Factory, the application of these computing approaches means that the previously closed environments, using only operation technology (OT), are now opened towards information technology (IT) thus providing opportunities regarding computing power but also challenges, e.g., from a security perspective.

Cloud Computing: Cloud Computing covers a computer science and hardware concept where central computing, network or software services are bought on service platforms. These are known as Platform as a Service (PaaS), if an organization is unable to establish its own data center or local data storage. Central indicates a single point of access to the service, which in turn may be within the cloud distributed transparently across multiple processing nodes. In the wake of Cloud Computing, different services emerged such as Software as a Service (SaaS), providing central, scalable software applications, which need not be installed locally but are accessible universally. Within the literature, Cloud Computing is strongly associated with Big Data as it offers the most benefits when huge amounts of data need to be analyzed with extensive, parallelized computing power which a local single-server unit may not be able to provide. In the context of Industry 4.0, Zhong et al. (2017) even speak of "*cloud manufacturing*" as a new model for an "*intelligent manufacturing system*" [4]. In this case, the company or the machines generating the data need to have a network connection to the cloud services to store their data inside the cloud environment and use the available processing power and tools for analysis. The results will be transmitted back to the company. One aspect discussed in this context is data privacy and intellectual property, following the externalization of company data into a booked cloud service. To meet these concerns, there are possibilities for using a public cloud, private cloud or a hybrid version (e.g., [3]). In the case of a public cloud, the data will not be public for all, but it indicates that different user groups or organizations are using this cloud or virtual environment together. The private cloud environment is available only for the organization itself. Hybrid clouds blend both concepts seamlessly.

Fog Computing: Next to Cloud Computing, Fog Computing, while not as prominent, still plays an important role in the implementation of computing resources in the Smart Factory. The OpenFog reference architecture of the OpenFog Consortium (2017) defines it as: "*A horizontal, system-level architecture that distributes computing, storage, control and networking functions closer to the users along a cloud-to-thing continuum*" [33]. Oftentimes it is also defined in relation to the other two types of computing: Cloud and Edge. Hu et al. (2017) summarize the concept of fog computing as "*Fog computing extends the cloud services to the edge of network, and makes computation, communication and storage closer to edge devices and end-users, which aims to enhance low-latency, mobility, network bandwidth, security and privacy*" [32]. The term "*edge of the network*" defines devices such as routers or gateways available for the edge devices to connect to the network. These devices are called fog nodes and offer the first central computing unit from the perspective of the edge device. Most of the time, fog nodes themselves are again connected to the cloud layer thus covering the computing unit in between [32]. Fei et al. (2019) pronounce the main difference towards Cloud Computing is that the available resources are more limited and require "*optimal management*", while on the other hand fog nodes are much closer to the edge devices, reducing network delays [22].

Edge Computing: Edge Computing targets a computing and storage unit that is directly attached to or embedded into an edge device (e.g., IoT Device, embedded sensor in a production unit). Edge computing describes a decentralized cloud computing architecture. The data generated, for example, from local computers or IoT devices, are processed immediately at or within the device itself or in the vicinity, i.e., at the outermost edge of the network of an IT infrastructure (e.g., [34]). This aspect has direct relations to stream processing, which will be discussed later in this section. This helps to offload network bandwidth and releases *"an important part of the computational load from the Cloud servers"* to a data center or cloud [34]. The most important feature of edge computing is the extension and trade-off of cloud services to the edge of the network. As a result, computation, communication, and storage are much closer to the end user and the vicinity of the creation of the data. These edge nodes may include, e.g., smart sensors, smartphones, smart vehicles, or dedicated edge servers [32]. Furthermore, the computing at the edge may introduce concepts of data separation where data are filtered, compacted, aggregated, and analyzed in the manufacturing environment while only resulting data with a potentially enriched information density is transmitted to other systems and services. Edge nodes may be used for different steps of the data analytics process [16]. This way it can also be scaled what information needs to be revealed to external services (e.g., public clouds) to receive a desired analytical feedback. The Edge Computing Task Group of the IIC concludes a vertical integration of edge computing over the whole technology stack of the *Industrial Internet Reference Architecture* (IIRA), while horizontally it may be used for *peer-to-peer networking, edge-device collaboration, distributed queries, distributes data management* or *data governance* [35].

3.3.2. Types of Data Analytics in Smart Factories

Data Analytics is a central factor for the implementation of automated or semi-automated Smart Factories. Depending on the literature reference, it may also be more specifically called *Industrial Analytics* (e.g., [35]). The analyzed data may be structured (e.g., sensor data), semi-structured (e.g., meta data or semi-structured feedback data) or even unstructured data in the form of varying texts or images. The ways of extracting knowledge from these data are diverse. *"Extracting practical knowledge from heterogeneous data is challenging and thus, determining the right methodologies and tools for querying and aggregating sensor data is crucial . . . "* [36].

Smart Factories emphasize the analysis of sensor data, due to the distributed sensor data streams created from singular or interconnected CPS or IoT devices inside the production plant or Smart Factory environment. *"Data is a key enabler for smart manufacturing"* [1]. While more data allow more insights, this holds true only if the fitting data analytics, statistics or machine learning methods are applied considering the purpose of why certain data streams or batches are being analyzed. From a knowledge integration perspective, it is important to define the use case (e.g., predictive maintenance of a machine, monitoring of the current CPS status, etc.) and fitting data analytics concepts before choosing a concrete implementation technology.

Data analytics in Smart Factories follow general lifecycle management models and dedicated analytics methods adapting them to the type of data available in the environment and the given task at hand. It is like any other data analytics application dependent on its lifecycle *"of data collection, transmission, storage, processing, visualization, and application"* [1]. The IIC (2017) summarizes the applicable data analytics method categories of *descriptive analytics*, *predictive analytics* and *prescriptive analytics*: For the descriptive category, both batch and stream processing may be included to create monitoring, diagnosis or reporting applications, but to support also the training of a machine learning model. Predictive analytics applies *"statistical and machine learning techniques"*, which may be used for, e.g., predictive maintenance or material consumption, while *"prescriptive analytics uses the results from predictive analytics"* to develop recommendations, process optimization and prevention of failures [16].

Next to the use case and envisioned category of analytics task, it is important to define how the specific data need to be prepared and what kind of pre-analysis steps should be executed (e.g., [1]), such as filtering or removal of invalid or irrelevant data [37], handling of data gaps [38] where, e.g., the sensor did not work, what patterns should be detected or which parameters are needed for the visualization of results, to name only a few. The definition and application of data quality criteria are essential at this step.

While talking about descriptive analytics, the terms stream and batch processing have been mentioned. A major deciding factor for the type of analysis is time. Is there a requirement for real-time analysis or not? Depending on the answer, stream analytics/real-time analytics or batch analytics may be implemented:

Stream Analytics: Stream Analytics or Stream Processing indicates that an incoming data stream, e.g., generated from one or more sensors attached to a CPS, is being analyzed immediately, utilizing "in place" algorithms which need a limited data window to operate and are computational sparse. Aggarwal (2013) highlights, "*with increasing volume of the data, it is no longer possible to process the data efficiently by using multiple passes. Rather, one can process a data item at most once*" [39]. To meet the requirements of real-time processing, the incoming data are being separated into small frames and immediately forwarded into an "analyzer module" to generate knowledge for reactive measures or monitoring inside the production plant. Training models may be created and trained at different computing points, e.g., edge nodes [16]. Turaga and van der Schaar (2017) summarize different forms of data streaming analytics features, such as *Streaming and In-Motion Analysis* for on the fly analytics, *Distributed Data Analysis*, *High-Performance and Scalable Analysis*, *Multi-Modal Analysis*, *Loss-Tolerant Analysis* or *Adaptive and Time-varying Analysis* [40]. Stream analytics require local processing units, which is why the concepts of fog or edge computing are prevalent for this type of processing. A transfer of stream data into a cloud environment before it is being analyzed might lead to too much latency and a delay in reaction time as the results oftentimes lead to an immediate action inside the production plant. Summarizing, Fei et al. (2019) proclaim "*data stream analytics [as] one of the core components of CPS*" thus being motivated to evaluate different machine learning (ML) algorithms regarding their applicability for analyzing CPS data streams [22].

Batch Analytics: Batch Analytics follow the principle of gathering data for some time before running a fitting analysis or ML method. Depending on the batch size, this may lead to big data analytics (e.g., [1]) and the necessity of applying technologies for handling huge amounts of data. A well explored and integrated, yet continuously developing set of methods for batch processing are deep learning networks, enabling the batch-wise processing of large volumes of data with a large input-feature space [41]. Production plants oftentimes run logging or monitoring applications, building up historians of accumulated data which might also be part of the analysis process. The purpose of data accumulation instead of immediate analysis is mostly due to the need to create reports (e.g., monthly) or visualize timeline developments or machine workload. Analyzed trends or what is part of predictive analytics also depends on past data instead of only ad hoc available data streams.

Overall, it is important to decide which data may be part of the stream processing and analysis and which ones are accumulated, e.g., with the help of cloud computing. This is not an either/or decision. Some data may also be analyzed immediately, while afterwards being stored in the batch/cloud to be part of another analysis, e.g., timeline analyses, or being used to offline-update deployed algorithms. These types of data are especially valuable as they offer a dual benefit and knowledge for ad hoc and long-term decisions.

3.3.3. Simulation and Decision Making in Smart Factories Using Digital Twins

Simulation in Smart Factories is directly associated with the concept of the "*Digital Twin*" (DT). Alternatively, the terms "*Digital Shadow*" (DS) or "*Digital Model*" (DM) are being used in the literature as well, but they do not necessarily have the same meaning [42]. The term digital twin indicates that the physical device, system or entity is being represented in a digital form which has identical characteristics and allows to interact, test or manipulate it in the digital space. Before explaining more about the DT, it is important to understand the differences to CPS or IoT as they are core building blocks of the Smart Factory and also bridge a gap between physical and cyber world. Tao et al. (2019) resume that *"CPS and DTs share the same essential concepts of an intensive cyber–physical connection, real-time interaction, organization integration, and in-depth collaboration"* but they are not the same concept [9]. *"CPS are multidimensional and complex systems that integrate the cyber world and the dynamic physical world"*, while DTs focus more on the virtual models, the matching of their behavior to the physical counterpart and feedback flow of data between both [9]. Harrison, Vera and Ahmad (2021) envision that DTs may support the whole lifecycle and engineering of CPPS [43]. Jacoby and Usländer compared the concepts of DT and IoT and related standards or representation models [44]. They resume that while they are similar and both *"center on resources"*, DTs aim for optimization and automation, applying computer science concepts such as AI or machine learning in an integrative manner [44].

DTs are being used to simulate situations or decision alternatives which may not be executed easily in the physical world during operation. The Industrial Internet Consortium and Plattform Industrie 4.0 (2020) summarized in a joint whitepaper, that the DT is *"adequate for communication, storage, interpretation, process and analysis of data pertaining to the entity in order to monitor and predict its states and behaviors within a certain context"* [45]. DT has advantages in the case of decision making, analysis of gathered data or optimization of the physical counterpart. It is important to define the scope and purpose as the DT may address various aspects next to simulation [46]. Martinez et al. (2021) recently conducted research where they reflected that DTs can be implemented on all levels of the classical automation pyramid while applying AI concepts, thus also indicating that a DT implementation is not only focused on one level but is a holistic approach [47]. Harrison, Vera and Ahmad (2021) also see an evolution of the automation pyramid due to *"data-driven manufacturing capabilities"* [43].

The creation of a DT requires key aspects such as the availability of underlying models, service interfaces for connecting to the DT, and data collected from the physical entity it is related to [45]. Kritzinger et al. (2017) conducted a literature study regarding DM, DS and DT and tried to differentiate between the terms based on the data flow between the physical and digital object. In the case of a Digital Model, they see only manual data flows; the Digital Shadow, however, also supports automated data flow from physical to digital object, while the Digital Twin supports a bidirectional automated data flow between the physical and digital object [42]. Schluse et al. (2018) combine these aspects and summarize that a DT *"integrates all knowledge resulting from modeling activities in engineering (digital model) and from working data captured during real-world operation (digital shadow)"* [46]. Hänel et al. (2021) reflect from a data and information perspective on the application possibilities of DTs *"for High-Tech Machining Applications"* and present a model targeting a *"shadow-twin transformation loop"* based on the acquired feedback data [48].

The models of a DT may cover different details such as geometrical, material or physical properties, behavior or rules as well as data models [9]. The complexity and possibilities for simulation applications in the Smart Factory based on DT depend highly on the level of detail of the DT and the available data gathered by or from the physical device and its integration and analysis in the DT. DTs may be independent/discrete or they are part of a composition of DTs thus allowing, e.g., the representation of a production line of different physical entities [45]. Tao et al. (2019) summarize this as the unit, system and system of system level of DTs [9]. From a practical perspective, Autiosalo et al. (2021) implemented different DTs with varying update times on the asset, fleet and strategic level for an overhead crane scenario, giving special focus to API standards and data flows [49].

From a knowledge integration perspective, the data flow is essential for the concept of the DT as it allows the creation of knowledge with the help of data analysis or machine learning methods, thus feeding the DT with new knowledge from its physical counterpart. At this stage there is an interrelation to the section of stream analytics (see Section 3.3.2) as the real-time simulation of the DT would require the application of real-time data processing and stream analytics as well. While real-time data may be a factor for the DT to show the current state of the physical entity with the help of a visualized DT, historical or accumulated data during, e.g., the operation process of the physical entity, allow more complex simulations for optimization, failure studies or predictive maintenance, thus building a bridge to batch analytics (see Section 3.3.2). To realize these complex applications, authors research how concepts from AI, machine learning or other data science aspects such as big data analytics, may be applied in an efficient manner, or for which use cases of the Smart Factory they are applicable (e.g., [50]).

3.3.4. Semi- and Unstructured Data Integration

In a Smart Factory, knowledge can be mined from a range of data types. Sensor readings, for example, represent a structured data type, where the meaning of each value is definite. However, process documentations in modern industries also hold a great potential to capture expert knowledge in Smart Factories. Such documentations include verbal descriptions of procedures, failures, maintenance, etc. and it reflects inherently the experience of the human who is authoring the document. These user-specific inputs from the experts, like textual formats in general, are only available in a semi-structured or unstructured form. The latter type represents a completely free user input, whose meaning is only determined by analyzing the written text. An example of this type can be a customer complaint report. The former type, semi-structured data, features some information about the free text that can help in determining its meaning. For example, a failure report can take the form of a table, in which certain fields are filled with free textual input. Here, although the cell content is free text, the cell itself is predefined by the table structure to contain information about a certain aspect of the failure, e.g., the description of a failure reason, which makes the overall data in the table semi-structured.

Extracting knowledge from semi- and unstructured data requires methods that can automatically analyze the textual content to mine the information and represent its meaning. For this reason, we will highlight in this section the role of text mining and semantic knowledge representation for extracting and modeling knowledge from these two types of data, in order to integrate this knowledge into the overall decision-making process of a Smart Factory.

Application of Text Mining and Text Generation in Smart Factories

Text mining is a collection of methods and algorithms intended for information extraction and knowledge discovery from structured, semi-structured and unstructured textual data [51,52]. The utilization of text mining methods and approaches has a wide spectrum in the state of the art. Its applications include:

- Information extraction and retrieval: The information extraction process can be described as turning the unstructured textual data into structured and usable information. Since textual data are prone to including inconsistencies and human errors or potentially irrelevant data, a pre-processing phase takes place before the information extraction. After the raw text is pre-processed, several approaches have been addressed in the literature for the extraction of knowledge from the text, including Named Entity Recognition (NER) [53] and Relation Extraction (RE) [54]. Both methods are directly relevant for a semantic representation of data, since entities and relations can be identified in a meaningful way to define and populate the nodes and edges, in graph based representations of texts, as, e.g., knowledge graphs, while utilizing additional pre-existing domain knowledge [55].
- Classification, clustering and topic modelling: Classification is one of the important methods that allow creating a connection between a document and a search query. It is utilized in several applications, including the medical, commercial and industrial design processes. In the latter, Jiang et al. (2017) have developed an algorithm that predicts the importance of a product feature based on written user reviews [56]. Traditional text classification methods include Naïve Bayes, Support Vector Machines (SVM), and k-Nearest Neighbor. All those algorithms depend on the features of a

document and the predefined labels that classify this document into a certain category. Unlike traditional approaches, which use Term Frequency–Inverse Document Frequency (TF-IDF), cosine similarities or probability functions, machine and deep learning algorithms learn to map document features to the available labels through creating a matching function that enables the system of categorizing a new document to one or more of the predefined classes based on its features.

In contrast to classification, clustering and topic modelling are unsupervised learning approaches. Clustering and topic modelling have been applied in the design and engineering fields. They, e.g., enable the analysis of previous textual data, collected from various process steps and activities, such as emails, change logs and regular reports, for knowledge extraction and reuse. Grieco et al. (2017) have applied clustering on free text written on the Engineering Change Request (ECR) [57]. This is a type of engineering design log record of frequently required changes during the design phase, which are collected for products and their components. This approach has been observed to significantly help to summarize the main features and changes to the product during the development process of previous projects.

Smart factories require architectures that integrate knowledge and focus on textual analysis of unstructured data. One approach to actively utilize unstructured data is the creation of an Intelligent Document and therefore the generation of texts, e.g., for reporting purposes, which are generated on demand and may include interactive elements [26]. An intelligent document collects and contains references to the identified process documentation and relevant extracted knowledge sources. This includes expert identifiers as explicit entities to indicate the source creator and therefore potentially more implicit process knowledge. Such linked resources can be either followed by the user through generated links or be imported into the intelligent document to be composed into one homogenous text. Besides reporting purposes, additional applications can be the analysis and integration of results from past incident analyses as a so-called "lessons learned" documentation. In the case of subsequent changes to product generations and processes, identified errors resulting from incident analyses (e.g., log analysis) and error causes during manufacturing (machine and production reports) can be considered. Further text generation techniques can be used to improve the blending of imported texts [58]. For the development of intelligent documents, Zenkert et al. (2018a) propose a four-step process [26]:

- Acquisition and pre-indexing: In the acquisition and indexing phase, all relevant knowledge assets—including experts and extracted content from the knowledge base —are considered for extraction. Domain specific information such as textual information from design and process documentation is considered and subsequently identified through text analysis.
- Initialization: In the initialization phase, relevant knowledge sources are collected on the basis of the acquisition and indexing results, and metadata is assigned to documents. To sustain the extraction context and metadata of extraction algorithms, such results are stored using a graph-based knowledge representation concept, called Multidimensional Knowledge Representation (MKR) [59]. Based on the graph-based MKR of each document, other entities and semantically related content from other documents can be referenced [59]. Each extracted document is processed using text mining techniques to feed into the meta-structure of the MKR.
- Process Enrichment: In the process enrichment phase, relevant content is collected from the MKR structure, based on the relationships between elements and the underlying texts, and put into the format of an intelligent document. Each generated, intelligent document is a composition of identified knowledge assets. Based on the interaction of users, links to further resources can be followed and the document can be modified on demand and further enriched by additional asset recommendations.
- Automated Update Cycle: The automatic enrichment of the intelligent documents runs iteratively and continuously and modifies documents or adds content to new assets

when new or changed source texts or extraction results are available (e.g., new metadata such as keywords, entity links, related process contexts, new incident reports).

Semantic Knowledge Representation in Smart Factories

Knowledge graphs (KG), or knowledge maps, are graphical representations of a knowledge base. Whether ontologies or other semantic representations that are based on relations between graph entities, such networks are well suitable for knowledge representation and decision support in complex environments. As a knowledge representation tool, knowledge graphs are constructed through analyzing available information and representing it and its relevant semantic relationships as a graph through using rule-based integration mechanisms or adhering to a formal representation structure. The added value of such semantic representations is rooted in their ability to describe the relationships between pieces of information. This interconnectivity and contextualization of information enables it to be used as a medium of stored knowledge. To extract information from a KG, knowledge querying procedures are used.

Yahya et al. (2021) surveyed the use of semantic technologies in the scope of industry 4.0 and smart manufacturing [60]. They point out the multiple domains, in which KGs are involved to solve production challenges, such as predictive maintenance, resource optimization, and information management. They also propose an enhanced and generalized model, namely the reference generalized ontological model (RGOM), based on RAMI 4.0. Another study by Beden et al. (2021) analyzes the role of semantic technologies in the asset administration shells in RAMI 4.0 [61]. The authors show that semantic approaches enhance the way physical assets are digitally represented through adding an element of formalism to the knowledge representation.

Interaction with the knowledge graph, in terms of their construction and querying, is essential for their utilization. Several approaches for this process have been investigated, including Triple Pattern Fragments [62] and Example Entity Tuples [63]. Knowledge graphs have been exploited in several applications. Their semantic structure was used alongside text mining in order to enrich a graph with extracted textual data. The result of this combination is also addressed by Dietz et al. (2018), who highlighted in their summary on the utilization of knowledge graphs with text mining, the rising role of knowledge graphs in text-centric retrieval, especially for search system applications [64]. This further highlights the interaction between knowledge graphs and natural language processing (NLP). Knowledge graphs have provided the opportunity to bridge the gap between the data-driven and symbolic approaches, which considerably influenced the research in this field [65–67]. On the other hand, knowledge graphs can be constructed in different ways, being tailored to the input data and the graph's domain of application. As a consequence, the information extracted from graphs with different semantic representations and tailoring can also have different relevance in the specific targeted application [26].

The ability of knowledge graphs to provide the foundation for search and matching tasks can be supported by several data mining techniques. For example, Tiwari et al. (2015) utilize RDF triples to better represent large, heterogeneous, and distributed databases [68]. Their approach relies on RDFs being a machine-understandable format, which can be integrated with human-understandable sentences using natural language processing and generation. Abu-Rasheed et al. (2022) construct a multidimensional knowledge graph from multiple documentation sources in the electronic chip design process [69]. They develop a domain-specific text mining workflow, in which expert knowledge of the domain vocabulary complements and enhances the intelligent model's predictions. The results from the text mining are then used to define the nodes and relations of the knowledge graph for different entity types, where each type is tailored to a certain information source. The knowledge graph is then used as a source and graphical representation for an explainable information retrieval task, which is based on a transitive graph-based search and relation-based reasoning [69].

Although the nature of knowledge graphs depends on describing the information in the form of subject–predicate–object triples, labeled graphs can also be represented in the form of adjacency tensors [70,71]. In this form, machine learning and deep neural networks can be utilized for the analysis process and thus for prediction and decision support.

Contemporary research suggests that not only machine learning and deep learning techniques are enhancing the construction of knowledge graphs, but also the graphs themselves are utilized for enhancing the performance of those techniques. Together, the combination of both approaches is investigated for empowering prediction processes. This is observed in the work of Qin et al. (2018), who suggest an approach for the anomaly detection problem based on a deep learning model that has been supported by a knowledge graph [72].

The previous approaches of semi- and unstructured data analysis and integration complement the other procedures in a Smart Factory to accomplish knowledge integration and enhance processes with decision support. Text mining and semantic technologies play an important role in this integration process. They provide tailored solutions to utilize semi- and unstructured data in addressing the needs and requirements of the factory's domain of application.

4. Conclusions

In this article, the relevant concepts of knowledge integration in a Smart Factory were presented. Different aspects of the Smart Factory environment and its architecture as well as the building blocks such as the Internet-of-Things (IoT), Cyber–Physical Systems (CPS) and Cyber–Physical Production Systems (CPPS) were explained and set into context. Horizontal and vertical knowledge integration in a Smart Factory were presented from the organizational, employee, and technology viewpoint as overarching perspectives of knowledge management in Smart factories.

From the organizational perspective, the horizontal integration of production and transportation processes along the value chain was explained. The vertical perspective shows how the Smart Factory manufacturing environment is changing towards a data-driven, decentralized environment and how this has to be connected and synchronized with other organizational units such as planning, logistics, sales and marketing. A key aspect for both integration perspectives is the data gathered along the value chain and the extracted and related knowledge to support offline and real-time decision-making.

From the employee perspective, the role of employees in Smart Factories was discussed and concepts such as the human–machine interaction (HMI) and the increasing need to interact intelligently with machines, to learn and work jointly were highlighted.

From the technology perspective, this article focuses on the fundamental concepts of Cloud Computing, Fog Computing, and Edge Computing, as well as intelligent methods to support and facilitate the implementation of and knowledge integration within a Smart Factory. In this context, the article addresses the processing variations of structured data in terms of stream analysis and batch analysis. Furthermore, the concept of the Digital Twin is explained and the possibilities for simulation and decision support are shown. In addition to the processing of structured data, the high potential of unstructured data in smart factories is increasingly being recognized but yet not fully unearthed for the Smart Factory of the future. Therefore, this article highlights a range of techniques for unstructured data processing, such as semantic analysis and text mining, with new forms of knowledge representation, such as knowledge graphs and applications such as intelligent documents.

Considering the presented state of the art, Smart Factories are a reality now. Data-driven techniques and the integration of information on all levels of a factory have working solutions in theory and practice. However, the leap towards knowledge integration is scarce and nuanced consideration of information or knowledge has not been made. To do the leap, strategies for a steady contextualization of information across the value chain are needed. Data-driven techniques are foremost quantitatively driven, deriving, and recognizing patterns from large amounts of data. Deriving conclusions from such patterns and stepping up from "being informed" to "knowing how to act", context and an actionable representation of the formed knowledge are needed. To do so, it needs contextualizing semantic techniques on the technical side and the integration of the human factor, as an informed and integrated partner, on the organizational side. Both needs do have working solutions independently, but the joint integration into the highly flexible, complex, digitized and rapidly changing environment of a Smart Factory still needs better solutions. Following the perspective of knowledge integration, as envisioned by this article, will create the fundament for this new vision of integration.

Author Contributions: Conceptualization, J.Z., C.W., M.D. and H.A.-R.; methodology, J.Z., C.W., M.D., H.A.-R. and M.F.; validation, J.Z., C.W., M.D. and H.A.-R.; formal analysis, J.Z., C.W., M.D., H.A.-R. and M.F.; investigation, J.Z., C.W., M.D., H.A.-R. and M.F.; resources, M.D.; writing—original draft preparation, J.Z., C.W., M.D. and H.A.-R.; writing—review and editing, J.Z., C.W., M.D., H.A.-R. and M.F.; visualization, M.D.; supervision, J.Z., M.D. and M.F.; project administration, J.Z. and M.F.; All authors have read and agreed to the published version of the manuscript.

Funding: This research received no external funding.

Acknowledgments: The authors would like to thank their students Hüseyin Turan and Sefa Colak for their support in the pre-study phase of this entry.

Conflicts of Interest: The authors declare no conflict of interest.

Entry Link on the Encyclopedia Platform: https://encyclopedia.pub/14253.

References

1. Tao, F.; Qi, Q.; Liu, A.; Kusiak, A. Data-driven smart manufacturing. *J. Manuf. Syst.* **2018**, *48*, 157–169. [CrossRef]
2. Alcácer, V.; Machado, V. Scanning the Industry 4.0: A Literature Review on Technologies for Manufacturing Systems. *Eng. Sci. Technol. Int. J.* **2019**, *22*, 899–919. [CrossRef]
3. Oztemel, E.; Gursev, S. Literature review of Industry 4.0 and related technologies. *J. Intell. Manuf.* **2020**, *31*, 127–182. [CrossRef]
4. Zhong, R.Y.; Xu, X.; Klotz, E.; Newman, S.T. Intelligent Manufacturing in the Context of Industry 4.0: A Review. *Engineering* **2017**, *3*, 616–630. [CrossRef]
5. Neumann, W.P.; Winkelhaus, S.; Grosse, E.H.; Glock, C.H. Industry 4.0 and the human factor—A systems framework and analysis methodology for successful development. *Int. J. Prod. Econ.* **2021**, *233*, 107992. [CrossRef]
6. Roth, A. (Ed.) *Einführung und Umsetzung von Industrie 4.0: Grundlagen, Vorgehensmodell und Use Cases aus der Praxis*; Springer Gabler: Berlin, Germany, 2016; ISBN 978-3-662-48504-0.
7. Shi, Z.; Xie, Y.; Xue, W.; Chen, Y.; Fu, L.; Xu, X. Smart factory in Industry 4.0. *Syst. Res. Behav. Sci.* **2020**, *37*, 607–617. [CrossRef]
8. Strozzi, F.; Colicchia, C.; Creazza, A.; Noè, C. Literature review on the 'Smart Factory' concept using bibliometric tools. *Int. J. Prod. Res.* **2017**, *55*, 6572–6591. [CrossRef]
9. Tao, F.; Qi, Q.; Wang, L.; Nee, A. Digital Twins and Cyber–Physical Systems toward Smart Manufacturing and Industry 4.0: Correlation and Comparison. *Engineering* **2019**, *5*, 653–661. [CrossRef]
10. Frey-Luxemburger, M. (Ed.) *Wissensmanagement-Grundlagen und Praktische Anwendung: Eine Einführung in das IT-Gestützte Management der Ressource Wissen*, 2nd ed.; IT im Unternehmen; Springer Vieweg: Wiesbaden, Germany, 2014; ISBN 978-3-658-04752-8.
11. Capestro, M.; Kinkel, S. Industry 4.0 and Knowledge Management: A Review of Empirical Studies. In *Knowledge Management and Industry 4.0: New Paradigms for Value Creation*; Bettiol, M., Di Maria, E., Micelli, S., Eds.; Knowledge Management and Organizational Learning; Springer International Publishing: Cham, Switzerland, 2020; pp. 19–52. ISBN 978-3-030-43589-9.
12. Brauckmann, O. *Smart Production: Wertschöpfung durch Geschäftsmodelle*; Springer Vieweg: Berlin, Germany, 2015; ISBN 978-3-662-45301-8.
13. Bettiol, M.; Maria, E.D.; Micelli, S. (Eds.) *Knowledge Management and Industry 4.0: New Paradigms for Value Creation*; Knowledge Management and Organizational Learning; Springer International Publishing: Cham, Switzerland, 2020; ISBN 978-3-030-43588-2.
14. Ansari, F.; Erol, S.; Sihn, W. Rethinking Human-Machine Learning in Industry 4.0: How Does the Paradigm Shift Treat the Role of Human Learning? *Procedia Manuf.* **2018**, *23*, 117–122. [CrossRef]
15. Seeber, I.; Bittner, E.; Briggs, R.O.; de Vreede, T.; de Vreede, G.-J.; Elkins, A.; Maier, R.; Merz, A.B.; Oeste-Reiß, S.; Randrup, N.; et al. Machines as teammates: A research agenda on AI in team collaboration. *Inf. Manag.* **2020**, *57*, 103174. [CrossRef]
16. Industrial Internet Consortium (IIC). The Industrial Internet of Things Volume T3: Analytics Framework. Available online: https://iiconsortium.org/pdf/IIC_Industrial_Analytics_Framework_Oct_2017.pdf (accessed on 6 August 2021).
17. Plattform-i4.0 Referenzarchitekturmodell (RAMI) 4.0 (Reference Architecture Model). Available online: https://www.plattform-i40.de/PI40/Redaktion/DE/Infografiken/referenzarchitekturmodell-4-0.html (accessed on 6 August 2021).
18. Bagheri, B.; Yang, S.; Kao, H.-A.; Lee, J. Cyber-physical Systems Architecture for Self-Aware Machines in Industry 4.0 Environment. *IFAC-PapersOnLine* **2015**, *48*, 1622–1627. [CrossRef]

19. Karnouskos, S.; Ribeiro, L.; Leitão, P.; Luder, A.; Vogel-Heuser, B. Key Directions for Industrial Agent Based Cyber-Physical Production Systems. In Proceedings of the 2019 IEEE International Conference on Industrial Cyber Physical Systems (ICPS), Taipei, Taiwan, 6–9 May 2019; pp. 17–22.
20. Dornhöfer, M.; Sack, S.; Zenkert, J.; Fathi, M. Simulation of Smart Factory Processes Applying Multi-Agent-Systems—A Knowledge Management Perspective. *J. Manuf. Mater. Process.* **2020**, *4*, 89. [CrossRef]
21. Soic, R.; Vukovic, M.; Skocir, P.; Jezic, G. Context-Aware Service Orchestration in Smart Environments. In *Agents and Multi-agent Systems: Technologies and Applications 2019*; Jezic, G., Chen-Burger, Y.-H.J., Kusek, M., Šperka, R., Howlett, R.J., Jain, L.C., Eds.; Springer: Singapore, 2020; pp. 35–45.
22. Fei, X.; Shah, N.; Verba, N.; Chao, K.-M.; Sanchez-Anguix, V.; Lewandowski, J.; James, A.; Usman, Z. CPS data streams analytics based on machine learning for Cloud and Fog Computing: A survey. *Futur. Gener. Comput. Syst.* **2019**, *90*, 435–450. [CrossRef]
23. Gorodetsky, V.I.; Kozhevnikov, S.S.; Novichkov, D.; Skobelev, P.O. The Framework for Designing Autonomous Cyber-Physical Multi-Agent Systems for Adaptive Resource Management. In *Industrial Applications of Holonic and Multi-Agent Systems*; Mařík, V., Kadera, P., Rzevski, G., Zoitl, A., Anderst-Kotsis, G., Tjoa, A.M., Khalil, I., Eds.; Springer International Publishing: Cham, Switzerland, 2019; pp. 52–64.
24. Industrial Internet Consortium (IIC). The Industrial Internet of Things Volume G1: Reference Architecture. Available online: https://www.iiconsortium.org/pdf/IIRA-v1.9.pdf (accessed on 6 August 2021).
25. Bullinger, H.-J.; Wörner, K.; Prieto, J. Wissensmanagement—Modelle und Strategien für die Praxis. In *Wissensmanagement: Schritte zum Intelligenten Unternehmen*; Bürgel, H.D., Ed.; Edition Alcatel SEL Stiftung; Springer: Berlin/Heidelberg, Germany, 1998; pp. 21–39. ISBN 978-3-642-71995-0.
26. Zenkert, J.; Weber, C.; Klahold, A.; Fathi, M.; Hahn, K. Knowledge-Based Production Documentation Analysis: An Integrat-ed Text Mining Architecture. In Proceedings of the 2018 IEEE 61st International Midwest Symposium on Circuits and Sys-tems (MWSCAS), Windsor, ON, Canada, 5–8 August 2018; pp. 717–720.
27. Plattform-i4.0 Infographic about Hierarchy in Industry 3.0. Available online: https://www.plattform-i40.de/PI40/Redaktion/DE/Infografiken/hierarchie-in-der-industrie-3-0.html (accessed on 6 August 2021).
28. Plattform-i4.0 Infographic about Hierarchy in Industry 4.0. Available online: https://www.plattform-i40.de/PI40/Redaktion/DE/Infografiken/hierarchie-in-der-industrie-4-0.html (accessed on 6 August 2021).
29. Vogel-Heuser, B.; Bauernhansl, T.; Hompel, M. (Eds.) *Handbuch Industrie 4.0 Bd.4: Allgemeine Grundlagen*, 2nd ed.; VDI Springer Reference; Springer Vieweg: Berlin, Germany, 2017; ISBN 978-3-662-53253-9.
30. North, K.; Maier, R.; Haas, O. Value Creation in the Digitally Enabled Knowledge Economy. In *Progress in IS*; Springer International Publishing: Berlin, Germany, 2018; pp. 1–29.
31. Gillespie, R.B.; Colgate, J.E.; Peshkin, M.A. A General Framework for Cobot Control. *IEEE Trans. Robot. Autom.* **2001**, *17*, 391–401. [CrossRef]
32. Hu, P.; Dhelim, S.; Ning, H.; Qiu, T. Survey on fog computing: Architecture, key technologies, applications and open issues. *J. Netw. Comput. Appl.* **2017**, *98*, 27–42. [CrossRef]
33. OpenFog Consortium OpenFog Reference Architecture for Fog Computing. Available online: https://www.iiconsortium.org/pdf/OpenFog_Reference_Architecture_2_09_17.pdf (accessed on 6 August 2021).
34. Sittón-Candanedo, I.; Alonso, R.S.; Rodríguez-González, S.; García Coria, J.A.; De La Prieta, F. Edge Computing Architec-tures in Industry 4.0: A General Survey and Comparison. In *International Workshop on Soft Computing Models in Industrial and Environmental Applications*; Martínez Álvarez, F., Troncoso Lora, A., Sáez Muñoz, J.A., Quintián, H., Corchado, E., Eds.; Springer International Publishing: Cham, Switzerland, 2020; pp. 121–131.
35. Industrial Internet Consortium Edge Computing Task Group Introduction to Edge Computing. Available online: https://hub.iiconsortium.org/intro-edge-computing (accessed on 6 August 2021).
36. Lim, K.Y.H.; Zheng, P.; Chen, C.-H. A state-of-the-art survey of Digital Twin: Techniques, engineering product lifecycle management and business innovation perspectives. *J. Intell. Manuf.* **2020**, *31*, 1313–1337. [CrossRef]
37. Papadimitriou, S.; Sun, J.; Faloutsos, C.; Yu, P.S. Dimensionality Reduction and Filtering on Time Series Sensor Streams. In *Managing and Mining Sensor Data*; Springer: Boston, MA, USA, 2012; pp. 103–141.
38. Enders, C.K. *Applied Missing Data Analysis*, 1st ed.; Guilford Publications: New York, NY, USA, 2010; ISBN 978-1-60623-639-0.
39. Aggarwal, C.C. Mining Sensor Data Streams. In *Managing and Mining Sensor Data*; Springer: Boston, MA, USA, 2012; pp. 143–171.
40. Turaga, D.S.; Van Der Schaar, M. Distributed Online Learning and Stream Processing for a Smarter Planet. In *Fog for 5G and IoT*; Chiang, M., Balasubramanian, B., Bonomi, F., Eds.; John Wiley & Sons, Inc.: Hoboken, NJ, USA, 2017; pp. 234–260. ISBN 978-1-119-18720-2.
41. Zhang, Q.; Yang, L.-T.; Chen, Z.; Li, P. A survey on deep learning for big data. *Inf. Fusion* **2018**, *42*, 146–157. [CrossRef]
42. Kritzinger, W.; Karner, M.; Traar, G.; Henjes, J.; Sihn, W. Digital Twin in manufacturing: A categorical literature review and classification. *IFAC-PapersOnLine* **2018**, *51*, 1016–1022. [CrossRef]
43. Harrison, R.; Vera, D.A.; Ahmad, B. A Connective Framework to Support the Lifecycle of Cyber-Physical Production Systems. *Proc. IEEE* **2021**, *109*, 568–581. [CrossRef]
44. Jacoby, M.; Usländer, T. Digital Twin and Internet of Things—Current Standards Landscape. *Appl. Sci.* **2020**, *10*, 6519. [CrossRef]

45. Industrial Internet Consortium and Plattform Industrie 4.0 Digital Twin and Asset Administration Shell Concepts and Application in the Industrial Internet and Industrie 4.0: An Industrial Internet Consortium and Plattform Industrie 4.0 Joint Whitepaper. Available online: https://www.iiconsortium.org/pdf/Digital-Twin-and-Asset-Administration-Shell-Concepts-and-Application-Joint-Whitepaper.pdf (accessed on 6 August 2021).
46. Schluse, M.; Priggemeyer, M.; Atorf, L.; Rossmann, J. Experimentable Digital Twins—Streamlining Simulation-Based Systems Engineering for Industry 4.0. *IEEE Trans. Ind. Inform.* **2018**, *14*, 1722–1731. [CrossRef]
47. Martinez, E.; Ponce, P.; Macias, I.; Molina, A. Automation Pyramid as Constructor for a Complete Digital Twin, Case Study: A Didactic Manufacturing System. *Sensors* **2021**, *21*, 4656. [CrossRef]
48. Hänel, A.; Seidel, A.; Frieß, U.; Teicher, U.; Wiemer, H.; Wang, D.; Wenkler, E.; Penter, L.; Hellmich, A.; Ihlenfeldt, S. Digital Twins for High-Tech Machining Applications—A Model-Based Analytics-Ready Approach. *J. Manuf. Mater. Process.* **2021**, *5*, 80. [CrossRef]
49. Autiosalo, J.; Ala-Laurinaho, R.; Mattila, J.; Valtonen, M.; Peltoranta, V.; Tammi, K. Towards Integrated Digital Twins for Industrial Products: Case Study on an Overhead Crane. *Appl. Sci.* **2021**, *11*, 683. [CrossRef]
50. Rathore, M.M.; Shah, S.A.; Shukla, D.; Bentafat, E.; Bakiras, S. The Role of AI, Machine Learning, and Big Data in Digital Twinning: A Systematic Literature Review, Challenges, and Opportunities. *IEEE Access* **2021**, *9*, 32030–32052. [CrossRef]
51. Allahyari, M.; Pouriyeh, S.; Assefi, M.; Safaei, S.; Trippe, E.D.; Gutierrez, J.B.; Kochut, K. A Brief Survey of Text Mining: Clas-sification, Clustering and Extraction Techniques. *arXiv* **2017**, arXiv:1707.02919.
52. Shi, F.; Chen, L.; Han, J.; Childs, P. A Data-Driven Text Mining and Semantic Network Analysis for Design Information Retrieval. *J. Mech. Des.* **2017**, *139*, 111402. [CrossRef]
53. Nadeau, D.; Sekine, S. A survey of named entity recognition and classification. *Lingvisticae Investig.* **2007**, *30*, 3–26. [CrossRef]
54. Pawar, S.; Palshikar, G.K.; Bhattacharyya, P. Relation Extraction: A Survey. *arXiv* **2017**, arXiv:1712.05191.
55. Abu Rasheed, H.; Weber, C.; Zenkert, J.; Czerner, P.; Krumm, R.; Fathi, M. A Text Extraction-Based Smart Knowledge Graph Composition for Integrating Lessons Learned during the Microchip Design. In *Advances in Intelligent Systems and Computing*; Springer International Publishing: Cham, Switzerland, 2020; pp. 594–610.
56. Jiang, H.; Kwong, C.K.; Yung, K.L. Predicting Future Importance of Product Features Based on Online Customer Reviews. *J. Mech. Des.* **2017**, *139*, 111413. [CrossRef]
57. Grieco, A.; Pacella, M.; Blaco, M. On the Application of Text Clustering in Engineering Change Process. *Procedia CIRP* **2017**, *62*, 187–192. [CrossRef]
58. Klahold, A.; Fathi, M. *Computer Aided Writing*; Springer International Publishing: Berlin, Germany, 2020.
59. Zenkert, J.; Klahold, A.; Fathi, M. Knowledge Discovery in Multidimensional Knowledge Representation Framework: An Integrative Approach for the Visualization of Text Analytics Results. *Iran J. Comput. Sci.* **2018**, *1*, 199–216. [CrossRef]
60. Yahya, M.; Breslin, J.; Ali, M. Semantic Web and Knowledge Graphs for Industry 4.0. *Appl. Sci.* **2021**, *11*, 5110. [CrossRef]
61. Beden, S.; Cao, Q.; Beckmann, A. Semantic Asset Administration Shells in Industry 4.0: A Survey. In Proceedings of the 2021 4th IEEE International Conference on Industrial Cyber-Physical Systems (ICPS), Victoria, BC, Canada, 10–12 May 2021; pp. 31–38.
62. Heling, L.; Acosta, M.; Maleshkova, M.; Sure-Vetter, Y. Querying Large Knowledge Graphs over Triple Pattern Fragments: An Empirical Study. In *The Semantic Web-ISWC 2018*; Vrandečić, D., Bontcheva, K., Suárez-Figueroa, M.C., Presutti, V., Ce-lino, I., Sabou, M., Kaffee, L.-A., Simperl, E., Eds.; Springer International Publishing: Cham, Switzerland, 2018; Volume 11137, pp. 86–102. ISBN 978-3-030-00667-9.
63. Jayaram, N.; Khan, A.; Li, C.; Yan, X.; Elmasri, R. Querying Knowledge Graphs by Example Entity Tuples. *IEEE Trans. Knowl. Data Eng.* **2015**, *27*, 2797–2811. [CrossRef]
64. Dietz, L.; Kotov, A.; Meij, E. Utilizing Knowledge Graphs for Text-Centric Information Retrieval. In Proceedings of the 41st International ACM SIGIR Conference on Research & Development in Information Retrieval, Ann Arbor, MI, USA, 8–12 July 2018; pp. 1387–1390.
65. Li, H.; Xiong, C.; Callan, J. Natural Language Supported Relation Matching for Question Answering with Knowledge Graphs. In *The First Workshop on Knowledge Graphs and Semantics for Text Retrieval and Analysis (KG4IR 2017)*; CEUR: Tokyo, Japan, 2017.
66. Shekarpour, S.; Marx, E.; Auer, S.; Sheth, A. RQUERY: Rewriting Natural Language Queries on Knowledge Graphs to Alle-viate the Vocabulary Mismatch Problem. In Proceedings of the Thirty-First AAAI Conference on Artificial Intelligence, San Francisco, CA, USA, 4–9 February 2017.
67. Hu, S.; Zou, L.; Yu, J.X.; Wang, H.; Zhao, D. Answering Natural Language Questions by Subgraph Matching over Knowledge Graphs. *IEEE Trans. Knowl. Data Eng.* **2018**, *30*, 824–837. [CrossRef]
68. Tiwari, A.; Rajesh, K.; Srujana, P. Semantically Enriched Knowledge Extraction With Data Mining. *Int. J. Comput. Appl. Technol. Res.* **2015**, *4*, 7–10. [CrossRef]
69. Abu-Rasheed, H.; Weber, C.; Zenkert, J.; Krumm, R.; Fathi, M. Explainable Graph-Based Search for Lessons-Learned Docu-ments in the Semiconductor Industry. In *Intelligent Computing*; Arai, K., Ed.; Springer International Publishing: Cham, Switzerland, 2022; pp. 1097–1106.
70. Nickel, M.; Murphy, K.; Tresp, V.; Gabrilovich, E. A Review of Relational Machine Learning for Knowledge Graphs. *Proc. IEEE* **2015**, *104*, 11–33. [CrossRef]

71. Liu, Z.; Han, X. Deep Learning in Knowledge Graph. In *Deep Learning in Natural Language Processing*; Deng, L., Liu, Y., Eds.; Springer: Singapore, 2018; pp. 117–145. ISBN 978-981-10-5208-8.
72. Qin, Z.; Cen, C.; Jie, W.; Gee, T.S.; Chandrasekhar, V.R.; Peng, Z.; Zeng, Z. Knowledge-Graph Based Multi-Target Deep-Learning Models for Train Anomaly Detection. In Proceedings of the 2018 International Conference on Intelligent Rail Transportation (ICIRT), Singapore, 12–14 December 2018; pp. 1–5.

encyclopedia

MDPI

Entry

Automobile Tires' High-Carbon Steel Wire

Marina Polyakova [1,*] and Alexey Stolyarov [2]

1 Department of Material Processing, Metallurgy, Mechanical Engineering and Materials Processing Institute, Nosov Magnitogorsk State Technical University, 455000 Magnitogorsk, Russia
2 OJSC Magnitogorsk Hardware and Sizing Plant "MMK-METIZ", 455002 Magnitogorsk, Russia; Stolyarov.AY@mmk-metiz.ru
* Correspondence: m.polyakova@magtu.ru; Tel.: +7-3519-29-84-81

Definition: It is a well-known fact that to manufacture an automobile tire more than 200 different materials are used, including high-carbon steel wire. In order to withstand the affecting forces, the tire tread is reinforced with steel wire or other products such as ropes or strands. These ropes are called steel cord. Steel cord can be of different constructions. To ensure a good adhesive bond between the rubber of the tire and the steel cord, the cord is either brass-plated or bronzed. The reason brass or bronze is used is because copper, which is a part of these alloys, makes a high-strength chemical composition with sulfur in rubber. For steel cord, the high carbon steel is usually used at 0.70–0.95% C. This amount of carbon ensures the high strength of the steel cord. This kind of high-quality, unalloyed steel has a pearlitic structure which is designed for multi-pass drawing. To ensure the specified technical characteristics, modern metal reinforcing materials for automobile tires, metal cord and bead wire, must withstand, first of all, a high breaking load with a minimum running meter weight. At present, reinforcing materials of the strength range 2800–3200 MPa are increasingly used, the manufacture of which requires high-strength wire. The production of such wire requires the use of a workpiece with high carbon content, changing the drawing regimes, patenting, and other operations. At the same time, it is necessary to achieve a reduction in the cost of wire manufacturing. In this context, the development and implementation of competitive processes for the manufacture of high-quality, high-strength wire as a reinforcing material for automobile tires is an urgent task.

Keywords: high carbon steel wire; reinforcing material; automobile tire; steel cord; bead wire; drawing; patenting; brass-plated wire; laying

Citation: Polyakova, M.; Stolyarov, A. Automobile Tires' High-Carbon Steel Wire. *Encyclopedia* **2021**, *1*, 859–870. https://doi.org/10.3390/encyclopedia1030066

Academic Editors: Raffaele Barretta, Ramesh Agarwal, Krzysztof Kamil Żur and Giuseppe Ruta

Received: 18 July 2021
Accepted: 20 August 2021
Published: 24 August 2021

1. Introduction

Over the past, relatively short, time, the range of reinforcing materials for car tires has undergone significant change. Firstly, this can be explained by the increased requirements for automobile tires, which are now more stringent for mileage, weight, imbalance (power non-uniformity), and so on (Figure 1) [1]. To ensure the elevated technical characteristics of tires, the modern metal cord and bead wire have to withstand a high breaking load with minimum mass of a running meter (linear density), have a sufficient level of bond strength with rubber, and have an increased resistance to fatigue failure under applied loads. To date, special attention is paid to such indicators as the level of residual torsion, straightness, and deflection arrow, which directly affect the manufacturability of the rubber cord sheets' (bead rings) technological processing on modern rubber lines.

The idea of increasing the strength of reinforcing materials for automobile tires and decreasing the volume weight was justified in 1979 through the experience of such leading manufacturers of metal cord and bead wire as Bekaert (Belgium), Goodyear, Firestone (USA), Michelin (France), Bridgestone (Japan), and Pirelli» (Italy) [2]. But the more active process of the substitution of normal-strength reinforced materials for high-strength materials started at the beginning of the 1990s.

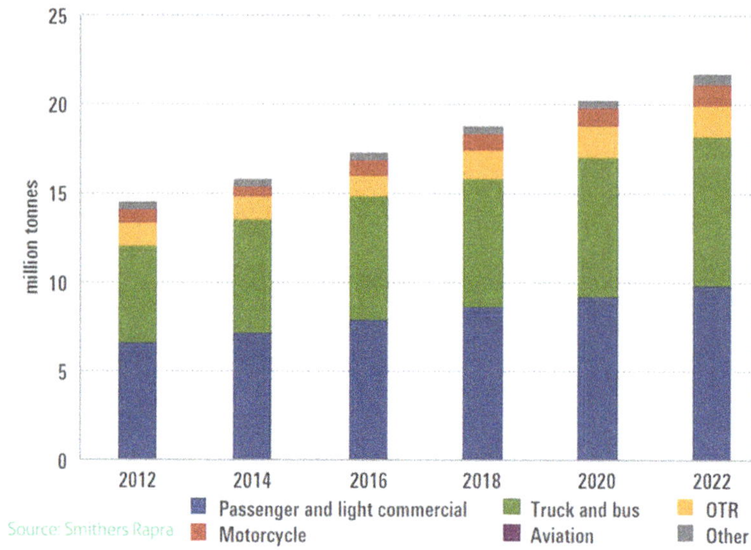

Figure 1. Manufacturing of tires for different applications. Reprinted from ref. [1].

While designing new constructions of high-strength steel cord and wires for reinforcing the bead rings of tires, different companies have developed technologies for production wire for high-strength reinforcing materials. The most progressive technologies of manufacturing bead wire were produced in Japan.

At the moment, reinforcing materials of high strength are increasingly used instead of materials with normal tensile (NT) 2400–2800 MPa. They are divided into the following groups: high tensile materials (HT) 2800–3200 MPa and super tensile (ST) materials 3200–3500 MPa. Furthermore, the increase of tensile strength promotes an increased endurance strength of brass-plated wire for metal cord, especially compact beam structures for reinforcing the car tire carcass (Figure 2) [3].

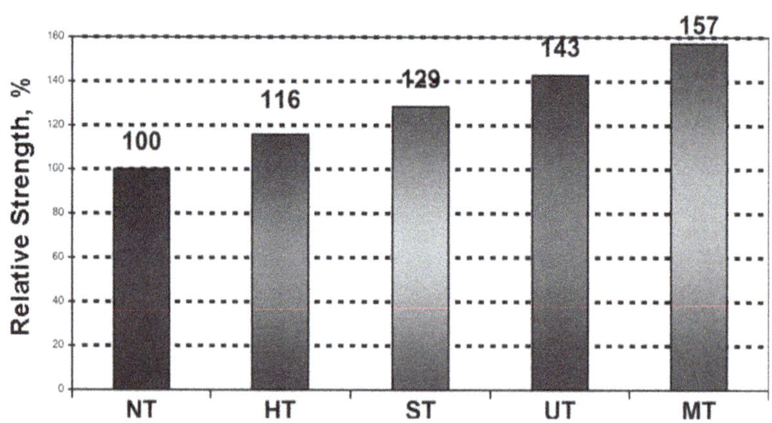

Figure 2. Change of wire strength depending on its group. Reprinted from ref. [3].

Figure 3 shows the level of tensile strength of materials which are used for metal cord manufacturing [4].

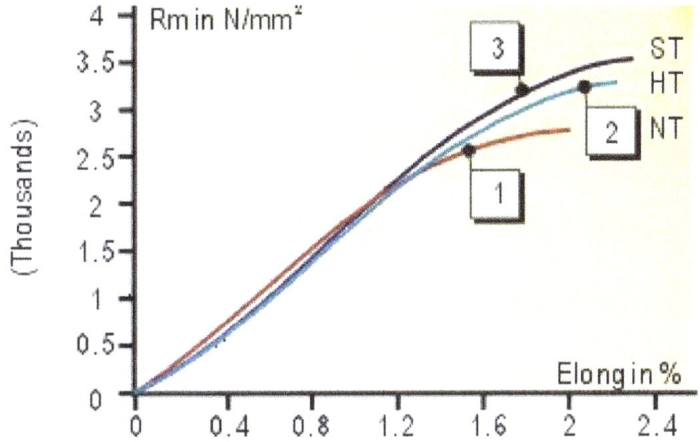

Figure 3. Tensile test data for 0.175 mm diameter filaments for (1) normal products, (2) high tensile grade, and (3) an experimental super-high tensile grade. Reprinted from ref. [4].

At present time the tendency to increase the tensile strength of steel cord is observed as shown in Figure 4 [5–8].

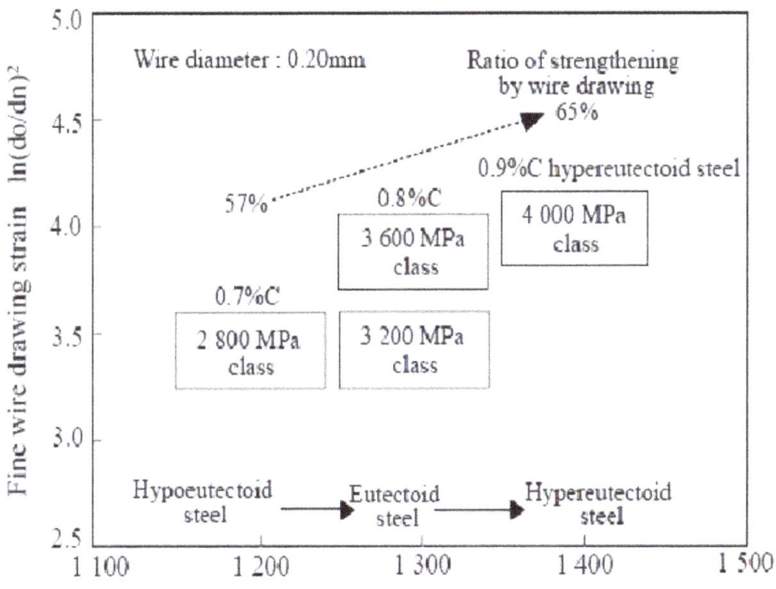

Figure 4. Trend of high tensile strength of steel cord. Reprinted from ref. [7].

The tensile strength of metal cord with a diameter of 0.20 mm was 2800 MPa in the 1970s, 3300 MPa in the 1980s, and reached a high strength of 3600 MPa in the early 1990s. The increase of speed and the increase in highway transport demanded a raise in the level of tensile strength to higher values [6,7].

In addition to high tensile strength, the wire for modern reinforcing materials must have a high range of ductile (fatigue) properties. As for metal cord, this condition is necessary to ensure the processability of the double twisting method of high-speed laying. At the level of the HT group (approximately 3000–3200 MPa) thin brass-plated wire with 0.2–0.35 mm in diameter must withstand a certain amount of forward and backward twists.

Otherwise, the quality of the finished steel cord (fatigue endurance) and the productivity of the lay process are sharply reduced.

The aim of this paper is to describe the peculiarities of the manufacturing process of the high-carbon steel wire which is used as the reinforcing material for automobile tires. A general description of the technological process is given in Section 2, which also contains information about special aspects of every technological operation of the manufacturing process. This overview can help the reader to learn about those technological techniques which are necessary in order to produce high-carbon steel wire with the desired exploitation properties. The main tendencies for the improvement of high-carbon steel manufacturing process are denoted in the conclusion.

2. Structure, Role, and Demands of the Technological Process of High-Carbon Steel Wire Manufacturing

The main direction of the perspective technological design and development of new technological processes in metallurgy is the creation of such technological systems which are based on low-operational, unmanned, and waste-free technology providing a multiple increase in labor productivity and a significant improvement in product quality and other indicators.

The technology for the manufacture of high-strength wire for automobile tires should be generally observed. For example, in Japan, there are conceptually two main directions to achieve the required level of steel-wire strength: strengthening in patenting and strengthening in drawing. Moreover, these two directions are each esteemed comprehensively in terms of the regularity of the pearlite structure refinement in the wire [9–11].

At the input stage of the technological process of manufacturing wire for reinforcing materials for automobile tires, there are main and auxiliary materials (high-carbon steel wire rod, copper and zinc anodes, etc.), and at the output of the process there is the cold-deformed (brass plated, bronzed) wire.

The structure of the actual technological process of manufacturing wire for the reinforcement of materials for automobile tires consists of the following main interrelated subprocesses:

- surface preparation and "rough-medium" drawing of sorbitized wire rod;
- patenting of cold deformed workpiece. Patenting can be a transitory operation used for the formation of the final properties of the wire;
- brass plating to ensure the adhesion of the metal cord to rubber;
- finishing drawing;
- laying of the metal cord;
- annealing, processing in alternative deformation (if necessary for bead rings of tires).

The technological scheme can be presented by means of blocks. Each block contains information about the name of the technological operation. The technological scheme for brass-plated, high-carbon steel wire with high strength for steel cord is presented in Figure 5, as are the range of diameters of the processed wire. For thin high-strength brass plated wire with 0.85% C the diameters of patented workpiece were chosen as shown in the blocks. Based on the experimental results, it was proved [12] that the intermediate operation of patenting was obligatory in the manufacturing process because it reduced wire breakage in drawing.

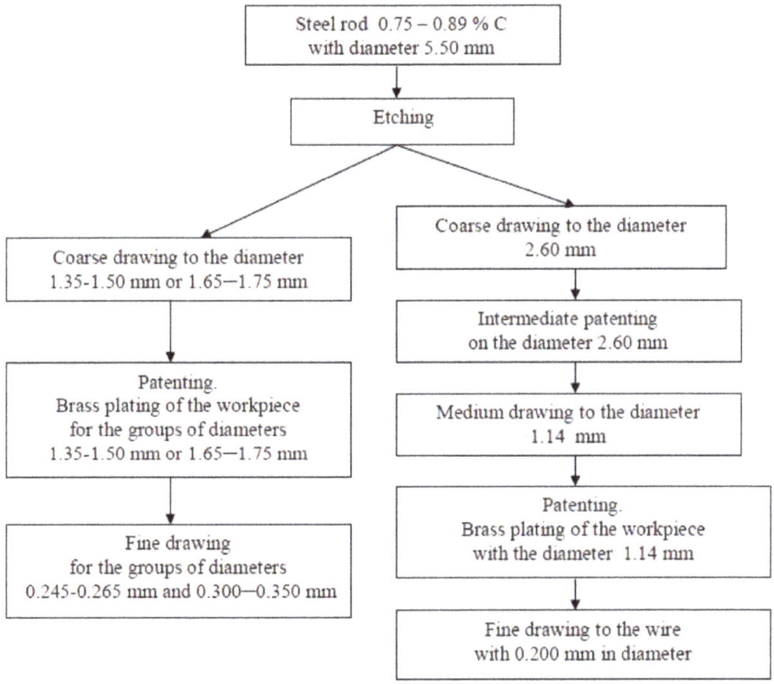

Figure 5. Technological scheme for the manufacture of brass-plated, high-carbon steel wire with high strength for steel cord.

At the present time high-strength, bronzed-steel wire with a diameter of 1.60 mm is highly requested for all-steel automobile tires. The current way to get the desired level of mechanical properties is to use a special kind of heat treatment in an air-fluidized bed with alternative bending as final operations. The application of these kinds of processes guarantees a ratio of yield strength to tensile strength of 75–85%. The technological scheme for bronzed, high-carbon steel wire is presented in Figure 6.

The implementation of these technological schemes (see Figures 5 and 6) at the industrial scale makes it possible to improve the competitiveness of the manufactured high-strength steel wire for cord [12].

2.1. Steel Rod for High-Strength Wire Manufacturing

The choice of steel rod for the manufacture of high-strength wire has a significant role in the technological process of the production of reinforcing materials for automobile tires [13]. One of the basic factors which affects the technological effectiveness of metal cord manufacturing, as well as its technical and exploitation characteristics, is the quality of the high carbon steel rod which is used as a reinforcing material in automobile tires. Demands on the steel rod for metal cord and bead wire are formulated, first of all, taking into consideration further regimes of its processing and the functions of the final product.

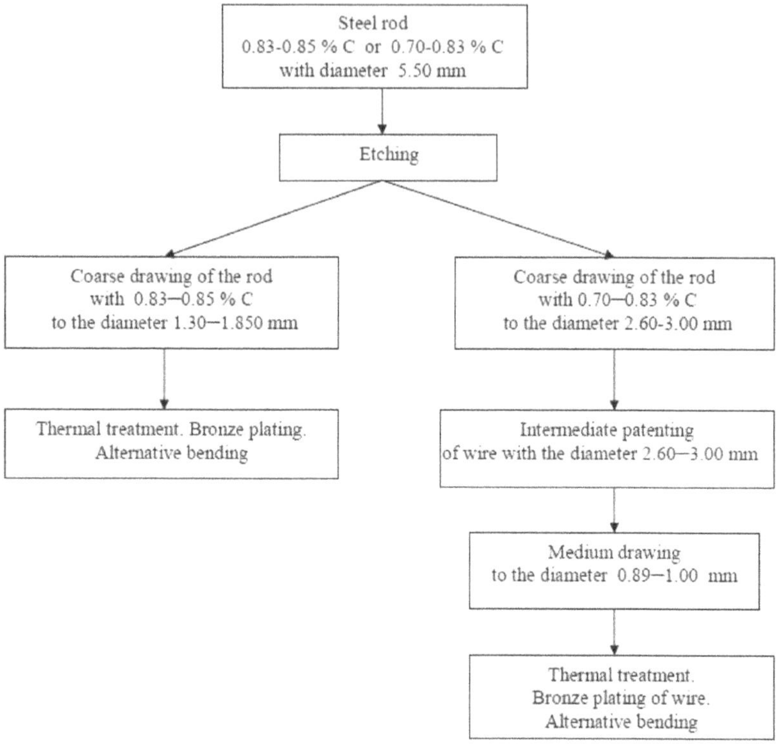

Figure 6. Technological scheme for the manufacture of bronzed, high-carbon steel wire with high strength.

To manufacture high-strength and ultra-high-strength metal cord, steel rod made from high-carbon steel with 0.70–0.95% C and 5.5 mm in diameter is used. The pearlitic microstructure is typical for steel with such an amount of carbon and consists of ferrite-carbide mixture (Figure 7).

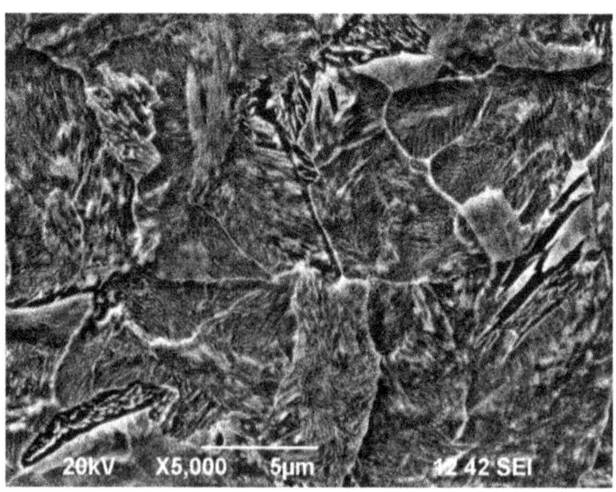

Figure 7. Microstructure of steel rod 0.70% C for metal cord manufacturing before drawing.

As shown in Figure 7, the microstructure consists of troostite with small amount of bainite, and ferrite which is located as a net around pearlite colonies

Special demands are exhibited to the chemical composition of steel, the quantity of impurities and imperfections in steel, and the macro and microstructure. To ensure the required level of properties such companies as Cobe Steel [14], Nippon Steel [15], Kawasaki Steel (Japan) [16], THYSSEN (Germany) [17], and others alloy their steel with chromium, copper, manganese, cobalt, etc.

It is stated in many papers [18–24] that chemical composition, pollution of steel by non-metallic inclusions, results of liquation processes, presence of scale on the surface of rod and its decarburization, and the peculiarities of macro and microstructure have a great influence on the processability of steel rod in the following operations of technology: rough drawing, patenting, drawing of brass-plated wire, and laying, as well as the quality of the final product.

2.2. Role of Drawing in the Technological Process of High-Strength Wire Manufacturing

In drawing, the cross section of a long rod or wire is reduced when it is pulled through a die. Tensile strain and compression strain are obvious in drawing. The major processing variables in drawing are reduction in cross-sectional area, die angle, friction along the die-workpiece interface, and drawing speed. Drawing is usually performed as a cold working operation. Drawing speeds are as high as 50 m/s for steel cord. In drawing, reductions in the cross-sectional area per pass range up to about 45%. Usually, the smaller the initial cross section, the smaller the reduction per pass. Fine wires for steel cord usually are drawn at 15 to 25% reduction per pass. In order to avoid the breakage of wire in high-speed drawing, the emulsion coolant "oil in water" is used.

In metal cord manufacturing it is impossible to produce high carbon steel wire with a diameter less than 1 mm directly from the rod because of the large amount of total reduction in drawing [25]. For this reason, the technological process «Rod—Wire for Metal Cord» is divided into several subprocesses and can be presented as the combination of basic operations of drawing in monolithic dies and thermal treatment (patenting).

Conditionally it can be determined as two variants:

1. rod—workpiece for the final wire (rough process stage)
2. brass-plated wire after patenting—thin bras-plated wire (final process stage).

As a matter of fact, the rough process stage «Rod—Workpiece for the Final Wire» is the shape-generating stage which ensures the necessary diameter of the workpiece for the further drawing of the rod so as to manufacture the final wire with the definite diameter. To lower costs for the rough process stage, it is necessary, on the one hand, to reduce the quantity of thermal treatments and, on the other, it is necessary to keep in mind that with the increase of the total deformation degree the probability of breakage of the wire in drawing also raises. In particular, cracks, tears, and other kinds of breakage are dangerous because these kinds of defects do not disappear during further heat treatment and decrease the wire quality as well as the metal cord laid from this wire.

In the manufacture of bead wire with a diameter of between 1.30–1.85 mm at present time both physical and chemical properties of the final product are dependent on the process stage «Rod—Workpiece for the Final Wire». For this reason, special attention is paid to the regimes of coarse drawing in the technological process of bead bronzed wire.

The role of the final process stage (fine drawing) in the manufacture of metal cord, besides shaping, is of ensuring the strength and ductile properties of the final wire. This is why the diameter of the workpiece for the final wire is chosen by taking into consideration the necessary degree of total deformation. The key points in this case are the steel composition (carbon content), the degree of total deformation (determination of the diameter of patented brass-plated workpiece), and the regimes of wet drawing. In drawing, pearlite colonies of the processed high-carbon steel wire elongate towards the drawing direction as shown in Figure 8. This kind of microstructure is characterized by a disposition of grains along the force applied.

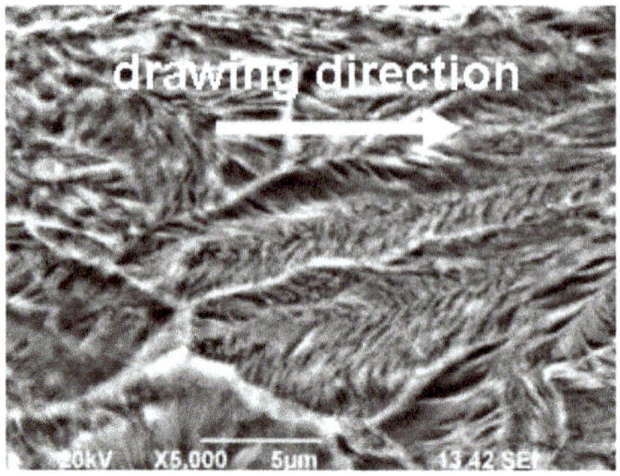

Figure 8. Microstructure of high-carbon steel wire 0.70% C after drawing.

As compared with coarse drawing, the fine drawing of brass-plated wire is characterized by tough friction conditions and a higher drawing rate. The approaches used for the designs associated with drawing of high-strength wire for metal cord are presented in [26–28]. Quality control in drawing is based on the distribution of hardness across the wire including fine brass-plated wire. The difference in hardness between outer and internal areas of wire should not be more than 7% [26]. This is why, besides the magnitude of reduction, the control factor to ensure the properties of cold-drawn wire is the angle of the drawing tool.

In the drawing of high-carbon steel wire, much attention is paid to the negative affect of the heat deformation warming-up. It is considered that the temperature of the wire on the finishing drum of the drawing mill should not be more than 150 °C. The negative affect of the temperature is proved during tribological analysis of the contact system «Brass-Lubricant-Drawing Tool» which was carried out by specialists of «Michelin» (France) [29].

For the coarse drawing of wire, the direct-flow drawing mills with intensive system for cooling drums and drawing tools are used. Well-known drawing mills are produced by «GCR EURODRAW SPA» (Italy), «MARIO FRIGERIO SPA COMPANY» (Italy), «ERNST KOCH GMBH @ CO.» (Germany), «SWARAJ TECHNOCRAD PVT. LTD.» (India). Special attention is paid to the quality of the surface of motoblocs.

For wet drawing of brass-plated wire, the drawing mills of higher deformation ratio produced by «M + E Macchine + Engineering S.p.a», «VVM», «Team Meccanica», and «Samp Steel» (Italy) are used. Drawing mills are equipped with a high pressure emulsion supply system [30] and cooling for the drawing tools, fine dies, and drawing drums [31]. The maximum drawing rate reaches 20–25 m/s. The drawing emulsion is fed to the group of mills through a closed loop, which make it possible to effectively control its parameters.

It is known [32] that an increase in drawing rate leads to a reduction of the viscosity of the lubricant and the thickness of its layer in the deformation zone. As a result, the wear of the drawing tool increases and the warming-up of the wire and pulling pulleys of the drawing mill intensify. This should be taken into consideration when designing the regimes of wet drawing for thin high-carbon steel wire on sliding drawing mills.

The increase of drawing rate also facilitates the localization of deformations on the outer layers of the wire and, eventually, an irregularity across the wire cross section. This fact enhances the influence of surface phenomena during the wet drawing of brass-plated wire, in other words, it enhances the influence of the scale factor.

It has been stated [25,33–35] that when drawing high-carbon steel wire, the development of dynamic and static deformation aging processes leads to a deterioration in the plastic and fatigue life of thin brass-plated wire. For this reason, in drawing thin brass-plated wire on drawing mills of wet drawing, lower deformation degrees are used as compared with coarse drawing. Wire slip on the pull pulleys of the drawing mill is the result of additional thermal effects on the wire. Taking into consideration the negative effect of temperature in drawing thin brass-plated wire it is necessary to reduce wire slip on the final passes as well as single reductions [36] and ensure the effective cooling of the wire in its exit from the finishing die.

2.3. Role of Thermal Treatment in the Technological Process of High-Strength Wire Manufacturing

There are two kinds of thermal treatment in the technological processes of metal cord and bead wire manufacturing. Patenting is used to recover ductility of cold-drawn wire and to ensure the necessary level of mechanical properties in the final product. Annealing of the final bead wire is used for stress relaxation which is necessary to match the requirements of normative and technical documentation to the relative elongation values of the finished wire. In both cases, a reliable and efficient implementation of the temperature regime is required, which provides not only the required complex of properties for the finished product, but also a minimal energy consumption for the operation.

Analysis of the applied technologies of patenting shows that to get the desired microstructure, air cooling, heating (cooling) in fluidized area of particles, quenching in water, keeping temperature by the direct transmission of electric power, etc. are used [26,37–42].

There are two variants of practice in patenting. In the first case, the cooling rate is regulated only by the difference of temperatures between the heating of the wire in a furnace and a bath of isothermal decomposition; while other parameters also affect the cooling rate, in particular the coefficient of forced convective heat transfer between the wire and the bath environment, they are not taken into account. The other way is to consider both the difference of temperatures between the heating of the wire in a furnace and a bath of isothermal decomposition and the coefficient of forced convective heat transfer between the wire and the bath environment. Special attention to this aspect is paid in [43–45] where different methods which ensure the reliable regime of wire cooling are described.

With regard to the process of patenting wire in lead, the efficiency of convective heat transfer during the decomposition of supercooled austenite can be increased by raising the speed of movement of the lead (wire).

More perspectives, from the point of view of energy saving, ecology, and harmful effects on the human body, for methods of wire heating and cooling can be used not only in patenting but also during annealing of finished bead wire. In particular, fluidized bed heating technology, which, with proper technical support, has a number of advantages over heating in lead is widely used to heat the wire to 450–500 °C.

2.4. Deposition of Adhesive Coatings on the Metal Cord Wire

To date, the assortment of wire for tire bead ring reinforcement has become wider with a consequent substitution of brass-plated wire for bronze-plated wire which is considered to be more competitive [46,47]. Technological schemes of bead brass-plated wire and bronze-plated wire are different. Brass-plated coating is deposited on the wire by the consistent electrochemical deposition of the copper layer and the zinc layer. In this case, it is necessary to heat the wire to initiate the diffusion process of copper and zinc. The technological process of bronze deposition is more efficient. Bronze is deposited chemically by means of simultaneous deposition of copper and tin in one bath. One of the disadvantages of bronze coating, as compared with brass-plated coating, is its low level of adhesion with rubber. However, the technology of preparation of rubber mixtures at tire manufacturing enterprises makes it possible to change this parameter through a correction of the compounding. As a result, adhesion of the bronzed wire increases which allows it to be used quite successfully for reinforcing the bead rings of tires.

Taking into consideration manufacturing costs together with the level of exploitation properties the perspective way is to carry out the industrial technology of deposition of bronze coating on bead wire instead of brass-plating.

2.5. Use of Setups for Alternative Bending to Increase the Ductility of Bead Wire

The application of enterprises for tire manufacturing using modern high-capacity bead-making units has led to the formulation of strict demands to the bead wire mechanical properties. As a result, the percentage ratio of yield strength to tensile strength should be equal to 75–85% in accordance with the demands in technical certificates to the bead wire. This can be ensured by alternative bending of cold drawn wire.

Alternative bending causes the appearance of stresses which lead to the breakup the unstable substructures in the processed wire [48,49]. This kind of processing promotes the increase of its ductile properties.

2.6. Laying

The laying of metal cord is basically carried out on single twisting machines when the wire is not exposed to alternating deformation. For this reason, the existing reserve of plasticity in the wire ensures a sufficient level of its manufacturability in laying. Breakage of the wire in laying can be predominantly explained by the presence of non-metallic inclusions in the steel [50].

Machines operating on the principle of double twisting, when metal cord is twisted in two pitches during one rotation of the rotor, are usually used for laying. Laying on double twisting machines is more efficient and effective as compared with the same operation when single twisting rotor type machines are used. But at the same time, thin brass-plated wire is exposed to high alternative deformation, hence it raises demands to the mechanical properties [51].

3. Conclusions and Prospects

The manufacturing process consists of complex technological actions on the workpiece. During any technological operation the workpiece changes its parameters. Furthermore, products made of modern materials can be processed technologically in a number of different ways. Under such conditions the manufacturer should have some algorithms and models to select the technological process considered to be optimal taking into consideration the peculiarities of the industrial enterprise. This technological process has to guarantee the production of the finished product with the related level of quality and exploitation properties.

Because of high strength and high corrosion resistance, the steel cord still remains the main reinforcing material for tires of different types of automobiles. New trends in steel cord manufacturing processes are presented in [3,52,53]. The necessity to decrease artificial pollutants in the use the gasoline engines put forward new tasks for engineers to find new ways to increase the steel cord tensile strength. One of the prospective ways is to use steel with a nanostructure which ensures high values of both tensile strength and ductility in the processed material [54–57]. At the present time, the implementation of methods of severe plastic deformation under industrial conditions is on the cutting edge of technological progress. However, considering that the diameter of steel wire for cord is less than 1 mm, it would be necessary to create alternative ways to achieve a similar nanostructure in the processed material.

The technological process of steel cord manufacturing consists of several operations of different physical natures. For this reason, the risk of breakage of the processed material increases. This is why one of the important problems of the manufacturing process is to decrease the quantity of non-metallic inclusions, segregation of alloyed elements in steel for cord, surface blemishes, etc. Engineering should be addressed to solve these issues.

Perspectives for the design of the manufacturing process for a competitive high-carbon steel wire for steel cord and bead wire should be based on the solution of the following tasks:

- development of a methodology for calculating the drawing modes of high-carbon wire based on the selection and use of the fracture criterion and assessment of the influence of the deformation zone shape factor on its destruction;

- investigation of the nature of metal flow in the near-surface layer during drawing, assessment of the influence of drawing factors on its depth, and development of practical recommendations for calculating deformation modes of thin high strength brass plated, and bronze coated wire for metal cord and bead wire.

Furthermore, the level of technology of every manufacturing process has a decisive influence on its economic performance. This is why the choice of the optimal variant of the technological process should be carried out on the basis of the most important indicators of its effectiveness: productivity, cost, and quality of products. The tendency to find new materials to substitute steel wire for automobile tires with the required level of exploitation properties remains a challenge for scientists and engineers.

Author Contributions: Conceptualization, M.P.; project administration, A.S. All authors have read and agreed to the published version of the manuscript.

Funding: This research received no external funding.

Conflicts of Interest: The authors declare no conflict of interest.

Entry Link on the Encyclopedia Platform: : https://encyclopedia.pub/14524.

References

1. Smithers Rapra. Available online: https://www.smithersrapra.com/resources/2017/april/tire-industry-in-boom-cycle (accessed on 30 June 2021).
2. Fleming, R.A.; Livinsto, D.I. *Tire Reinforcement and Tire Performance*; American Society for Testing and Materials: Philadelphia, PA, USA, 1979; pp. 112–134.
3. Vedeneev, A.V.; Bobarikin, Y.A.; Zalewski, V.P. Analysis of the development of consumption and production of steel cord. *Foundry Prod. Metall.* **2019**, 48–59. [CrossRef]
4. AZoM Materials. Available online: https://www.azom.com/ (accessed on 1 August 2021).
5. Tarui, T.; Takahashi, T.; Tashiro, H.; Nishida, S. Metallurgical design of ultra high strength steel wires for bridge cable and tire cord. In *Metallurgy Processing and Applications of Metal Wires*; Paris, H.G., Kim, D.K., Eds.; TMS: Warrendale, PA, USA, 1996; pp. 87–96.
6. Minamida, T.; Hiraga, N.; Shibata, T. R&D Kobe Steel Engineering Reports. 2000, p. 32. Available online: https://www.kobelco.co.jp/technology-review/pdf/50_3/031-035.pdf (accessed on 20 August 2021).
7. Tashiro, H.; Tarui, T. Nippon Steel Technical Report. 2003, p. 87. Available online: https://www.nipponsteel.com/en/tech/report/nsc/pdf/n8818.pdf (accessed on 20 August 2021).
8. Kirihara, K. Production technology of wire rod for high tensile strength steel cord. *Kobelco Technol. Rev.* **2011**, *30*, 62–65.
9. Kanao, M. Improvement of the technology to get fine pearlite structure in wire. *News Ferr. Metall. Abroad* **2004**, *1*, 61–62.
10. Dailoh, V.; Hamada, T. Microstructures of heavily deformed high steel wires. *J. Iron Steel Inst. Jpn.* **2003**, *2*, 105–110.
11. Tashiro, H.; Tarui, T. State of the Art for High Tensile Strength Steel Cord: Technical Report. *Nippon. Steel Tech. Rep.* **2003**, *88*, 77–80. Available online: http://nsc.co.jp (accessed on 30 June 2021).
12. Kharitonov, V.A.; Stolyarov, A.Y. Development of a competitive technology to make wire for metal cord. *Metallurgist* **2013**, *57*, 320–325. [CrossRef]
13. Parusov, V.; Derevyanchenko, I.V.; Sychkov, A.B.; Nesterenko, A.M.; Zhigarev, M.A. Ensuring high quality indices for the wire rod used to make metal cord. *Metallurgist* **2005**, *49*, 439–448. [CrossRef]
14. KOBELCO. Available online: https://www.kobelco.co.jp/english/ (accessed on 1 August 2021).
15. Nippon Steel Corporation. Available online: https://www.nipponsteel.com/ (accessed on 1 August 2021).
16. JFE Steel Corporation. Available online: https://www.jfe-steel.co.jp/en/ (accessed on 1 August 2021).
17. Thyssenkrupp. Engineering. Tomorrow. Together. Available online: https://www.thyssenkrupp-steel.com/en/? (accessed on 1 August 2021).
18. Wang, L.; Xi, Z.; Li, C. Modification of type B inclusions by calcium treatment in high-carbon hard-wire steel. *Metals* **2021**, *11*, 676. [CrossRef]
19. Dey, I.; Chandra, S.; Saha, R.; Ghosh, S. Effect of Nb micro-alloying on microstructure and properties of thermo-mechanically processed high carbon pearlitic steel. *Mater. Charact.* **2018**, *140*, 45–54. [CrossRef]
20. Han, K.; Edmonds, D.V.; Smith, G.D.W. Optimization of mechanical properties of high-carbon pearlitic steels with Si and V additions. *Met. Mater. Trans. A* **2001**, *32*, 1313–1324. [CrossRef]
21. Han, K.; Smith, G.; Edmonds, D. Developments in ultra-high-carbon steels for wire rod production achieved through microalloying additions. *Mater. Des.* **1993**, *14*, 79–82. [CrossRef]

22. Tarui, T.; Maruyama, N. Microstructure Control and Strengthening of High-Carbon Steel Wires Nippon Steel Technical Report No. 91. January 2005. Available online: https://www.nipponsteel.com/en/tech/report/nsc/pdf/n9112.pdf (accessed on 18 July 2021).
23. Jiang, Y.; Lei, J.; Zhang, J.; Xiong, R.; Zou, F.; Xue, Z. Effect of carbon content on ti inclusion precipitated in tire cord steel. *J. Surf. Eng. Mater. Adv. Technol.* **2013**, *3*, 283–286. [CrossRef]
24. Lei, J.; Xue, Z. Study on tin precipitation during solidification for hypereutectoid tire cord steel. *Metal. Int.* **2012**, *17*, 10–15.
25. Vereshchagin, M.N.; Bobarikin, Y.L.; Savenok, A.N.; Vedeneev, A.V.; Tseluev, M.Y.; Ignatenko, O.I. Influence of the drawing rate on the temperature and stress-strain state of high-carbon wire. *Steel Transl.* **2007**, *37*, 1036–1041. [CrossRef]
26. Kazama, H.; Ishimoto, K. Super High Tensile Steel Wire for Rubber Product Reinforcement, Steel Cord Using for Steel Wire and Radial Tire Using This Steel Cord. U.S. Patent 5888321, 15 May 1996.
27. Person, L.E. Drawing wire with a smaller die angle. *Wire Ind.* **2004**, *71*, 437–441.
28. Mamoru, N.; Takeshi, K. Suppression of delamination in hyper-eutectoid steel wires by multi-skin pass drawing. *Tetsu to Hagane* **2004**, *90*, 588–592.
29. Hollinger, S.; Depraetere, E.; Giroux, O. Wear mechanism of tungsten carbide dies during wet drawing of steel tire cords. *Wear* **2003**, *255*, 1291–1299. [CrossRef]
30. Matsuyama, K. Method for Manufacturing Brass-Plated Steel Wire and Apparatus for Drawing Brass Plated Steel Wire. U.S. Patent 20100294013A, 26 January 2009.
31. Demidov, A.V.; Muraveiko, I.A. The experience of modernization of mills of wet wire drawing for the metal wire cord. *Foundry Prod. Metall.* **2019**, *4*, 97–102. [CrossRef]
32. Haddi, A.; Imad, A.; Vega, G. Analysis of temperature and speed effects on the drawing stress for improving the wire drawing process. *Mater. Des.* **2011**, *32*, 4310–4315. [CrossRef]
33. Fetisov, V.P. Change of deformational aging rate of brass plated wire under high reductions. *Steel Transl.* **1998**, *11*, 55–57.
34. Zelin, M. Microstructure evolution in pearlitic steels during wire drawing. *Acta Mater.* **2002**, *50*, 4431–4447. [CrossRef]
35. Kemp, I.P.; Pollard, G.; Bramley, A.N. Static strain aging in high carbon steel wire. *Mater. Sci. Technol.* **1990**, *6*, 331–337. [CrossRef]
36. Lee, S.-K.; Lee, S.-B.; Kim, B.-M. Process design of multi-stag wet drawing for improving the drawing speed for 0.72 wt% steel wire. *J. Mater. Process. Technol.* **2010**, *210*, 776–783. [CrossRef]
37. Process for Heat Treating a Carbon Steel Wire. U.S. Patent 4830684.122113, 18 November 1987.
38. Process for Producing Patented Steel Wire. U.S. Patent 5749981.767467, 16 December 1996.
39. High Flexibility Steel Wire and Method of Treating Same. U.S. Patent 3574000.675522, 17 February 1969.
40. High Flexibility Steel Wire and Method of Treating Same. U.S. Patent 3584494.826229, 20 May 1969.
41. Heat Treatment of Steel Wire. U.S. Patent 6228188.08/278910, 22 July 1994.
42. Process for Manufacturing Pearlitic Steel and Product Made Thereby. U.S. Patent 4759806.948077, 31 December 1986.
43. Tyl, T.W. Thermodynamic wire transformation process in the manufacture of steel tire cord. *Wire J. Int.* **2008**, *7*, 80–86.
44. Tyl, T.W. Steel patenting technology in the manufacture of steel tire cord. *Wire J. Int.* **2008**, *10*, 80–87.
45. Kruzel, R.; Suliga, M. The impact of the heat treatment parameters on patenting line on mechanical-technological properties of steel cord wires. *Mater. Sci.* 2016. Available online: https://www.semanticscholar.org/paper/The-impact-of-the-heat-treatment-parameters-on-line-Kruzel-Suliga/29650620f2e28463c4b1a3133a021d9d2f738657#citing-papers (accessed on 18 July 2021).
46. Buytaert, G.; Coornaert, F.; DeKeyser, W. Characterization of the steel tire cord—Rubber interface. *Rubber Chem. Technol.* **2009**, *82*, 430–441. [CrossRef]
47. Khrol, O.N. Influence of the quality of compounds on the adhesion properties of metal cord. *Foundry Prod. Metall.* **2020**, *2*, 36–41. [CrossRef]
48. Su, Y.-Y.; Shemenski, R.M. Investigation the parameters for torsion ductility of bead wire. *Mater. Des.* **2010**, *31*, 1423–1430. [CrossRef]
49. Zhou, L.; Fang, F.; Wang, L.; Hu, X.; Xie, Z.; Jiang, J. Torsion performance of pearlitic steel wires: Effects of morphology and crystallinity of cementite. *Mater. Sci. Eng. A* **2018**, *743*, 425–435. [CrossRef]
50. Sychkov, A.B.; Stolyarov, A.Y.; Kamalova, G.Y.; Efimova, Y.Y.; Gulin, A.E.; Selivanov, V.N. Peculiarities of structure formation in thin wire. *Vestn. Nosov Magnitogorsk State Tech. Univ.* **2017**, *15*, 75–84. (In Russian) [CrossRef]
51. Parusov, V.; Starov, R.V.; Derevyanchenko, I.V.; Sychkov, A.B.; Kucherenko, O.L. A decade of quality steel production: Developing a production technology for steel used in metal-cord manufacture. *Steel Transl.* **2010**, *40*, 82–87. [CrossRef]
52. Qiao, L.; Wang, Z.; Wang, Y.; Zhu, J. Mechanical performance-based optimum design of high carbon pearlitic steel by particle swarm optimization. *Steel Res. Int.* **2020**, *92*, 2000252. [CrossRef]
53. Shmurak, I.I. Improvement of metal cord for tyres. *Int. Polym. Sci. Technol.* **2006**, *33*, 1–4. [CrossRef]
54. Valiev, R.; Islamgaliev, R.; Alexandrov, I. Bulk nanostructured materials from severe plastic deformation. *Prog. Mater. Sci.* **2000**, *45*, 103–189. [CrossRef]
55. Valiev, R.Z.; Estrin, Y.; Horita, Z.; Langdon, T.G.; Zechetbauer, M.J.; Zhu, Y. Producing bulk ultrafine-grained materials by severe plastic deformation. *JOM* **2006**, *58*, 33–39. [CrossRef]
56. Langdon, T.G. Processing by severe plastic deformation: Historical development and current input. *Mater. Sci. Forum* **2011**, *667–669*, 9–14.
57. Segal, V. Review: Modes and processes of severe plastic deformation (SPD). *Materials* **2018**, *11*, 1175. [CrossRef]

encyclopedia

MDPI

Entry

Low-Pressure Turbine Cooling Systems

Krzysztof Marzec

Faculty of Mechaincal Engineering and Aviation, Rzeszow University of Technology, 35-959 Rzeszów, Poland; k_marzec@prz.edu.pl

Definition: Modern low-pressure turbine engines are equipped with casings impingement cooling systems. Those systems (called Active Clearance Control) are composed of an array of air nozzles, which are directed to strike turbine casing to absorb generated heat. As a result, the casing starts to shrink, reducing the radial gap between the sealing and rotating tip of the blade. Cooling air is delivered to the nozzles through distribution channels and collector boxes, which are connected to the main air supply duct. The application of low-pressure turbine cooling systems increases its efficiency and reduces engine fuel consumption.

Keywords: cooling systems; turbine casings; Active Clearance Control

Citation: Marzec, K. Low-Pressure Turbine Cooling Systems. *Encyclopedia* **2021**, *1*, 893–904. https://doi.org/10.3390/encyclopedia1030068

Academic Editors: Raffaele Barretta, Ramesh Agarwal, Krzysztof Kamil Żur and Giuseppe Ruta

Received: 19 July 2021
Accepted: 27 August 2021
Published: 31 August 2021

1. Introduction

Gas path sealing is a challenging problem of aircraft gas turbine engine design. It is caused because the clearance between the blade tip (rotating structure) and casing with sealing (static structure) tends to vary during engine operation due to various mechanical and thermal loads. What is more, inertial (maneuver) and aerodynamic (pressure) loads during flight have to be taken into consideration. More factors, which also have a negative influence on tip clearance control, are manufacturing and assembly limitations such as case ovalization effects, tolerance stack-ups, shaft deflections, etc. Additionally, the clearances between blade tip and sealing vary along the lifespan of the whole engine as well as the part itself, as the wear and thermal erosion on all of the parts occur.

Despite the large number of limitations that have to be considered, low-pressure turbines are commonly equipped with impingement cooling systems, called Active Clearance Control (ACC), which help to control gas path sealing and therefore reduce gas path leakages. The main role of the impingement cooling system (to provide efficient gap control) is gap reduction between the tip of the blades and sealing during engine operation in the cruise phase. The benefits of active clearance control are, among others, increased engine efficiency, reduced specific fuel consumption (SFC), and reduced NOx and CO emissions.

An impingement cooling system comprises an array of nozzles, which direct jets of high-velocity fluid at a target surface, thus securing proper operating conditions of attached hardware (e.g., the blade tip and sealing surface) through convective heat transfer between fluid and target surfaces.

Lowering air leakage in the area of blade tips boosts turbine efficiency, making it possible for the engine to fulfill thrust and performance targets utilizing less fuel and with the lower temperature at the rotor inlet. The lifecycle of the hot section components may be increased by operating the turbine at lower temperatures, which consequently will increase the engine's service life as a result of the greater interval between overhauls. Lattime and Steinetz [1] give overviews of multiple advantages of advanced active clearance control systems. Taking fuel savings into consideration, a reduction of tip clearance by 0.010 in. decreases specific fuel consumption by ~0.8% to 1%. A significant cut down in NOx, CO, and CO_2 emissions are also achievable by reduction of fuel consumption. Exhaust gas temperature (EGT) may be lowered by ~10 °C through the reduction of tip clearances by 0.010 in. The main reason for the removal of an aircraft engine from service is the deterioration of the EGT margin. Up to 1000 extra cycles of engine on-wing time may be achieved by operating it at lower temperatures, thus increasing the life of hot sections parts. Other advantages are an increase in payload and mission range.

A certain amount of technical issues has to be focused on to field an efficient active clearance control system. Two main challenges encompass the high-temperature environment and the necessity for precise control.

2. Basic Design Principle

The impingement cooling systems comprise one or more distribution channels, every one of which comprises an adequate number of cooling nozzles directed at the target surface at a specified angle, see Figure 1. Air is supplied to the distribution channels via an inlet tube. The valve regulates air flow, which feeds the ACC system. The aim of an impingement cooling system is to decrease the radial gap between the tips of rotating blades and the sealing.

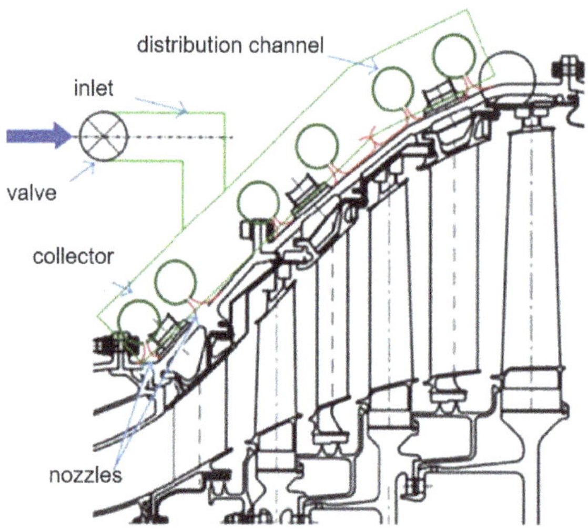

Figure 1. Schematic section of low-pressure turbine equipped with Active Clearance Control (ACC) [2].

Figure 2 depicts the typical cooling system of low and high-pressure turbine casing of turbofan engine.

The depicted low-pressure turbine cooling system is built of a number of tubular distribution channels at a distance to the casing surface. Each of the distribution channels comprises an appropriate number of holes, through which the air flows at high velocity and onto the surface of the casing. The role of the cooling medium is ensuring a proper heat exchange through convection with the casing heated up as a result of hot gas flow occurring in it. A various number of nozzles and, therefore, a various amount of coolant is employed in different circumferential regions along the engine axis according to the local casing temperature.

Higher temperature gas flows through the front part of the casing (with a smaller diameter) due to the proximity of the combustion chamber.

The decrease of temperature of casing, to which a number of rows of sealing are attached, results in a decrease of tip clearance between rotating and static components.

Figure 2. Typical tube design of the ACC Cooling system of low-pressure (LP) and high-pressure (HP) turbines [3].

The air is delivered to distribution channels through collector channels, which are connected to the main air supply duct, which in turn provides air to the cooling system of a low-pressure turbine. In the typical low-pressure turbine casing cooling systems, the distribution channels are connected with collector channels by welding. The whole system is mounted on brackets distributed circumferentially, which in turn are fixed to the casing.

The cross-sections of the flow channels are selected in a way that minimizes the linear pressure losses in the flow of the cooling medium. Additionally, the geometrical shape of flow channels limits the generation of local pressure losses during the operation of the system.

The crucial aspect of impingement cooling systems design is ensuring a proper clearance between casing's and cooling systems' components to guarantee that thermal stresses, which occur during engine operation, are compensated for. These stresses are a result of, among others, various operating temperatures of particular components and various heat expansion coefficients of applied materials.

The next decisive feature of impingement cooling systems design is the utilization of materials, which guarantee proper operation of components in high temperature and an increased pressure related to the flow of cooling fluid.

The dynamic stresses generated, e.g., due to the turbine shaft's imbalance, are also significant factors. These stresses may lead to damaging of the parts of a cooling system in the areas of increased concentration of stress (e.g., weld seam areas).

The air, which is fed to the cooling systems, is transported from the turbofan engine's bypass airflow. The mass output of air is governed by a valve. In cooling systems' construction, it is crucial that the cooling nozzles are positioned at an adequate distance to the cooled target surface and that they are aimed at an appropriate angle, which ensures an effective heat exchange between the cooling medium and the casing.

The result of the operation of a cooling system is an increase of kinetic energy of gases being burnt, which is converted into mechanical energy of a rotating shaft. Impingement cooling systems are applied both to the cooling of high and low-pressure turbines of turbofan engines [1]. Cylindrical [4,5] and slot [6,7] nozzles are often used in the construction of cooling systems. The main parameter that is characteristic for the cylindrical nozzle is its diameter D, and for the slot nozzle—the length of slot H. The primary parameters, which are characteristic for the impingement cooling systems, are dimensionless coefficients Y/D, H/B, and S/D. They are depicted in Figure 3.

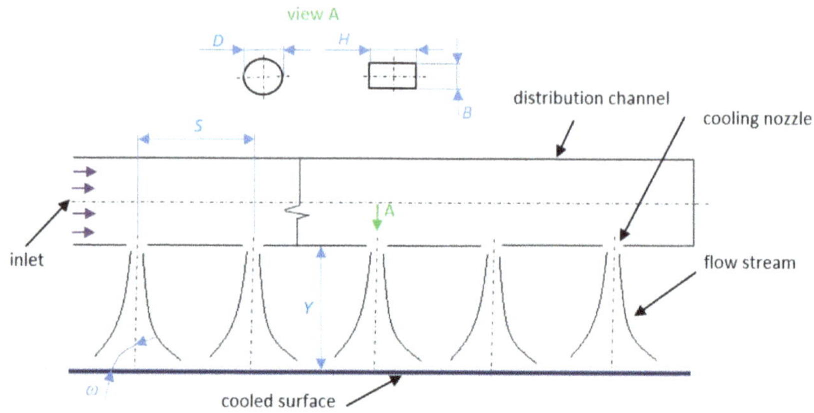

Figure 3. Impingement cooling systems distribution channel's axis cross-section.

Relative distance Y/D determines the ratio between the distance of the nozzle and a target cooled surface Y and the nozzle diameter D, H/B determines the relationship between the slot lengths H to the sloth width B. The relative distribution of nozzles S/D determines the ratio between the distances between the neighboring nozzles' axes S to the nozzle diameter D. The cooling efficiency is additionally affected by, among others, the angle ω between the nozzle and the cooled target surface, the sort of cooling medium and Reynolds number *Re*.

The impingement cooling systems are also divided according to the possibility of fluid flow between the cooled surface and the surrounding environment. There are cooling systems in which the cooling medium flows parallel to the cooled surface before leaving the system [5,7], and systems in which the stream of cooling medium flows out of the system immediately after coming in contact with the cooled target surface [8,9].

In impingement cooling systems, a "fountain effect" [10], which is cooling medium recirculation, occurs due to neighboring streams' collisions. Figure 4 depicts an example of velocity vectors distribution in the x-y plane in the area of four nozzles of the cooling system, which shows the effect of fountain generation as a result of collisions of neighboring fluid streams.

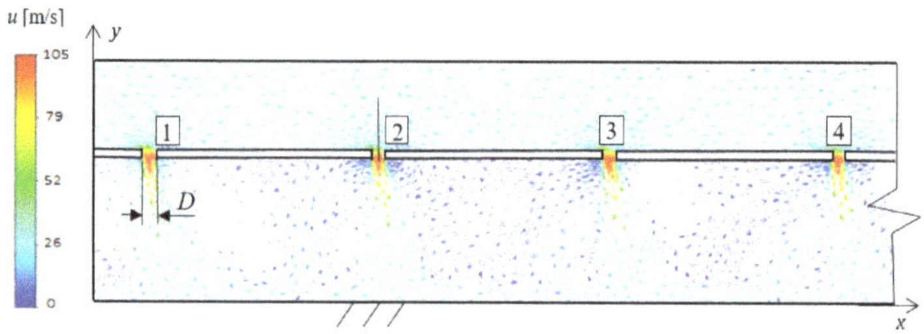

Figure 4. Distribution of velocity vectors in the x-y plane in the area of four cooling nozzles depicting the generation of fountain effect [10].

The flow field through a single nozzle is characterized by the existence of the nozzle area, stagnation area, and wall area (Figure 5). The flow of impinging jet goes through a number of specific areas, as pictured in Figure 5. The jet comes out of a nozzle or aperture at a temperature and velocity profile and turbulence characteristics determined by the upstream flow. Considering a pipe-shaped nozzle, also known as a tube or cylindrical nozzle, the flow is established into a parabolic velocity profile characteristic to pipe flow, and a mild amount of turbulence is generated upstream. On the contrary, a flow conveyed by the application of differential pressure traversing a space of a thin, flat orifice will generate a starting flow with a quite flat velocity profile, lower turbulence, and a downstream flow contraction (vena contracta) [10].

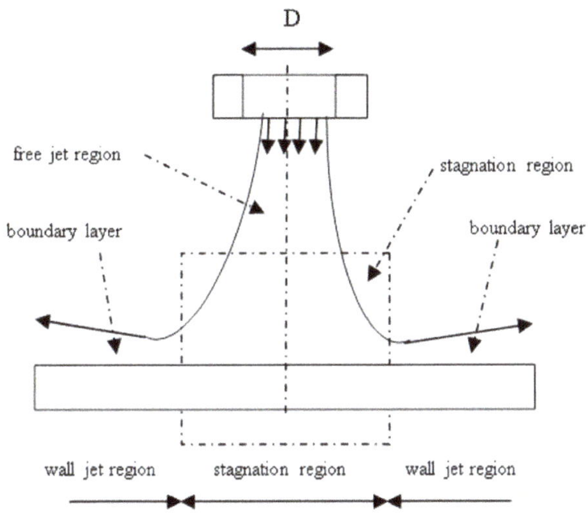

Figure 5. Characteristics of flow through a single nozzle.

3. ACC Operation Rationales

The function of an ACC is to provide proper clearance between the tip of the blade and sealing during the entire engine operation. There are several distinct stages in the flight cycle, which translate directly to engine conditions—the schematic cycle is shown in Figure 6.

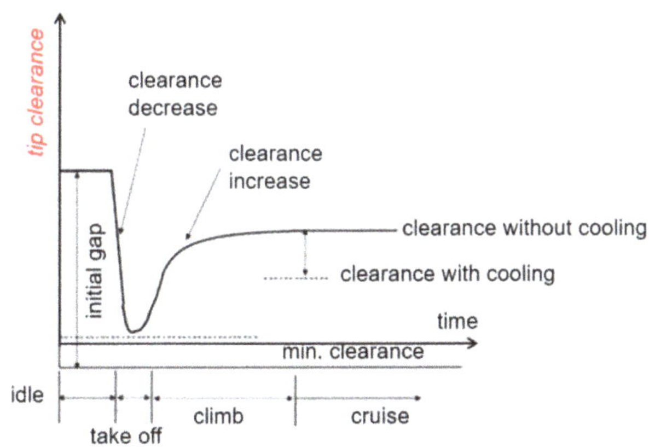

Figure 6. Tip clearance control [2].

The tip clearance at the ground idle condition, in which the engine is free spinning on the ground, is constant as no significant centrifugal force acts on the rotating part, and thermal loads also do not appear.

After taxing to take-off condition, the throttle is applied. The temperature change ratio is highest in this condition. Core flow (the air, which flows through all stages of a compressor into the combustion chamber, is then mixed with fuel and burnt to flow through the turbine and into the exhaust area) heats the inner (core) part of the engine at a very high rate. Additionally, the centrifugal force increases with the rising engine's rotational speed and starts acting on the rotating components. The combination of these factors causes the blade tip with sealing fins to displace outwards. On the other hand, the casing is heating up at a much slower rate, as it is much further away from the hottest core flow. The sum of these two displacements results in gap closure and leads to achieving a minimum clearance value. In this condition, the flow efficiency is at its best, as the leakages of hot core flow are lowest, thus transforming the maximum amount of aerodynamic energy acting on turbine blades into the rotational speed of the shaft.

Following is the climb condition, during which the casing heats up continually to reach its ADP (aero design point) condition—the state designed for the cruise of an aircraft. Tip clearance in these conditions increases as the casing expands due to rising temperature. After reaching cruise condition, a gap is established at ADP. Active clearance control executed through impingement cooling system allows reducing the tip clearance by cooling down the casing (using bypass air), which is corresponding to the rightmost section of the graph in Figure 6.

4. Heat Transfer between Casing and Air Stream from ACC

The heat transfer process between the hot casing and cooling air distributed via impingement cooling nozzles is characterized by dimensionless Nusselt number Nu. This number is a basic parameter, which determines the cooling efficiency of the wall by the air stream. Its value characterizes heat flow intensity on the boundary between fluid and wall [9,11]. The Nusselt number is directly proportional to heat transfer coefficient h, which in turn is dependent on the temperature difference between the wall (T_w) and fluid temperature (T_j). The value of Nusselt number Nu is given by the equation:

$$\mathrm{Nu} = \frac{\mathrm{hD}}{\mathrm{k}} \tag{1}$$

where:

h—heat transfer coefficient on the fluid-wall boundary;

D—cooling nozzle diameter;

k—heat conduction coefficient of the fluid;

Heat transfer coefficient h is given by the equation [12]:

$$\mathrm{h} = \frac{-\mathrm{k}\frac{\partial \mathrm{T}}{\partial \mathrm{y}}}{(\mathrm{T_j} - \mathrm{T_w})} \tag{2}$$

where:

T_w—wall temperature;

T_j—nozzle temperature;

$\partial T/\partial y$—derivative of temperature in the direction perpendicular to the cooled surface.

To define heat transfer rates between the target plate and the coolant line, the averaged Nusselt number (3) and area-averaged Nusselt number (4) have to be evaluated.

Line averaged Nusselt number [13]:

$$\overline{\mathrm{Nu}} = \frac{1}{\mathrm{L}} \int_{\mathrm{L}} \mathrm{Nu}(\mathrm{x})\mathrm{dL} \tag{3}$$

where:

L—averaging line parallel to the target plate;

Nu (x)—local Nusselt number along line L;

Area averaged Nusselt number [13]:

$$\overline{\overline{\mathrm{Nu}}} = \frac{1}{\mathrm{A}} \int_{\mathrm{A}} \mathrm{Nu}(\mathrm{x},\mathrm{y})\mathrm{dA} \tag{4}$$

where:

A—averaging area parallel to the target plate;

Nu (x,y)—local Nusselt number on area A;

The other dimensionless parameters describing heat transfer between target plate and cooling fluid are:

- nozzle height to nozzle diameter (H/D);
- distance between nozzle and target plate to nozzle diameter (Y/D);
- distance between neighboring nozzles to nozzle diameter (S/D);
- distance from jet axis to nozzle diameter (r/D);
- Reynolds number according to the following expressions:

$$Re = \frac{U_o D}{\upsilon} \quad (5)$$

where:

U_o—mean nozzle velocity at the exit area;

υ—kinematic viscosity of the cooling fluid;

- Mach number based on nozzle exit average velocity (for the Ma < 0.3, there is a negligible impact on heat transfer rates)

The heat transfer rates between the target surface and cooling fluid and thus cooling efficiency also depends on the shape of the nozzles. Paper [14] shows numerical analyses of the influence of various nozzle shapes (cylindrical, convergent, divergent, and cylindrical elongated) on Nusselt number distribution along plate cooled with one row of ten nozzles. The highest average value of the Nusselt number was obtained for the cylindrical ones.

In turn, Marzec, K. [15] showed numerical analyses of the influence of various nozzle positions on the Nusselt number distribution of an array of ten cooling orifices placed along the non-planar target surface. Results presented in this paper determine the most optimum position of the cooling nozzles (up to two orifice diameters) to provide a high rate of heat transfer for seven various dimensionless jet positions.

Besides presented parameters, heat transfer between target plate and cooling fluid depends mostly on turbulence intensity, Reynolds number, and Mach number (negligible influence for Ma < 0.3).

To evaluate the stream of nozzle mass flow, the discharge coefficient has to be considered. Discharge coefficient (Cd) is a ratio of the real mass flow rate through the orifice in relation to the isentropic flow rate. This coefficient includes losses (pressure, friction), which reduces the mass flow rate through a nozzle. The ideal mass flow is calculated assuming a one-dimensional isentropic expansion through an orifice from coolant pipe (secondary flow) total pressure (Pt) to the main flow (primary flow) static pressure (Pd) with the obtained expression:

$$C_d = \frac{\dot{m}}{P_t\left(\frac{P_d}{P_t}\right)^{\frac{\gamma+1}{2\gamma}} \sqrt{\frac{2\gamma}{(\gamma-1)RT_t}\left(\left(\frac{P_t}{P_d}\right)^{\frac{\gamma-1}{\gamma}} - 1\right)} \frac{\pi}{4}D^2} \quad (6)$$

where:

$\dot{m}$—real mass flow rate;

γ—heat coefficient ratio;

R—gas constant;

P_t—total pressure;

P_d—static pressure;

T_t—total temperature.

Active clearance control systems consist of a number of nozzles placed in an array. For low values of Y/D and low values of S/D, the flow delivered by each nozzle (due to limited space to leave impingement region) forms a crossflow with the neighboring flow. The presence of the crossflow causes asymmetric nozzle flow area, moves stagnation points, and results in a thicker boundary layer. These effects have a negative impact on average heat transfer rates. The presented nozzle interference does not have a significant influence on the peak of the Nusselt number value; however, the averaged $\overline{Nu}$ value shows a decrease [13].

Another behavior affecting heat transfer rate between coolant and target surface is thermal interaction of the hot spent impingement flow coming out from the upstream (after striking target surface) with the downstream nozzle fresh coolant. This phenomenon has a negative impact on heat transfer rate and depends mostly on Y/D and S/D factors.

5. Review of Basic Geometrical and Physical Cooling Systems' Parameters

The research described in the literature takes into consideration both geometrical and physical parameters, which characterize the operation of impingement cooling systems.

Various cooling systems nozzles diameters D are considered [2,16,17]. The paper [2] includes a numerical analysis of impingement cooling system with seven different values of cooling nozzles' diameters: D = 0.45 mm; D = 0.5 mm; D = 0.64 mm; D = 0.8 mm; D = 1.6 mm; D = 2.4 mm; D = 3.2 mm. The highest mean values of Nusslet number (Nu) were achieved in the case of a nozzle with diameter D = 3.2 mm and Mach number Ma = 0.1. For comparison, the paper [2] describes the research on cylindrical nozzles with larger diameters D = 6 mm, D = 7.3 mm, D = 10 mm, and D = 15 mm respectively. The analysis of incompressible fluid flow shows that the highest values of Nusselt number (Nu) were achieved for a value of coefficient Y/D = 6 and for a nozzle diameter D = 15mm. The performed research showed that the application of nozzles with higher diameter D in cooling systems has an influence on the increase of the amount of heat being transferred on the fluid-wall boundary.

San et al. [4] performed research with an aim to determine the influence of relative distance Y/D and Reynolds number (Re) on the distribution of Nusselt number along the cooled surface. The values taken for the research were: Re = 10,000, Re = 15,000, Re = 30,000 and Y/D = 1, Y/D = 2, Y/D = 4, and Y/D = 6. Results of the research indicated cooled areas characterized by the occurrence of local maximum of the Nusselt number, among others, for Y/D = 1 and Reynolds number Re = 30,000. This research demonstrates a major influence of both relative distance Y/D and Reynolds number Re of the flowing fluid on the cooling efficiency.

Trinh et al. [18] showed experimental research of three kinds of nozzles: tube-shaped, cylindrical, and cross-shaped nozzles placed on a hemisphere. In the results of the research presented, the highest value of Nusselt number on the boundary of the fluid-wall area was achieved when the cylindrical nozzle was applied, and the lowest values were observed for the cross-shaped nozzle.

The cooling efficiency is additionally affected by the angle of inclination of the nozzle to the cooled surface ω [19–21]. Taking a three-dimensional problem into consideration, the change of angle of inclination of the nozzle to the cooled surface results in forming of elliptical distribution of Nusselt number on the cooled surface [9]. For comparison, in the case of the nozzles directed perpendicularly to the cooled surface, the distribution of Nusselt number on the cooled surface is cylindrical. Afroz and Sharif [17] showed various inclination angles of nozzle ω to the cooled surface have been examined at two values of Reynolds number: Re = 23,000 and Re = 50,000. The value of Nusselt number decreased by 25–50% with a decrease of angle from $\omega = 90^\circ$ *to* $\omega = 45^\circ$ for the given values of Reynolds number (Re). Moreover, when angle $\omega > 50^\circ$ and the Re = 50,000, the occurrence of the "Coand effect" was observed. This phenomenon is characterized by the tendency of the fluid stream to adhere to the neighboring surface. When the angle of inclination of the nozzle to the cooled surface is low, the fluid stream starts to adhere to the adiabatic wall of the fluid distribution channel positioned nearby. At a Re of 23,000, it was possible to observe the Coand effect for an angle $\omega > 30^\circ$. This effect leads to a drastic decrease in surface cooling efficiency. In the literature, the phenomenon of fountain effect created as a result of the collision of neighboring streams of cooling fluid is also described [22]. In such a case, recirculation of fluid occurs. The paper [22] shows, that when the relative distribution of nozzles acquires a value $S/D < 4$, the fluid recirculation as a result of collisions of neighboring streams occurs. The authors prove, that if $Y/D = 2$, then the collisions of neighboring streams occur when $S/D \leq 10$. The maximum value of Nusselt number along the cooled surface has been achieved at $S/D = 8$. San and Lai [23] confirmed that at $S/D = 14$, no interaction between neighboring streams occurs.

The presented results of the literature research show the broad differentiation of operation parameters of impingement cooling systems. The increase of flow velocity of fluid stream represented by Re number causes the increase of heat exchange efficiency between the fluid and cooled surface.

The temperature difference between the cooled surface and a cooling fluid stream, nozzle inclination angle, and geometrical shape of it is also of high significance.

6. Numerical and Experimental Methods of ACC Researches

Researches focused on flow behavior and the heat transfer process of the impingement cooling systems are divided into two categories: numerical modeling [24–26] and experimental researches [27,28]. Both categories are analyzing possible ways of ACC efficiency improvement. Experimental results are often used to validate numerical methods used for the calculations of ACC performance.

Experimental researches on impingement cooling systems are focused mostly on measuring the surface heat transfer coefficients. In such experiments, arrays of nozzles drilled in distribution tubes are installed above a target surface (flat, cylindrical, or conical) with spacers. The air supply setup consists of a pump, air reservoir, dryer, filter, and a control valve to regulate pressure and mass flow. The target surface is often supplied with constant heat flux ($\dot{q}$) using thin steel strips bonded to the target plate. To reduce lateral heat transfer within the target plate, low thermal conductivity materials are chosen for the target plate. The idea of this approach is to ensure that electrically generated energy is taken over by the fluid in the direction perpendicular to the target surface with constant heat flux. The temperature distribution on the backside of the target plate (T_w) is measured with non-contact optical devices such as infrared (IR) thermography cameras or thermochromic liquid crystals that change their color with the temperature. The surface heat transfer coefficient (h) is calculated with the equation [9]:

$$h = \frac{\dot{q}}{T_w - T_j} \tag{7}$$

The flow temperature is often measured using thermocouples inside the distribution tubes in a location corresponding to the central nozzle, while the ambient temperature is also recorded on both sides of the target plate in order to evaluate heat losses [29].

Experimental measurements allow investigating the influence of various geometrical (nozzle diameter, shape, Y/D, S/D, etc.) and thermo-flow parameters (heat flux, Reynolds number, etc.) on the heat transfer performance of the impingement cooling system. Numerical researches of impinging cooling systems allow the assessment of the heat transfer coefficient, Nusselt number distribution, pressure drops, velocity fields, and temperature distribution for a given flow in advance of manufacturing the hardware.

Impingement cooling systems applications (ACC) to reduce the radial thermal growth of a casing involve turbulent flow downstream of the nozzles. Accurate prediction of the behavior of turbulent flow downstream the nozzles is a big challenge in numerical modeling. Finite volume and finite element computational fluid dynamics (CFD) solvers (which solve Navier–Stokes equations) are often used to evaluate the behavior of the flow and heat transfer of the impinging cooling systems. The accurate predictions of velocity fields, pressure drops, or heat transfer coefficients using CFD methods depend strongly on the modeling of turbulence and the interaction of the turbulent flow field with the target surface. The numerical approach (validated by the experimental procedure in advance) gives a chance to evaluate flow behavior and heat transfer rates for various geometries of the cooling system and target plate without limitations that can occur for the experimental procedures.

Numerical methods give an opportunity for further development of the impingement cooling systems efficiency in terms of very precise adjustment of tip clearance across the different engine operating conditions and mainly during cruise operation. The current trend in increasing the engine bypass ratio to enhance the propulsive system efficiency pushes the limits of ACC traditional design performance. It is caused in most designs by feeding air for the active clearance control (ACC) system of the LPT comes from secondary bypass flow [30]. In fact, the fan pressure ratio tends to fall, thus reducing the pressure ratio on the ACC inlet piping. Therefore impingement cooling systems require more efficient heat transfer between the casing and cooling liquid to ensure adequate tip control between rotating structure and sealing. Reduced inlet pressure, thus reduced Reynolds number requires to be compensated by modification of geometrical parameters (dimensionless nozzle-target distance Y/D, smaller pitch distance S/D, etc.) to keep the efficiency of the system on the proper level. In turn, reducing nozzle-target distance brings the risk of cooling system interference with casing during engine operation due to vibratory stresses, dynamic loads, etc. This fact has to be taken into consideration during ACC components design, especially when manufacturing tolerances are defined. Reduced nozzle-target distance with a defined chain of tolerances has to ensure that in the worst-case scenario, there is no interference between the casing and cooling system. In turn, this approach requires very precise machining and measuring tools, which makes ACC design more expensive.

7. Conclusions

Impingement cooling systems play a significant role in many technical applications, especially in the aero industry (Active Clearance Control systems). Due to the adjusted amount of cooling air directed onto the turbine casing surface, they are able to control clearance between the blade tip (rotating structure) and casing with sealing (static structure). This clearance tends to vary during engine operation due to various mechanical, thermal, inertial (maneuver), and aerodynamic (pressure) loads.

The radial clearance closure is a function of physical parameters like impingement flow, Reynolds number in the orifice region, or turbulence intensity. Nozzle diameter, dimensionless factors like relative distance Y/D, relative nozzle position S/D, the shape of the orifices, or inclination angle also have a significant influence on heat transfer rates along the cooled surface

The surrounding environment also has an impact on the heat transfer rate. Cooling systems in which the cooling medium flows parallel to the cooled surface before leaving the system (confined) are less efficient (due to the heat accumulation) in comparison to the systems in which the stream of cooling medium flows out of the system immediately after coming in contact with cooled target surface (unconfined).

Manufacturing feasibility and assembly limitations such as case ovalization effects, tolerance stack-ups, or shaft deflections also have to be considered during the design of a dedicated Active Clearance Control system.

The application of turbine cooling systems (ACC) helps to increase engine efficiency, reduce specific fuel consumption (SFC), and reduce NOx and CO emissions.

Funding: This research received no external funding.

Conflicts of Interest: The author declares no conflict of interest.

Entry Link on the Encyclopedia Platform: https://encyclopedia.pub/14765.

References

1. Lattime, S.B.; Steinetz, B.M. Turbine Engine Clearance Control System: Current Practices and Future Directions. NASA/TM-2002-211794. 2002. Available online: http://ntrs.nasa.gov/archive/nasa/c-asi.ntrs.nasa.gov/20030003697.pdf (accessed on 19 July 2021).
2. Ahmed, F.B.; Weigand, B.; Meier, K. Heat transfer and pressure drop characteristics for a turbine casing impingement cooling system. In Proceedings of the 14th International Heat Transfer Conference IHTC, Washington, DC, USA, 8–13 August 2010; p. 14.
3. da Soghe, R.; Bianchini, C.; D'Errico, J.; Tarchi, L. Effect of Temperature Ratio on Jet Impingement Heat Transfer in Active Clearance Control Systems. *J. Turbomach.* **2019**, *141*, 081009. [CrossRef]
4. San, Y.-Y.; Shiao, W.-Z. Effects of jet plate size and plate spacing on the stagnation Nusselt number for a confined circular air jet impinging on a flat surface. *Int. J. Heat Mass Transf.* **2006**, *49*, 3477–3486. [CrossRef]
5. Shu, J.-J. Heat Transfer on a Hot Surface Impinged by a Cold Circular Liquid Jet. V European Conference on CFD, ECCOMAS CFD. 2011. Available online: http://congress.cimne.com/-eccomas/cfd2010/papers/01620.pdf (accessed on 19 July 2021).
6. Nirmalkumar, M.; Katti, V.; Prabhu, S.V. Local heat transfer distribution on a smooth flat plate impinged by a slot jet. *Int. J. Heat Mass Transf.* **2011**, *54*, 727–738. [CrossRef]
7. Zukowski, M. Heat transfer performance of a confined single slot jet if air impinging on a flat surface. *Int. J. Heat Mass Transf.* **2013**, *57*, 484–490. [CrossRef]
8. Andreini, A.; da Soghe, R.; Facchini, B.; Maiuolo, F.; Tarchi, L.; Coutandin, D. Experimental and numerical analysis of multiple impingement jet arrays for an active clearance control system. *J. Turbomach.* **2013**, *135*, 031016. Available online: http://turbomachinery.asmedigitalcollection.asme.org/article.aspx?articleID=1672752 (accessed on 19 July 2021). [CrossRef]
9. Zuckerman, N.; Lior, N. Jet impingement heat Transfer: Physics, Correlations and Numerical Modeling. *Adv. Heat Transf.* **2006**, *39*, 565–631.
10. Marzec, K. Influence of impingement cooling nozzle geometry on heat transfer. *Modelowanie Inżynierskie* **2018**, *68*. Available online: http://www.kms.polsl.pl/mi/pelne_37/10_37_68.pdf (accessed on 19 July 2021).
11. Limaye, M.D.; Vedula, R.P.; Prabhu, S.V. Local heat transfer distribution on a flat plate impinged by a compressible round air jet. *Int. J. Therm. Sci.* **2010**, *49*, 2157–2168. [CrossRef]
12. Katti, V.; Prabhu, S.V. Influence of spanwise pitch on local heat transfer distribution for in-line arrays of circular jets with spent air flow in two opposite directions. *Exp. Therm. Fluid Sci.* **2008**, *33*, 84–95. [CrossRef]
13. Aldabbagh, L.B.Y.; Sezai, I. Numerical simulation of three-dimensional laminar multiple impinging square jets. *Int. J. Heat Fluid Flow* **2002**, *23*, 509–518. [CrossRef]
14. Marzec, K.; Kucaba-Pięta1, A. Numerical investigation of local heat transfer distribution on surfaces with a non-uniform temperature under an array of impinging jets with various nozzle shapes. *J. Theor. Appl. Mech.* **2017**, *4*, 1313–1324. [CrossRef]
15. Marzec, K. Influence of a jet position on local heat transfer distribution under an array of impinging nozzles with non-planar contour of the cooled surface. *Heat Transf. Eng.* **2021**, *42*, 1506–1521. [CrossRef]
16. Goodro, M.; Park, J.; Ligrani, P.; Fox, M.; Moon, H.-K. Effects of Mach number and Reynolds number on jet array impingement heat transfer. *Int. J. Heat Mass Transf.* **2007**, *50*, 367–380. [CrossRef]
17. Lee, J.; Ren, Z.; Ligrani, P.; Fox, M.D.; Moon, H.-K. Crossflows from jet array impingement cooling: Hole spacing, target plate distance, Reynolds number effects. *Int. J. Therm. Sci.* **2015**, 7–18. [CrossRef]
18. Trinh, X.T.; Fénot, M.; Dorignac, E. The effect of nozzle geometry on local convective heat transfer to unconfined impinging air jets. *Exp. Therm. Fluid Sci.* **2016**, *70*, 1–16. [CrossRef]
19. Afroz, F.; Sharif, M.A.R. Numerical study of heat transfer from an isothermally heated flat surface due to turbulent twin oblique confined slot-jet impingement. *Int. J. Therm. Sci.* **2013**, *74*, 1–13. [CrossRef]
20. Beitelmal, A.H.; Saad, M.A.; Patel, C.D. The effect of inclination on the heat transfer between a flat surface and an impinging two-dimensional air jet. *Int. J. Heat Fluid Flow* **2000**, *21*, 156–163. [CrossRef]
21. Choo, K.; Kang, T.Y.; Kim, S.J. The effect of inclination on impinging jets at small nozzle to plate spacing. *Int. J. Heat Mass Transf.* **2012**, *55*, 3327–3334. [CrossRef]
22. San, J.Y.; Lai, M. Optimum jet-to-jet spacing of heat transfer for staggered arrays of impinging air jets. *Int. J. Heat Mass Transf.* **2001**, *44*, 3997–4007. [CrossRef]
23. Hosaina, M.L.; Fdhilaa, R.B.; Daneryda, A. Multi-Jet Impingement Cooling of a Hot Flat Steel Plate, The 6th International Conference on Applied Energy—ICAE2014. *Energy Procedia* **2014**, *61*, 1835–1839. [CrossRef]
24. da Soghe, R.; Bianchini, C. Aero-Thermal Investigation of Convective and Radiative Heat Transfer on Active Clearance Control Manifolds. In Proceedings of the ASME Turbo Expo 2019: Turbomachinery Technical Conference and Exposition, Phoenix, AZ, USA, 17–21 June 2019; ASME: New York, NY, USA, 2019.
25. Ahmed, F.B.; Tucholke, R.; Weigand, B.; Meier, K. Numerical Investigation of Heat Transfer and Pressure Drop Characteristics for Different Hole Geometries of a Turbine Casing Impingement Cooling System. In Proceedings of the ASME 2011 Turbo Expo: Turbine Technical Conference and Exposition, Vancouver, BC, Canada, 6–10 June 2011; ASME: New York, NY, USA, 2012; pp. 95–1108.
26. da Soghe, R.; Bianchini, C.; Andreini, A.; Facchini, B.; Mazzei, L. Heat transfer augmentation due to coolant extraction on the cold side of active clearance control manifold. In Proceedings of the ASME Turbo Expo 2015: Turbine Technical Conference and Exposition, Montreal, QC, Canada, 15–19 June 2015.

27. Otero-Peérez, J.; Sandberg, R.D.; Mizukami, S.; Tanimoto, K. High-Fidelity Simulations of Multi-Jet Impingement Cooling Flows. *J. Turbomach.* **2021**, *143*, 081011. [CrossRef]
28. Xiao-ming, T.; Jing-zhou, Z.; Hua-sheng, X. Experimental investigation on impingement/effusion cooling with short normal injection holes. *Int. Commun. Heat Mass Transf.* **2015**, *69*, 1–10. [CrossRef]
29. Bu, X.; Peng, L.; Lin, G.; Bai, L.; Wen, D. Experimental study of jet impingement heat transfer on a variable-curvature concave surface in a wing leading edge. *Int. J. Heat Mass Transf.* **2015**, *90*, 92–101. [CrossRef]
30. da Soghe, R.; Mazzei, L.; Tarchi, L.; Cocchi, L.; Picchi, A.; Facchini, B.; Simon, M. Development of Experimental and Numerical Methods for the Analysis of Active Clearance Control Systems. *J. Eng. Gas Turbines Power* **2021**, *143*, 021018. [CrossRef]

encyclopedia

MDPI

Entry

Silicon Micro-Strip Detectors

Gregorio Landi [1,*] and Giovanni E. Landi [2]

1 Dipartimento di Fisica e Astronomia, Universita' di Firenze and INFN Sezione di Firenze, Largo E. Fermi 2 (Arcetri), 50125 Firenze, Italy
2 ArchonVR S.a.g.l., Via Cisieri 3, 6900 Lugano, Switzerland; giovannilandi@yahoo.it
* Correspondence: gregorio.landi@fi.infn.it

Definition: Silicon micro-strip detectors are fundamental tools for the high energy physics. Each detector is formed by a large set of parallel narrow strips of special surface treatments (diode junctions) on a slab of very high quality silicon crystals. Their development and use required a large amount of work and research. A very synthetic view is given of these important components and of their applications. Some details are devoted to the basic subject of the track reconstruction in silicon strip trackers. Recent demonstrations substantially modified the usual understanding of this argument.

Keywords: silicon micro-strips; tracker detectors; positioning algorithms; least-squares method; track reconstructions

Citation: Landi, G.; Landi, G.E. Silicon Micro-Strip Detectors. *Encyclopedia* **2021**, *1*, 1076–1083. https://doi.org/10.3390/encyclopedia1040082

Academic Editors: Raffaele Barretta, Ramesh Agarwal, Krzysztof Kamil Żur and Giuseppe Ruta

Received: 20 August 2021
Accepted: 22 October 2021
Published: 25 October 2021

1. Properties and Operation of Silicon Micro-Strip Detectors

The study of matter at extreme conditions requires collisions of elementary particles at the maximum energies allowed by the actual accelerators. Detailed analysis of the reaction products of those collisions enable the extraction of the physical parameters relevant for this study. The reaction products are a large set of other particles, some stable or of sufficiently long half-life to cross the nearby detectors. Charged particles are easily detected for their streams of ionization released in solid or gaseous matter. The neutral components of the reaction products require other very special detectors, able to transform their energy in detectable signals. The stream of ionization of the charged particles in solid or gaseous materials is a fundamental source of information about the properties of the incident particles, therefore, large efforts are dedicated to the observation and measurement of their ionization streams. Evidently, the amount of ionization, on the unit of path length, is an essential parameter for the quality of the detectors. The average ionization released in a solid silicon slab is around ten times that released in a gaseous detector. The production of an electron-hole pair, in a silicon detector, requires an average of 3.6 eV. Instead, the average ionization is around 30 eV in a gaseous detector. However, the collection of the charges released in silicon slabs is a much more elaborate operation than the equivalent operation in gaseous detectors. Thus, only in recent years, silicon detectors acquired extensive applications. The huge use of silicon crystals, in all the electronic devices, drastically reduced the high production costs of silicon crystals of the best quality (detector grade) required by the detector construction. Further details and plots on this subject can be found in Refs. [1,2].

For its electronic composition, a silicon crystal is a semiconductor, having a resistivity of 400 kΩ cm at 300 K, intermediate between that of a conductor and that of an insulator. Impurities and crystal imperfections give an effective resistivity around 10 kΩ cm [2], for mass production crystals. The effects of this reduction in resistivity, of the higher quality crystals, can be compensated resorting to the properties of reverse-biased diodes. In fact, in a semiconductor, the density and the type of current carriers can be tuned with the addition of appropriate impurities, able to add electrons to the conduction band or holes to the valence band, giving n-type or p-type material. The surface treatments of an n-type or p-type materials with the impurity of opposite property produce the diode structures (diode-junctions). The geometrical dispositions of those impurities on a surface can transform a slab 300 μm thick (the typical thickness of a detector) in an array of parallel lines of diode-junctions: a micro-strip silicon detector. The anisotropy of the impurity distributions has a drastic effect on the current conduction, allowing only the conduction for a given sign of the applied tension. Applying a reverse-bias (i.e., the tension that can not conduct the current in the diode), the density of the charge carriers is reduced in the thickness of the slab. The depletion from free charge carriers, of the detector active region, eliminates the recombination with the charges released by the ionizing particle, keeping intact the incoming ionization distribution.

The reverse-bias is increased to obtain the largest allowed depleted region. The reverse-bias has a maximum beyond which the electric field can accelerate the charge carriers (always present, even if in negligible concentrations) above the ionization of the material, generating a rapid rise of the diode current. Therefore, to have a detection, the maximum bias can not be reached. As an ionizing particle crosses the depleted region, the applied bias guides the two types of produced charges toward the corresponding collecting electrodes. In the actual experiments, the energies of the charge particles move their ionization release in regions with a relative minimum. Charged particles of this type are indicated as minimum ionizing particles (MIPs). Evidently, the particle detectors must efficiently operate for the MIPs. The charge released is proportional to the thickness of the depleted region. However, this region, and the thickness of the detector, must be limited to avoid that the hard scattering of the MIP, with the constituents of the crystal (multiple Coulomb scattering), randomly deviates the particle from its path. For a MIP in a silicon detector, 300 μm thick, the most probable number of electron-hole pairs is $\approx$ 23,000, collected by the corresponding electrodes in a few tens of nanoseconds. To handle such low charges, consistent amplifications with very low noise amplifiers are required. Hence, the strips of diode junctions are connected to the charge amplifiers thought decoupling capacitors (AC connection). These capacitors are easily produced with a thin layer of silicon oxide. The array of charge preamplifiers are realized in very large integration device (custom produced) with a separation, of the component amplifiers, equal to that of the strips; they are optimized for the corresponding applications [2] (Refs. [3,4] for those used in the CMS and ALICE trackers). The preamplifiers and their service electronics are contained in a different component (often called "hybrid"). Each preamplifier, of the array, is wire-bounded to the corresponding strip. A schematic set-up of a silicon micro-strip detector is represented in Figure 1 on the junction side.

Double sided strip detector. The structure of the micro-strip detectors discussed above, contains only a parallel array of linear diodes, evidently this structure allows the measurement of a single coordinate of the path of a MIP relative to the silicon surface (another coordinate is always given by the position of the sensor plane in the supporting mechanical structure). To measure the other coordinate, an additional detecting layer is required, with its strips orthogonal to the previous one. The combination of the two detections increases the random disturbance of the multiple Coulomb scattering to the particle tracks and augments the complexity of the trackers, and the weight for the satellite experiments [5] (Figure 2, array of double sided micro strips for a the PAMELA satellite). To eliminate a second detector, the other surface of the silicon slab, opposed to that with the array of diode-junctions, is armed with a set collecting electrodes oriented orthogonal to the direction of diode junctions [5–7]. Various specialized implants are required for the proper functioning of this second detecting layer (conventionally called ohmic side, Figure 1), among them, this side must be instrumented with preamplifiers and service electronics whose reference is not the ground, as usual, but the depleting bias. Special types of power supplies are required to distribute the electric currents necessary for the operation of the preamplifiers and the service electronics.

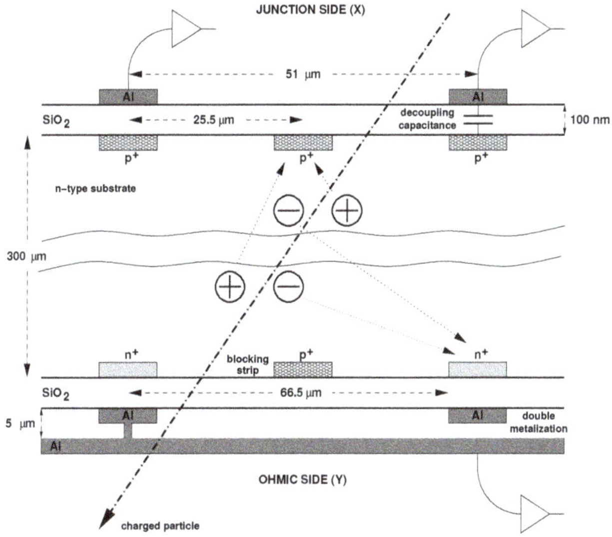

Figure 1. Schematic set-up of a double-sided micro-strip detector, the junction side is the set-up typical of the single side micro-strip detector.

Figure 2. Array of double sided micro-strip detectors for the payload for antimatter matter exploration and light-nuclei astrophysics (PAMELA) satellite experiment. The right side of the detector layers shows the connections to the service electronics.

3D micro strips. Recent developments of silicon detectors explore the possibility to confine the diode structure in deep holes orthogonal to the layer surface. As usual, a depleting bias is provided and now the electric field collects the released charges toward the nearest hole. The signals from an array of holes are collected by an external electrode (strip). The 3D micro-strips [8] are conceived to better sustain the huge radiation damage of the high luminosity large hadron collider (LHC).

Forms of the charge distributions. The charges of each sign, released by the ionizing particles drift toward the collecting electrodes with a constant velocity (different for the electrons and the holes) proportional to the local electric field. During their drifting times the electron and the holes are subject to a diffusion due to the thermal fluctuation of the medium (further mathematical details in Ref. [9]).

Assuming a continuous charge distribution along the particle track, the forms of the charge density at the detection plane can be calculated as in Ref. [9]. The projected distributions on a plane orthogonal to the strips are reported in Figure 3. The combination of drift and diffusion contributes to the form of the charge distributions arriving to the collecting electrodes. The sole drift in the electric field gives rectangular distributions with a side equal to the projection of the track on a direction orthogonal to the strips.

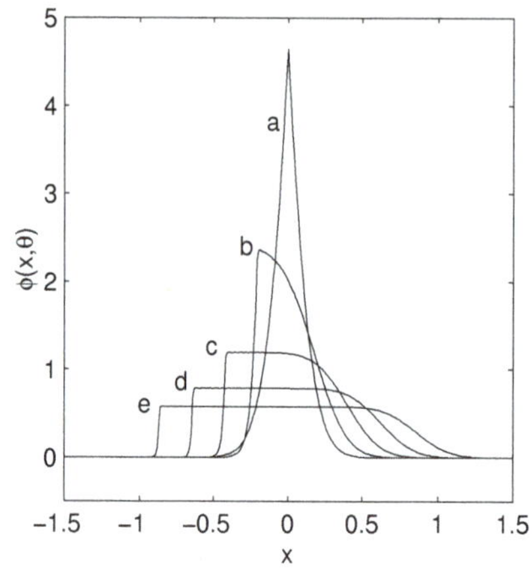

Figure 3. Forms of the charge distributions (all normalized to one) at various incidence angles calculated with a continuity assumption. The curves *a*, *b*, *c*, *d*, and *e* are, respectively, for $\theta = 0°$, $\theta = 5°$, $\theta = 10°$, $\theta = 15°$, and $\theta = 20°$ (from [9]).

The forms of Figure 3 can be considered averages over very large numbers of realistic distributions. In fact, the actual distributions are produced by a finite number of charges released along the particle track. This number fluctuates as a Landau distribution with the parameters corresponding to the crossed material. In addition, each segment of any track has itself a Landau distribution of released charges, according to the segment length. Each charge follows a random walk resulting from the combination of drifting and diffusion or scattering by the thermal fluctuations of the crossed medium. Along the track, some electrons acquire an energy well above the mean value of the other charges (δ-rays) and produce the tails of the Landau distributions above their most probable values. Thus, a large set of signal fluctuations must be expected as output of the strip preamplifiers.

Strip calibrations. The reconstruction of the particle signals in the detector requires some detailed operation on each strip. The outputs of the each strip-preamplifier are converted in a number by the analog digital converter (ADC). These numerical data require corrections with some strip parameters. The pedestals and strip noise must be calculated in absence of particle signals. The pedestals and the strip noise are constants and are calculated (or recalculated) when the status of the detector is supposed to be changed. Another parameter (defined as common noise) must be determined on event by event basis, it is particularly relevant for double sided detectors. The common noise is produced by the fluctuations of the power supply and other electromagnetic interferences, and it is constant for all the preamplifiers contained in the same integrated device. An iterative procedure allows the definition of these parameters (also discarding possible particle hits). A set of random triggers is generated, the first iteration supposes the absence of the common noise and the output of each strip is averaged for all the triggers. This average eliminates the noise of the read-out electronics and defines the effective zero (pedestal) of the strip signal. With the initial set of pedestals, the common noise for each trigger is calculated as the difference of the strip output and its pedestal averaged over all the strips connected to the same integrated amplifier device. This first set of common noise is subtracted from the output of each random trigger and a new set of pedestals is generated. These iterations are repeated until convergence. The strip noise is given by the root mean square of the differences of the strip outputs minus the pedestal and minus the common noise. The large majority of the strips have very similar noises, a small fraction (usually less than 5%) have higher noise. The noisy strips require a special attention in handling their data.

Cluster detection. The signals released by the charged particles (the MIP in particular) require careful criteria to distinguish them from the background noise. The forms of Figure 3 suggest that the particle signal distribution is spread on a small group of nearby strips: a cluster. To reduce the probability of false detections, a strip is inserted in a cluster if its signal is at least few times the strip noise. This threshold is tuned on the characteristic of the tracker detector and to efficiency required for the detection. Some experiments defines the seed of the cluster to be up to 7 times the corresponding strip noise. The other component of cluster are selected with a lower threshold. If several adjacent strips are classified as seed, the strip with the highest signal-to-noise ratio is the cluster seed. Additionally, the global quality of the cluster can be tested exploring its global signal-to-noise ratio. The cluster detection requires a due care to the inclusion (or exclusion) of the noisy strips.

Hit positioning. Typically, silicon micro-strip detectors are composed in regular arrays (called trackers) in a mechanical support structure finalized to reconstruct the paths of the incoming charged particles. The addition of a magnetic field allows the extraction of the particle momentum and sign of its charge. Thus, a dedicated algorithm must be applied on the detected clusters to extract the position where the particle hits the micro-strip detector. Evidently, for its geometry, a micro-strip detector can only give a single coordinate of the track path, that orthogonal to the strip direction (another coordinate is fixed by the mechanical structure of the sensor plane and is given by other measuring systems). Even if the dimensions of the clusters are in the range of few hundred microns, at most, much more precise positions can be extracted. The quality of the hit position has a fundamental effect on the track reconstruction and on the extraction of the track parameters. The center-of-gravity (COG) of the cluster is the most used positioning algorithms. The COG (sometimes called barycenter or weighted average) is defined by $X_g = (\sum_i C_i Y_i)/(\sum C_i)$, where C_i is the charge released in the strip i whose center (orthogonal to the strip direction) is in Y_i. This definition must be used with the due care. In fact, as discussed in Refs. [9,10], a set of systematic errors and anomalies are contained in this very simple definition. The first set of anomalies are gaps in the COG distributions that depends from the size of the charge distribution and the number of strips inserted in the COG algorithm. In general, an even number of strips generates gaps around $X_g = 0$, instead, an odd number has gaps around $X_g = \pm 1/2$ (the strip length is put to be one). As consequence, the corresponding probability density functions (PDFs) of positioning errors turn out to be very different. The PDFs, calculated in Refs. [11,12] illustrates these differences. To limit the influence of different PDFs, the same number of strips should be inserted in the COG calculation. This implies to neglect part of the information contained in the cluster, for clusters larger than the average, or insert strip-signals discarded by the algorithm of the cluster construction. Each case has a negligible effect on the hit position reconstruction. In fact, large clusters are produced by the high side of the Landau PDF and each of its strips has a good signal-to-noise ratio, thus the elimination of a lateral strip has a small effect on the result. Similarly, for the clusters with a lower number of strips, they are given by the low charge side of the Landau PDF, with a low signal-to-noise ratio, the addition of another strip add a small part of signal. In any case, the hit quality is low and the addition of another strip-signal has a negligible effect on the hit resolution. This recipe is useful for clusters formed by a single strip where the COG algorithm can not be used.

Another very important weakness of the COG algorithm is a systematic error contained in its position determination [9,10]. An effective algorithm (called η-algorithm) was defined in Ref. [13], further refinements are contained in Refs. [9,14], converging toward the elimination of this error. With η-correction, the hit positioning has only a random noise, becoming an unbiased position estimator. Its variance can be further reduced by a best fit with other hits of the particle in the detecting layers of the tracker.

Track fitting. The final use of the silicon micro-strips is in the track reconstruction in tracker detectors. These detectors are always immersed in homogeneous magnetic fields that impose to the incoming particles circular paths in the direction orthogonal to the field direction (helical paths in three dimensions). The radius of each circle is proportional to the particle momentum orthogonal to the magnetic field (transverse momentum). The arrays of micro-strips are arranged to optimize the measurement of the track bending and the extraction of the transverse particle-momentum. For example, the micro-strips are arranged in cylindrical surfaces if the magnetic field is directed parallel to the beam pipe of the accelerator (as in CMS [3] or ALICE [4]). The track fitting is always done with the standard least squares (sometimes called linear regression) even if for various necessity is implemented as the Kalman filter. The key point of these fitting methods is the assumption of an identity of the variance of the measurements (homoscedasticity). Rare deviations from this assumption are allowed, the observations with variances larger than the rest of the observations are called outliers. The Kalman filter is supposed to be able to detect the rare outliers, and their elimination restores the consistency of the approach. The assumption of homoscedasticity implies an enormous simplification to the fitting algorithms. The algorithms are independent from the details of the detectors. At most, to account for substantial different technology of a given detector layer, a single parameter is introduced for all the hits of that layer.

In addition, the homoscedasticity gives equations for the extraction of the observation variance (a single value) from the mean square of the fitting residuals; equations inconsistent for real systems. Often the variance, obtained with the homoscedastic equations, is improperly indicated as an experimental determination. Another assumption is the Gaussian PDF of the hit errors. If the inconsistency of homoscedasticity is proved, the optimality of all the fitting procedure breaks down.

Toward optimal fittings. Ref. [15,16] proved that for non-homoscedastic (heteroscedastic) systems, the use of homoscedastic algorithms implies an increase in the variance (often very large) of the estimators. Due to the definition of optimal algorithm, as that with minimum variance, each fit on heteroscedastic systems with homoscedastic fitting algorithms turns out to be non-optimal with a consistent loss of resolution. The very different analytical forms of the COG PDF of Refs. [11,12] prove that the simple neglect of the strip number differences in the calculation of the COGs introduces an evident trivial heteroscedasticity also in homoscedastic systems or increases the hit-differences in the heteroscedastic ones. Another source of heteroscedasticity is the different signal-to-noise ratio of each hit given by Landau PDF of the charge released along the track. Furthermore, the simple observation of the scatter plot of any simulation of the two (or three) strip COG in function of the impact point (as the left side of Figure 4) shows immediately the impossibility of a single variance for each COG value. It must be recalled that also the histogram of the two strip COG (or three strip) shows substantial variations. These can only be generated by large variations of the probability, for the noise, to produce a given COG value for each position on the strip width (uniform population of hits on the strip is always assumed). This correlation is sufficiently strong the be directly used in the track fitting as weight of the hit with the η positioning (the lucky model of Ref. [17]) giving a good improvement of the estimator resolutions. Better results can be obtained if effective variances, for each hit, is extracted by the two-dimensional PDFs for COG error in function of the impact points for the charge released by the MIP in the seed strip and the two lateral ones. A sample of those PDFs is reported in the right side of Figure 4. The calculated effective variances are inserted in the weighted least squares (the schematic model of Ref. [17]). Similar approximate uses of the full PDFs were suggested also by Gauss [18] in his book of 1821 to avoid the complexity of the full maximization of products of PDFs (a century later the method acquired the name of maximum likelihood). However, the mean variance of the estimators is further reduced with the search of the maximum likelihood (just the procedure not-recommended by Gauss for its complexity). The hit PDFs are extracted from surfaces similar to that of Figure 4 for constant values of the (two strip) COG and with the charge collected by three strips of the cluster (ref. [17] and therein references). Again these variances could not be the minima, other improvements of minor details could produce further reductions. However, the hit-error PDFs substantially differs from Gaussian PDFs, hence, the linear equations of the least squares no longer coincide with the likelihood maximization. The hit-error PDFs have heavy tails, similar to Agnesi–Cauchy PDFs, and the maximization of the likelihood can only be done with numerical methods.

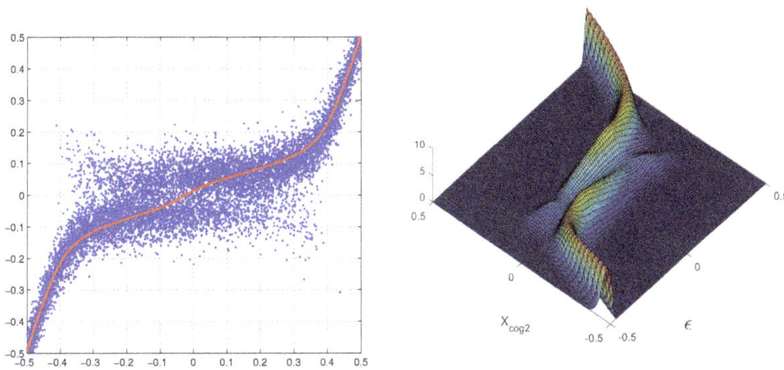

Figure 4. Left plot: Scatter plot of a two strip COG in function of the impact point for silicon strip detector of general type. Right plot: Calculated probability density function for a two strip COG (X_{cog2}) in function of the impact point (ϵ) with the most probable signal released by a MIP in three strips.

2. Summary and Conclusions

This synthetic discussion of the silicon micro-strip detectors illustrates only essential features of these instruments. Some details are reported of their set-ups, applications and the fitting methods, specialized for high energy physics. Evidently, the use of ionizing radiations is extended well beyond high energy physics, therefore these devices find applications in a large set of industrial, medical and technological environments with dramatic increases of performance compared to the previous detection systems. However, the next upgrade of LHC, with its large increase in the reaction rates, requires additional developments of these detectors to survive to the intense radiations produced by the beam collisions and fast acquisition times to faithful reconstruct the signals released by the reaction products.

Author Contributions: Conceptualization, G.L. and G.E.L.; software, G.E.L. All authors have read and agreed to the published version of the manuscript.

Funding: This research received no external funding.

Conflicts of Interest: The authors declare no conflicts of interest.

Entry Link on the Encyclopedia Platform: https://encyclopedia.pub/16469.

Abbreviations

The following abbreviations are used in this manuscript:

MIP	Minimum ionizing particle
COG	Center of gravity
PDF	Probability density function

References

1. Hartmann F. Silicon tracking detectors in high-energy physics. *Nucl. Instrum. Meth. A* **2012**, *666*, 25. [CrossRef]
2. Zyla, P.A.; Particle Data Group. Review of Particle Physics. *Prog. Theor. Exp. Phys.* **2020**, *083C01*, 569–573. [CrossRef]
3. The CMS Collaboration. The CMS experiment at the CERN LHC. *JINST* **2008**, *3*, S08004.

4. The ALICE Collaboration. The ALICE experiment at the CERN LHC. *JINST* **2008**, *3*, S08002.
5. Adriani, O.; Bongi M.; Bonechi, L.; Bottai, S.; Castellini, G.; Fedele, D.; Landi, G.; Grandi, M.;, Papini, P.; Vannuccini, E.; et al. In-flight performance of the PAMELA magnetic spectrometer. In Proceedings of the 16th International Workshop on Vertex Detectors, Lake Placid, NY, USA, 23–28 September 2007. [CrossRef]
6. Batignani, G.; Bosi, F.; Bosisio, L.; Conti, A.; Focardi, E.; Forti, F.; Giorgi, M.A.; Parrini, G.; Scarlini, E.; Tempesta, P.; et al. Double sided read out silicon strip detectors for the ALEPH minivertex. *Nucl. Instrum. Methods Phys. Res. A* **1989**, *277*, 147–153. [CrossRef]
7. Adriani, O.; Alcaraz, J.; Ahlen, S.; Ambrosi, G.; Baschirotto, A.; Battiston, R.; Bay, A.; Bencze, G.; Bertucci, B.; Biasini, M.; et al. The New double sided silicon microvertex detector for the L3 experiment. *Nucl. Instrum. Methods Phys. Res. A* **1994**, *348*, 431–435. [CrossRef]
8. Meschini, M.; Cassese, A.; Ceccarelli, R.; Viliani, L.; Dinardo, M.; Gennai, S.; Zuolo, D.; Messineo, A.; Parolia, S.; Ebrahimi, A.; et al. Radiation resistant innovative 3D pixel sensors for the CMS upgrade at the High Luminosity LHC. *Nucl. Instrum. Methods Phys. Res. A* **2020**, *978*, 164429. [CrossRef]
9. Landi G. Problems of position reconstruction in silicon microstrip detectors. *Nucl. Instrum. Methods Phys. Res. A* **2005**, *554*, 226. [CrossRef]
10. Landi G. The center of gravity as an algorithm for position measurements. *Nucl. Instrum. Methods Phys. Res. A* **2002**, *485*, 698. [CrossRef]
11. Landi, G.; Landi, G.E. Probability Distributions of Positioning Errors for Some Forms of Center-of-Gravity Algorithms. *arXiv* **2020**, arXiv:2004.08975.
12. Landi, G.; Landi, G.E. Probability Distributions of Positioning Errors for Some Forms of Center-of-Gravity Algorithms. Part II *arXiv* **2020**, arXiv:2011.14474.
13. Belau, E.; Klanner, R.; Lutz, G.; Neugebauer, E.; Seebrunner, H.J.; Wylie, A. Charge collection in silicon strip detector. *Nucl. Instrum. Methods Phys. Res. A* **1983**, *214*, 253–260. [CrossRef]
14. Ebrahimi, A; Feind, F.; Fretwurst, E.; Garutti, E.: Hajheidari, M.; Klanner, R.; Pitzl, D.; Schwandt, J.; Steinbrück, G.; Zoi, I. Position reconstruction for segmented detectors. *Nucl. Instrum. Methods Phys. Res. A* **2021**, *1014*, 165744.
15. Landi, G.; Landi, G.E. Generalized inequelities to optimizing the fitting method for track reconstructions. *Physics* **2020**, *2*, 608–623.
16. Frühwirth, R. Regression with Gaussian mixture models applied to track fitting. *Instruments* **2020**, *4*, 25. [CrossRef]
17. Landi, G.; Landi, G.E. Beyond the $\sqrt{N}$-limit of the least squares resolution and the lucky-model. *arXiv* **2018**, arXiv:1808.06708.
18. Gauss, C.F. Méthode des Moindres Carrés. Mémoires sur la Combination des Observations. Franch Translation by J. Bertrand; Revised by the Author; Mallet-Bachelier: Paris, France, 1855. Available online: https://books.google.it/books?id=_qzpB3QqQkQC (accessed on 1 September 2018).

encyclopedia

Entry

Natural Disasters—Origins, Impacts, Management

Muhammad T. Chaudhary [1] **and Awais Piracha** [2,*]

[1] Department of Civil Engineering, Kuwait University, Shadadiya Campus, P.O. Box 5969, Safat 13060, Kuwait; tariq.chaudhary@ku.edu.kw
[2] School of Social Sciences and Psychology, Western Sydney University, Penrith, NSW 2751, Australia
* Correspondence: a.piracha@westernsydney.edu.au

Definition: Natural hazards are processes that serve as triggers for natural disasters. Natural hazards can be classified into six categories. Geophysical or geological hazards relate to movement in solid earth. Their examples include earthquakes and volcanic activity. Hydrological hazards relate to the movement of water and include floods, landslides, and wave action. Meteorological hazards are storms, extreme temperatures, and fog. Climatological hazards are increasingly related to climate change and include droughts and wildfires. Biological hazards are caused by exposure to living organisms and/or their toxic substances. The COVID-19 virus is an example of a biological hazard. Extraterrestrial hazards are caused by asteroids, meteoroids, and comets as they pass near earth or strike earth. In addition to local damage, they can change earth inter planetary conditions that can affect the Earth's magnetosphere, ionosphere, and thermosphere. This entry presents an overview of origins, impacts, and management of natural disasters. It describes processes that have potential to cause natural disasters. It outlines a brief history of impacts of natural hazards on the human built environment and the common techniques adopted for natural disaster preparedness. It also lays out challenges in dealing with disasters caused by natural hazards and points to new directions in warding off the adverse impact of such disasters.

Keywords: natural hazards; disasters; global impacts; disaster management; built environment

Citation: Chaudhary, M.T.; Piracha, A. Natural Disasters—Origins, Impacts, Management. *Encyclopedia* **2021**, *1*, 1101–1131. https://doi.org/10.3390/encyclopedia1040084

Academic Editors: Raffaele Barretta, Ramesh Agarwal, Krzysztof Kamil Żur and Giuseppe Ruta

Received: 22 September 2021
Accepted: 28 October 2021
Published: 30 October 2021

1. Introduction

Earthquakes, floods, cyclones, storms, wildfires, volcanic eruptions, and landslides are natural processes that have sculptured the landscape of the earth for millenniums. These natural processes can cause natural disasters on interaction with human-made features such as settlements, agriculture, and infrastructure. This article begins with an overview of the various natural processes that have potential to cause natural disasters. After that, a brief history of impacts of natural hazards on the human built environment is provided, followed by a description of the common techniques adopted for natural disaster management. The chapter concludes with a review of challenges in dealing with disasters caused by natural hazards and points to new directions in building the capacity to ward off the adverse impact of natural disasters on vulnerable sections of society.

2. Natural Processes or Natural Hazards

The natural processes (or hazards) that are the triggers for natural disasters are broadly classified into six categories [1,2]. The definitions and descriptions of each hazard are as follows:

1. Geophysical: This is also termed as geological hazard and originates from the solid crust of the Earth. The events associated with this hazard include earthquakes, volcanic activity, and dry mass movement.
2. Hydrological: This hazard is associated with the occurrence, movement, and distribution of fresh and saltwater over or beneath the Earth's surface. The events created by this hazard include floods, landslides, and scour and wave action.

3. Meteorological: This hazard constitutes short-lived events having a time-span of minutes to a few days and are caused by micro- (<1 km) to meso-scale (2~2000 km) atmospheric conditions that can be exacerbated by global climate change. Convective storms (or tornadoes), extra-tropical storms (occurring in the middle, i.e., 30° to 60° latitude), tropical storms (occurring up to 30° latitude), fog, and sudden extreme temperature variation are included in this category of hazard.
4. Climatological: A hazard linked with variability in climate over a broad time-span ranging from intra-season to multi-decade at a meso- to macro- (>2000 km) scale. Droughts, wildfires, glacial movement, and glacial lake outburst are some of the events associated with this hazard.
5. Biological: A hazard originating from a biological substance, e.g., venom, mold, or a vector carrying disease-causing organisms, exposure to which poses a threat to other living beings or humans. Locust swarms, algal blooms, venomous wildlife infestation, and vector-borne diseases such as plague, malaria, dengue, and COVID-19 are some examples of this hazard.
6. Extraterrestrial: A hazard originating outside the Earth's atmosphere that may be caused by residues of asteroids, meteorites, comets, or human space debris, when these enter Earth's atmosphere, or the impact caused by these objects on Earth's surface. This hazard may also be caused by interplanetary conditions such as solar flares that can cause disruption in the Earth's magnetosphere, thermosphere, or ionosphere.

Figure 1 depicts these categories of hazards along with main events and pertinent examples of peril/harm for each type of natural hazard. This article will explore the origin, impact, and management of natural disaster in the context of geophysical, hydrological, and climatological hazards only. These hazards were chosen in this article due to their pronounced impact on the human-built infrastructure and the socio-economic consequences posed by disasters caused by these natural hazards.

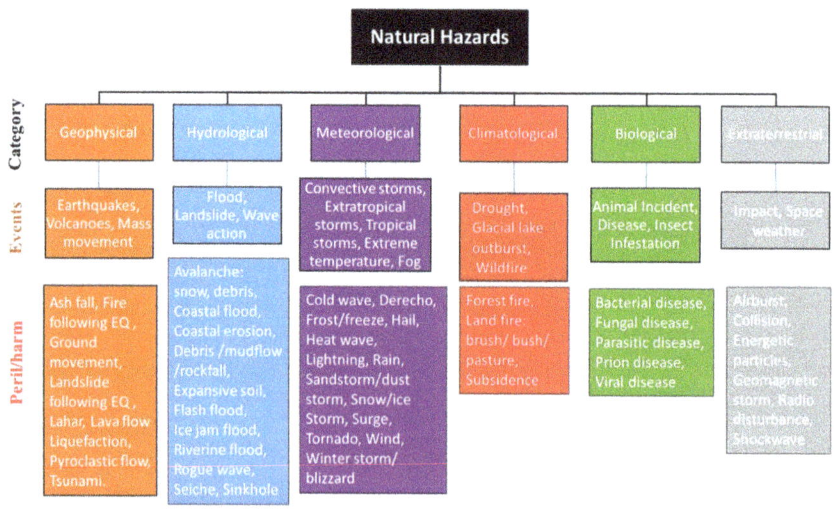

Figure 1. Classification of natural hazards with examples of events and peril/harm for each category.

3. Definitions and Terminologies

The field of natural disaster preparedness and hazard mitigation has been dynamically evolving for the past six decades. It is therefore necessary to obtain a clear understanding of various terms used in the context of natural disaster planning, preparedness, and mitigation.

3.1. Hazard

According to Cutter [3], *"A hazard, in the broadest term, is a threat to people and the things they value. Hazards have a potentiality to them (they could happen), but they also include the actual impact of an event on people or places. Hazards arise from the interaction between social, technological, and natural systems"*. This definition of hazard implies that the interaction between the natural and the social systems is the key element, which transforms a natural process to a hazard. It is also to be understood that 'hazard' by itself is harmless, as it is only a 'threat' that has the potential to cause harm. Therefore, Federal Emergency Management Agency (FEMA) [4] portrays hazards as *"events or physical conditions that have potential to fatalities, injuries, property damage, infrastructure damage, agricultural loss, damage to the environment, interruption of business, or other types of harm or loss"*. In the same vein, the United Nations International Strategy for Disaster Reduction (UNISDR) [5] defines a natural hazard as *"any natural process or phenomenon that may cause loss of life, injury or other health impacts, property damage, loss of livelihoods and services, social and economic disruption or environmental damage"*.

3.2. Exposure

According to the Cambridge Dictionary, exposure is *"the fact of experiencing something or being affected by it because of being in a particular situation or place"*. Therefore, in the context of natural disaster management, exposure refers to the inventory of elements (i.e., people, property, systems, or functions) in an area in which hazard events may occur [6,7]. Hence, if human or capital resources are not located in an area that is exposed to natural hazard(s), then there is no risk of a natural disaster. Exposure to a hazard is a necessary but not a sufficient requirement for a disaster situation to develop. For example, an asset could be exposed to a hazard but may possess sufficient capacity to withstand the hazard without damage and thus avoiding a disaster.

3.3. Vulnerability

Vulnerability refers to the susceptibility to loss of human life, physical injury, or economic loss of livelihoods and assets when exposed to hazard events [6,7]. The extent of vulnerability depends on the construction, predisposition, fragilities, inherent capacity, or weakness of the exposed elements [8].

3.4. Disaster

According to the Merriam Webster dictionary, a disaster is defined as *"a sudden calamitous event bringing great damage, loss, or destruction"*. Therefore, a disaster is an actual event having unfavorable consequences, unlike a hazard or risk, which is a potential threat. In the lexicon of natural disaster management community, a disaster is *"a serious disruption of the functioning of a community or a society involving widespread human, material, economic or environmental losses or impacts which exceed the ability of the affected community or society to cope using its own resources"* [5]. On the other hand, according to the Center for Research on the Epidemiology of Disasters (CRED) [9], a disaster is *"a situation or event which overwhelms local capacity, necessitating a request to a national or international level for external assistance; an unforeseen and often sudden event that causes great damage, destruction and human suffering"*.

Natural hazards have their origins in natural processes, but disasters affect a community and have social consequences that disrupt societal functioning and cause human and/or material loss. A hazardous process (event) occurring in an uninhabited region is not termed as a disaster, as it does not influence people (society) and their possessions (infrastructure). Similarly, occurrence of a hazardous process in a community that has built sufficient protection against such an event may also avoid a disaster. The following mnemonic expression eloquently describes this relationship:

$$\text{Disaster} = (\text{Hazard} + \text{Vulnerability})/\text{Capacity} \quad (1)$$

The degree of exposure to a hazard and the level of vulnerability is directly related to the magnitude of a disaster, whereas disaster magnitude is inversely proportional to capacity.

3.5. Risk

The Oxford English Dictionary defines risk as "*(Exposure to) the possibility of loss, injury, or other adverse or unwelcome circumstance; a chance or situation involving such a possibility*". Ansell and Wharton [10] argue: "*Risk is the likelihood of an event's occurrence multiplied by the consequences of that event, if it occurs*" and can be stated as the following mnemonic:

$$\text{Risk (R)} = \text{Hazard (H)} \times \text{Vulnerability (V)} \quad (2)$$

Risk depends on the combination of hazard, vulnerability, and exposure. Risk is the estimated impact that a hazard would have on people, services, infrastructure, and physical assets in a community. It refers to the likelihood of a hazard event becoming a disaster [7].

Wisner et al. [11] modified the relationship presented in (2) by including personal protection capacity (C) and larger scale risk mitigation measures at the societal level (M) and proposed the following mnemonic relationship between these variables:

$$R = H \times [(V/C) - M] \quad (3)$$

It is to be noted that the expressions given by the mnemonics above are not exact mathematical relationships but are merely attempts to correlate various factors in the complex phenomenon.

4. Theories of Natural Disasters

Theories of origin of disasters have evolved over time, showing advancements in human understanding of the physical natural phenomena and their interaction with the social systems and infrastructure built by humankind [12]. An understanding of these theories is necessary for natural disaster planning, preparedness, and mitigation. Four theories of disaster are briefly reviewed in the following.

4.1. Disaster as a Retribution—An Act of God

Earliest usage, with continued acceptance in some communities, suggests that disasters are acts of God, which happen as "*a divine retribution for human misdeeds and failings*" [13]. A recent study found that the concept of disasters as act of God is still prevalent worldwide, and such a belief is strengthened after occurrence of a major natural disaster [14]. This fatalistic viewpoint encourages accepting the negative consequences of such event(s) as part of one's fate and proposes that mitigation of a disaster's impact is beyond human capacity [15]. Such fatalistic attitude could be one of the reasons for lack of disaster preparedness and adoption of better land-use planning and disaster mitigation measures in many parts of the world [16–18]. However, it is to be noted that the disaster risk management (DRM) community has moved away from this theory of disasters since the 18th century.

4.2. Disaster as a Physical Phenomenon—An Act of Nature

Progress in scientific thinking and knowledge after the Renaissance started to alter the perception of disasters from the supernatural paradigm to the natural physical realities. The Lisbon earthquake of 1755 was probably the first natural disaster that shaped the viewpoint of natural and geophysical phenomena as the agents responsible for a natural disaster [19]. According to Dynes [19]:

> "*Prior to that, earthquakes traditionally had been interpreted as a dramatic means of communication between gods and humans. In particular, such events previously had been explained as indicating some disturbance between earthly and heavenly spheres. The Lisbon earthquake can be identified as a turning point in human history which moved the*

consideration of such physical events as supernatural signals toward a more neutral or even a secular, proto-scientific causation".

This theory of disasters became widely accepted by the early 20th century. However, the fatalism associated with disasters remained to some extent, especially for the geophysical hazards of earthquakes and volcanic activity. The only difference was the change in the causative agent, from God to Mother Nature.

This theory was instrumental in the adoption of engineering measures to 'tame' the natural forces that cause disasters in human settlements. The earliest examples of such attempts can be found in the building of river dams in the Middle East about 4000 years ago and earthquake-resistant dwellings in China about 2000 years ago [20]. Great strides were made in understanding the origin, physical causative mechanism, and future prediction of natural hazards (e.g., floods, earthquakes, storms, volcanic activity, etc.) after the advent of the industrial age in the late 18th century. Continuous discovery and innovation in this field continue today. This scientific knowledge was then utilized for engineering solutions that can either 'tame' the forces of nature (as is the case with flood control dykes, dams, embankments, and related irrigation works) or withstand the impact of brutal forces unleashed by natural phenomena such as earthquakes, windstorms, or volcanic activity by building strong, ductile, and integrated structures.

However, despite the adoptions of these engineered solutions, continuously increasing human life and economic losses stemming from natural disasters in the early half of the 20th century led to the realization that natural phenomena alone are not the only cause of disasters, and the problem cannot be adequately solved by adopting hard scientific and engineering methods alone. This led to the third theory, that disasters happen due to interaction between natural phenomena and societal systems.

4.3. Disaster as an Act of Nature–Human Interplay

Carr [21] was the first to propose that disasters occur due to the interaction between a geophysical (natural) system and a human-use system. Absence of either one will not result in a disaster. For example, a powerful earthquake happening in a remote uninhibited area is a natural hazard but will not result in a disaster.

After observing the limits of flood protection works to reduce economic losses in the USA, White [22] introduced his theory that disasters have a societal dimension, in addition to the presence of a geophysical hazard agent and the human-use system. He noted with dismay that reliance placed on the engineered solutions of flood protection works encouraged the social behavior of development of flood-prone lands for short-term economic gains. However, such actions resulted in greater economic loss after failure of the flood protection system. He advocated the 'human ecology' concept of Barrows [23], which calls for judicious land use planning and interconnectivity between the natural and the human systems for betterment of the society as well as the natural environment. This concept was applied to more complex interactions in subsequent studies by various authors [24–26].

A similar viewpoint of ecological design was championed by McHarg [27] for urban planning, which called for modification in the natural face of the earth for human use with due consideration to the ecology of the landscape. He argued that such planning will reduce the impact of natural hazards on human settlements. Recent studies [28,29] applied principles advocated by McHarg to the 2011 Fukushima Nuclear Power Plant disaster in Japan and to the settlements in Staten Islands subjected to Hurricane Sandy in 2012, respectively, and concluded that the economic impact of these disasters could have been considerably reduced by implementing the ecological design principles proposed by McHarg.

4.4. Disaster as a Complex Nexus of Natural-Human-Social-Economic Factors

By the late 20th century, it was clear that certain nations and segments of population were more vulnerable to the impact of natural disasters than others. Researchers and international donor agencies tried to uncover links between under-development and effects of natural disasters [30]. Two disturbing facts were noted: (1) disaster fatalities were disproportionately higher in the least developed countries (LDCs), and (2) although the absolute economic loss in LDCs was lower than that of the more developed countries (MDCs), the per capita cost of natural disasters in terms of GDP was more than 20 times higher in LDCs than in MDCs [20,31]. LDCs were noted to be caught up in a viscous circle of under-development, which was exacerbated by recurrence of natural disasters at regular intervals, which diverted the scarce and often borrowed human and infrastructure development funds to relief and reconstruction activities [32].

It is worthwhile to point out that O'Keefe et al. [33] questioned the use of the term 'natural disasters', as it appears to wrongly attribute disasters to nature and masks the role of decision makers responsible for underdevelopment, which is the main cause of people's increased vulnerability to natural hazards. Recently, Chmutina and von Meding [34] carried out an extensive analysis of the usage of the term 'natural disasters' by researchers and professionals involved in disaster risk management (DRM) and concluded that most of the authors see its usage as a 'convenience term' while being fully aware that non-natural factors are mainly responsible for turning a natural hazard to a disaster. Recently, a shift in discontinuing the usage of this misleading term can be noticed in academia [35–38] and DRM organizations [39–42]. However, it is anticipated that the usage of the term will continue among disaster academicians, professionals, journalists, and mass media in the near future. Therefore, the term 'natural disaster' is used in this article to point to the trigger or the hazard that has its origin in the 'natural' physical phenomena and not to people's vulnerabilities associated with 'acts of nature'.

Cultural and social aspects, political instability, lack of will, civil unrest, fatalistic beliefs, and other anthropologic dimensions present a myriad of challenges in implementing capacity building and vulnerability mitigation measures against natural disasters [43]. Therefore, various holistic approaches to natural disaster mitigation have been proposed, which strive to integrate the diverse physical triggers of natural hazards, associated engineering solutions, and socio-economic–political–cultural dilemmas [44–47]. Increasingly, humans are no longer seen as the victims of natural disasters but as contributors to the misery caused by a hazardous natural process through irrational human exploitation of natural resources, contribution to climate change, and inefficient functioning of the political and economic systems [11,48,49]. Therefore, in order to address this complex paradigm, a long-term focus is needed on capacity building that ensures equitable distribution of economic resources, reduction in poverty, and participation of local communities in incorporating local knowledge and practices in the proposed solution to natural hazards [50–52].

5. Global Impact of Natural Hazards

Natural hazards have interacted with human settlements since the dawn of civilization. Accounts of such encounters are preserved in ancient religious texts, historical accounts, and local folklore around the globe. It is reasonable to assume that the number of reported natural disasters, as well their impact on human life and property, has historically increased over the course of time with growth in human population as well as inhibition of hazard-prone areas. In more recent times, evidence has been documented that disaster risk and the occurrence of disasters have significantly increased over the last six decades [53]. Boudreau [54] estimates that about 85% of the world's population has been affected by at least one natural disaster in the past 30 years. Economic impact of routine natural disasters (i.e., without counting a significantly large event) is estimated to be around USD 100–200 billion/year worldwide since the 1990s [53,55]. The grim reality related to the economic development of countries is amply reflected in the facts that 90% of the fatalities attributed to natural disasters occur in developing countries, while 90% of the economic loss is borne by the developed nations [56].

This section examines the distribution and impact disasters caused by three types of natural hazards, i.e., geophysical, hydrological, and climatological, which happened around the world since 1900. The reason for focusing on this type of disasters is given in Section 2. Data used to compile the presented information were obtained from the Emergency Events Database (EM-DAT) maintained by the Centre for Research on the Epidemiology of Disasters (CRED) at the Catholic University of Louvain, Belgium [57]. For a disaster event to be listed in this database, at least one of the following criteria is to be met:

1. Deaths: 10 or more people;
2. Affected: 100 or more people affected/injured/homeless;
3. Declaration/international appeal: Declaration by the country of a state of emergency and/or an appeal for international assistance.

Total estimated damages and direct or indirect economic losses in the EM-DAT are based on equivalent of 2017 US dollars.

A knowledge of global disaster hotspots resulting from various natural hazards is the first step in mitigating the impact of these disasters. Therefore, subsequent sections are devoted to identification of these hotspots based on the available natural hazard and disaster loss data.

5.1. Geophysical Disasters

Geophysical disasters are the result of earthquakes, dry mass movement, or volcanic activity. Figure 2 presents the distribution of such events, the death toll, persons affected, and the economic loss in various regions of the world from 1900–2020 AD. It is to be noted that more than 99% of the fatalities, adverse effects, and economic loss has been due to earthquakes, and the share of dry mass movement and volcanic activity is miniscule. More than 2.4 million people have lost their lives, and more than 206 million people have been injured or rendered homeless/jobless in these disasters. The economic toll of these events for the noted period is estimated to be more than USD 1.3 trillion. It is to be noted that the Asia–Pacific region is significantly affected by these disasters, as more than 50% of the entries in all the four categories were in this region, which can be attributed to the geomorphology, geophysical location, and larger area and population of this region.

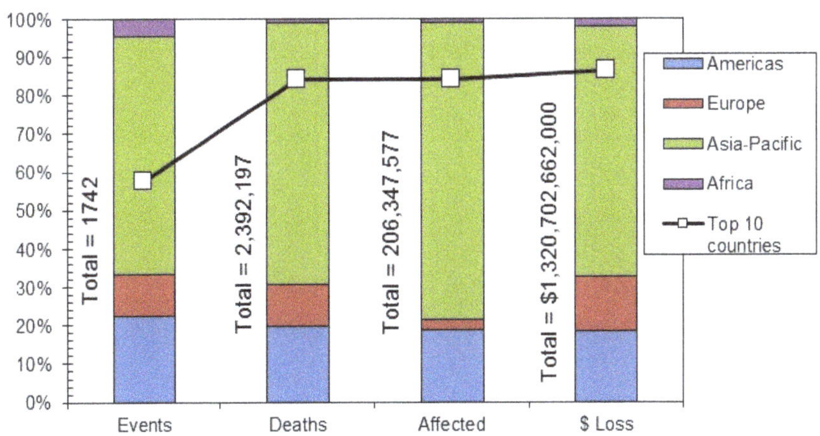

Figure 2. Distribution of events, deaths, persons affected, and economic loss due to geophysical disasters across the globe during the period 1900–2020 AD (Data Source: [57]).

It is also noteworthy in Figure 2 that almost half of these events occurred in the top 10 countries, while the burden of human and economic loss shared by the top 10 countries is more than 80% of the overall total. This finding is in line with the earlier observation that, since the year 1900 AD, only about 100 major earthquakes have resulted in more than 93% of the total fatalities [58] and more than 95% of the total economic loss [59].

The distribution of number of geophysical disaster events, fatalities, number of affected persons, and economic loss for the top 10 countries in each category for the period 1900–2020 is depicted on the world map in Figure 3 and listed in Table 1. Although the numbers of events are distributed more or less evenly among the top five countries, the fatalities and number of affected are disproportionately higher in China (32.5% and 44.9%, respectively), while Japan bore the brunt of the economic loss (46.4%). Italy is the only European country that is among the top 10 countries in three categories, while no country from the African continent is among the pool of top 10 countries for any category. Haiti tops the fatality list in the Americas, while Peru is the only country in this region listed in three categories. More detailed maps of these hotspots are presented in [60].

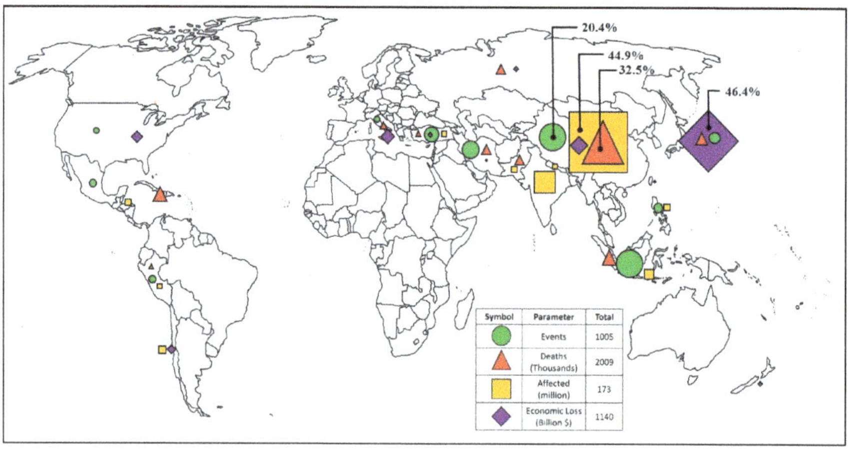

Figure 3. Global hotspots for geophysical disasters based on data for the top 10 countries in each category from 1900 to 2020 AD (Data Source: [57]).

Table 1. Distribution of events, fatalities, affected people, and economic loss in top 10 countries due to geophysical disasters from 1900 to 2020 AD (Data Source: [57]).

Country	Events (%)	Country	Fatalities (%)	Country	Affected (%)	Country	Economic Loss (%)
China	20.40	China	32.50	China	44.96	Japan	46.37
Indonesia	19.40	Haiti	11.08	India	16.49	China	13.31
Iran	13.03	Indonesia	10.79	Indonesia	6.96	Italy	10.91
Turkey	11.04	Japan	9.39	Chile	5.78	USA	8.88
Japan	8.46	Iran	7.68	Philippines	5.34	Chile	6.54
Philippines	7.16	Russia	7.28	Guatemala	4.73	Turkey	3.47
Peru	5.77	Pakistan	7.18	Pakistan	4.29	Russia	3.42
Mexico	5.17	Italy	5.79	Turkey	4.14	New Zealand	2.91
Italy	4.88	Turkey	4.69	Nepal	3.68	Iran	2.12
USA	4.68	Peru	3.62	Peru	3.65	Taiwan	2.08
Total	1005		2009 (thousand)		173 (million)		1140 (billion USD)

5.2. Hydrological Disasters

Hydrological disasters are caused by floods, landslide, and wave action, with riverine flooding being the most dominant cause for such disasters. The distribution of these events in various regions of the world from 1900–2020 AD, the associated death toll, persons affected, and economic loss are depicted in Figure 4. More than 7 million people lost their lives in these events, and more than 3.8 billion people were injured or displaced in these disasters. The economic toll of these events for the noted period is estimated to be around 1.3 trillion dollars. China was the most affected country, losing more than 6.6 million people in these deluges that also affected more than 2 billion Chinese people. Economic loss in China was also the highest, at more than USD 413 billion. It can be noted in Figure 4 that the Asia–Pacific region was the most affected by these disasters, as more than 50% of events happened here that caused more than 95% of the total global fatalities and displacement of people, as well as more than 60% of the total global economic losses. Social impact of these disasters, in terms of fatalities and people affected, in the Americas and Europe was very low, whereas the economic losses were close to 40% of the global total. This trend of human and economic loss is similar to the one noted for the geophysical disasters across various regions of the world.

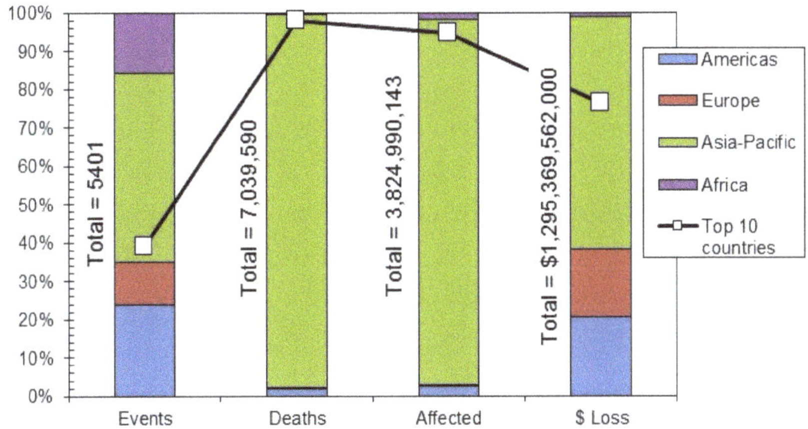

Figure 4. Distribution of events, deaths, persons affected, and economic loss due to hydrological disasters across the globe during the period 1900–2020 AD (Source: [57]).

The distribution of number of hydrological disaster events, fatalities, number of affected persons, and economic loss for the top 10 countries in each category for the period 1900–2020 AD are depicted on the world map of Figure 5, while the percentage distribution is listed in Table 2. Although the numbers of events are distributed evenly among the top five countries, the fatalities, number of affected, and economic loss are disproportionately higher in China (96.02%, 58.22%, and 41.93%, respectively). China lost more than 6.2 million people in just three events. Figure 5 reveals that China and the Indian sub-continent are the hotspots for these disaster events as well as social and economic losses. Italy, Germany, and the UK are the European countries that are among the top 10 countries for economic loss. No country from the African continent is among the pool of top 10 countries for any category. The USA tops the economic loss list in the Americas. It is noteworthy that no country outside the Asia–Pacific region is listed in the top 10 for three or more categories.

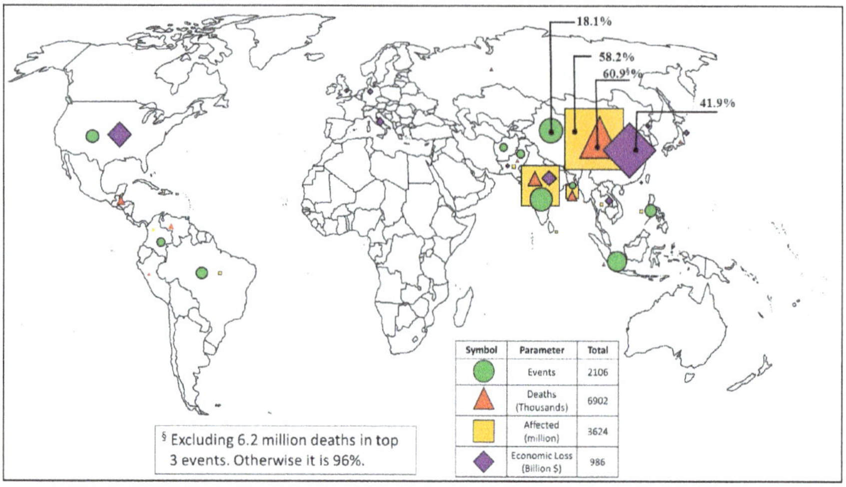

Figure 5. Global hotspots for hydrological disasters based on data for the top 10 countries in each category from 1900 to 2020 AD (Data Source: [57]).

Table 2. Distribution of events, fatalities, affected people, and economic loss in top 10 countries due to hydrological disasters from 1900 to 2020 AD (Data Source: [57]).

Country	Events (%)	Country	Fatalities (%)	Country	Affected (%)	Country	Economic Loss (%)
China	18.14	China	96.02	China	58.22	China	41.93
India	17.05	India	1.17	India	24.83	USA	17.97
Indonesia	14.58	Bangladesh	0.77	Bangladesh	9.53	India	11.44
USA	9.54	Guatemala	0.60	Pakistan	2.23	Italy	6.05
Philippines	8.78	Venezuela	0.44	Thailand	1.69	Thailand	5.77
Brazil	8.50	Pakistan	0.27	Philippines	0.94	Germany	3.90
Colombia	6.46	Japan	0.21	Vietnam	0.92	Japan	3.88
Pakistan	6.17	Russia	0.20	Brazil	0.69	Pakistan	3.16
Afghanistan	5.94	Peru	0.16	Sri Lanka	0.47	North Korea	2.97
Bangladesh	4.84	Indonesia	0.15	Colombia	0.46	UK	2.92
Total	2106		6902 (thousand)		3624 (million)		986 (billion USD)

5.3. Meteorological Disasters

Disasters caused by a meteorological hazard can be due to convective storms, extratropical storms, tropical storms, extreme temperature, and fog. In this article, only the effect of disasters caused by storms is discussed. Figure 6 presents the distribution of these storm events in various regions of the world for the period of 1900 to 2020 AD in terms of fatalities, persons affected, and the associated economic loss. Again, more than 50% of the events and more than 90% of the social cost in terms of lives lost and person affected were in the Asia–Pacific region. However, the lion's share of economic loss was suffered by the Americas, with USA having the largest share in this loss. Similar to geophysical and hydrological disasters, the top 10 countries accounted for more than 90% of the total human and economic losses while subjected to about 60% of these events. The share of African and European continents in social losses was relatively small. However, Europe suffered economic damage roughly proportional to the number of reported events. Overall, about 1.4 million lives were lost worldwide to disasters caused by storms, and more than 1.2 billion people were affected. Total economic loss due to disasters in this category was more than 2.1 trillion dollars, which is the largest among the three types of disasters considered in this article.

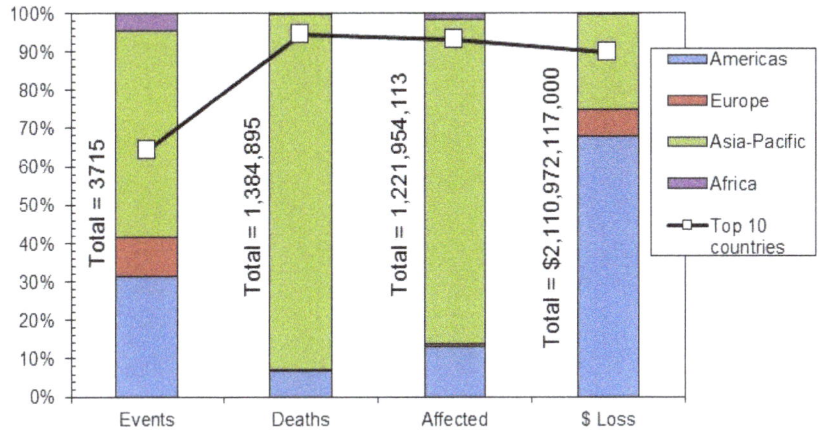

Figure 6. Distribution of events, deaths, persons affected, and economic loss due to metrological disasters (storms only) across the globe during the period 1900–2020 AD (Data Source: [57]).

Global hotspots for disasters caused by storms in the metrological hazard category for the four metrics of number of events, fatalities, person affected, and economic loss between 1900 AD and 2020 AD are depicted on the world map in Figure 7. The corresponding data for the top 10 countries in each category are presented in Table 3. The USA accounted for the greatest number of events (28.4%) and suffered the most economically, i.e., 65.3% of the total loss for the top 10 countries. On the other hand, the most fatalities occurred in Bangladesh (49.63%), and the most people were displaced in China (43.86%). Madagascar was the only African country that was listed among the top 10 in any category (i.e., number of people affected). Four countries from the Asia–Pacific region were among the top 10 in all four categories, while the USA was the only non-Asia–Pacific country to be listed among the top 10 in all four categories for any of the three types of natural disasters studied in this chapter (i.e., geophysical, hydrological, and metrological).

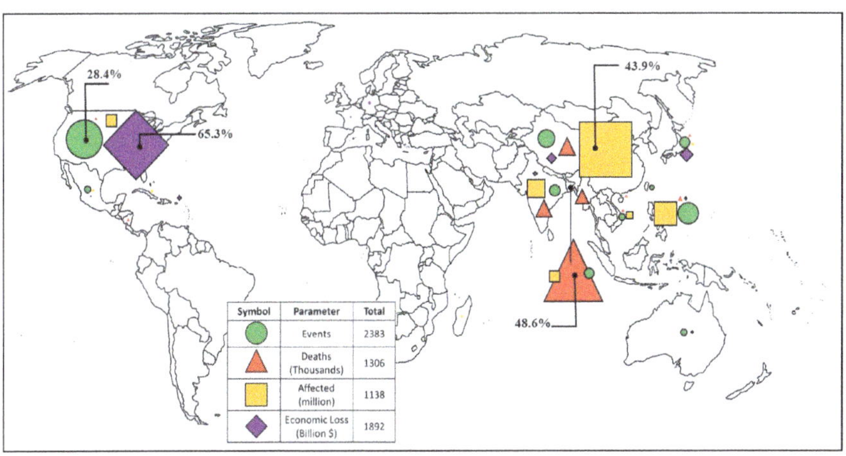

Figure 7. Global hotspots for meteorological disasters (storms only) based on data for top 10 countries in each category from 1900 to 2020 AD. (Data Source: [57]).

Table 3. Distribution of events, fatalities, affected people, and economic loss in top 10 countries due to meteorological disasters (storms only) from 1900 to 2020 AD. (Data Source: [57]).

Country	Events (%)	Country	Fatalities (%)	Country	Affected (%)	Country	Economic Loss (%)
USA	28.41	Bangladesh	48.63	China	43.86	USA	65.34
Philippines	15.69	China	13.41	Philippines	16.89	Japan	8.42
China	13.34	India	12.79	India	13.13	China	7.73
India	8.60	Myanmar	11.08	USA	8.92	Puerto Rico	4.08
Japan	7.85	Philippines	3.82	Bangladesh	7.92	India	3.02
Bangladesh	7.64	Japan	2.68	Vietnam	4.90	Germany	2.91
Vietnam	5.25	USA	2.38	Cuba	1.95	Mexico	2.33
Mexico	4.95	Honduras	1.90	Madagascar	0.92	Australia	2.19
Australia	4.66	Hong Kong	1.82	Mexico	0.76	France	2.18
Taiwan	3.61	Vietnam	1.51	Japan	0.76	Philippines	1.80
Total	2383		1306 (thousand)		1138 (million)		1892 (billion USD)

6. Natural Disaster Management

Disaster management comprises actions taken before and after the occurrence of a natural disaster to manage the negative consequences of the event. Disaster management is a cyclic process and comprises two stages and four phases, as depicted in Figure 8. Two phases (i.e., prevention and preparedness) precede a disaster event and are categorized as the development stage activities. The other two phases, i.e., response and relief, that immediately follow a disaster event constitute the humanitarian aid stage activities. This comprehensive model of disaster management developed in the USA is termed as PPRR [61]. Risk-driven and vulnerability-focused disaster management paradigms have also been proposed as competing models to PPRR [62–64]. However, the PPRR model is still widely used, and disaster management concepts in this article will be explained based on PPRR.

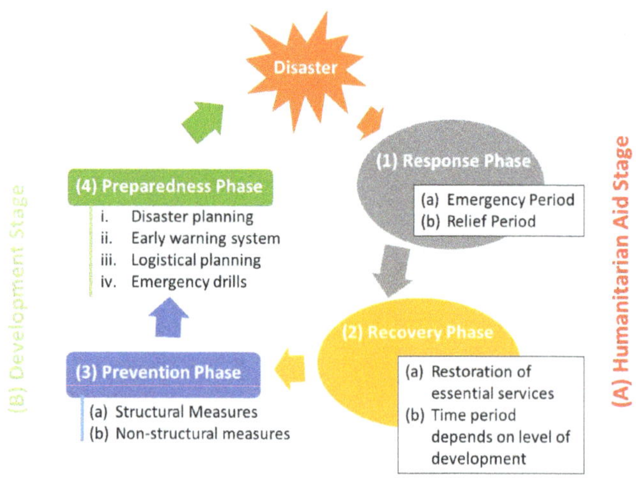

Figure 8. RRPP/PPRR model of disaster management.

Due to its cyclic nature, explanation of the disaster management cycle can start from any phase. Herein, the discussion on the disaster management cycle will begin with the phase immediately following the disaster, i.e., response. Therefore, the model can be renamed to RRPP with no change in the tasks performed in each phase. It is to be noted that

RR and PP phases belong to the humanitarian aid and development stages, respectively. Detailed discussion on these phases in the two stages is presented in the following.

6.1. Humanitarian Aid Stage

The humanitarian aid stage comprises the 'RR', i.e., response and recovery phases. Discussion on each of these phases is undertaken in the following.

6.1.1. Response Phase

This is the phase immediately following a disaster event that is part of the humanitarian aid stage. The focus of this phase is on search and rescue, preventing further mortalities and meeting the basic needs (healthcare, subsistence, and shelter) of the survivors. This phase can be further divided into two periods, viz., emergency response and relief periods.

(a) Emergency response period

This is a time-sensitive phase that can last up to a couple of weeks after the disaster event. The main task during the first 72 'golden hours' is search and rescue of disaster victims. Disaster survivors from the community and local medical teams play the most vital role in rescuing and treating the victims trapped in destroyed buildings or other infrastructure during this short period of time in which outside help is unable to arrive and timely join rescue activities. Smith [20] notes that almost 90% of the victims brought out alive from damaged buildings during earthquakes were rescued within the first 24 h. The numbers of survivors rescued from collapsed buildings usually decrease with each passing day after the disaster along with their chances of survival [65,66].

Effective communication is essential for rapid response and rescue efforts. Recently, a new informal communication channel for coordinating disaster rescue operation between affected citizens has emerged, which has utilized various social media platforms to play the role of rescuer, dispatcher, or information compiler [67]. Members of such platforms share information about alternate routes to reach/avoid a disaster area, location of rescuees, condition of damage to infrastructure, health and rescuee needs of affected persons, etc. Academics, disaster management professionals, and governmental agencies have provided policy frameworks and guidelines for the responsible use of social media for disaster response [68–71]. It is also noted with concern that this new form of communication can also be used to misguide, hamper, or even thwart rescue operations [69]. However, recently, social media platforms have been positively employed for rescue operations during hurricanes, rains, and earthquakes/tsunamis [72–75].

Other equally important tasks during this period include provision of healthcare facilities to the survivors, burial of the deceased, provision of shelter and food, and prevention of communicable diseases.

(b) Relief period

The relief period could last up to 5–6 months after the end of the emergency period. During this period, outside assistance in terms of trained personnel or materials is generally available. The focus is on prevention of additional mortalities and comforting the survivors to restart their lives. Typical tasks undertaken in this period include debris removal, demolishing of unsafe structures, restoration of lifelines (i.e., water, power, sanitation, and transportation networks), provision of temporary shelters and healthcare facilities, post-trauma counseling, and assessment of inflicted damage.

Each type of natural disaster requires a specific kind of medical help during this period. For example, earthquake victims need attention for fractures, respiratory tract infections, trauma to internal organs, and psychological stress, while control and treatment of water submersion (near drowning), snakebites, diarrhea, and communicable diseases is the focus during floods [76].

Although community involvement and tapping into the social network of the affected community are important, a centralized command and control approach to provision of relief efforts has been found to work satisfactorily during this period [77].

6.1.2. Recovery Phase

The recovery phase part of the humanitarian aid stage overlaps the relief period and can continue for many years. During this phase, the emergency situation created by the disaster event no longer exists, and focus is on restoration of daily life and economic activities of the affected population. Priority is given to restoration of essential services such as housing, subsistence, utilities, mobility, and healthcare.

The duration of the recovery phase is strongly affected by the level of development of the affected community [78]. Developing countries may experience slow recovery due to the viscous cycle of underdevelopment, as funds dedicated for development activities will be spent on emergency response and relief [79]. Effective recovery depends on the level of community involvement as well as appropriate and equitable allocation of external funds and material resources. Corruption in allocation and utilization of disaster relief assistance is another prevalent problem [80,81]. The recovery phase slowly blends into the long-term development phases of prevention and preparedness.

6.2. Development Stage

The phases in the disaster management cycle included in this stage, which precedes a disaster event, are the two Ps, i.e., prevention and preparedness. The development stage is the time-period in which a community prepares for a future disaster event. Due to the knowledge accumulated over the past two centuries about the mechanism of natural hazards and occurrence of these hazards in various parts of the world, most communities and nations are aware of the natural hazards frequently faced by them. However, lack of planning, community involvement, political will, or financial resources leave a number of communities unprepared for a future disaster. Therefore, disaster mitigation has been aptly termed as a social rather than a biophysical process [82].

Steps taken and plans implemented during the development stage have a profound impact on the post-disaster phases. This is the stage in which a community builds the necessary resilience to withstand a natural disaster and quickly navigate through the humanitarian aid stage of the disaster management cycle. Building resilience requires involvement of various stakeholders for improving the physical infrastructure, social engagement, disaster planning, and establishing a warning system to reduce the negative consequences when struck with a natural disaster. Such efforts are rewarded with significant payback, as evident from substantially less loss of life and property in recent natural disasters in China [83–85] and other countries [86,87].

A clear relationship between the level of development and the misery caused by a natural disaster in terms of lost lives and economic damage has emerged after careful analysis of natural disasters in the 20th century [88]. Developing countries are stuck in a viscous cycle of under-development, where emergency response and relief efforts after a natural disaster eats up funds dedicated for development activities [89]. Procrastination regarding building the necessary safeguards against future natural disasters is evident in almost all countries around the globe [90,91]. Preventive measures are delayed, even though it is well known that the benefit-to-cost ratio (BCR) of disaster prevention measures is around 60 for flood hazard and varies between 3 and 15 for all other natural hazards across different countries [92].

It was noted by the donor agencies that the situation of natural disaster risk in developing countries has not improved over the years despite spending billions of dollars. For example, World Bank loans for natural disaster relief totaled more than USD 14 billion for the last two decades of the 20th century [79]. Therefore, the focus has now shifted from disaster relief assistance to programs that are more directed towards reducing poverty, building resilient infrastructure, ensuring community participation, and creating social support mechanisms.

As noted in the beginning, the PP phases, i.e., prevention and preparedness, comprise the development stage of the disaster management cycle and are the focus of the discussion presented below.

6.2.1. Prevention Phase

Activities done in this phase are meant to prevent the negative consequences of a disaster event. However, it has been observed that fallout from a natural disaster cannot be fully prevented due to the uncertainties in frequency and magnitude of the hazard(s) as well as capacity of the prevention measure(s). Therefore, the phrase 'prevention' is replaced with 'mitigation' in recent disaster management literature that focuses on mitigating the harmful impacts of the disaster within a specified probabilistic tolerance. Mitigation measures can be classified as either structural or non-structural, as explained in the following for the three types of natural hazards discussed in this paper.

(a) Structural measures

Structural measures are the ones that change the characteristics of a natural hazard or improve the strength of an infrastructure component to withstand the impact of the forces unleashed by a natural hazard. Due to the inherent uncertainty of the magnitude and frequency of natural hazards, there is always a non-zero probability of failure of a structural measure. Much effort has been devoted and considerable progress has been made by scientist and engineers over the past 150 years to understand the physical process underlying various natural hazard phenomena as well as to devise cost-effective engineering solutions to improve the strength of the infrastructure components. A brief overview of these efforts, related to the geophysical, hydrological, and meteorological hazards, is presented below.

(i) Geophysical hazards

Probably the first documented account of adopting engineered structural measures in buildings to withstand the forces generated by earthquakes was in Japan in 1895 [93], while earlier seismic design codes date back to 1927 in USA [94] and in early 1930s in Chile [95] and India/Pakistan [96]. Ever since these early efforts, the field of earthquake resistant design has advanced to the use of various forms of traditional seismic lateral force resistance systems such as ductile moment frames, braced frames, and shear walls, as well as the use of base-isolation, supplemental damping devices, and active control of lateral forces induced by an earthquake in a variety of structures ranging from residential houses to skyscrapers to bridges and nuclear power plants [97].

Despite these impressive strides in the scientific knowledge and engineering applications of earthquake resistant design, the sad fact is that earthquakes are the leading cause of natural disaster mortality, as detailed in Section 2. A further examination [98] revealed that 90% of these fatalities were caused by collapse of non-engineered or semi-engineered structures—mostly houses built of adobe or unreinforced masonry (URM). This seismically vulnerable building stock is present in both developed and developing countries. However, such existing buildings in the developed countries have largely been replaced or retrofitted, and the inventory of the seismically deficient structures is continuously decreasing. Contrarily, in the developing countries, stock of such non-engineered or semi-engineered buildings is steadily increasing, and so is the risk to life and property.

It was noted that the reason for this increased vulnerability is not due to the absence of technical knowledge to cost-effectively build such buildings but an ignorance among homeowners about the available engineering solutions, lax code enforcement, outdated construction practices, and a lack of appreciation for the life-cycle cost and benefit of an improved construction methodology [99]. Detailed guidelines for cost-effective design and retrofit of non-engineered structures was originally published by the International Association for Earthquake Engineering (IAEE) in 1986 and has recently been adopted by UNESCO for widespread distribution [100]. An example of confined masonry for seismic resistance of ordinary houses and simple structures is depicted in Figure 9.

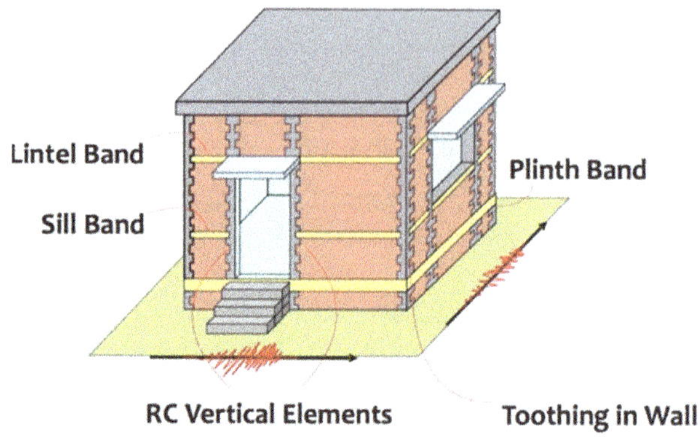

Figure 9. Concept of confined masonry for seismic resistance.

It is relatively easy to enforce improved construction practice in new construction when it is associated with an incentive, as was the case of construction of about 400,000 seismically designed houses in Pakistan after the 2005 Kashmir earthquake [101]. However, it has proven to be a hard task to convince owners to strengthen and retrofit their existing houses [102].

(ii) Hydrological hazards

Two types of structural measures are commonly adopted to mitigate hydrological hazards. One intends to 'tame' the flood, while the other provides capacity to withstand the flood waters. Use of the first measure in the form of levees (also called embankment or dykes) to control riverine flooding dates back to early civilizations in China, Mesopotamia, Egypt, and Pakistan almost 4000 years ago [20]. Over the centuries, other means to 'tame' the rivers in the form of barrages, dams, and weirs have been devised. The hydrological and engineering aspects of flood control measures are rather well understood. Therefore, the use of structural measures for its control is the most widespread compared to other natural hazards.

The 'levee effect', i.e., the sense of security provided by the levee, is the driver for floodplain development and hence the losses associated with the flood when the levee is breached due to a construction defect or poor maintenance or overtopped due to an exceptional flood. Despite the widespread use of levees, flood losses are on the rise [103].

The second form of structural measure against riverine floods is 'flood proof' construction. This involves building the structure above the predicted flood level by either building it on an embankment or supporting it on stilts and providing adequate strength in the structural members to withstand the hydraulic forces generated by flowing flood water [104].

In urban areas, the structural measures to mitigate the effect of pluvial or surface flooding caused by rainfall independent of an overflowing water body consist of storm water drainage system, comprising catch pits, manholes, storm sewers, culverts, detention ponds, and pumping stations [105]. Figure 10 details some of the measures that can be adopted to mitigate effects of flood hazard for new and existing developments.

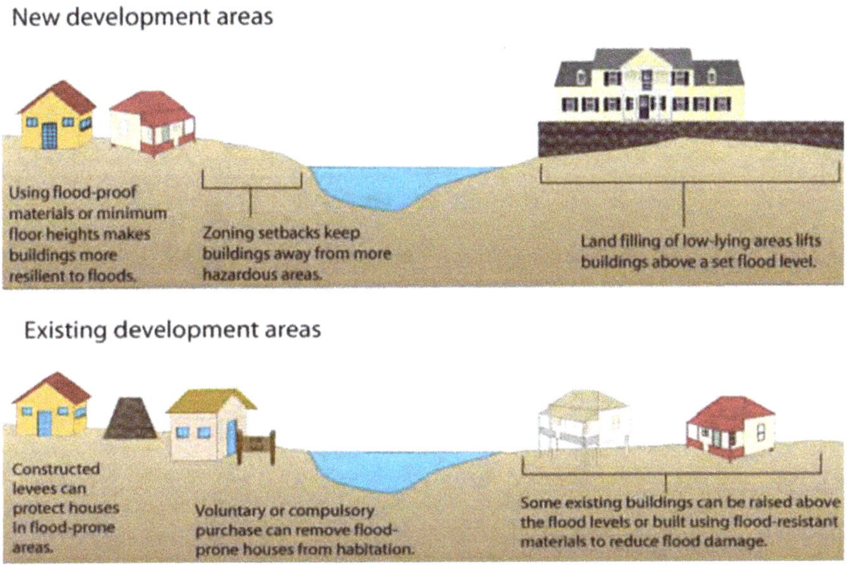

Figure 10. Flood mitigation measures for new and existing developments.

(iii) Meteorological hazards

Meteorological hazards are caused by tropical storms termed as cyclones, hurricanes, or typhoons in different parts of the world. These storms are characterized by strong winds accompanied by rain and storm surge (i.e., an increase in the sea level). A tropical storm has 1 min sustained wind speed between 17.4 and 33.1 m/s, and when this speed exceeds 33.1 m/s, it is termed a hurricane [106]. Structural measures to mitigate the effect of this hazard need to cater for two very different forces of nature, i.e., wind and flowing water. There is a possibility of structural damage due to lateral wind load on the walls as well as uplift of the roof in addition to flooding caused by rainwater or storm surge.

Nearly all types of buildings (wood frame, steel, unreinforced masonry, and reinforced concrete) and their components are at risk of damage or failure due to high winds [107]. As with the seismic design, the main source of failure of these structures is a lack of continuity in the load path from roof/floors to walls to foundations [108]. Foundations could also be compromised due to flooding and erosion caused by flowing water, loss of soil bearing capacity due to submerged water conditions, or loss of stability due to buoyancy [109]. Proper appraisal of wind forces and adequate design of connections against uplift is the key to avoiding structural failure during a hurricane. Most often, the cost of these retrofit and mitigation measures is relatively small compared to the incurred damage [110].

(b) Non-structural measures

It was noted earlier that the non-structural measures do not rely on physical construction to mitigate the risk associated with a natural hazard but depend on enforcement of building codes and land-use regulations and community awareness to break the disaster–rebuild–disaster cycle. Use of insurance as a collaborative tool is encouraged to share economic losses with wider global financial markets that would otherwise have to be borne by the individuals or governments alone. This section presents an overview of commonly adopted non-structural measures to mitigate the risk of three types of natural hazards (i.e., geophysical, hydrological, and meteorological) that are the focus of this article.

(i) Geophysical hazards

Earthquakes happen without any warning. Unlike hydrological and meteorological hazard events, there is no reliable scientific method to forecast the location, time, and magnitude of seismic events. Therefore, being prepared and alert is the best defense to reduce human and material losses resulting from seismic hazards. Some of the non-structural measures to reduce risk of seismic disasters are reviewed below:

Land-use planning

Moving out of the seismic-hazard-prone areas seems to be the most logical and effective measure to reduce the risk of this disaster, as the locations of most of the active seismic regions are known through historical accounts or scientific investigations. Enforcing of seismic zoning laws may be possible only for new developments, as adopted in the Alquist-Priolo Earthquake Fault Zoning Act in California that limits any development near active faults [111]. However, such measures cannot be adopted for existing cities such as Tokyo, San Francisco, Tehran, Christchurch, etc., that have witnessed many recorded earthquakes but have surprisingly evolved into even bigger metropolises after successive disaster events.

Relocation of an entire town or a city is not only an expensive logistic nightmare but also a socially sensitive issue. For example, Yungay, a town in the Peruvian Andes, was completely destroyed by mudslides triggered by the Ancash earthquake in 1970, and the surviving residents forced the government to rebuild instead of relocating [112]. Many communities have other compelling reasons to inhibit places of known seismic hazard. In the case of many towns and cities in Iran (Tehran included), it is the lure of life-sustaining water that is brought up from the depths of the earth by seismic faulting associated with earthquakes, which forces these communities to live in the shadow of persistent seismic hazard [113]. These examples illustrate some serious limitations of land-use planning as a non-structural measure to mitigate seismic hazard.

Personal safety measures

Personal safety measures are important to prevent physical injuries, fires, loss of utilities, and damage to non-structural components inside buildings and houses. These measures include securing heavy and movable objects, such as cabinets, furniture, computers, etc., fuel (gas) cylinders, keeping building exits clear of any obstructions, etc. [114]. Injuries and deaths caused by loose objects and fires after an earthquake pose equal hazard to life and property as structural damage to buildings. Education and community awareness play a key part in achieving seismic hazard mitigation goals.

Insurance

Insurance is a disaster management tool that is most beneficial during the recovery phase. However, in order to reap its benefit, it should have been initiated before the disaster, i.e., in the prevention phase of the development stage. The majority of seismic risk around the world is uninsured due to the limited capacity of private insurers to cover the cost of potential damages [115]. This means that, inevitably, the government has to be involved as a provider, reinsurer, or regulator to assist the affected citizens. Setting the right insurance premium is the trickiest part. If it is too low, then the tendency is to have excessive assets built in the exposed area, which may also be of low quality, and if it is too high, then few buy the insurance. Both scenarios increase people's vulnerability [116].

Access to seismic insurance is available mostly in developed countries and few developing countries. However, even with government assistance, seismic insurance uptake is generally low in most of the seismic hazard prone countries. For example, it is around 20% in California, Mexico, Turkey, and Italy, 30% in Japan and Chile, and 80% in New Zealand [117]. Paleari [118] found that higher seismic insurance penetration in the European Union countries is more, due to government involvement, than its voluntary or mandatory nature. With government support and compulsory offer or purchase, very high (90–100%) bundled seismic and flood insurance penetration is noted in Spain, France, and Belgium. On the other hand, Italy, which has observed the most seismic activity and damage in Europe, as noted in Figure 3, has seen a steady rise in the uptake of seismic insurance since 2009, with a nearly 0% insurance rate of residential properties to about 20% in 2019 [119]. This low uptake ratio in the majority of these countries could be due to high premiums, owner's perception of risk, insurance history, governmental participation, uncertainties in the seismic loss models, involvement of insurers in development and promotion of seismic research and code development, and other issues [120,121].

The concept of seismic insurance is almost non-existent in developing countries, and its adaptation is a big challenge [122]. The active government role in enacting pertinent legislation and assurance to assist with loss coverage are essential for insurance uptake, as was the case in Turkey, where the Turkish Catastrophe Insurance Pool (TCIP) was established after the 1999 Marmara earthquake, and in Taiwan, where the Taiwanese Residential Earthquake Insurance program was initiated after the 1999 Chi-Chi earthquake. It is estimated that less than 1% of losses due to natural hazards are insured in the developing countries, which constitute a significant portion (almost 13%) of their GNP [123].

Natural hazard awareness and education

Preparedness through natural hazard perception, awareness, and education are of utmost importance to minimize loss of life and property when natural disasters occur. Everyone in a community must know the nature of natural hazards in their environments and how to prepare for them [124]. The National Research Council [125] compiled a very comprehensive set of recommendations for natural hazards education and awareness. NRC proposes national education and awareness campaigns for various levels and forums in society. They include home, community, schools, and the workplaces. All homes should have basic information about natural hazards in that area, emergency supplies, and escape and evacuation plans. Community centers, churches, and hospitals should all participate in natural hazard awareness and education. In addition to evacuation plans and emergency procedures, they should have information on shelter and treatment they would be able to provide in case of a natural disaster.

Disaster training and education is part of the curriculum at school and university levels in many countries all over the world. The Building Research Institute and the National Graduate Institute for Policy Studies in Japan took stock of disaster education at primary, secondary, and tertiary levels in Japan, Fiji, Indonesia, Uzbekistan, India, and Nepal [126]. It traced best practices and guided how countries can learn from each other to improve disaster education. Boon and Pagliano [127] looked into the disaster education in Australian schools and pointed to the need to significantly improve the same. Similar work on disaster education undertaken in parts of world has been reported in articles and reports (e.g., [128–131]).

Situation awareness of natural hazards depends on the type of hazard encountered. The Centre for Disease Control (CDC) provides biological hazard type specific awareness information [132]. Resources for public disaster education and awareness are made available through disaster related agencies such as meteorological departments, wildfire safety agencies, earthquake and tsunami warning agencies, and others. These agencies maintain situation awareness webpages for hurricanes, wildfires, floods, earthquakes, volcanos, and winter weather. The Rural Fire Service (RFS) NSW (Australia) is an example of such agencies. RFS (https://www.rfs.nsw.gov.au/ (accessed on 29 October 2021)) provides real time information about the bushfires (wildfires) on their website and via their smart phone app. The RFS places and manages bushfire warning signs that indicate the level of bushfire threat in real time. The RFS website provides guidance to property owners on how to prepare their properties to avoid damage from bushfires. They also educate the public on how to prepare bushfire survival plans. While such agencies may exist in developing countries, these are often not very active and/or effective in public awareness and education of disasters due to lack of resources and/or poor governance [133].

(ii) Hydrological hazards

Flood risk is not only created by the combination of the flood hazard and the inadequacy of the engineering (or technical) solution. It is a combination of natural, social, cultural, and technological aspects that need to be ethically considered in a holistic manner as opposed to the stand-alone decisions based on technological, materialistic (economical), or political benefits. White [22] was perhaps the first to realize the ineffectiveness of structural measures in reducing flood damages in the US and strongly advocated adaptation of land-use planning as a solution. Ever since, various attempts have been made to include non-structural measures, such as land-use planning, flood plain management, social equity and inclusion, public awareness, and loss sharing through insurance, to mitigate the impact of flood disasters [134].

Structural measures endeavor to keep flood water away from people and property, whereas non-structural strategies strive to keep people and property away from the flood water. Non-structural measures fall into two broad categories: (1) those that modify the susceptibility to damage and (2) those that modify the loss. Flood plain management, flood proofing of structures, flood awareness, and societal participation are examples of the first type of measures while flood insurance represents the second category. Additional details can be found in [135,136].

Land-use planning

Land-use planning as an effective solution for reducing flood related economic loss has been advocated in the USA since 1930s [22]. It was demonstrated through a case study in Austria [137] that implementation of stricter land use control and/or flood proofing measures by the owners reduced the flood risk by about 30%. In contrast, ignoring such measures increased the losses by an additional 17%.

Social justice

It is also argued to include 'social justice' in the metric of flood mitigation measures. The authors of a study on this topic state [138]:

> *" … … whatever risk mitigation measures are taken they will never be able to bring equal benefits to all members of the society, and even if they do so for the present generation they may not do the same for future generations. Consequently, on one hand, there will be members of the society who will benefit from such measures in one way or the other, and on the other hand, there will be other members of the society who will be more burdened by the same measures".*

Inclusion of the 'social justice' element is especially critical in the case of implementation of extensive structural measures in developing countries that are funded by borrowed funds through international donors. There is a propensity among the aid donors to push for solutions that are better suited to serve the political and technical establishments of their countries with little regard to the social and cultural norms of the recipient country. A flood control mega-project in Bangladesh can be cited as one such example where various governments and international aid agencies influenced different aspects of the project [139].

(iii) Meteorological hazards

Regions prone to hurricanes are well known. However, exposure of human populations and related assets to hurricane hazard has different reasons in developed and the developing countries. In the US, the population of these coastal regions has almost doubled since 1960 due to the lure of beach, blue waters, and sunny weather. As a result, costly building assets are willingly exposed to hurricane risk. However, for millions of people living in the Caribbean islands and coastal areas of Bangladesh and India, there is no choice other than to live with the consistent threat of hurricane winds and storm surge. Community preparedness, storm warning system, and storm shelters are effective means to mitigate threat to life and property [140].

Meteorological hazards are characterized by high wind, moderate-to-intense rainfall, and storm surge. Non-structural measures can be effective against flooding and storm surge, while structural measures need to be adopted for protection against high winds. Land-use planning provides basic guidelines for development in coastal zones that caters for risk caused by the combination of high winds, urban flash flooding, and storm surge. The adoption of updated coastal hazard risk maps often faces challenges from developers and owners due to potential devaluation of the properties and increase in insurance premiums.

Coastal zone management

Utilization of stabilized sand dunes, coastal mangrove forests, man-made breakwaters, and sea walls are important defense mechanisms against storm surge, as depicted in Figure 11. These mitigation measures may not fully protect coastal structures from flooding but help in reducing coastal erosion. A mix of structural and non-structural measures, depending on the local geophysical conditions and community needs, usually result in a sustainable solution.

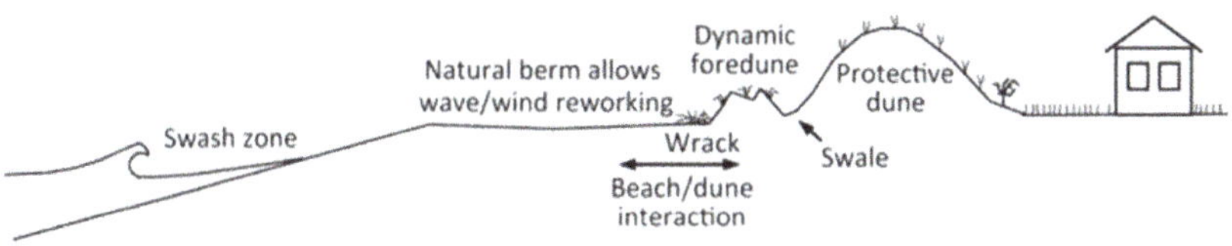

Figure 11. Use of sand dunes for protection against storm surge.

Loss sharing through insurance

Storm and flood insurance is almost non-existent in developing countries. Therefore, the losses are borne mostly by individuals, accompanied by some contribution by the national government. However, efforts are being made to extend insurance cover in these countries through subsidized schemes [141,142]. In the USA, the problem is complex, as almost 80% of at-risk properties are insured, and in the event of a major hurricane disaster, the insurance industry is faced with huge liabilities.

6.2.2. Preparedness Phase

This phase in the development stage of the disaster management cycle immediately precedes a disaster event and critically affects emergency response during the initial phases of the disaster. It is imperative that disaster preparedness plans are in place for various scenarios of anticipated natural disasters and various dimensions of vulnerability so that prompt actions can be taken to save lives and lessen the impact of economic loss.

It is important to recognize various dimensions of vulnerability to natural disasters for effective preparation and response. Vulnerability is the propensity of an asset or people or an institution to suffer damage and lose functionality as the result of a disaster. The various dimensions of vulnerability are: physical (risk of damage to physical infrastructure such as buildings, transportation networks, flood protection works, etc.); economic (impact on agriculture, fishing, commerce, and manufacturing); institutional (providing effective leadership, support, response and protection to the affectees in a streamlined manner without a chaos); and social (awareness and knowledge of disaster, ability to act on warnings and follow competent authorities, social bonding and cooperation, and insurance to cover the economic losses). Important dimensions of vulnerability of a community need to be properly assessed during the disaster preparedness phase, and appropriate contingency plans are to be prepared for effective response during a disaster.

The aim of this phase is to achieve a readiness level that is appropriate for the type of encountered disaster by building leadership, organizational, management, and technical capabilities at institutional, societal, and individual levels. Specific measures that can be taken during this phase include the following:

(a) Disaster planning

This includes drawing and following up on long- and short-term disaster emergency response mechanisms and procedures that detail responsibilities and actions of various organizations and individuals and clearly establish chain of command and communication protocols. Such disaster planning is recommended for communities and nations that are historically prone to natural disasters. Other activities include construction and regular maintenance of disaster shelters, publicizing of evacuation routes, provision of transport for evacuation, and safety and well-being of evacuees.

(b) Early warning system

Establishment of a reliable early warning system can save numerous lives for slow-onset disasters such as floods or storms. It is also important that the warning message is conveyed to the targeted population segment in a clear way; preferably, according to a well-established protocol, using a well-publicized or mutually agreed mechanism (e.g., public alarms, radio, TV, social media, etc.). Technological advancements and utilization of scientific knowledge is a key component of such warning systems.

With the widespread use of mobile telephones, social media has emerged as a new form of communication, through which warning messages can be reached out to people who are potentially in the harm's way. Use of social media is no longer limited only to the search and rescue phases. It is being actively used to educate people about the risks associated with natural hazards, to adopt proactive measures for reducing the impact of a future disaster, and to create awareness [143–146]. Citizens expect disaster management agencies to engage with them through social media [147]. This is the reason that Red Cross, FEMA, CDC, NOAA, many city fire and police departments in USA, and other organizations have active and updated social media accounts on various platforms. As a case study, Tagliacozzo and Magni [148] analyzed how social media was effectively used for the post-disaster recovery (PDR) phase following the Emilia Romangna (northern Italy) earthquake in 2012. They found social media helpful at a grassroots level, enabling peer-to-peer communication for mobilizing public opinion about the reconstruction efforts.

(c) Logistical planning

This includes ensuring a sufficient stockpile of life-saving medicines, food, water, emergency power source, transportation, rescue equipment, and trained emergency response personnel. Mobilization and deployment of the army to assist the civilian administration can also be a part of such planning.

(d) Emergency drills

Emergency drills ensure the level of preparedness and can identify any lapses in planning and response. These drills include individual response during an emergency situation caused by the natural disaster, as well as societal and organizational responses. These rehearsal drills are effective defense mechanisms against rapid onset natural disasters such as earthquakes or fires.

(e) Knowledge and awareness

The role of natural hazard awareness and knowledge has already been emphasized in Section 6.2.1. Being knowledgeable of the nature of the natural hazard and various dimensions of vulnerabilities associated with it takes the 'surprise factor' out of disaster response. It prevents chaos and confusion and helps in coping with the disaster situation with focus and calm. Awareness and knowledge of individuals is a critical factor in rescue operations during the first few hours of a disaster occurrence, when limited outside help is available.

7. Challenges and New Directions in Natural Hazard Preparedness

Climate change is increasing extreme climatological events in both intensity and frequency. As a result, storm- and flood-related disasters are increasing and becoming more devastating. Continued population and economic growth are leading to ever-increasing natural resource consumption. Humans are increasingly encroaching on natural areas and coming into contact with natural processes. That is increasing disaster vulnerability.

In the following accounts, both the challenges of and new directions in natural hazard preparedness are elaborated.

7.1. Population and Economic Growth

The world human population has exploded to about eight billion from close to one billion one hundred years ago [149]. It is forecasted to continue to grow in the future. The UN estimates that the human population will increase to ten billion by 2050 [149]. Seven billion people that were added in the last hundred years had to be provided with shelter, food, water, clothing, household objects, and means of transportation. That need was fulfilled by enormously increasing exploitation of natural resources, including land. Vast swaths of natural forested land had to be converted to human settlements, agriculture, aquaculture, and other anthropogenic land uses [150]. The loss of natural lands was also due to adoption of low-density car-reliant suburban development. Greek planner Constantinos Doxiadis pointed out that average densities in several major cities decreased by two-thirds in the 40 years to 1968 [151]. TNC [152] forecasts that, at the current rate of urban expansion, the world will encroach into natural lands equal to the size of London every seven weeks.

Expansion of human settlements and activities into natural lands has increased human exposure to natural disasters in multiple ways. Human settlements have had to be built in more precarious lands in river catchments, on unstable grounds, in steep hills, closer to fire-prone forests, etc. In addition, loss of natural wooded land reduced the ability of land to slow down the rainfall runoffs, increasing risk of flooding. Replacing natural wooded and grassed surfaces with less pervious agriculture and impervious concrete and bitumen surfaces, as well as replacement of meandering natural streams with straightened and concrete lined canals and drainage channels, leads to excessively high and fast runoff [153]. Expansion of land under human habitat thus has led to increased exposure to floods, landslides, droughts, and wildfires.

7.2. Climate Change Related Weather Extremes

In July 2021, Germany and Belgium experienced intense rainfall and floods that had never been experienced before. Unprecedently intense rainfall and resulting devastation is linked with climate change [154]. The floods caused 184 deaths and property damage of more than EUR 2.5 billion in Germany [155]. Unprecedented extreme weather events are being observed in all parts of world with increasing frequency. In early September 2021, New York (NY) and New Jersey (NJ) experienced more than 100 mm/h of rainfall for extended periods. That intensity of rain was never seen before [156]. The rain flooded NY subways and transformed Manhattan streets into rivers.

Climate change-induced heat in the Earth's atmosphere is causing extended wildfire season and more extensive damage all over the world [157]. Unprecedentedly extensive and prolonged wildfires in California and in Australia are examples of that. In the Southern Hemisphere summer of 2019–2020, Australia experienced a mega bushfire that burnt about 20 million hectares of land and displaced or killed nearly 3 billion animals [158,159].

The world continues to release very high volumes of GHG (greenhouse gases) resulting from population and economic growth and a consumptive lifestyle. The Earth's atmosphere is thus likely to continue to heat up for many years to come. The Intergovernmental Panel on Climate Change (IPCC) anticipates that the world will see an increasing frequency and intensity of weather patterns, causing ever-bigger disasters [157]. It also warns that the poor and dense areas of the world are the most vulnerable from climate-induced extreme weather events of rainstorms, cyclones, heatwaves, and droughts [157].

7.3. Better Weather and Climate Change Modelling

Tradition hydrological modeling is based on probability analysis of occurrence of rainfall events of a certain intensity and related forecast of flooding. This analysis entirely relies on historic rainfall data to estimate the probability of floods likely to occur every certain number of years—such as one in one hundred years. Related to those estimates is the demarcation of areas that would come under water when a flood, say, one in one hundred years, occurs. Hydrological modeling has assisted in locating new human settlements outside one in one hundred years flood zones [153] and thus largely out of the way of harm of the frequent floods.

Hydrological modelling relies on historic rainfall data that goes back to the latter half of the 19th century, when measurement and recording of rainfall began [160]. However, due to climate change, historic data is no longer a good predictor of the rainfall intensity and flood frequency [161]. One way of dealing with the uncertainty of rainfall in the future is to extend the rainfall records of past 150 years to a much longer period in the past, with estimation of older rainfalls from thickness of annual tree rings and annual alluvial deposits [162]. A second way of dealing with uncertainties introduced by the changes in rainfall patterns is to integrate hydrological modeling with the climate change modelling [163]. However, significant improvements in climate change models are required to achieve this integration. It is also reported that the clouds and precipitation forecast part of climate change models need significant improvements as well [164].

7.4. Compact and Sustainable Living

First and foremost, the biggest contributors to climate change and hence increasing risk of disasters must be tackled through modifications in the way people live. Resource- and energy-intensive suburban life in large houses and mobility reliant on cars will have to change. Living in smaller dwellings and compact neighborhood with access to transit for travel would reduce both encroachment into the natural lands as well as reduce GHG emissions. City planners have, for decades, advocated higher densities through concepts of smart growth, new urbanism, and transit-oriented development (TOD).

The concept of smart growth arose from the negative impacts of the low-density car-reliant urban development. smart growth, a USEPA collaboration, promotes compact and walkable, self-sustaining, and attractive urban development [165,166]. New urbanism recommends a gridiron street pattern, narrow streets, smaller lots, and shallow setbacks for achieving walkability and compactness [167]. Transit-oriented development (TOD) recommends higher densities and mixed-use urban nodes that are connected to each other by transit bus, light rail, or metro lines.

8. Conclusions

The Earth's landscape is a product of natural processes of earthquakes, floods, cyclones, storms, wildfires, volcanic eruptions, and landslides. Disasters take place when these processes interact with human settlements and/or human economic activities. This article presents a broad and in-depth analysis of natural disasters, focusing on their origin, impact, and management. It describes a brief history of impacts of natural hazards on the human built environment. It lays out an account of natural processes and theories related to natural disasters. Theoretical knowledge in this field has evolved over the years. While in the past, they were attributed to acts of gods or acts of nature, they are now understood as a complex nexus of natural–human–social–economic factors.

A very elaborate discussion on global impacts of natural disasters is included in the paper. The paper provides detailed and spatially-differentiated statistics on harm from the natural disasters for various categories of natural hazards. About 85% of the world's population has been affected by at least one natural disaster in the past 30 years [54]. A significant loss of human life is also associated with natural disasters that kill 60,000 people on average yearly [168]. Ninety percent of these fatalities take place in developing countries [56].

The impact of natural disasters is growing due to their increasing frequency due to climate change and increasing human encroachment and encounter with nature. In the twenty-year period from 1980 to 1999, 4212 disasters were recorded all over the world. In the same period, 1.19 million lives were lost, 3.25 billion people were affected, and

USD 1.63 trillion in economic losses was incurred from these disasters. In the 20 years that followed (2000 to 2019), a large increase was recorded in number of disasters (7348), number of people affected (4.2 billion), and economic loss (USD 2.97 trillion) [169]. That increase in disaster frequency and damage is attributed to climate change, which is making climatological events more frequent and more extreme.

The paper covered a thorough discussion on natural disaster management. The four-phase PPRR—preparedness, prevention, response, and recovery—model is described in great detail. Various common techniques adopted for natural disaster preparedness, including structural measures for preparedness, are elaborated. Details of various responses adopted at various stages of disaster are also described.

Awareness and education play a vital role in disaster preparedness and mitigation. Natural hazard awareness and education is to be tailored for different demographics. The situational awareness would also be different depending on the type of natural hazards. The education and awareness can and should take place at all administrative levels, i.e., local, regional, national, and international. The paper also presented other aspects and dimensions of awareness, of which social media and new tools such as smartphone apps are the most noteworthy.

Huge advancements have been made in the past few decades in the forecast and modelling of natural hazards, disaster communication and warning systems, transport and mobility for quick evacuation and arrival disaster relief, and technology to build safer structures. However, lack of resources, unchecked population growth, political instability, dysfunction and fatalism in poor countries, and continuation and expansion of highly consumptive and unsustainable lifestyles in richer countries remain significant challenges.

Author Contributions: Conceptualization, M.T.C. and A.P.; methodology, A.P. and M.T.C.; investigation, M.T.C.; resources, A.P.; data curation, M.T.C.; writing—original draft preparation, M.T.C. and A.P.; writing—review and editing, A.P. and M.T.C.; visualization, A.P. and M.T.C. All authors have read and agreed to the published version of the manuscript.

Funding: This research received no external funding.

Conflicts of Interest: The authors declare no conflict of interest.

Entry Link on the Encyclopedia Platform: https://encyclopedia.pub/16699.

References

1. Below, R.; Wirtz, A.; Guha-Sapir, D. *Disaster Category Classification and Peril Terminology for Operational Purposes*; Report No. 264; Université Catholique de Louvain: Louvain-la-Neuve, Belgium; Centre for Research on the Epidemiology of Disasters and Munich Reinsurance Company: Munich, Germany, 2009; 20p, Available online: http://hdl.handle.net/2078.1/178845 (accessed on 28 June 2021).
2. Integrated Research on Disaster Risk (IRDR). Peril Classification and Hazard Glossary. IRDR DATA Publication No. 1. Beijing: Integrated Research on Disaster Risk. 2014. Available online: https://www.irdrinternational.org/uploads/files/2020/08/2h6G5J59fs7nFgoj2zt7hNAQgLCgL55evtT8jBNi/IRDR_DATA-Project-Report-No.-1.pdf (accessed on 28 June 2021).
3. Cutter, S.L. Chapter 1—The Changing Nature of Risks and Hazards. In *American Hazardscapes: The Regionalization of Hazards and Disasters*; Joseph Henry Press: Washington, DC, USA, 2001.
4. FEMA. *Multi Hazard Identification and Assessment*; FEMA: Washington, DC, USA, 1997.
5. UN/ISDR (UN/International Strategy for Disaster Reduction). *UNISDR Terminology on Disaster Risk Reduction*; United Nations: Geneva, Switzerland, 2009.
6. Cardona, O.D.; van Aalst, M.K.; Birkmann, J.; Fordham, M.; McGregor, G.; Perez, R.; Pulwarty, R.S.; Schipper, E.L.F.; Sinh, B.T. Determinants of risk: Exposure and vulnerability. In *Managing the Risks of Extreme Events and Disasters to Advance Climate Change Adaptation*; Field, C.B., Barros, V., Stocker, T.F., Qin, D., Dokken, D.J., Ebi, K.L., Mastrandrea, M.D., Mach, K.J., Plattner, G.-K., Allen, S.K., et al., Eds.; A Special Report of Working Groups I and II of the Intergovernmental Panel on Climate Change (IPCC); Cambridge University Press: Cambridge, UK; New York, NY, USA, 2012; pp. 65–108.
7. FEMA. *The Disaster Dictionary—Common Terms and Definitions Used in Disaster Operations (9071.1-JA Job Aid)*; Federal Emergency Management Agency (FEMA): Washington, DC, USA, 2001.
8. Thywissen, K. Core terminology of disaster risk reduction: A comparative glossary. In *Measuring Vulnerability to Natural Hazards*; Birkmann, J., Ed.; UNU Press: Tokyo, Japan, 2006; pp. 448–496.

9. Guha-Sapir, D.; Hoyois, P.; Wallemacq, P.; Below, R. *Annual Disaster Statistical Review 2016: The Numbers and Trends*; Center for Research on the Epidemiology of Disasters (CRED): Brussels, Belgium, 2016; 91p.
10. Ansell, J.; Wharton, F. (Eds.) *Risk: Analysis, Assessment and Management*; John Wiley: Chichester, UK, 1992; 220p.
11. Wisner, B.; Blaikie, P.; Cannon, T.; Davis, I. *At Risk: Natural Hazards, People's Vulnerability and Disasters*; Routledge: New York, NY, USA, 2004; 496p.
12. Furedi, F. The changing meaning of disaster. *Area* **2007**, *39*, 482–489. [CrossRef]
13. White, G.F.; Kates, R.W.; Burton, I. Knowing Better and Losing Even More: The Use of Knowledge in Hazards Management. *Environ. Hazards* **2001**, *3*, 81–92.
14. Bentzen, S.J. Acts of God? Religiosity and natural disasters across subnational world districts. *Econ. J.* **2019**, *129*, 2295–2321. [CrossRef]
15. Ruiu, G. *Is Fatalism a Cultural Belief? An Empirical Analysis of the Origin of Fatalistic Tendencies*; Munich Personal RePEc Archive, No. 41705; University Library of Munich: Munich, Germany, 2012; pp. 1–27. Available online: http://mpra.ub.uni-muenchen.de/41705/ (accessed on 5 August 2021).
16. Paradise, T.R. Perception of earthquake risk in Agadir, Morocco: A case study from a Muslim community. *Environ. Hazards* **2005**, *6*, 167–180. [CrossRef]
17. Baytiyeh, H.; Naja, M. The effects of fatalism and denial on earthquake preparedness levels. *Disaster Prev. Manag. Int. J.* **2016**, *25*, 154–167. [CrossRef]
18. Yari, A.; Zarezadeh, Y.; Ostadtaghizadeh, A. Prevalence of fatalistic attitudes toward earthquake disaster risk management in citizens of Tehran, Iran. *Int. J. Disaster Risk Reduct.* **2019**, *38*, 101181. [CrossRef]
19. Dynes, R.R. Disaster Reduction: The Importance of Adequate Assumptions about Social Organization. *Sociol. Spectr.* **1993**, *13*, 175–192. [CrossRef]
20. Smith, K. *Environmental Hazards: Assessing Risk and Reducing Disaster*, 6th ed.; Routledge: Oxford, UK, 2013; p. 470.
21. Carr, L.J. Disaster and the Sequence-Pattern Concept of Social Change. *Am. J. Sociol.* **1932**, *38*, 207–218. [CrossRef]
22. White, G.F. The limit of economic justification for flood protection. *J. Land Public Util. Econ.* **1936**, *12*, 133–148. [CrossRef]
23. Barrows, H.H. Geography as Human Ecology. *Ann. Assoc. Am. Geogr.* **1923**, *13*, 1–14. [CrossRef]
24. Burton, I.; Kates, R.W.; White, G.F. *The Human Ecology of Extreme Geophysical Events*; Natural Hazards Research, Working Paper 1; University of Chicago: Chicago, IL, USA, 1968; 38p.
25. White, G.F. *Natural Hazards, Local, National, Global*; Oxford University Press: Oxford, UK, 1974.
26. Burby, R.J. Natural hazards and land use: An introduction. In *Cooperating with Nature: Confronting Natural Hazards with Land-Use Planning for Sustainable Communities*; Burby, R.J., Ed.; Joseph Henry Press: Washington, DC, USA, 1998; pp. 1–26.
27. McHarg, I.L. *Design with Nature*; Natural History Press: New York, NY, USA, 1969; 198p.
28. Uehara, M. The Long Term Economic Value of Holistic Ecological Planning for Disaster Risk. In *Rethinking Resilience, Adaptation and Transformation in a Time of Change*; Yan, W., Galloway, W., Eds.; Springer: Cham, Switzerland, 2017. [CrossRef]
29. Wagner, M.; Merson, J.; Wentz, E.A. Design with Nature: Key lessons from McHarg's intrinsic suitability in the wake of Hurricane Sandy. *Landsc. Urban Plan.* **2016**, *155*, 33–46. [CrossRef]
30. UNDP. *An Overview of Disaster Management*; United Nations Development Program: New York, NY, USA, 1992; 136p.
31. Linnerooth-Bayer, J.; Amendola, A. Global change, natural disasters and loss-sharing: Issues of efficiency and equity. *Geneva Pap. Risk Insur.-Issues Pract.* **2000**, *25*, 203–219. [CrossRef]
32. Cuny, F.C. *Disasters and Development*; Oxfam: Dallas, TX, UDA, 1994.
33. O'Keefe, P.; Westgate, K.; Wisner, B. Taking the naturalness out of natural disasters. *Nature* **1976**, *260*, 566–567. [CrossRef]
34. Chmutina, K.; Von Meding, J. A Dilemma of language: "Natural disasters" in academic literature. *Int. J. Disaster Risk Sci.* **2019**, *10*, 283–292. [CrossRef]
35. Wijkman, A.; Timberlake, L. *Natural Disasters: Acts of God or Acts of Man?* Routledge: London, UK, 1984; 150p.
36. Kelman, I. Natural Disasters Do Not Exist (Natural Hazards Do Not Exist Either) Version 3. 9 July 2010 (Version 1 Was 26 July 2007). Available online: http://www.ilankelman.org/miscellany/NaturalDisasters.doc (accessed on 20 October 2021).
37. Bosher, L.; Chmutina, K. *Disaster Risk Reduction for the Built Environment*; John Wiley & Sons: Hoboken, NJ, USA, 2017; 304p.
38. Chmutina, K.; von Meding, J.; Bosher, L. Language Matters: Dangers of the "Natural Disaster" Misnomer. Contributing Paper to the Global Assessment Report (GAR) 2019. UNDRR, 2019. Available online: https://www.preventionweb.net/publications/view/65974 (accessed on 20 October 2021).
39. UNDRR (United Nations Office for Disaster Risk Reduction). Global Assessment Report 2019. Available online: https://gar.unisdr.org/ (accessed on 13 October 2021).
40. Reliefweb. Why Disasters Are Not Natural? 2021. Available online: https://reliefweb.int/report/world/why-disasters-are-not-natural (accessed on 20 October 2021).
41. Preventionweb. Changing the Discourse of Disaster Reporting. 2021. Available online: https://www.preventionweb.net/blog/nonaturaldisasters-changing-discourse-disaster-reporting (accessed on 20 October 2021).
42. Global Network of Civil Society Organisations for Disaster Reduction (GNDR). Disasters Are Not Natural. 2021. Available online: https://www.gndr.org/news/item/1499-disasters-are-not-natural.html (accessed on 20 October 2021).
43. Weichselgartner, J. Disaster mitigation: The concept of vulnerability revisited. *Disaster Prev. Manag. Int. J.* **2001**, *10*, 85–95. [CrossRef]

44. Burton, I.; Kates, R.W.; White, G.F. *The Environment as Hazard*; The Guilford Press: New York, NY, USA, 1993.
45. Quarantelli, E.L. *What Is a Disaster? Perspectives on the Question*; Routledge: New York, NY, USA, 1998.
46. Mileti, D. *Disasters by Design: A Reassessment of Natural Hazards in the United States*; Joseph Henry Press: Washington, DC, USA, 1999.
47. McEntire, D.A. Triggering agents, vulnerabilities and disaster reduction: Towards a holistic paradigm. *Disaster Prev. Manag. Int. J.* **2001**, *10*, 189–196. [CrossRef]
48. Messerli, B.; Grosjean, M.; Hofer, T.; Nunez, L.; Pfister, C. From nature-dominated to human-dominated environmental changes. *Quat. Sci. Rev.* **2000**, *19*, 459–479. [CrossRef]
49. Van Aalst, M.K. The impacts of climate change on the risk of natural disasters. *Disasters* **2006**, *30*, 5–18. [CrossRef]
50. UN/ISDR. *Living with Risk: A global Review of Disaster Reduction Initiatives*; United Nations: Geneva, Switzerland, 2004; 429p, Available online: https://www.undrr.org/publication/living-risk-global-review-disaster-reduction-initiatives (accessed on 2 July 2021).
51. Wenger, D. Hazards and disaster research, how would the past 40 years rate. *Nat. Hazards Obs.* **2006**, *31*, 1–28.
52. World Bank. *Natural Hazards, Unnatural Disasters: The Economics of Effective Prevention*; The World Bank: Washington, DC, USA, 2010.
53. Munich, Re. *TOPICS GEO, Natural Catastrophes 2019, Analyses, Assessments, Positions*; Munich Re: Munich, Germany, 2020.
54. Boudreau, T. *Solving the Risk Equation: People-Centred Disaster Risk Assessment in Ethiopia*; HPN Network Paper 66; Humanitarian Practice Network, Overseas Development Institute: London, UK, 2009; 40p.
55. Clarke, C.L.; Munasinghe, M. Economic aspects of disasters and sustainable development: An introduction. In *Disaster Prevention for Sustainable Development: Economic and Policy Issues*; International Decade for Natural Disaster Reduction (IDNDR): Washington, DC, USA, 1995; pp. 1–10.
56. Ranke, U. *Natural Disaster Risk Management: Geosciences and Social Responsibility*; Springer International Publishing: Cham, Switzerland, 2016; p. 514.
57. EM-DAT. Emergency Management Database, CRED, Catholic University of Louvain. 2021. Available online: http://www.public.emdat.be (accessed on 3 July 2021).
58. Daniell, J.E.; Schaefer, A.M.; Wenzel, F. Losses Associated with Secondary Effects in Earthquakes. *Front. Built Environ.* **2017**, 3, 30. [CrossRef]
59. Daniell, J.E.; Khazai, B.; Wenzel, F.; Vervaeck, A. The CATDAT damaging earthquakes database. *Nat. Hazards Earth Syst. Sci.* **2011**, *11*, 2235–2251. [CrossRef]
60. Dilley, M.; Chen, R.S.; Deichmann, U.; Lerner-Lam, A.L.; Arnold, M.; Agwe, J.; Buys, P.; Kjekstad, O.; Lyon, B.; Yetman, G. *Natural Disaster Hotspots: A Global Risk Analysis*; Disaster Risk Management Series 5; The World Bank: Washington, DC, USA, 2005; p. 132.
61. National Governors' Association. *Comprehensive Emergency Management: A Governor's Guide, Center for Policy Research*; U.S. Defense Civil Preparedness Agency: Washington, DC, USA, 1979; 65p.
62. Salter, J. Risk management in a disaster management context. *J. Contingencies Crisis Manag.* **1997**, *5*, 60–65. [CrossRef]
63. Cronstedt, M. Prevention, preparedness, response, recovery-an outdated concept? *Aust. J. Emerg. Manag.* **2002**, *17*, 10–13.
64. Rogers, P. Development of resilient Australia: Enhancing the PPRR approach with anticipation, assessment and registration of risks. *Aust. J. Emerg. Manag.* **2011**, *26*, 54–58.
65. Pretto, E.A.; Ricci, E.; Klain, M.; Safar, P.; Semenov, V.; Abrams, J.; Tisherman, S.; Crippen, D.; Comfort, L. Disaster reanimatology potentials: A structured interview study in Armenia. III. Results, conclusions, and recommendations. *Prehospital Disaster Med.* **1992**, *7*, 327–337. [CrossRef]
66. Comfort, L.K. *Self-Organization in Disaster Response: The Great Hanshin, Japan, Earthquake of 17 January 1995*; Quick Response Report 78; Natural Hazards Research and Applications Information Center: Boulder, CO, USA, 1996; 29p, Available online: http://cidbimena.desastres.hn/docum/crid/Abril2006/CD2/pdf/eng/doc8691/doc8691-a.pdf (accessed on 16 August 2021).
67. Smith, W.R.; Stephens, K.K.; Robertson, B.R.; Li, J.; Murthy, D. Social media in citizen-led disaster response: Rescuer roles, coordination challenges, and untapped potential. In Proceedings of the 15th ISCRAM Conference, Rochester, NY, USA, 20–23 May 2018.
68. Houston, J.B.; Hawthorne, J.; Perreault, M.; Park, E.H.; Hode, M.G.; Halliwell, M.R.; McGowen, S.E.T.; Davis, R.; Vaid, S.; McElderry, J.A.; et al. Social media and disasters: A functional framework for social media use in disaster planning, response, and research. *Disasters* **2014**, *39*, 1–22. [CrossRef]
69. Lindsay, B.R. *Social Media and Disasters: Current Uses, Future Options, and Policy Considerations*; Report Number: R41987; Congressional Research Service: Washington, DC, USA, 2011.
70. Manso, M.; Manso, B. The role of social media in crisis: A European holistic approach to the adoption of online and mobile communications in crisis response and search and rescue efforts. In *Strategic Intelligence Management*; Butterworth-Heinemann: Oxford, UK, 2013; pp. 93–107.
71. Jayathilaka, G.S.; Siriwardana, C.; Amaratunga, D.; Haigh, R.; Dias, N. A Conceptual Framework for Social Media Use During Disasters. In *Multi-Hazard Early Warning and Disaster Risks*; Amaratunga, D., Haigh, R., Dias, N., Eds.; Springer: Cham, Switzerland, 2021; pp. 659–684.
72. Finau, G.; Tarai, J.; Varea, R.; Titifanue, J.; Kant, R.; Cox, J. Social media and disaster communication: A case study of Cyclone Winston. *Pac. J. Rev.* **2018**, *24*, 123–137. [CrossRef]

73. Takahashi, B.; Tandoc, E.C., Jr.; Carmichael, C. Communicating on Twitter during a disaster: An analysis of tweets during Typhoon Haiyan in the Philippines. *Comput. Hum. Behav.* **2015**, *50*, 392–398. [CrossRef]
74. Yadav, M.; Zillur, R. The social role of social media: The case of Chennai rains—2015. *Soc. Netw. Anal. Min.* **2016**, *6*, 101. [CrossRef]
75. Peary, B.D.M.; Shaw, R.; Takeuchi, Y. Utilization of social media in the east Japan earthquake and tsunami and its effectiveness. *J. Nat. Disaster Sci.* **2012**, *34*, 3–18. [CrossRef]
76. Noji, E.K. The Public Health Consequences of Disasters. *Prehospital Disaster Med.* **2000**, *15*, 21–31. [CrossRef]
77. Nowell, B.; Steelman, T.; Velez, A.L.K.; Yang, Z. The structure of effective governance of disaster response networks: Insights from the field. *Am. Rev. Public Adm.* **2018**, *48*, 699–715. [CrossRef]
78. Platt, S. Factors affecting the speed and quality of post-disaster recovery and resilience. In *International Conference on Earthquake Engineering and Structural Dynamics*; Springer: Cham, Switzerland, 2017; pp. 369–403.
79. Freeman, P.K. Infrastructure, natural disasters, and poverty. In *Managing Disaster Risk in Emerging Economies*; Kreimer, A., Arnold, M., Eds.; Disaster Risk Management Series No. 2; World Bank: Washington, DC, USA, 2000; pp. 55–61.
80. Leeson, P.T.; Sobel, R.S. Weathering corruption. *J. Law Econ.* **2008**, *51*, 667–681. [CrossRef]
81. Yamamura, E. Impact of natural disaster on public sector corruption. *Public Choice* **2014**, *161*, 385–405. [CrossRef]
82. Weichselgartner, J.; Bertens, J. Natural disasters: Acts of God, nature or society? On the social relation to natural hazards. *WIT Trans. Ecol. Environ.* **2000**, *45*, 10.
83. Zhang, H. What has China learnt from disasters? Evolution of the emergency management system after SARS, Southern Snowstorm, and Wenchuan Earthquake. *J. Comp. Policy Anal. Res. Pract.* **2012**, *14*, 234–244. [CrossRef]
84. Lu, Y.; Xu, J. The progress of emergency response and rescue in China: A comparative analysis of Wenchuan and Lushan earthquakes. *Nat. Hazards* **2014**, *74*, 421–444. [CrossRef]
85. Yang, J.; Chen, J.; Liu, H.; Zheng, J. Comparison of two large earthquakes in China: The 2008 Sichuan Wenchuan Earthquake and the 2013 Sichuan Lushan Earthquake. *Nat. Hazards* **2014**, *73*, 1127–1136. [CrossRef]
86. Pelling, M. *Making Disaster Risk Reduction Work. A Review and Discussion of Key Themes, Challenges and Potential Contributions to Be Made by ProVention in Promoting Disaster Risk Reduction*; Provention Consortium: Geneva, Switzerland, 2007; 24p, Available online: https://www.preventionweb.net/files/11114_Forum2007report1.pdf (accessed on 30 June 2021).
87. Basher, R. *Science and Technology for Disaster Risk Reduction: A Review of Application and Coordination Needs*; UNISDR Report; 2013; Available online: https://www.preventionweb.net/posthfa/documents/Science-and-Technology-for-Disaster-Risk-Reduction.pdf (accessed on 22 June 2021).
88. Kreimer, A.; Arnold, M. Introduction. In *Managing Disaster Risk in Emerging Economies*; Kreimer, A., Arnold, M., Eds.; Disaster Risk Management Series No. 2; The World Bank: Washington, DC, USA, 2000; pp. 1–7.
89. Alcántara-Ayala, I. Geomorphology, natural hazards, vulnerability and prevention of natural disasters in developing countries. *Geomorphology* **2002**, *47*, 107–124. [CrossRef]
90. Tekeli-Yeşil, S.; Dedeoğlu, N.; Braun-Fahrlaender, C.; Tanner, M. Factors motivating individuals to take precautionary action for an expected earthquake in Istanbul. *Risk Anal. Int. J.* **2010**, *30*, 1181–1195. [CrossRef]
91. Hochrainer-Stigler, S.; Patnaik, U.; Kull, D.; Singh, P.; Wajih, S. Disaster financing and poverty traps for poor households: Realities in Northern India. *Int. J. Mass Emerg. Disasters* **2011**, *29*, 57–82.
92. Shreve, C.M.; Kelman, I. Does mitigation save? Reviewing cost-benefit analyses of disaster risk reduction. *Int. J. Disaster Risk Reduct.* **2014**, *10*, 213–235. [CrossRef]
93. Otani, S. The dawn of structural earthquake engineering in Japan. In Proceedings of the 14th World Conference on Earthquake Engineering, Beijing, China, 12–17 October 2008.
94. Beavers, J.E. A review of seismic hazard description in US design codes and procedures. *Prog. Struct. Eng. Mater.* **2002**, *4*, 46–63. [CrossRef]
95. Reitherman, R.K. *Earthquakes and Engineers: An International History*; American Society of Civil Engineers: Reston, VA, USA, 2012; p. 474.
96. Jain, S.K.; Nigam, N.C. Historical developments and current status of earthquake engineering in India. In Proceedings of the Twelfth World Conference on Earthquake Engineering, Auckland, New Zealand, 30 January–4 February 2000; Volume 30.
97. Scawthorn, C.; Chen, W.F. *Earthquake Engineering Handbook*; CRC Press: Boca Ratan, FL, USA, 2002; 1480p.
98. Spence, R.; So, E. *Estimating Shaking-Induced Casualties and Building Damage for Global Earthquake Events*; Cambridge Architectural Research Ltd.: Cambridge, UK, 2009.
99. Chaudhary, M.T.A. Seismic Vulnerability Mitigation of Non-Engineered Buildings. *Int. J. Civ. Environ. Struct. Constr. Archit. Eng.* **2014**, *8*, 37–41.
100. Arya, A.S.; Boen, T.; Ishiyama, Y. *Guidelines for Earthquake Resistant Non-Engineered Construction*; UNESCO: Paris, France, 2014; 191p.
101. Piracha, A.; Chaudhary, M.T. Better awareness for better natural hazards preparedness in Pakistan. *Int. J. Built Environ. Sustain.* **2015**, *2*, 252–259. [CrossRef]
102. Hochrainer-Stigler, S.; Kunreuther, H.; Linnerooth-Bayer, J.; Mechler, R.; Michel-Kerjan, E.; Muir-Wood, R.; Ranger, N.; Vaziri, P.; Young, M. *The Costs and Benefits of Reducing Risk from Natural Hazards to Residential Structures in Developing Countries*; University of Pennsylvania: Philadelphia, PA, USA, 2010.

103. Montz, B.; Gruntfest, E.C. Changes in American urban floodplain occupancy since 1958: The experiences of nine cities. *Appl. Geogr.* **1986**, *6*, 325–338. [CrossRef]
104. FEMA. *Engineering Principles and Practices for Retrofitting Flood-Prone Residential Structures*, 3rd ed.; FEMA P-259; Federal Emergency Management Agency (FEMA): San Francisco, CA, USA, 2012.
105. Guo, J.C. *Urban Storm Water Design*; Water Resources Publication: Littleton, CO, USA, 2003.
106. NOAA. *Glossary of Terms, National Oceanic and Atmospheric Administration*; NOAA: Silver Spring, MD, USA, 2021. Available online: https://w1.weather.gov/glossary/ (accessed on 2 July 2021).
107. Amini, M.; Memari, A.M. Review of literature on performance of coastal residential buildings under hurricane conditions and lessons learned. *J. Perform. Constr. Facil.* **2020**, *34*, 04020102. [CrossRef]
108. Taher, R. *General Recommendations for Improved Building Practices in Earthquake and Hurricane Prone Areas*; Prepared in the Aftermath of the Haiti Earthquake of 2010 for Architecture for Humanity; New Jersey Institute of Technology, College of Architecture & Design, University Heights: Newark, NJ, USA, 2010; Available online: www.njit.edu (accessed on 6 August 2021).
109. Nadal, N.C.; Zapata, R.E.; Pagan, I.; Lopez, R.; Agudelo, J. Building damage due to riverine and coastal floods. *J. Water Resour. Plan. Manag.* **2010**, *136*, 327–336. [CrossRef]
110. Li, Y. Assessment of damage risks to residential buildings and cost–benefit of mitigation strategies considering hurricane and earthquake hazards. *J. Perform. Constr. Facil.* **2012**, *26*, 7–16. [CrossRef]
111. Bryant, W.A.; Hart, E.W. *Fault-Rupture Hazard Zones in California*; Special Publication 42; California Department of Conservation: Sacramento, CA, USA, 2007; 50p.
112. Oliver-Smith, A. *The Martyred City: Death and Rebirth in the Andes*; University of New Mexico Press: Albuquerque, NM, USA, 1986; 280p.
113. Jackson, J. Fatal attraction: Living with earthquakes, the growth of villages into megacities, and earthquake vulnerability in the modern world. *Philos. Trans. R. Soc. A Math. Phys. Eng. Sci.* **2006**, *364*, 1911–1925. [CrossRef]
114. Whitney, D.J.; Dickerson, A.; Lindell, M.K. Nonstructural seismic preparedness of Southern California hospitals. *Earthq. Spectra* **2001**, *17*, 153–171. [CrossRef]
115. Kunreuther, H.; Ginsberg, R.; Miller, L.; Sagi, P.; Slovic, P.; Borkan, B.; Katz, N. *Disaster Insurance Protection: Public Policy Lessons*; Wiley: New York, NY, USA, 1978.
116. Walker, G.R. Insurance as a tool for reducing natural hazard impact. In *Insurance Viability & Loss Mitigation*; Britton, N.R., McDonald, J., Oliver, J., Eds.; Alexander Howden Reinsurance Brokers (Australia) Ltd.: Sydney, Australia, 1995; pp. 211–223.
117. Swiss Re. *Lessons from Recent Major Earthquakes*; Swiss Re Institute: Zurich, Switzerland, 2012; p. 18. Available online: https://www.swissre.com/dam/jcr:dbdf9660-9061-4eca-a5b3-ed48af267bfe/Exp_Pub_Lessons_from_recent_major_earthquakes.pdf (accessed on 15 July 2021).
118. Paleari, S. Disaster risk insurance: A comparison of national schemes in the EU-28. *Int. J. Disaster Risk Reduct.* **2018**, *35*, 101059. [CrossRef]
119. Gizzi, F.T.; Kam, J.; Porrini, D. Time windows of opportunities to fight earthquake under-insurance: Evidence from Google Trends. *Humanit. Soc. Sci. Commun.* **2020**, *7*, 61. [CrossRef]
120. Walker, G.R. Earthquake engineering and insurance: Past, present and future. In Proceedings of the 12th World Conference on Earthquake Engineering, Auckland, New Zealand, 30 January–4 February 2000. Paper No. 1472.
121. Gkimprixis, A.; Douglas, J.; Tubaldi, E. Seismic risk management through insurance and its sensitivity to uncertainty in the hazard model. *Nat. Hazards* **2021**, *108*, 1629–1657. [CrossRef]
122. Linnerooth-Bayer, J.; Mechler, R.; Hochrainer-Stigler, S. Insurance against losses from natural disasters in developing countries: Evidence, gaps and the way forward. *J. Integr. Disaster Risk Manag.* **2011**, *1*. [CrossRef]
123. World Bank. *Damage, Loss, and Needs Assessment for Disaster Recovery and Reconstruction after the 2008 Cyclone Season in Madagascar*; Government and Madagascar, United Nations and World Bank: Washington, DC, USA, 2008.
124. Agrawal, N. Disaster Perceptions. *Nat. Disasters Risk Manag. Can.* **2018**, *49*, 193–217. [CrossRef]
125. NRC. *A Safer Future: Reducing the Impacts of Natural Disasters*; National Research Council, The National Academies Press: Washington, DC, USA, 1991. [CrossRef]
126. BRI and GRIPS. *Disaster Education*; Building Research Institute and National Graduate Institute for Policy Studies: Tokyo, Japan, 2007.
127. Boon, H.; Pagliano, P. Disaster Education in Australian Schools. *Aust. J. Environ. Educ.* **2014**, *30*, 187–1977. [CrossRef]
128. Unicef. *Disaster Risk Reduction in Education: Building Resilience with and for Children in Central Asia and South Caucasus*; The United Nations International Children's Emergency Fund: Almaty, Kazakhstan, 2013.
129. Thayaparan, M.; Malalgoda, C.; Keraminiyage, K.; Amaratunga, D. Disaster Management Education through Higher Education—Industry Collaboration in the Built Environment. *Procedia Econ. Finance* **2014**, *18*, 651–658. [CrossRef]
130. Kitagawa, K. Conceptualising 'Disaster Education'. *Educ. Sci.* **2021**, *11*, 233. [CrossRef]
131. Shaw, R.; Shiwaku, K.; Takeuchi, Y. *Disaster Education*; Emerald: Bingley, UK, 2011.
132. CDC. Emergency Preparedness and Response: CDC Situation Awareness—2021 Natural Hazards and Severe Weather. 2021. Available online: https://emergency.cdc.gov/situationawareness/naturalhazards/ (accessed on 18 October 2021).
133. The World Bank. *Disaster Risk Management in South Asia: A Regional Overview*; The World Bank: Washington, DC, USA, 2012.

134. Ruangpan, L.; Vojinovic, Z.; Di Sabatino, S.; Leo, L.S.; Capobianco, V.; Oen, A.M.P.; McClain, M.E.; Lopez-Gunn, E. Nature-based solutions for hydro-meteorological risk reduction: A state-of-the-art review of the research area. *Nat. Hazards Earth Syst. Sci.* **2020**, *20*, 243–270. [CrossRef]
135. Andjelkovic, I. *Guidelines on Non-Structural Measures in Urban Flood Management*; International Hydrological Programme (IHP), United Nations Educational, Scientific and Cultural Organization (UNESCO): Paris, France, 2001.
136. Jha, A.K.; Bloch, R.; Lamond, J. *Cities and Flooding: A Guide to Integrated Urban Flood Risk Management for the 21st Century*; World Bank: Washington, DC, USA, 2012; pp. 46–49.
137. Thieken, A.H.; Cammerer, H.; Dobler, C.; Lammel, J.; Schöberl, F. Estimating changes in flood risks and benefits of non-structural adaptation strategies—A case study from Tyrol, Austria. *Mitig. Adapt. Strat. Glob. Chang.* **2014**, *21*, 343–376. [CrossRef] [PubMed]
138. Vojinovic, Z.; Abbott, M.B. *Flood Risk and Social Justice*; IWA Publishing: London, UK, 2012.
139. Boyce, J.K. Birth of a megaproject: Political economy of flood control in Bangladesh. *Environ. Manag.* **1990**, *14*, 419–428. [CrossRef]
140. Hallegatte, S. *A Cost Effective Solution to Reduce Disaster Losses in Developing Countries: Hydro-Meteorological Services, Early Warning, and Evacuation*; World Bank Policy Research Working Paper 6058; World Bank: Washington, DC, USA, 2012.
141. Linnerooth-Bayer, J.; Mechler, R.; Pflug, G. Refocusing disaster aid. *Science* **2005**, *309*, 1044–1046. [CrossRef]
142. Michel-Kerjan, E.; Hochrainer-Stigler, S.; Kunreuther, H.; Linnerooth-Bayer, J.; Mechler, R.; Muir-Wood, R.; Ranger, N.; Vaziri, P.; Young, M. Catastrophe risk models for evaluating disaster risk reduction investments in developing countries. *Risk Anal.* **2012**, *33*, 984–999. [CrossRef]
143. Kankanamge, N.; Yigitcanlar, T.; Goonetilleke, A. How engaging are disaster management related social media channels? The case of Australian state emergency organisations. *Int. J. Disaster Risk Reduct.* **2020**, *48*, 101571. [CrossRef]
144. Zhang, C.; Fan, C.; Yao, W.; Hu, X.; Mostafavi, A. Social media for intelligent public information and warning in disasters: An interdisciplinary review. *Int. J. Inf. Manag.* **2019**, *49*, 190–207. [CrossRef]
145. Department of Homeland Security. Social Media Strategy. Washington, DC, USA, 2012. Available online: https://www.dhs.gov/sites/default/files/publications/Virtual%20Social%20Media%20Working%20Group%20VSMWG%20Social%20Media%20Strategy.pdf (accessed on 15 October 2021).
146. Kaminska, K.; Rutten, B. *Social Media in Emergency Management: Capability Assessment*; Defence Research and Development Canada-Centre for Security Science (CSSP): Ottawa, ON, Canada, 2014.
147. American Red Cross Crisis Data Summit 2010. Available online: http://www.scribd.com/doc/40080608/The-Path-Forward-ARC-CrisisData-Summit-Wrap-Up (accessed on 15 October 2021).
148. Tagliacozzo, S.; Magni, M. Government to Citizens (G2C) communication and use of social media in the post-disaster reconstruction phase. *Environ. Hazards* **2017**, *17*, 1–20. [CrossRef]
149. DESA (Department of Economic and Social Affairs). World Population Prospects, United Nations. 2021. Available online: https://population.un.org/wpp/ (accessed on 4 September 2021).
150. Veach, V.; Moilanen, A.; Di Minin, E. Threats from urban expansion, agricultural transformation and forest loss on global conservation priority areas. *PLoS ONE* **2017**, *12*, e0188397. [CrossRef] [PubMed]
151. Giovanni, G.; Mokhtarian, P. Impacts of Information and Communication Technology. In *The Geography of Urban Transportation*; Giuliano, G., Hanson, S., Eds.; Guilford Press: New York, NY, USA, 2017.
152. TNC (The Nature Conservancy). *Nature in the Urban Century*; Stockholm Reliance Centre, Stockholm University: Stockholm, Sweden, 2018.
153. Packman, J. *The Effects of Urbanisation on Flood Magnitude and Frequency*; Report #63; Institute of Hydrology: Wallingford, UK, 1980.
154. Cornwall, W. Europe's Deadly Floods Leave Scientists Stunned. Science. 20 July 2021. Available online: https://www.science.org/news/2021/07/europe-s-deadly-floods-leave-scientists-stunned (accessed on 5 September 2021).
155. Knolle, K.; Alkousaa, R. Germany Counts Cost of Floods as Hopes of Finding Survivors Fade. Reuters. 22 July 2021. Available online: https://www.reuters.com/world/europe/no-more-survivors-likely-be-found-german-flood-zone-relief-official-2021-07-21/ (accessed on 5 September 2021).
156. The Weather Channel. Available online: https://weather.com/news/weather/news/2021-09-02-northeast-historic-rainfall-flooding-stats (accessed on 5 September 2021).
157. IPCC. *Climate Change 2021: The Physical Science Basis. Contribution of Working Group I to the Sixth Assessment Report of the Intergovernmental Panel on Climate Change*; Cambridge University Press: Cambridge, UK, 2021.
158. UNEP. Ten Impacts of the Australian Bushfires, United Nations Environmental Program. 2020. Available online: https://www.unep.org/news-and-stories/story/ten-impacts-australian-bushfires (accessed on 5 September 2021).
159. BBC. Australia's Fires 'Killed or Harmed Three Billion Animals'. 2020. Available online: https://www.bbc.com/news/world-australia-53549936 (accessed on 5 September 2021).
160. MetOffice. Recording Observations for over 100 Years, Meteorological Office of the UK. 2020. Available online: https://www.metoffice.gov.uk/weather/learn-about/how-forecasts-are-made/observations/recording-observations-for-over-100-years (accessed on 3 September 2021).
161. Her, Y.; Yoo, S.-H.; Cho, J.; Hwang, S.; Jeong, J.; Seong, C. Uncertainty in hydrological analysis of climate change: Multi-parameter vs. multi-GCM ensemble predictions. *Sci. Rep.* **2019**, *9*, 4974. [CrossRef]
162. Pechtl, J.; Land, A. Tree rings as a proxy for seasonal precipitation variability and Early Neolithic settlement dynamics in Bavaria, Germany. *PLoS ONE* **2019**, *14*, e0210438. [CrossRef]

163. Kleinen, T.; Petschel-Held, G. Integrated assessment of changes in flooding probabilities due to climate change. *Clim. Chang.* **2007**, *81*, 283–312. [CrossRef]
164. Flato, G.; Marotzke, J.; Abiodun, B.; Braconnot, P.; Chou, S.C.; Collins, W.; Cox, P.; Driouech, F.; Emori, S.; Eyring, V.; et al. Evaluation of Climate Models. In *Climate Change 2013: The Physical Science Basis. Contribution of Working Group I to the Fifth Assessment Report of the Intergovernmental Panel on Climate Change*; Stocker, T.F., Qin, D., Plattner, G.-K., Tignor, M., Eds.; Cambridge University Press: Cambridge, UK, 2013.
165. Smart Growth Network. This is Smart Growth. 2013. Available online: SmartGrowth.Org (accessed on 8 September 2021).
166. USEPA. Smart Growth in your Community. 2021. Available online: https://www.epa.gov/smartgrowth/smart-growth-your-community (accessed on 22 August 2021).
167. Berke, P.; Godschalk, D.; Kaiser, E.; Rodriguez, D. *Urban Land Use Planning*, 5th ed.; University of Illinois Press: Champaign, IL, USA, 2006.
168. Ritchie, H.; Roser, M. Natural Disasters. 2014. Available online: https://ourworldindata.org/natural-disasters (accessed on 18 October 2021).
169. Centre for Research on the Epidemiology of Disasters (CERD); United Nations Office for Disaster Risk Reduction (UNDRR). The Human Cost of Disasters: An Overview of the Last 20 Years (2000–2019). Available online: https://www.undrr.org/media/48008/download (accessed on 18 October 2021).

encyclopedia

Entry

Ionospheric Remote Sensing with GNSS

YuXiang Peng [1,2,3,*] and Wayne A. Scales [1,2]

1 Center for Space Science and Engineering Research, Virginia Tech, Blacksburg, VA 24061, USA; wscales@vt.edu
2 Bradley Department of Electrical and Computer Engineering, Virginia Tech, Blacksburg, VA 24061, USA
3 Qualcomm Technologies Inc., Santa Clara, CA 95051, USA
* Correspondence: yuxiang7@vt.edu

Definition: The Global Navigation Satellite System (GNSS) plays a pivotal role in our modern positioning, navigation and timing (PNT) technologies. GNSS satellites fly at altitudes of approximately 20,000 km or higher. This altitude is above an ionized layer of the Earth's upper atmosphere, the so called "ionosphere". Before reaching a typical GNSS receiver on the ground, GNSS satellite signals penetrate through the Earth's ionosphere. The ionosphere is a plasma medium consisting of free charged particles that can slow down, attenuate, refract, or scatter the GNSS signals. Ionospheric density structures (also known as irregularities) can cause GNSS signal scintillations (phase and intensity fluctuations). These ionospheric impacts on GNSS signals can be utilized to observe and study physical processes in the ionosphere and is referred to ionospheric remote sensing. This entry introduces some fundamentals of ionospheric remote sensing using GNSS.

Keywords: GNSS; ionosphere; remote sensing

Citation: Peng, Y.; Scales, W.A. Ionospheric Remote Sensing with GNSS. *Encyclopedia* **2021**, *1*, 1246–1256. https://doi.org/10.3390/encyclopedia1040094

Academic Editors: Raffaele Barretta, Ramesh Agarwal, Krzysztof Kamil Żur and Giuseppe Ruta

Received: 11 October 2021
Accepted: 17 November 2021
Published: 22 November 2021

1. GNSS Introduction

The positioning, navigation, and timing (PNT) provided by the Global Navigation Satellite System (GNSS) forms a ubiquitous technological infrastructure in modern society. The high-level GNSS positioning is simply based on the principle of trilateration. To determine the unknown location (x, y, z) of a GNSS receiver as shown in Figure 1, for simplicity let us assume the locations of three beacon GNSS satellites in the sky are known beforehand (transmitted by the GNSS satellites to the receiver via navigation messages). When the receiver acquires and tracks the incoming GNSS signals from the three satellites, it can determine the signal propagation time Δt (transmission time minus reception time). Assume the GNSS signals propagate at the speed of light (c), the distances from the receiver to the 3 beacon satellites (R_1, R_2, and R_3) can be estimated by multiplying c with Δt. Then, a set of trilateration equations can be established as:

$$c(\Delta t^m) = \sqrt{(x - x^m)^2 + (y - y^m)^2 + (z - z^m)^2} \text{ where } m = 1, 2, 3. \quad (1)$$

Given x^m, y^m, z^m, Δt^m, and c are known, the three unknowns x, y, and z can be determined by solving the three equations simultaneously and the solutions will give two positions (one outside of the Earth, one on the surface on the Earth). It is important to note in reality, there is an unknown bias in the signal propagation time from every beacon satellite due to a common time error from the inaccurate receiver clock (δt). Therefore, an additional clock bias term must be introduced as the fourth unknown, implying in reality that four satellites are needed to determine the receiver position. Consequently, an additional GNSS beacon satellite needs to be tracked to obtain a fourth sphere equation.

$$c(\Delta t^m + \delta t) = \sqrt{(x - x^m)^2 + (y - y^m)^2 + (z - z^m)^2} \text{ where } m = 1, 2, 3, 4. \quad (2)$$

This set of four equations, involving reception of at least four GNSS satellite signals, forms the underlying algorithm to solve a simple static positioning problem in the 3D space including the receiver clock bias.

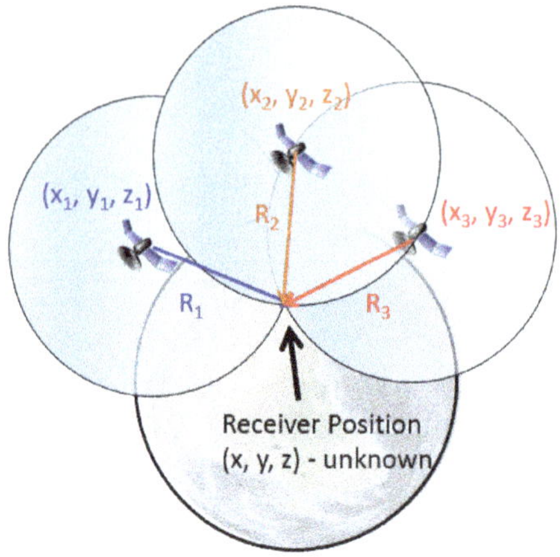

Figure 1. Trilateration principle of GNSS positioning.

By definition, GNSS are satellite navigation systems with global signal coverage. Currently, there are four operational GNSS constellations: USA's Global Positioning System (GPS), Russia's Global'naya Navigatsionnaya Sputnikovaya Sistema (GLONASS), European Union's Galileo, and China's BeiDou Navigation Satellite System (BDS, formerly known as COMPASS). As of October 2021, the GPS, GLONASS, and Beidou constellations are fully operational. The Galileo constellation is expected to reach a full operational capability (FOC) stage soon. A brief status summary of four GNSS constellations is given in Table 1.

Table 1. Current status of GNSS constellations (* n stands for GLONASS frequency channel number).

	GPS	GLONASS	Galileo	Beidou
Affiliation	U.S.A.	Russia	European Union	P.R. China
Orbital Altitude	MEO (∼20,200 km)	MEO (∼19,100 km)	MEO (∼23,222 km)	MEO/GEO/IGSO
Number of Operational Satellites (Scheduled total)	31 (32)	23 (29)	22 (30)	56 (≥ 61)
Signal and Frequency (in MHz)	L1 (1575.42), L2 (1227.6), L5 (1176.45)	G1 (1602 + n × 0.5625), G2 (1246 + n × 0.4375), G3 (1201 + n × 0.4375) *	L1 (1575.42), E5 (1191.795), E5a (1176.45), E5b (1207.14), E6 (1278.75)	B1 (1561.10/1575.42), B2 (1207.14/1176.45), B3 (1268.52)

The GPS satellites are located within six different orbital planes of medium Earth orbit (MEO) with an altitude of ∼20,200 km. Each two neighboring orbital planes are separated by 60 degrees in Ω (longitude of the ascending node). The inclination angle of all GPS satellites is approximately 55 degrees. The orbital period of all GPS satellites is approximately 12 h. By design, a GPS receiver at any place on the Earth's open surface should be able to track at least six line-of-sight (LOS) direction satellites. The GPS constellation is designed with a total number of 32 satellites in orbit. Currently among the 31 operational GPS satellites, 11 satellites broadcast the L1 (1575.42 MHz) signal only, 7 satellites broadcast the L1 and L2 (1227.6 MHz) signals, and 13 satellites broadcast the L1, L2 and L5 (1176.45 MHz) signals. The transmission of these GPS civilian radio-frequency (RF) signals is based on the Code Division Multiple Access (CDMA) spread-spectrum technology. The details of GPS signal structure can be found in the Interface Control Documents (ICD) [1]. The latest status of the GPS constellation can be found at the U.S. Coast Guard Navigation Center website [2].

Currently, 24 operational GLONASS satellites are located within three different MEO orbital planes with an altitude of $\sim$19,100 km, which indicates GLONASS satellites' orbital period is $\sim$11 h 15 min. Each two neighboring GLONASS orbital planes are separated by an Ω of 120 degrees, and all satellite inclination angles are approximately 64.8 degrees. In contrast to the GPS constellation, the transmission of GLONASS civilian RF signals is based on the frequency division multiple access (FDMA) technique. Therefore, the RF transmitting frequencies under the same frequency band on different GLONASS satellites are different. 22 out of the 24 GLONASS satellites transmit dual frequency bands, where their G1 center frequency is $1602 + n \times 0.5625$ MHz and G2 center frequency is $1246 + n \times 0.4375$ MHz (where n is the satellite frequency channel number). The other two GLONASS satellites transmit an additional frequency band, G3, with a center frequency of $1201 + n \times 0.4375$ MHz. The details of GLONASS signal structure can be found in their ICDs [3]. The latest status of the GLONASS constellation can be found at the Russian Information and Analysis Center for Positioning, Navigation and Timing website [4].

Currently, 22 operational Galileo satellites are located within three different MEO orbital planes with an altitude of $\sim$23,200 km, which gives an orbital period $\sim$14 h 7 min for the Galileo satellites. The FOC stage of Galileo is expected to be composed of 30 satellites in total. Each neighboring Galileo orbital plane is separated by an Ω of 120 degrees, and all satellite inclination angles are approximately 56 degrees. The Galileo constellation utilizes the CDMA technique for RF signal transmission. All the Galileo satellites broadcast E1 (1575.42 MHz), E5a (1176.45 MHz), E5b (1207.14 MHz) and E6 (1278.75 MHz) civilian signals. Galileo receivers may receive the E5 AltBOC modulation signal, a modified version of a Binary Offset Carrier (BOC) with a center frequecy of 1191.795 MHz. The details of Galileo signal structure can be found in their ICDs [5]. The latest status of the Galileo can be found at the European GNSS Service Centre website [6].

The orbits of the Beidou constellation are more complicated than the other three constellations. Out of the 43 currently operational Beidou satellites, 5 satellites are in the geostationary orbit (GEO) over the Asian sector; 10 satellites are in inclined geosynchronous orbits (IGSO) with an altitude of $\sim$35,786 km and an inclination angle of 55°; 28 satellites are in MEO with a nominal altitude of $\sim$21,528 km. The MEO satellites are separated within three orbital planes (equally divided by 120 degrees) with an orbital period of $\sim$12 h and 53 min. Compared to other GNSS constellations, the more complicated orbit geometry of the BDS can increase positioning accuracy by providing lower horizontal, vertical and temporal dilution of precision (DOP) particularly in a large portion of the eastern hemisphere region (60° S–60° N and 50° E–170° E) [7]. More BDS satellites will be launched to complete the full constellation with at least 49 satellites in total. Beidou second generation satellites transmit public RF signals in three different frequency bands: B1 (1561.098 MHz), B2 (1207.14 MHz) and B3 (1268.52 MHz). Beidou third generation satellites also transmit three different frequency bands of public signals: B1 (1575.42 MHz), B2 (1176.45 MHz) and B3 (1268.52 MHz). The details of BDS signal structure can be found in their ICDs [8]. The latest status of the Beidou constellation can be found at the Test and Assessment Research Center of China Satellite Navigation Office website [9].

2. GNSS Observables and Ranging Errors

In order to achieve PNT, a standard GNSS receiver's measurement module must generate GNSS observables and the receiver's positioning module must properly model each ranging error for GNSS observables. GNSS observables refer to the basic measurements produced from a GNSS receiver, including pseudorange, carrier-phase, and Doppler shift. Together with the epoch time and the carrier-to-noise density ratio (often denoted as C/N_0), these data are conventionally stored in a observation file based on the receiver independent exchange format (RINEX) [10].

Pseudorange (also known as code-range) refers to the sum of the real range and the range equivalent of various ranging errors or offsets. Pseudorange (P) can be defined as:

$$P^m = R^m + c(\delta T^m + \delta t) + I^m + T^m + M^m + D^m + W^m \tag{3}$$

where m denotes a specific GNSS satellite number. R is the true geometric range from the satellite to receiver's antenna phase center in meters, c is the speed of light in vacuum in meters per second, δT is the satellite clock bias in seconds, δt is the receiver clock bias in seconds, I is the ionospheric signal delay in meters, T is the troposphere signal delay in meters, M is the multipath effect delay in meters, D is the error caused by the geometric dilution effect in meters, W refers to all other noise effects (e.g., receiver noise, relativity effects). The standard unit of P is meters.

Doppler shift (f_D or simply Doppler) is the frequency shift caused by the relative motion between a GNSS satellite (transmitter) and receiver (antenna). f_D is related to the satellite radial velocity (v_r), which is equal to the pseudorange rate:

$$f_D^m = \frac{v_r^m}{\lambda} = \frac{\dot{P}^m}{\lambda} \tag{4}$$

where λ is the wavelength of the GNSS signal being used. f_D is typically not used to compute navigation solutions, but instead used to determine the receiver velocity. The standard unit of f_D is Hz.

Carrier-phase (or beat carrier-phase) refers to the time integral of the carrier Doppler shift. Carrier-phase (CP) can be defined as:

$$CP^m = \frac{1}{\lambda}[R^m + c(\delta T^m + \delta t) - I^m + T^m + M^m + D^m + W^m] + N^m \tag{5}$$

where N is the carrier-phase measurement ambiguity (sometimes known as integer ambiguity). CP typically is measured in number of cycles of the carrier signals being received and tracked. The relationship between f_D and CP can be expressed as:

$$CP^m = N^m + \int_0^t (f_D^m) dt_\Delta \tag{6}$$

where t_Δ is the elapsed time during a carrier-phase measurement.

3. Ionospheric Characteristics and Phenomena

The ionosphere is an ionized layer of the Earth's upper atmosphere. Due to the solar radiation, the neutral particles in the Earth's atmosphere are converted to electrons and ions. Overall, these electrons are concentrated in the altitude range from ~60 km to 1000 km but their characteristics are spatially and temporally dynamic [11]. In equatorial and low-latitude regions (latitude < 30 degrees North/South), the magnetic field is nearly horizontal and ionospheric electron density is typically higher than other regions due to the high solar angle effect. Equatorial Spread F (ESF) or Equatorial plasma bubbles (EPB) form density structures in the ionosphere and have a typical scale size between 1° (~115 km) and 4° (~460 km) [12]. Additionally, plasma irregularities driven by plasma instabilities can be detected to smaller scale (~10^{-1} to 10^3 m) as well. The high-latitude (or polar) ionosphere is the region above 60° magnetic latitude (e.g., auroral zone, polar cap), where plasma instabilities and other dynamic processes (e.g., coupling physics between the magnetosphere, ionosphere and thermosphere) cause ionospheric structures and irregularities [11]. Under the influence of the nearly vertical geomagnetic field as well as the horizontal variation of plasma density and electric fields driven by plasma instabilities, various multi-scale ($\sim 10^{-2}$ to 10^5 m) ionospheric structures lead to phenomena such as aurora (and the associated arcs), sub-auroral polarization streams (SAPS), as well as polar tongues of ionization (TOI). A broad range of observation techniques must be used to study these multi-scale space weather phenomena which span seven orders of magnitude or more in space and time scales.

Aurora, one of the most famous and important space weather phenomena, is typically seen as a visual phenomenon caused by charged particle precipitation along polar geomagnetic field lines and subsequent interaction with the neutral particles in the upper atmosphere. The energetic charged particles are often driven by the solar wind [11]. The length of auroral arcs can range from 100 to 1000 km, the width can range from 50 m to 10 km, and the altitude (maximum energy of particles in the primary beam) is typically from 80m to 400 km [13]. SAPS often refers to a sunward plasma drift/convection in the sub-auroral region with an approximate spatial span of $\sim 3°$–$5°$ latitudinally and temporal duration of several hours in the evening sector [14]. TOI is a continuous stream of cold plasma enhancement with an entrainment pattern of high-latitude convection. The spatial scale of TOI can range from about 100 to 1000 km [15]. A hardware-in-the-loop simulation of TOI and its impact on GNSS was reported by [16].

4. Ionospheric Remote Sensing

GNSS is not only the ubiquitous modern technology for PNT, but also a versatile remote sensing tool for many areas (e.g., space weather, geodesy, geophysics, and oceanography). For example, GNSS is widely applied to ionospheric remote sensing. Plasma physics describes the basic science of the ionosphere. An important parameter—plasma frequency (ω_P) can be defined as:

$$\omega_P = q\sqrt{\frac{n_e}{\varepsilon_0 m_e}} \tag{7}$$

where q is elementary charge ($\approx 1.6 \times 10^{-19}$ C); n_e is electron density; ε_0 is the permittivity of free space; and m_e is the electron rest mass. For radio waves with frequencies below ω_P, typically 10's of MHz, the ionosphere can reflect the RF waves (e.g., amplitude modulated radios) and enables long distance over the horizon radio communications globally. For radio waves frequencies above ω_P, such as GNSS, the signals penetrate through the ionosphere. Due to the difference in index of refraction in the ionosphere compared to a vacuum (free space), the ionosphere can delay, attenuate, disturb or induce Faraday rotation on GNSS signal propagation. The index of refraction depends on electron density and RF wave frequency [17]. These ionospheric effects can dramatically degrade the PNT accuracy, precision, and integrity of GNSS. Conversely, GNSS can be (and has been) utilized to monitor and study the ionosphere.

Associated with the ionospheric delay effect, the total electron content (TEC) of the ionosphere can be measured by multi-frequency GNSS receivers on the ground or in space. TEC is defined as the total number of electrons within a cross-sectional volume along the LOS between two points (e.g., a GNSS satellite and a ground-based GNSS receiver): $\text{TEC} = \int_0^R n_e dr$, where R is the LOS's distance. The TEC is measured in units of number of electrons per m^2, but more often expressed in units of TECU (10^{16} electrons/m^2). The ionospheric delay I (measured in meters) on a ranging Equation (Equations (4) and (6)) with a specific frequency band can be expressed in terms of TEC as [17]:

$$I = \frac{40.308}{f^2} \times \text{TEC} \tag{8}$$

where f is the GNSS signal frequency. Using the pseudorange from two GNSS different frequency bands (e.g., GPS L1 and L2), the TEC can be measured based on the following formula:

$$\mathrm{TEC}_{P_{L1}-P_{L2}} = \frac{1}{40.308}\left(\frac{f_{L1}^2 f_{L2}^2}{f_{L1}^2 - f_{L2}^2}\right)(P_{L2} - P_{L1}) \tag{9}$$

where f_{L1} is GPS L1 frequency, f_{L2} is GPS L2 frequency, P_{L1} is GPS L1 pseudorange, and P_{L2} is GPS L2 pseudorange. Due to the differential code bias (DCB), an offset caused by different hardware delays on GNSS code/pseudorange observations with different signal frequencies, an additional DCB bias term needs to be accounted for when implementing Equation (10). More details about DCB estimations can be found at [18]. The MIT Madrigal database gathers thousands of multi-frequency GPS/GNSS receiver data to create global TEC maps [19] as shown in Figure 2, which is advantageous for global ionospheric weather monitoring and studies. Note, vertical TEC (VTEC) is an integration of the electron density along the direction perpendicular to the ground standing. There are several other global TEC monitoring systems/institutes, such as NASA Jet Propulsion Laboratory (JPL) [20], International GNSS Service (IGS) [21], and United States' National Oceanic and Atmospheric Administration (NOAA) [22].

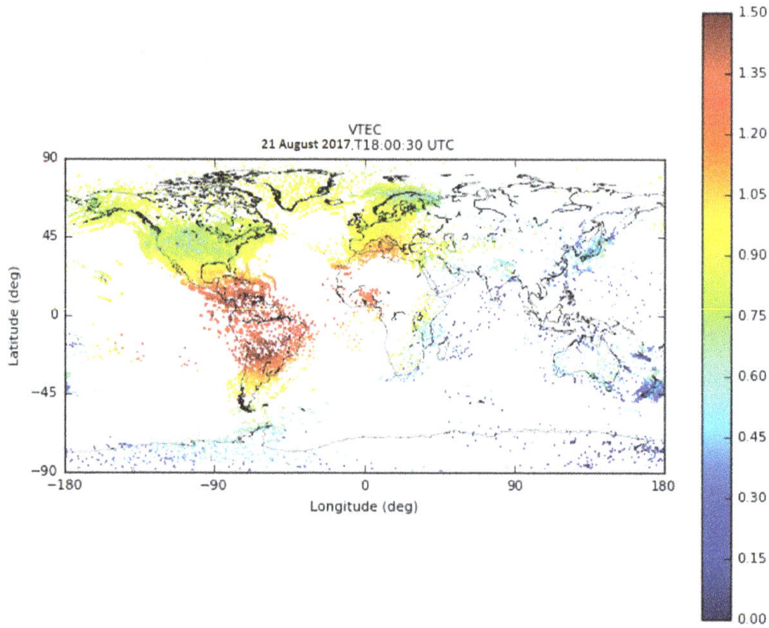

Figure 2. A global VTEC map from MIT Madrigal world-wide GPS receiver network [23].

Other than ionospheric delay, ionospheric irregularities can cause GNSS scintillations (rapid fluctuations in signal's intensity or/and phase). According to [24], ionospheric irregularities are "small-scale structures in the ionospheric plasma density generally oriented so that the plasma density variations occur rapidly across the geomagnetic field but slowly (or not at all) along the geomagnetic field". Two ionospheric scintillation indices are used to quantify the scintillation severity:

(a) S4 (amplitude scintillation index): the ratio of the standard deviation of the signal power to the average signal power computed over a period of time (typically 1 min) as defined by [25]:

$$S4 = \sqrt{\frac{\langle A^2 \rangle - \langle A \rangle^2}{\langle A \rangle^2}} \tag{10}$$

where A is the signal intensity or power, and $\langle\rangle$ denotes ensemble averaging ($\approx$time averaging).
(b) sigma phi or σ_ϕ (phase scintillation index): the standard deviation of σ in radians, where σ is the refractive component of the GNSS signal phase as defined by [25] and ϕ_{detr} is the detrended phase.

$$\sigma_\phi = \sqrt{\langle \phi_{detr}^2 \rangle - \langle \phi_{detr} \rangle^2} \tag{11}$$

Ionospheric irregularities may simultaneously lead to GNSS ranging errors (TEC delay) and GNSS signal phase scintillations. The relationship between the GPS phase scintillations and TEC variation can be expressed as (described in detail in [26]):

$$\Delta\text{TEC} = \frac{0.75 f_{\text{L1}} \Delta\phi_{\text{L1L2}}}{f_{\text{L1}}^2 / f_{\text{L2}}^2 - 1} \tag{12}$$

where the differential carrier phase between L1 and L2 frequency bands ($\Delta\phi_{L1L2}$) is proportional to TEC variation (ΔTEC).

As shown in Figure 3, an example GPS scintillation event observed at the Antarctic McMurdo scintillation Station from MIT Madrigal contains both high amplitude and phase scintillation measurements. The background color represents VTEC level in a similar way in Figure 2, and the red circles represent S4 in Figure 3a and sigma phi (σ_ϕ) in Figure 3b. On 7 January 2014, the polar ionospheric irregularities and density structures in the southern polar region induced by an incoming solar storm caused an observation of this scintillation event (with relatively high S4 and σ_ϕ) using ground-based GPS receivers.

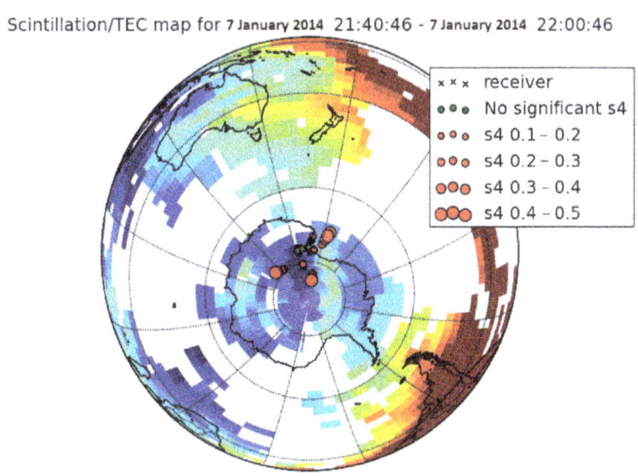

(a)

Figure 3. *Cont.*

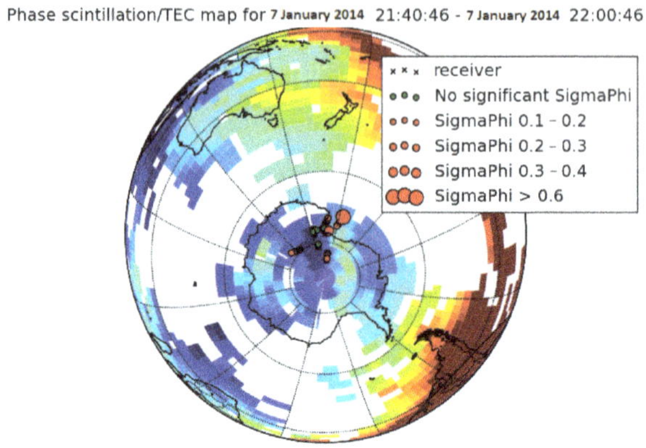

(**b**)

Figure 3. An example GPS scintillation event observed at the Antarctic McMurdo scintillation Station from MIT Madrigal. Adapted from [27] (**a**) S4 measurement; (**b**) SigmaPhi (σ_ϕ) measurement.

GNSS is widely used to measure S4 and σ_ϕ in order to observe and study the associated ionospheric irregularities. GNSS phase scintillations can cause cycle slips in carrier-phase and put pressure on the tracking loops of GNSS receivers. Severe GNSS scintillations can even lead to GNSS receiver loss-of-track and thus reduce positioning accuracy and availability. A great number of ground-based receivers are deployed in different regions around the world to detect and measure ionospheric space weather including the plasma irregularities that disturb GNSS signals. For instance, the chain of autonomous adaptive low-power instrument platforms (AAL-PIP) [28] on the East Antarctic Plateau has been used to observe ionospheric activity in the South Polar region. Together with six ground-based magnetometers, four dual frequency GPS receivers of the AAL-PIP project have been used to capture ionospheric irregularities and ultra-low frequency (ULF) waves associated with geomagnetic storms by analyzing the GPS TEC and scintillation data collected in Antarctica [29]. Furthermore, the ESA Space Weather Service Network is hosting several ionospheric scintillation monitoring systems developed by the German Aerospace Center (DLR), Norwegian Mapping Authority (NMA), and Collecte Localisation Satellites (CLS) [30]. Figure 4 gives a high-level illustration of two ionospheric impacts on GNSS—ranging errors and scintillation.

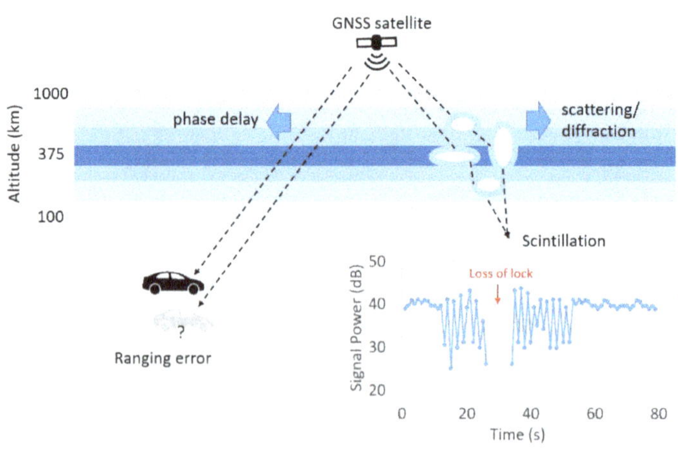

Figure 4. An illustration of ionospheric impacts on GNSS.

Besides ground-based GNSS ionospheric remote sensing, there are space-based approaches that utilize the spaceborne GNSS receivers on satellites for ionospheric radio soundings. For example, the Constellation Observing System for Meteorology, Ionosphere, and Climate (COSMIC) mission uses the radio occultation technique (a bending effect on the GNSS signals propagating through the Earth's upper atmosphere) to measure space-based TEC and scintillations, detect ionospheric irregularities, and reconstruct global electron density profiles using ionospheric tomography techniques [31]. Using low-Earth-orbit GNSS receivers sensors in proximity together with spacecraft formation flying techniques, the ionospheric TEC, electron density, and scintillation index can also be measured globally with high flexibility [32–34].

5. Conclusions and Prospects

Fundamental physics and engineering of GNSS and ionospheric remote sensing are introduced in this entry. It is important to monitor and understand the ionospheric impact on GNSS, because the ionosphere can cause delays or scintillation of GNSS signals which eventually degrade the PNT solutions from GNSS. As a reflection of ionospheric ionization level, TEC is an integration of the electron density along the LOS between two points. The larger the TEC, the larger ranging offset in the GNSS observable caused by the ionosphere. S4 and σ_ϕ are the two commonly used ionospheric scintillation indexes to quantify the GNSS signal fluctuation level in the amplitude and phase domain, respectively. Ionospheric irregularities can cause scintillations of GNSS signals, which may lead to signal attenuation, carrier phase cycle slip or even loss of lock. The ubiquitous GNSS is a powerful engineering tool for ionospheric remote sensing. Ionospheric remote sensing studies using ground-based GNSS receivers have been conduced over the past several decades, while ionospheric measurement using space-based GNSS techniques is emerging rapidly and providing much higher coverage and flexibility.

Author Contributions: Conceptualization, Y.P. and W.A.S.; methodology, Y.P.; software, Y.P.; validation, Y.P. and W.A.S.; formal analysis, Y.P.; investigation, Y.P.; resources, Y.P. and W.A.S.; writing—original draft preparation, Y.P.; writing—review and editing, W.A.S.; visualization, Y.P.; supervision, W.A.S. All authors have read and agreed to the published version of the manuscript.

Funding: This work is supported by the AFOSR (Grant No. 13-0658-09) and Virginia Tech.

Data Availability Statement: GPS TEC data products and access through the Madrigal distributed data system are provided to the community by the Massachusetts Institute of Technology under support from US National Science Foundation grant AGS-1952737. Data for the TEC processing is provided from the following organizations: UNAVCO, Scripps Orbit and Permanent Array Center, Institut Geographique National, France, International GNSS Service, The Crustal Dynamics Data Information System (CDDIS), National Geodetic Survey, Instituto Brasileiro de Geografia e Estatística, RAMSAC CORS of Instituto Geográfico Nacional de la República Argentina, Arecibo Observatory, Low-Latitude Ionospheric Sensor Network (LISN), Topcon Positioning Systems, Inc., Canadian High Arctic Ionospheric Network, Institute of Geology and Geophysics, Chinese Academy of Sciences, China Meteorology Administration, Centro di Ricerche Sismologiche, Système d'Observation du Niveau des Eaux Littorales (SONEL), RENAG : REseau NAtional GPS permanent, GeoNet—the official source of geological hazard information for New Zealand, GNSS Reference Networks, Finnish Meteorological Institute, SWEPOS—Sweden, Hartebeesthoek Radio Astronomy Observatory, TrigNet Web Application, South Africa, Australian Space Weather Services, RETE INTEGRATA NAZIONALE GPS, Estonian Land Board, Virginia Tech Center for Space Science and Engineering Research, and Korea Astronomy and Space Science Institute.

Acknowledgments: The authors would like to thank Anthea Coster, Greg Earle, Michael Ruohoniemi, Robert Clauer, and Jonathan Black for their comments and inputs on our work.

Conflicts of Interest: The authors declare no conflict of interest.

Entry Link on the Encyclopedia Platform: https://encyclopedia.pub/17411.

Abbreviations

The following abbreviations are used in this manuscript:

GNSS	Global navigation satellite system
PNT	Positioning, Navigation and Timing
GPS	Global Positioning System
GLONASS	Global'naya Navigatsionnaya Sputnikovaya Sistema
BDS	BeiDou Navigation Satellite System
FOC	Full Operational Capability
MEO	medium Earth orbit
LOS	line-of-sight
RF	radio-frequency
CDMA	Code Division Multiple Access
FDMA	Frequency Division Multiple Access
BOC	Binary Offset Carrier
GEO	geostationary orbit
IGSO	inclined geosynchronous orbits
DOP	dilution of precision
RINEX	Receiver Independent Exchange Format
P	Pseudorange
CP	Carrier-phase
ESF	Equatorial Spread F
EPB	Equatorial plasma bubbles
SAPS	sub-auroral polarization streams
TOI	tongues of ionization
TEC	total electron content
DCB	differential code bias
MIT	Massachusetts Institute of Technology
AAL-PIP	Autonomous Adaptive Low-Power Instrument Platforms
ULF	ultra-low frequency
COSMIC	Constellation Observing System for Meteorology, Ionosphere,and Climate

References

1. GPS Interface Control Documents. Available online: https://www.gps.gov/technical/icwg/#current (accessed on 11 October 2021).
2. GPS Constellation Status (U.S. Coast Guard Navigation Center). Available online: https://www.navcen.uscg.gov/?Do=constellationstatus (accessed on 10 May 2021).
3. GLONASS Interface Control Documents. Available online: https://www.glonass-iac.ru/en/documents/ (accessed on 11 October 2021).
4. GLONASS Constellation Status (IAC; Information and Analysis Center for Positioning, Navigation and Timing). Available online: https://www.glonass-iac.ru/en/GLONASS/ (accessed on 10 May 2021).
5. European GNSS Service Center Programme Reference Documents. Available online: https://www.gsc-europa.eu/electronic-library/programme-reference-documents (accessed on 11 October 2021).
6. Galileo Constellation Information (European GNSS Service Centre). Available online: https://www.gsc-europa.eu/system-service-status/constellation-information (accessed on 10 May 2021).
7. Wang, M.; Wang, J.; Dong, D.; Meng, L.; Chen, J.; Wang, A.; Cui, H. Performance of BDS-3: Satellite visibility and dilution of precision. *GPS Solut.* **2019**, *23*, 56. [CrossRef]
8. BeiDou Navigation Satellite System Interface Control Documents. Available online: http://en.beidou.gov.cn/SYSTEMS/ICD/ (accessed on 11 October 2021).
9. BeiDou Constellation Status (CSNO-TARC; Test and Assessment Research Center of China Satellite Navigation Office). Available online: http://www.csno-tarc.cn/en/system/constellation (accessed on 10 May 2021).
10. Rinex Version 3.05. Available online: https://files.igs.org/pub/data/format/rinex305.pdf (accessed on 10 May 2021).
11. Kelley, M.C. *The Earth's Ionosphere: Plasma Physics and Electrodynamics*; Academic Press: Cambridge, MA, USA, 2009.
12. Smith, J.; Heelis, R.A. Equatorial plasma bubbles: Variations of occurrence and spatial scale in local time, longitude, season, and solar activity. *J. Geophys. Res. Space Phys.* **2017**, *122*, 5743–5755. [CrossRef]
13. Davis, T.N. Observed characteristics of auroral forms. *Space Sci. Rev.* **1978**, *22*, 77–113. [CrossRef]
14. Coster, A.; Foster, J. Space-weather impacts of the sub-auroral polarization stream. *URSI Radio Sci. Bull.* **2007**, *321*, 28–36.

15. Carlson, H.C. Sharpening our thinking about polar cap ionospheric patch morphology, research, and mitigation techniques. *Radio Sci.* **2012**, *47*, 1–16. [CrossRef]
16. Peng, Y.; Scales, W.A.; Lin, D. GNSS-based hardware-in-the-loop simulations of spacecraft formation flying with the global ionospheric model TIEGCM. *GPS Solut.* **2021**, *25*, 1–14. [CrossRef]
17. Kintner, P.M.; Mark, L.P. *Global Positioning System Theory and Design*, 3rd ed.; Cornell University: Ithaca, NY, USA, 2015.
18. Rideout, W.; Coster, A. Automated GPS processing for global total electron content data. *GPS Solut.* **2006**, *10, 3*, 219–228. [CrossRef]
19. Vierinen, J.; Coster, A.J.; Rideout, W.C.; Erickson, P.J.; Norberg, J. Statistical framework for estimating GNSS bias. *Atmos. Meas. Tech.* **2016**, *9*, 1303–1312. [CrossRef]
20. NASA JPL Global Maps of Ionospheric Total Electron Content. Available online: https://iono.jpl.nasa.gov/latest_rti_global.html (accessed on 11 August 2021).
21. IGS Ionospheric Products. Available online: https://igs.org/products/#ionospheric_products (accessed on 11 August 2021).
22. NOAA US Total Electron Content. Available online: https://www.swpc.noaa.gov/products/us-total-electron-content (accessed on 11 August 2021).
23. Coster, A. MIT/Haystack Observatory. Data from the CEDAR Madrigal Database. 2017. Available online: cedar.openmadrigal.org (accessed on 10 September 2021).
24. Perkins, F.W. Ionospheric irregularities. *Rev. Geophys.* **1975**, *13, 3*, 884–884. [CrossRef]
25. Kintner, P.M.; Ledvina, B.M.; De Paula, E.R. GPS and ionospheric scintillations. *Space Weather* **2007**, *5*, 09003. [CrossRef]
26. Najmi, A.; Milikh, G.; Secan, J.; Chiang, K.; Psiaki, M.; Bernhardt, P.; Briczinski, S.; Siefring, C.; Chang, C.L.; Papadopoulos, K. Generation and detection of super small striations by F region HF heating. *J. Geophys. Res. Space Phys.* **2014**, *119*, 6000–6011. [CrossRef]
27. Weatherwax, A.T. Department of Physics. Data from the CEDAR Madrigal Database. 2014. Available online: https://w3id.org/cedar?experiment_list=experiments/2014/mcs/07jan14&file_list=1774_2_00.gps_all.out.hdf5 (accessed on 11 July 2021).
28. Xu, Z.; Hartinger, M.D.; Clauer, R.; Weimer, D.; Deshpande, K.; Kim, H.; Musko, S.; Willer, A.; Edwards, T.; Coyle, S.; et al. Newly established autonomous adaptive low-power instrument platform (AAL-PIP) chain on East Antarctic Plateau and operations. *Adv. Polar Sci.* **2019**, *30*, 362–374.
29. Peng, Y., Scales, W.A., Hartinger, M.D., Xu, Z, Coyle, S. Characterization of multi-scale ionospheric irregularities using ground-based and space-based GNSS observations. *Satell. Navig.* **2021**, *2*, 14. [CrossRef]
30. ESA Space Weather Service Network Near-Real-Time Ionospheric Scintillation Maps. Available online: https://swe.ssa.esa.int/tio_sci# (accessed on 11 August 2021).
31. Pinto Jayawardena, T.S.; Chartier, A.T.; Spencer, P.; Mitchell, C.N. Imaging the topside ionosphere and plasmasphere with ionospheric tomography using COSMIC GPS TEC. *J. Geophys. Res. Space Phys.* **2016**, *121*, 817–831. [CrossRef]
32. Peng, Y.; Scales, W.A.; Edwards, T.R. GPS-based satellite formation flight simulation and applications to ionospheric remote sensing. *Navigation* **2020**, *67*, 3–21. [CrossRef]
33. Peng, Y.; Scales, W.A. Satellite Formation Flight Simulation Using Multi-Constellation GNSS and Applications to Ionospheric Remote Sensing. *Remote Sens.* **2019**, *11*, 2851. [CrossRef]
34. Peng, Y.; Scales, W.A.; Esswein, C.E.; Hartinger, D.H. Small Satellite Formation Flying Simulation with Multi-Constellation GNSS and Applications to Future Multi-Scale Space Weather Observations. In Proceedings of the 32nd International Technical Meeting of the Satellite Division of the Institute of Navigation (ION GNSS+ 2019), Miami, FL, USA, 16–20 September 2019; pp. 2035–2047. [CrossRef]

encyclopedia MDPI

Entry

Opportunities for Catalytic Reactions and Materials in Buildings †

Praveen Cheekatamarla

Buildings and Transportation Sciences Division, Oak Ridge National Laboratory, Oak Ridge, TN 37831, USA; cheekatamapk@ornl.gov; Tel.: +1-865-341-0417

† This manuscript has been authored by UT-Battelle, LLC, under contract DE-AC05-00OR22725 with the US Department of Energy (DOE). The US government retains and the publisher, by accepting the article for publication, acknowledges that the US government retains a nonexclusive, paid-up, irrevocable, worldwide license to publish or reproduce the published form of this manuscript, or allow others to do so, for US government purposes. DOE will provide public access to these results of federally sponsored research in accordance with the DOE Public Access Plan (http://energy.gov/downloads/doe-public-access-plan).

Definition: Residential and commercial buildings are responsible for over 30% of global final energy consumption and accounts for ~40% of annual direct and indirect greenhouse gas emissions. Energy efficient and sustainable technologies are necessary to not only lower the energy footprint but also lower the environmental burden. Many proven and emerging technologies are being pursued to meet the ever-increasing energy demand. Catalytic science has a significant new role to play in helping address sustainable energy challenges, particularly in buildings, compared to transportation and industrial sectors. Thermally driven heat pumps, dehumidification, cogeneration, thermal energy storage, carbon capture and utilization, emissions suppression, waste-to-energy conversion, and corrosion prevention technologies can tap into the advantages of catalytic science in realizing the full potential of such approaches, quickly, efficiently, and reliably. Catalysts can help increase energy conversion efficiency in building related technologies but must utilize low cost, easily available and easy-to-manufacture materials for large scale deployment. This entry presents a comprehensive overview of the impact of each building technology area on energy demand and environmental burden, state-of-the-art of catalytic solutions, research, and development opportunities for catalysis in building technologies, while identifying requirements, opportunities, and challenges.

Keywords: catalysis; buildings; heat pumps; dehumidification; carbon capture; emissions; indoor air quality; cogeneration; non-precious metals; photo-catalysis; electrocatalysis

Citation: Cheekatamarla, P. Opportunities for Catalytic Reactions and Materials in Buildings. *Encyclopedia* **2022**, 2, 36–55. https://doi.org/10.3390/encyclopedia2010004

Academic Editors: Raffaele Barretta, Ramesh Agarwal, Krzysztof Kamil Żur and Giuseppe Ruta

Received: 8 November 2021
Accepted: 22 December 2021
Published: 28 December 2021

Publisher's Note: MDPI stays neutral with regard to jurisdictional claims in published maps and institutional affiliations.

1. Introduction

Energy plays a vital role in modern society; however, it is also responsible for greenhouse gas emissions. Global energy consumption is on the rise driven by economic and population growth. As shown in Figure 1, although renewables are expected to become the primary resource, fossil fuels continue to increase, given the global demand increase [1].

By year 2050, energy information administration (EIA) projects a 50% increase in global energy demand driven by a 65% increase in building energy consumption, a 79% increase in electrical power generation, and a 40% increase in natural gas consumption [2]. A continuous upsurge in energy demand due to increased affordability, economic growth, and new energy consumers poses a significant challenge to the energy infrastructure and the environment. Three major energy-consuming sectors in any economy are buildings, transportation, and industry. Amongst these energy consumers, residential and commercial buildings account for more than 30% of primary energy consumption. Combined direct and indirect carbon footprint of the buildings sector approached almost 40% of total carbon dioxide emissions [3]. According to energy information administration statistics [4], residential and commercial buildings in the USA consumed 41 exajoules (1.3 terawatt-year) of energy in the year 2019, accounting for almost 39% of total primary energy. As shown in Figure 2, the electricity supply to the end consumer in the USA comes at a premium value since ~65% of the primary energy is lost in the process of production and distribution.

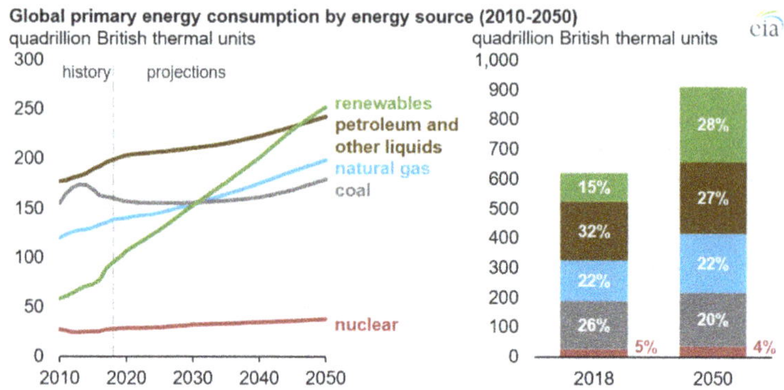

Figure 1. Global primary energy consumption by energy source, quadrillion Btu [2].

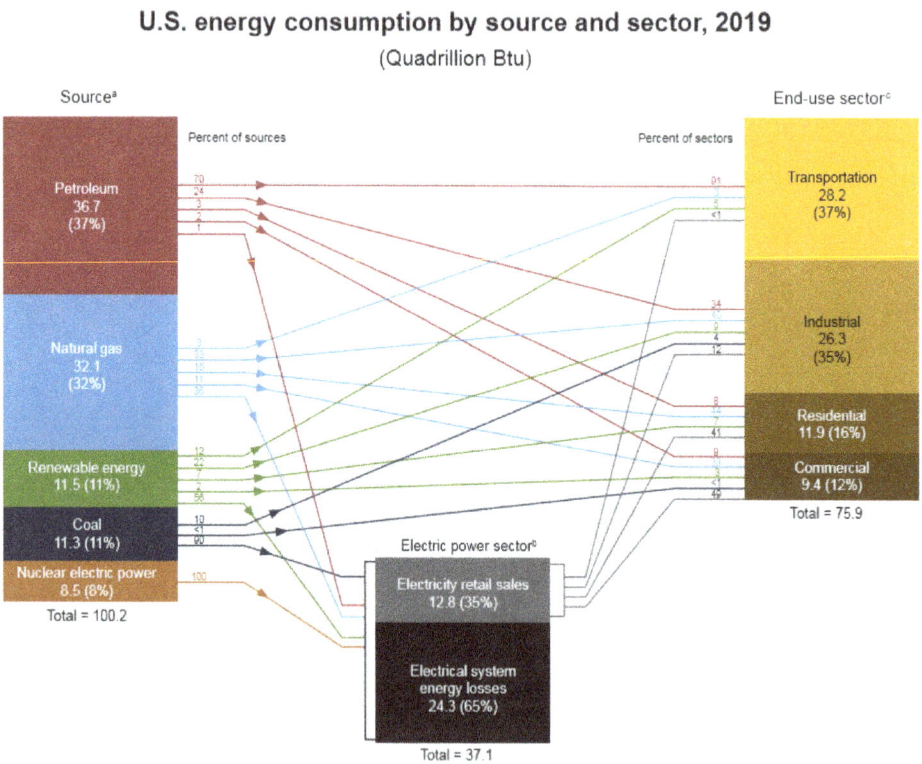

Figure 2. U.S. Energy consumption by source and sector [4]. [a] Primary energy consumption. Each energy source is measured in different physical units and converted to common British thermal units (Btu). [b] The electric power sector includes electricity-only and combined-heat-and-power plants. [c] End-use sector consumption of primary energy and electricity retail sales, excluding electrical system energy losses from electricity retail sales.

Given this high premium, it is essential that the electric energy supplied to buildings is utilized to its full potential at the highest possible conversion and utilization efficiency. From an energy efficiency perspective, heating and cooling equipment in a residential or commercial building consumes up to 50% of the total energy supply [5,6] in the USA. Heating systems, for instance, utilize ~60% fossil fuels as the primary energy resource [7] in the USA, as shown in Figure 3. As a result, building energy usage accounts for ~35% of total annual carbon dioxide emissions in the USA [8], as shown in Figure 4.

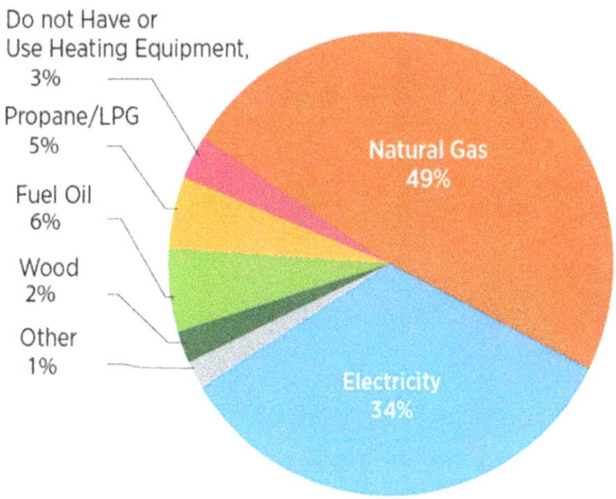

Figure 3. U.S. household heating systems' energy source [7].

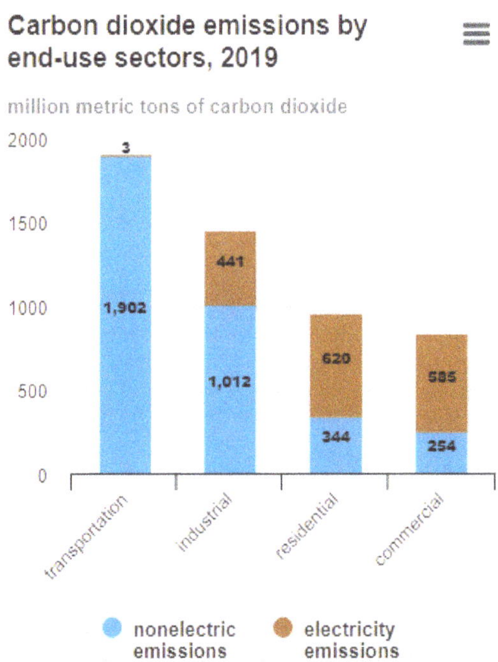

Figure 4. U.S. carbon dioxide emissions from energy consumption by sector [8].

Given these energy and environmental impacts and projected statistics, technology improvements to address high energy demand, carbon footprint, and emissions associated with building energy consumption is of enormous value for a sustainable energy future. Additionally, buildings offer great potential to improve quality of life, health, and workforce productivity (defined as an assessment of the efficacy of worker(s)). For instance, a recent study by national institute of health [9] focusing on the relationship between carbon dioxide, volatile organic compounds (VOC) concentrations, and associated the impact of air quality on cognitive function scores in office environments. The authors looked at nine different cognitive function domains as a function of carbon dioxide and VOC concentrations and found that clean and green buildings improve the scores by 50% to 200% depending on the activity domain.

Similarly, indoor air quality improvements are highly essential for the improved health of occupants in buildings. Various studies have established the significance of pollutant concentrations in homes and their impact on health and productivity of residents [10]. A review of interactions between energy performance of the buildings, outdoor air pollution and the indoor air quality was recently reported [11,12]. These studies investigated the impact of building energy efficiency improvements on pollutant concentrations and their health risk.

Increase in energy demand coupled with intermittent renewable energy resources in the grid infrastructure also emphasizes the importance of energy management and flexibility. For instance, efficient use of energy resources focusing on grid resiliency and environmental security supported by cogeneration and trigeneration technologies is a highly impactful approach. Numerous studies have already established the significance of such systems [13,14]. These behind-the-meter assets have a significant potential in enabling sustainable energy technologies and can be deployed across the grid. Considering these factors, many opportunities exist in improving the health, productivity, energy efficiency, carbon footprint, and energy management for a sustainable future. Building energy resources such as heating, cooling, dehumidification, and thermal storage equipment, along with carbon capture/conversion, cogeneration and biomass conversion technologies play a vital role in realizing energy sustainability, occupant health and productivity, and environmental responsibility. In addition, energy efficiency, affordability, and retrofittability are some of the key attributes necessary for successful transition towards new technologies.

All these technologies utilize a vast array of chemical transformations, including (i) oxidation—for instance, pollutants, fuel conversion to syngas, and thermal storage, (ii) reduction—carbon dioxide conversion, (iii) hydration—thermal storage, (iv) dehydration—dehumidification, (v) absorption/chemisorption—heating and cooling, carbon capture, and (vi) hydrolysis and methanogenesis—biomass conversion to biogas. Given these chemical reactions and their potential for enabling sustainable societies, catalysis has a new significant role to play in buildings. In this context, the objective of this entry is to provide a comprehensive overview of the state of the art of catalysts in building applications, their impact, and future research and development opportunities. The next section introduces individual technology areas followed by a discussion on future directions for catalysts in enabling clean and sustainable energy in buildings.

2. Catalysts in Building Technologies

As discussed above, energy and environmental challenges associated with buildings can be overcome through the application of catalytic science. Many energy consumers within building envelope can exploit the advantages offered by innovations in the field of catalysis. Application of various catalytic reaction schemes shown in Figure 5 include electrochemical, photochemical, photoelectrochemical, and thermochemical can aid in lowering the environmental and energy burden of the buildings. The rest of this section provides an overview of this.

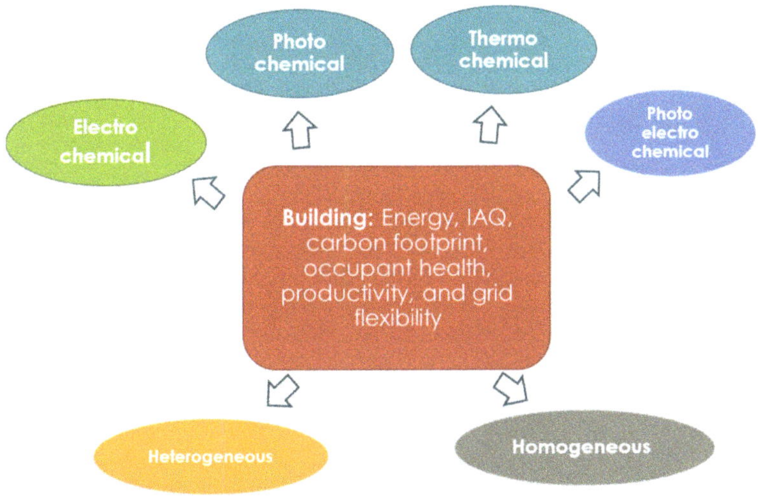

Figure 5. Opportunities for catalysis in buildings—sustainable energy, environment, and health.

2.1. Indoor Air Quality and Emissions

Common indoor air pollutants in buildings include volatile organic compounds (VOC), formaldehyde, benzene emanating from furniture, decoration, and personal care products [15], along with combustion byproducts such as carbon monoxide, nitrogen dioxide, and sulfur dioxide from cooking and heating equipment. Energy efficiency improvements leading to tight sealing of the building envelope and lack of fresh makeup air in buildings can easily increase the concentration of these toxic compounds beyond what is generally observed outdoors [16]. A review of standards and guidelines set by international bodies for the parameters of indoor air quality was recently published [17]. Some of these threshold values are: 7 mg/m^3 over 24 h for carbon monoxide (CO), 0.1 mg/m^3 over 30 min for formaldehyde, 200 $\mu g/m^3$ over 1 h for nitrogen dioxide, and 0.5 mg/m^3 for total VOCs [18].

Application of catalysts for mitigation of toxic indoor air pollutants has been extensively studied and reported [19]. More recently, Gomez et al. applied a novel photocatalytic lime render for improving the indoor and outdoor air quality. This research team applied advanced lime-based binders with titanium dioxide (TiO_2) nano particles for the breakdown of formaldehyde via visible light and NOx pollutants via UV light [20]. Similarly, photocatalytic binder-based mortars for renders and panels were applied to improve indoor air quality [21]. Catalysts offer higher efficiencies at low operating temperatures for a wide variety of pollutants without producing any undesired byproducts. Photocatalytic oxidation using photosensitive semiconductors such as TiO_2 produces strong oxidizing agents to convert VOCs and formaldehyde into carbon dioxide and water [22]. Supported metal oxide catalysts with enhanced catalytic activity was demonstrated via modified preparation procedure [23,24]. Composite materials consisting of bioactive char and carbon nitride (C_3N_4) were shown to have a high formaldehyde conversion efficiency of 85% [25]. Palladium (Pd) doped TiO_2 was shown to convert formaldehyde at ambient conditions [26] and modified TiO_2/zeolite catalysts were highly effective in complete oxidation of benzene. Among different catalysts studied for the oxidation of VOCs and formaldehyde, TiO_2, cobalt (II, III) oxide (CO_3O_4), and alumina supported transition metal (Mn, nickel oxide (NiO_x), Fe) and precious metal (Pd, Ag) catalysts performed at high conversion rates exceeding 90% [19]. Zeng et al. reported a multifunctional catalyst capable of oxidizing VOCs at high gas hourly space velocities of 594,000 h^{-1} [27]. More recently, a review of catalytic oxidation of indoor air pollutants was reported [28]. This study provided a comprehensive review of advances in oxidation catalysis for addressing harmful indoor air pollutants. Similarly, Boyjoo et al. provided a broad overview of photocatalysis, catalyst development as well as reactor design for implementing these catalysts for commercial air purification. Gas phase photodegradation of both VOCs and inorganic gases including ozone was reviewed in this study [29]. Malayeri et al. conducted a thorough review of modeling approaches used in photocatalytic oxidation of VOCs. This study provided a broad overview of reaction mechanisms and kinetic models [30]. Usage of photocatalysts in concrete as an environmentally friendly photocatalyst treatment methodology was reported by Nath et al. [31]. Several studies looked at applying nonprecious metals as formaldehyde oxidation catalysts [32–34].

Primary polluting and toxic constituents from gas fired equipment in buildings include carbon monoxide, nitrogen dioxide, unconverted hydrocarbons, and NOx emissions. Natural gas fired cooking and heating equipment in buildings can significantly impact indoor air quality if proper ventilation is not provided [35]. Additionally, life-time performance degradation may also lead to undesired pollutants. Utilization of catalysts in building heating equipment to suppress these emissions is underexplored and is yet to enter the buildings market. NOx suppression via burner design has been heavily investigated although application of catalysts to tackle all criterion pollutants is sparse.

2.2. Dehumidification

Humidity control in buildings is an important research area with direct implications related to energy efficiency. Relative humidity (RH) in building environment plays a significant role in the health of occupants, building performance, and energy efficiency. The built environment can be severely impacted by high humidity. For instance, as shown in Figure 6, high humidity levels beyond 65% can lead to condensation, mold/mildew growth, corrosion, equipment damage, building fabric deterioration, loss of insulation properties, discoloration, and potentially slip hazards [36]. Dry environment, for instance RH below 20% can thrive atmospheres suitable for viruses and infection [37]. Temperature of course plays a role in compounding the above issues further. Building energy performance, occupant comfort and health impact is well documented in several research studies. Vellei et al. studied the influence of relative humidity on thermal comfort [38], where the authors analyzed and summarized designer friendly RH inclusive adaptive model to extend the range of indoor conditions for low energy naturally conditioned buildings globally. The impact of RH and temperature on structural properties of buildings was also reported [39]. Similarly, a comprehensive review of indoor environment on mold contamination and hygrothermal effect was reported. The study analyzed the growth models in relation to temperature and humidity within the building envelope with a focus towards walls.

Air dehumidification, on the other hand, also has a significant potential in lowering the energy consumption during the space cooling processes. A reduction in latent cooling load via dehumidification helps decrease the energy consumption as well as it enhances the cooling comfort temperature range. In vapor compression systems (e.g., HVAC), latent cooling load could be as high as 50% of the total cooling load which is typically accomplished via cooling the process air below its dew point. Dehumidification can certainly lower or eliminate this load if it is accomplished via physisorption, chemisorption or electrolytic processes.

Electrolytic dehumidification using enhanced oxygen evolution reaction (OER) catalysts were employed in recent studies. For instance, Gao et al. utilized manganese (II, III) oxide (Mn_3O_4) supported on cobalt selenide ($CoSe_2$) catalyst for alkaline OER catalyst exhibiting superior performance [40]. Stabilization of commercial iridium oxide (IrO_2) catalysts via usage of transparent oxides, such as tin doped indium oxide and fluorine doped tin oxide was employed [41,42]. Antimony pentoxide (Sb_2O_5):Tin oxide (SnO_2) was also employed as a support material in improving the stability and conductivity of OER catalysts. Li et al. studied different structurally modified OER catalysts via nanodendrites as supports for IrO_2 catalyst where the Polymer electrolyte membrane-based dehumidification was enhanced by 45% [43].

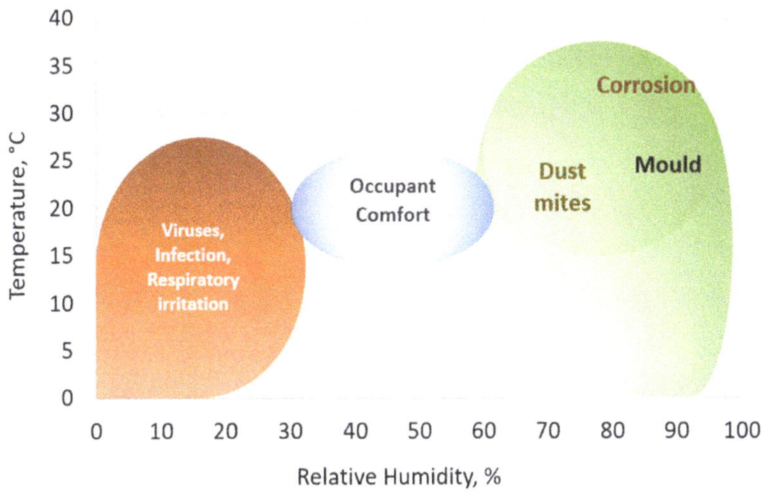

Figure 6. Relative humidity and temperature relationship in closed building environment—Comfortable vs. undesired ranges.

Photocatalysts for dehumidification of ambient air is another approach employed by few researchers. Carbon nitride catalysts were studied for dehumidification and hydrogen evolution reaction to develop metal free catalysts for photoelectrocatalysis and electrocatalysis [44]. Similarly, tantalum nitride (Ta_3N_5) nanorod crystals grown on potassium tantalate ($KTaO_3$) particles were also studied for generating hydrogen via photocatalytic splitting of water [45]. More recently, Yang et al. [46] developed a novel hybrid system consisting of Zn/Co based superhygroscopic hydrogel incorporated with copper oxide (Cu_2O) and barium titanate ($BaTiO_3$) nanoparticles for dehumidifying air.

Improved performance of silica gel by employing promoters such as Al, Co, and Ti ions was also reported [47]. This study showed improved adsorption capacity, stability, and longevity due to the formation of Si-O-M (M = promoter element) linkages in the doped silica gel. Similarly, several researchers also enhanced the performance of silica gels by adding promoters such as calcium chloride and other halide complexes [48–50].

2.3. Thermal Energy Storage

Energy storage is critical to enable integration of intermittent renewable energy into grid infrastructure. A significant portion of peak demand consumption in buildings is associated with thermal energy (as high as 40%). Storage and utilization of thermal energy is of great value in peak shaving applications. Thermal energy storage via utilizing sensible and latent storage technologies is very well studied. However, thermochemical energy storage offers higher energy density due to high amount of enthalpy change associated with reversible chemical reactions [51,52], as shown in Figure 7.

In reversible thermochemical reactions, heat is stored during endothermic reaction and released during the reverse exothermic reaction. A review of high temperature thermochemical reactions was recently published where the concepts and reaction mechanisms of several classes of reversible chemical reactions are discussed [53]. Thermal storage with metal hydrides is an attractive option for moderate thermal energy storage. Several reaction and catalytic strategies to enhance thermal storage capacity and stability of metal hydrides was recently published [54]. Magnesium hydride (MgH_2) coated with iron, vanadium, chromium catalysts was shown to improve the reaction kinetics and thermal conductivity of MgH_2 [55]. Additionally, Titanium based catalysts were shown to enhance the reversible hydrogen release and uptake in sodium-aluminum hydride complex [56].

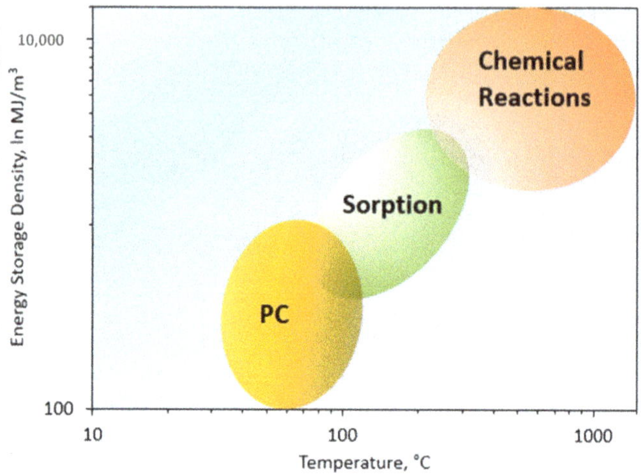

Figure 7. Comparison of different thermal energy storage technologies' energy storage density and reaction temperature.

Dry reforming of methane using solar thermal power plants is an attractive option for thermochemical energy storage due to its high endothermicity (259 kJ/mol). Several studies provided a comprehensive review of catalysts for enabling this reaction [57–59]. Alumina-supported bimetallic Pt-Ru catalyst was recently utilized in dry reforming of methane as a thermochemical energy storage medium where the authors showed a stable performance for continuous cycling in a closed loop application [60].

Iron (III) oxide was utilized as a catalyst and heat transfer medium in enabling a sulfur trioxide (SO_3) to sulfur dioxide (SO_2) reversible thermochemical dissociation reaction for storing concentrated solar thermal power at 850 °C [61]. This study showed the significance of sintering temperature of the catalytic particles in enhancing the stability and cyclability. Similarly, calcium oxide redox couple with carbon dioxide has been extensively researched for applications in thermal energy storage owed to its high endothermicity (178 kJ/mol). Enhancement of the material's optical properties, catalytic properties, and cycling stability via promoter elements such as iron and manganese was shown in a recent study, achieving 2.5 MJ/kg of energy storage density [62]. Ammonia based thermochemical thermal energy storage with a reaction enthalpy of 66.8 kJ/mol is another area of high interest when utilizing concentrated solar thermal power.

Although several reaction schemes are possible for thermal energy storage, moderate-to-low temperature reactions are highly desirable for applications in buildings. Salt hydrates, hydration, and hydrogenation of metal oxides with promoters such as rare-earth metals and ternary compounds with transition metals or alkaline earth metals have been shown to enable these reactions, while improving stability and cyclability. Ideally, reactions capable of utilizing low grade waste heat while supplemented with renewable solar thermal energy all potential candidates for applications in certain climatic regions.

2.4. Carbon Dioxide Capture and Conversion

As described in the introduction section, buildings account for 40% of total carbon dioxide emissions globally (including electricity generation and onsite fuel consumption). The capturing and storage or conversion of this carbon dioxide to useful products is an underexplored area in the context of buildings. Post-combustion carbon capture technologies are suitable for removing carbon dioxide from flue gas generated by fuel combustion in furnaces, boilers, and water heaters. Additionally, direct air capture technologies are suitable for removing the carbon dioxide from air stream recirculated within the building. For instance, a residential building consuming 12 kWh/day of thermal energy can produce up to 2.1–2.5 kg/day of carbon dioxide, depending on the thermal efficiency of the heating equipment. Various materials have been explored to capture carbon dioxide, including metal oxide frameworks (MOF) [63,64], carbon fiber [65,66], ionic liquids [67,68], activated carbon [69,70], non-carbon solid sorbents [71], aqueous sorbents [72], and organic membranes [73].

Mukherjee et al. reviewed the application of activated carbon for post-combustion removal of carbon dioxide from flue gases [74]. The application of photocatalysts based on carbon nitrides (CN) and their enhancement with inorganic semiconductors, carbon materials, ruthenium catalysts, MOFs, and porous materials to promote gas adsorption was reviewed by Liu et al. [75]. In the same study, phosphorus doped carbon nitride-based nanotubes were also investigated for the photocatalytic reduction in CO_2 to enhance electrical conductivity and photo-reactivity. This study concluded that phosphorus doping enhances charge separation, surface area, adsorption capability, and morphology to improve the overall carbon dioxide reduction activity.

Conversion of carbon dioxide into useful products via many chemical transformation routes including electrocatalysis, photothermal catalysis, and thermocatalysis has gained significant momentum in the last two decades, owing to the climate change urgency. Such mechanisms will not only suppress the direct carbon footprint, but also negate the indirect carbon dioxide emissions associated with producing chemicals at the production source (e.g., methanol—industrial chemical; potassium carbonate—glass, soap industry; etc.).

Amplification of photothermal heterogeneous catalysts for valorization of carbon dioxide has been recently reviewed [76]. A comprehensive interpretation of the heat-light interaction and directions for CO_2 utilization via novel photothermal catalyst design was provided. CO_2 hydrogenation via black indium oxide as a photothermal catalyst was also reported [77]. Enhancement of light absorption by making yellow indium oxide into black color was synthesized and reported, showing 100% selectivity towards the hydrogenation of CO_2 to carbon monoxide with a turnover frequency of 2.44 s^{-1}. Similarly, Jantarang et al. investigated the role of support in photothermal hydrogenation of carbon dioxide using ceria-titania composite supported nickel catalyst [78].

Thermocatalytic conversion of CO_2 to useful products is a highly attractive consideration if the net CO_2 output is not increased as a result of the conversion process. Usage of renewable energy to enable these reactions would be the desirable pathway, as shown in Figure 8. Some of the possible reaction schemes are listed below:

$$2CO_2 \rightarrow 2CO + O_2 \ \Delta H^0 = +293 \text{ kJ/mol} \quad (1)$$

$$CO_2 + H_2 \longleftrightarrow CO + H_2O \ \Delta H_{298K} = +41 \text{ kJ/mol} \quad (2)$$

$$CO_2 + 3H_2 \longleftrightarrow CH_3OH + H_2O \ \Delta H_{298K} = -49.1 \text{ kJ/mol} \quad (3)$$

$$CO_2 + 4H_2 \longleftrightarrow CH_4 + 2H_2O \ \Delta H_{298K} = -165 \text{ kJ/mol} \quad (4)$$

$$CO_2 + CH_4 \longleftrightarrow 2CO + 2H_2 \ \Delta H_{298K} = +247 \text{ kJ/mol} \quad (5)$$

Developments in the application of heterogeneous catalysis for CO_2 utilization as a feedstock to produce fine chemicals and renewable fuels was reviewed by De et al. [79]. This is an excellent study providing a very broad overview of thermodynamic considerations in designing a suitable catalyst for CO_2 activation and reaction on several classes of heterogeneous catalytic compositions. Additionally, a literature review of all possible reaction schemes and catalyst options was conducted and reported. Reactions schemes in this study included the production of CO via hydrogenation and the production of alcohols and hydrocarbon chains. Some of the catalyst options for CO_2 hydrogenation reported include (i) noble metal catalysts and their combinations as bimetallic catalysts with Co, Ni, Au, K (ii) supported Cu, Fe-Ni, Ru-Fe, La-Fe, Fe-Ni-Zn catalysts, (iii) support materials including zeolites, molybdenum carbide (Mo_2C), ceria, alumina, ceria-alumina (CeAl), SiO_2, doped ceria, etc. GHSV varied from 30,000 to 300,000 mL.$g_{ca}{}^{-1}h^{-1}$ depending on the catalysts employed, while the selectivity and conversion rates varied significantly. Ethanol synthesis catalysts included precious group metals (PGM) and non-PGM catalysts supported on iron(II, III) oxide (Fe_3O_4), TiO_2, Co_3O_4, MCM-41, etc. and the reported GHSV was in the range of 6000 to 30,000 $mL.g_{ca}{}^{-1}h^{-1}$. A non-exhaustive list of catalysts for the production of multi-carbon hydrocarbon chains from CO_2 was also reported. These reactions were enabled by non-PGM, transition metal, rare earth metal catalysts including Fe, In, K, Zr, Zn, Na, Ga, etc. supported mostly on nano tubes, ZSM, and SAPO.

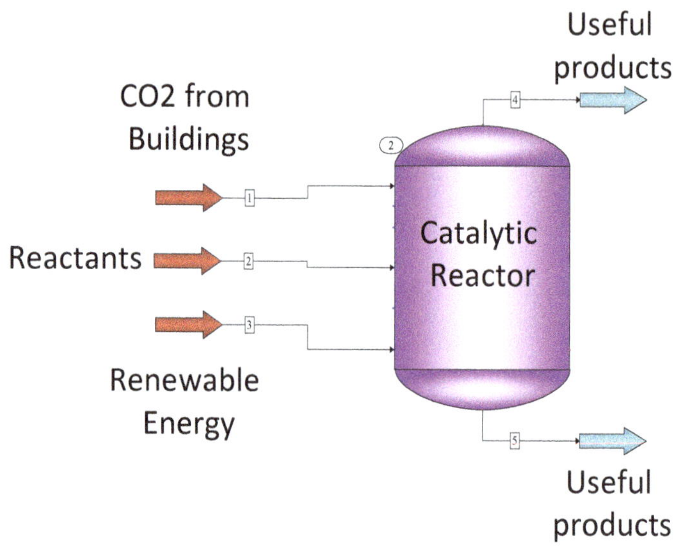

Figure 8. Catalytic reactor scheme to convert CO_2 from buildings to produce useful products. 1—carbon dioxide source, 2—other reactants, 3—renewable energy source, 4—gaseous products, 5—liquid products.

The electrochemical reduction in CO_2 using various electrocatalysts has been recently reviewed in a comprehensive manner. Possible ways of converting waste CO_2 into useful products and thus contributing to the control of global warming while also producing new materials was thoroughly reviewed. Conversion of CO_2 to diverse products such as formic acid, carbon monoxide, methane, and methanol using flow cells, high pressure strategies, molecular catalysis, high temperature solid oxide electrolysis, and non-aqueous electrolytes was reviewed. The authors emphasized the significance of understanding the mechanism including that of poisoning for maturing these technologies in realizing carbon neutral cycles [80,81].

Similarly, evolution of the electrocatalytic reduction in CO_2 from 2004 to 2018 was reviewed by Lee et al. The authors identified the bottlenecks for large scale implementation of the technology and concluded that the gas phase electrolysis offers a beneficial approach [82]. Application of two-dimensional materials as electrocatalysts in reducing CO_2 as opposed to 3D structures was reviewed by Zhu et al. where the authors provided theoretical insights into the advantages of 2D approach such as improved electrical conductivity, larger surface area and the ability to access abundant surface active sites, etc. [82]. Another prominent approach in electrocatalytic conversion of CO_2 via ionic liquids was recently reviewed by Lim et al. A review of experimental and theoretical investigations in to room temperature ionic liquids (RTILs) was provided, owing to their high selectivity and efficiency benefits [83].

2.5. Heating and Cooling

Thermal energy in buildings is a significant energy burden. Considering all the energy needs in a typical residential or commercial building, heating and cooling equipment consume up to 40% of the total energy supply. Utilization of low-quality thermal resources in upgrading the heat to higher temperatures suitable for buildings, via heat pump technology is a highly effective method.

Electrically driven heat pumps use a vapor compression cycle to upgrade the heat for occupant comfort in a building, by harvesting ambient thermal energy. Energy resources for such heat pumps include geothermal, ambient air, and surrounding water bodies which act as heat sources to enhance the thermodynamic cycle performance in providing both heating and cooling energy. Other heat pump approaches involve absorption or solid–gas adsorption cycles to lower primary energy consumption and greenhouse gas emissions by utilizing waste heat or solar thermal energy. Fossil fuel combustion-driven heat pumps (thermally driven heat pumps) can also achieve coefficient of performance (defined as useful thermal energy produced vs. primary energy consumed) above 1 (1.2–2.5 in heating mode and 0.8–1.6 in cooling mode) but are lower than electrically driven vapor compression cycles.

Waste heat as the primary energy resource can offset the overall carbon footprint associated with thermal production. The working principle involves evaporation and consecutive absorption or adsorption of a refrigerant/coolant on the medium, as shown in Figure 9. Absorption cycles typically employ lithium halide/water and water/ammonia as the working fluid pair. Adsorption cycles utilize physisorption on a solid surface where the working fluid evaporates producing cold for cooling or heat is extracted from a low temperature source in heat pump application. Adsorption heat pumps utilize microporous materials such as zeolites [84,85], aluminophosphates [86], porous polymers [87], and metal oxide frameworks (MOF) [88].

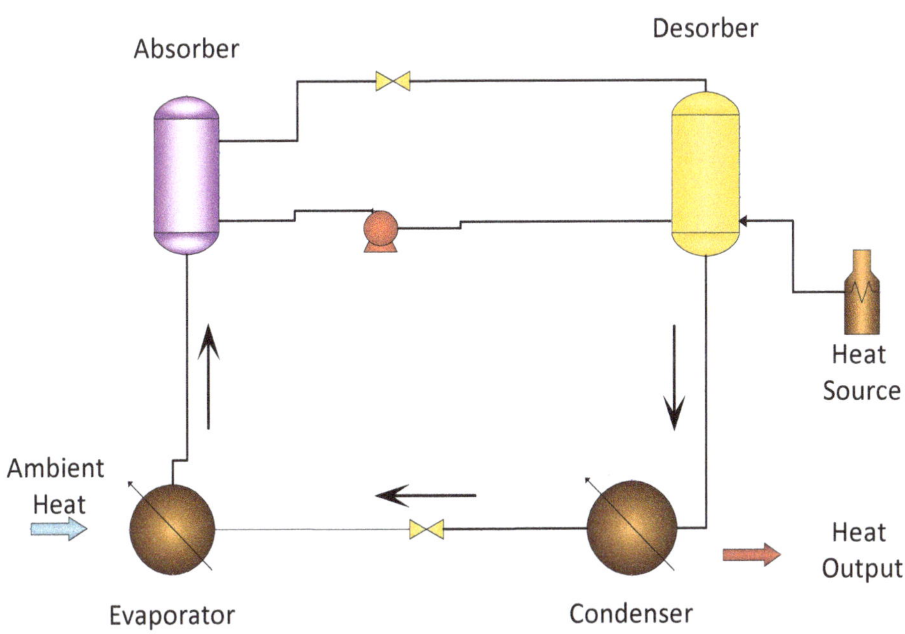

Figure 9. Depiction of a thermally driven heat pump cycle providing thermal comfort in a building.

Compared to the above sensible heat technologies, chemical heat pumps offer higher energy density due to high heat of reaction, as discussed in the thermochemical thermal energy storage topic 2.3 above. The chemical heat pump composes of an endothermic reactor absorbing low temperature heat and an exothermic reactor releasing high temperature heat by utilizing reversible thermochemical reactions. Such a system typically consists of a condenser, an evaporator, and a reactor or adsorber coupled with a generator. It is used for upgrading and storing low grade heat via reversible chemical reactions.

Wongsuwan reviewed chemical heat pumps and their applications in using low grade waste heat [89]. Dehydrogenation of isopropanol over Raney nickel (RN) catalyst [90] was extensively studied and reported for a rapidly quenched and standard RN catalysts. Advances in organic liquid-gas based chemical heat pumps assisted by catalysts was reviewed by Cai et al. [91]. Some of the reactions and catalysts reviewed included dehydrogenation of isopropanol and hydrogenation of acetone (carbon-supported PGM, Raney nickel, metal oxide-supported Cu), dehydrogenation of cyclohexane (Pt/Al_2O_3), dehydration of tert-Butanol (ion exchange resins), depolymerization of paraldehyde (solid acid). This study also highlighted the advantages of chemical heat pumps compared to vapor compression cycles. Additionally, a detailed overview of problems and future directions concerning chemical heat pumps was provided.

2.6. *Cogeneration*

The simultaneous production of electrical and thermal energy (heating or cooling) to offset energy consumption in a building is of significant value in the context of primary energy consumption reduction, grid flexibility, carbon footprint reduction and operational cost savings [13]. Thermal energy and electrical energy generation from the prime mover can be stored and utilized via hydronic systems (to provide heating) and dehumidification (to reduce the latent cooling load, cooling). The produced electrical power can directly support the load profile and/or stored for later consumption via batteries or exported to the grid. An approach to utilizing a cogeneration system in a building is displayed in Figure 10.

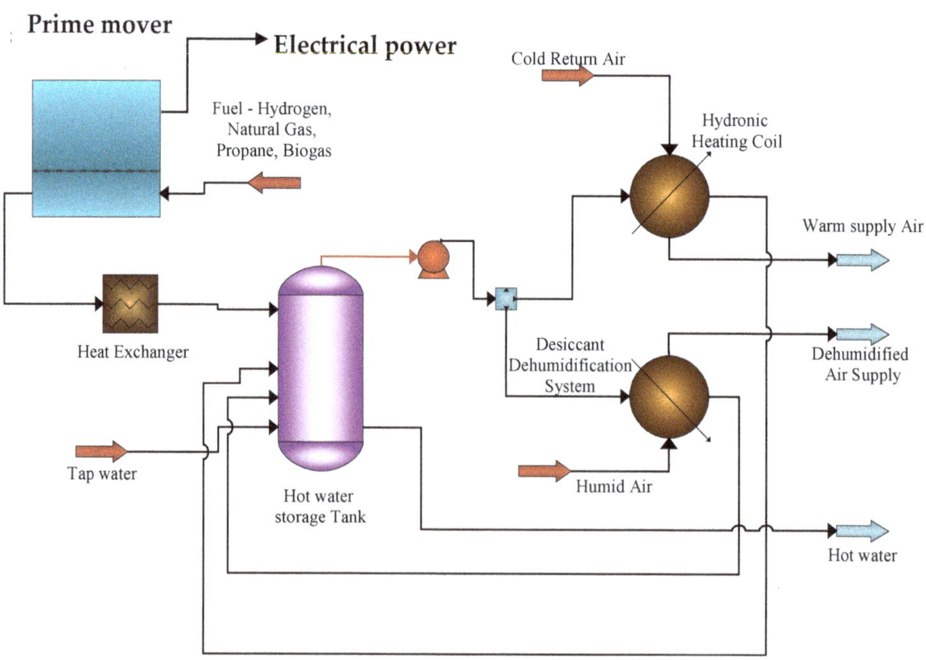

Figure 10. Cogeneration system using a prime mover to provide electrical and thermal energy in a building.

Fuel-driven prime movers suitable for building energy include polymer electrolyte fuel cells (PEM), solid oxide fuel cells (SOFC), reciprocating engines, thermoelectric, Stirling engines, etc. Some of these technologies may not be suitable for building applications due to the physical footprint and other integration challenges. These technologies also act as bridging solutions for enabling low carbon building energy while the electric power grid transitions towards higher efficiencies and composes mainly of clean renewable resources, such as solar and wind power.

The role of catalysts in prime mover technologies is well established. Three main classes of catalysts are utilized in these systems viz. fuel processing (including reforming, purification, and poison removal), emissions suppression, and electrode catalysts. Such catalysts are either employed in thermo-catalytic or electrochemical or electro-thermo-catalytic reactions.

Fuel processing catalysts enable the reforming of hydrocarbon fuels to generate syngas via partial oxidation, steam reforming, dry reforming, autothermal reforming for SOFC applications and downstream purification of hydrogen via water-gas shift reaction, and preferential oxidation for hydrogen generation in PEM fuel cell applications. This catalyst class also includes the removal of poisonous compounds from fuel (for instance, sulfur compounds from fuel). Emissions suppression catalysts help remove undesired toxic components such as carbon monoxide, nitrogen oxides, sulfur dioxide, etc. from the flue gas to enable clean fuel oxidation and release into the atmosphere. The primary function of an electrocatalyst is to oxidize the fuel (anode half-cell reaction) and reduce oxygen (cathode half-cell reaction).

A comprehensive review of reforming catalysts for hydrogen generation in fuel cell applications was provided where the authors reviewed reactions and catalyst systems for syngas generation [92]. Several review articles focusing on reforming catalysts were published in the last five years and some of them are listed here, as follows: methanol steam reforming catalysts [93,94], non-precious metal-based ethanol steam reforming catalysts [95], dry methane reforming catalysts [58], bio-oil steam reforming catalysts [96], dry reforming catalysts [97], biomass pyrolysis oil steam reforming catalysts [98], partial oxidation reforming catalysts [99], and nickel-based methane steam reforming catalysts [100].

Lehtoranta et al. recently reported the use of two different catalysts in natural gas emission reduction strategies from engines [101]. Another study investigated the catalytic oxidation of methane fuel in the exhaust gas of a lean-burn natural gas engine [102]. Similarly, Gremminger et al. investigated PGM catalysts for the oxidation of methane and formaldehyde in the exhaust gas, where the authors concluded that Pd-based catalysts were suitable for methane oxidation in lean conditions, while Pt catalysts were suitable for stoichiometric conditions. Additionally, an alumina-supported Pt catalyst was shown to provide excellent formaldehyde oxidation activity but high sensitivity to CO poisoning. Several research studies also investigated and reviewed the application of novel catalysts for the removal of NO_x and SO_2 from flue gas [103,104].

The field of electrocatalysts is a heavily investigated area due to its implications in carbon dioxide conversion, automotive and stationary power generation, distributed power solutions, etc. Sui et al. recently provided a comprehensive review of Pt electrocatalysts for oxygen reduction reaction (ORR) [105], while Ioroi et al. and Yang et al. reviewed a wide range of electrocatalysts for PEM fuel cell applications [106,107]. Several studies also looked at recent developments in electrocatalyst design with and without noble metals [108,109]. An interesting study by Ma et al. recently concluded that novel ORR electrocatalysts in PEM fuel cells via introduction of heteroatoms and defects into carbon materials (e.g., N-doped carbon nano tubes CNT) offer significant promise in enhancing the reaction kinetics [110].

Development of porous SOFC electrode structures with enhanced oxygen reduction reaction at lower operating temperatures has taken a front seat in the last decade. The primary objective of these research efforts is to extend triple-phase boundaries while enhancing the ionic and electronic conductivities. Detailed reviews of several classes of cathode materials have been published: core–shell structures [111], composite materials [112], Ruddlesden-Popper perovskites [113], perovskite oxides [114].

Similarly, several researchers investigated anode electrode modification for direct internal reforming and enhancing carbon and sulfur tolerance by incorporating dopants and catalytic functional layers [115–118]. A comprehensive review of recent advances in the development of anode materials for direct hydrocarbon fuel conversion was provided by Shabri et al. [119]. This study focused on carbon deposition issues related to direct reforming on Ni and Ni alloys with several transition metals and made a recommendation for a hybrid mixed oxygen carrier perovskite and Ni/Ni alloy as a suitable anode material for enhanced carbon tolerance.

3. Discussion

In the field of indoor air quality and emissions, there is an absolute need for highly active, non-PGM catalytic materials which are tolerant to poisoning species. Incorporation of these materials into reactor structures offering negligible pressure drop with high capacity (lifetime enhancement) and ability to integrate in existing building stock is a critical requirement. Catalytic materials capable of addressing multiple reactions with high selectivity would be ideally suited for building air quality improvements. Such catalysts must be capable of suppressing CH_4, CO, NO_x, formaldehyde emissions from the flue gas generated from gas fired heating, and cooking equipment as well as from prime movers in cogeneration systems. Photocatalyst based solutions could play a vital role in successful implementation of such solutions. Buildings offer a variety of surfaces where such systems can be incorporated, for instance, selective surfaces within the building envelope. Hybrid materials with adsorbents and catalysts are necessary for improved conversion and selectivity.

Primary energy consumption reduction by lowering the latent cooling load via dehumidification of return air or fresh air supply in building cooling system also relies on some of the above-mentioned features necessary for successful application in residential equipment. Economic and population growth coupled with affordability are poised to drive the energy consumption associated with cooling needs in a building higher. The ability to incorporate dehumidification strategies in the air conditioning unit requires high capacity, selectivity, and lower regeneration energy. Another novel area with high potential for future deployment in buildings include dehumidification coupled with power production using advanced hybrid photocatalytic materials.

Thermal energy storage is another high impact area where the beneficial features of catalytic materials can be employed in enhancing the ability to use different heat sources, including low grade waste heat and renewable solar thermal energy, more effectively. Intermittent renewable energy sources, such as wind and solar integration at a grid scale, require energy storage technologies to offset the unpredictability by storing and using the energy at low cost. Successful deployment of high energy density thermal energy storage technologies enabled by heat of reaction from a reversible chemical reaction can help address grid flexibility and energy management. Catalytic promoters and additives to address the cyclability and stability of thermal energy storage materials can certainly help bring these solutions in to the real world.

Carbon dioxide emissions suppression from existing buildings' gas fired heating/cooking equipment requires affordable, retrofittable, and easy to maintain solutions in capturing CO_2 and converting to useful products. Products from carbon conversion however requires a logistical chain to handle the derivatives appropriately. Ideally, a locally consumable product (e.g., fertilizer) would be beneficial to lock in the carbon. Co-adsorption of CO_2 and steam from the flue gas followed by in situ chemical transformation on a photocatalyst is a highly desirable route for carbon management in a building. Fu et al. recently published an article focusing on such an approach [120]. Another useful approach could be the incorporation of catalytically enhanced CO_2 sorbents into the duct work of buildings as a long-term storage solution; however, the adsorption–desorption kinetics will have to be considered carefully. Metal support interaction phenomenon observed in heterogeneous catalysis may help address and tailor the reaction schemes such that CO_2 concentration does not pose an issue during transient cycles. Similarly, the indoor emissions and air quality improvements could be accomplished with the same duct with targeted materials placed strategically in each zone of the building (e.g., kitchen vs. rest of the building). Additionally, carbon dioxide capture should not impose further energy burden in a building. Hence, the usage of renewable energy and application of photocatalysts is desirable in enabling such chemical transformations without energy penalty. Song et al. recently proposed a hybrid approach for CO_2 capture [121].

Non-precious metal-based, earth abundant, durable, easy-to-process-and-manufacture catalysts are needed for integration with cogeneration technologies. Fuel processing catalysts which can suppress carbon deposition along with being tolerant towards sulfur species are necessary. SOFC-based electrocatalysts capable of operating at lower temperatures in the range of 500 °C can enable affordable power generators with low capital expenditure costs.

In conclusion, implementation of catalytic technologies in buildings is needed to improve our health and workforce efficiency or productivity, lower the building energy consumption, and lower its carbon footprint. Catalysts have a new role to play in enabling sustainable energy future to meet the growing energy needs of the population without impacting the climate. Although the ideal technical requirements of a catalyst in providing different solutions in buildings is well understood, the realistic factors which will enable impactful deployment are exclusively associated with the cost, reliability, longevity, retrofittability, manufacturability, and ready availability of materials.

Funding: This research received no external funding.

Acknowledgments: This research was supported by the Department of Energy (DOE) Office of Energy Efficiency and Renewable Energy (EERE), Building Technologies Office and used resources at the Building Technologies Research and Integration Center, a DOE-EERE User Facility at Oak Ridge National Laboratory.

Conflicts of Interest: The author declare that they have no known competing financial interest or personal relationships that could have appeared to influence the work reported in this entry.

Entry Link on the Encyclopedia Platform: https://encyclopedia.pub/18811.

Nomenclature & Abbreviations

Btu	British Thermal Units
EJ	Exajoules
g	gram
h	Hour
kg	Kilogram
kJ	Kilo Joules
mL	Milliliter
MJ	Mega Joules
TWh	Terawatt-hours
CNT	Carbon Nano Tubes
EIA	Energy Information Administration
EPA	Environment Protection Agency
HVAC	Heating, Ventilation, and Air Conditioning
IAQ	Indoor Air Quality
IEA	International Energy Agency
MOF	Metal Oxide Framework
NIH	National Institute of Health

OER	Oxygen Evolution Reaction
ORR	Oxygen Reduction Reaction
Quad	Quadrillion
PEM	Polymer Electrolyte Fuel Cells
PGM	Precious Group Metals
RH	Relative Humidity
RTIL	Room Temperature Ionic Liquids
SAPO	Silicoalumino Phosphate Zeolite
SOFC	Solid Oxide Fuel Cell
USA	United States of America
VOC	Volatile Organic Compounds

References

1. U.S. Energy Information Administration. EIA Projects Nearly 50% Increase in World Energy Use by 2050, Led by Growth in Renewables. Available online: https://www.eia.gov/todayinenergy/detail.php?id=49876 (accessed on 21 December 2021).
2. U.S. Energy Information Administration. Global Primary Energy Consumption by Energy Source (2010–2050). Available online: https://www.eia.gov/todayinenergy/detail.php?id=41433 (accessed on 21 December 2021).
3. International Energy Agency. Tracking Buildings. Available online: https://www.iea.org/reports/tracking-buildings-2020 (accessed on 21 December 2021).
4. U.S. Energy Information Administration. How the United States Uses Energy. Available online: https://www.eia.gov/energyexplained/use-of-energy/ (accessed on 30 September 2021).
5. U.S. Energy Information Administration. Residential Energy Consumption Survey (RECS). Available online: https://www.eia.gov/consumption/residential/ (accessed on 30 September 2021).
6. U.S. Energy Information Administration. Commercial Buildings Energy Consumption Survey (CBECS). Available online: https://www.eia.gov/consumption/commercial/reports/ (accessed on 31 January 2021).
7. U.S. Department of Energy. Home Heating Systems. Available online: https://www.energy.gov/energysaver/heat-and-cool/home-heating-systems (accessed on 28 February 2021).
8. U.S. Energy Information Administration. Energy and the Environment Explained, Where Greenhouse Gases Come from. Available online: https://www.eia.gov/energyexplained/energy-and-the-environment/where-greenhouse-gases-come-from.php (accessed on 31 January 2021).
9. Allen, J.G.; MacNaughton, P.; Satish, U.; Santanam, S.; Vallarino, J.; Spengler, J.D. Associations of cognitive function scores with carbon dioxide, ventilation, and volatile organic compound exposures in office workers: A controlled exposure study of green and conventional office environments. *Environ. Health Perspect.* **2016**, *124*, 805–812. [CrossRef] [PubMed]
10. Huang, K.; Sun, W.; Feng, G.; Wang, J.; Song, J. Indoor air quality analysis of 8 mechanically ventilated residential buildings in northeast China based on long-term monitoring. *Sustain. Cities Soc.* **2020**, *54*, 101947. [CrossRef]
11. Spiru, P.; Simona, P.L. A review on interactions between energy performance of the buildings, outdoor air pollution and the indoor air quality. *Energy Procedia* **2017**, *128*, 179–186. [CrossRef]
12. Ma, N.; Aviv, D.; Guo, H.; Braham, W.W. Measuring the right factors: A review of variables and models for thermal comfort and indoor air quality. *Renew. Sustain. Energy Rev.* **2021**, *135*, 110436. [CrossRef]
13. Cheekatamarla, P. Performance analysis of hybrid power configurations: Impact on primary energy intensity, carbon dioxide emissions, and life cycle costs. *Int. J. Hydrog. Energy* **2020**, *45*, 34089–34098. [CrossRef]
14. Cheekatamarla, P.; Abu-Heiba, A. A Comprehensive Review and Qualitative Analysis of Micro-Combined Heat and Power Modeling Approaches. *Energies* **2020**, *13*, 3581. [CrossRef]
15. Sun, D.; Le, Y.; Jiang, C.; Cheng, B. Ultrathin Bi_2WO_6 nanosheet decorated with Pt nanoparticles for efficient formaldehyde removal at room temperature. *Appl. Surf. Sci.* **2018**, *441*, 429–437. [CrossRef]
16. Wallace, L.A. *Total Exposure Assessment Methodology (TEAM) Study: Summary and Analysis*; Environmental Protection Agency: Washington, DC, USA, 1987; Volume 1.
17. Abdul-Wahab, S.A.; En, S.C.F.; Elkamel, A.; Ahmadi, L.; Yetilmezsoy, K. A review of standards and guidelines set by international bodies for the parameters of indoor air quality. *Atmos. Pollut. Res.* **2015**, *6*, 751–767. [CrossRef]
18. World Health Organization. WHO Guildelines for Indoor Air Quality. Available online: https://www.euro.who.int/__data/assets/pdf_file/0009/128169/e94535.pdf (accessed on 31 January 2021).
19. Yue, X.; Ma, N.L.; Sonne, C.; Guan, R.; Lam, S.S.; Van Le, Q.; Chen, X.; Yang, Y.; Gu, H.; Rinklebe, J. Mitigation of indoor air pollution: A review of recent advances in adsorption materials and catalytic oxidation. *J. Hazard. Mater.* **2020**, *405*, 124138. [CrossRef] [PubMed]
20. Ibáñez Gómez, J.A.; Giampiccolo, A.; Tobaldi, D.M.; Mair, S.; da Silva, C.F.; Barrasa, A.M.C.; Maskell, D.; Ansell, M.P.; Kurchania, R.; Mayer, F. Photocatalytic lime render for indoor and outdoor air quality improvement. *Catalysts* **2021**, *11*, 296. [CrossRef]
21. Giosuè, C.; Pierpaoli, M.; Mobili, A.; Ruello, M.L.; Tittarelli, F. Influence of binders and lightweight aggregates on the properties of cementitious mortars: From traditional requirements to indoor air quality improvement. *Materials* **2017**, *10*, 978. [CrossRef]

22. Shayegan, Z.; Lee, C.-S.; Haghighat, F. TiO_2 photocatalyst for removal of volatile organic compounds in gas phase–A review. *Chem. Eng. J.* **2018**, *334*, 2408–2439. [CrossRef]
23. Rezaee, A.; Rangkooy, H.; Khavanin, A.; Jafari, A.J. High photocatalytic decomposition of the air pollutant formaldehyde using nano-ZnO on bone char. *Environ. Chem. Lett.* **2014**, *12*, 353–357. [CrossRef]
24. Li, Y.; Zhang, C.; Ma, J.; Chen, M.; Deng, H.; He, H. High temperature reduction dramatically promotes Pd/TiO_2 catalyst for ambient formaldehyde oxidation. *Appl. Catal. B Environ.* **2017**, *217*, 560–569. [CrossRef]
25. Li, X.; Qian, X.; An, X.; Huang, J. Preparation of a novel composite comprising biochar skeleton and "chrysanthemum" g-C3N4 for enhanced visible light photocatalytic degradation of formaldehyde. *Appl. Surf. Sci.* **2019**, *487*, 1262–1270. [CrossRef]
26. Nasriddinov, A.; Rumyantseva, M.; Marikutsa, A.; Gaskov, A.; Lee, J.-H.; Kim, J.-H.; Kim, J.-Y.; Kim, S.S.; Kim, H.W. Sub-ppm formaldehyde detection by nn TiO_2@ SnO_2 nanocomposites. *Sensors* **2019**, *19*, 3182. [CrossRef] [PubMed]
27. Zeng, Y.; Xie, R.; Cao, J.; Chen, Z.; Fan, Q.; Liu, B.; Lian, X.; Huang, H. Simultaneous removal of multiple indoor-air pollutants using a combined process of electrostatic precipitation and catalytic decomposition. *Chem. Eng. J.* **2020**, *388*, 124219. [CrossRef]
28. Zhao, G.; Zou, J.; Zhang, T.; Li, C.; Zhou, S.; Jiao, F. Recent progress on removal of indoor air pollutants by catalytic oxidation. *Rev. Environ. Health* **2020**, *35*, 311–321. [CrossRef]
29. Boyjoo, Y.; Sun, H.; Liu, J.; Pareek, V.K.; Wang, S. A review on photocatalysis for air treatment: From catalyst development to reactor design. *Chem. Eng. J.* **2017**, *310*, 537–559. [CrossRef]
30. Malayeri, M.; Haghighat, F.; Lee, C.-S. Modeling of volatile organic compounds degradation by photocatalytic oxidation reactor in indoor air: A review. *Build. Environ.* **2019**, *154*, 309–323. [CrossRef]
31. Nath, R.K.; Zain, M.; Jamil, M. An environment-friendly solution for indoor air purification by using renewable photocatalysts in concrete: A review. *Renew. Sustain. Energy Rev.* **2016**, *62*, 1184–1194. [CrossRef]
32. Zang, M.; Zhao, C.; Wang, Y.; Chen, S. A review of recent advances in catalytic combustion of VOCs on perovskite-type catalysts. *J. Saudi Chem. Soc.* **2019**, *23*, 645–654. [CrossRef]
33. Miao, L.; Wang, J.; Zhang, P. Review on manganese dioxide for catalytic oxidation of airborne formaldehyde. *Appl. Surf. Sci.* **2019**, *466*, 441–453. [CrossRef]
34. Guo, J.; Lin, C.; Jiang, C.; Zhang, P. Review on noble metal-based catalysts for formaldehyde oxidation at room temperature. *Appl. Surf. Sci.* **2019**, *475*, 237–255. [CrossRef]
35. Singer, B.C.; Pass, R.Z.; Delp, W.W.; Lorenzetti, D.M.; Maddalena, R.L. Pollutant concentrations and emission rates from natural gas cooking burners without and with range hood exhaust in nine California homes. *Build. Environ.* **2017**, *122*, 215–229. [CrossRef]
36. U.S. Energy Information Administration. Why and Where Mold Grows? Available online: https://www.epa.gov/mold/mold-course-chapter-2 (accessed on 28 February 2021).
37. Yang, W.; Marr, L.C. Mechanisms by Which Ambient Humidity May Affect Viruses in Aerosols. *Appl Environ Microbiol.* **2012**, *78*, 6781–6788. [CrossRef] [PubMed]
38. Vellei, M.; Herrera, M.; Fosas, D.; Natarajan, S. The influence of relative humidity on adaptive thermal comfort. *Build. Environ.* **2017**, *124*, 171–185. [CrossRef]
39. Zonno, G.; Aguilar, R.; Boroschek, R.; Lourenço, P.B. Analysis of the long and short-term effects of temperature and humidity on the structural properties of adobe buildings using continuous monitoring. *Eng. Struct.* **2019**, *196*, 109299. [CrossRef]
40. Gao, M.-R.; Xu, Y.-F.; Jiang, J.; Zheng, Y.-R.; Yu, S.-H. Water oxidation electrocatalyzed by an efficient Mn3O4/CoSe2 nanocomposite. *J. Am. Chem. Soc.* **2012**, *134*, 2930–2933. [CrossRef]
41. Liu, G.; Xu, J.; Wang, Y.; Wang, X. An oxygen evolution catalyst on an antimony doped tin oxide nanowire structured support for proton exchange membrane liquid water electrolysis. *J. Mater. Chem. A* **2015**, *3*, 20791–20800. [CrossRef]
42. Oh, H.S.; Nong, H.N.; Strasser, P. Preparation of mesoporous Sb-, F-, and In-doped SnO_2 bulk powder with high surface area for use as catalyst supports in electrolytic cells. *Adv. Funct. Mater.* **2015**, *25*, 1074–1081. [CrossRef]
43. Li, D.; Yan, W.; Qi, R.; Zhang, L.-Z. Development of structurally modified OER catalysts with enhanced performance and longevity for PEM-based electrolytic air dehumidification. *Int. J. Hydrog. Energy* **2021**, *46*, 9267–9279. [CrossRef]
44. Zhou, G.; Shan, Y.; Hu, Y.; Xu, X.; Long, L.; Zhang, J.; Dai, J.; Guo, J.; Shen, J.; Li, S. Half-metallic carbon nitride nanosheets with micro grid mode resonance structure for efficient photocatalytic hydrogen evolution. *Nat. Commun.* **2018**, *9*, 1–9. [CrossRef] [PubMed]
45. Wang, Z.; Inoue, Y.; Hisatomi, T.; Ishikawa, R.; Wang, Q.; Takata, T.; Chen, S.; Shibata, N.; Ikuhara, Y.; Domen, K. Overall water splitting by Ta_3N_5 nanorod single crystals grown on the edges of $KTaO_3$ particles. *Nat. Catal.* **2018**, *1*, 756–763. [CrossRef]
46. Yang, L.; Ravi, S.K.; Nandakumar, D.K.; Alzakia, F.I.; Lu, W.; Zhang, Y.; Yang, J.; Zhang, Q.; Zhang, X.; Tan, S.C. A hybrid artificial photocatalysis system splits atmospheric water for simultaneous dehumidification and power generation. *Adv. Mater.* **2019**, *31*, 1902963. [CrossRef]
47. Fang, Y.; Liu, T.; Zhang, Z.; Gao, X. Silica gel adsorbents doped with Al, Ti, and Co ions improved adsorption capacity, thermal stability and aging resistance. *Renew. Energy* **2014**, *63*, 755–761. [CrossRef]
48. Aristov, Y.I.; Tokarev, M.; Gordeeva, L.; Snytnikov, V.; Parmon, V. New composite sorbents for solar-driven technology of fresh water production from the atmosphere. *Sol. Energy* **1999**, *66*, 165–168. [CrossRef]
49. Ge, T.; Ziegler, F.; Wang, R. A mathematical model for predicting the performance of a compound desiccant wheel (A model of compound desiccant wheel). *Appl. Therm. Eng.* **2010**, *30*, 1005–1015. [CrossRef]

50. Zhang, X.; Sumathy, K.; Dai, Y.; Wang, R. Dynamic hygroscopic effect of the composite material used in desiccant rotary wheel. *Sol. Energy* **2006**, *80*, 1058–1061. [CrossRef]
51. André, L.; Abanades, S.; Flamant, G. Screening of thermochemical systems based on solid-gas reversible reactions for high temperature solar thermal energy storage. *Renew. Sustain. Energy Rev.* **2016**, *64*, 703–715. [CrossRef]
52. Bauer, T.; Steinmann, W.-D.; Laing, D.; Tamme, R. Thermal energy storage materials and systems. *Annu. Rev. Heat Transf.* **2012**, *15*, 131–177. [CrossRef]
53. Prasad, J.S.; Muthukumar, P.; Desai, F.; Basu, D.N.; Rahman, M.M. A critical review of high-temperature reversible thermochemical energy storage systems. *Appl. Energy* **2019**, *254*, 113733. [CrossRef]
54. Manickam, K.; Mistry, P.; Walker, G.; Grant, D.; Buckley, C.E.; Humphries, T.D.; Paskevicius, M.; Jensen, T.; Albert, R.; Peinecke, K. Future perspectives of thermal energy storage with metal hydrides. *Int. J. Hydrog. Energy* **2019**, *44*, 7738–7745. [CrossRef]
55. Mistry, P.C.; Grant, D.M.; Stuart, A.D.; Manickam, K.; Walker, G.S. Evolution of catalyst coated atomised magnesium spheres–An alternative thermal storage medium for concentrated solar power applications. *Int. J. Hydrog. Energy* **2017**, *42*, 28453–28463. [CrossRef]
56. Bogdanović, B.; Schwickardi, M. Ti-doped alkali metal aluminium hydrides as potential novel reversible hydrogen storage materials. *J. Alloys Compd.* **1997**, *253*, 1–9. [CrossRef]
57. Aramouni, N.A.K.; Touma, J.G.; Tarboush, B.A.; Zeaiter, J.; Ahmad, M.N. Catalyst design for dry reforming of methane: Analysis review. *Renew. Sustain. Energy Rev.* **2018**, *82*, 2570–2585. [CrossRef]
58. Abdulrasheed, A.; Jalil, A.A.; Gambo, Y.; Ibrahim, M.; Hambali, H.U.; Hamid, M.Y.S. A review on catalyst development for dry reforming of methane to syngas: Recent advances. *Renew. Sustain. Energy Rev.* **2019**, *108*, 175–193. [CrossRef]
59. Jang, W.-J.; Shim, J.-O.; Kim, H.-M.; Yoo, S.-Y.; Roh, H.-S. A review on dry reforming of methane in aspect of catalytic properties. *Catal. Today* **2019**, *324*, 15–26. [CrossRef]
60. Du, J.; Yang, X.-x.; Ding, J.; Wei, X.-l.; Yang, J.-p.; Wang, W.-l.; Yang, M.-l. Carbon dioxide reforming of methane over bimetallic catalysts of Pt-Ru/γ-Al 2 O 3 for thermochemical energy storage. *J. Cent. South Univ.* **2013**, *20*, 1307–1313. [CrossRef]
61. Tsongidis, N.I.; Karagiannakis, G.; Sakellariou, K.G.; Pagkoura, C.; Konstandopoulos, A.G. Iron oxide-based particles for high temperature thermochemical energy storage via the elemental sulfur thermochemical cycle. In *Proceedings of AIP Conference Proceedings*; AIP Publishing: Long Island, NY, USA, 2019; p. 210009.
62. Yang, L.; Huang, Z.; Huang, G. Fe-and Mn-Doped Ca-Based Materials for Thermochemical Energy Storage Systems. *Energy Fuels* **2020**, *34*, 11479–11488. [CrossRef]
63. Samanta, A.; Zhao, A.; Shimizu, G.K.; Sarkar, P.; Gupta, R. Post-combustion CO_2 capture using solid sorbents: A review. *Ind. Eng. Chem. Res.* **2012**, *51*, 1438–1463. [CrossRef]
64. Ghanbari, T.; Abnisa, F.; Daud, W.M.A.W. A review on production of metal organic frameworks (MOF) for CO_2 adsorption. *Sci. Total Environ.* **2020**, *707*, 135090. [CrossRef]
65. Mondal, M.K.; Balsora, H.K.; Varshney, P. Progress and trends in CO_2 capture/separation technologies: A review. *Energy* **2012**, *46*, 431–441. [CrossRef]
66. Li, B.; Duan, Y.; Luebke, D.; Morreale, B. Advances in CO_2 capture technology: A patent review. *Appl. Energy* **2013**, *102*, 1439–1447. [CrossRef]
67. Aghaie, M.; Rezaei, N.; Zendehboudi, S. A systematic review on CO_2 capture with ionic liquids: Current status and future prospects. *Renew. Sustain. Energy Rev.* **2018**, *96*, 502–525. [CrossRef]
68. Zhang, X.; Zhang, X.; Dong, H.; Zhao, Z.; Zhang, S.; Huang, Y. Carbon capture with ionic liquids: Overview and progress. *Energy Environ. Sci.* **2012**, *5*, 6668–6681. [CrossRef]
69. Creamer, A.E.; Gao, B. Carbon-based adsorbents for postcombustion CO_2 capture: A critical review. *Environ. Sci. Technol.* **2016**, *50*, 7276–7289. [CrossRef]
70. Shafeeyan, M.S.; Daud, W.M.A.W.; Houshmand, A.; Shamiri, A. A review on surface modification of activated carbon for carbon dioxide adsorption. *J. Anal. Appl. Pyrolysis* **2010**, *89*, 143–151. [CrossRef]
71. Lee, S.-Y.; Park, S.-J. A review on solid adsorbents for carbon dioxide capture. *J. Ind. Eng. Chem.* **2015**, *23*, 1–11. [CrossRef]
72. Dutcher, B.; Fan, M.; Russell, A.G. Amine-based CO_2 capture technology development from the beginning of 2013–A Review. *ACS Appl. Mater. Interfaces* **2015**, *7*, 2137–2148. [CrossRef]
73. Prasetya, N.; Himma, N.F.; Sutrisna, P.D.; Wenten, I.G.; Ladewig, B.P. A review on emerging organic-containing microporous material membranes for carbon capture and separation. *Chem. Eng. J.* **2019**, *391*, 123575. [CrossRef]
74. Mukherjee, A.; Okolie, J.A.; Abdelrasoul, A.; Niu, C.; Dalai, A.K. Review of post-combustion carbon dioxide capture technologies using activated carbon. *J. Environ. Sci.* **2019**, *83*, 46–63. [CrossRef] [PubMed]
75. Liu, B.; Ye, L.; Wang, R.; Yang, J.; Zhang, Y.; Guan, R.; Tian, L.; Chen, X. Phosphorus-doped graphitic carbon nitride nanotubes with amino-rich surface for efficient CO_2 capture, enhanced photocatalytic activity, and product selectivity. *ACS Appl. Mater. Interfaces* **2018**, *10*, 4001–4009. [CrossRef]
76. Xie, B.; Lovell, E.; Tan, T.H.; Jantarang, S.; Yu, M.; Scott, J.; Amal, R. Emerging material engineering strategies for amplifying photothermal heterogeneous CO_2 catalysis. *J. Energy Chem.* **2020**, *59*, 108–125. [CrossRef]
77. Wang, L.; Dong, Y.; Yan, T.; Hu, Z.; Jelle, A.A.; Meira, D.M.; Duchesne, P.N.; Loh, J.Y.Y.; Qiu, C.; Storey, E.E. Black indium oxide a photothermal CO_2 hydrogenation catalyst. *Nat. Commun.* **2020**, *11*, 1–8. [CrossRef]

78. Jantarang, S.; Lovell, E.C.; Tan, T.H.; Scott, J.; Amal, R. Role of support in photothermal carbon dioxide hydrogenation catalysed by Ni/CexTiyO_2. *Prog. Nat. Sci. Mater. Int.* **2018**, *28*, 168–177. [CrossRef]
79. De, S.; Dokania, A.; Ramirez, A.; Gascon, J. Advances in the Design of Heterogeneous Catalysts and Thermocatalytic Processes for CO_2 Utilization. *ACS Catal.* **2020**, *10*, 14147–14185. [CrossRef]
80. Jones, J.P.; Prakash, G.S.; Olah, G.A. Electrochemical CO_2 reduction: Recent advances and current trends. *Isr. J. Chem.* **2014**, *54*, 1451–1466. [CrossRef]
81. Kumar, S.; Wani, M.Y.; Arranja, C.T.; e Silva, J.d.A.; Avula, B.; Sobral, A.J. Porphyrins as nanoreactors in the carbon dioxide capture and conversion: A review. *J. Mater. Chem. A* **2015**, *3*, 19615–19637. [CrossRef]
82. Lee, M.-Y.; Park, K.T.; Lee, W.; Lim, H.; Kwon, Y.; Kang, S. Current achievements and the future direction of electrochemical CO_2 reduction: A short review. *Crit. Rev. Environ. Sci. Technol.* **2020**, *50*, 769–815. [CrossRef]
83. Lim, H.-K.; Kim, H. The mechanism of room-temperature ionic-liquid-based electrochemical CO_2 reduction: A review. *Molecules* **2017**, *22*, 536. [CrossRef]
84. Calabrese, L.; Bonaccorsi, L.; Freni, A.; Proverbio, E. Synthesis of SAPO-34 zeolite filled macrocellular foams for adsorption heat pump applications: A preliminary study. *Appl. Therm. Eng.* **2017**, *124*, 1312–1318. [CrossRef]
85. Restuccia, G.; Recupero, V.; Cacciola, G.; Rothmeyer, M. Zeolite heat pump for domestic heating. *Energy* **1988**, *13*, 333–342. [CrossRef]
86. Henninger, S.K.; Jeremias, F.; Kummer, H.; Schossig, P.; Henning, H.-M. Novel sorption materials for solar heating and cooling. *Energy Procedia* **2012**, *30*, 279–288. [CrossRef]
87. Mandal, J.; Fu, Y.; Overvig, A.C.; Jia, M.; Sun, K.; Shi, N.N.; Zhou, H.; Xiao, X.; Yu, N.; Yang, Y. Hierarchically porous polymer coatings for highly efficient passive daytime radiative cooling. *Science* **2018**, *362*, 315–319. [CrossRef] [PubMed]
88. Cui, S.; Marandi, A.; Lebourleux, G.; Thimon, M.; Bourdon, M.; Chen, C.; Severino, M.I.; Steggles, V.; Nouar, F.; Serre, C. Heat properties of a hydrophilic carboxylate-based MOF for water adsorption applications. *Appl. Therm. Eng.* **2019**, *161*, 114135. [CrossRef]
89. Wongsuwan, W.; Kumar, S.; Neveu, P.; Meunier, F. A review of chemical heat pump technology and applications. *Appl. Therm. Eng.* **2001**, *21*, 1489–1519. [CrossRef]
90. Xin, F.; Xu, M.; Huai, X.-L.; Li, X.-F. Characteristic and kinetic of liquid-phase isopropanol dehydrogenation over Raney nickel catalysts for chemical heat pump. *Appl. Therm. Eng.* **2014**, *70*, 580–585. [CrossRef]
91. Cai, J.; Li, X.; Tao, Y.; Huai, X.; Guo, Z. Advances in Organic Liquid-Gas Chemical Heat Pumps. *Chem. Eng. Technol.* **2011**, *34*, 1603–1613. [CrossRef]
92. Cheekatamarla, P.K.; Finnerty, C. Reforming catalysts for hydrogen generation in fuel cell applications. *J. Power Sources* **2006**, *160*, 490–499. [CrossRef]
93. Xu, X.; Shuai, K.; Xu, B. Review on copper and palladium based catalysts for methanol steam reforming to produce hydrogen. *Catalysts* **2017**, *7*, 183. [CrossRef]
94. Bepari, S.; Kuila, D. Steam reforming of methanol, ethanol and glycerol over nickel-based catalysts—A review. *Int. J. Hydrog. Energy* **2020**, *45*, 18090–18113. [CrossRef]
95. Ogo, S.; Sekine, Y. Recent progress in ethanol steam reforming using non-noble transition metal catalysts: A review. *Fuel Process. Technol.* **2020**, *199*, 106238. [CrossRef]
96. Chen, J.; Sun, J.; Wang, Y. Catalysts for steam reforming of bio-oil: A review. *Ind. Eng. Chem. Res.* **2017**, *56*, 4627–4637. [CrossRef]
97. Aziz, M.; Setiabudi, H.; Teh, L.; Annuar, N.; Jalil, A. A review of heterogeneous catalysts for syngas production via dry reforming. *J. Taiwan Inst. Chem. Eng.* **2019**, *101*, 139–158. [CrossRef]
98. Setiabudi, H.; Aziz, M.; Abdullah, S.; Teh, L.; Jusoh, R. Hydrogen production from catalytic steam reforming of biomass pyrolysis oil or bio-oil derivatives: A review. *Int. J. Hydrog. Energy* **2020**, *45*, 18376–18397. [CrossRef]
99. Arku, P.; Regmi, B.; Dutta, A. A review of catalytic partial oxidation of fossil fuels and biofuels: Recent advances in catalyst development and kinetic modelling. *Chem. Eng. Res. Des.* **2018**, *136*, 385–402. [CrossRef]
100. Meloni, E.; Martino, M.; Palma, V. A short review on Ni based catalysts and related engineering issues for methane steam reforming. *Catalysts* **2020**, *10*, 352. [CrossRef]
101. Lehtoranta, K.; Murtonen, T.; Vesala, H.; Koponen, P.; Alanen, J.; Simonen, P.; Rönkkö, T.; Timonen, H.; Saarikoski, S.; Maunula, T. Natural gas engine emission reduction by catalysts. *Emiss. Control Sci. Technol.* **2017**, *3*, 142–152. [CrossRef]
102. Hutter, R.; De Libero, L.; Elbert, P.; Onder, C.H. Catalytic methane oxidation in the exhaust gas aftertreatment of a lean-burn natural gas engine. *Chem. Eng. J.* **2018**, *349*, 156–167. [CrossRef]
103. Silas, K.; Ghani, W.A.W.A.K.; Choong, T.S.; Rashid, U. Carbonaceous materials modified catalysts for simultaneous SO_2/NOx removal from flue gas: A review. *Catal. Rev.* **2019**, *61*, 134–161. [CrossRef]
104. Gholami, Z.; Luo, G.; Gholami, F.; Yang, F. Recent advances in selective catalytic reduction of NOx by carbon monoxide for flue gas cleaning process: A review. *Catal. Rev.* **2020**, *63*, 68–119. [CrossRef]
105. Sui, S.; Wang, X.; Zhou, X.; Su, Y.; Riffat, S.; Liu, C.-j. A comprehensive review of Pt electrocatalysts for the oxygen reduction reaction: Nanostructure, activity, mechanism and carbon support in PEM fuel cells. *J. Mater. Chem. A* **2017**, *5*, 1808–1825. [CrossRef]
106. Ioroi, T.; Siroma, Z.; Yamazaki, S.i.; Yasuda, K. Electrocatalysts for PEM fuel cells. *Adv. Energy Mater.* **2019**, *9*, 1801284. [CrossRef]

107. Yang, L.; Shui, J.; Du, L.; Shao, Y.; Liu, J.; Dai, L.; Hu, Z. Carbon-based metal-free ORR electrocatalysts for fuel cells: Past, present, and future. *Adv. Mater.* **2019**, *31*, 1804799. [CrossRef]
108. Ren, X.; Lv, Q.; Liu, L.; Liu, B.; Wang, Y.; Liu, A.; Wu, G. Current progress of Pt and Pt-based electrocatalysts used for fuel cells. *Sustain. Energy Fuels* **2020**, *4*, 15–30. [CrossRef]
109. Ercolano, G.; Cavaliere, S.; Roziere, J.; Jones, D.J. Recent developments in electrocatalyst design thrifting noble metals in fuel cells. *Curr. Opin. Electrochem.* **2018**, *9*, 271–277. [CrossRef]
110. Ma, R.; Lin, G.; Zhou, Y.; Liu, Q.; Zhang, T.; Shan, G.; Yang, M.; Wang, J. A review of oxygen reduction mechanisms for metal-free carbon-based electrocatalysts. *NPJ Comput. Mater.* **2019**, *5*, 1–15. [CrossRef]
111. Qiu, P.; Yang, X.; Zhu, T.; Sun, S.; Jia, L.; Li, J. Review on core-shell structured cathode for intermediate temperature solid oxide fuel cells. *Int. J. Hydrog. Energy* **2020**, *45*, 23160–23173. [CrossRef]
112. Abd Aziz, A.J.; Baharuddin, N.A.; Somalu, M.R.; Muchtar, A. Review of composite cathodes for intermediate-temperature solid oxide fuel cell applications. *Ceram. Int.* **2020**, *46*, 23314–23325. [CrossRef]
113. Ding, P.; Li, W.; Zhao, H.; Wu, C.; Zhao, L.; Dong, B.; Wang, S. Review on Ruddlesden-Popper perovskites as cathode for solid oxide fuel cells. *J. Phys. Mater.* **2021**, *4*, 022002. [CrossRef]
114. Sunarso, J.; Hashim, S.S.; Zhu, N.; Zhou, W. Perovskite oxides applications in high temperature oxygen separation, solid oxide fuel cell and membrane reactor: A review. *Prog. Energy Combust. Sci.* **2017**, *61*, 57–77. [CrossRef]
115. Niakolas, D.K.; Neofytidis, C.S.; Neophytides, S.G. Effect of au and/or Mo doping on the development of carbon and sulfur tolerant anodes for SOFCs—A short review. *Front. Environ. Sci.* **2017**, *5*, 78. [CrossRef]
116. Cheekatamarla, P.K.; Finnerty, C.M.; Du, Y.; Jiang, J.; Dong, J.; Dewald, P.; Robinson, C. Advanced tubular solid oxide fuel cells with high efficiency for internal reforming of hydrocarbon fuels. *J. Power Sources* **2009**, *188*, 521–526. [CrossRef]
117. Cheekatamarla, P. *Performance and Reliability Advancements in a Durable Low Temperature Tubular SOFC*; Atrex Energy Inc.: Walpole, MA, USA, 2019.
118. Lyu, Z.; Li, H.; Han, M. Electrochemical properties and thermal neutral state of solid oxide fuel cells with direct internal reforming of methane. *Int. J. Hydrog. Energy* **2019**, *44*, 12151–12162. [CrossRef]
119. Shabri, H.A.; Othman, M.H.D.; Mohamed, M.A.; Kurniawan, T.A.; Jamil, S.M. Recent progress in metal-ceramic anode of solid oxide fuel cell for direct hydrocarbon fuel utilization: A review. *Fuel Process. Technol.* **2021**, *212*, 106626. [CrossRef]
120. Fu, H.-C.; You, F.; Li, H.-R.; He, L.-N. CO_2 Capture and in situ Catalytic Transformation. *Front. Chem.* **2019**, *7*, 525. [CrossRef]
121. Song, C.; Liu, Q.; Ji, N.; Deng, S.; Zhao, J.; Li, Y.; Song, Y.; Li, H. Alternative pathways for efficient CO_2 capture by hybrid processes—A review. *Renew. Sustain. Energy Rev.* **2018**, *82*, 215–231. [CrossRef]

encyclopedia

Entry

Aircraft Icing Severity Evaluation

Sibo Li [1,*] and Roberto Paoli [1,2]

1 Department of Mechanical and Industrial Engineering, University of Illinois at Chicago, Chicago, IL 60607, USA; robpaoli@uic.edu
2 Computational Science Division and Leadership Computing Facility, Argonne National Laboratory, Lemont, IL 60439, USA
* Correspondence: sli218@uic.edu

Definition: Aircraft icing refers to the ice buildup on the surface of an aircraft flying in icing conditions. The ice accretion on the aircraft alters the original aerodynamic configuration and degrades the aerodynamic performances and may lead to unsafe flight conditions. Evaluating the flow structure, icing mechanism and consequences is of great importance to the development of an anti/deicing technique. Studies have shown computational fluid dynamics (CFD) and machine learning (ML) to be effective in predicting the ice shape and icing severity under different flight conditions. CFD solves a set of partial differential equations to obtain the air flow fields, water droplets trajectories and ice shape. ML is a branch of artificial intelligence and, based on the data, the self-improved computer algorithms can be effective in finding the nonlinear mapping relationship between the input flight conditions and the output aircraft icing severity features.

Keywords: aircraft icing; aircraft safety; computational fluid dynamics; OpenFOAM; machine learning; data-driven modeling

Citation: Li, S.; Paoli, R. Aircraft Icing Severity Evaluation. *Encyclopedia* **2022**, 2, 56–69. https://doi.org/10.3390/encyclopedia2010005

Academic Editors: Raffaele Barretta, Ramesh Agarwal, Krzysztof Kamil Żur and Giuseppe Ruta

Received: 27 November 2021
Accepted: 4 January 2022
Published: 6 January 2022

1. Introduction

Aircraft icing represents a serious hazard in aviation and has been the principal cause of several flight accidents in the past [1]. According to the International Civil Aviation Organization (ICAO), 42 plane accidents caused by icing are reported from 1986 to 1996, and 39% of them were fatal for at least one person [2]. When an aircraft encounters the supercooled water droplets that are naturally present in humid and cold atmosphere, a fraction of the supercooled droplets freezes upon the impact on the aircraft surface. The ice accretion on the wing's leading edge changes the original wing's shape and affects the aerodynamic performances. For example, the ice buildup on the wing decreases the maximum lift coefficient and increases the drag, which may cause instability and further lead to a crash [3]. Additionally, the ice accretion position is extremely important in evaluating the icing severity. For example, a small amount of ice at a key location might cause more severe performance degradation than a large amount of ice at a less important location. Therefore, evaluating the icing mechanism, ice shape and severity is of great importance to improving the flight safety.

Aircraft icing is an active research area and several approaches have been developed to investigate the ice accretion, including experimental study, numerical simulation and data-driven modeling. In terms of experimental study, NASA conducted a test flight and the testing data show that the effect of aircraft icing on the stability increases with the increasing angle of attack [4]. Papadakis et al. conducted experiments to study the effect of ice accretion on the aircraft aerodynamic performance and handling qualities at different icing times [5]. Wind tunnel tests have also been conducted to study the ice accretion process on aircraft, which provides valuable data of icing effects on aircraft stability [6].

Although experiments provide direct results and valuable information for icing mechanism investigation, carrying out the experimental study can be expensive and time-consuming; thus, with the building up of the theoretical icing models, more research has been focusing on the numerical simulation approach. To conduct the numerical simulation, the program that implements a mathematical model for the aircraft icing needs to be established. Then, the program can be run on a computer to obtain the icing results. Since the aircraft icing mathematical model is too complex to obtain the analytical solution, numerical simulation is essential to study the ice accretion process. For example, the LEWICE code [7,8], developed by the NASA Glenn Research Center, applied the Messinger icing model [9] to study the ice accretion for different flight conditions. FENSAP-ICE [10] implements a three-dimensional ice accretion solver which solves the Reynolds-Averaged Navier–Stokes (RANS) equation for airflow field and Messinger model for ice accretion. MULTI-ICE [11] achieves the functionality to compute ice accretion on multi-element airfoils. It applies a panel method for solving the aerodynamic field and Messinger model for icing computation. Cao et al. [12,13] established a numerical simulation method to predict the ice accretions based on the Eulerian two-phase flow theory. The permeable wall was proposed to simulate the droplet impingement on the iced surface effectively. Li et al. [14,15] developed the icing solver based on the OpenFOAM framework [16] to investigate the ice accretion process in a multi-shot manner; the icing solver is able to predict the ice shape as well as the effect of the ice accretion on the aerodynamic performance. Moreover, due to the highly modular structure of OpenFOAM, more features can be easily implemented into the solver. For example, the PoliMIce ice accretion modeling framework [17] was coupled with OpenFOAM to enable more accurate aerodynamics computation. Based on the computed airflow field, a generalized mass balance was introduced in PoliMIce to conserve the liquid fraction at the interface between the glaze and the rime ice types to achieve smooth transition between the two types of ice. In addition, surface roughness caused by ice accretion is also an important factor due to its effect on the heat transfer characteristics. For example, Fortin et al. [18] developed a thermodynamic model that combines mass and heat balance equations to the water states analytical representation to calculate the airfoil surface roughness caused by ice accretion. Han et al. [19] conducted experimental and analytical studies on airfoils roughened by natural ice accretion to improve the accuracy of current aircraft ice-accretion prediction tools. Recently, there have been studies focusing on predicting the flow field around the iced airfoil by using time-accurate methods such as detached eddy simulation (DES) [20]. Xiao et al. [21] improved the DES prediction of flow around airfoils with leading edge horn ice, which is important in studying the effect of ice on the aerodynamics.

In recent years, there has been growing interest in applying machine learning methods to aircraft icing research. It is motivated, on one hand, by the progress of artificial intelligence (AI) incorporating richer and/or more complex algorithms and, on the other hand, by the need of limiting the high computational cost of carrying out the numerical simulation [22]. AI is intelligence demonstrated by a computer program which has the ability to perform tasks associated with intelligence displayed by human beings. Machine learning (ML) is a branch of AI. Based on the training data, ML models are capable of addressing strong nonlinearity with the aid of constructing black-box input–output mapping [23]. Due to the complex interaction of multiple flight conditions, the mapping relationship between the input flight conditions and the output aircraft icing severity features is likely to be strongly nonlinear [24]; thus, ML has been implemented in several applications in aircraft icing to predict ice shape [25], icing area, maximum ice thickness, icing severity level [24,26] and the effect of ice on the aircraft aerodynamic performance [27]. The details will be given in Section 4. The accuracy of the ML models' predictions needs to be evaluated quantitatively by the error analysis method containing multiple statistical measures [23]. The trained ML models can make predictions based on any given flight conditions at a very fast pace. With reasonable accuracy, the built ML model has the potential to be an attractive alternative to the numerical simulation approach. Specifically, in aircraft icing, ML has a significant impact at three levels: for fast evaluating icing severity under different flight conditions, for estimating degradation of the aircraft aerodynamic performance by coupling with other computational fluid dynamics (CFD) codes and for increasing the flight safety by incorporating ice protection systems [26].

2. Aircraft Icing

2.1. Aircraft Icing Type

Based on the aerodynamic and meteorological factors, three types of ice can be generated in aircraft icing: rime ice, glaze ice and mixed ice [28]. The main characteristics and forming conditions for the three types of ice are summarized. It should be noted that the ice accretion depends on many factors and due to the complexity of icing mechanism, only the main features are discussed here. A detailed explanation can be found in the Aircraft Icing Handbook [29].

2.1.1. Rime Ice

When the supercooled water droplets become fully frozen immediately upon the impact on the aircraft surface, rime ice is formed [30]. It usually occurs in an environment of low flight speed and low temperature; the freezing is very fast and there exists no liquid water film. The shape of rime ice is relatively smooth and usually seen as a spear-like shape on the leading edge.

2.1.2. Glaze Ice

Glaze ice usually occurs in relatively warm temperatures and high flight speed. In such conditions, only a fraction of the supercooled water droplets become frozen upon the impact on the aircraft surface and the rest still remain in liquid state. The formed liquid film moves along the aircraft surface, which might be blown away by the aerodynamic forces or become frozen when its energy is deprived. Due to the movement of the liquid film, the shape of glaze ice is often characterized by the formation of one or two horns. Additionally, glaze ice often has a greater density and is usually tightly attached to the aircraft surface, and thus is more difficult to remove [3]. Mikkelsen et al. [31] demonstrated that due to the irregular shape, glaze ice might affect the aircraft performance far more seriously than rime or mixed ice.

2.1.3. Mixed Ice

During flight, it is possible to form different mixtures of the rime ice and glaze ice, both in time and space. The water droplets diameter and concentration vary widely in the atmosphere; in certain temperature range, the ice might characterize both the glaze ice and rime ice features. Additionally, rime ice may occur in the beginning of the icing process; however, as the icing process continues, the thickness of the ice layer increases and the heat loss due to conduction becomes weaker, which may lead to the generation of a liquid water layer. Therefore, glaze ice might be formed in the later icing stage.

2.2. Aircraft Icing Parameters

Ice accretion process is a complex interaction of aerodynamic and environmental variables, including flight speed, attack angle, exposure time, liquid water content (LWC), droplet diameter and environmental temperature.

The faster the flight speed, the greater the mass of water droplets that will impact on the aircraft surface, and hence the greater the amount of ice accretion. Additionally, the higher flight speed led to higher aerodynamic heat. Aerodynamic heat might cause the temperature increase near the stagnation point of the wing leading edge, which affects the type and shape of the formed ice. Indeed, based on the US military standard (MIL-A9482), when the flight speed is above 530 knots (sea level), the ice can be melted completely by the aerodynamic heat [28].

Changing the aircraft angle of attack affects the water droplets impingement location, collection efficiency distribution and the ice shape. Li et al. [14] developed a simulation framework to effectively study the ice accretion under a different angle of attack. It was found that when the wing is at a 4° angle of attack, the main impingement region is the lower surface of the wing. The same phenomenon is also observed in the ice height distribution, as shown in Figure 1 [14], with more ice accreted in the lower region. The black line represents the clean airfoil shape, the green dots represent the ice shape obtained in the experiment [32] and the blue and red solid lines represent the numerical predicted ice shape in Cao [12] and Li's [14] work, respectively.

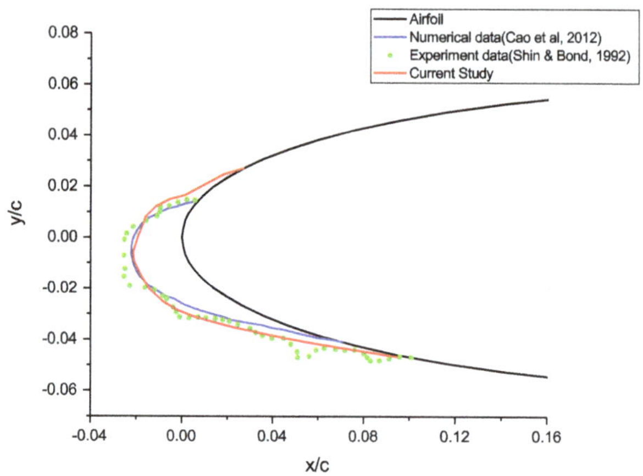

Figure 1. Visualization of the ice shape comparison on the NACA0012 airfoil [14].

Exposure time is the time that an aircraft spent in the icing condition. It directly affects the ice shape and the longer a flight stays in icing condition, the more severe the icing severity is. During flight, the exposure time is directly related to the cloud size [33].

Liquid water content (LWC) is normally expressed as the number of grams of liquid water per cubic meter of air. It represents the amount of supercooled water droplets that can impact on the aircraft surface in a given air mass. Therefore, LWC is an important factor in aircraft icing. As LWC increases, the amount of ice and ice thickness also increase, which cause more severe damage to the flight safety.

A water droplet's mass is directly proportional to the cube of the droplet diameter. Due to higher inertia, the droplets with higher diameter will be more likely to impact on the aircraft surface. On the other hand, smaller droplets tend to follow the air streamlines and avoid impacting on the surface. Therefore, the droplet diameter can affect the ice shape and ice layer thickness. The water droplet diameter is usually characterized as median volumetric diameter (MVD). According to the FAR-25 [33], the LWC and MVD is closely related; the relationship depends on the cloud type and environmental temperature. Therefore, at different temperatures, the distribution of LWC and MVD varies. The environmental temperature also affects the ice type. For example, it is more likely to form rime ice when the environmental temperature is low enough to make the entire water droplet to become frozen immediately upon the impact. On the other hand, glaze ice is more likely to be created when the water droplets only partially freeze [3].

2.3. Aircraft Icing Severity Levels

Defining the aircraft icing severity levels is important to give pilots a good idea how hazardous the icing is. The Aeronautical Information Manual [34] introduces four icing severity levels: trace, light, moderate and severe. However, since it simply serves as a reference for the pilots to give icing severity to the control tower, its classification is qualitative and vague. Later on, the NCCAM icing standard [35] defines the classification based on the accretion rate on a small probe (Table 1). Similarly, a icing severity level, as shown in Table 2, is established based on the maximum ice thickness [32]. Four levels are introduced to describe the icing severity, including light, moderate, heavy and severe. The pilots could use the standard as a reference to assess the severity of the flight condition [36]. It is reasonable to establish the standard based on the maximum ice thickness instead of the ice accretion rate because, during flight, the aircraft safety will only be a little affected if the time spent in severe icing state is limited.

Table 1. 1964 NCCAM icing standards [35].

Icing Severity Level	Accretion Rate on a Small Probe
Trace	1/2 inch in 80 miles
Light	1/2 inch in 40 miles
Moderate	1/2 inch in 20 miles
Heavy	1/2 inch in 10 miles

Table 2. Icing severity level based on icing thickness [37].

Icing Severity Level	Maximum Ice Thickness (mm)
Light	0.1–5.0
Moderate	5.1–15
Heavy	15.1–30
Severe	>30

3. Numerical Simulation for Aircraft Icing

Since the flight test and experimental simulations are expensive to carry out, numerical simulation is adopted widely. There have been numerous discussions about the numerical tools to predict the ice accretion under different flight conditions [7,13–15]. The computational method proposed by Li et al. [14] for aircraft icing includes four main steps:

1. Solve the air flow field around the aircraft.
2. Simulate the droplet impingement on the aircraft surface.
3. Solve the ice accretion model to compute the ice shape.
4. Apply mesh morphing algorithm to account for the shape change caused by ice accretion.

The four steps construct the conventional icing simulation framework. This numerical tool is able to predict the ice shape as well as the effect of ice accretion on the aerodynamics performance. The framework is presented in Figure 2.

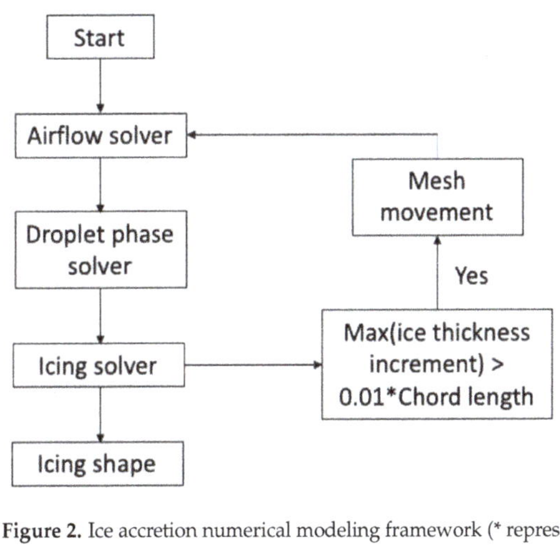

Figure 2. Ice accretion numerical modeling framework (* represents product of 0.01 and chord length).

3.1. Airflow Field

Solving the airflow field is the first step in the ice accretion numerical modeling framework. The airflow field will affect the movement of the supercooled water droplets via drag force. The airflow field can be obtained by solving the potential flow equation [38] or the Euler equations [39] or the Navier–Stokes equations [14]. For example, the potential flow equation is solved in LEWICE code [7]. Cao et al. [13] solve the Eulerian equations to obtain the airflow field. Li et al. [14] solve the compressible Navier–Stokes equations based on the OpenFOAM framework [16,40]. Generally speaking, the Navier–Stokes is more complex to solve; however, the airflow field obtained by solving Navier–Stokes is more accurate, especially in the icing simulation where complex flow behavior often occurs. By solving the airflow field, the air flow velocity, pressure and temperature distribution is obtained and passed to the droplet impingement simulation.

3.2. Droplet Impingement

In the droplet impingement simulation, the crucial quantity that needs to be determined is the water droplet collection efficiency which reflects how often the droplets impact on the aircraft surface. The calculation of droplet collection efficiency on the airfoil/wing surface is crucial in numerically simulating ice accretions. The most natural technique to track the droplet motion is to individually compute the trajectory of each droplet, which is referred as Lagrangian method [41]. The Lagrangian method is easy to implement; however, it has a severe drawback in icing simulations, which is the enormous computational cost required to model the droplets. Another computational method available is Eulerian two-phase flow method [14]. Eulerian two-phase model considers the droplets in the airflow as continuous field which interpenetrates with the air. The collection efficiency is obtained through solving the droplet velocity and droplet volume fraction. The computational grid used in the airflow field simulation can be used for the Eulerian two-phase model, which further improved the efficiency. Eulerian two-phase model has been applied in many studies to simulate the droplet impingement [12,14,15].

When applying the Eulerian two-phase model, different levels of interaction between air and droplets can be defined, as shown in Figure 3. The simplest model is one-way interaction in which only the airflow field affects the droplet phase. The most complex one is the four-way interaction where not only two-way interaction between air and droplets exist but the collision effect between droplets is considered as well. In many icing conditions, the droplet volume fraction is below 10^{-6}; thus, one-way interaction is accepted [42].

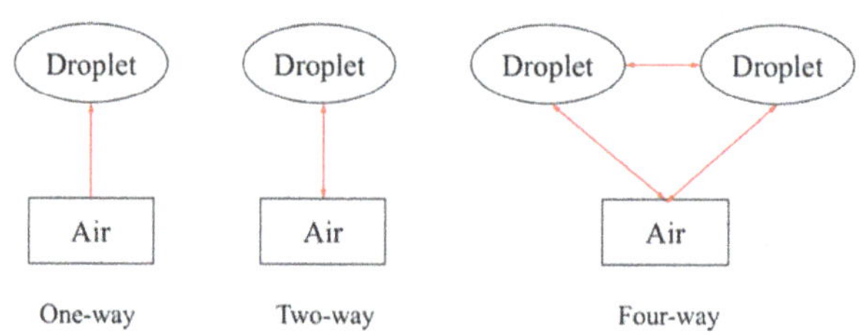

Figure 3. Different levels of interaction between air and droplet.

In constructing the governing equations for the Eulerian two-phase model, the following assumptions [14] need to be made.

1. The distribution of the water droplets is uniform, and they are simplified as sphere with a median volumetric diameter.
2. The physical parameters of the droplets do not change by assuming that there is no heat or mass transfer between the droplets and air.
3. The droplet collision, splashing and bouncing effects are neglected.
4. The airflow viscosity has no effect on the droplets.

3.3. Ice Accretion

Many numerical approaches [7,12–14] apply the Messinger model [9] to solve the ice accretion process. The Messinger model constructs the mass balance and energy balance equations in the control volume on the aircraft surface based on the following assumptions [14].

1. There is no runback water in the control volume at the stagnation point, and any runback water flowing out of the control volume flows along the direction away from the stagnation point.
2. The heat and mass transfer only happens in the direction normal to the wing's surface.
3. In the mixture of water and ice, a balance temperature is reached.

As shown in Figure 4 [14], the mass coming into the control volume includes impinging water droplets, $\dot{m}_{imp}$, and the water flow into the control volume from the upstream adjacent cell, $\dot{m}_{flowin}$. The mass going out of the control volume consists of the generated ice, $\dot{m}_{ice}$, the evaporation or sublimation, $\dot{m}_{es}$, and the water flow out of the control volume to the downstream adjacent cell, $\dot{m}_{flowout}$. The mass balance equation can be written as

$$\dot{m}_{imp} + \dot{m}_{flowin} = \dot{m}_{ice} + \dot{m}_{es} + \dot{m}_{flowout} \quad (1)$$

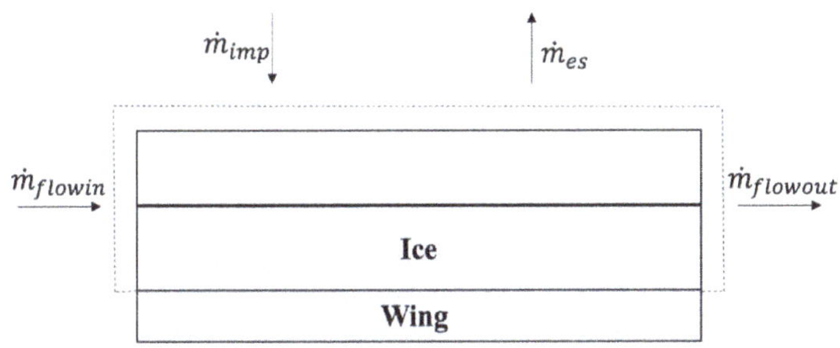

Figure 4. Mass balance in the control volume [14].

For the energy conservation, as shown in Figure 5 [14], the contributing energy terms are the convective heat, $\dot{Q}_{ca}$, kinetic energy of impinging water droplets, $\dot{Q}_{imp}$, latent heat, $\dot{Q}_{latent}$, and sensible heat, $\dot{Q}_{sensible}$. The energy balance equation can be written as

$$\dot{Q}_{ca} + \dot{Q}_{imp} + \dot{Q}_{latent} + \dot{Q}_{sensible} = 0 \quad (2)$$

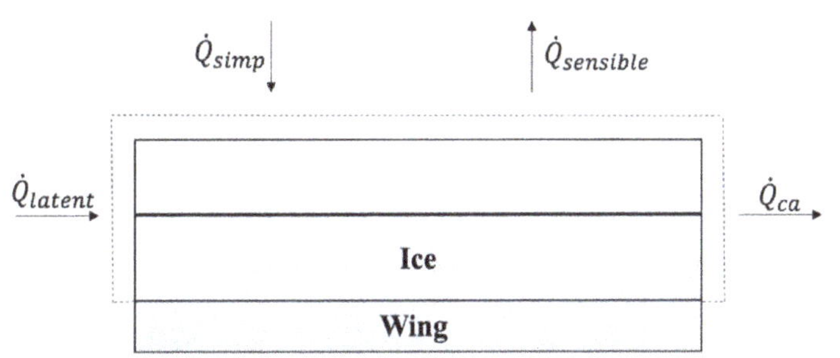

Figure 5. Energy balance in the control volume [14].

By solving the mass balance and energy balance equations, the ice layer thickness distribution can be obtained. Many studies [12–14] have shown that this approach can give good agreement with the experimental data [32].

3.4. Mesh Morphing

Ice accretion changes the wing's shape, which affects the previously solved airflow field and droplet impingement in step 1 and step 2 of the ice accretion numerical modeling framework. Therefore, it is necessary to re-calculate the airflow field and droplet flow field in order to obtain the new ice shape accurately based on the updated mesh. The irregular ice shape represents a major challenge in numerical simulation of long-time icing because manually remeshing is a time-consuming procedure. Kinzel et al. [43] use the mesh generation tool, AFLR3, to achieve the automated re-gridding procedure. Li et al. [14] build a mesh morphing algorithm to move the internal mesh nodes to achieve a smooth transition and maintain the mesh quality after constructing the ice shape.

4. Data-Driven Modeling for Aircraft Icing

This section details the different types of ML and then presents prominent research and findings of ML applications in aircraft icing. As described in the previous section, the aircraft icing process is a complex interaction of multiple variables and explicitly modeling the icing formation process often requires computationally expensive and/or cumbersome treatments to calculate the ice displacement and accretion along the wing, such as remeshing [22]. Data-driven methods can help alleviate this constraint by applying regression analysis and machine learning models to predict the aircraft icing based on icing data collected in experimental campaigns and/or numerical simulations [24]. Within the domain of aircraft icing, ML is applied in two main areas: ice shape prediction and icing severity evaluation.

4.1. Machine Learning

Since the ice accretion process shows strong nonlinearity, linear algorithms such as linear regression [44] and logistic regression [45] are not suitable [24]. Common nonlinear algorithms include the classification and regression trees (CART) [46], Naïve Bayes (NB) [47], k-Nearest Neighbors (KNN) [48] and Support Vector Machines (SVM) [49]. CART is referred to as decision tree (DT) algorithms that can be used for classification or regression predictive modeling problems. DT is a supervised learning method which can be used to make predictions in both regression and classification problems. DT constructs a binary tree from the training data, and the goal is to predict values by learning simple decision rules inferred from the data features. A tree can be seen as a piecewise constant approximation, and the split points are chosen greedily to minimize a cost function, such as Gini index [46]. The recursive binary splitting procedure needs to know when to stop splitting. The most common stopping criteria is the number of training instances assigned to each leaf node, whose value should be carefully tuned during the training process to avoid overfitting. Naïve Bayes calculates the probability of each class based on the Bayes theorem. It computes the conditional probability of each class given each input value based on the assumption of conditional independence between every pair of features given the value of the class variable. KNN is implemented through the instance bases learning with parameter k, which uses a majority voting mechanism [48] to make predictions in both regression and classification problems. One critical step in KNN is to determine which of the k instances in the training dataset are most similar to a new input. The common way is to use a distance measure, such as Euclidean distance and Hamming Distance. SVM seeks a line that best separates two classes. The optimal line will have the largest margin, which is the distance between the line and the closest data points. In practice, the SVM algorithm is implemented using a kernel, such as linear kernel and polynomial kernel, which defines the similarity or a distance measure between new data and the support vectors.

Besides the conventional ML models mentioned above, ensemble machine learning algorithms are also widely used in aircraft icing applications [24,26]. Ensemble ML methods create multiple models and then combine them to generate improved prediction results [50]. Li and Paoli [51] investigated the effectiveness of different conventional and ensemble ML methods on the icing applications. Ensemble methods usually have stronger prediction power and produce more accurate solutions than a single model would. Random Forest (RF) [52] is a type of ensemble method. RF constructs a set of decision trees during training, each individual tree gives a class prediction and the class that has the most votes becomes the model's prediction. XGBoost is another ensemble model, which shows excellent performance on structured or tabular datasets on classification and regression predictive modeling problems [53]. In XGBoost, models are added sequentially, and new models are added to correct the errors made by existing models. Additionally, to avoid overfitting, XGBoost adds the regularization factor to the loss, which represents the complexity of the trees. It has been shown that XGBoost has promising performance in exploring the complex pattern between different types of icing conditions [26].

4.2. Machine Learning for Ice Shape Prediction

In aircraft icing, one of the applications of ML is to predict the ice shape. Predicting ice shape by using numerical simulation approach generally requires solving the partial different equations, which requires high computational cost. Many studies have been carried out to develop ML data-driven models as an alternative to the traditional numerical simulation approach. Ogretim et al. [25] incorporate the Fourier series expansion of an ice shape following a conformal mapping, which suppresses the effect of airfoil geometry, and then utilize neural networks to model the Fourier coefficients and the downstream extent of the ice shape. A set of 20 Fourier terms is given to the network. The training data for the NNs were generated at the NASA Icing Research Tunnel at NASA Glenn, which was reported in the LEWICE validation report [54]. Neural networks (NNs) have been widely used in supervised learning; an input space is mapped onto an output space through the constructed hidden layers [55]. The small computational resource requirement and reasonable accuracy make this method a promising alternative to the numerical simulation approach. NNs can also be combined with wavelet packet transform (WRT) to predict the ice shape [56]. Chang et al. [56] selected five variables (velocity, temperature, liquid water content, median volumetric diameter and exposure time) as input data and WRT is applied to reduce the number of input vectors to increase the convergence efficiency. The number of training samples is 43. Then, a back propagation network that consists of 1 hidden layer with 39 nodes is established, and the output coefficients are reconstructed to generate ice shape. The neural network schematic diagram is shown in Figure 6. The hyperbolic tangent sigmoid transfer function is used in the hidden layer, and the linear function is used in the output layer.

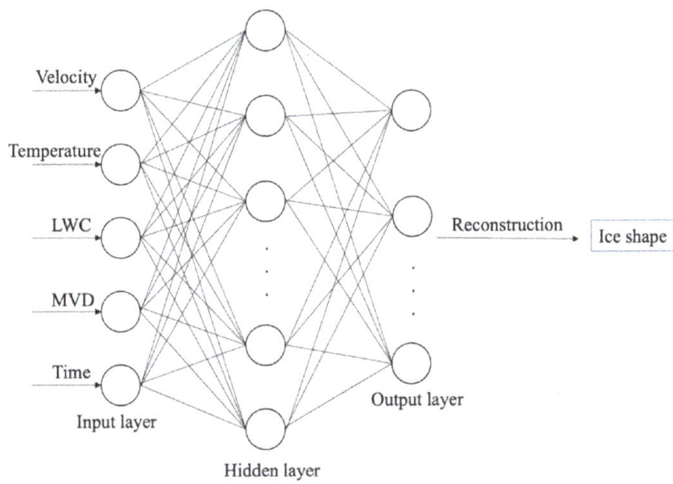

Figure 6. Schematic diagram of neural network structure [55].

4.3. *Machine Learning for Icing Severity Prediction*

In aircraft icing severity evaluation, the use of ML methods has been shown to aid in reducing the computational time, enabling aircraft safety improvement. Li et al. [24] proposed a method for aircraft icing severity prediction at different flight conditions based on machine learning model XGBoost. Based on the numerical modeling, six flight conditions (flight speed, angle of attack, exposure time, LWC, MVD and freestream temperature) are considered as input to the ML model. A total of 1890 samples are selected to form the dataset. The model is trained to predict three icing severity features: the size of the area on the airfoil covered by ice, maximum ice thickness and icing severity level (Table 2). During the training process, an important step is to find the optimal parameter settings for the ML models. A scikit-learn class called "GridSearchCV" [57] can be applied to identify the optimal hyperparameter set to improve the prediction accuracy of the ML models. For example, in predicting the icing severity level using XGBoost [24], the number of trees is 80, interaction depth is 10, shrinkage factor is 0.1, subsample ratio is 1 and minimum child weight is 0.1. In order to evaluate the models' performance, multiple statistical measures can be applied. For regression problems, the Root Mean Square Error (RMSE) [26], coefficient of determination R^2 [26] and Median Absolute Error (MAE) [24] can be applied.

$$RMSE = \sqrt{\frac{1}{N_{data}} \sum_{m=1}^{N_{data}} \left(Y_{m,predicted} - Y_{m,true}\right)^2} \tag{3}$$

$$R^2 = 1 - \frac{\sum_m \left(Y_{m,predicted} - Y_{m,true}\right)^2}{\sum_m \left(\overline{Y} - Y_{m,true}\right)^2} \tag{4}$$

$$MAE = \frac{1}{N_{data}} \sum_{m=1}^{N_{data}} \left|Y_{m,predicted} - Y_{m,true}\right| \tag{5}$$

where N_{data} is the number of data samples, $Y_{predicted}$ and Y_{true} represent the value predicted by the model and the value prepared in the dataset, respectively.

For classification problems, several model evaluation indicators, such as precision, recall rate, F1 score and confusion matrix, can be applied [26]. Figure 7 shows the comparison between the observed results and predicted results. The red line has an intercept of zero and a slope of one, which represents a perfect prediction. It can be observed that the sufficient agreement is achieved between the predicted and observed results in predicting the maximum ice thickness and icing area. The R^2 computed from Figure 6 is 0.995 for both cases. Due to the low computational cost requirement and skillful prediction, some potential uses for this prediction model include the following:

1. The proposed method has the potential to be coupled with other ice protection systems to further increase the flight safety, such as being an ice protection mechanism trigger.
2. The hybrid machine learning and CFD methods can be applied to the estimation of the degradation of the aircraft performance.

The built model can also output feature importance to indicate how valuable each flight condition is toward the three icing severity features. As an example, Figure 8 presents the feature importance of LWC, exposure time and droplet diameter in regard to the icing severity level. The F score value indicates how useful each feature is in the model building process. The higher the F score is, the more important the feature is. It can be seen that exposure time and droplet diameter have a comparable level of importance, while LWC has the lowest importance with regards to icing severity level.

Cao et al. [27] developed a methodology for predicting the effects of ice shape on airfoil aerodynamic performance based on a feed-forward neural network (NN). The model considers multiple ice geometry features, including ice horn leading-edge radius, ice height and ice horn position on airfoil surface. The model is trained to predict the lift coefficient, drag coefficient and moment coefficient, which are critical aerodynamic coefficients toward flight safety. Due to the sufficient agreement with the tunnel test data, the model can be further developed as a research tool to evaluate airfoil performance in different ice cloud conditions. McCann [37] built a pair of neural networks (NNICE) to recognize vertical atmospheric patterns associated to different icing intensities. A total of 398 samples are prepared for training the network. NNICE includes not only temperature and relative humidity at the flight level but also humidity data above and below flight level and a profile of the potential instability; therefore, it can make icing forecasts of all intensities with reasonable accuracy.

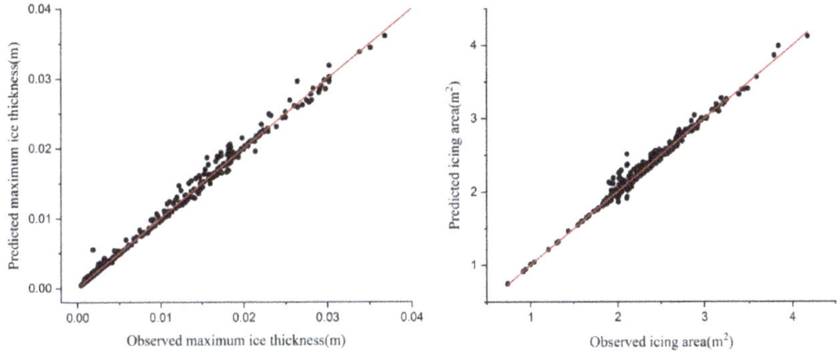

Figure 7. Scatter plot of observed results vs. predicted results. Left panel: maximum ice thickness; right panel: icing area [24].

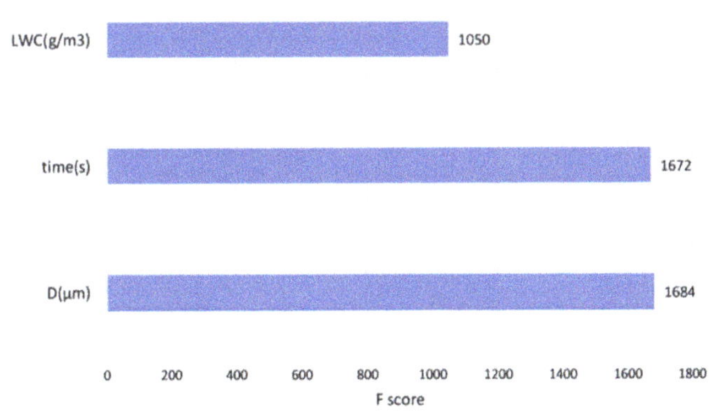

Figure 8. Bar chart of XGBoost feature importance in regard to icing severity level prediction [26].

5. Conclusions and Prospects

Aircraft icing has been studied by numerical modeling and machine learning methods. In research, numerical simulation has been shown to be effective in calculating the ice shape, analyzing the detailed flow fields and predicting the effect of ice shape on aerodynamics. The highly modular structure of the numerical simulation framework can easily incorporate different CFD and ice accretion models to provide more in-depth analysis. ML has been successfully used to accelerate the ice shape prediction process and enable fast evaluation of aircraft icing severity. The ability of ML models to generate the feature importance can help to study the effect of different aerodynamic and meteorological factors on the icing severity results. Due to the small computational resource requirement, fast performance and reasonable accuracy, the ML models can provide an attractive alternative to the traditional numerical simulation approach. Additionally, the ML models have the potential to be coupled with CFD codes and ice protection system to further increase the flight safety. Accurate data, such as those from high-fidelity CFD simulation, would help build useful training datasets for ML developments, especially for unsteady flows that occur in certain phases and/or specific wing configurations (flap deployment, retraction).

Author Contributions: Conceptualization, S.L. and R.P.; methodology, S.L. and R.P.; software, S.L.; validation, S.L.; formal analysis, S.L.; investigation, S.L. and R.P.; resources, R.P.; data curation, S.L.; writing—original draft preparation, S.L.; writing—review and editing, S.L. and R.P.; visualization, S.L.; supervision, R.P.; project administration, R.P.; funding acquisition, R.P. All authors have read and agreed to the published version of the manuscript.

Funding: This work was supported by Argonne National Laboratory through grant number ANL 0J-60008-0019A, titled "High-performance computing and physics-informed machine learning for multiscale flows", and by National Science Foundation through grant number 1854815, titled "High-Performance Computing and Data-Driven Modeling of Aircraft Contrails," awarded to R. Paoli.

Data Availability Statement: Data sharing not applicable.

Conflicts of Interest: The authors declare no conflict of interest.

Entry Link on the Encyclopedia Platform: https://encyclopedia.pub/19038.

References

1. Mclean, J.J. Determining the effects of weather in aircraft accident investigations. In Proceedings of the 24th Aerospace Sciences Meeting, Reno, NV, USA, 6–9 January 1986. [CrossRef]
2. Chang, L. Aircraft icing and aviation safety. *Aeronaut. Sci. Technol.* **2010**, *5*, 12–14.
3. Cao, Y.; Tan, W.; Wu, Z. Aircraft icing: An ongoing threat to aviation safety. *Aerosp. Sci. Technol.* **2018**, *75*, 353–385. [CrossRef]
4. Ratvasky, T.; Van Zante, J.; Riley, J. Thomas Ratvasky, "NASA/FAA Tailplane Icing Program overview". In Proceedings of the 37th Aerospace Sciences Meeting and Exhibit, Reno, NV, USA, 11–14 January 1999.
5. Papadakis, M.; Yeong, H.W.; Vargas, M.; Potapczuk, M. Aerodynamic Performance of a Swept Wing with Ice Accretions. In Proceedings of the 41st Aerospace Sciences Meeting and Exhibit, Reno, NV, USA, 6–9 January 2003. [CrossRef]
6. Ratvasky, T.P.; Ranaudo, R.J. Icing Effects on Aircraft Stability and Control Determined from Flight Data, Preliminary Results. In Proceedings of the 31st Aerospace Sciences Meeting, Reno, NV, USA, 11–14 January 1993; No. NASA TM 105977.
7. Ruff, G.; Berkowitz, B. *User's Manual for the NASA Lewis Ice Accretion Prediction Code (LEWICE)*; NASA: Washington, DC, USA, 1990.
8. Wright, W.B. *User Manual for the NASA Glenn Ice Accretion Code LEWICE, Ver. 2.2.2*; NASA/CR-2002-211793; NASA: Washington, DC, USA, 2002.
9. Messinger, B. Equilibrium temperature of an unheated icing surface as a functionof air speed. *J. Aeronaut. Sci.* **1953**, *20*, 29. [CrossRef]
10. Beaugendre, H.; Morency, F.; Habashi, W.G. FENSAP-ICE's three-dimensional inflightice accretion module: ICE3D. *J. Aircr.* **2003**, *40*, 239. [CrossRef]
11. Mingione, G.; Brandi, V. Ice Accretion Prediction on Multielement Airfoils. *J. Aircr.* **1998**, *35*, 240–246. [CrossRef]
12. Cao, Y.; Ma, C.; Zhang, Q.; Sheridan, J. Numerical simulation of ice accretions on an aircraft wing. *Aerosp. Sci. Technol.* **2011**, *23*, 296–304. [CrossRef]
13. Cao, Y.; Huang, J.; Yin, J. Numerical simulation of three-dimensional ice accretion on an aircraft wing. *Int. J. Heat Mass Transf.* **2016**, *92*, 34–54. [CrossRef]
14. Li, S.; Paoli, R. Modeling of Ice Accretion over Aircraft Wings Using a Compressible OpenFOAM Solver. *Int. J. Aerosp. Eng.* **2019**, *2019*, 4864927. [CrossRef]
15. Li, S.; Paoli, R. Numerical Study of Ice Accretion over Aircraft Wings Using Delayed Detached Eddy Simulation. *Bull. Am. Phys. Soc.* **2019**, *64*, Q23-009.
16. Weller, H.G.; Tabor, G.; Jasak, H.; Fureby, C. A tensorial approach to computational continuum mechanics using object-oriented techniques. *Comput. Phys.* **1998**, *12*, 620–631. [CrossRef]
17. Gori, G.; Zocca, M.; Garabelli, M.; Guardone, A.; Quaranta, G. PoliMIce: A simulation framework for three-dimensional ice accretion. *Appl. Math. Comput.* **2015**, *267*, 96–107. [CrossRef]
18. Fortin, G.; Laforte, J.-L.; Ilinca, A. Heat and mass transfer during ice accretion on aircraft wings with an improved roughness model. *Int. J. Therm. Sci.* **2006**, *45*, 595–606. [CrossRef]
19. Han, Y.; Palacios, J. Surface Roughness and Heat Transfer Improved Predictions for Aircraft Ice-Accretion Modeling. *AIAA J.* **2017**, *55*, 1318–1331. [CrossRef]

20. Spalart, P.R.; Deck, S.; Shur, M.L.; Squires, K.D.; Strelets, M.K.; Travin, A. A New Version of Detached-eddy Simulation, Resistant to Ambiguous Grid Densities. *Theor. Comput. Fluid Dyn.* **2006**, *20*, 181–195. [CrossRef]
21. Xiao, M.; Zhang, Y. Improved Prediction of Flow Around Airfoil Accreted with Horn or Ridge Ice. *AIAA J.* **2021**, *59*, 2318–2327. [CrossRef]
22. Li, S.; Paoli, R.; D'Mello, M. Scalability of OpenFOAM Density-Based Solver with Runge–Kutta Temporal Discretization Scheme. *Sci. Program.* **2020**, *2020*, 9083620. [CrossRef]
23. Moacir, R.F.; Ponti, A. *Machine Learning: A Practical Approach on the Statistical Learning Theory*; Springer: Cham, Switzerland, 2018.
24. Li, S.; Qin, J.; Paoli, R. Data-Driven Machine Learning Model for Aircraft Icing Severity Evaluation. *J. Aerosp. Inf. Syst.* **2021**, *18*, 876–880. [CrossRef]
25. Ogretim, E.; Huebsch, W.; Shinn, A. Aircraft Ice Accretion Prediction Based on Neural Networks. *J. Aircr.* **2006**, *43*, 233–240. [CrossRef]
26. Li, S.; Qin, J.; He, M.; Paoli, R. Fast Evaluation of Aircraft Icing Severity Using Machine Learning Based on XGBoost. *Aerospace* **2020**, *7*, 36. [CrossRef]
27. Cao, Y.; Yuan, K.; Li, G. Effects of ice geometry on airfoil performance using neural networks prediction. *Aircr. Eng. Aerosp. Technol.* **2011**, *83*, 266–274. [CrossRef]
28. Vukits, T. Overview and risk assessment of icing for transport category aircraft and components. In Proceedings of the 40th AIAA Aerospace Sciences Meeting & Exhibit, Reno, NV, USA, 14–17 January 2002. [CrossRef]
29. Heinrich, A.; Ross, R.; Zumwalt, G.; Provorse, J.; Padmanabhan, V. *Aircraft Icing Handbook*; Federal Aviation Administration: Washington, DC, USA, 1993; Volume 1–3.
30. Bragg, M.B. Rime Ice Accretion and Its Effect on Airfoil Performance. Ph.D. Thesis, Ohio State Univ., Columbus, OH, USA, September 1982. Available online: https://ntrs.nasa.gov/citations/19820016290 (accessed on 26 November 2021).
31. Mikkelsen, K.; Mcknight, R.; Ranaudo, R.; Perkins, J.P. Icing flight research—Aerodynamic effects of ice and ice shape documentation with stereo photography. *AIAA Pap.* **1985**, *85*, 468. [CrossRef]
32. Shin, J.W.; Bond, T.H. *Experimental and Computational Ice Shapes and Resulting Drag Increase for a NACA 0012 Airfoil*; National Aeronautics and Space Administration: Washington, DC, USA, 1992.
33. *F.A. Regulations, Part 25—Airworthiness Standards: Transport Category Air-planes*; Federal Aviation Administration: Washington, DC, USA, 2013.
34. Federal Aviation Administration. *Aeronautical Information Manual (AIM), Updated Annually*; Federal Aviation Administration: Washington, DC, USA, 2021.
35. Mitchell, L.V. Aircraft Icing—A New Look. *Aerosp. Saf.* **1964**, 9–11.
36. McCann, D.W. NNICE—A neural network aircraft icing algorithm. *Environ. Model. Softw.* **2005**, *20*, 1335–1342. [CrossRef]
37. Zhang, C. *Flight Meteorology*; Meteorological Press: Beijing, China, 2000.
38. Chanson, H. *Applied Hydrodynamics: An Introduction to Ideal and Real Fluid Flows*; CRC Press Taylor & Francis Group: Lei-den, The Netherlands, 2009.
39. Gibbon, J.D.; Moore, D.R.; Stuart, J.T. Exact, infinite energy, blow-up solutions of the three-dimensional Euler equa-tions. *Nonlinearity* **2003**, *16*, 1823–1831. [CrossRef]
40. Li, S.; Qiao, H. Development of a Fast Fluid Dynamics Model Based on PISO Algorithm for Simulating Indoor Airflow. In Proceedings of the ASME 2021 Heat Transfer Summer Conference, HT 2021, Virtual, Online, 16–18 June 2021. [CrossRef]
41. Hoffmann, L.D.; Bradley, G.L. *Calculus for Business, Economics, and the Social and Life Sciences*, 8th ed.; McGraw-Hill: New York City, NY, USA, 2004.
42. Elghobashi, S. On predicting particle-laden turbulent flows. *Flow Turbul. Combust.* **1994**, *52*, 309–329. [CrossRef]
43. Kinzel, M.P.; Noack, R.W.; Sarofeen, C.M.; Boger, D.; Kreeger, R.E. A CFD Approach for Predicting 3D Ice Accretion on Aircraft. In Proceedings of the SAE 2011 International Conference on Aircraft and Engine Icing and Ground Deicing, Chicago, IL, USA, 13–17 June 2011. [CrossRef]
44. Darlington, R.B.; Hayes, A.F. *"Multicategorical Regressors", Regression Analysis and Linear Models: Concepts, Applications, and Implementation*; Guilford Publications: New York, NY, USA, 2016.
45. Harrell, F.E. Ordinal Logistic Regression. In *Regression Modeling Strategies*; Springer: Cham, Switzerland, 2015; pp. 311–325.
46. Kamiński, B.; Jakubczyk, M.; Szufel, P. A framework for sensitivity analysis of decision trees. *Central Eur. J. Oper. Res.* **2017**, *26*, 135–159. [CrossRef] [PubMed]
47. Webb, G.I.; Boughton, J.R.; Wang, Z. Not So Naive Bayes: Aggregating One-Dependence Estimators. *Mach. Learn.* **2005**, *58*, 5–24. [CrossRef]
48. Altman, N.S. An Introduction to Kernel and Nearest-Neighbor Nonparametric Regression. *Am. Stat.* **1992**, *46*, 175. [CrossRef]
49. Cortes, C.; Vapnik, V. Support-vector networks. *Mach. Learn.* **1995**, *20*, 273–297. [CrossRef]
50. He, H.; Yang, Y.; Pan, Y. Machine learning for continuous liquid interface production: Printing speed modelling. *J. Manuf. Syst.* **2019**, *50*, 236–246. [CrossRef]
51. Li, S.; Paoli, R. Comparison of Machine Learning Models for Data-Driven Aircraft Icing Severity Evaluation. *J. Aerosp. Inf. Syst.* **2021**, *18*, 973–977. [CrossRef]
52. Breiman, L. Random Forests. *Mach. Learn.* **2001**, *45*, 5–32. [CrossRef]

53. Chen, T.; Guestrin, C. XGBoost: A Scalable Tree Boosting System. In Proceedings of the 22nd ACM SIGKDD International Conference on Knowledge Discovery and Data Mining, San Francisco, CA, USA, 13–17 August 2016; pp. 785–794.
54. Wright, W.B.; Rutkowski, A. *Validation Results for LEWICE 2.0*; NASA: Washington, DC, USA, 1999.
55. Sosnovik, I.; Oseledets, I. Neural networks for topology optimization. *Russ. J. Numer. Anal. Math. Model.* **2019**, *34*, 215–223. [CrossRef]
56. Chang, S.; Leng, M.; Wu, H.; Thompson, J. Aircraft ice accretion prediction using neural network and wavelet packet transform. *Aircr. Eng. Aerosp. Technol.* **2016**, *88*, 128–136. [CrossRef]
57. Pedregosa, F.; Varoquaux, G.; Gramfort, A.; Michel, V.; Thirion, B.; Grisel, O.; Blondel, M.; Prettenhofer, P.; Weiss, R.; Dubourg, V. Scikit-Learn: Machine Learning in Python. *J. Mach. Learn. Res.* **2011**, *12*, 2825–2830.

encyclopedia

Entry

Prefabricated Building Systems—Design and Construction

Tharaka Gunawardena * and Priyan Mendis

Faculty of Engineering and Information Technology, The University of Melbourne, Parkville, VIC 3010, Australia; pamendis@unimelb.edu.au
* Correspondence: tgu@unimelb.edu.au

Definition: Modern Methods of Construction with Offsite Manufacturing is an advancement from prefabricated technologies that existed for decades in the construction industry, and is a platform to integrate various disciplines into providing a more holistic solution. Due to the rapid speed of construction, reduced requirement of labour and minimised work on site, offsite manufacturing and prefabricated building systems are becoming more popular, and perhaps a necessity for the future of the global construction industry. The approach to the design and construction of prefab building systems demands a thorough understanding of their unique characteristics.

Keywords: offsite manufacturing; inter-modular connections; Design for Manufacturing and Assembly (DfMA); structural design; modularisation; modular construction; panelised construction; connection design—worked example; design for transportation; lifting and handling

Citation: Gunawardena, T.; Mendis, P. Prefabricated Building Systems—Design and Construction. *Encyclopedia* **2022**, 2, 70–95. https://doi.org/10.3390/encyclopedia2010006

Academic Editors: Raffaele Barretta, Ramesh Agarwal, Krzysztof Kamil Żur and Giuseppe Ruta

Received: 20 November 2021
Accepted: 28 December 2021
Published: 6 January 2022

1. Introduction

A prefabricated (prefab) building, by definition, is where an entire building or an assembly of its components is manufactured at an offsite facility and assembled onsite from self-sustained volumetric modules or separate panels. Prefabrication has existed in construction for many decades in various forms such as dry wall systems, structural insulated panels (SIP), prestressed beams, prefabricated roof trusses, prefabricated reinforcement cages, etc. [1–3]. Modern Methods of Construction (MMC) with Offsite Manufacturing (OSM) have arisen to integrate these various technologies into a more holistic and systematic solution, and most modern prefab manufacturers will cater for various architectural designs with prefab units (modules or panels) of creative geometries and innovative connection systems. Such prefab units are mass produced in factories with upskilled specialist workmanship, and at times with automation and robotics in an offsite manufacturing facility, transforming the traditional site-based and labour-intensive approach to construction. Prefab building units have been widely used for residential, commercial and public infrastructure, post-disaster structures and many other applications around the world. Particularly in Australia, the use of modular construction in public infrastructure is a highlight in applications such as railway stations, schools, hospitals, police stations, childcare facilities, etc. [4].

Offsite-manufactured prefabricated building systems are built in three main types of construction, as listed below:

1. Modular (volumetric) construction: manufacturing of fully self-contained units in an offsite facility to be transported to site to be assembled to form a complete structure.
2. Panelised construction: manufacturing of flat panel units in an offsite facility to be transported to site to be assembled to form a complete structure.
3. Hybrid prefab construction (semi-volumetric): combining both panelised and modular methods. Compact modular units (pods) are used for the highly serviced and more repeatable areas such as kitchens and bathrooms, with the remainder of the building constructed using panels or modules.

As shown in Figure 1, prefab structures can be built out of steel, timber (including engineered wood products), concrete or a combination of these. The two main benefits of prefab construction are speed of construction and reduced labour compared to traditional methods of construction. Some of the other benefits and features of prefab construction are as follows:

- The reduced need for labour, especially for skilled trades such as welding, enables the construction process to move away from being a labour-oriented operation to a more process-oriented manufacturing and assembly process. This allows construction projects to achieve reduced on-site congestion, waste and pollution.
- The prefab units, especially volumetric modules, can be removed from the main structure for future reuse or relocation. The reusability of prefab units contributes significantly to prefab buildings having a much lower life cycle energy [5].
- Enables construction to commence earlier than usual, as prefab panels or modules can be manufactured in the factory while onsite preparation and foundations works get underway concurrently.
- Prefab construction at present reduces construction time by at least 50% from a site-intensive building [6]. This ensures that the client of the project starts generating revenue much earlier than usual.
- The construction process is significantly less vulnerable to adverse weather, resulting in an even faster construction.

Figure 1. A steel modular unit with softwood wall frames (**left**) and a steel corner supported modular unit with a cold formed light-gauge steel wall frame (**right**), showing various materials used in prefabricated construction.

Panelised units are loadbearing in nature, and most of these are integrated with insulation, cladding and internal plasterboards. Structural insulated panels (SIP) are a popular prefabricated panel type in use, especially in countries like Australia. Structural systems of modular buildings are developed using two main types of modules according to their load transfer mechanisms [7], namely:

1. **Loadbearing modular structural systems**—The perimeter wall structure of the module carries adequate stiffness and continuous connectivity to transfer the gravity loads to the modules below. The roof beams and floor beams are often assisted by stud columns to add stiffness to the walls. This wall stiffness contributes to resistance against racking actions that are experienced during temporary loading scenarios such as moving and handling inside factories, loading and unloading to and from trucks during transportation and during onsite lifting and handling. This system is only feasible for low-rise applications due to the nature of loadbearing systems.
2. **Corner-supported modular structural systems**—Corner columns in the module take up the floor loads and transfer them to the columns below. This system is also capable of resisting horizontal loads such as earthquake and wind forces—as long as the connections are designed properly, with adequate stiffness, and complying to

design standards—thus ideal for multi-storey applications. In most current modular applications, a structural system such as this, with columns and structural connections, is made from steel.

As discussed above, most multi-storey modular buildings found around the world are assemblies of corner-supported modules that are laterally connected to a cast in situ concrete core. This in situ core effectively acts as the primary lateral load resisting element and, in many of these, the floors are poured with concrete after installing the modules. Although these methods are innovative and do save construction time initially, they do not characterise the structures as purely modular nor provide them with the previously mentioned benefits of modular construction. Gunawardena et al. [8] introduced a new concept of an advanced corner-supported structural system, where the shear core of the building could be formed by special modules with infill concrete walls. This would allow a multi-storey building to be constructed using corner-supported modules that could later be dismantled to construct other buildings elsewhere. Since then, similar concepts have realised into prefab structural systems in high-rise modular buildings such as the La Trobe tower in Melbourne, Australia [2].

2. Approach to Structural Design in a DfMA (Design for Manufacturing and Assembly) Format

As shown in Figure 2, the design stage of a traditional construction project would follow a path where initially, a conceptual structural design, also known as a schematic design, would be drafted. At this conceptual design phase, the design loads on the structure are determined; locations and arrangement of the structural elements are decided; and preliminary sizing of all structural elements are carried out. This conceptual design would produce structural GAs (general arrangements) that contain a preliminary arrangement and the dimensions of all structural elements with a construction detail of foundations (as foundations need to be constructed first). These schematic designs would be adequate to be produced for approvals from councils (or a similar regulatory body), and even to call for tenders from potential builders. The take-offs for bills of quantities (BOQs) are usually obtained from these schematic GAs using rules of thumb to estimate quantities where no detailed designs are provided (such as the quantities of steel reinforcement). Thereafter, a detailed structural design would be produced, with all the necessary details needed to fully construct the structure (for example, reinforcement details of concrete beams, columns, etc. and connection details of steel and timber structures).

The design process of a prefab building takes a somewhat different and an arguably improved approach (Figure 3). This is mainly due to the necessity of completing the project in a much shorter time and also because the structural design takes place centred around the particular prefab builder, while the specific design task may be subcontracted to an outside structural engineering firm. It is significant how the collaboration of both the builder and the structural engineer is critically important in achieving the design of a prefab building. An efficient structural design of a prefab building will only be achieved with continuous consultation with the builder, and particularly their in-house teams that look after construction logistics and procurement. Mainly, two different types of prefab projects can be identified as they reach the office of a prefab builder or a designer:

Type A—Projects that were conceptualised from the early stages to be built as a prefab building.

Type B—Projects that were initially conceptualised to be built with traditional construction methods (in situ) but are subsequently required to be built using prefab methods.

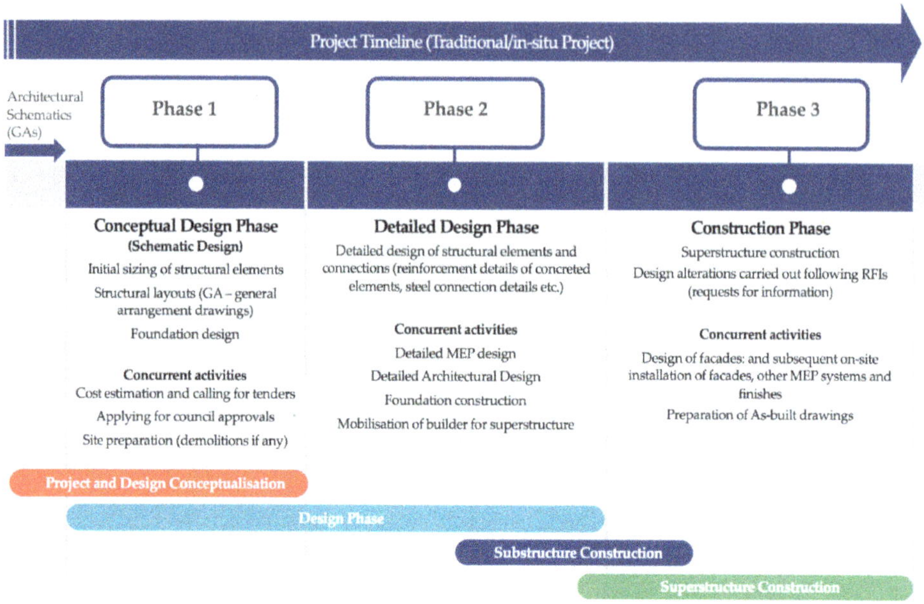

Figure 2. Phases in a traditional structural design process for an in-situ construction alongside other related activities that occur concurrently with each phase of the project.

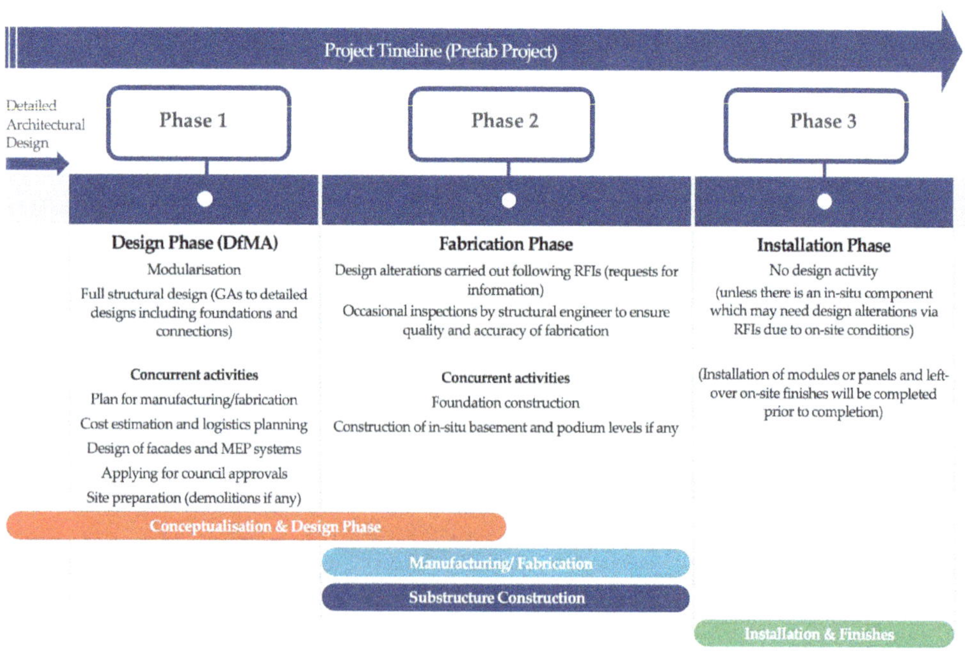

Figure 3. A typical DfMA (Design for Manufacturing and Assembly) process for a prefab construction alongside other related activities that occur concurrently with each phase of the DfMA process.

Out of these, a Type A project would be designed initially by an architect who is quite knowledgeable and experienced in prefab construction methods and would result in a rather smooth process thereafter for the structural designers and the prefab builders. On the contrary, converting a design that was initially conceptualised to be built as an in-situ building (Type B) to then be a prefab building is a rather tedious and at times a very complex activity.

In a nutshell, the design process of an offsite manufactured building has one single design phase (Figure 3), and it takes place centred around the prefab builder (contrary to a traditional design, which is centred around architectural and structural design firms). This design phase, when structured properly, takes the form of a Design for Manufacturing and Assembly (DfMA) format. In this DfMA setup, all relevant faculties of the design, i.e., architectural, structural, MEP and interior designs, are integrated with the necessary detailing and instructions for its in-factory manufacturing, transportation and on-site handling and assembly. A well-organised Building Information Modelling (BIM) framework would ideally integrate of all such specialities. The quantity surveying activity for cost estimation of a traditional construction project gets enriched into a much broader process involving logistics and procurement. Principles of 'Lean Manufacturing' have been adopted to incorporate logistics management of both in-factory and on-site work in recent research [9,10] and are gradually being adopted by prefab builders, especially the ones that have a degree of automation within their facilities. Concepts from blockchain and cryptocurrency have also been considered in recent research for developing smart procurement methods and smart contracts to add further efficiencies to prefab building processes [11–13].

2.1. Modularisation

Once an architectural concept or a developed design reaches a prefab builder's office, the first design activity that occurs is the modularisation of the given floor plans. For a rectangular or a similarly simple floor plan, it could be as straightforward as dividing up a given plan into segments, of which the sizes represent the viable modular or panel dimensions. However, every modularisation activity requires the builder to consider a number of key parameters prior to arriving at his viable modular or panel dimensions and their arrangement. Some of these key parameters are listed below:

1. **Architectural constraints**—The architectural space arrangements will dictate where possible modular splits can be drawn for a modular or flatpack design. Similarly, the locations and dimensions of architectural features such as windows, openings, facades, etc. will dictate where the panel boundaries can be located for a panelised design. A 'Type A' design, as discussed previously, will result in a fairly straightforward modularisation process, since an architect who has experience in prefab designs will have prepared architectural general arrangements (GAs) to allow for more efficient modular or panel splits. However, if prefabrication was not envisaged from the beginning (Type B), the splits would need to be carefully drawn around spaces and architectural features, considering how they might impact the rest of the manufacturing, transport, handling and installation processes.
2. **Transportation limitations**—The capability of the fleet of trucks that a particular prefab builder has access to will dictate the maximum length of a module or panel. While, for example, in Australia, the maximum allowable length for transportation is 30 m (Table 1), most trucks can carry modules and panels up to about 16 m in length. The lane widths of the relevant arterial road network will reflect upon the allowed maximum width for transportation. This may not be a great concern for panels, since they are usually transported vertically. However, for modules, this is one of the main design limitations. A 5 m maximum is allowed in Australia (Table 1), and similar limits apply in other countries as well. Depending on a particular project, other dimensional and weight limits could very possibly apply to the design. Issues such as the condition of access roads, nearby obstructions and other applicable regulations and constraints

(for example, heritage structures that cannot be demolished) can impose conditions on the dimensions of prefab units. Similarly, if the prefab units are to be shipped, the volumetric and weight constraints of the vessel will also apply, in addition to the transportation limitations that apply to the embarking and disembarking countries or regions.

3 **Lifting and handling limitations**—Prefab units need to be designed to dimensions and weights that are practically feasible when it comes to lifting and handling both within the manufacturing facility as well as at the onsite installation. If modules are to be shipped, the lifting and handling limitations at ports will also apply. Depending on the fleet of lifting vehicles such as mobile cranes that are accessible to the prefab builder, and the site access conditions of the particular project site (which dictates the lifting angles and distances), the weight of each module and panel will need to be limited. At least a preliminary understanding of the lifting and handling strategy of a particular project needs to be developed for an efficient modularisation.

4 **Onsite installation**—The sequence of installation and locations and access to structural connections are important parameters to be considered when dividing up a structure into modules or panels. One of the main intentions of a prefab structure is to reduce the necessity for skilled trades to be on site. Therefore, the onsite installation needs to reduce skilled activities such as welding to a minimum. The connections also need to be safely and conveniently accessible. The arrangement of prefab units needs to be carried out with these in mind.

5 **Manufacturing process**—The modularisation activity needs to consider the dimensions of various activity stations of the manufacturing plant and its sequence of operations. The prefab units need to move around the facility from one station to the other without the need to restructure any of their geometries and without hindering the normal operations of the factory.

6 **Preliminary locations of structural columns and connections**—A preliminary idea of the required structural design needs to be considered during the modularisation activity. The location of columns and their structural connections is a critical aspect of efficient modularisation. In contrast to a traditional structural design, the location of structural columns will not only relate to the eventual structure, but also to the structural integrity of a given prefab unit during its transportation, lifting and handling stages, since the lifting connectors should ideally be located on the structural columns.

7 **Preliminary locations and dimensions of service spaces (MEP)**—A preliminary idea of the required spaces for mechanical, electrical and plumbing (MEP) services such as service ducts and ceilings and heavily serviced areas such as toilets and kitchens need to be known at the modularisation stage. Modular and panel splits will need to be arranged accordingly, and any necessary alterations could be sorted via a well-formed BIM platform or by communicating with the architects and building services engineers.

Table 1. Dimension and weight limits to be followed in Victoria, Australia by vehicles transporting houses and prefabricated buildings. Adapted from ref. [14].

Dimension	Maximum Limit
Width	5.0 m
Height [1]	5.0 m
Length	30.0 m
Weight [2]	43 tonnes

[1] The total allowable height includes 1.2 m of trailer deck height (i.e., the height of modules or panels need to be less than 3.8 m). [2] The total allowable weight includes the weight of the steer axle concession, which varies according to the vehicle.

The abovementioned parameters in a modularisation activity will affect key decision variables that depend on these parameters. Cost and logistical items such as labour and equipment requirements related to the manufacturing process, lifting and handling, transportation and any other part of the project will depend on the outcomes of the modularisation activity.

2.2. Structural Design

As identified in the Australian design standards, and similarly stated in many international codes of practice and guidelines that use a 'limit state design' approach, any structural design should comply with three main design criteria, namely:

1 Stability;
2 Serviceability; and
3 Strength.

Structural design standards found elsewhere in the world follow the same principles but may identify these criteria under various different terms (for example, the term 'Ultimate limit' is used instead of 'strength design' in many standards). None of these would and should change in designing a prefabricated structure. While prefabrication and offsite manufacturing are the methods used to replace traditional construction, the end product is still a building used to serve the same traditional functions. Therefore, the authors, in principle, see no particular need to have separate modular or prefab design standards. However, the authors are quite supportive of modifying the already existing structural design standards to incorporate modular and other prefab construction methods. A good example for such an inclusion of prefab concepts into a traditional design standard is the latest version of the Australian concrete design code, AS3600:2018 [15], where new clauses were added to provide guidelines on minimum levels of strength and safety to prefabricated concrete structures, especially in the design of connections [16].

As previously discussed, the structural design of a prefab project is carried out in one single phase, although the specific tasks related to conceptual design and detailed design are still mostly distinct.

In terms of identifying the loads acting on a given prefab structure, as in any traditional structural design, the usage and importance level need to be determined according to its architectural design and the relevant codes of practice (for example, for structures built in Australia, AS1170.0:2002 [17] provides guidance on choosing importance levels according to their intended purpose and usage, with further guidance on load combinations and return periods (probability of exceedance); AS1170.1:2002 [18] provides guidance on minimum live loads; AS1170.2:2021 [19] provides guidance on minimum wind loads to consider and AS1170.4:2007 [20] provides guidance on minimum earthquake loads to consider).

The selection of most construction materials for facades and other finishes will be decided according to the architectural design. However, the structural design being centred around the prefab builder becomes a significant factor affecting the choice of material for the main structural elements. Depending on the normal practice of the particular prefab builder, the main structural form could be either concrete, timber, steel or a mix of these, and this extends to partition wall frames and the building envelope as well. The decision to go for cold-formed steel frames or softwood frames, or another form, such as SIP (structurally insulated panels), mostly depends on the normal practice of the prefab builder. Having prior knowledge of the finishing materials and their weights adds an extra advantage to the overall structural design, since there would be minimum uncertainty on how heavy some of the finishes and cladding would be (especially since there is minimal opportunity for these choices to be changed later, as the project timeline is very short). The exact values of their weights with a lower factor of safety can be used in determining some of the superimposed dead loads such as cladding, partitions and finishes.

The general arrangement of structural elements will take a different approach to a traditional conceptual structural design. The location of structural columns, especially for a modular building design, would already be decided during the modularisation activity.

The floor and roof elements would then be designed within the frame of each module. There could still be many structural components that need to be built in a traditional form, such as foundations, basements, shear walls and podiums, and they would follow the traditional format of design and would usually be built by separate contractors. A panelised building could have even more structural components that need to be built in situ. However, as far as the structural design of the panelised parts is concerned, the design approach would be similar to that of a modular building.

The abovementioned steps of a conceptual structural design of a prefab building are still principally similar to those of a traditional in situ building. However, there are a few further unique steps required for a prefab building design. These unique requirements arise as a result of the structural design being centred around the prefab builder, as discussed previously. Firstly, similar to the modularisation activity, the conceptual structural design needs to consider the strategy for the transportation, lifting and handling of the prefab units within the manufacturing facility and afterwards, following a DfMA format. DfMA consists of two components, namely, Design for Manufacturing (DfM) and Design for Assembly (DfA) [21–23]. The arrangement of structural elements and frames need to be compatible with the existing factory setup and manufacturing (fabrication) process of the prefab builder. Some activities, such as framing of steel or timber modular floor and ceiling frames, could be fully or partly automated within an offsite manufacturing facility. The structural design should be carried out in a way that assists the smooth functioning of the fabrication process to obtain optimum outputs with greater accuracy, while ensuring minimised waste generation, energy usage and emissions.

Similarly, certain structural elements would need to be strategically placed in a way to assist the planned lifting process. Most lifting connectors (lifting eyes) can be fitted into the structural columns of modules and panels using welded nuts. This is a much safer method of lifting heavy prefab units compared to lifting from beams and stub columns that are temporarily welded to the prefab units. Similarly, certain limitations and requirements are imposed on the structural design by the method of transportation. While each case would need to be studied individually, a common requirement for modular units would be to have a level floor bed as much as possible (without drops), to allow the modules to sit stably on the trailer bed of the transport vehicle.

The structural design concepts followed in 'serviceability design' and 'strength design' along with specific detailing requirements are discussed in the following sections.

3. Serviceability Design and Temporary Conditions

3.1. Deflections

As with any other structural design, serviceability deflections (long-term) of a modular structure would still need to follow the maximum deflection criteria of the relevant code of practice. For example, a steel-framed module built for a structure in Australia would need to follow the serviceability criteria as per AS 4100:2020 [24] where a span/250 is the maximum allowed deflection. In practice, for in situ buildings, many engineers keep a more stringent deflection limit of span/300 or span/400 to allow for uncertainties such as subsequent changes to the design, especially in its superimposed dead loads such as services, finishes, partitions and cladding.

However, a prefab structure is built in a significantly shorter time period, and the structural engineer can be more certain about many of the superimposed dead loads, as mentioned above, as they are fitted to the actual modules within a few weeks on many occasions. Therefore, due to the reduced level of uncertainty, the authors in practice have used span/250 (the minimum allowable limit as per the code of practice without any further factors of safety) for steel modules, with a maximum total deflection at any point of 25 mm, to prevent the cracking of any brittle finishes or fittings such as glass curtain walls or plasterboard partitions. A similar approach is recommended for establishing allowable deflections for prefabricated buildings built out of other materials.

Footfall vibration is another serviceability criteria that needs to be checked for prefab units. While the criteria for footfall vibrations is not different to that from a traditional building, the advantage of knowing the finishes with more certainty could result in more accurate predictions.

3.2. Design for Transportation, Lifting and Handling

The entire design process of a prefabricated building needs to be comprehensively aligned with the transportation, lifting and handling strategy of a given project and the relevant capabilities of the builder and its subcontractors. As far as the structural design is concerned, there are two main temporary conditions that the structure undergoes, which are:

- Temporary loads; and
- Temporary support conditions.

3.2.1. Temporary Loads

Temporary loads act on structures in any form of construction. In an in-situ construction, temporary loads are usually in the form of temporary live loads such as the weight of equipment, formwork and stored materials, in addition to the live load of construction workers. However, since on-site time is minimal in a prefab construction, temporary loads act on fully or partially finished modules or panels during their transport, lifting and handling. These loads that act on a structure during its transportation can be estimated according to the findings of some recent studies [25–27]. Although the stiffness of temporary bracing and the permanent structural frame should be checked against these loads, proper guidance on transportation-induced loading has not yet adequately appeared in the design standards of buildings. Much more groundwork in both research and practice needs to be carried out in achieving this. For the time being, it is prudent to overdesign the temporary bracings to encounter any unpredictable forces that the structure may undergo during all stages of its movement, since temporary bracings can be reused many times. As a result, as shown in Figure 4, temporary bracings and temporary columns are a common feature in modular construction at present.

Figure 4. A temporary cross bracing applied between two temporary columns in a prefabricated volumetric module (all to be removed after installation).

3.2.2. Temporary Support Conditions

Similar to temporary loads, temporary support conditions also exist in any form of construction. In an in-situ construction, temporary support conditions can be commonly seen in situations such as a temporarily unbraced column or lift core that is usually cast ahead of the rest of the building. The transportation, lifting and handling stages and even the installation stage of a prefab construction would similarly impose temporary support conditions on a prefab structure.

Figure 5 shows a situation where temporary cantilevers are imposed on a module while it is being transported due to its width being larger than the width of the truck bed. Such situations need to be foreseen by the structural engineer, especially since the details of the fleet of trucks accessible to the prefab builder can easily be known.

Figure 5. Temporary support conditions—An example of a situation where cantilevers which would not be there in the final state of a structure are imposed on a module during transportation.

Similarly, the lifting and handling of prefab units impose temporary loads and support conditions on them. **Top lifts** and **bottom lifts** are the two main methods of lifting modular and panelised units. A top lift is commonly arranged with adjustable spreaders and chains that lift a unit from the top. These top lifts should ideally be carried out by connecting the lifting mechanism to structural columns which are adequately stiff to carry the tensile force applied on them when the full weight of the prefab unit acts upon them. Bottom lifts are carried out by connecting the lifting mechanism to the bottom of the module or panel, usually arranged with a single spreader beam on top.

Unfortunately, certain unsafe practices are seen in lifting modules and panels, mainly due to the lack of adequate standards to provide guidance. Only precast concrete panels have well-established standards and guidelines, and these are not always directly applicable to modern modular or panellised units. Some contractors and even designers allow lifting points to be on roof beams or stub columns that are at times only temporarily welded to roof frames or floor frames. This is a practice that should be avoided, since temporarily welded elements are much less stiff than a full-length structural column, and can fail, especially at the welds, if one or two of the lifting chains are not in full tension. Such failures have occurred in practice, and they can be easily avoided by applying engineering common sense and proper preparation for the lift. The preparation should come from the structural design itself. The structural engineer must design and specify the complete lifting strategy and should not leave this to be a decision for the lifting contractor or the builder. Figure 6 shows an extract from a structural drawing that specifies the lifting strategy and how the lifting capacities are worked out using the capacities of the crane and the available lifting angles according to site conditions. Figure 7 shows an example of 'good practice' when it comes to lifting heavy modules from their full-length structural columns.

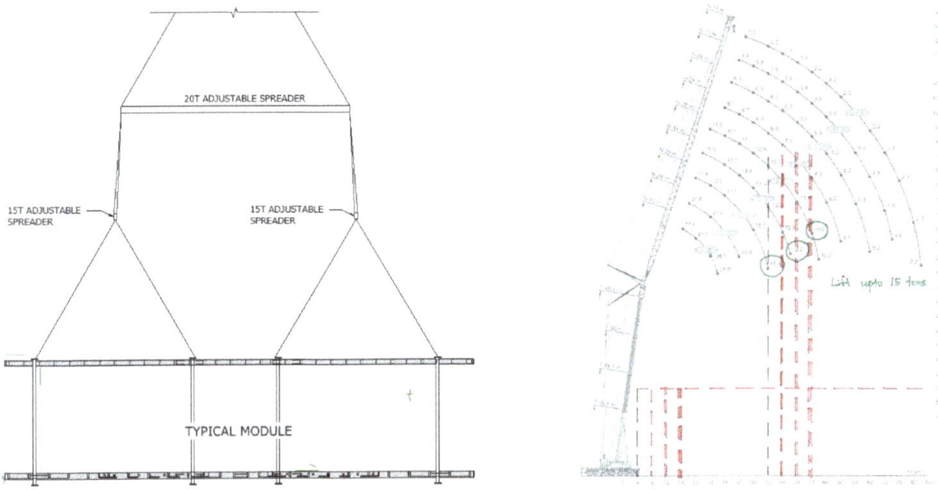

Figure 6. The lifting design of a prefabricated volumetric module where a top lift with an adjustable spreader is designed (**left**) and the lifting capacity for a mobile crane is worked out using lifting angles and distances (**right**).

Figure 7. A partly finished structural module being lifted into place using a top-lift method where the lifting mechanism is properly connected to full-length structural columns.

Lifting stability can be provided by designing for suitable lifting capacity from the mobile crane (or the tower crane). Most lifts end up being unstable and shaky due to the selection of cranes with capacities with an allowable lifting weight just above the actual weight of the prefab unit. It is observed in practice that a fairly stable lift can be achieved by selecting a mobile crane with a capacity that results in an allowable lifting weight which is around 2 to 3 times the actual weight of the prefab unit being lifted. The extra cost for hiring a larger capacity crane would be soon offset by achieving a faster installation due to having a more stable lift. It also helps, especially in lifting modules, if the structural design ensures an even weight distribution in the horizontal plane of the module (on floors and ceilings), so that the module can be lowered without being tilted to a side.

The handling of prefab units within the factory as well as onsite can impose loads and unforeseen support conditions. Depending on the capability of the prefab builder, various handling methods could be in place in an offsite manufacturing facility. Figure 8 shows a modular unit being handled with forklifts within a factory. Since manually operated forklifts cannot be fully synchronised, forces will be induced on the partially finished panels or modules when being handled in such a manner. More sophisticated facilities would at times have hydraulic mechanisms to handle prefab frames and they would have more damping to prevent any undue jerking forces being applied on the frames.

Figure 8. Forklifts handling partly finished modules within an offsite manufacturing facility.

4. Strength Design for Ultimate Limit State

The strength design for ultimate limit state in a prefab structure would not be very different to that of an in situ built traditional structure. Eventually, the demand imposed by various combined vertical and lateral loads on each structural column, beam, floor system, ceiling frame, wall, roof member and connection are to be checked against their respective structural capacities (in bending, compression, tension, shear, torsion, etc.). To formulate the strength design of structural systems for prefab buildings, it is essential to understand the load paths taken by each type of load from their origin to their eventual transfer down to the foundation. Not understanding load paths adequately will lead to very inefficient structural designs in prefab buildings.

4.1. Gravity Loads (Vertical Loads)

In any structural form, gravity loads always take the path from the thinnest member (least stiff) to the thickest member (most stiff) in the direction of gravity. Gravity-load-resisting structural systems have developed over the years for traditional in-situ built structures, and very little needs to change for prefabricated structures, although some prefab designers seem to have over-complicated this. It is most advantageous for a structural system to be repeatable to allow for a more efficient offsite fabrication process following DfMA principles, as previously discussed. While there may be many architectural customisations, the structural system could still be designed to be repeatable (typification). The key to such a design is a minimalistic and efficient gravity load resisting structural system. This can then integrate into the larger superstructure with any additions such as building services, to form a wholistic structure (unification).

Due to the nature of many steel modular and flatpack building units, one-way spanning floor systems commonly provide the most cost-effective and structurally efficient floor designs. Their repeatability and simplicity in terms of the arrangement of structural members and lower number of welded or bolted joints results in a more efficient fabrication process compared to a two-way spanning floor frame. The same applies to ceiling or roof frames, resulting in a holistic and efficient DfMA design. Figure 9 shows the load paths in a typical module within a corner-supported modular structural system. Once the load paths are properly designed, each structural member can be checked for their structural capacities according to the relevant structural codes of practice (for example, AS 4100:2020 [24] for the design of steel members of structures in Australia).

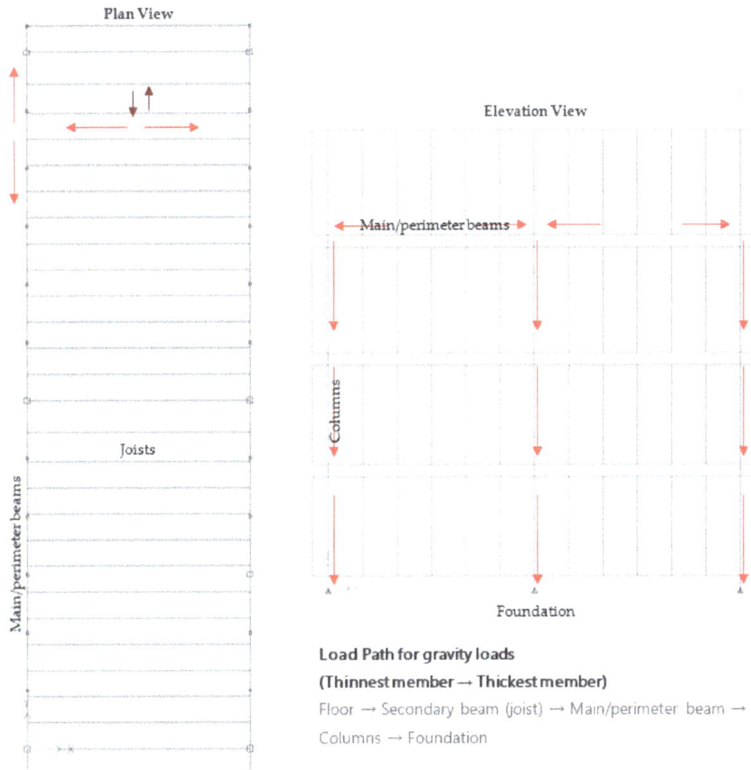

Figure 9. Load path of gravity loads in a corner-supported modular structure shown in plan and elevation views.

4.2. Lateral Loads (Wind and Earthquake Loads)

Wind loads first need to be estimated according to the relevant loading standard (for example, AS 1170.2:2021 [19] for structures in Australia) and applied to a global structural analysis model to observe overall effects such as top deflection, inter-storey drifts and base moments. The design criteria of strength, stability and serviceability for prefabricated buildings against wind loads apply as follows:

Stability—against overturning, uplift and/or sliding of the structure as a whole.

Strength—capacity of the structural members in bending, shear, torsion, etc. are required to withstand without failure against wind loads applied under long return period winds (typically, 500- or 1000-year return period, decided according to the importance level and design life).

Serviceability—where inter-storey and overall deflections are within acceptable limits. Wind accelerations also need to be checked for taller structures to ensure that the acceleration limits are within acceptable criteria for human perception of motion (typically, 25- or 50-year return period for deflections and 1- to 10-year return period for accelerations, decided according to the importance level and design life).

For a complex prefab structure, such as a tall building or a building that covers a large area, a load evaluation merely according to the loading standard may not be adequate. In such cases, a complete set of wind studies using a physical or virtual wind tunnel (using computational fluid dynamics) will be necessary to estimate the complete wind effects on the particular building [28,29]. The types of wind studies to consider are:

1 Environmental wind studies—to study the wind effects on the surrounding environment caused by erecting a tall prefab building or a building with complex geometry (this study may also include a pedestrian wind comfort study for ground level and podium levels or terraces).

2 Wind loads for the façade—to assess design wind pressures throughout the surface area of the building to design the cladding and façade system.

3 Wind loads for the structure—to determine the design wind loads to design the lateral load-resisting structural system.

The earthquake performance of prefab buildings and their connections have been studied in detail in many recent studies [2,7,30,31]. As explained in these studies, the stiffness of the connections and how and where they connect to the main structural elements is critical in how they perform in transferring earthquake loads. Earthquake forces impact a building at its base, and propagates to higher levels as an inertial force according to the mass and stiffness of each level (or each module). The connections would transfer these loads in the form of shear forces to main structural elements. The connection to the foundation is also critical, since the mass of the building (dead load) acting on the columns would be the largest at ground level. As a result, the largest component of the earthquake shear would be felt at the ground level. The Australian standard AS1170.4:2007 [20] provides comprehensive guidance in estimating both static and dynamic earthquake forces for each type of structure.

The code-based earthquake design is commonly known as a 'force-based' design approach. The force-based approach follows the commonly practiced limit state analysis principles, which are practiced through many design standards around the world. In addition, for the design of prefab buildings, a performance-based design approach may also be adopted. Performance-based design (also known as displacement-based design) focuses more on the performance aspects of the structure, such as its functionality and operability after an earthquake event, and is checked in relation to its target displacements or drifts under a given earthquake force. As one of the pioneering steps into performance-based seismic design (PBSD), the Structural Engineers Association of California (SEAOC) [32] proposed a framework (Figure 10) with target performance levels for the design and verification of new building constructions. Kappos and Panagopoulos [33] introduced more clarity into how PBSD could be employed through inelastic static and dynamic analyses, and check the design against the following target performance objectives:

- Two distinct performance objectives were 'serviceability' (damage limitation), which is typically associated with an earthquake probability of 50% in 50 years, and 'life safety', which is typically associated with an earthquake probability of 10% in 50 years, are explicitly considered. The basic strength level of the design of the structure does not relate to the 'life safety' criteria as in most existing codes, but to the serviceability criteria. A third performance objective in the form of 'collapse prevention', which is typically associated with an earthquake probability of 2% in 50 years is considered in the design against shear, and for the seismic detailing of elements.

- Inelasticity is also explicitly considered in the analysis, but is restricted only to those members that are considered as seismic energy dissipating zones (plastic hinge zones—see Figure 11) by the structural engineer. The well-established capacity design principles are followed in selecting these zones [34].
- The confinement detailing of critical structural members is also performance-based, i.e., it is determined according to the post-yield deformation requirements predicted from the inelastic analysis [35].

Earthquake Design Levels	Earthquake Performance Level				
		Fully Operational	Immediate Occupancy	Life Safety	Collapse Prevention
	Frequent (50 Years)	Basic Objective	Unacceptable	Unacceptable	Unacceptable
	Occasional (72 Years)	Essential/Hazardous Objective	Basic Objective	Unacceptable	Unacceptable
	Rare (475 Years)	Safety Critical Objective	Essential/Hazardous Objective	Basic Objective	Unacceptable
	Very Rare (2475 Years)	Uneconomical	Safety Critical Objective	Essential/Hazardous Objective	Basic Objective

Figure 10. Framework proposed by SEAOC for performance-based seismic design (PBSD) [7].

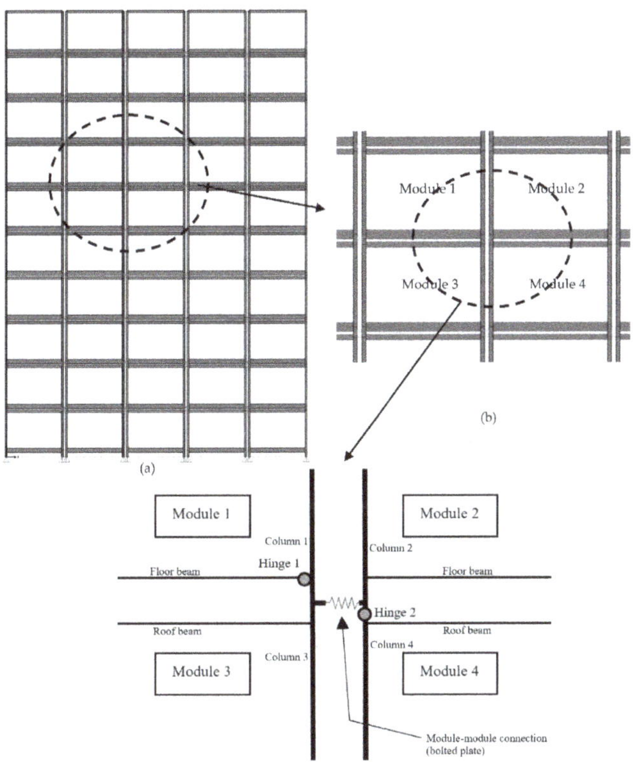

Figure 11. An illustration of the location of module–module connections in a multi-storey building with advanced corner-supported modular system as introduced by Gunawardena et al. [7,8] and possible hinge locations (**c**) zoomed-in from the front elevation, (**a**) and a close-up of the elevation view of neighbouring modules (**b**) connected by this module–module connection.

It is very important to check inter-storey drift against both wind and earthquake loads. Every modular floor in a corner-supported system has two different structural frames, as the roof or ceiling frame of the lower storey, and the floor frame of the upper storey. This makes the estimation of inter-storey drift uniquely different for corner-supported modular structural systems compared to any other structural form. For example, a traditional in situ building would have both these frames represented by a single slab or floor frame. Therefore, the inter-storey height can be defined as the height between two consecutive inter-modular connections, as shown in Figure 12, in estimating inter-storey drift.

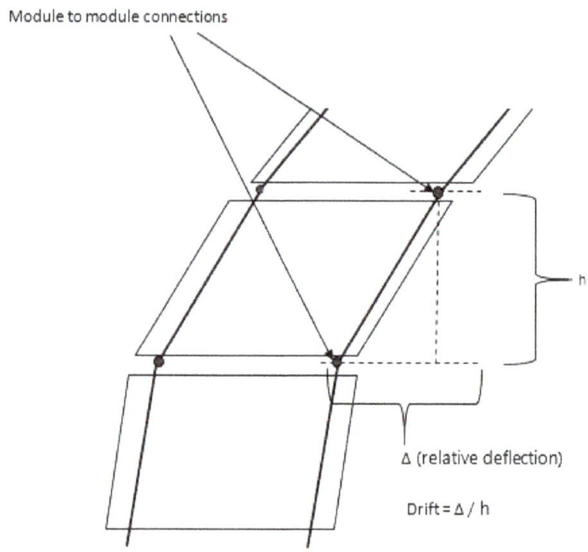

Figure 12. Illustration on how inter-storey drift needs to be considered for a corner-supported modular structural system against lateral loads.

5. Design of Connections (with a Worked Example)

Structural connections are the most uniquely different aspect of prefabricated structures compared to an in situ built structure. They are also the most critical structural component, which provides a prefabricated structure its buildability, speed of construction and overall effectiveness.

In the advanced corner-supported modular structural system introduced by Gunawardena et al. [7,8], the connections were designed in a way to connect neighbouring modules both laterally and vertically via the corner columns (Figure 12). The transfer of vertical and lateral loads is designed to occur via these bolted steel connections that connect two adjacent corner columns of two neighbouring modules. The details of this connection's design are shown in Figure 13. The conceptual make-up of the connection was intended to be as generic as possible, so that any engineer is at liberty to follow the findings and recommendations of this study to create their own unique connection design.

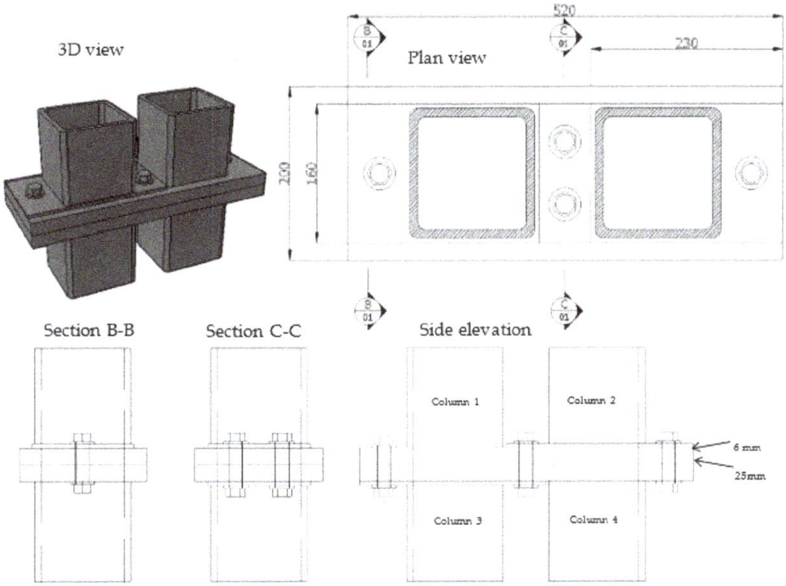

Figure 13. The inter-modular connection designed for the advanced corner-supported structural system [7,8]; the connection will connect four adjacent corner-supporting columns of four volumetric modules, as shown in Figure 12.

The strength and serviceability limit states for the structural design of this connection were determined following the Australian steel design standard, AS4100:2020 [24]. The steps to its complete design calculation are shown in Section 5.1. The supporting guidelines provided by Gorenc et al. [36] were also followed in the design calculations in addition to the requirements of AS 4100:2020 [24]. It was assumed that the column sizes were all 150 mm × 150 mm × 9 mm SHS (square hollow section) members; and the material properties of the other components of this proposed connection design are listed in Table 2. Both gravity and lateral loads are transferred via this connection, thus it needs to resist the bearing and shear actions that are generated as a result. The lateral forces extracted from a global structural analysis can be used to estimate the design actions on such an inter-modular connection or a connection of a panelised structure.

Table 2. Material properties of the components in the modular connection.

Dimension	Maximum Limit
Bottom and top plates of Columns 1 and 4	25 mm thick steel plates with a yield strength of 350 MPa
Bottom and top plates of Columns 2 and 3	6 mm thick steel plates with a yield strength of 350 MPa
Grade of bolts	Grade 8.8 with a yield strength of 660 MPa
Minimum tensile strength of grade 8.8 bolts	800 MPa (AJAX, 1999) [37]
Washers	4 mm thick stainless-steel (SS) washers

5.1. Serviceability Limit State Design of the Connection

The design calculations for the serviceability limit state of the inter-modular connection shown in Figure 13 are explained in this section. This connection needs to be treated as a 'slip critical' connection due its geometric position and orientation in the overall structure and the type of actions that it would undergo during an earthquake. The Australian standard AS 4100:2020 [24] addresses slip failure as a serviceability limit sate criterion and the slip resistance (V_{sf}) is calculated as follows:

$$V_{sf} = \mu n_{ei} N_{ti} k_b \tag{1}$$

where,

μ = Coefficient of friction between plies
n_{ei} = Number of shear planes
N_{ti} = Minimum pretension imparted on the bolts during installation
k_b = Factor for hole type (1.0 for standard holes, 0.85 for oversized holes and short slotted holes, and 0.70 for long slotted holes)

Following the AJAX Fastener Handbook [37], a value of 31.8 kN is considered for N_{ti} and the calculation for slip resistance is as follows:

$$V_{sf} = 0.2 \times 1 \times 31.8 \text{ kN} \times 1 = 6.36 \text{ kN} \tag{2}$$

Therefore, the slip would occur at a load of 6.36 kN for one bolt in the given connection. The total slip resistance of the entire inter-modular connection is 25.44 kN (6.36 × 4).

To be prudent, a low coefficient of friction (μ), such as 0.2 or 0.15, can be considered to anticipate the worst-case slip scenario. In reality, the connection would have protective coatings for durability and fire resistance applied on its components. Such coatings or paints could also result in low coefficients of friction in all relevant slip planes. As a conservative yet practical solution, a larger bolt size can be used to achieve a higher slip resistance, since it would require a higher bolt pre-tension.

5.2. Strength Limit State Design of the Connection

The design calculations for the strength (ultimate) limit state of the inter-modular connection shown in Figure 13 are explained in this section. As per AS4100:2020 [24], the strength limit state design of a bolted steel plate would need to be checked for its shear capacity, bearing capacity and tear-out failure capacity.

Considering no shear planes in the threaded region, the shear capacity (V_f) is calculated as follows:

$$V_f = 0.62 k_r f_{uf} n_x A_0 \tag{3}$$

where,

k_r = Reduction factor for length of bolt line (bears a value of 1.0 for connections other than lap connections)
f_{uf} = Minimum tensile strength of the bolt
n_x = Number of shear planes in the unthreaded region
A_0 = Bolt shank area

The calculation for one bolt is as follows:

$$V_f = 0.62 \times 1 \times 800 \frac{\text{N}}{\text{mm}^2} \times 1 \times 113.1 \text{ mm}^2 \times 10^{-3} = 56.1 \text{ kN}$$

Therefore, the shear capacity of all four bolts would be 224.4 kN (56.1 kN × 4). A further capacity reduction factor of 0.8 applies over this value for the design shear capacity of the connection as per AS 4100:2020 [24] resulting in a design shear capacity of 179.5 kN.

Check for bearing (V_b) is as follows:

$$V_b = 3.2 t_p d_f f_{up} \tag{4}$$

where,

t_p = Thickness of the ply
f_{up} = Minimum tensile strength of the ply
d_f = Bolt diameter

The calculation for one of the 6 mm plies for crushing under the pre-tension of one of the bolts is as follows:

$$V_b = 3.2 \times 6 \text{ mm} \times 12 \text{ mm} \times 450 \text{ N/mm}^2 = 103.7 \text{ kN}$$

Therefore, all bolt pre-tensions applied need to be less than 103.7 kN each.

Check for tear-out failure (V_p) is as follows:

$$V_p = a_e t_p f_{up} \tag{5}$$

where,

a_e = Minimum distance from the ply edge to the centre of the hole in the direction of the bearing load

The calculation for one of the 6 mm plies for tear-out near one of the bolts is as follows:

$$V_p = 35 \text{ mm} \times 6 \text{ mm} \times 450 \text{ N/mm}^2 = 94.5 \text{ kN}$$

Therefore, the tear-out capacity of the entire inter-modular connection would be 378 kN (94.5 kN × 4). This value is higher than the shear capacity of the connection. Therefore, it can be deduced here that checking for shear capacity would satisfy the other two failure criteria as well.

Following an experimental investigation on this inter-modular connection, Gunawardena [7] observed that subsequent to slip failure, the bolts transfer into a stage where they resist a combined shear and bearing action. Therefore, the critical design criterion is the check for combined shear and tension.

The combined capacity for shear and tensile bearing can be checked for the design values as follows:

$$\left(\frac{V_f^*}{\varnothing V_f}\right)^2 + \left(\frac{N_{tf}^*}{\varnothing N_{tf}}\right)^2 \leq 1 \tag{6}$$

where,

N_{tf} = nominal tension capacity of a bolt

N_{tf}^* = design tensile force on a bolt

$\varnothing$ = 0.8 for combined actions

Once the capacities of the connection design are calculated as above, they can be compared with the forces obtained from a global structural analysis. With a consideration on maintaining the slip resistance, it would be prudent to increase bolt size prior to increasing the number of bolts where the design capacities are not adequate.

6. Importance of Structural Detailing

Clear and precise structural detailing is essential in effectively communicating a structural design to the builder and other operators of a construction. This requirement is even more critical in prefab construction, as the onsite tolerances are quite small in high-precision assembly of prefabricated panelised and modular units. In a DfMA setup, precise detailing contributes significantly to ensuring the smooth-functioning of the manufacturing and assembly line of prefab units. Due to the high demand for producing prefab units in a short period of time, there is no time go back and forth with traditional RFIs (requests for information) within a prefab facility. Therefore, clear and precise detailing is key to an efficient and error-free prefab production line.

In addition to the abovementioned requirement of providing precision and adequate detail for the manufacturing stage, structural detailing becomes quite necessary for other stages of a prefab construction as well. Figure 14 shows how a structural floor design of a modular unit is detailed with strategically placed steel noggins to enable the tying down of the module to the trailer bed of the particular transport vehicle. This in turn implies the necessity to know the fleet of trucks at the builder's disposal during the design stage, as discussed previously. Figure 15a shows how the structural detailing allows for the easy installation of structural modules (this method is also applicable for panels) by leaving a post-installation strip in the façade or cladding. The external skin of a building needs to be watertight and connect seamlessly at all joints. Unless a special gasket connection is designed, for the façade joint, a watertight connection would result in a very time-consuming installation process. Therefore, to allow for a fast installation, most builders leave a post-installation strip for the façade or external cladding which could be installed after all the modules are fully installed. This detail needs to be provided in the design. Figure 15b shows how structural detailing needs to be provided for internal finishes such as plasterboard joints. It is prudent to leave a movement gap between brittle plasterboards at ceiling to wall and floor to wall joints, since they may produce cracks during transportation, lifting and handling if joined rigidly. A good practice solution to this would be to leave a gap at these joints to behave as a movement joint and cover them up with cornices (at ceilings) and skirtings (at floorings) afterwards.

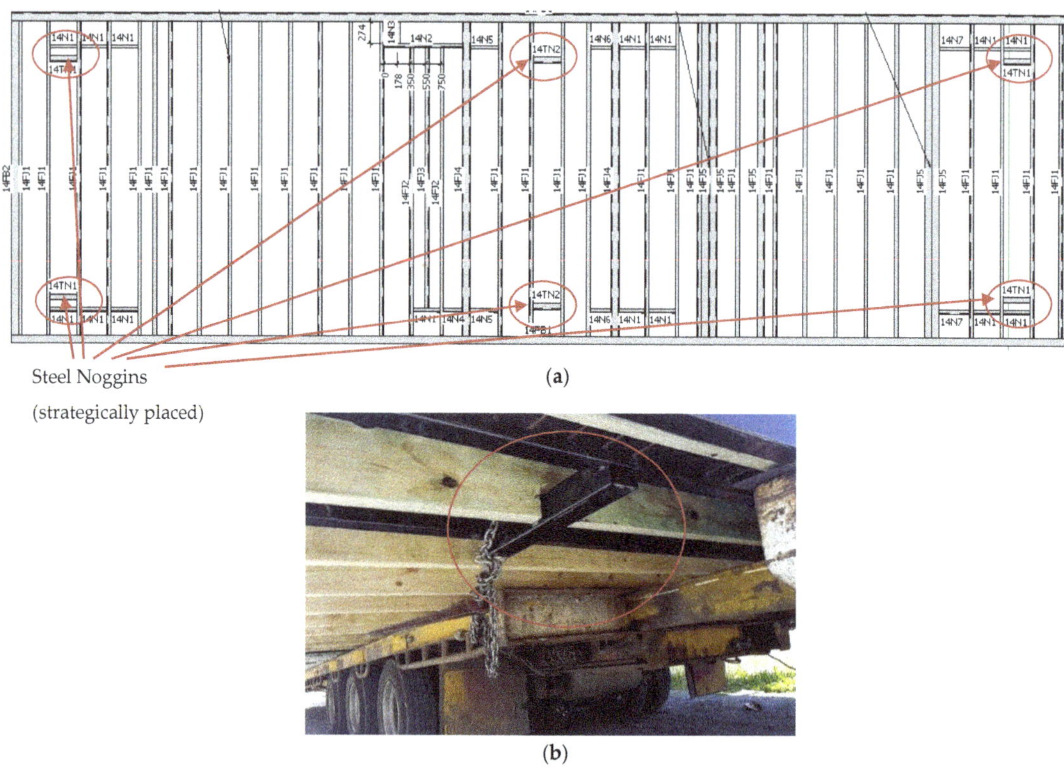

Figure 14. Detailing structural floor frames of volumetric modules to enable safe transportation where, (**a**) shows the placement of steel noggins on structural general arrangement drawings in between steel joists to provide elements to tie the module down to the trailer's bed and (**b**) shows how the module is tied down to the trailer's bed in reality.

(a)

(b)

Figure 15. Detailing of Finishes: (**a**) External finishes—A module split detailed so that a certain strip of the facade can be installed on site, allowing more work to be completed offsite; (**b**) Internal finishes—A gap left between the plasterboards of a ceiling and a partition wall (usually covered by a cornice) to act as a movement joint during transportation.

7. Construction Technology

7.1. Factory-Based Fabrication (Offsite Manufacturing)

Offsite manufacturing transforms many of the more site-intensive construction activities of traditional construction to in-house activities within a prefab manufacturing facility. This not only ensures a greater overall quality of the product, since it is under more organised supervision and monitoring, but also significantly improves the occupational health and safety of construction workers.

A great degree of automation is commonly seen in the manufacturing of panelised units, with automated assembly lines including a fair amount of robotics. However, such a level of automation is seldom seen in the manufacturing of modular units in most parts of the world. Regardless of the level of automation, the level of accuracy must be maintained at very high levels at all times to make sure that on-site installation can proceed without unnecessary delays due to manufacturing defects. Due to this reason, manufacturing tolerances must ideally be kept lower than the traditionally allowed construction tolerances. Design and practice standards and guidelines will need to change to take this into consideration.

In ensuring manufacturing accuracy, the movement within the assembly line or each stage of manufacturing needs to be planned carefully. Especially for modular units, it must be noted that a module needs to achieve its full rigidity before moving from its structural framing stage to a different stage. If a module is not rigid enough before being moved, the handling itself could alter its geometry and location of connections and cause issues during installation if the error goes undetected, and worse if the error accumulates.

7.2. Importance of Prototyping

Since there is no room for error once the prefab units reach the construction site, all measures must be taken to ensure accuracy while the units are still within the builder's manufacturing facility. Prototyping is an essential activity that provides an opportunity to a builder to inspect the installation process beforehand. All aspects related to buildability, such as access to connections, safety of installers, accuracy of installing sequence and adequacy of tolerances can all be checked during a well-organised prototyping activity. Figure 16 shows a full-scale prototyping of a double-storey modular building carried out at an offsite manufacturing facility. While this activity would point out all practical issues as previously explained, it also stands as a testament to the full reusability and relocatability of such prefab modules, since the modules would need be fully disassembled from their inter-modular connections before being transported to site to be assembled again.

Figure 16. Example of a full-scale prototype carried out within the manufacturing facility of a prefab builder for a double storey-steel modular building to check issues in buildability and installation sequence.

While most builders would prefer a full-scale prototyping of the real structure, the continuous increase in the demand for prefabricated buildings may not warrant this in the future. Rapid prototyping methods such as 3D printing provide the solution to carry out accurate, small-scale and quick prototype checks. Figure 17 shows a small-scale 3D printed rapid prototype of a façade extrusion against a full-scale prototype of an inter-modular connection. Both were prepared for similar reasons and provided similar inputs to the builder. However, the latter used less labour, time and cost in preparation. This highlights the advantage of using rapid prototyping methods for the naturally fast-paced prefab industry.

Figure 17. Examples of prototyping for prefabricated structures—A crude prototype with accurate full-scale dimensions of an inter-modular connection to check accessibility to the connection during installation (**left**) and a 3D printed rapid prototype of an extrusion frame of a façade to check buildability and fabrication issues (**right**).

7.3. Prefab Foundations

Most multi-storey modular and panelised structures are still largely built on top of in-situ built foundations and basements. Prefabrication is yet to develop to a level where underground structures, such as carparks, with retaining walls and proper waterproofing can be constructed with fully prefab units. However, there are prefab foundation methods that have been successfully used in low-rise prefab buildings, as shown in Figure 18, and some initial research work has been carried out on their performance [38]. These foundations are usually built with driven piles where the prefab structure transfers its weight to the piles via steel bearers (ground beams) and base plates.

Figure 18. A partially finished modular unit being installed on top of a prefab foundation (driven in piles with a base plate and steel bearers on top).

The technology for prefabricated foundations, especially with regard to deep foundations with basements for high-rise structures, stands out as one of the critical areas for future improvement demanding a considerable amount of research. This may need to be looked into together with prefabricating structural cores, since together, these activities form the backbone of the critical path in a typical high-rise construction. The 'advanced corner supported prefab structural system' proposed by Gunawardena et al. [7,8] could prove to be an initial step for such concepts.

7.4. Modular Building Services (MEP)

Similar to prefab foundations, technologies for prefabricating mechanical, electrical and plumbing (MEP) works, commonly known as building services, is also a less-explored area in prefab construction. Only a handful of recent research work has captured the need for modular MEP systems and introduced concepts to practically implement them [39–41]. However, the industry uptake of these concepts seems to be slow. There is a critical need to modularise MEP systems in prefab buildings, since the installation of building services currently takes up a considerable amount of time and remains a bottleneck in most prefab design and construction schedules. Modular MEP systems need to be adopted more frequently for them to be recognised and regulated by design standards and for them to develop further to fill a critical gap in the prefab industry. Design for building acoustics and thermal and fire performance could also be included in this category, and similar need for more detailed research as well as industry uptake exists in making these systems integrated more efficiently into the modular or prefab setup [42–44].

8. Conclusions and Prospects

Prefab technology has initiated a remarkable transformation in modern construction by reducing the amount of on-site activities and the time spent, with the help of efficient offsite manufacturing. The inclusion of Design for Manufacturing and Assembly (DfMA) is enabling a rapid evolution in the construction industry and design practice by creating better-performing prefab systems that reduce factory overheads and enable more automation. A combination of these technologies will realise projects even faster and generate better profit margins for investors. In addition to fast constructions, offsite manufacturing will also create a more quality-oriented practice and a safer working environment for upskilled building technicians.

The structural design of prefab systems and their connections still need to follow the basic principles of structural design and satisfy strength, serviceability and stability criteria. In a fundamental sense, prefab structural systems should not differ by any means from traditional structural systems in achieving the required performance. Designers need to provide the capability of easy installation on site, and easy accessibility to connections even after installation. The overall geometry, rigidity and weight of modules would need to satisfy lifting, transportability and assembly criteria to make any prefabricated system viable.

Where skilled labour is diminishing as a resource globally, it would be prudent to invest more in prefabrication technologies, supported by more detailed research. Further, prefab builders would also benefit from various economies of scale by having a factory-oriented process with minimum on-site costs. The disassembly and reusability aspects add another dimension into the value, by improving the sustainability and reducing the overall carbon footprint of a construction. Automation in manufacturing, with the use of robotics, and enhanced productivity, with the use of artificial intelligence, can further enhance the future value of offsite manufacturing.

Author Contributions: Conceptualisation, T.G. and P.M., methodology, T.G., formal analysis, T.G., investigation, T.G., resources, T.G. and P.M., writing—original draft preparation, T.G, writing-review and editing, T.G. and P.M. All authors have read and agreed to the published version of the manuscript.

Funding: This research was funded by the Australian Research Council Industrial Transformation Research Programme, IC150100023: ARC Training Centre for Advanced Manufacturing of Prefabricated Housing.

Acknowledgments: The authors acknowledge with gratitude the support given by industry partner Prebuilt Pty Ltd., Australia by providing access to their manufacturing facility and projects to which the authors have contributed.

Conflicts of Interest: The authors declare no conflict of interest.

Entry Link on the Encyclopedia Platform: https://encyclopedia.pub/19039.

References

1. Gunawardena, T.; Ngo, T.; Mendis, P.; Aye, L.; Crawford, R. Time-Efficient Post-Disaster Housing Reconstruction with Prefabricated Modular Structures. *Open House Int.* **2014**, *39*, 59–69. Available online: https://www.scopus.com/inward/record.uri?eid=2-s2.0-84929162675&partnerID=40&md5=b34c27beef2b3323975d693c4aa48046 (accessed on 17 December 2021). [CrossRef]
2. Gunawardena, T.; Ngo, T.; Mendis, P.; Kumar, S. Performance of multi-storey prefabricated modular buildings with infill concrete walls subjected to earthquake loads. *Concr. Aust.* **2017**, *43*, 51–58.
3. Navaratnam, S.; Ngo, T.; Gunawardena, T.; Henderson, D. Performance Review of Prefabricated Building Systems and Future Research in Australia. *Buildings* **2019**, *9*, 38. [CrossRef]
4. Gunawardena, T.; Mendis, P.; Ngo, T.; Rismanchi, B.; Aye, L. Effective Use of Offsite Manufacturing for Public Infrastructure Projects in Australia. Presented at the International Conference on Smart Infrastructure and Construction 2019 (ICSIC), Cambridge, UK, 8–10 July 2019; Available online: https://www.icevirtuallibrary.com/doi/abs/10.1680/icsic.64669.267 (accessed on 17 December 2021).

5. Aye, L.; Ngo, T.; Crawford, R.H.; Gammampila, R.; Mendis, P. Life cycle greenhouse gas emissions and energy analysis of prefabricated reusable building modules. *Energy Build.* **2012**, *47*, 159–168. [CrossRef]
6. Lawson, R.M.; Ogden, R.G.; Bergin, R. Application of Modular Construction in High-Rise Buildings. *J. Arch. Eng.* **2012**, *18*, 148–154. [CrossRef]
7. Gunawardena, T. Behaviour of Prefabricated Modular Buildings Subjected to Lateral Loads. 2016. Available online: http://hdl.handle.net/11343/123961 (accessed on 17 December 2021).
8. Gunawardena, T.; Ngo, T.; Mendis, P.; Alfano, J. Innovative Flexible Structural System Using Prefabricated Modules. *J. Arch. Eng.* **2016**, *22*, 05016003. [CrossRef]
9. Peiris, P.A.N.; Hui, F.K.P.; Ngo, T.; Duffield, C.; Garcia, M.G. A Case Study on Early Stage Adoption of Lean Practices in Prefabricated Construction Industry. In *ICSECM 2019*; Dissanayake, R., Mendis, P., Weerasekera, K., De Silva, S., Fernando, S., Eds.; Springer: Singapore, 2021; pp. 589–600.
10. Sepasgozar, S.; Hui, F.; Shirowzhan, S.; Foroozanfar, M.; Yang, L.; Aye, L. Lean Practices Using Building Information Modeling (BIM) and Digital Twinning for Sustainable Construction. *Sustainability* **2020**, *13*, 161. [CrossRef]
11. Hijazi, A.A.; Perera, S.; Calheiros, R.N.; Alashwal, A. Rationale for the Integration of BIM and Blockchain for the Construction Supply Chain Data Delivery: A Systematic Literature Review and Validation through Focus Group. *J. Constr. Eng. Manag.* **2021**, *147*, 03121005. [CrossRef]
12. Nanayakkara, S.; Perera, S.; Senaratne, S.; Weerasuriya, G.T.; Bandara, H.M.N.D. Blockchain and Smart Contracts: A Solution for Payment Issues in Construction Supply Chains. *Informatics* **2021**, *8*, 36. Available online: https://www.mdpi.com/2227-9709/8/2/36 (accessed on 17 December 2021). [CrossRef]
13. Rodrigo, M.N.N.; Perera, S.; Senaratne, S.; Jin, X. Potential Application of Blockchain Technology for Embodied Carbon Estimating in Construction Supply Chains. *Buildings* **2020**, *10*, 140. Available online: https://www.mdpi.com/2075-5309/10/8/140 (accessed on 17 December 2021). [CrossRef]
14. Transporting Houses and Prefabricated Buildings. 2018. Available online: https://www.vicroads.vic.gov.au/business-and-industry/heavy-vehicle-industry/heavy-vehicle-road-safety/transporting-houses-and-prefabricated-buildings (accessed on 17 December 2021).
15. Standards Australia. *AS 3600:2018 Concrete Structures*; Standards Australia: Sydney, NSW, Australia, 2018.
16. Jayasinghe, T.; Gunawardena, T.; Mendis, P. Advancements in Concrete Technology for Offsite Manufactured Buildings and Infrastructure Systems. In Proceedings of the CONCRETE 2021, Concrete Institute of Australia's Biennial National Conference, Perth, WA, Australia, 5–8 September 2021.
17. Standards Australia. *AS 1170.0:2002 Structural Design Actions General Principles*; Standards Australia: Sydney, NSW, Australia, 2002.
18. Standards Australia. *AS 1170.1:2002 Structural Design Actions Permanent, Imposed and Other Actions*; Standards Australia: Sydney, NSW, Australia, 2002.
19. Standards Australia. *AS 1170.2:2021 Structural Design Actions Wind Actions*; Standards Australia: Sydney, NSW, Australia, 2021.
20. Standards Australia. *AS 1170.4:2007 Structural Design Actions, Part 4: Earthquake Actions in Australia*; Standards Australia: Sydney, NSW, Australia, 2007.
21. Annamalai, K.; Naiju, C.; Karthik, S.; Prashanth, M.M. Early Cost Estimate of Product during Design Stage Using Design for Manufacturing and Assembly (DFMA) Principles. *Adv. Mater. Res.* **2012**, *622–623*, 540–544. Available online: https://search.ebscohost.com/login.aspx?direct=true&AuthType=sso&db=edo&AN=ejs29232997&site=eds-live&scope=site&custid=s2775460 (accessed on 17 December 2021). [CrossRef]
22. Jovicic, G.; Lukic, D.; Todic, V.; Milosevic, M.; Vukman, J. Design for manufacturing and assembly within design for excellence: Approaches, methods and methodologies. *Tehnika* **2014**, *69*, 233–242. [CrossRef]
23. Leaney, P.; Wittenberg, G. Design for Assembling. *Assem. Autom.* **1992**, *12*, 8–17. [CrossRef]
24. Standards Australia. *AS 4100:2020 Steel Structures*; Standards Australia: Sydney, NSW, Australia, 2020.
25. Godbole, S.; Lam, N.; Mafas, M.; Fernando, S.; Gad, E.; Hashemi, J. Dynamic loading on a prefabricated modular unit of a building during road transportation. *J. Build. Eng.* **2018**, *18*, 260–269. [CrossRef]
26. Godbole, S.; Lam, N.; Mafas, M.; Gad, E. Pounding of a modular building unit during road transportation. *J. Build. Eng.* **2020**, *36*, 102120. [CrossRef]
27. Godbole, S.; Lam, N.; Muhinadeen, M.M.M. Vehicle–road interaction analysis for pounding between cargo and trailer-bed. *Veh. Syst. Dyn.* **2019**, *59*, 547–567. [CrossRef]
28. Abu-Zidan, Y.; Mendis, P.; Gunawardena, T. Optimising the computational domain size in CFD simulations of tall buildings. *Heliyon* **2021**, *7*, e06723. [CrossRef] [PubMed]
29. Abu-Zidan, Y.; Mendis, P.; Gunawardena, T. Impact of atmospheric boundary layer inhomogeneity in CFD simulations of tall buildings. *Heliyon* **2020**, *6*, e04274. [CrossRef]
30. Gunawardena, T.; Ngo, T.; Mendis, P. Behaviour of Multi-Storey Prefabricated Modular Buildings under seismic loads. *Earthq. Struct.* **2016**, *11*, 1061–1076. [CrossRef]
31. Gunawardena, T.; Ngo, T.; Mendis, P.; Aye, L.; Alfano, J. Structural performance under lateral loads of innovative prefabricated modular structures. In *From Materials to Structures: Advancement through Innovation. In Proceedings of the 22nd Australasian Conference on the Mechanics of Structures and Materials, ACMSM 2012, Sydney, NSW, Australia, 11–14 December 2012*; Taylor and Francis: London, UK, 2013; pp. 717–722. Available online: http://www.scopus.com/inward/record.url?eid=2-s2.0-84881164688&partnerID=MN8TOARS (accessed on 17 December 2021).

32. Structural Engineers Association of California. *Recommended Lateral Force Requirements and Commentary*, 7th ed.; Structural Engineers Association of California: Sacramento, CA, USA, 1999.
33. Kappos, A.; Panagopoulos, G. Performance-Based Seismic Design of 3D R/C Buildings Using Inelastic Static and Dynamic Analysis Procedures. *ISET J. Earthq. Technol.* **2004**, *41*, 141–158.
34. Paulay, T.; Priestly, M.J.N. *Seismic Design of Reinforced Concrete and Masonry Buildings*; Wiley: New York, NY, USA, 1992.
35. Menegon, S.J.; Wilson, J.L.; Lam, N.T.K.; McBean, P. RC walls in Australia: Seismic design and detailing to AS 1170.4 and AS 3600. *Aust. J. Struct. Eng.* **2018**, *19*, 67–84. [CrossRef]
36. Gorenc, B.; Tinyou, R.; Syam, A. *Steel Designers' Handbook*, 8th ed.; USNW Press: Kensington, NSW, Australia, 2012.
37. AJAX. *Fastener Handbook: Bolt Products*; AJAX Fasteners: Braeside, VIC, Australia, 1999.
38. Teodosio, B.; Baduge, K.S.K.; Mendis, P.; Heath, D.J. Prefabrication of substructures for single-detached dwellings on reactive soils: A review of existing systems and design challenges. *Aust. J. Civ. Eng.* **2019**, *17*, 120–133. [CrossRef]
39. Samarasinghe, T.; Gunawardena, T.; Mendis, P.; Sofi, M.; Aye, L. Dependency Structure Matrix and Hierarchical Clustering based algorithm for optimum module identification in MEP systems. *Autom. Constr.* **2019**, *104*, 153–178. [CrossRef]
40. Samarasinghe, T.; Mendis, P.; Aye, L.; Gunawardena, D.; Karunaratne, R. BIM and Modular MEP Systems for Super-Tall and Mega-Tall Buildings. 2017. Available online: http://hdl.handle.net/11343/197525 (accessed on 17 December 2021).
41. Samarasinghe, T.; Mendis, P.; Aye, L.; Vassos, T. Applications of Design for Excellence in Prefabricated Building Services Systems. In Proceedings of the 7th International Conference on Sustainable Built Environment (ICSBE), Kandy, Sri Lanka, 16–18 December 2016.
42. Jayalath, A.; Navaratnam, S.; Gunawardena, T.; Mendis, P.; Aye, L. Airborne and impact sound performance of modern lightweight timber buildings in the Australian construction industry. *Case Stud. Constr. Mater.* **2021**, *15*, e00632. [CrossRef]
43. Caniato, M.; Bettarello, F.; Ferluga, A.; Marsich, L.; Schmid, C.; Fausti, P. Acoustic of lightweight timber buildings: A review. *Renew. Sustain. Energy Rev.* **2017**, *80*, 585–596. [CrossRef]
44. Caniato, M.; Bettarello, F.; Ferluga, A.; Marsich, L.; Schmid, C.; Fausti, P. Thermal and acoustic performance expectations on timber buildings. *Build. Acoust.* **2017**, *24*, 219–237. [CrossRef]

encyclopedia

Entry
Mechanics and Mathematics in Ancient Greece

Danilo Capecchi [1] and Giuseppe Ruta [1,2,*]

[1] Department of Structural and Geotechnical Engineering, University "La Sapienza", I-00184 Rome, Italy; danilo.capecchi@fondazione.uniroma1.it
[2] National Group for Mathematical Physics, I-00185 Rome, Italy
* Correspondence: giuseppe.ruta@uniroma1.it; Tel.: +39-(0)-644585085

Definition: This entry presents an overview on how mechanics in Greece was linked to geometry. In ancient Greece, mechanics was about lifting heavy bodies, and mathematics almost coincided with geometry. Mathematics interconnected with mechanics at least from the 5th century BCE and became dominant in the Hellenistic period. The contributions by thinkers such as Aristotle, Euclid, and Archytas on fundamental problems such as that of the lever are sketched. This entry can be the starting point for a deeper investigation on the connections of the two disciplines through the ages until our present day.

Keywords: mechanics; mathematics; natural philosophy; Aristotle; Euclid

Citation: Capecchi, D.; Ruta, G. Mechanics and Mathematics in Ancient Greece. *Encyclopedia* **2022**, 2, 140–150. https://doi.org/10.3390/encyclopedia2010010

Academic Editor: César M. A. Vasques

Received: 3 December 2021
Accepted: 11 January 2022
Published: 14 January 2022

1. Introduction

The link of mathematics with mechanics is ancient and complex enough to be given a thorough and neat explanation. The same terms 'mechanics' and 'mathematics' today have a very different meaning with respect to that in ancient times. In particular, herein, we focus on their meaning and relations in classical and Hellenistic Greece.

Actually, the ancient Greek term μεχανη ('mechane') indicated the discipline that dealt with equipments, in particular those for lifting weights. However, the meaning is not limited to this and in general also encompasses other inventions. As for mathematics, the question is more complex: as a matter of fact, it is not even clear how Pythagoras and Thales, for example, intended their discipline; moreover, there are no mathematical treatises before Plato (4th century BCE), even though during the last half of the 5th century BCE, there were meaningful testimonials of a handful of mathematicians intensely concerned with geometrical problems. This period is named the *Heroic Age of Mathematics* in [1] (p. 42.): its main characters were Archytas of Tarentum (b. c. 428 BCE), Hippasus of Metapontum (f. c. 480 BCE), Democritus of Abdera (b. c. 460 BCE), and Hippocrates of Chios (b. c. 470 BCE). In any case, the mathematicians of the 4th century that attended Plato's Academy also left no written epistemological considerations: Taethetus of Athen (c. 314-c. 369 BCE), Theodorus of Cyrene (b. c. 390 BCE), Eudoxus of Cnidus (d. c. 355 BCE), Menaechmus (f. c. 350 BCE), and Autolycus of Pitane (f. c. 330 BCE), see for instance [1,2].

On the other hand, Aristotle's position is clear, probably interpreting the thought of the important mathematicians of his time:

> [...] as the mathematician investigates abstractions (for before beginning his investigation he strips off all the sensible qualities, e.g., weight and lightness, hardness and its contrary, and also heat and cold and the other sensible contrarieties, and leaves only the quantitative and continuous, sometimes in one, sometimes in two, sometimes in three dimensions, and the attributes of these *qua* [i.e., inasmuch] quantitative and continuous, and does not consider them in any other respect, and examines the relative positions of some and the attributes of these, and the commensurabilities and incommensurabilities of others, and the ratios of others; but yet we say there is one and the same science of all these things—geometry), the same is true with regard to being [3] (11, 3. 27).

Here, mathematics seems to be identified with geometry, but in other passages of Aristotle's writings, the meaning of the term is broader and also applies to optics, astronomy, and music, which Aristotle considered subordinate to geometry (optics and astronomy) and arithmetic (music):

> Similar evidence is supplied by the more physical of the branches of mathematics, such as optics, harmonics, and astronomy. These are to some extent the converse of geometry. While geometry investigates physical lines but not *qua* physical, optics investigates mathematical lines, but *qua* physical, not *qua* mathematical [4] (II, 194a, 7 ff.).

In another point of Aristotle's writings mechanics as well appears as a subordinate science, thus another form of mathematics:

> The same account may be given of harmonics and optics; for neither considers its objects *qua* sight or *qua* voice, but *qua* lines and numbers; but the latter are attributes proper to the former. And mechanics too proceeds in the same way [3] (M, 3, 1078a).

It is not known why Aristotle exposed this thesis, as the mathematical treatment of mechanics was for sure more recent to him than that of the other subordinate sciences. It is probable that Archytas of Tarentum was the first to introduce a geometric study of mechanics, perhaps limited to the lever. This is what Diogenes Laertius (c. 200 CE) wrote on the subject much time later:

> He was the first to bring mechanics to a system by applying mathematical principles; he also first employed mechanical motion in a geometrical construction, namely, when he tried, by means of a section of a half-cylinder, to find two mean proportionals in order to duplicate the cube. In geometry, too, he was the first to discover the cube, as Plato says in the *Republic* [5] (volume 2, book 8, 83, pp. 395–396).

It is unlikely that Aristotle was referring to his own studies as the *Mechanica problemata* (see below): indeed, it is true that this treatise employs geometry, but it is also of dubious attribution and of uncertain dating.

2. Aristotle's *Mechanica Problemata*

The first mathematical treatise on mechanics known today is precisely the *Mechanica Problemata* [6]. It has been proved that the Arab world knew the treatise or at least a part of it [7]. During the Middle Ages and the Renaissance, the attribution to Aristotle was substantially undisputed; an investigation on the Renaissance is in [8], while for more recent periods see [9]. It is worth noticing that Fritz Kraft considers the *Mechanica Problemata* to be an early work by Aristotle, when he had not yet fully developed his physical concepts [10], while a recent paper by Thomas Nelson Winter considers Archytas of Tarentum as the author of the *Mechanica Problemata* [11]; some more investigations are in [12]. This last attribution is suggestive, though not very convincing. A possibility, not exploited indeed in the literature, is that the *Mechanica Problemata* is based on a previous text written by Archytas of Tarentum and subsequently elaborated in an Aristotelian environment. Here, we do not enter into the merit of this attribution, and for the sake of simplicity, we consider the *Mechanica Problemata* as an Aristotelian work instead of a pseudo-Aristotelian one, as it is frequently seen. For the same reasons of simplicity, we do not comment the great relevance assumed by the Aristotelian treatise in the Renaissance and the role it had in the history of mechanics; for this, reference can be made to [9,13].

The treatise can be divided into two parts: one theoretical and one applicative. In the theoretical initial part, the law of the lever is proved, and geometry openly insinuates into Aristotle's physics, leading to a proof based on intuition and analogy more than on a rigorous argumentation, as was customary among the mathematicians of the time. Starting from the law of the lever, the second part presents the qualitative and quantitative behavior of simple machines: the screw, the pulley, the wedge, and the winch (according to how they have been called subsequently in history).

In the following section, we go into some detail in the demonstration of the law of the lever, trying to understand its weaknesses and strengths without being influenced by modern notions. However, before commenting the treatise, we consider noteworthy to reproduce its *incipit* in the form we know:

> One marvels at things that happen according to nature, to the extent the cause is unknown, and at things happening contrary to nature, done through art for the advantage of humanity. Nature, so far as our benefit is concerned, often works just the opposite to it. For nature always has the same bent, simple, while use gets complex. So whenever it is necessary to do something counter to nature, it presents perplexity on account of the difficulty, and art [τεχνη, *techne*] is required. We call that part of art solving such perplexity a *mechane* [11] (p. 1).

This *incipit* is quite 'exotic' for a modern mathematician, and we suppose also for mathematicians contemporary to Aristotle, who is much more sober in his philosophical writings. Indeed, Aristotle's basic idea on moving bodies is:

> Now if of two objects moving under the influence of the same force one suffers more interference, and the other less; it is reasonable to suppose that the one suffering the greater interference should move more slowly than that suffering less [6] (p. 341).

In this argument, the two terms *force* and *interference* have a vague meaning, which is fine only for a natural philosopher wishing to provide qualitative explanations but not for a mathematician. The locution *more slowly* refers to a motion along the arc of circle and means that in a given finite time the arc of circle passed is shorter. To make this argument more precise and susceptible of a quantitative treatment, Aristotle, or whoever for him, introduced geometry to explain the meaning of the term 'interference'. Note that this is the first written testimonial in which mathematics was used in mechanics.

The proof is based on Figure 1, where we report two representations of Aristotle's argument, since we do not have original drawings but only graphical interpretations provided by translators and commentators, see for instance [14]. Figure 1a is the more interesting one for us, as commented hereinafter; Figure 1b is the one more commonly found in the Renaissance editions of the *Mechanica Problemata*. Aristotle assumes a clockwise motion on the circumference starting from B and labels the motions ZΘ and KH, parallel to the tangent to the circumference, as 'according to nature'; he labels the motions XZ, BY along the radius of the same circumference, as 'against nature'. This denomination makes it natural in referring to Figure 1a, where the 'natural' motion is vertical and directed downward, just like that of a falling weight, to which Aristotle referred constantly in his philosophy of nature.

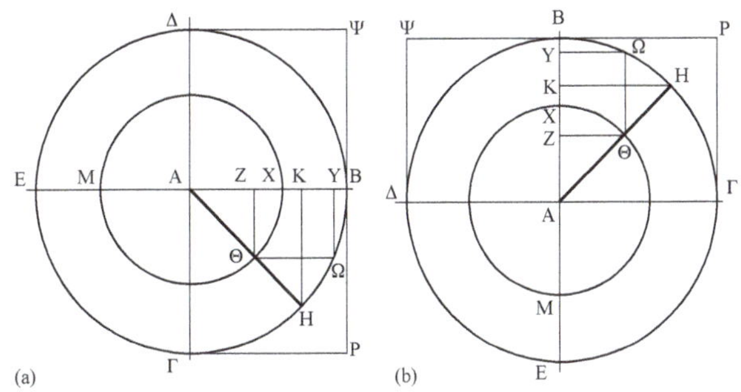

Figure 1. Vertical (**a**) and horizontal (**b**) motion according to nature.

Geometry proves that, in correspondence with the same natural motion ZΘ, the motion against nature decreases as the radius of the circle increases, since YB < ZX. Basically, what Aristotle said is that a body under a given force and constrained to move along a circumference moves faster the greater the radius of the circumference, because the deviation (the interference to motion) is smaller.

The other term that needs to be justified is force (ισχύος). The term used by Aristotle seems to refer to muscular force but on some occasions, more or less explicitly, identifies weight, which, like muscular force, exerts a dragging action. One can then paraphrase Aristotle saying that "a body of given weight constrained to move along a circumference moves faster the greater the radius of the circumference".

So far the reasoning is quite rigorous, but rigor fails when Aristotle tried to apply this result for a quantitative law of the lever, proposed in Problem 3 (the numbering of problems is different from edition to edition; here, we chose in [6]), where he states:

> [...] the ratio of the weight moved to the weight moving it is the inverse ratio of the distances from the centre [6] (p. 353).

This is the quantitative formulation of the law of equilibrium for the lever; however, Aristotle added nothing to common man knowledge, since that law was for sure known since a long time, based on simple observations and the knowledge of a little mathematics. More interesting is the explanation:

> The reason has been given before that the point further from the centre describes the greater circle, so that by the use of the same force, when the motive force is farther from the lever [sic! correct: fulcrum], it will cause a greater movement [6] (p. 353).

To a modern reader this does not seem like a rigourous explanation but simply a kinematical description of the phenomenon, which becomes causal and dynamical if one admits that there is a dragging effect of the weight that increases with its distance from the fulcrum. This can be suggested by the use of the term 'motive force' for one of the two weights of the lever; more explicitly, Winter translated the last part of the previous quotation as: "So by the same force, the mover will manage more the farther from the fulcrum" [11] (p. 11).

However, also accepting this view, one can correctly formulate only this rule: "a weight p located at a distance $d > D$ from the fulcrum can raise a weight $P > p$ located at a distance D", which is an indeterminate law, as remarked by Bernardino Baldi (1553-1617 CE).

> Thus when Aristotle discloses the reason for which the lever moves a weight more easily, he says that this happens because of the greater length on the side of the power that moves; and this [accords] quite well with his first principle, in which he assumes that things at the greater distance from the centre are moved more easily and with greater force, from which he finds the principal cause in the velocity with which the greater circle overpowers the lesser. So the cause is correct, but it is indeterminate; for I still do not know, given a weight and a lever and a force, how I must divide the lever at the fulcrum so that the given force may balance the given weight. Therefore Archimedes, assuming the principle of Aristotle, went on beyond him; nor was he content that the force be on the longer side of the lever, but he determined how much [longer] it must be, that is, with what proportion it must answer the shorter side so that the given force should balance the given weight [15] (pp. 54–55. Translation by [16], p. 14).

For the indeterminate law to become determinate, one needs to replace the relations of greater or lesser with an equality and to allow the variation of the value of the moving weight, so as to give: "a heavy body P brings about an action on another body Q, located on the opposite side of a lever, which is directly proportional to its distance d from the fulcrum and to its weight p"; in modern terms, it is the product dp that actually counts.

It is worth remarking that in the explanation of the law of the lever the circumstance that "the mover will manage more the farther from the fulcrum" is stressed more than the balance between the tendency to go down of the two weights on the lever. This is the same occurring in the Middle Age treatise *Ratione ponderis* attributed to Jordanus Nemorarius (13th century CE) [13] (p. 84).

Assuming the lever of Figure 2, one can accept that there is an equilibrium when two weights Γ are located at the same distance from the fulcrum A (that is, $AB = AM$) for symmetry reasons. However, the weight Γ at M can be replaced by a weight Δ at a distance AE such that $AE \times \Delta = AM \times \Gamma = AB \times \Gamma$, which is the law of the lever in its standard formulation.

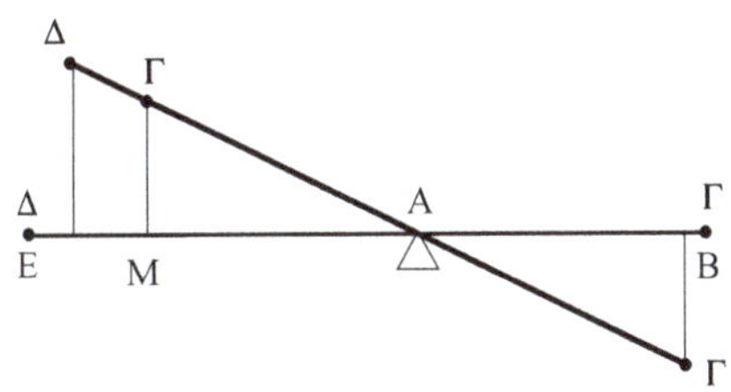

Figure 2. The law of the lever.

3. Euclid's Book on the Balance

We present the proof of the law of the lever contained in *The Book of the Balance* which can be attributed, with much reserve, to Euclid of Alexandria [17,18]. Doubts arise as no work by Euclid on mechanics exists in Greek, nor is he credited with any mechanical work by ancient writers. In 1851, Woepcke published an Arabic fragment that he had discovered in Paris under the title *Le livre d'Euclide sur la balance* [19], attributing it to Euclid. If that were a treatise of Euclid, it would be the first on mechanics after the *Mechanica Problemata*, less than one century later, and would be worthy to be commented on here, as we do.

Euclid's treatise has received little attention from the historians, most probably because of the dubious attribution; when speaking about the Greek proof of the law of the lever, reference is made only to the *Equilibrium of Planes* by Archimedes [20]. A recent discussion on *The Book of the Balance* can be found in [21], but the most interesting account is still due to a Pierre Duhem at the beginning of the 20th century [22] (volume 1, pp. 62–67). Duhem criticized Euclid's proof bitterly for his confusing argumentations; however, we do not agree with this position and consider the proof as interesting and smart as that proposed by Archimedes at least. In addition, it has the advantage of not requiring notions that cannot be found in the treatise, as it is instead the case for the notion of center of gravity, necessary in the *Equilibrium of Planes*.

The treatise is very terse, completely different from the wordy *Mechanica Problemata*. As usual in Greek writings on geometry, there are Definition 1, axioms (2), and propositions (4), not present in the Aristotelian treatise. Apart from the logical structure, the main difference with the *Mechanica Problemata* is that here there are no references to principles of a pre-established philosophy of nature, as it occurs also for a modern treatise on physical mathematics, after two thousand years. Moreover, the axioms in *The Book of the Balance*, although not necessarily shared by everyone at a first reading, are verifiable with simple experiments, even mental ones.

The definition and the axioms are referred to below:

Definition 1. *Weight is the measure of the heaviness and lightness of one thing compared to another by means of a balance.*

> *Axiom 1.* When there is a straight beam of uniform thickness, and there are suspended on its extremities two equal weights, and the beam is suspended

on an axis at the middle point between the two weights, then the beam will be parallel to the plane of the horizon.
Axiom 2. When two weights—either equal or unequal—are placed on the extremities of a beam, and the beam is suspended by an axis on some position of it such that the two weights keep the beam on the plane of the horizon, then if one of the two weights is left in its position on the extremity of the beam and from the other extremity of the beam a straight line is drawn at a right angle to the beam in any direction at all, and the other weight is suspended on any point at all of this line, then the beam will be parallel to the plane of the horizon as before [...] [19] (p. 220. Translation into English in [18]).

At least two other axioms, very intuitive indeed, are implicit:

Axiom 3'. Suspending any weight at the fulcrum, if the balance is horizontal it will remain horizontal.
Axiom 4'. Weight is an additive measure. That is, given two heavy bodies with equal weight, the heavy body composed by the two bodies has a double weight.

The lack of Axiom 3' is also noticed in [22] (volume 1, p. 65).

Before moving on to the analysis of the propositions, a short comment is necessary. For instance, it should be noted that the axioms refer explicitly to weights and they do not contain the term *force*, which, on the other hand, was used frequently in the *Mechanica problemata*, thus reducing the ambiguity of the Aristotelian treatise. Then, at first sight for a modern reader, Axiom 1 seems to be a simple reformulation of the Definition, because it can be immediately derived from it; however, this is not the case. Actually, in the Definition, the equality of the measure of the two weights is verified operationally and can be carried out with whatever a balance, even with one having different lever arms: two heavy bodies are considered to be of equal weight (that is, of the same degree of heaviness) if and only if when placed on the same plate of a scale they are balanced by the same heavy body placed on the other plate.

Moreover, Axiom 1 is not self-evident as it might seem at first sight; in fact, although the arms of the balance are equal, the two heavy bodies at the extremities can be of different shape and material, even if with the same weight: therefore, the principle of symmetry cannot be invoked. It could be said that Axiom 1 has the function of establishing that as far as equilibrium is concerned, the only relevant characteristic of a body is its weight.

Axiom 2 is somewhat complex and unpredictable: it states that if the axis of a balance is in a horizontal position under the action of some weights, it remains in that position even if the weights it supports are moved in any direction perpendicular to it. Figure 3 illustrates two possibilities of displacing the weights: in Figure 3a, the displacement is vertical, and the statement of Axiom 2 can be easily accepted. The case of Figure 3b is different; here, the fact that the axis-weight system is not in equilibrium, since there is a moment that tends to make the axis of the scale rotate around itself, is disturbing somehow. However, Axiom 2 does not say that the balance is in equilibrium but only that it remains horizontal.

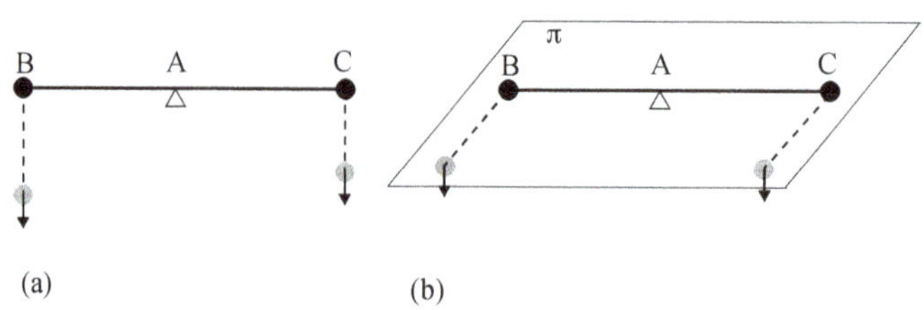

Figure 3. Displacement of weights orthogonal to the beam in a vertical (**a**) and a horizontal (**b**) plane.

As for the propositions, we focus mainly on the first:

> *Proposition 1.* This being assumed, we pose a straight line AB (see Figure 4) as a beam of a balance whose axis is at point C, and we draw CE at a right angle to line AB, and we extend it in a straight line to point D, and we make line CD equal to CE, and we complete the square CH by drawing parallels. Then we place equal weights at points A, H and E. And so I say that these three weights keep lines AB, ED parallel to the horizon.
> The proof is this: tne weight has been placed on one of the extremities of line AB at point A. From the other extremity we have drawn a line at right angles, the line BH, and we have placed on it a weight equal to the weight which is at point A. And so the two weights maintain the line AB parallel to the horizon [by Axiom II]. For the same reason it is necessary that the two weights which are at points E, H keep line ED parallel to the horizon. Thus weights A, E, H will keep lines AB, ED parallel to the horizon.
> It is clear that if the weight which is at point H is removed to point B from which the line BH was drawn at right angles, then with weight A it maintains line AB parallel to the horizon, just as it was necessary when the weight was at point H. The line ED will accordingly not be in equilibrium, since the weight E will make it incline on its side. But if weight E is moved to point C, or if weight E is left on its place and a weight equal to it is placed at point D, then the weight E balances the line ED and it will be parallel to the horizon. We conclude from this that the weight which is at point H was one weight which stood in place of two weights at points B, D, each of which was equal to it. [19] (p. 221. Translation into English in [18]).

Figure 4 illustrates the configuration of the lines ACB, CDE, BH, and DH drawn on a horizontal plane, so that the direction of gravity is perpendicular to the same plane. The proof of Proposition 1 is simple, even if it is not clear what its role is in the proof of the law of the lever. This approach is different from that in Aristotle's text and is typical of Greek geometry, in which without an initial comment explaining the strategy, a series of propositions are shown, the purpose of which is revealed only at the end.

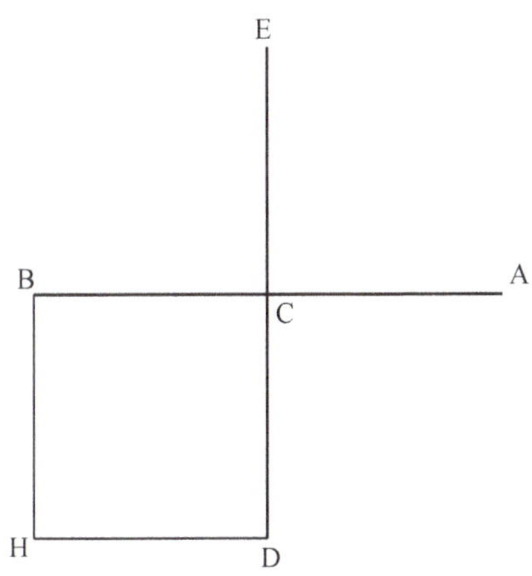

Figure 4. Equilibrium of a beam in the space. Redrawn from [19], p. 221.

Proposition 2 follows immediately from Proposition 1 and states that the equal arm balance of Figure 5a, with the fulcrum in C, is in equilibrium under three equal weights, one placed at the free end H, one at a distance CZ from the fulcrum, and the last at a distance TW from the free end H, so that CZ = WT. Then, since CZ and WT are arbitrary, the proposition allows us to affirm more generally that in a balance in equilibrium under three equal weights such as that in Figure 5a, the two weights hanging from the same arm can be shifted in opposite directions by the same amount (notice that if there is equilibrium for CZ = WT, there cannot be equilibrium for CZ $\neq$ WT).

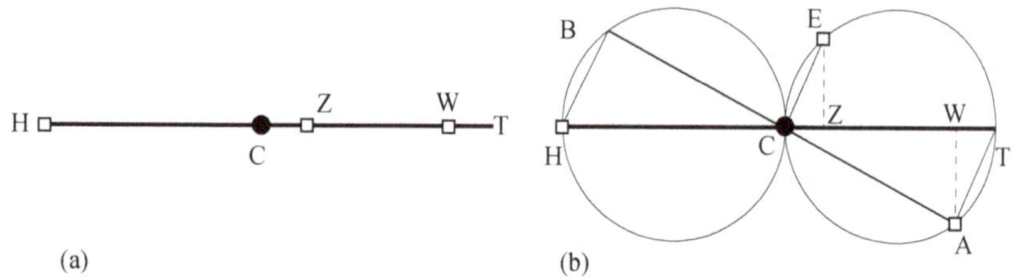

Figure 5. Equilibrium of a beam under three weights: (**a**) a modern interpretation; (**b**) appealing to the law of the lever; (**b**) is redrawn from [19], p. 222.

The proof refers to Figure 5b, in which the lines are supposed as drawn on a horizontal plane. Three equal weights are hung at points A, E, and H, where A can be thought as belonging to the equal arm balance ACB. The circles serve only to facilitate the reading of the drawing. According to Proposition 1, the three weights are in equilibrium: indeed, even if Proposition 1 was proved when the quadrilateral BCDH of Figure 4 is a square, it is easy to prove it holds also valid for a rectangle such as ATEC in Figure 5b. If the three equal weights are in equilibrium with respect to the balance ACB, the plane to which they belong is also in equilibrium. Thus, the straight line HT can be considered the axis of a balance with fulcrum in C to which the weights in A, E, and H are somehow hung, and is also in equilibrium. By Axiom 2, the two weights in A and E can be translated orthogonally to HCT, respectively, to Z and W, and HT remains horizontal: thus, the balance HCT with weights in H, Z, and W is in equilibrium.

Proposition 2 is not completely sufficient to prove the law of the lever; however, it suggests the following generalization, which can be assumed as a further axiom:

> *Axiom 5′*. If any number of weights keep a balance in the horizontal position and two weights hanging at one of the two arms are moved in the opposite direction of the same amount, the balance still remains in the horizontal position.

This should be considered as an axiom because Proposition 2 proves a similar statement for the case of three weights only, two of them on one arm of the balance. Duhem considered Propositions 1 and 2 useless, referring to them as "parasitic and vicious" [22] (p. 65), and assumed only Axiom 5′. Actually, Propositions 1 and 2 are very interesting, because they show how a mathematician can tackle a mechanical proof: without them, Axiom 5′ could be hardly imagined. Moreover, Proposition 2 let the law of the lever be proved at least for a ratio 1 : 2 of weights at the ends. Indeed, if the straight lines ACB and TCH are at $\pi/4$, the weights in Z and W coalesce in the middle of the arm CD, providing a weight double than that in A.

At this point, the law of the lever can be easily proved in Proposition 4, with reference to Figure 6a. The balance HCE is in equilibrium by Axiom 1 if two equal weights p are hung at H and E and two weights are suspended below the fulcrum C.

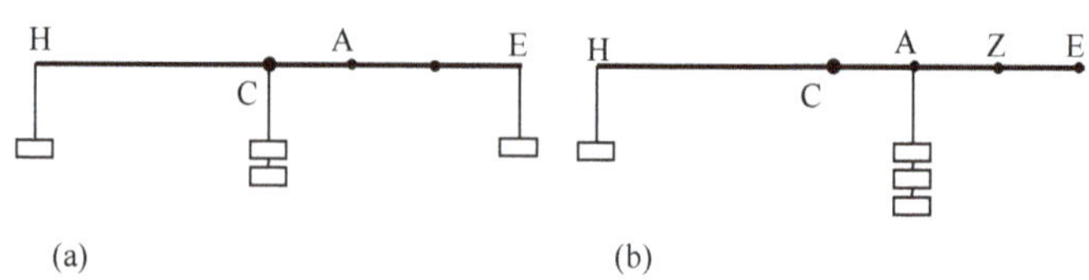

Figure 6. The law of the lever for a particular case: (**a**) first step of the proof; (**b**) second step of the proof; (**b**) is redrawn from [18], p. 27.

Due to Proposition 2, we can move the weight p in C to position A and the weight p in E to position Z without altering equilibrium (we could already start from the position with the weights in A and Z, in equilibrium according to Proposition 2). Then, one can repeat the operation moving another weight p from C to A and the weight p from Z to A; this can be performed in accordance with Axiom 5′ but not with Proposition 2, because on the arm CE, there are more than two weights. Thus, we obtain the configuration of Figure 6b with three equal weights in A, which for Axiom 4′ are equivalent to a weight having a triple value of the weight in B. The law is thus proved for a weight ratio of 3 : 1, and Euclid leaves to the reader the generalization of the law of the lever to the more general case of the ratio of weights expressed by any pair of integers.

A separate discussion is needed for Proposition 3, which is enunciated and proven by means of Figure 7, illustrating a balance BCA of equal arms with fulcrum in C, in equilibrium under three equal weights applied in B, Z, and D. The 10 segments into which BA is divided are all equal.

Figure 7. The force of weight. Redrawn from [19], p. 223.

Proposition 3 states that one can shift the weight at B onto T and the weight at D onto E, still keeping the balance horizontal. It is in fact a shorter balance than the previous one, for which Proposition 2 still holds. The same holds good if one displaces the weight at T onto H and the weight at D into E. The proposition expresses the fact, which is proved for the particular case of Figure 7, that in a balance with three weights it is possible to move each of two equal weights hanging from different arms in opposite directions along the arms, without altering the horizontality of the balance.

At first sight, Proposition 3 seems uninteresting because it concerns only a particular case and is not necessary for the proof of the law of the lever. Its usefulness can be seen from what is written later: in its essence, indeed, it wants to provide a causal justification of the efficiency of a weight on a balance, by introducing the term *force of weight* to indicate the tendency that a weight has to make the arm of a scale move, recalling to the modern concept of moment of a force; see this significative quotation:

> It is then clear that the diminution of force of weight when the weight is moved from B to T is equal to the diminution that occurs when a weight is moved from T to H. The same reasoning applies to all the quantities of equal lengths taken from CB [19] (p. 225. Translation into English in [18]).

This statement suggests to generalize Axiom 5′, by removing the clause "hung at one of the two arms", to become: "If any number of weights keep a balance in the horizontal position, if any two weights are moved along the balance in opposite directions of the same amount, the balance still remains in the horizontal position".

4. Concluding Remarks

Although this work has the main purpose of illustrating the way in which mathematics was used in ancient Greek mechanics, we believe it appropriate to make some considerations on the merits of the treatment of the law of the lever. In the interpretation provided above of the *Mechanica Problemata*, one starts from the idea that a force is the more effective the faster the movement of its point of application. In other words, to establish whether the lever is in a state of rest, reference is made to its possible motion, that is, to its kinematics. The formulations of the Hellenistic mathematicians concerning mechanics, of Euclid and Archimedes in primis, do not refer to kinematics; for this, they are labeled as purely static approaches.

Sometimes, it is improperly spoken of the Aristotelian way and the Archimedean way to equilibrium. While the attribution to Archimedes of the purely static approach can be accepted, that to Aristotle is much more problematic. Meanwhile, there are doubts about the attribution, but mainly, the ideas reported therein are very confused. The so-called Aristotelian way is actually the medieval way that was made explicit in the 13th century treatises [13]. In the Renaissance, this way was considered not very rigorous by mathematicians with a humanistic background.

Apart from the correct criticism regarding the lack of rigor of medieval treatises, there is also a criticism that lasted at least until the 19th century: what could be the meaning of investigating rest, that is, the absence of motion, by checking what motion is possible? Here is what Bernardino Baldi writes about this:

> The power acquires forces from the length of the arm and therefore from the consequent speed; in fact, the longer the arms, the more they are fast at their ends [...].
> This assertion is certainly true and abundantly substantiated. But we do not agree that the cause of this wonderful effect is the speed resulting from the length of the arm. What speed in fact can an immobile thing have? In fact, the lever and the balance are immobile as long as they are kept in balance and nothing but a small one power bears great weight [23] (p. 126).

It is worth remarking that Baldi attributed the idea of a dragging effect due to the speed of the moving weight to Aristotle, while such idea is actually difficult to find in the *Mechanica problemata*. The so-called Aristotelian way to statics began to revive in the 17th and 18th century with Johann Bernoulli and with Lagrange, and in the 19th century, it took on a sufficiently rigorous form accepted by all mathematicians.

Author Contributions: Conceptualization, D.C. and G.R.; resources, D.C. and G.R.; writing—original draft preparation, D.C.; writing—review and editing, G.R.; funding acquisition, G.R. All authors have read and agreed to the published version of the manuscript.

Funding: This research was funded by University "La Sapienza", Rome, Italy, grant numbers RM11916B7ECCFCBF, RM12017294D1B7EF.

Conflicts of Interest: The authors declare no conflict of interest.

Entry Link on the Encyclopedia Platform: https://encyclopedia.pub/19484.

References

1. Merzbach, U.; Boyer, C. *A History of Mathematics;* Wiley: Hoboken, NJ, USA, 1969.
2. Heath, T.L. *A History of Greek Mathematics: From Aristarchus to Diophantus;* Clarendon Press: Oxford, UK, 1921.
3. Aristotle. *Metaphysica;* Smith, J.A.; Ross, W.D., Translators; Clarendon Press: Oxford, UK, 1908.
4. Aristotle. Physica. In *The Internet Classical Archive;* Hardie, R.P.; Gaye, R.K., Translators. Available online: http://classics.mit.edu/Aristotle/physics.mb.txt (accessed on 27 December 2021).
5. Laertius, D. *Lives of the Eminent Philosophers (2 Vols);* Hicks, R.D., Translator; Heinemann: London, UK, 1925.
6. Aristotle. Mechanical problems. In *Aristotle. Minor Works;* Hett, W., Ed.; Heinemann: London, UK, 1955.
7. Abattouy, M. Nutaf Min Al-HiYal: A partial Arabic version of pseudo-Aristotle's Problemata mechanica. *Early Sci. Med.* **2001**, *6*, 96–122. [CrossRef]
8. Rose, P.; Drake, S. The pseudo-Aristotelian questions of mechanics in Renaissance culture. *Stud. Renaiss.* **1971**, *18*, 65–104. [CrossRef]
9. Aristotle. *Problemi Meccanici;* Bottecchia Dheò, M.E., Ed.; Rubbettino: Catanzaro, Italy, 2000.
10. Kraft, F. *Dynamische und Statische Betrachtungsweise in der Antiken Mechanik;* Steiner: Wiesbaden, Germany, 1970.
11. Winter, T. *The Mechanical Problems in the Corpus of Aristotle;* University of Nebraska: Lincoln, NE, USA, 2007.
12. McLaughlin, P. The Question of the Authenticity of the *Mechanical Problems*. 2013. Available online: https://www.uni-heidelberg.de/md/philsem/personal/mclaughlin_authenticity_2013_2.pdf (accessed on 27 December 2021).
13. Capecchi, D. *History of Virtual Work Laws;* Birkhäuser: Milan, Italy, 2012.
14. Leeuwen, J. *The Aristotelian Mechanics. Text and Diagrams;* Springer: Cham, Switzerland, 2012.
15. Baldi, B. Vita di Archimede. In *Vite Inedite dei Matematici Italiani Scritte da Bernardino Baldi;* Narducci, E., Ed.; Tipografia Delle Scienze Matematiche e Fisiche: Rome, Italy, 1887.
16. Drake, S.; Drabkin, I.E. *Mechanics in Sixteenth-Century Italy;* University of Wisconsin Press: Madison, WI, USA, 1969.

17. Bulmer-Thomas, I.; Murdoch, J. Euclid. In *Complete Dictionary of Scientific Biography*; Gillispie, C.C., Ed.; Scribner: New York, NY, USA, 1971; Volume 4.
18. Clagett, M. *The Science of Mechanics in the Middle Ages*; The University of Wisconsin Press: Madison, WI, USA, 1959.
19. Woepcke, F. Notice sur des traductions de deux ouvrages perdus d'Euclide. *Philos. Trans. R. Soc. Lond.* **1851**, *18*, 217–232.
20. Archimedes. On the equilibrium of planes. In *The Works of Archimedes*; Heath, T., Ed.; Dover: New York, NY, USA, 2002.
21. Renn, J.; Damerow, P.; McLaughlin, P. Aristotle, Archimedes, Euclid, and the origin of mechanics: The perspective of historical epistemology. In *Symposium Arquímedes. Fundación Canaria Orotava de Historia de la Ciencia*; Montesino-Sirera, J., Ed.; Preprint 239; Max-Planck-Institut für Wissenschaftsgeschichte: Berlin, Germany, 2003; pp. 43–59.
22. Duhem, P. *Études sur Leonard de Vinci (3 Vols)*; Hermann: Paris, France, 1906.
23. Baldi, B. *Bernardini Baldi Urbinatis Guastallae Abbatis in Mechanica Aristotelis Problemata Exercitationes*; Albini: Mainz, Germany, 1621.

encyclopedia MDPI

Entry

Energy Storage Flywheel Rotors—Mechanical Design

Miles Skinner and Pierre Mertiny *

Department of Mechanical Engineering, University of Alberta, 9211-116 St., Edmonton, AB T6G 1H9, Canada; maskinne@ualberta.ca
* Correspondence: pmertiny@ualberta.ca

Definition: Energy storage flywheel systems are mechanical devices that typically utilize an electrical machine (motor/generator unit) to convert electrical energy in mechanical energy and vice versa. Energy is stored in a fast-rotating mass known as the flywheel rotor. The rotor is subject to high centripetal forces requiring careful design, analysis, and fabrication to ensure the safe operation of the storage device.

Keywords: flywheel energy storage; high-speed rotors; mechanical design; manufacturing; analytical modeling; failure prediction

Citation: Skinner, M.; Mertiny, P. Energy Storage Flywheel Rotors—Mechanical Design. *Encyclopedia* **2022**, *2*, 301–324. https://doi.org/10.3390/encyclopedia2010019

Academic Editors: Raffaele Barretta, Ramesh Agarwal, Krzysztof Kamil Żur and Giuseppe Ruta

Received: 10 December 2021
Accepted: 25 January 2022
Published: 28 January 2022

1. Introduction

Between 2019 and 2020, the generation of solar energy grew by 26.0 TWh (24.1%) and 37.1 TWh (16.6%) for the two largest global consumers of energy, the Unites States of America and the People's Republic of China, respectively. Over the same timeframe, the growth in energy generation from wind for these two countries was correspondingly 42.0 TWh (14.1%) and 61.2 TWh (15.1%) [1]. For perspective, the total electricity generation of Canada was 643.9 TWh in 2020. Renewable energy generation capacity is expected to continue to increase rapidly as energy demands and pressure to reduce environmental impacts grow [2]. Additionally, the cost of renewable energy production has been falling dramatically over the last half decade [3], which further increases demand. However, as renewable energy production increases the intermittency from these sources necessitates significant energy storage capacity to meet demand at any particular moment [4].

Compounding the intermittency issue is the separation between peak power demands from residences and businesses and peak power production from renewable sources [5]. What is now recognized as the "Duck Curve" shows the difference between hourly demand and renewable energy production [6]. Energy consumption has been shown to peak in the mornings and evening while energy production typically peaks around midday, especially for solar photovoltaic systems.

Energy storage is among the largest obstacles facing modern energy grids as they transition to new renewable sources of energy while attempting to maintain both power supply and power quality. As the demand for renewable energy sources increases and the costs of that energy decrease, the economic and environmental benefits of maintaining large-scale energy storage systems increase [7]. The plethora of energy storage options [8] includes flywheel energy storage systems (FESS). FESS are among the oldest forms of energy storage, having been used to regulate power output in stone drills as early as 1000 BCE [9]. While the principal concept of flywheel energy storage, i.e., a large mass spinning on an axis, has changed little in the intervening millennia, the materials, control systems, and applications have continually evolved.

Modern high-speed flywheel energy storage systems have a wide range of applications in renewable energy storage, uninterrupted power supplies, transportation, electric vehicle charging, energy grid regulation, and peak shaving. They are recognized for a number of advantageous characteristics, including high charge/discharge rates, expected lifetimes of greater than 20 years, and specific energies in excess of 100 Wh/kg [5]. They are also unaffected by cyclic degradation or depth of discharge effects common to traditional electrochemical batteries, and their cycle efficiency can be up to 95% [10,11]. As can be inferred from the above applications, the advantage of FESS over more common energy storage technologies, such as electrochemical batteries and pumped hydro storage, is that FESS facilitate applications requiring high power and high specific energy [12,13]. FESS have faster response times than both electrochemical batteries or pumped hydro. Compared to batteries, FESS do not require the same level of delicate control over power and temperature, and, due to their high cycle lifetime and deep depth of discharge, FESS require less installed capacity than batteries while still meeting demand [7].

This is not to say FESS are an ideal solution to address all energy storage challenges. FESS experience high passive discharge losses [10], comparatively high initial investment costs [14], and ongoing efforts to understand long-term behavior of rotor materials and failure [15,16]. In an effort to understand and improve flywheel rotor performance and safe operating limits, analytical models have been developed that consider material selection, rotor construction, and operating conditions.

This entry focuses on the design and analysis of the flywheel rotor itself. It will begin by highlighting some FESS applications and performance, followed by the design and manufacturing approach commonly used for flywheel rotors. Analytical modeling approaches for typical flywheel rotors will be discussed, including the effects of variable angular velocity, viscoelastic stress relaxation, and acceleration. Finally, rotor failure criteria will be discussed.

2. Applications and Performance

FESS have a wide range of applications for uninterruptible power supplies, energy grid regulation for frequency and power quality, and electric vehicle and rail transportation. A general range of FESS performance characteristics is given in Table 1.

Table 1. General range of FESS performance characteristics.

Parameter	Value
Lifetime [years]	>20 [5]
Charge/discharge cycles	<10^7 [5]
Energy density [Wh/kg]	<130 [17]
Price [(USD)/kWh]	400–6960 [5,17]
Power density [W/kg]	~1000 [5]

Implementations of FESS are plentiful, so only a few examples are given here. An early application of FESS was the Gyrobus, which began operation in Switzerland and Belgium in 1952 with the goal of servicing low traffic public transport routes where installing overhead electrical catenary wire was deemed too costly [18]. In the late 1990s, Rosen Motors designed a hybrid power train for a vehicle with a gas turbine engine and a high-speed FESS supplementing acceleration in short bursts [19]. Later, Volvo developed a recumbent braking system for their S60 sedan, which recovered and stored energy during braking and subsequent use powering the vehicle [20]. Most recently, Porsche integrated a flywheel into their 911 GT3R race car to extend its range and achieve performance enhancements for long-distance racing [21]. FESS can also be installed on light rail transit systems, either in the cars or along the rail line, as a recumbent braking system to reduce operating costs [22]. Trials for these systems have been conducted in London, New York, Lyon, and Tokyo, among others [23]. Furthermore, utility-scale FESS installations have been implemented as temporary backup power for energy grids in Minto, Ontario [24], Stephentown, New York [25], and De La Salle, Philippines [26].

3. Manufacture

The primary components of FESS are the electrical machine (motor/generator unit), housing, flywheel rotor, and bearing assembly. As an illustration, Figure 1 depicts a cu-away schematic of a scaled-down FESS that was designed for short-term energy storage from regenerative braking in light-rail transit applications. The shown unit features a rotor with a full-size 400 mm outer diameter but axial height scaled to 24% of the full-scale design with 1.0 kWh nominal capacity.

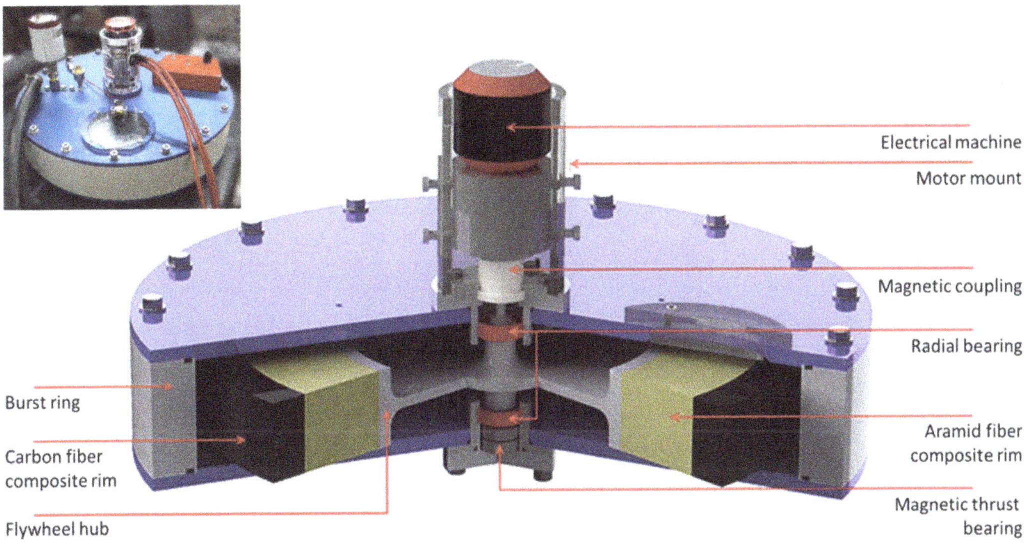

Figure 1. Cutaway schematic of a flywheel energy storage system for experimental research. Inset shows the actual device [16].

In FESS, the electrical machine is responsible for controlling the energy flow into and out of the system. Notably, the electrical machine can be selected independently from the desired energy capacity to meet the demands of a specific application. The housing, bearings, and rotor work in unison; however, while they have clear interactions with each other, changes to one do not necessarily impact the others. For example, any bearing assembly capable of supporting the rotor is acceptable, and different assemblies can be substituted provided they adequately support the rotor. In this way, FESS are highly modular, allowing the system to be finely tuned for optimized performance in a given application. Being the focus of the present entry, the construction of flywheel rotors can be broken down into the two main rotor components—the hub and the rotor rims—and their assembly.

3.1. Hub Construction

The hub of a flywheel rotor is responsible for supporting the rims and transferring torque from the electrical machine to the rest of the rotor. Rotor hubs are commonly constructed from either high-strength steel, aluminum, or fiber-reinforced polymer (FRP) composites. A metallic hub can be forged or machined into a variety of complex shapes. These shapes have been characterized in detail in a number of different works [13,27]. The advantages of various metallic hub geometries are discussed in greater detail below. Limited studies have been conducted on composite hubs that have been shown to be more compliant than metallic hubs, thus providing advantages supporting the rotor rims [28].

3.2. Rim Construction

Flywheel rotor rims can also be constructed from metals or FRP composites. Metallic flywheels are a well understood and comparatively low-cost option that can be forged or machined into rather complicated shapes to maximize performance. Additionally, the hub can be integrated with the rim into a single component, simplifying the manufacturing process. Kale et al. [29] developed an optimization method to maximize kinetic energy of a metal flywheels by varying the cross-section, speed, and size of the flywheel.

FRP rims are fabricated by either filament winding, as shown in Figure 2, or weaving [30,31]. Rectilinear fabric layup techniques have also been studied for constructing rotating disks [32]; however, fabric-based methods are uncommon, as they have not proven to be advantageous compared other techniques such as filament winding. Filament winding is a highly efficient method for fabricating FRP rotor rims due to the accurate control over fiber placement and orientation, axisymmetry of the finished product, and high fiber volume fraction [33] regardless of fiber material—carbon, glass, aramid, etc. Rim geometries are usually a simple thick-walled cylinder with rectangular cross section. The process involves passing long filaments through a resin bath to impregnate the dry fibers with a prepolymer. The fibers are then wound onto a mandrel by passing through the deposition head of the filament winding machine, which allows for precise control of fiber positioning and orientation, i.e., winding angle, of the fibers [34]. Filament winding is an additive manufacturing technique which is often automated to produce parts rapidly and efficiently while minimizing cost. After winding and curing, FRP rotors often require machining to their final dimensions, particularly on the outer surface where excess resin tends to accumulate during the winding process.

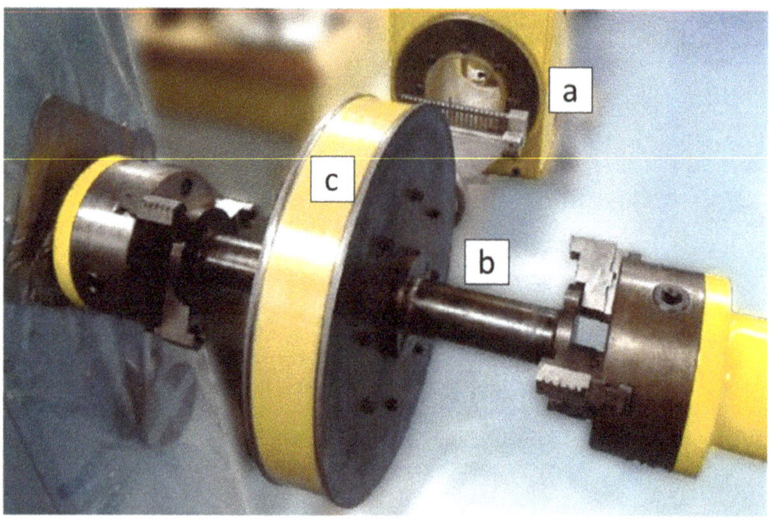

Figure 2. Composite flywheel rotor rim at the end of filament winding manufacturing process; (**a**) fiber payout eye and deposition head on winding machine carriage arm, (**b**) winding mandrel, and (**c**) completed aramid fiber/epoxy composite rim.

The majority of FRP composite rims are constructed with winding angles approaching 90 degrees, typically larger than 88 degrees, relative to the axis of rotation, as this maximizes circumferential strength in the rotor. However, investigations into the effects of variable winding angles have shown to improve rotor performance. Wild et al. [35] showed that periodically increasing the winding angle from the inner to outer radius increased compliance of the FRP at inner radii relative to outer radii allowing the inner portion of the rim to move disproportionately outward, preventing the buildup of large tensile radial stress, which is the driver for a primary failure mode. Recognizing the significance of radial tensile stress, Uddin et al. [36] conducted finite element analysis on FRP composite rotors filament-wound with a mosaic pattern. These complicated patterns were created by significantly changing the fiber angle between layers during the winding process. Results showed that radial stress could be significantly reduced, possibly leading to greater rotor energy storage capacities; however, effects on manufacturing cost have not been determined, and further research is, therefore, needed.

Wang et al. [30] discussed the possibility of creating woven FRP rims with fibers perpendicular to each other radially and circumferentially. They successfully created thin composite disks and conducted finite element analysis on the structures. Their results indicate the radially oriented fibers provide greater support when compared to unidirectional filament-wound rotors. Similar to the mosaic patter it is not clear if this technique improves specific energy, nor has the effect on manufacturing cost been clearly assessed.

3.3. Assembly

Assembly of a flywheel rotor is only necessary when it is constructed from multiple components, typically a hub and one or more FRP composite rims. For metallic flywheels, assembly is typically not required as they can be manufactured as a single part. For flywheel rotors constructed from a metallic hub and a single FRP rim, the composite can be wound directly onto the hub as discussed by Tzeng et al. [37] or joined with a press-fit [18]. An example of a thermal press-fit is shown in Figure 3.

Figure 3. Thermal press-fit accomplished by cooling the aluminum hub with liquid nitrogen before pressing into the composite rims.

While there is no consensus on the optimal method for assembling flywheel rotors, press-fitting is often considered for the construction of flywheel rotors with more than a single rim. When press-fitting FRP rims onto a hub or other FRP rims, they can be manufactured with a slight taper to reduce the required pressing force and minimize the risk of damaging the fibers [38]. When dissimilar materials are adjacent to each other it is often expedient to create a thermal press-fit by taking advantage of the different thermal expansion coefficients. This is especially true when assembling an FRP rim and a metallic hub [38]. The final step in flywheel rotor assembly is typically balancing to minimize vibrations and oscillations by ensuring mass is evenly distributed around the axis of rotation.

4. Analytical Modeling

4.1. Energy Storage and Power Capacity

Flywheel energy storage systems have often been described as 'mechanical batteries' where energy is converted from electrical to kinetic and vice versa. The rate of energy conversion is the power capacity of the system, which is chiefly determined by the electrical machine connected to the rotor [13,39]. The capacity of the FESS is determined by the size, shape, materials, and construction of the flywheel rotor [15]. As indicated above, modern high-speed flywheel rotors are typically constructed from a hub, responsible for torque transfer and structural support, and one or more rims [39]. Here, for the sake of explanation, a monolithic rotor geometry is considered to consist only of a hub without any added rims around its perimeter. Hub and rims can be constructed from either metals, ceramics, or composites [40,41] to maximize rotor performance. The kinetic energy of a rotor, as a rotating body, is defined as:

$$E_{\mathrm{K}} = E_{\mathrm{hub}} + \sum_{n=1}^{N} E_{\mathrm{rim}}^{n} = \frac{1}{2} I_{\mathrm{r}} \omega^2, \tag{1}$$

where E_{K} is the total kinetic energy of the rotor, I_{r} is the total moment of inertia for the rotor, ω is the angular velocity in units rad/s, and N is the number of rims such that $n = 1, 2, \ldots N$. The moment of inertia for the entire rotor is a superposition of the moment of inertia for the hub and all rims:

$$I_{\mathrm{r}} = I_{\mathrm{hub}} + \sum_{n=1}^{N} I_{\mathrm{rim}}^{n}, \tag{2}$$

where I_{hub} and I_{rim}^{n} is the moment of inertia for the hub and the n-th rim, respectively.

Considering the flywheel hub, defining the moment of inertia for simple geometries is straightforward, i.e., for rectangular cross sections of a solid or hollow disk, the moment of inertia can be defined as:

$$I_{\mathrm{hub}} = \frac{1}{2} m \left(r_o^2 + r_i^2\right) = \frac{1}{2} \rho \pi h \left(r_o^4 - r_i^4\right), \tag{3}$$

where ρ is the density of the hub material, h is the height of the hub (with respect to the axis of rotation), and r is the radius with the inner and outer dimension defined by subscripts 'i' and 'o'. In analytical modeling, the mass of the hub is calculated using the volume and density. A common approach for handling complex geometries and functionally graded materials is to discretize the shape into a series of uniform disks of arbitrary width and varying height [42], in which case, Equation (3) can be generalized by manipulating r_o, r_i, ρ, and h. As the hub cross section increases in complexity it is common to define the energy density (ratio of energy to mass) [13,27,43] of the hub as:

$$\frac{E_{\mathrm{hub}}}{m} = \frac{k\sigma}{\rho}, \tag{4}$$

where k is the shape factor of the hub and σ is the stress in the hub. When σ is equal to the ultimate tensile strength in the hub, energy density is maximized and can be used to find the maximum energy capacity of the flywheel rotor. Shape factors for common hub geometries are presented in Table 2; additional cross sections k-values are given in [13,43]. It has been noted [27] that the choice of material for the hub will strongly influence cross sectional geometries. Hub shape factors above 0.5 induce bidirectional stress states, which negatively impact composite materials, especially unidirectional composites, because transverse strength is typically significantly lower than strength in the fiber direction. For this reason, isotropic materials are more appropriate for cross sections with large shape factors. Discontinuous hub geometries, such as the split type hub [44], are either treated as continuous and analyzed as described above, or determined through numerical methods [45].

Focusing attention now on rotor rims, calculating the energy capacity is analogous to Equations (1)–(4). The vast majority of industrial and academic work focusing on flywheel rotors uses rims with rectangular cross sections [46–49]. While it has been shown that variable thickness flywheel rotors can produce a more favorable stress state [50], the energy capacity typically stuffers due to the reduction of mass at the largest radial coordinates and limited maximum angular velocity to minimize transverse loading. Variable thickness flywheel rotors with mass concentrated on the outer edges have been presented [45]; however, these have not proven to produce higher energy density or a more favorable stress state than traditional rotor designs, such as the Laval disk, with rims discussed in [43].

Table 2. Shape factor values (k) for various flywheel rotor cross sections.

Shape	Cross Section	k-Value
Laval disk		1.00
Laval disk real	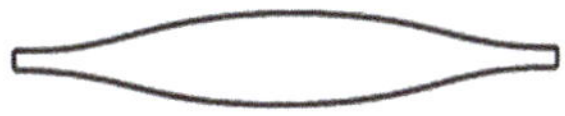	0.70–0.90
Conical disk		0.70–0.85
Solid disk		0.606
Thin ring		0.50
Thick rim		0.303

4.2. *Material Characterization*

Flywheel rotor material selection depends on a large variety of constraints, including system requirements, cost, operating conditions, and expected lifetime. Equation (1) indicates that energy capacity is quadratically related to angular velocity and radius. Therefore, increasing either one or both values is the most effective method to increase energy capacity. Moreover, Equation (4) shows that the energy density of a rotating rotor is proportional to the ratio of its material's strength and density. This suggest that high strength, low density materials such as carbon FRP composites are an ideal material for flywheel rotor construction. However, the stress state is also quadratically related to angular velocity and radius. Compounding this issue is the typically limited transverse strength of highly anisotropic materials [27], such as carbon FRP, suggesting that additional design features are required for achieving full energy capacity potential (e.g., press-fit assembly of multiple rotor rims). These considerations lead to the conclusion that the most suitable choice of material and geometry depends heavily on the application requirements and design constraints such as system geometry and cost.

The most common choices for modern flywheel rotors are either metals, such as aluminum and steel, or FRP composites [51]. With respect to single and multi-rim flywheel rotors, it has been shown that the optimal choice depends on the design criteria. When optimizing for specific energy, i.e., energy per unit mass, then FRP composites are usually the ideal choice, whereas metal flywheels are often superior when optimizing for energy per cost [40]. Another consideration is that isotropic materials are also better understood than advanced composite materials, which increases confidence in modeling and failure prediction, especially in design cases aiming for long lifetimes and operation near maximum energy capacity.

Regardless of material selection, it is necessary to describe the stress strain relationship for all materials in the rotor. Assuming time-independent linear elastic behavior [52], Hooke's law in cylindrical coordinates states:

$$\begin{bmatrix} \sigma_{11} \\ \sigma_{22} \\ \sigma_{23} \\ \sigma_{12} \\ \sigma_{13} \\ \sigma_{23} \end{bmatrix} = \begin{bmatrix} C_{11} & C_{12} & C_{13} & C_{14} & C_{15} & C_{16} \\ C_{21} & C_{22} & C_{23} & C_{24} & C_{25} & C_{26} \\ C_{31} & C_{32} & C_{33} & C_{34} & C_{35} & C_{36} \\ C_{41} & C_{42} & C_{43} & C_{44} & C_{45} & C_{46} \\ C_{51} & C_{52} & C_{53} & C_{54} & C_{55} & C_{56} \\ C_{61} & C_{62} & C_{63} & C_{64} & C_{65} & C_{66} \end{bmatrix} \begin{bmatrix} \varepsilon_{11} \\ \varepsilon_{22} \\ \varepsilon_{33} \\ \gamma_{12} \\ \gamma_{13} \\ \gamma_{23} \end{bmatrix}, \tag{5}$$

where σ is stress, C is an elastic modulus of elasticity, ε is linear strain, and γ the shear strain. The subscripts 1, 2, and 3 in the stress and strain terms indicate the rotor's radial, circumferential, and axial directions, respectively. The stiffness matrix, [C], given above, assumes a fully anisotropic material and has 36 independent moduli. However, materials used in flywheel rotor display varying levels of symmetry, so this matrix can be simplified based on the materials selection. Orthotropic carbon FRP flywheel rotors have been constructed by stacking woven carbon fiber laminates [30] or developing unique fabric layup patterns [36], discussed in Section 4.2, in which case the stiffness matrix becomes:

$$[C] = \begin{bmatrix} C_{11} & C_{12} & C_{13} & 0 & 0 & 0 \\ C_{12} & C_{22} & C_{23} & 0 & 0 & 0 \\ C_{13} & C_{23} & C_{33} & 0 & 0 & 0 \\ 0 & 0 & 0 & C_{44} & 0 & 0 \\ 0 & 0 & 0 & 0 & C_{55} & 0 \\ 0 & 0 & 0 & 0 & 0 & C_{66} \end{bmatrix}. \tag{6}$$

Further simplifying assumptions can be made for unidirectional FRP composites where the rotor is made by continuously winding long polymer resin impregnated filaments onto a mandrel before polymer solidification [28,38]. In this case, the fibers are all oriented circumferentially with the radial and axial directions both being transverse to the fibers. In this case, the material is considered transversely isotropic [53]:

$$C_{22} = C_{33}; \quad C_{12} = C_{13}; \quad C_{44} = C_{55}. \tag{7}$$

For fully isotropic materials, such as steel, the stiffness matrix simplifies significantly [54]:

$$C_{11} = C_{22} = C_{33}; \quad C_{12} = C_{13} = C_{23}; \quad C_{44} = C_{55} = C_{66} \tag{8}$$

Transversely isotropic and fully isotropic materials are most common in modern flywheel rotor construction due to their comparatively low cost, high strength, and ease of manufacturing.

A description of elasticity is sufficient to determine the instantaneous or time-independent rotor response to loading; however, this approach does not necessarily reflect the realistic material response to loading. Therefore, it is necessary to develop a description of the materials that depends on time, t. All engineering materials exhibit some viscoelastic response, meaning they have characteristics of elastic solids and viscous fluids [55]. However, at typical FESS operating temperatures, below 50 °C [56], metals display negligible viscoelastic behavior [57]; therefore, this discussion will focus on FRP composites.

The time-dependent compliance of a material is defined as the inverse of the stiffness matrix, such that $[S(t)] = [C(t)]^{-1}$. Then, the time-dependent compliance matrix for an orthotropic linearly elastic material is as follows:

$$[S(t)] = \begin{bmatrix} S_{11}(t) & S_{12}(t) & S_{13}(t) & 0 & 0 & 0 \\ S_{12}(t) & S_{22}(t) & S_{23}(t) & 0 & 0 & 0 \\ S_{13}(t) & S_{23}(t) & S_{33}(t) & 0 & 0 & 0 \\ 0 & 0 & 0 & S_{44}(t) & 0 & 0 \\ 0 & 0 & 0 & 0 & S_{55}(t) & 0 \\ 0 & 0 & 0 & 0 & 0 & S_{66}(t) \end{bmatrix}. \quad (9)$$

At this juncture it is worth taking a moment to define the S_{ij} terms with respect to moduli of elasticity, E, and Poison's ratios, ν:

$$[S(t)] = \begin{bmatrix} 1/E_1(t) & -\nu_{12}/E_2(t) & -\nu_{13}/E_3(t) & 0 & 0 & 0 \\ -\nu_{12}/E_1(t) & 1/E_2(t) & -\nu_{23}/E_3(t) & 0 & 0 & 0 \\ -\nu_{13}/E_1(t) & -\nu_{23}/E_2(t) & 1/E_3(t) & 0 & 0 & 0 \\ 0 & 0 & 0 & 1/G_{23} & 0 & 0 \\ 0 & 0 & 0 & 0 & 1/G_{23} & 0 \\ 0 & 0 & 0 & 0 & 0 & 1/G_{12} \end{bmatrix} \quad (10)$$

As shown earlier, the time-independent compliance matrix for transversely and fully isotropic materials can be found using Equations (7) and (8). For viscoelastic materials, the sustained imposition of a stress causes increasing strain, called creep. Conversely, subjecting a viscoelastic material to constant strain leads to decreasing stress, called relaxation. Creep occurs in three phases characterized by the linearity of the strain response as a function of time. Primary, or phase I, creep is characterized by logarithmic growth. In secondary, phase II, creep, deformation increases linearly with time. Finally, tertiary, phase III, creep is characterized by exponential growth until failure [55]. Methods for calculating the compliance from stress-strain data is well documented [58–61]. These methods typically involve applying a known stress to material samples while measuring strain and time data. From these data, stress-strain curves are constructed and functions are fit to the curves to define the time-dependent change in elastic modulus. It is worth noting that a number of phenomena affect the viscoelastic response of materials, including stress magnitude and direction, temperature, moisture, and age [62].

4.2.1. Hygroscopic Effects

The effects of moisture, also known as hygroscopic effects, on material properties have been documented for a both elastic and viscoelastic FRP composite materials [63]. However, hygroscopic effects are not expected to significantly affect the operation of flywheel rotors. FESS commonly comprise a vacuum enclosure designed to contain the flywheel and limit the aerodynamic drag acting on the rotor and bearing surfaces [39]. Hence, hygroscopic instability is not expected to affect the rotor material during operation, provided the vacuum environment is maintained. Consequently, viscoelastic material characterization should be performed on suitably dry specimens to most accurately describe the material in situ. If necessary, this can be accomplished conditioning specimens, e.g., by gently heating specimens to approximately 90 °C for up to 24 h [62].

4.2.2. Temperature Effects

Similar to hygroscopic effects, the vacuum condition in the FESS enclosure minimizes the influence of environmental temperature changes on the flywheel rotor during operation. On the other hand, a vacuum environment prevents convective heat transfer and, thus, impedes the removal of parasitic heat that is generated by energy losses, such as friction in bearings and eddy currents in the electrical machine. Hence, a flywheel rotor may still experience considerable temperature fluctuations depending on the FESS design configuration and operation, and hence, the study of temperature on flywheel rotor creep and relaxation should be considered in FESS design.

Challenges with assessing the creep behavior of FRP composite rotors arise from the projected long lifetimes of FESS. As a solution, time-temperature superposition principle (TTSP) can be used to predict long-term behavior using short-term viscoelastic test data. FRP composites are highly sensitive to temperature fluctuations with linear viscoelastic behavior being observed below the polymer matrix glass transition temperature, T_g, and non-linear viscoelasticity above. Elevated temperatures facilitate polymer chain mobility, causing a decrease in both moduli and strength [60]. For the TTSP, a trade-off is seen where increasing temperature increases the rate of viscoelastic response, and decreasing temperature decreases this response. By conducting short-term experiments at elevated temperatures, it is possible to predict the long-term behavior of the material at low temperatures. The basic procedure for the TTSP is discussed in [64]. First, material specimens are subjected to constant load at various temperatures in conventional creep testing. These data generate a series of compliance curves when plotted over time in logarithmic scale (log(time)). Second, an arbitrary reference temperature is selected. Third, all compliance curves are shifted along the time axis onto the reference temperature compliance curve to construct a master curve. As a demonstration, consider the data series of tensile experiments in Figure 4. Short-term tensile experiments were conducted on an FRP composite material at various temperatures to collect the viscoelastic data [65]. Data for all temperatures but the reference temperature were shifted along the time axis to construct the master curve at a reference temperature, T_r, of 40 °C.

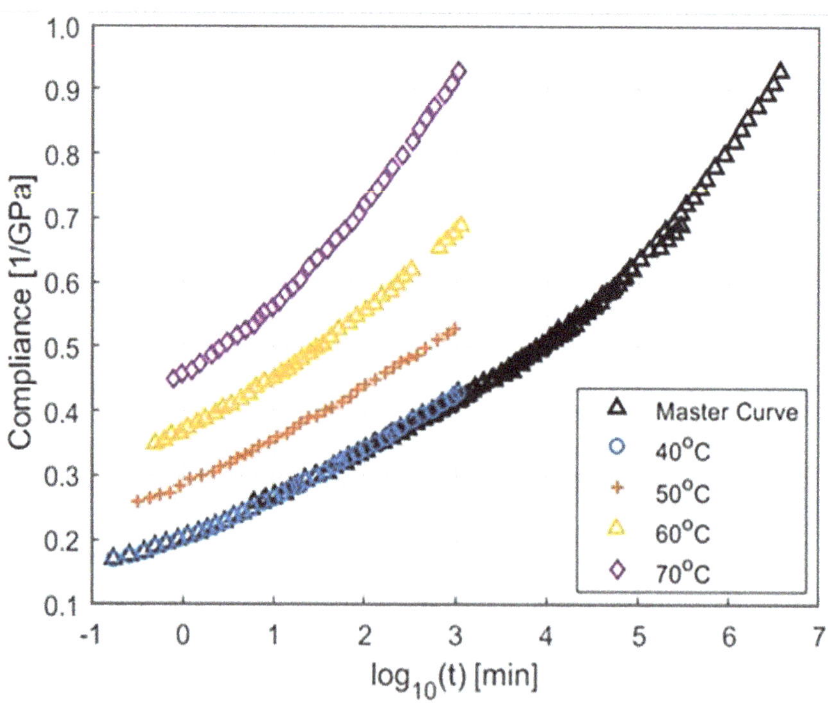

Figure 4. Time-temperature superposition experimental data, reproduced from [65]. Data were collected from tensile tests for an FRP composite at various temperatures and shifted along the time axis to create a master curve for a reference temperature of 40 °C.

An underlying assumption for the TTSP is that creep is controlled by the same mechanisms under the different temperatures. Therefore, the master curve is expected to be smooth throughout. Since it is constructed on a log(time) axis, the predicted compliance is sensitive to the shift factor, where a small discontinuity could result in errors of years or decades. If a smooth master curve exists by using only horizontal shift factors, then the material is considered thermorheologically simple. The need for vertical shift factors has been identified under some conditions [64], in which case materials are referred to as thermorheologically complex. The majority of materials, including FRP composites under normal conditions, are considered thermorheologically simple [64]. Notably, even though TTSP has been employed to characterize the linear viscoelastic behavior of epoxy polymers since at least the 1960s [66], there is still no established convention defining the optimal method to determine shift factors for each curve.

The distance each curve is shifted along the time axis is called the shift factor, a_T. There are several ways to determine the shift factor for each curve, all of which are designed to create a smooth master curve. Brinson [67] studied the time temperature response of Hysol 4290, a common contemporary two-part epoxy. Brinson conducted tensile tests on samples of the material at temperatures between 90 °C and 130 °C and, thus, constructed a master curve covering creep at 90 °C over approximately 6 months. The shift factor was determined using the William-Landel-Ferry (WLF) equation [68], which requires a knowledge of T_g and a set of experimentally determined material constants. While WLF can create a smooth master curve, it is limited to temperatures above T_g, so it may not be suitable for all applications. Another common method is using an Arrhenius' equation [69,70], which requires knowledge of the activation energy and gas constant. The activation energy is typically determined using dynamic mechanical analysis [71].

Both of the above mechanistic methods attempt to define a relationship between certain material properties and the creep response. However, Gergesova et al. [72] recognized that a smooth master curve can be constructed without this mechanistic relationship by mathematically minimizing the horizontal distance between two adjacent curves. His algorithm considers overlapping region of data between adjacent curves. Before shifting these regions, one defines an area that is delineated on either side by the experimental data and on top and bottom by the height of the overlap. This area can be minimized by applying a shift factor to one or both curves depending on the chosen reference temperature. Using this method, the shift factor and master curve can be found without the need for additional experiments or prior knowledge of the activation energy. It is worth noting that Sihn and Tsai [65] used an Arrhenius equation, while the master curve in Figure 4 was created using the algorithm from Gergesova et al. [72].

Applying a best fit curve to the compliance master curve defines a function used to determine the material's stiffness at any time throughout its lifetime:

$$S_{ij}(t, T_r) = S_{ij}(a_T, t, T) \tag{11}$$

where S_{ij} is the compliance and T is the experimental temperature. Tensile experiments must be conducted to determine $[S]$ for each independent modulus in Equation (10), i.e., E_1, E_2, E_3, etc., and will vary depending on whether the material is isotropic, transversely isotropic, orthotropic, or fully anisotropic.

4.2.3. Aging Effects

Aging is a continuous process which occurs at all temperatures and is caused by polymer chains evolving toward equilibrium. This is ultimately a densification process which results in a decreased chain mobility and compliance. The effect of aging is similar to temperature in that it is continuous; however, aging always results in a decrease in compliance, whereas temperature can result in either an increase or decrease. Aging effects can be included in directional compliance similarly to temperature effects. Compliance is measured from material specimens at various ages and resulting curves are shifted to define the age shift factor, a_{te}. Then, S_{ij} becomes the following:

$$S_{ij}(t_e, T_r) = S_{ij}(a_T, a_{te}, t, T). \quad (12)$$

where t_e is the age for which the master curve is created. Under isothermal conditions, the aging shift factor can be calculated as a ratio between a reference aging time and an experimental aging time raised to an experimentally determined thermal shift rate [73]. While it is possible to experimentally determine and account for material aging when modeling flywheel rotors, it is more practical to thoroughly stabilize the flywheel rotor by aging at an elevated temperature under no load conditions until the rotor reaches equilibrium before operation. This stiffens the material, minimizes creep, and provides a more repeatable starting point for designing flywheel rotors. Sullivan [74] showed equilibrium can be achieved by aging epoxy polymers at 115 °C for 1000 h. It is recommended that flywheel rotors be aged to minimize material evolution during operation, which will improve rotor response to applied loads and increase confidence in any simulation or modeling conducted during the design of the rotor.

4.2.4. Stress Magnitude

Akin to temperature, the viscoelastic material response is closely linked to the stress magnitude. At low magnitudes, FRP composite materials typically display linear viscoelastic behavior. As the stress magnitude increases, the material begins displaying non-linear viscoelastic behavior. Experimental findings on different material systems indicate significant variation in the stress magnitude and temperature levels necessary to predict linear viscoelastic response [62]. Currently, there is no conclusive method for determining at what temperature and stress the material will transition from a linear to non-linear response. However, it has been shown that linear response, necessary for TTSP, and fatigue resistance, necessary for flywheel operation, can be ensured by limiting the temperature to below T_g [75] and stress to below 50% of the failure strength [76].

4.3. *Quasi-Static Analysis*

In 1957, Lekhnitskiy [77] defined the stress equilibrium equations for an arbitrary homogeneous anisotropic plate in cylindrical coordinates. These equations define the radial, circumferential, axial, and tangential (shear) equilibrium for an anisotropic body with applied forces, such as rotation, and the resulting internal stresses. Leknitskiy worked with thin plates assuming a plane stress state for the body. If a thin uniform circular disk is in equilibrium, axisymmetric, neither accelerating nor decelerating, and not experiencing out of plane forces, it means the only the radial equilibrium equation is non-trivial.

Leknitskiy's original analysis have been expanded upon with focus specifically on multi-rim FRP composite flywheel rotors. Chamis and Kiraly [78] applied analytical modeling to determine the stress and vibration induced in thin FRP flywheel rotors. They found that high aspect ratio flywheel rotors were the most weight efficient elements of a rotor, and that a flywheel can efficiently provide power in excess of 10 kW for several days when needed.

By the 1990s, analytical analysis of flywheel rotors had been generalized to predict the stress and displacement of multi rim flywheel rotors through work such as Gabrys and Bakis [79], Ha et al. [80], and Wild and Vickers [35]. Gabrys and Bakis developed a complete method for designing composite flywheel rotors from one or more FRP rims press-fitted together. Their method relied on defining an optimization routine that maximizes angular velocity, while ensuring radial and circumferential failures occur simultaneously. Through their method, the thickness of each rim in a press-fit rotor can be found, thus defining an optimal rotor design. They also state that rim materials should decrease in density and increase in stiffness as rims are positioned further from the axis of rotation. In other words, the densest and least stiff material should be used for the innermost rim, while the least dense and most stiff material should form the outer most rim. This recommendation is reasonable considering largest radial positions will experience the greatest loading from centripetal forces due to rotation and reaction forces from other rims deforming outward. At the same time, this design approach alleviates the buildup of radial tensile stress that acts transverse to the fibers, i.e., the direction with greatest susceptibility to failure.

Ha et al. [80] recognized that solving the analytical equations for multi-rim rotors results in a series of non-linear equations, which led them to develop a unique method for solving all the equation simultaneously, thus minimizing the time and computational effort needed to analyzed flywheel rotors. They then went on to apply a similar optimization routine as Gabrys and Bakis [79] to optimize the radial thickness of each rim for multi-rim rotors constructed of various materials. Ha et al. considered rotors with an embedded permanent magnet at the inner surface and up to four different rims: glass/epoxy, aramid/epoxy, and two different carbon/epoxy variants, i.e., AS/H3501, T300/5208, and IM6/epoxy. They showed that no multi-rim solution exists when density and stiffness decrease with radius, contrary to typical construction. The optimization algorithm always trended toward eliminating (i.e., zero radial rim thickness) all but the innermost rim.

Methods for solving Equation (13) to find radial displacement, radial stress, and circumferential stress have been described extensively in literature [16,80,81] so only a brief description is provided here. The radial equilibrium equation is as follows:

$$\frac{\partial \sigma_r}{\partial r} + \frac{\sigma_r - \sigma_\theta}{r} + \rho r \omega^2 = 0, \tag{13}$$

where σ is the internal stress in either the radial, subscript r, or circumferential, subscript θ, direction, ρ is the density of the material, and ω is the angular velocity. The stresses are defined by Hooke's law, Equation (5), and the stiffness matrix is defined with any of the Equations (6)–(8), depending on the material response. Fundamentally, a two-dimensional assumption can be made which is suitable for high aspect ratio flywheel rotors, i.e., thin rotors with radial dimensions significantly larger than axial dimensions. The directional strains are defined as:

$$\varepsilon_\theta = \frac{u_r}{r}; \quad \varepsilon_r = \frac{\partial u_r}{\partial r}; \quad \varepsilon_z = \varepsilon_{\theta z} = 0 \tag{14}$$

where u_r is the radial displacement and the subscript z signifies the rotor axial direction. Then, Equation (14) can be substituted into Hooke's law which is further substituted into Equation (13). This yields a second order inhomogeneous ordinary differential equation, which can be solved for the radial displacement and radial stress, yielding the following:

$$\begin{aligned} u_r &= -\rho\omega^2\varphi_0 r^3 + C_1\varphi_1 r^{\kappa} + C_2\varphi_2 r^{-\kappa} \\ \sigma_r &= -\rho\omega^2\varphi_3 r^2 + C_1 r^{\kappa-1} + C_2 r^{-\kappa-1} \end{aligned} \tag{15}$$

where φ and κ are constants based on the material properties of the rim, and C_1 and C_2 are integration constants, detailed in [80], which must be determined by the boundary conditions, see [81].

All research mentioned up to this point, and in fact the majority of flywheel research, has been conducted on relatively thin disks. Such rotor geometries tend to minimize material and fabrication costs and simplify analytical modeling by allowing for a two-dimensional or plane stress assumption. Additionally, axial stress arises merely due to Poisson's effects from the combination of radial and circumferential stress. Moreover, for typical rotor configurations, it is challenging to measure radial deformation experimentally. For these reasons, a thin composite disk is beneficial especially for research purposes.

While Ha et al. [82] has extensively explored modeling under plane stress, work by this group of researchers also involved two alternate assumptions: plane strain (PS) and modified generalized plane strain (MGPS). The PS assumption is true for a thick rotor where the axial dimension is significantly larger than the radial dimension, and defines the axial strain as zero while the axial stress is allowed to vary [81]. Generalized PS and MGPS allow the axial strain to vary according to a constant and a linear relation, respectively. Ha et al. compared the axial stress results for single-, two-, and three-rim rotor simulations conducted with PS, MGPS, and finite element modeling (FEM). They found axial stress results to have the best correlation between MGPS and FEM. For the two-dimensional case, such as the one solved using the model by Lekhnitskiy, plane stress and PS are identical because there is no third dimension for stress or strain. As the flywheel rotor increases in thickness, PS was shown to be more appropriate than plane stress approximately when the rotor radial dimension equals the axial dimension. While MGPS is relatively uncommon in modern flywheel research due to its complexity, PS and generalized PS are still part of contemporary research.

A number of studies have been published discussing analyses that specifically target flywheel rotor design for energy storage applications [14,46,47]. Much of recent research into FRP composite flywheels has focused on optimizing the design to minimize cost, in an effort to make the technology a more attractive alternative to other conventional storage technologies, primarily electrochemical batteries. Hearn et al. [83] and Rupp et al. [22] focused on minimizing FESS cost for public transportation. Both studies found rotors with rectangular cross sections and no more than three rims to be ideal for maximizing storage capacity while minimizing cost; a storage capacity of approximately 3 to 5 kWh was considered appropriate for public transportation. Recalling Equations (2) and (15), rectangular cross sections maximize the volume of material at a given radius while providing in-plane support for material at smaller radial locations. Rectangular cross section rotors are also comparatively easy to manufacture. Recent efforts [84] have employed advanced multi-factor optimization algorithms to develop methods for designing FESS appropriate for a wide range of application include grid storage, grid regulation [85], and energy storage in addition to public transport.

In the most recent decade, research has shown a trend to move away from either the PS or plane stress assumptions to include full three-dimensional analyses. Pérez-Aparicio and Ripoll [86] described exact solutions for the analytical equations in the radial, circumferential, axial, and tangential (shear) directions. They also compare two failure criteria, discussed later. Zheng et al. and Eraslan and Akis [41,87] discussed the instantaneous stresses induced in functionally graded rotating disks of variable thickness. A functionally graded rotor is one where the material properties smoothly vary as a function of radius, in contrast to a multi-rim rotor, where material properties change discretely. These results show carefully controlling rotor thickness and material properties can significantly reduce induced stress and minimizing the risk of failure due to crack initiation and propagation. The methods discussed in these studies are valuable tools in understanding rotor mechanics; however, they fail to consider aspects such as energy storage capacity and manufacturing costs.

While there has been significant development in the understanding and optimization of quasi-static composite rotor stress responses, there has been comparatively little development in the understanding of viscoelastic and dynamic behavior of composite rotors, which is the subject matter of the following two sections. This is especially surprising given one of the primary advantages of FESS over other storage systems is the expected long lifetimes of these systems.

4.4. Viscoelastic Analysis

Viscoelastic creep and stress relaxation continuously evolve over the operation of an FRP composite flywheel rotor. Viscoelasticity has been suggested to significantly affect the interface pressure at either the hub-rim or rim-rim interfaces, depending on rotor construction, which is critical for the integrity of rotors assembled via press-fitting. Creep rupture in the composite materials is an additional concern [88]. Trufanov and Smetannikov [89] investigated a flywheel rotor constructed from a variable thickness filament-wound composite wrapped in an organic plastic shell. They tracked the change in radial and circumferential stress at several key points over a simulated period of 10,000 h. Depending on the location in the shell, their results showed that circumferential tensile stresses can increase between 4% and 15% and radial compressive stresses could increase by up to 40%. In the composite rim, the maximum circumferential stress increased by 7.5%. At the same time, the maximum radial stress decreased by 33%. The construction of this flywheel is unusual for modern high-speed flywheel rotor; however, these results demonstrate that radial and circumferential stresses are highly variable and the potential for creep rupture or loss of interfacial pressure between rotor components exists.

Portnov and Bakis [90] presented complete solutions for the analytical equilibrium equations including creep. They studied a thick unidirectional FRP composite rim with rectangular cross section filament-wound around a small metallic hub. Their results showed that after complete relaxation, radial strain was maximized at the outer radius of the rotor, with strains being predicted to be approximately three times larger than the circumferential strain at the same position. This further supports the conclusion that creep rupture may be of significant concern.

Subsequent studies by Tzeng et al. [91,92] simulated arbitrarily long composite flywheel rotors press-fit or wound onto metallic hubs similar to those seen in industry [93,94]. They employed the generalized PS assumption due to the assumed length of the rotor and predicted stress and displacement in the radial and circumferential direction after 1 year, 10 years, and infinite time (10^{10} years). Similar to previous work, Tzeng showed that radial stress could decrease by as much as 35%, while circumferential stress could increase by up to 9%. Tzeng also studied flywheels with variable winding angles and found similar though slightly improved results.

While this body of work is compelling, the majority of it has been conducted analytically with relatively little available experimental data. Emerson [62] attempted to resolve this issue by, first, measuring the transverse strength and modulus of a glass fiber composite used in flywheel rotor construction, to improve simulation reliability, and, second, by taking in situ strain measurements using optoelectronic strain measurements. The material testing was conducted according to the methods described in Section 4.2. The flywheel measurements were to be conducted using a custom-built test apparatus. Unfortunately, this testing was inconclusive due to a series of mechanical failures and was not able to eliminate the possibility of creep, significantly impacting rotor structural health.

While some studies suggest that over extremely long times of operation, e.g., 10^{10} years or the time required to reach full relaxation, viscoelastic behavior of the composite can significantly impact rotor structural health by facilitating either creep rupture, the loss of rotor integrity by the loss of interfacial pressures between hub and rims, or both. However, the expected lifetime for flywheel rotors, as discussed, is between 10 and 20 years [5]. Furthermore, many of these studies occurred on either thick composite disks or arbitrarily long flywheel rotors. Skinner and Mertiny addressed this issue in [16], where a carbon FRP composite flywheel rotor was simulated for up to 10 years. The analytical process they followed to simulate the rotor behavior is similar to that pursued by previous researchers, so it is worth taking a brief aside to discuss this work here.

The analytical methodology used for viscoelastic simulations is fundamentally a quasi-static analysis; therefore, the viscoelastic solution procedure requires approximating time-varying behavior through a number of discrete time and load steps. The response at each step is used to calculate stress for the flywheel rotor throughout the simulation. First, the rotor dimensions, material properties, and simulation parameters—time and velocity vectors of interest—are defined as inputs to the algorithm. Then, beginning at the first time and velocity of interest, the material stiffness matrix is calculated for each rim of the flywheel rotor. Next, the boundary conditions at each interface and at the inner and outer surface of the rotor are calculated. Through these steps, the rotor response is calculated for the current time and velocity iteration. Finally, the algorithm proceeds to the next time and velocity of interest. Iterations continue for all discrete times and velocities of interest, which yields the induced stress for all points in the flywheel rotor at all times and velocities of interest.

The results from Skinner and Mertiny, Figure 5, showed that during operation, radial and circumferential stresses in the carbon FRP composite rotor were predicted to decrease by 1% and 5%, respectively. Additionally, as was seen by other researchers, interfacial pressure was predicted to have the most significant variation with an overall decrease of up to 36%. Despite these changes, viscoelastic stress relaxation is not expected to cause complete loss of interfacial pressure between hub and rim during the expected lifetime, nor is it expected to be a primary cause of failure. It was postulated that viscoelastic behavior of the material may play a role in other failure modes, such as fatigue damage and matrix cracking, but is ultimately unlikely to be the dominant cause for rotor failure.

4.5. Shear Stress

The presence of shear stresses in FRP composite flywheel rotors has not been studied extensively. Nevertheless, the analytical equilibrium equations have been defined for rotating anisotropic disks, and extensive work has been completed in this field for isotropic and functionally graded rotating disks of constant and variable thickness. An exact solution for the tangential (shear) equilibrium equation of a rotating disk was presented by Pérez Aparicio and Ripoll [86]. The equilibrium equation, given by Equation (16), has a similar form to the radial equilibrium equation, Equation (13):

$$\frac{d\tau_{r\theta}}{dr} + \frac{2}{r}\tau_{r\theta} + \rho\alpha r = 0 \tag{16}$$

where $\tau_{r\theta}$ is the in-plane shear stress and α is angular acceleration. Shear strain is defined as:

$$\gamma_{r\theta} = \frac{dv}{dr} - \frac{v}{r} \tag{17}$$

Solving the resulting second order inhomogeneous ordinary differential equation, in the same manner as previously discussed, yields the tangential stress and displacement equations:

$$v = C_1 r^{-1} + C_2 r + \frac{\rho\alpha}{8G_{r\theta}} r^3; \quad \tau_{r\theta} = G_{r\theta}\left[-\frac{2C_1}{r^2} + \frac{\rho\alpha}{4G_{r\theta}} r^2\right] \tag{18}$$

where v is the tangential displacement and C_1 and C_2 are integration constants. Notice that tangential stress is dependent on a single integration constant because when the strain, Equation (17), is substituted into tangential displacement, the second integration constant, C_2, is eliminated. The integration constants can be found through the boundary conditions as functions of the rotor geometry, density, shear modulus, and angular acceleration. Pérez Aparicio and Ripoll considered a worst-case scenario where peak shear stress is caused by a severe acceleration of 3.6×10^5 rad/s^2. For this considered worst-case scenario, resulting stress states were described as possibly critical for the hub rather than the rotor.

Tang [95] conducted an early study on shear stress in accelerating disks mounted to a ridged shaft. They showed that shear stress was dependent on the acceleration and the ratio between the inner and outer rotor radius. When this ratio is greater than 0.15, the shear stress will increase drastically and may need to be considered when designing structural components.

Much of the studies on shear stress in rotating disks focuses on variable thickness and functionally graded materials for applications in turbines and engines. Reddy and Srinath [96] presented a method to study acceleration in high-temperature rotating disks with variable thickness. They showed that the cross section of the disk may have a significant impact on shear stress and should, therefore, not be discounted. Continuing with rotating disks for turbine applications, Eraslan and Akais [87] and Zheng et al. [41] presented a method to analyze instantaneous shear stress in rotating disks. They showed that carefully controlling the rotor cross section and properties produces an optimum stress profile. Zheng et al. also showed that the presence of shear stress can shift the maximum stress location from the inner radius to near the mid-radius, depending on shear stress magnitude and direction. Note, shear stress directionality is relative to the rotating direction, where accelerating the rotor causes positive shear stress and decelerating the rotor causes negative shear stress. Shear direction is important, for example, for predicting failure such as using the Tsai-Wu criteria discussed below.

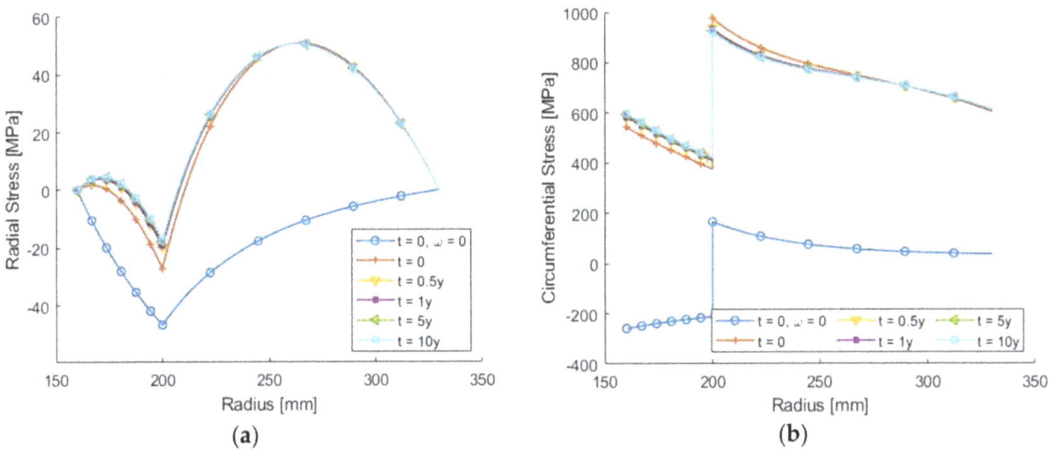

Figure 5. Evolution of (**a**) radial and (**b**) circumferential stresses at different times of operation (0–10 years) of a flywheel rotor with an aluminum hub and carbon FRP composite rim due to viscoelastic stress relaxation [16].

Salehian et al. [97] investigated instantaneous shear stress in functionally graded constant and variable thickness rotating disks. They conducted both analytical and numerical analyses. The functionally graded flywheels they studied featured increasing material density as a function of radius. They also showed that both methods are equally accurate and that shear stress can be significant for functionally graded materials.

Previous studies were conducted assuming an essentially instantaneous event subjecting a rotating disk to angular acceleration. However, in the context of FESS, shear stress created by accelerating or decelerating the flywheel rotor should be considered for typical FESS energy transfer, i.e., the supply or demand of power. The relationship between power and acceleration is found through the applied torque, such that:

$$P = T\omega; T = I_r\alpha \tag{19}$$

where P is power and T is torque. From Equation (19), it is clear that power is related linearly to angular acceleration and velocity at a given instant. Furthermore, from Equation (18), shear stress is linearly related to angular acceleration. Therefore, even for constant acceleration, power varies over time, and so do radial and circumferential stresses as the velocity changes due to angular acceleration. Considering the opposite case of constant power, acceleration necessarily needs to vary. For example, at an initially low angular velocity and constant power supply, the flywheel rotor acceleration and shear stresses would be much larger than at a later time when velocity has increased due to the imposed acceleration.

Combining Equations (18) and (19), it is possible to determine the stress state as a result of a given power supply or demand, and vice versa. Recalling the work by Pérez Aparicio and Ripoll [86] mentioned above, a flywheel rotor was simulated with an inner radius, outer radius, height, and density of 0.08 m, 0.2 m, 0.06 m, and 1800 kg/m^3, respectively. For an angular velocity of 17,425 rpm (1827.6 s^{-1}), a supplied power of 1.67 GW is associated with an angular acceleration of 3.6 × 10^5 s^{-2} for 0.005 s. Pérez Aparicio and Ripoll explained that power supplied at this magnitude would occur in specific applications, such as military artillery; however, it is atypical for energy storage systems.

The shear stress investigations discussed above presented solutions to analytical equilibrium equations and described instantaneous behavior of variable thickness FRP and functionally graded rotating disks. Moreover, shear stress resulting from a given peak acceleration of a flywheel rotor was discussed. However, the technical literature is ambiguous regarding time-dependent behavior, evolution of the rotor stress states, and possible damage events resulting from typical operating conditions, i.e., repeated energy transfer cycles over the flywheel lifetime.

5. Failure Analysis

5.1. Failure Criteria

Several criteria have been applied to predicting failure of FRP composite flywheel rotors. A large body of the available research considers rotor failure a quasi-static process caused by excessive loading from centripetal forces due to rotation exceeding material ultimate strengths [45]. The most common failure models are the maximum stress or strain [98], von Mises [41], and Tsai-Wu failure criteria [16,99]. Additionally, attempts have been made to predict rotor failure with progressive damage models [100]. Other less common methods, such as the Christensen model [86], have been used to a limited extent for predicting the failure of composite flywheel rotors.

5.2. Maximum Stress Criterion

The maximum stress and maximum strain failure criteria are the most widely used due to their simple application and analysis. The maximum stress failure criterion defines the failure ratio in each material direction to be the ratio of the applied stress to the failure strength. Consider the failure stress in the fiber direction of the material in the tensile or compressive direction to be σ_{1t} or σ_{1c}, respectively. In the transverse directions, the material is assumed to be transversely isotropic such that the 2 and 3 directions are congruent; thus, $\sigma_{2t} = \sigma_{3t}$ and $\sigma_{2c} = \sigma_{3c}$. Shear stress is dominated by matrix deformation τ_{12} and τ_{23}. With the applied stress tensor as $[\sigma_\theta, \sigma_z, \sigma_r, \tau_{r\theta}]$, the maximum stress failure criterion is defined as:

$$\begin{gathered}\frac{\sigma_\theta}{\sigma_{1t}} \leq 1 \text{ if } \sigma_\theta \geq 0 \text{ or} \frac{|\sigma_\theta|}{\sigma_{1c}} \leq 1 \text{ if } \sigma_\theta \leq 0, \\ \frac{\sigma_z}{\sigma_{2t}} \leq 1 \text{ if } \sigma_z \geq 0 \text{ or} \frac{|\sigma_z|}{\sigma_{2c}} \leq 1 \text{ if } \sigma_z \leq 0, \\ \frac{\sigma_3}{\sigma_{3t}} \leq 1 \text{ if } \sigma_r \geq 0 \text{ or} \frac{|\sigma_r|}{\sigma_{3c}} \leq 1 \text{ if } \sigma_r \leq 0, \\ \frac{|\tau_{r\theta}|}{\tau_{12}} \leq 0 \text{ and } \frac{|\tau_{rz}|}{\tau_{23}} \leq 0\end{gathered} \tag{20}$$

Failure occurs when any of the above ratios is larger than unity. Similar inequalities can be written for the maximum strain criteria to find the ratio between applied strain and failure strain. While these criteria are well suited to predict failure when the primary failure mode is uniaxial loading, they neglect load interactions in a rotor.

5.3. Tsai-Wu Criterion

To address multiaxial loading conditions present in flywheel rotors, the Tsai-Wu failure criterion is frequently employed. The Tsai-Wu failure criterion involves independent interaction terms, considers strength parameters both for tension and compression, and enables treating different classes of materials, multi-axial stress, and multi-dimensional space [101]. As presented by Tsai and Wu, this method considers 27 independent terms which normalize the applied stress in a particular direction with the strength parameter in that direction. If the sum of these terms, called the failure index F, is equal to unity, failure is predicted. When applied to FRP flywheel rotors, the analysis problem is often simplified using material symmetry and certain modeling assumptions. For example, consider a thin, transversely isotropic FRP rotor operating at constant velocity, axial stress terms can be neglected, and all out-of-plane and shear terms vanish. Therefore, the Tsai-Wu criterion can be reduced to six terms. Depending on the material and modeling assumptions, the exact number of terms that must be considered will vary. It is worth noting, when applied to an isotropic material with equal tensile and compressive strengths, the Tsai Wu criteria will simplify to the von Mises failure criterion [102]. Therefore, the Tsai-Wu criterion can expediently be applied to multi-material flywheel rotors where the hub and rims may be constructed from materials that are either isotropic, e.g., metals, or anisotropic, e.g., FRP composites. The Tsai-Wu failure criterion, which has widely been applied for the failure prediction in FRP flywheel rotors for decades [16,32,99,103], is given for a three-dimensional transversely isotropic material as:

$$F = F_{11}\sigma_1^2 + F_{22}\left(\sigma_2^2 + \sigma_3^2\right) + (2F_{22} - F_{44})\sigma_2\sigma_3 + 2F_{12}\sigma_1(\sigma_3 + \sigma_2) + F_1(\sigma_1 + \sigma_2) + F_2\sigma_3 + F_{44}\tau_{23}^2 + 2F_{66}\left(\tau_{12}^2\right) \quad (21)$$

where F_{ij} are material coefficients dependent on the tensile and compressive strengths in each direction. A complete list of coefficients is available in [102]. The Tsai-Wu failure criterion can be modified to find the strength ratio (SR), which is the ratio between the applied stress and the failure stress [16,80,100]. Failure is predicted when SR is greater than or equal to unity. This approach provides an intuitive and easily represented term, which facilitates the comparison of combined stresses across the entire flywheel rotor.

5.4. Progressive Failure Analysis

Progressive failure analysis (PFA) has been applied to composite rotors and other structures in a number of studies in the proceeding decade [30,100,104,105]. The premise underlying this approach is that composite materials may initially experience benign failure modes, e.g., matrix micro-cracking and interlaminar fracture, without complete loss of structural integrity. In this case, the structure can continue to support applied loads until the accumulation of damage causes ultimate (catastrophic) failure. As applied to flywheel rotors, matrix damage such as cracking, delamination, and interlaminar fracture can be classified as benign failure modes while fiber rupture is considered catastrophic. This type of failure analysis is iterative. First, rotor simulations are conducted as discussed above to determine the maximum rotor velocity, and failure mode and location. In case of a benign failure mode, a knockdown factor that depends on the failure mode and the material characteristics is applied to the material properties at that location. This process is repeated until catastrophic failure is predicted [99].

PFA has been shown to accurately predict failure dynamics in woven composite disks [30]; however, only limited studies have been conducted on filament-wound flywheel rotors [100]. In contrast to radially oriented fibers, in the woven disk designs, the fibers provide the majority of radial support for the rotor to resist the centripetal forces. However, this is not the case for filament-wound flywheel rotors where radial stresses are borne chiefly by the matrix. Notably, circumferential matrix fracture in a filament-wound rotor would result in practically complete loss of radial integrity. Furthermore, analytical methods described above assume the rotor to be continuous; however, progressive damage events may introduce discontinuities which may or may not violate this assumption. For example, if a damage location, such as a circumferential matrix fracture, is under compressive stress, then crack closure may ensue, and hence, a continuity assumption could be upheld. In such as case, the fractured structure could be considered as two separate rims of the same material that are press-fitted together. However, under tensile stress, the crack is forced open, violating the continuity assumption. Situation like these have not been addressed in the technical literature, so further studies into PFA are needed to better understand its applicability to predicting FRP flywheel rotor failure.

6. Conclusions and Prospects

The present entry has presented an overview of the mechanical design of flywheel energy storage systems with discussions of manufacturing techniques for flywheel rotors, analytical modeling of flywheel rotors including multi-rim configurations, and contemporary failure criteria. Flywheel construction employing metallic hubs and rotors was also considered, as was the assembly of components by either filament-winding or press-fitting. Analytical techniques for modeling multi-rim flywheel rotors constructed from either metallic or FRP composite materials were described for quasi-static, viscoelastic, and variable angular velocity operating conditions. Finally, contemporary failure criteria were discussed along with their advantages and limitations. Clearly, the understanding of flywheel rotor construction, analysis, and failure prediction has advanced significantly in the last several decades. Nevertheless, despite flywheel energy storage being a maturing field, some gaps in understanding still exist. For example, further investigations into the cost of manufacturing and the efficacy of variable winding angle flywheel rotors seems warranted. Further studies on the effects of shear stress and time-dependent effects, including cyclic loading and fatigue, in FRP composite rotors may be warranted to better understand behavior and improve failure predictions for flywheel rotors for long-term operation. Additionally, experimental data characterizing long-term behavior of FRP composite materials, especially in the transverse direction, would be valuable for improving the accuracy of long-term modeling of stress and failure predictions. Finally, progressive damage failure analysis, while compelling, would benefit substantially from experimental validation of modeling results to clearly discern its merit compared to other failure predictions.

Author Contributions: Conceptualization and methodology, M.S. and P.M.; validation, formal analysis, and investigation, M.S.; resources, P.M.; data curation, M.S.; writing—original draft preparation and visualization, M.S.; writing—review and editing, M.S. and P.M.; supervision, project administration, and funding acquisition, P.M. All authors have read and agreed to the published version of the manuscript.

Funding: This work was funded by the Canada First Research Excellence Fund with grant number Future Energy Systems T06-P03.

Conflicts of Interest: The authors declare no conflict of interest.

Entry Link on the Encyclopedia Platform: https://encyclopedia.pub/20231.

References

1. British Petroleum Statistical Review of World Energy. *Globally Consistent Data on World Energy Markets and Authoritative Publications in the Field of Energy*; British Petroleum: London, UK, 2021; Volume 70.
2. Chen, H.; Cong, T.N.; Yang, W.; Tan, C.; Li, Y.; Ding, Y. Progress in electrical energy storage system: A critical review. *Prog. Nat. Sci.* **2009**, *19*, 291–312. [CrossRef]
3. Kåberger, T. Progress of renewable electricity replacing fossil fuels. *Glob. Energy Interconnect.* **2018**, *1*, 48–52. [CrossRef]
4. Moriarty, P.; Honnery, D. Can renewable energy power the future? *Energy Policy* **2016**, *93*, 3–7. [CrossRef]
5. Hadjipaschalis, I.; Poullikkas, A.; Efthimiou, V. Overview of current and future energy storage technologies for electric power applications. *Renew. Sustain. Energy Rev.* **2009**, *13*, 1513–1522. [CrossRef]
6. Denholm, P.; O'Connell, M.; Brinkman, G.; Jorgenson, J. *Overgeneration from Solar Energy in California: A Field Guide to the Duck Chart*; National Renewable Energy Lab. (NREL): Golden, CO, USA, 2015.
7. Amiryar, M.E.; Pullen, K.R. A review of flywheel energy storage system technologies and their applications. *Appl. Sci.* **2017**, *7*, 286. [CrossRef]
8. Sabihuddin, S.; Kiprakis, A.E.; Mueller, M. A numerical and graphical review of energy storage technologies. *Energies* **2015**, *8*, 172–216. [CrossRef]
9. Ilan, D. The ground stone components of drills in the ancient Near East: Sockets, flywheels, cobble weights, and drill bits. *J. Lithic Stud.* **2016**, *3*, 261–277. [CrossRef]
10. Skinner, M. Characterization of Passibe Sischarge Losses in a Flywheel Energy Storage System. Masters's Thesis, University of Alberta, Edmonton, AB, Canada, 2017. [CrossRef]
11. Luo, X.; Wang, J.; Dooner, M.; Clarke, J. Overview of current development in electrical energy storage technologies and the application potential in power system operation. *Appl. Energy* **2015**, *137*, 511–536. [CrossRef]
12. Hebner, R.; Beno, J.; Walls, A. Flywheel batteries come around again. *IEEE Spectr.* **2002**, *39*, 46–51. [CrossRef]

13. Bolund, B.; Bernhoff, H.; Leijon, M. Flywheel energy and power storage systems. *Renew. Sustain. Energy Rev.* **2007**, *11*, 235–258. [CrossRef]
14. Krack, M.; Secanell, M.; Mertiny, P. Rotor Design for High-Speed Flywheel Energy Storage Systems. In *Energy Storage in the Emerging Era of Smart Grids*; InTech: London, UK, 2011.
15. Skinner, M.; Suess, M.; Secanell, M.; Mertiny, P. Design of a Composite Flywheel Rotor For Long-Term Energy Storage in Residential Applications. In Proceedings of the The Canadian Society of Mechanical Engineering International Congress, Kelowna, BC, Canada, 26–29 June 2016; pp. 1–5.
16. Skinner, M.; Mertiny, P. Effects of Viscoelasticity on the Stress Evolution over the Lifetime of Filament-Wound Composite Flywheel Rotors for Energy Storage. *Appl. Sci.* **2021**, *11*, 9544. [CrossRef]
17. Pullen, K. The Status and Future of Flywheel Energy Storage. *Joule* **2019**, *3*, 1394–1399. [CrossRef]
18. The GYROBUS: Something New Under the Sun? *Motor Trend*, January 1952; 37.
19. Wakefield, E. *History of the Electric Automobile: Hybrid Electric Vehicles*; Society of Automotive Engineers: Warrendale, PA, USA, 1998; ISBN 978-0-7680-0125-9.
20. Weiss, C.C. Volvo Confirms Fuel Savings of 25 Percent with Flywheel KERS. Available online: https://newatlas.com/volvo-flywheel-kers-testing/27273/ (accessed on 9 December 2021).
21. Porche GT3R Technical Specs. Available online: https://www.porsche.com/international/_iceland_/motorsportandevents/motorsport/customerracing/racingcars/991-2nd-gt3-r/ (accessed on 9 December 2021).
22. Rupp, A.; Baier, H.; Mertiny, P.; Secanell, M. Analysis of a Flywheel Energy Storage System for Light Rail Transit. *Energy* **2016**, *107*, 625–638. [CrossRef]
23. Tarrant, C. Kinetic Energy Storage Wins Acceptance. Available online: https://www.railwaygazette.com/kinetic-energy-storage-wins-acceptance/27250.article (accessed on 9 December 2021).
24. NRStor Inc. 2 MW Minto Flywheel Facility: A Fast-Ramping Resource for Grid Regulation and Other Electricity Services. Available online: http://nrstor.com/2019/11/21/2-mw-minto-flywheel-facility-market-impact-case-study-power-advisory/ (accessed on 9 December 2021).
25. Beacon Power, Operating Plants Stephenton, New York. Available online: https://beaconpower.com/stephentown-new-york/ (accessed on 9 December 2021).
26. Amber Kinetics. The World's Only Flywheel Innovation Hub. Available online: https://amberkinetics.com/installation/the-worlds-only-flywheel-innovation-hub/ (accessed on 9 December 2021).
27. Genta, G. *Kinetic Energy Storage: Theory and Practice of Advanced Flywheel Sysstems*; Butterworth-Heinemann: London, UK, 2014; ISBN 0-408-01396-6.
28. Ha, S.K.; Han, H.H.; Han, Y.H. Design and manufacture of a composite flywheel press-fit multi-rim rotor. *J. Reinf. Plast. Compos.* **2008**, *27*, 953–965. [CrossRef]
29. Kale, V.; Thomas, M.; Secanell, M. On determining the optimal shape, speed, and size of metal flywheel rotors with maximum kinetic energy. *Struct. Multidiscip. Optim.* **2021**, *64*, 1481–1499. [CrossRef]
30. Wang, Y.; Dai, X.; Wei, K.; Guo, X. Progressive failure behavior of composite flywheels stacked from annular plain profiling woven fabric for energy storage. *Compos. Struct.* **2018**, *194*, 377–387. [CrossRef]
31. Ornaghi, H.L.; Neves, R.M.; Monticeli, F.M.; Almeida, J.H.S. Viscoelastic characteristics of carbon fiber-reinforced epoxy filament wound laminates. *Compos. Commun.* **2020**, *21*, 100418. [CrossRef]
32. Takkar, S.; Gupta, K.; Tiwari, V.; Singh, S.P. Dynamics of Rotating Composite Disc. *J. Vib. Eng. Technol.* **2019**, *7*, 629–637. [CrossRef]
33. Eggers, F.; Almeida, J.H.S.; Azevedo, C.B.; Amico, S.C. Mechanical response of filament wound composite rings under tension and compression. *Polym. Test.* **2019**, *78*, 105951. [CrossRef]
34. Rejab, R.; Kumar, N.M.; Ma, Q.; Idris, M.S.; Zhang, B.; Merzuki, M.N.M. Wireless technology applied in 3-axis filament winding machine control system using MIT app inventor Wireless technology applied in 3-axis filament winding machine control system using MIT app inventor. *IOP Conf. Ser. Mater. Sci. Eng.* **2019**, *469*, 012030.
35. Wild, P.M.; Vickers, G.W. Analysis of filament-wound cylindrical shells under combined centrifugal, pressure and axial loading. *Compos. Part A Appl. Sci. Manuf.* **1997**, *28*, 47–55. [CrossRef]
36. Sayem Uddin, M.; Morozov, E.V.; Shankar, K. The effect of filament winding mosaic pattern on the stress state of filament wound composite flywheel disk. *Compos. Struct.* **2014**, *107*, 260–275. [CrossRef]
37. Tzeng, J.T.; Emerson, R.P.; O'Brien, D.J. Viscoelasticity Analysis and Experimental Validation of Anisotropic Composite Overwrap Cylinders. In Proceedings of the ASME 2012 International Mechanical Engineering Congress and Exposition, Houston, TX, USA, 9–15 November 2012; p. 429. [CrossRef]
38. Ertz, G. Development, Manufacturing and Testing of a Multi-Rim (Hybrid) Flywheel Rotor. Diploma Thesis, Leibniz Universität Hannover, Hannover, Germany, 2014.
39. Skinner, M.; Mertiny, P. Experimental Characterization of Low-Speed Passive Discharge Losses of a Flywheel Energy Storage System. *Appl. Mech.* **2021**, *2*, 1–15. [CrossRef]
40. Kale, V.; Secanell, M. A comparative study between optimal metal and composite rotors for flywheel energy storage systems. *Energy Rep.* **2018**, *4*, 576–585. [CrossRef]
41. Zheng, Y.; Bahaloo, H.; Mousanezhad, D.; Mahdi, E.; Vaziri, A.; Nayeb-Hashemi, H. Stress analysis in functionally graded rotating disks with non-uniform thickness and variable angular velocity. *Int. J. Mech. Sci.* **2016**, *119*, 283–293. [CrossRef]

42. Yeh, K.Y.; Han, R.P.S. Analysis of high-speed rotating disks with variable thickness and inhomogeneity. *J. Appl. Mech. Trans. ASME* **1994**, *61*, 186–191. [CrossRef]
43. Ertz, G.; Twiefel, J.; Krack, M. Feasibility Study for Small Scaling Flywheel-Energy-Storage Systems in Energy Harvesting Systems. *Energy Harvest. Syst.* **2014**, *1*, 233–241. [CrossRef]
44. Ha, S.K.; Kim, M.H.; Han, S.C.; Sung, T.H. Design and spin test of a hybrid composite flywheel rotor with a split type hub. *J. Compos. Mater.* **2006**, *40*, 2113–2130. [CrossRef]
45. Hartl, S.; Schulz, A.; Sima, H.; Koch, T.; Kaltenbacher, M. A Static Burst Test for Composite Flywheel Rotors. *Appl. Compos. Mater.* **2016**, *23*, 271–288. [CrossRef]
46. Han, Y.; Ren, Z.; Tong, Y. General Design Method of Flywheel Rotor for Energy Storage System. *Energy Procedia* **2012**, *16*, 359–364. [CrossRef]
47. Mittelstedt, M.; Hansen, C.; Mertiny, P. Design and multi-objective optimization of fiber-reinforced polymer composite flywheel rotors. *Appl. Sci.* **2018**, *8*, 1256. [CrossRef]
48. Krack, M.; Secanell, M.; Mertiny, P. Cost optimization of hybrid composite flywheel rotors for energy storage. *Struct. Multidiscip. Optim.* **2010**, *41*, 779–795. [CrossRef]
49. Skinner, M.; Secanell Gallart, M.; Mertiny, P. Observed Effects of Vibrationally Induced Fretting on Bearing–Shaft Systems in Flywheel Energy Storage Systems. *J. Fail. Anal. Prev.* **2018**, *18*, 837–845. [CrossRef]
50. Allam, M.N.M.; Tantawy, R.; Yousof, A.; Zenkour, A.M. Elastic and viscoelastic stresses of nonlinear rotating functionally graded solid and annular disks with gradually varying thickness. *Arch. Mech. Eng.* **2017**, *64*, 423–440. [CrossRef]
51. Long, Z.; Zhiping, Q. Review of Flywheel Energy Storage System. In *Proceedings of ISES World Congress 2007 (Vol. I–Vol. V)*; Springer: Berlin/Heidelberg, Germany, 2008; pp. 2815–2819.
52. Lai, W.M.; Rubin, D.; Krempl, E. Chapter 5: The Elastic Solid. In *Introduction to Continuum Mechanics*; Elsevier: Amsterdam, The Netherlands, 2010; pp. 201–352.
53. Ding, H.; Chen, W.L.Z. *Elasticity of Transversely Isotropic Materials*; Gladwell, G.M.L., Ed.; Springer: Dordrecht, The Netherlands, 2006; ISBN 9781119130536.
54. Zhao, J.; Song, X.; Liu, B. Standardized compliance matrixes for general anisotropic materials and a simple measure of anisotropic degree based on shear extension coupling coefficient. *Int. J. Appl. Mech.* **2011**, *8*, 1–28.
55. Lakes, R. *Viscoelastic Materials*; Cambridge University Press: Cambridge, UK, 2009; Volume 1, ISBN 9780511626722.
56. Buchroithner, A.; Haidl, P.; Birgel, C.; Zarl, T.; Wegleiter, H. Design and experimental evaluation of a low-cost test rig for flywheel energy storage burst containment investigation. *Appl. Sci.* **2018**, *8*, 2622. [CrossRef]
57. Rojas, J.I.; Nicolás, J.; Crespo, D. Study on mechanical relaxations of 7075 (Al-Zn-Mg) and 2024 (Al-Cu-Mg) alloys by application of the time-temperature superposition principle. *Adv. Mater. Sci. Eng.* **2017**, *2017*, 2602953. [CrossRef]
58. Mahdavi, R.; Goodarzi, V.; Jafari, S.H.; Saeb, M.R.; Shojaei, S.; Khonakdar, H.A. Experimental analysis and prediction of viscoelastic creep properties of PP/EVA/LDH nanocomposites using master curves based on time—Temperature superposition. *J. Appl. Polym. Sci.* **2018**, *135*, 46725. [CrossRef]
59. Barbero, E.J. *Time-Temperature-Age Superposition Principle for Predicting Long-Term Response of Linear Viscoelastic Materials*, 2nd ed.; Elsevier: Amsterdam, The Netherlands, 2019; ISBN 9780081026014.
60. Yeow, Y.; Morris, D.; Brinson, H. Time-Temperature Behavior of a Unidirectional Graphite/Epoxy Composite. In *Fifth Conference on Composite Materials: Testing and Design*; ASTM International: West Conshohocken, PA, USA, 1979; pp. 1–37.
61. Koyanagi, J.; Sato, M. *Time and Temperature Dependence of Transverse Tensile Failure of Unidirectional Carbon Fiber-Reinforced Polymer Matrix Composites*, 2nd ed.; Elsevier: Amsterdam, The Netherlands, 2019; ISBN 9780081026014.
62. Emerson, R.P. Viscoelastic Flywheel Rotors: Modeling and Measurement. Ph.D. Thesis, Pennsylvania State University, State College, PA, USA, 2002.
63. Aniskevich, A.; Glaskova-Kuzmina, T. *Effect of Moisture on Elastic And Viscoelastic Properties of Fiber Reinforced Plastics: Retrospective and Current Trends*, 2nd ed.; Elsevier: Amsterdam, The Netherlands, 2019; ISBN 9780081026014.
64. Alwis, K.G.N.C.; Burgoyne, C.J. Time-temperature superposition to determine the stress-rupture of aramid fibres. *Appl. Compos. Mater.* **2006**, *13*, 249–264. [CrossRef]
65. Sihn, S.; Tsai, S. Automated shift for time-temperature superposition. In Proceedings of the 12th International Comittee on Composite Materials, Paris, France, 5–9 July 1999; Volume 51.
66. Brinson, H.F.; Griffith, W.I.; Morris, D.H. Creep Rupture of Polymer-matrix Composites. *Exp. Mech.* **1981**, *57*, 329–335. [CrossRef]
67. Brinson, H.F. Mechanical and optical viscoelastic characterization of Hysol 4290. *Exp. Mech.* **1968**, *8*, 561–566. [CrossRef]
68. Williams, M.L.; Landel, R.F.; Ferry, J.D. The Temperature Dependence of Relaxation Mechanisms in Amorphous Polymers and Other Glass-forming Liquids. *J. Am. Chem. Soc.* **1955**, *77*, 3701–3707. [CrossRef]
69. Krauklis, A.E.; Akulichev, A.G.; Gagani, A.I.; Echtermeyer, A.T. Time-temperature-plasticization superposition principle: Predicting creep of a plasticized epoxy. *Polymers* **2019**, *11*, 1848. [CrossRef] [PubMed]
70. Ganß, M.; Satapathy, B.K.; Thunga, M.; Weidisch, R.; Pötschke, P.; Janke, A. Temperature dependence of creep behavior of PP-MWNT nanocomposites. *Macromol. Rapid Commun.* **2007**, *28*, 1624–1633. [CrossRef]
71. Jain, N.; Verma, A.; Singh, V.K. Dynamic Mechanical Analysis and Creep-recovery behaviour of Polyvinyl Alcohol based cross-linked Biocomposite reinforced with Basalt fiber. *Mater. Res. Express* **2019**, *6*, 105373. [CrossRef]

72. Gergesova, M.; Zupančič, B.; Saprunov, I.; Emri, I. The closed form t-T-P shifting (CFS) algorithm. *J. Rheol.* **2011**, *55*, 1–16. [CrossRef]
73. Bradshaw, R.D.; Brinson, L.C. Recovering nonisothermal physical aging shift factors via continuous test data: Theory and experimental results. *J. Eng. Mater. Technol. Trans. ASME* **1997**, *119*, 233–241. [CrossRef]
74. Sullivan, J.L. Creep and physical aging of composites. *Compos. Sci. Technol.* **1990**, *39*, 207–232. [CrossRef]
75. Lou, Y.C. Viscoelastic Characterization of Nonlinear Fiber-Reinforced Plastic. *J. Compos. Mater.* **1971**, *5*, 208–234. [CrossRef]
76. Stinchcomb, W.W.; Bakis, C.E. Fatigue Behavior of Composite Laminates. *Compos. Mater. Ser.* **1991**, *4*, 105–180. [CrossRef]
77. Lekhnitskiy, S.G. *Anisotropic Plates*; Tekhniko-Teoreticheskoy Literatury; Air Force Systems Command: Moscow, Russia, 1957.
78. Chdnis, C.C.; Kiruly, L.J. *Rim-Spoke Composite Flywheels—Stress a N D Vibration Analysis*; NASA Lewis Research Center: Cleveland, OH, USA, 1976.
79. Gabrys, C.W.; Bakis, C.E. Design and Testing of Composite Flywheel Rotors. In *Composite Materials: Testing and Design, Thirteenth Volume*; ASTM International: West Conshohocken, PA, USA, 1997; pp. 1–22.
80. Ha, S.K.; Yang, H.-I.; Kim, D.-J. Optimum design of a hybrid composite flywheel with permanent magnet rotor. *J. Compos. Mater.* **1999**, *33*, 1544–1575. [CrossRef]
81. Kelly, P. Mechanics Lecture Notes. Available online: http://homepages.engineering.auckland.ac.nz/~{}pkel015/SolidMechanicsBooks/index.html (accessed on 15 November 2021).
82. Ha, S.K.; Kim, D.J.; Sung, T.H. Optimum design of multi-ring composite flywheel rotor using a modified generalized plane strain assumption. *Int. J. Mech. Sci.* **2001**, *43*, 993–1007. [CrossRef]
83. Hearn, C.S.; Flynn, M.M.; Lewis, M.C.; Thompson, R.C.; Murphy, B.T.; Longoria, R.G. Low cost flywheel energy storage for a fuel cell powered transit bus. In Proceedings of the IEEE Vehicle Power and Propulsion Conference, Arlington, TX, USA, 9–12 September 2007; pp. 829–836. [CrossRef]
84. Krack, M.; Secanell, M.; Mertiny, P. Advanced optimization strategies for cost-sensitive design of energy storage flywheel rotors. *J. Adv. Mater.* **2011**, *43*, 65–78. [CrossRef]
85. Kheawcum, M.; Sangwongwanich, S. A Case Study on Flywheel Energy Storage System Application for Frequency Regulation of Islanded Amphoe Mueang Mae Hong Son Microgrid. In Proceedings of the 17th International Conference on Electrical Engineering/Electronics, Computer, Telecommunications and Information Technology, ECTI-CON 2020, Phuket, Thailand, 24–27 June 2020; pp. 421–426.
86. Pérez-Aparicio, J.L.; Ripoll, L. Exact, integrated and complete solutions for composite flywheels. *Compos. Struct.* **2011**, *93*, 1404–1415. [CrossRef]
87. Eraslan, A.N.; Akis, T. On the plane strain and plane stress solutions of functionally graded rotating solid shaft and solid disk problems. *Acta Mech.* **2006**, *181*, 43–63. [CrossRef]
88. Saleeb, A.F.; Arnold, S.M.; Al-Zoubi, N.R. A study of time-dependent and anisotropic effects on the deformation response of two flywheel designs. In Proceedings of the 14th Symposium on Composite Materials: Testing and Design, Pittsburgh, PA, USA, 11–12 March 2003.
89. Trufanov, N.A.; Smetannikov, O.Y. Creep of Composite Energy Accumulators. *Strength Mater.* **1991**, *23*, 671–675.
90. Portnov, G.G. Estimation of Limit Strains in Disk-Type Flywheels Made of Compliant Elastomeric Matrix Composite Undergoing Radial Creep. *Mech. Compos. Mater.* **2000**, *36*, 55–58. [CrossRef]
91. Tzeng, J.T. *Viscoelastic Analysis of Composite Flywheel for Energy Storage*; Army Research Laboratory: Aberdeen Proving Ground, Aberdeen, MD, USA, 2001.
92. Ghosh, T.N.; Tzeng, J.T.; Emerson, R.P.; O'Brien, D.J.; Ghosh, T.N. Viscoelasticity Analysis and Experimental Validation of Anisotropic Composite Overwrap Cylinders. *Trans. Ophthalmol. Soc. UK* **2012**, *101*, 200–202. [CrossRef]
93. Levistor Boosting Forecourt Grid Power for the Next Generation of Fast Charging Electric Vehicles. Available online: https://levistor.com/#about (accessed on 9 December 2021).
94. Stornetic GmbH Powerful Storage System for Grid Services. Available online: https://stornetic.com/assets/downloads/stornetic_general_presentation.pdf (accessed on 9 December 2021).
95. Tang, S. Note on acceleration stress in a rotating disk. *Int. J. Mech. Sci.* **1970**, *12*, 205–207. [CrossRef]
96. Reddy, T.Y.; Srinath, H. Effects of acceleration stresses on the yielding of rotating disks. *Int. J. Mech. Sci.* **1974**, *16*, 593–596. [CrossRef]
97. Salehian, M.; Shahriari, B.; Yousefi, M. Investigating the effect of angular acceleration of the rotating disk having variable thickness and density function on shear stress and tangential displacement. *J. Braz. Soc. Mech. Sci. Eng.* **2019**, *41*, 1–11. [CrossRef]
98. Li, S. The Maximum Stress Failure Criterion and the Maximum Strain Failure Criterion: Their Unification and Rationalization. *J. Compos. Sci.* **2020**, *4*, 157. [CrossRef]
99. Ha, S.K.; Lee, D.G.; Kim, D.J. Optimization of hybrid composite rotor in flywheel battery. *SAE Tech. Pap.* **1998**. [CrossRef]
100. Corbin, C.K. Burst Failure Prediction of Composite Flywheel Rotors: A Progressive Damage Approach via Stiffness Degredation. Ph.D. Thesis, Stanford University, Stanford, CA, USA, 2005.
101. Tsai, S.W.; Wu, E.M. A General Theory of Strength for Anisotropic Materials. *J. Comp. Mater.* **1971**, *5*, 58–80. [CrossRef]
102. Li, S.; Sitnikova, E.; Liang, Y.; Kaddour, A.S. The Tsai-Wu failure criterion rationalised in the context of UD composites. *Compos. Part A Appl. Sci. Manuf.* **2017**, *102*, 207–217. [CrossRef]

103. Roy, A.K.; Tsai, S.W. Design of Thick Composite Cylinders. *J. Press. Vessel Technol.* **2009**, *110*, 255. [CrossRef]
104. Chen, Q.; Li, C.; Tie, Y.; Liu, K. Progressive Failure Analysis of Composite Flywheel Rotor Based on Progressive Damage Theory. *J. Mech. Eng.* **2013**, *49*, 1–6. [CrossRef]
105. Kotelnikova-Weiler, N.; Baverel, O.; Ducoulombier, N.; Caron, J.F. Progressive damage of a unidirectional composite with a viscoelastic matrix, observations and modelling. *Compos. Struct.* **2018**, *188*, 297–312. [CrossRef]

encyclopedia

Entry

Tsunami Alert Efficiency

Amir Yahav [1,*,†] and Amos Salamon [2,*,†]

1 National Steering Committee for Earthquake Preparedness, 6th Habaal Shem-tov, Lod 7128906, Israel
2 Geological Survey of Israel, 32 Yeshayhu Leibowitz Street, Jerusalem 9692100, Israel
* Correspondence: av_y@hotmail.com (A.Y.); salamon@gsi.gov.il (A.S.); Tel.: +972-5062-99024 (A.Y.); +972-5062-35281 (A.S.)
† These authors contributed equally to this work.

Definition: "Tsunami Alert Efficiency" is the rapid, accurate and reliable conduct of tsunami warning messaging, from the detection of potential tsunamigenic earthquakes to dissemination to all people under threat, and the successful survival of every person at risk on the basis of prior awareness and preparedness.

Keywords: decision matrix; tsunami alert; tsunami awareness; tsunami efficiency; tsunami hazard; tsunami messages; tsunami preparedness; tsunami ready; tsunami risk; tsunami warning

1. Introduction

Lessons learnt from recent disastrous tsunamis point towards significant gaps between the science behind tsunami warning and the practice of saving lives and minimizing risk [1–3]. Most notable was the identification of the 26 December 2004 Sumatra Mw 9.1 tsunamigenic earthquake in near real time, and due to the lack of communication means and unpreparedness there was no way to alert the circum-Indian Ocean inhabitants. Consequently, a quarter of million people lost their lives [4]. This catastrophe was considered an "eye-opener" [5], showing that, clearly, although tsunamis cannot be prevented, the massive loss of lives was avoidable and the scope of damages was mitigable.

About 7 years later, on 11 March 2011 the world faced another deadly tsunami event caused by the Mw 9.0 tsunamigenic Tohoku-Oki earthquake east of Honshu Island in Japan. This calamity cost the lives of about 18,500 people [6].

"Recognizing the increasing impact of disasters and their complexity in many parts of the world" [7], the third UN World Conference on Disaster Risk Reduction met on 18 March 2015 in Sendai, Japan, and decided to adopt the "Sendai Framework for Disaster Risk Reduction 2015–2030" [8]. The Sendai Framework presented four priorities for action: (1) understanding disaster risk; (2) strengthening disaster risk governance to manage disaster risk; (3) investing in disaster risk reduction for resilience; and (4) enhancing disaster preparedness for effective responses and to "Build Back Better" in recovery, rehabilitation and reconstruction. In addition, the Sendai declaration urged stakeholders to take actions in order to "... enhance our efforts to strengthen disaster risk reduction to reduce disaster losses of lives and assets worldwide" [7].

The disasters motivated the Intergovernmental Oceanographic Commission (IOC) of United Nations Educational, Scientific and Cultural Organization (UNESCO) to establish Intergovernmental Coordination Groups (ICGs) for tsunami early warning and mitigation systems (TWS) in the Indian Ocean [9]; the north-eastern Atlantic, Mediterranean and Connected Seas (NEAMTWS) [10]; and the Caribbean (ICG/CARIBE EWS) [11]; in addition to the already existing Pacific Tsunami Warning Center (PTWC) [12] in Hawaii and the Japanese Meteorological Agency (JMA) [13].

In fact, ICG/PTWC is a new name for the existing International Coordination Group for the Tsunami Warning System in the Pacific (ICG/ITSU), that was established in 1965

Citation: Yahav, A.; Salamon, A. Tsunami Alert Efficiency. *Encyclopedia* **2022**, 2, 383–399. https://doi.org/10.3390/encyclopedia2010023

Academic Editors: Raffaele Barretta, Ramesh Agarwal, Krzysztof Kamil Żur and Giuseppe Ruta

Received: 29 October 2021
Accepted: 26 January 2022
Published: 1 February 2022

after several decades of deadly tsunami catastrophes in the Pacific Ocean by a joint international effort under the umbrella of the IOC/UNESCO, which was thus the pioneer of the ICG/TWC groups [14]. Nowadays, " ... the (PTWC) provides warnings of tsunamis to the public and to organizations responsible for public safety in coastal areas of Hawai'i (since 1949), the Pacific Ocean (since 1965), the Indian Ocean (since 2005), and the Caribbean Sea (since 2006)." [15].

Thus, the space between tsunami generation at the one end, and the civil and public response at the other end, is nowadays covered by a systematic architecture of organizations that transfer tsunami alerts from end to end rapidly, accurately and reliably, on the basis of systematic Standard Operational Procedures (SOP) [16]. Yet the array of various bodies may complicate and delay the timely arrival of warning messages up to the very last threatened citizen, and therefore the alerting process should be conducted efficiently [17]. Orderly SOPs are of course required, and usually they are taken care of within the organizations [18], yet there is a need for efficient communication, because the chain is no stronger than its weakest link. Moreover, receiving the warning messages on time does not assure successful lifesaving conduct. Appropriate awareness [19] and preparedness [20] are necessary requirements for effective lifesaving behavior and must be integrated in the alerting process.

Here we describe the leading concepts behind the tsunami alerting process, emphasizing the importance of the corresponding awareness and preparedness, and discuss the difficulties and uncertainties that may downgrade its efficiency, because the effective conduct of the alerting process is the ultimate key to saving lives under threat. We aim not to rephrase existing SOPs or user guides, but bring to mind some thoughts on making tsunami alerts more efficient and effective.

2. The Alerting Process—End to End Organizing Architecture

Tsunami alert efficiency is defined here as the rapid, accurate and reliable conduct of tsunami warning messaging, from the detection of a potential tsunamigenic earthquake to the dissemination to all people under threat, and to the successful survival of every person at risk on the basis of prior tsunami readiness, awareness and preparedness. Accordingly, successful lifesaving shows good conduct and high alert efficiency, while a loss of lives reflects a failure of conduct and poor efficiency. The key to success or reasons for failure originate in the alerting process, and thus we center the following overview on this issue.

The alerting process is performed through several channels in an end-to-end (ETE) architecture (Figure 1) which is a chain that aims to transmit reliable data and tsunami warnings, urgently and fast, with the ultimate goal of saving the lives of people at risk. Here we follow the UNESCO/IOC [18] recommendations on NEAMTWS ETE architecture, and elaborate on:

- The ETE architecture;
- The elements that compose the ETE architecture;
- The regional, national and local domains under which the chain operates;
- The communication that bonds the elements;
- The last mile: lifesaving and self-evacuation;
- Factors behind the alert—awareness and preparedness;
- The fact that not all tsunamis are equal—regional, local and spontaneous events.

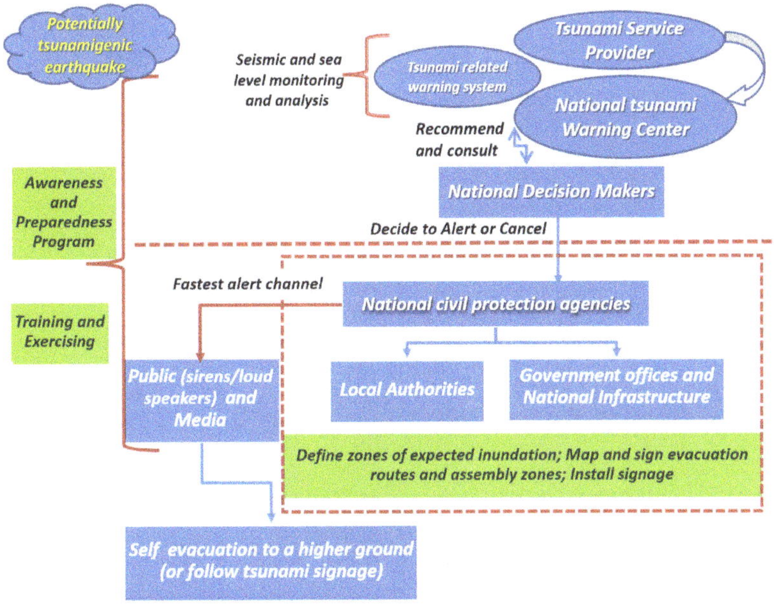

Figure 1. Generic end-to-end organizing architecture and the expected real-time flow of tsunami alerting and information.

The rationale is that proper application of all the components of the ETE architecture is the key to efficient tsunami alerting. "Efficiency" in this context means a reliable, accurate, fast and successful reach of the warning messages, from potential tsunami generation up to the last citizen, with an emphasis on the clear understanding and proper lifesaving behavior of the very last recipients.

We do not intend to present tsunami alert SOPs, detailed user guides or manuals; those can be found elsewhere [21]. Rather, we focus on the ideas and concepts behind the ETE architecture, with the understanding that the efficient conduct of the alerting process in which awareness and preparedness are integral parts, is the key to successful lifesaving. We further describe several examples experienced by Israel from its perspective as a member state of the ICG/NEAMTWS, and also mention other examples from elsewhere in the world.

3. The Elements of the ETE Architecture

The ETE architecture consists of several independent elements that all together enable the flow of information and warning messages along this chain (Figure 1). There is no restriction on the exact configuration of the elements, but they must bridge the whole ETE span, leaving no link open. Here we refer to a generic architecture, noticing that each setting is unique to its geographical, political and social environment; some elements can be merged into a single unit, while others can be sub-divided. Detailed information on contemporary tsunami early warning systems appear on the website http://www.ioc-tsunami.org/ (accessed on 25 January 2022) [22].

3.1. Tsunami Service Provider (TSP)

The first and leading element, the tsunami service provider (TSP), is a center responsible 24/7 for identifying tsunami generation in the geographical area it oversees, and if needed it issues appropriate tsunami threat information [23]. TSPs should monitor, detect, collect, record, process and analyze all relevant earthquake data—mainly the epicenter, depth, magnitude and origin time—that indicate the occurrence of a potential tsunamigenic earthquake. If they do, the TSP calculates the estimated arrival time and severity of the threat to a predefined list of coastal forecast points, and distributes the warning messages to its list of recipients. "To respect country sovereign rights, TSPs cannot issue warnings for another country, but can of course act as the NTWC issuing warnings for its own country" [24].

In parallel, the TSPs monitor in real-time sea-level indicators such as the tide, wave heights and sea-floor pressure fluctuations, in order to verify whether a tsunami was generated or not, reevaluate the potential threat of tsunami and update the warning messages. The TSPs follow the event closely as long as required, and update the warning messages according to the level of threat until it is over. If no tsunami is observed or tsunami threat no longer exists, the TSP cancels the alert and issues a cancelation message in a prescribed format [23].

As tsunami-meter and tide gauge systems are expanded around the world [25], highly sophisticated arrays are installed near potential tsunamigenic sources in Japan [26], cheaper and easier to operate tsunami alerting devices [27] are developed and set up [28], and new techniques are introduced [29], the verification of tsunami generation becomes faster and more reliable.

The main TSP challenge is to apply raw scientific data to operational use. On the one hand, the TSP is a scientifically based research institute, while on the other hand it should work under strict and high-level operational procedures. In doing so, the TSP must cope in near real time with preliminary, partial sets of seismological and sea level data, large uncertainties originating from the monitoring systems, calculation procedures and empirical decision matrix, under time pressure, and yet still issue timely and reliable warning messages [30,31]. A TSP that maintains and performs successfully the SOP requirements approved by the ICG it belongs to is accredited officially.

3.2. *National Tsunami Warning Center (NTWC)*

States who do not operate TSPs are still required to evaluate and deliver tsunami threat warnings to its governed coasts. Whether such National Tsunami Warning Centers (NTWCs) manage "in-house" analysis capabilities for the independent assessment of tsunami threat, or receive ready-made tsunami warning messages from TSPs, they are responsible for verifying and issuing official alerts. Preferably, an NTWC should be a scientific unit that is able to compile a large body of information in a short time and come up with a clear decision on the level of tsunami threat to its mandated coasts [32].

The challenge NTWCs and TSPs face is to evaluate the tsunami threat on the basis of limited seismological and sea level data within a short time window, and still carry full responsibility for its administrated population at risk. The need to issue a rapid evaluation before the tsunamis hit the coasts does not allow enough time for a definite determination of tsunami generation and providing an unequivocal true and reliable alarm. Thus, at present tsunami warning is associated with large uncertainties [33], and false alarms are unavoidable. To cope with these shortcomings, warning centers adopt a probabilistic tsunami forecasting (PTF) approach in order to quantify and reduce those uncertainties in real-time [1,34]. Without explaining this to the general public, warning messages can be perceived as "cry wolf" [35], lead to warning fatigue [36], and TSPs or NTWCs may lose trust and credibility.

Furthermore, NTWCs can receive several different TSP messages at the same time and should arrive at a clear, unequivocal decision on the immediate threat posed to its controlled coasts. Usually, TSPs rely on their own data sets, computer systems, calculation programs, mathematical models, and SOPs, and thus often determine different levels of threat posed by the same event on a given forecast point. This issue is further analyzed and discussed in a case in which different tsunami messages were received at the same time, for the same event and the same coasts in Israel (Section 9 below).

Overall, TSPs and NTWCs should balance their decisions between the need to inform the public of any slight chance of a coming tsunami and avoid missed alarms, while at the same time maintaining the credibility and trust of the public in the reliability of the warning system.

3.3. National Decision Makers (NDM)

While TSPs and NTWCs are required to maintain scientific capabilities, they lack the civilian and social perspectives and experience necessary for far reaching decision-making at the national level. National decision makers (NDM) must understand and accept that the information distributed by the TSPs and NTWCs is based on preliminary scientific data associated with large uncertainties and is therefore not definite. In fact, the first issue is that tsunami alerts present the "potential threat of a not yet verified tsunami", and it is up to the NDM to decide whether to alert the country or not. Their complex, far reaching decision must consider a sequence of numerous and drastic actions that should be taken once the tsunami alert is activated, while it is not yet clear whether a tsunami is actually on its way or not: for example, the immediate evacuation of hundreds of thousands of people away from the expected inundation areas to higher ground; shutting down of coastal power plants, which means blacking out its service area; evacuating sea ports and coastal infrastructure facilities such as desalination plants, oil and gas refineries, etc. From a state point of view, this is a massive operation that distract daily life activities on a national scale. This is of course well justified in case of a real alarm, yet the cost of a false alert can be tremendous.

The challenge is obvious: is there a real need to alert the country or not? *True alarms* are necessary and must not be missed, *false alarms* are forgivable if the public is well educated to live with them, yet *missed alarms* are not bearable. In order to make the alert efficient and avoid confusing the chain of operation and eventually the public, these issues should be elaborated between the scientific unit and the national decision makers ahead of time, and even in real time when needed [32].

3.4. National Civil Protection Agencies and Disaster Management Offices

The alerts distributed by the NDM should address, rapidly and simultaneously, all levels of recipients under threat within the national civil protection Agencies (CPAs) and disaster management offices (DMOs). Included are all kinds of national units, governmental offices, infrastructure facilities, local authorities, municipalities etc., with the aim of reaching all the relevant public at risk. This is the "last mile" of the alerting process, but it is not short at all.

The challenge of the CPAs and DMOs is to stand by 24/7, while the repeat time of the tsunami hazard may last decades or even centuries. They must fully take responsibility to consider and perform all aspects needed for an emergency response beforehand [18], understand the content of the warning messages in real time and be able to distribute the alert down-stream to all people at risk. Clearly, the efficient conduct of the alerting process depends on proper planning and detailed preparations ahead of time, so that at the moment of truth, every person in need gets the warning, knows what to do and is able to save their life [37].

3.5. The Public

The general public, including all persons under threat and at risk, in any given place, moment and circumstances, is the end recipient of the alerting messages. The challenge to every person is not only to receive the message, but to fully understand its meaning, know what to do and how to survive. This is not simple at all, because proper conduct under life threatening circumstances depends on prior awareness and proper preparedness. Nonetheless, it is the responsibility of the national and district civil protection agencies, as well as the local authorities, to guide and prepare the public for such a catastrophe [37]. These issues are elaborated further in Section 7.1.

4. The Domains

The great destruction caused by the May 1960 Chilean and March 1964 Alaskan Tsunamis around the Pacific emphasized the need for an International Tsunami Warning System [38], thus bringing the management of the initial stages of tsunami warning to the international sphere. As a result, the ETE architecture operates under three domains: regional, national and local.

The regional is the uppermost sphere in which the ICGs operate, and each of them involves a group of several nations that share the same tsunami-threatened basins [39]. The national domain operates under the regional one, and it consists of the relevant ICG member states. The local is the lowest domain; it operates under the national level, and could be of any kind of DMO, CPA, governmental or ministerial office; municipality or authority; or any other group of citizens, such as a "Tsunami Ready" community [37]. The local domain, which is the last unit of the ETE chain, is responsible for distributing the warning to whomever is under threat.

The need to cooperate and coordinate the transfer of warning messages from top to bottom across the three domains, including lifesaving operations, raises obvious and trivial difficulties to cope with, particularly technical means of communication, a common language and terminology, and even the need to bridge cultural gaps [40]. In order to achieve the efficient conduct of alerting, all three domains must recognize and be familiar with each other and coordinate the exchange of messages.

5. Communication—The Key to the Successful Transmission of an Alert

The warning messages must be transmitted across the ETE architectural chain, fast and efficiently, without losing their level of severity, urgency and certainty. This is not trivial at all, because one end reads and speaks a professional scientific language and the other end expects to receive a simple, layman's explanation and instruction. Furthermore, the standard TSP language is English, while the end recipients speak a national tongue, if not a specific, local dialect [40].

Thus, while communication is the bridge to transfer tsunami alerts, it poses potential barriers. The challenge, then, is to construct a common language and terminology, clear and understandable to both the sender and the receiver, built within a structure defined and accepted by the two sides ahead of time. The content of the sent messages must convey the necessary information required by the recipient. TSPs should rephrase scientific jargon into decision-makers terms, which in turn need to reshape it into clear instructions to civil protection agencies, which in turn have to distribute it to the public in a daily spoken language.

Overall, there are several preconditions for an efficient conduct and transfer of the warning messages along the ETE chain. Each element should be familiar with the architecture and standard of practice of its "neighbor"; agree and be familiar with the common language, terminology, structure and content of messages; and coordinate means of transmission and communication, including the simultaneous use of several channels (e.g., GTS, e-mail, SMS and facsimile) for backing up the alert. Eventually, the credibility of the system is no stronger than its weakest link.

Another obstacle is the poor ability of contemporary science and technology to forecast quantitatively the expected level of tsunami threat in real time, shortly after the earthquake. Empirical data and experience allow the construction of tsunami decision matrixes that are limited to distinguishing negligible, low and high levels of threat only [18], with a significant degree of uncertainty. The challenge is thus to rapidly, accurately and reliably distribute tsunami warning messages on the basis of preliminary data and large uncertainty, and still warn the public with a short, clear and simple message, such as "Tsunami alert, get away from the coast, go to high ground".

6. The Last Mile

Distributing the official alert to the general public is the ultimate goal of the alerting process, because this is what drives lifesaving and damage-minimizing actions. The challenge is reaching every person at risk and making sure they know what to do and are able to do so [41].

6.1. Reaching the Public

In the case of far field tsunamis, where no natural signal of earthquake shaking is available, efficient alerting requires that all means of communication should be involved. The fastest and most efficient way to alert the public is to establish a national or regional public address (PA) system (of loudspeakers for example) that is controlled from a single operational center and distributes the alert within seconds of the NDM decision. Likewise, a direct alert to critical infrastructure and industry facilities can be disseminated from the same operation center. In parallel, and of no less importance as a relatively fast and effective course of action, the alert can be broadcast by bursting the media (radio and television) and dispersing it across social web networks. Farther along, CPAs (e.g., police and coast guards) and their personnel should target the alert to those within the zone of expected inundation by all available means, such as sirens and loudspeakers. A further advantage is that these professional forces are closely familiar with the alerted area and can react fast and effectively to any developing situation in real time. Even helicopters can be recruited by flying along the coastline and alerting vacationers on the beach and inside the sea. This is a slower way of action due to the long chain of command required to initiate the operation and the time needed to reach the threatened areas, but it is an effective way to reach remote and isolated places.

Near-field tsunamis impose automatic reaction and shortcut procedures in order to cope with the short arrival time to the coast, and this is again the responsibility of the CPAs and DMOs to educate and direct the public on how to conduct self-lifesaving actions.

6.2. Saving Lives

Successful ETE architecture and efficient data and messaging flow are not enough, because the people should know what to do, and if needed do it. The arrival of a tsunami alert to the end member recipients should trigger lifesaving actions, and the challenge is to drive the crowd to act properly as soon as the alert is accepted. It requires the public to trust the alerting system, understand the message, know what to do and just do it, even though at some point the alert can be canceled. The public should be informed and educated that the warning process improves as more information arrives and therefore be ready to accept a cancelation of the alert without losing trust in the system. The public should be taught how to react according to where they are during the event and trained in doing the right actions, which basically are self-saving acts.

The key to the successful completion of the lifesaving process is thus proper awareness and preparedness of the public ahead of time. This is further elaborated below.

7. Beyond the Alert—Awareness and Preparedness

Alerting a public that is not aware of the hazard and not prepared to take lifesaving actions is hopeless, and this is the role of awareness and preparedness. Thus, while the ETE architecture aims to convey the warning messages to the public in need, the end recipients must be attentive and ready to take action, otherwise the alert is wasted.

The essence of awareness is bringing both the authorities and the public to be familiar with the tsunami phenomenon; learn what its typical characteristics are and how to identify them; why it is a hazard and its possible consequences; be aware that it can really happen—and where and when it can do so [42]. The challenge is to teach the civil authorities, coastal management bodies and the public to remain on standby for a long time and take all necessary actions at the right moment, even though it is a rare event and most of them have not experienced such an event before.

Preparedness in the present context refers to actions necessary to prevent loss of life and minimize damages, and it includes all needed actions that enable a country to better overcome a devastating emergency event. Effective and suitable preparedness requires a good understanding of the specific socio-cultural context of the local community, and taking those into account when programing suitable activities needed to target all levels, from the governmental all the way down to every household at the community level [43]. The responsibility to realize preparedness should be authorized at the state level and directed from top CPAs and DMOs levels down to all national infrastructure, civilian authorities, private sectors and eventually the general public.

In fact, establishing and maintaining all the components of the ETE architecture within the national and local domains (as already described above) is the foremost and elementary stage of preparedness that a state should have. It demands a great deal of resources to organize and specific focus to implement, but an efficient alert can only be achieved when this suite of activities is coordinated and activated simultaneously at all levels and times.

Even environmental and urban planning, as well as formulating proper codes for making buildings resistant to tsunami impacts [44], and particularly locating and constructing vertical evacuation structures [45,46], should be considered integral parts of tsunami lifesaving preparedness activities. The key is in bringing to mind tsunami awareness at all levels from the beginning, including planners and engineers as pointed above, so as to avoid unfortunate actions, such as placing schools in zones vulnerable to tsunami inundation. The ultimate goal is making all different clients and recipients be familiar and capable of the right lifesaving and damage minimizing actions ahead of time and particularly at the right moment [40].

7.1. *Evacuation, Signage and Route Mapping*

Experience shows that the probability of surviving a 1 m tsunami inundation wave height is about 50% and above 2 m high it is hopeless [47]. This is the reason why evacuation far from or above the inundation zone is the most important lifesaving action. Nevertheless, it is the responsibility of the civil authorities to identify ahead of time the zones of expected inundation; plan and sign evacuation routes and assembly points in a safe zone; and educate and train the public in how to react if needed. Posting road signs of tsunami hazard symbols is the most common way to notify the public whether they are in a zone of expected inundation, and where to evacuate and assemble if a tsunami alert is triggered and effective lifesaving actions are needed. The tendency is to follow a common, international signage language recommended by ISO.org. [48], and so people become familiar with tsunami hazard signs, no matter where they are coming from and what language they speak. There are many examples of tsunami signage from around the world, such as those guided by the New Zealand Ministry of Civil Defense & Emergency Management [49], or Indonesia, Chile, Japan, Hawaii (USA) and Washington (USA), as are presented at itic.ioc-unesco.org [50]. Figure 2 illustrates tsunami signage recently posted along the Mediterranean coast of Israel, including the signs that indicate the entrance to a tsunami hazard zone (Figure 2a), the fastest and safest escape route and distance to the assembly zone (Figure 2b) and the location of the assembly zone (Figure 2c).

Much effort is invested in developing innovative portal and cellphone apps in order to ensure a proper response and enhance traditional evacuation drills of the public [51] or allow the public to learn their evacuation route ahead of time [52]. CWarn [53], for example, a not-for-profit humanitarian initiative website, allows its members to register for a free tsunami warning wherever they are around the world by SMS text message on their mobile phones. As much as such initiatives are highly acknowledged, it is the sole responsibility of any sovereign country to provide each of its citizens professional, reliable and timely 24/7 tsunami early warning.

(a) (b) (c)

Figure 2. Tsunami hazard road signs installed along the Mediterranean coast of Tel-Aviv, Israel: (**a**) hazard zone; (**b**) escape route and distance to the assembly zone; and (**c**) assembly point. Photographs by A.Y., Tel Aviv, 2016.

To increase the effect and visibility of the signage system, the Israeli National Steering Committee for Earthquake Preparedness is developing a location map that shows each and every hazard, escape and assembly sign. The maps will be shared with all the local authorities; emergency and rescue forces; and the public, and assist the emergency forces in reaching more directly and faster the public in need, save crucial time and improve lifesaving actions [54].

The map symbology reflects the exact location of every single sign and its purpose. For example, a red flag denotes a tsunami hazard zone (Figure 3a), a yellow flag denotes an escape route (Figure 3b), and a green flag indicates an assembly zone (Figure 3c). This way the public will get a comprehensive view and better understanding of the entire emergency plan.

(a) (b) (c)

Figure 3. Flag symbols of the tsunami road signs used for the self-evacuation map: (**a**) red for the hazard zone; (**b**) yellow for evacuation routes; and (**c**) green for assembly points (see also Figure 4). Signage symbols by Dr. Orna Ido-Lichtman, Israel's National Steering Committee for Earthquake Preparedness.

Better yet is to upload the maps on an interactive cellphone application (under development) and allow every person in need to identify in real time their present position in regard to the expected inundation area [55]. Furthermore, such an app can allow one to find one's way to the closest assembly zone along a signed and safe route, as is being developed today in Israel (Figure 4a,b).

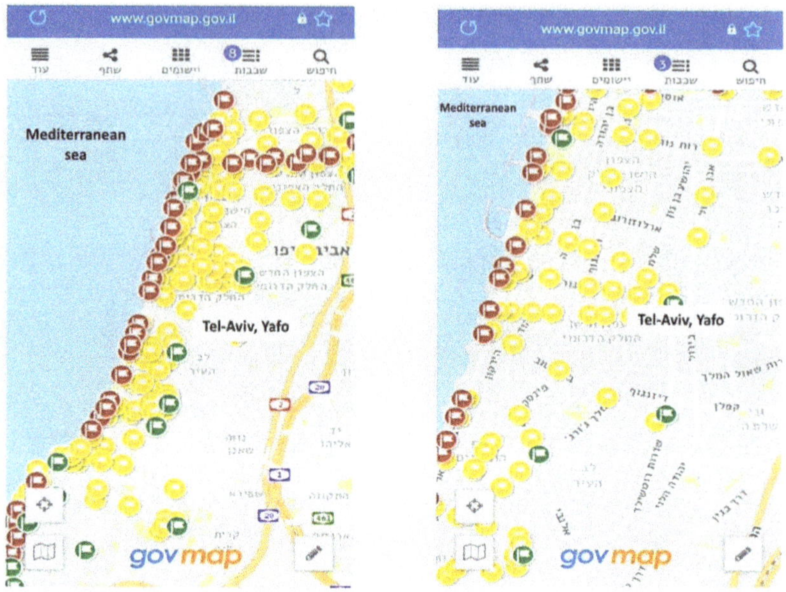

(a) (b)

Figure 4. Tsunami signage self-evacuation map of: (**a**) greater Tel Aviv; (**b**) central Tel Aviv. Color flag symbols denote the location of the tsunami road signs as follows: red for a hazard zone, yellow for evacuation routes and green for assembly points (see also Figure 3). Background map by the Survey of Israel, https://www.govmap.gov.il (accessed on 25 January 2022); signage of self-evacuation plan by the author A. Y.; and courtesy of Dr. Orna Ido-Lichtman and Mr. Daniel Lanza, Israel's National Steering Committee for Earthquake Preparedness.

7.2. *Educate, Train and Drill*

The time between tsunami events can last years, and tsunami preparedness can be perceived as a waste of time and resources. However, there are no shortcuts: "If you think education is expensive, try ignorance", especially because the potential unbearable cost of tsunami casualties is avoidable by very simple measures—"go away and high from the inundation zone".

The best way, though not necessarily the only one, to convey this notion is education [42]. The formal, national education system is the most efficient alternative, because young kindergarten and school children can bring home to their parents what they have learned at school, carry this knowledge with them through their lives and tell their children the story. Nonetheless, sustaining public education over generations remains one of the biggest challenges, but it is also arguably a keystone activity for saving lives from tsunamis [14].

Training is an integral part of learning, especially in lifesaving actions, without which the acquired knowledge and understanding fades away. The need is to exercise routinely the whole process by both the authorities who are in charge of the practice and the public, and particularly drill the actual evacuation and walking along the escape routes to the assembly zones [56].

7.3. *Care of the Disabled—No One Is Left Behind*

Often there is not enough time left for evacuation before the tsunami arrives, especially for slow moving and disabled people. Therefore, evacuation planning, education and training must consider solutions for those who cannot practice self-saving actions alone [57]. For example, mutual assistance should be an integral part of the practice, taken as the duty of young students. Other creative solutions such as vertical evacuation are needed as well.

8. Not All Tsunamis Are Generated Equal

While planning the tsunami emergency response, one should bear in mind that tsunamis are different from each other and much flexibility is required to cope with future scenarios. It is not only the unexpected timing and magnitude of the event, but also the source and mechanism that may surprise and catch us "not prepared".

The most significant difference is between tsunamis coming from a far distance—in which case the warning may arrive by telecommunication and there can be some time to react—and local tsunamis in which the earthquake shaking or abnormal behavior of the sea are the only warning signals and immediate evacuation is necessary. Awareness and preparedness need to consider both scenarios [58]. Yet further complexities should be expected, because all types of earthquakes can generate submarine and/or subaerial tsunamigenic landslides—even on-land earthquakes near the coast [59,60]. Furthermore, the common notion is that thrust mechanism earthquakes, mostly along subduction zones, pose the main threat and thus strike–slip events may appear as a "surprise" [61–63]; tsunamigenic volcanoes also go underrated [64], and so on. What remain unexpected indeed, and so far lack preceding signals, are spontaneous tsunamigenic submarine and subaerial landslides. Real time sea-level monitoring systems can bridge this gap.

The challenge is thus this: while there are numerous plausible tsunami scenarios that ETE architecture needs to be aware of and prepare for, the public should get clear and simple instructions on how to cope with this natural hazard, regardless of its generating mechanism.

9. A Case Study of a Decision-Making Dilemma

Real events present numerous dilemmas to coping with and arriving at a practical, fast and efficient solution. Behrens [33] proposed how to handle the problematics of the inherent uncertainties and differences that arise in real time from TSP warning messages. A classic example is the tsunami warning messages issued by the already accredited NEAMTWS TSPs regarding the 25 October 2018, 16.8 km depth, mb 6.3, MS 6.9 [65] tsunamigenic earthquake (Figure 5) in the Mediterranean Sea southwest of Greece. Israel, a NEAMTWS member state, received at that time two different, legitimate tsunami-warning messages. The first one arrived from the INGV (Italy TSP) at 23:02, sending Israel a "Tsunami Information" (Figure 6a). The second message, arriving from NOA (Greece TSP) one minute later, notified a "Tsunami Advisory" level of alert to Israel (Figure 6b). The analyses, the best practice of the two TSPs, were regarded reliable, and the differences may have originated from the SOPs and mathematical and decision modules each TSP was using.

(Map courtesy of Google maps)

Figure 5. Location map of the 25 October 2018 earthquake south west of Greece. Origin time 22:54:51.80, lat. 37.47, long. 20.72, depth 16.8 km, mb 6.3, MS 6.9 [65,66].

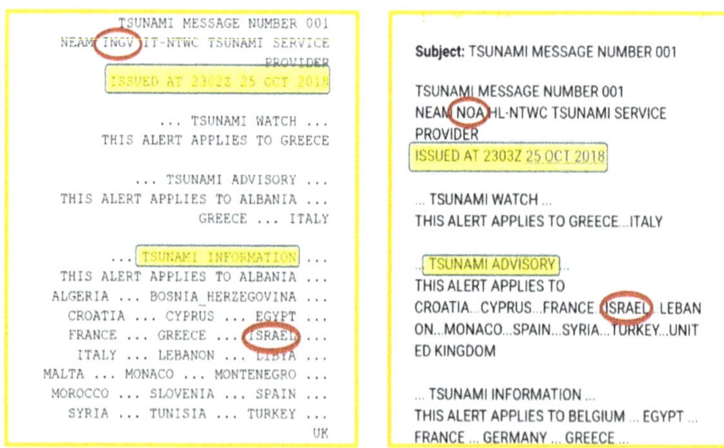

(a) (b)

Figure 6. Tsunami alert messages received in Israel regarding the 25 October 2018 earthquake southwest of Greece: (**a**) INGV message issued on 23:02 "Tsunami Information" to Israel; (**b**) NOA message issued on 23:03 "Tsunami Advisory" to Israel.

From Israel's stand point, the near real time source parameters determined by the two TSPs indicated a "Tsunami Advisory" on the basis of the decision matrix recommended for use in Israel [67]. However, as further information regarding sea level measurements in Crete and Cyprus arrived with no indication of clear abnormal behavior, the Israel

NDM (named "Migdalor" in Hebrew) consulted its NTWC (named "Nachshol Nitzpeh" in Hebrew) and concluded a "Tsunami Information" only, and no official alert was issued.

10. Concluding Words

10.1. Comprehensive and Flexible End-to-End Architecture

The leading concept behind efficient tsunami alerting is lifesaving. It requires the good conduct of transmitting the warning from one end—the detection of a potential tsunamigenic earthquake—to the other end—the people at threat. Yet without proper awareness and preparedness, the alert may just confuse the public and disorient rescue actions. The current NEAMTWS ETE architecture [18] presents a sound management of the alerting process. Here we proposed complementing this data and messaging flow by optional shortcuts and reinforcing the alerting process with appropriate awareness and preparedness (Figure 7).

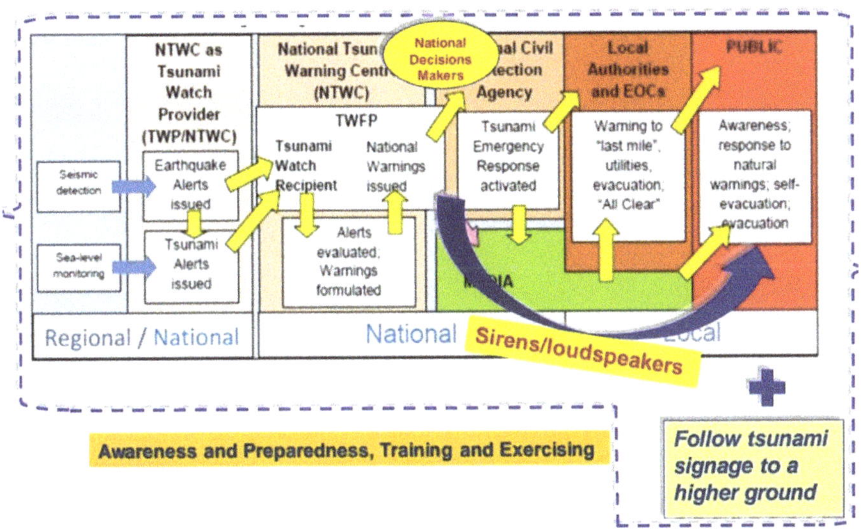

Figure 7. Suggested modifications to the NEAMTWS end-to-end alerting architecture [18]. The focus is on adding national decision makers (NDM) components and complementary shortcuts that allow more direct and faster flows of messaging from the NDM to the public in case it is clear that a tsunami was generated and arrival time is short. In addition, there is a need to emphasize tsunami awareness and preparedness across all the participating units.

A trivial yet crucial shortcut is a must if a significant tsunami is generated and the arrival time is short. In such events, NDMs must alert and activate the public directly, while in parallel roll the alerting commands along the formal chain of responsible authorities, and thus save precious time. Similarly, an NTWC can alert simultaneously both the NDM and the local authorities. Clearly, these actions should be formulated and agreed in advance by all participating units.

Nonetheless, even an efficient and fast flow of messages would be ineffective if the public is not tsunami ready. The alerting process must be supported by proper education and training, complemented by the delineation of inundation zones and evacuation routes, so as that in the moment of truth people under threat will indeed receive the alert and save their lives [68].

10.2. The Alerting Chain Is as Strong as Its Weakest Unit

Much has been published on orderly SOPs; manuals; and user guides on how to manage tsunami warnings, awareness and preparedness, and many such examples were mentioned in this entry. Altogether and above the technicalities, the alerting procedure seeks flexibility in coping with unexpected scenarios and the efficient conduct of lifesaving as its ultimate goal.

Beyond the formalities, alerting would be as effective as its most incompetent unit, the flow of warning messages would be as fast as its slowest segment, the ETE architecture would be as efficient as its most inexpert component and thus the alerting chain would be as strong as its weakest unit.

Author Contributions: Both authors, A.Y. and A.S., contributed equally and shared conceptualization, writing and editing of the text. All authors have read and agreed to the published version of the manuscript.

Funding: This research received no external funding.

Acknowledgments: We thank Orna Ido-Lichtman and Daniel Lanza, Israel's National Steering Committee for Earthquake Preparedness, for their assistance in planning the self-evacuation map. We highly appreciate the four anonymous reviewers for their critical readings of the manuscript and constructive comments.

Conflicts of Interest: The authors declare no conflict of interest.

Entry Link on the Encyclopedia Platform: https://encyclopedia.pub/20356.

References

1. Behrens, J.; Løvholt, F.; Jalayer, F.; Lorito, S.; Salgado-Gálvez, M.A.; Sørensen, M.; Abadie, S.; Aguirre-Ayerbe, I.; Aniel-Quiroga, I.; Babeyko, A.; et al. Probabilistic Tsunami Hazard and Risk Analysis: A Review of Research Gaps. *Front. Earth Sci.* **2021**, *9*, 628772. [CrossRef]
2. Lorito, S.; Behrens, J.; Løvholt, F.; Rossetto, T.; Selva, J. Editorial: From Tsunami Science to Hazard and Risk Assessment: Methods and Models. *Front. Earth Sci.* **2021**, *9*, 764922. [CrossRef]
3. Okal, E.A. Twenty-Five Years of Progress in the Science of "Geological" Tsunamis Following the 1992 Nicaragua and Flores Events. *Pure Appl. Geophys.* **2019**, *176*, 2771–2793. [CrossRef]
4. Marris, E. Inadequate warning system left Asia at the mercy of tsunami. *Nature* **2005**, *433*, 3–5. [CrossRef] [PubMed]
5. Okal, E.A. The Quest for Wisdom: Lessons from 17 Tsunamis, 2004–2014. *Phil. Trans. R. Soc. A* **2015**, *373*, 20140370. [CrossRef] [PubMed]
6. Uchida, N.; Bürgmann, R. A decade of lessons learned from the 2011 Tohoku-Oki earthquake. *Rev. Geophys.* **2021**, *59*, e2020RG000713. [CrossRef]
7. WCDRR (World Conference on Disaster Risk Reduction). The Sendai Framework for Disaster Risk Reduction 2015–2030. In Proceedings of the Third United Nations World Conference on Disaster Risk Reduction, Sendai, Miyagi, Japan, 14–18 March 2015. Available online: https://www.preventionweb.net/files/43291_sendaiframeworkfordrren.pdf (accessed on 25 January 2022).
8. WCDRR (World Conference on Disaster Risk Reduction). The Sendai Declaration. In Proceedings of the Third United Nations World Conference on Disaster Risk Reduction, Sendai, Miyagi, Japan, 14–18 March 2015. Available online: https://www.preventionweb.net/files/43300_sendaideclaration.pdf (accessed on 25 January 2022).
9. UNESCO/IOC (Intergovernmental Oceanographic Commission of UNESCO). *Tsunami Risk Assessment and Mitigation for the Indian Ocean; Knowing Your Tsunami Risk–and What to Do About It. IOC Manuals and Guides No. 52*, 2nd ed.; UNESCO: Paris, France, 2015. (In English)
10. North-Eastern Atlantic, Mediterranean and connected seas Tsunami Information Centre. NEAMTIC-Home. Available online: http://neamtic.ioc-unesco.org/ (accessed on 25 October 2021).
11. ICO. Proceedings of the Eleventh Session of the Intergovernmental Coordination Group for the Tsunami and Other Coastal Hazards Warning System for the Caribbean and Adjacent Regions, Cartagena de Indias, Colombia, 5–7 April 2016. Available online: http://www.ioc-tsunami.org/index.php?option=com_oe&task=viewEventRecord&eventID=1784 (accessed on 25 January 2022).
12. Tsunami Warning Centers. Available online: https://tsunami.gov/ (accessed on 24 October 2021).
13. Japan Meteorological Agency. Available online: https://www.jma.go.jp/jma/indexe.html (accessed on 24 October 2021).
14. Kong, L.S.L.; Paula, K.; Dunbar, A.N. (Eds.) *Pacific Tsunami Warning System: A Half Century of Protecting the Pacific, 1965–2015*; International Tsunami Information Center: Honolulu, HI, USA, 2015.
15. PTM (Pacific Tsunami Museum). Available online: http://tsunami.org/the-warning-system/ (accessed on 24 December 2021).
16. ICG/NEAMTWS-VIII. *Interim Operational Users Guide for the Tsunami Early Warning and Mitigation System in the North-eastern Atlantic, the Mediterranean and Connected Seas (NEAMTWS)*; Version 2.00; ICG/NEAMTWS-VIII: Rome, Italy, 2011.

17. Papadopoulos, G.; Lekkas, E.; Katsetsiadou, K.-N.; Rovithakis, E.; Yahav, A. Tsunami Alert Efficiency in the Eastern Mediterranean Sea: The 2 May 2020 Earthquake (Mw6.6) and Near-Field Tsunami South of Crete (Greece). *GeoHazards* **2020**, *1*, 44–60. [CrossRef]
18. UNESCO/IOC (Intergovernmental Oceanographic Commission of UNESCO). In *Reducing and Managing the Risk of Tsunamis*; IOC Manuals and Guides; United Nations Educational, Scientific and Cultural Organization: Paris, France, 2011; Volume 57.
19. Fukuji, T. ITIC Tsunami Awareness and Education Materials-International Tsunami Information Center. 2021. Available online: http://itic.ioc-unesco.org/index.php?option=com_content&view=article&id=1349&Itemid=+1075&lang=en (accessed on 25 October 2021).
20. NWS JetStream-Tsunami Preparedness and Mitigation: Individuals (You!). Available online: https://www.weather.gov/jetstream/prep_you (accessed on 25 October 2021).
21. UNESCO/IOC (Intergovernmental Oceanographic Commission of UNESCO). Users Guide for the Pacific Tsunami Warning Center Enhanced Products for the Pacific Tsunami Warning System. In *IOC Technical Series No 105*; Revised edition; 2014. Available online: http://itic.ioc-unesco.org/images/stories/about_warnings/what_are_they/ts105-Rev2_eo_220368E.pdf (accessed on 25 January 2022).
22. UNESCO/IOC (Intergovernmental Oceanographic Commission of UNESCO). Available online: http://www.ioc-tsunami.org/ (accessed on 26 December 2021).
23. UNESCO/IOC (Intergovernmental Oceanographic Commission of UNESCO). Tsunami Service Providers (TSPs). Available online: https://ioc.unesco.org/our-work/global-tsunami-early-warning-and-mitigation-programme (accessed on 24 December 2021).
24. IOC (Intergovernmental Oceanographic Commission of UNESCO). *Plans and Procedures for Tsunami Warning and Emergency Management*; IOC Manuals and Guides; United Nations Educational, Scientific and Cultural Organization: Paris, France, 2017; Volume 76, p. 72.
25. Michaels, M. *Buoy Data and Tsunami Preparedness*; WeatherNation: Centennial, CO, USA, 2021. Available online: https://www.weathernationtv.com/news/buoy-data-and-tsunami-preparedness/ (accessed on 26 October 2021).
26. Mulia, I.E.; Satake, K. Developments of Tsunami Observing Systems in Japan. *Front. Earth Sci.* **2020**, *8*, 145. [CrossRef]
27. Annunziato, A. The inexpensive device for sea level measurements. *Sci. Tsunami Hazards J. Tsunami Soc. Int.* **2015**, *34*, 199–211. Available online: https://webcritech.jrc.ec.europa.eu/TAD_server/Data/Documents/IDSL%20Description.pdf (accessed on 25 January 2022).
28. European Comission. *TAD Server*. Available online: https://webcritech.jrc.ec.europa.eu/TAD_server/Home (accessed on 26 October 2021).
29. Lipa, B.; Barrick, D.; Isaacson, J. Evaluating HF Coastal Radar Site Performance for Tsunami Warning. *Remote Sens.* **2019**, *11*, 2773. [CrossRef]
30. Amato, A. Some reflections on tsunami Early Warning Systems and their impact, with a look at the NEAMTWS. *Bollettino di Geofisica Teorica ed Applicata* **2020**, *61*, 403–420. [CrossRef]
31. Necmioğlu, Ö.; Turhan, F.; Özer Sözdinler, C.; Yılmazer, M.; Güneş, Y.; Cambaz, M.D.; Altuncu Poyraz, S.; Ergün, T.; Kalafat, D.; Özener, H. KOERI's Tsunami Warning System in the Eastern Mediterranean and Its Connected Seas: A Decade of Achievements and Challenges. *Appl. Sci.* **2021**, *11*, 11247. [CrossRef]
32. Salamon, A.; Rosen, D.S.; Gitterman, Y.; Yahav, A.; Ben-Arieh, S.; Debutton, Y.; Dover, D.; Vatenmacher, M.; Miloh, T. Recommendations on the Policy, Warning Principles and Frame of Preparedness for Tsunamis in Israel. The Scientific Committee for Formulating Policy of Early Warning Principles for Tsunami Hazards in Israel; Geological Survey of Israel Report GSI/16/2014; Geophysical Institute of Israel, Report 30/797/14; Israel Oceanographic and Limnological Research, Report H36/2014 (In Hebrew, English Abstract). 2014. Available online: https://www.gov.il/BlobFolder/reports/salamon-et-al-report-2014-16/he/report_2014_Salamon-Amos-Policy-Principles-Preparedness-Tsunamies-Israel-GSI-16-2014-GII-030-797-14-H36-2014.pdf (accessed on 25 January 2022).
33. Behrens, J. How to Handle Diverging TSP Warning Messages and Know about the Inherent Uncertainty. In Proceedings of the 14th Session of ICG/NEAMTWS, Lisbon, Portugal, 21–23 November 2017. Available online: http://legacy.iocunesco.org/index.php?option=com_oe&task=viewDocumentRecord&docID=20623 (accessed on 25 January 2022).
34. Selva, J.; Lorito, S.; Volpe, M.; Romano, F.; Tonini, R.; Perfetti, P.; Bernardi, F.; Taroni, M.; Scala, A.; Babeyko, A.; et al. Probabilistic tsunami forecasting for early warning. *Nat. Commun.* **2021**, *12*, 5677. [CrossRef] [PubMed]
35. Shepardson, D. FCC Says Appears Hawaii Had No Safeguard to Stop Missile Scare. Reuters. 14 January 2018. Available online: https://www.reuters.com/article/usa-missiles-falsealarm-fcc-idINKBN1F30TP (accessed on 25 January 2022).
36. Samarajiva, S.; Gunawardene, N. Crying Wolf Over Disasters Undermines Future Warnings. Available online: https://www.scidev.net/global/opinions/crying-wolf-over-disasters-undermines-future-warnings/ (accessed on 25 January 2022).
37. IOC Tsunami Ready-International Tsunami Information Center. Available online: http://itic.ioc-unesco.org/index.php?option=com_content&view=category&id=2234&Itemid=2758 (accessed on 24 October 2021).
38. Pararas-Carayannis, G. Tsunami Warning System in the Pacific: An Example of International Cooperation. In *Natural and Man-Made Hazards*; El-Sabh, M.I., Murty, T.S., Eds.; Springer: Dordrecht, The Netherlands, 1988. [CrossRef]
39. Intergovernmental-International Tsunami Information Center. Available online: http://itic.ioc-unesco.org/index.php?option=com_content&view=category&layout=blog&id=2002&Itemid=2002 (accessed on 26 October 2021).

40. Tsunami Kit-Developing Early Warning and Community Preparedness in Indonesia. Available online: https://www.gitews.org/tsunami-kit/index_en.html (accessed on 26 October 2021).
41. GITEWS: "Last Mile". Available online: https://www.gitews.org/en/capacity-building/last-mile/ (accessed on 26 October 2021).
42. Fukuji, T. TsunamiTeacher (International)-International Tsunami Information Center. 2021. Available online: http://itic.ioc-unesco.org/index.php?option=com_content&view=article&id=1441:tsunamiteacher-international&catid=2102&Itemid=2420 (accessed on 27 October 2021).
43. Kong, L. Guarding against Tsunamis: The Challenge of Building Preparedness at the National and Local Levels. In *Liaison*; 2006. Available online: https://nctr.pmel.noaa.gov/education/ITTI/usgs/seismic-tsunami-training-maldives/tsunami_preparedness_COE122905_3.pdf (accessed on 25 January 2022).
44. Chock, G. Design for tsunami loads and effects in the ASCE 7-16 standard. *J. Struct. Eng.* **2016**, *142*, 04016093. [CrossRef]
45. FEMA. Guidelines for Design of Structures for Vertical. Evacuation from Tsunamis. Available online: http://www.fema.gov/media-library-data/20130726-1641-20490-9063/femap646.pdf (accessed on 25 January 2022).
46. FEMA. Vertical Evacuation from Tsunamis: A Guide for Community Officials. Available online: https://damfailures.org/wp-content/uploads/2019/12/fema_p646a.pdf (accessed on 25 January 2022).
47. FEMA. Hazus Tsunami Model User Guidance. Available online: https://www.fema.gov/media-library-data/1511284000292-4f18206fb0c7bab3c5ecbbbdf504b9fd/Hazus_40_Tsunami_User_Manual.pdf (accessed on 25 January 2022).
48. ISO. Available online: https://www.iso.org/obp/ui#iso:grs:7010:W056 (accessed on 26 October 2021).
49. New Zealand Ministry of Civil Defence & Emergency Management. *National Tsunami Signage Technical Standard for the CDEM Sector [TS 01/08]*; Ministry of Civil Defence & Emergency Management: Wellington, New Zealand, 2008; ISBN 0-478-25480-6.
50. Signs & Symbols. Available online: http://itic.ioc-unesco.org/index.php?option=com_content&view=category&id=1406&Itemid=2280 (accessed on 25 December 2021).
51. Leelawat, N.; Suppasri, A.; Latcharote, P.; Abe, Y.; Sugiyasu, K.; Imamura, F. Tsunami evacuation experiment using a mobile application: A design science approach. *Int. J. Disaster Risk Reduct.* **2018**, *29*, 63–72. [CrossRef]
52. NVS Tsunami Evacuation Zones. Available online: http://nvs.nanoos.org/TsunamiEvac (accessed on 25 December 2021).
53. CWarn Tsunami Early Warning System. Available online: https://cwarn.org/ (accessed on 8 January 2022).
54. UNESCO/IOC (Intergovernmental Oceanographic Commission of UNESCO); ICG/PTWS Working Group 3 (Awareness and Response). Defining Tsunami Hazard Areas: Concepts for Tsunami Evacuation Planning and Hazard Mapping. 2012. Available online: http://www.ioc-tsunami.org/index.php?option=com_oe&task=viewDocumentRecord&docID=12617 (accessed on 25 January 2022).
55. NVS Tsunami Evacuation, NANOOS. Available online: https://apps.apple.com/us/app/nvs-tsunami-evacuation/id478984841 (accessed on 25 January 2022).
56. UNDP (United Nations Development Programme). Feel. See. Hear. Run. Don't Panic This Is Not Real, It's Just a Drill. Available online: https://feature.undp.org/lessons-for-life/ (accessed on 26 December 2021).
57. Scheer, S.; Gardi, A.; Guillande, R.; Eftichidis, G.; Varela, V.; de Vanssay, B.; Colbeau-Justin, L. *Handbook of Tsunami Evacuation Planning, SCHEMA (Scenarios for Hazard-Induced Emergencies Management), Project n°, 030963*; Publications Office of the European Union: Ispra, Italy, 2011.
58. American Red Cross. Tsunami Preparedness, Learn How, Why and Where to Evacuate if Conditions Suggest a Tsunami Risk in Your Area. 2021. Available online: https://www.redcross.org/get-help/how-to-prepare-for-emergencies/types-of-emergencies/tsunami.html (accessed on 24 December 2021).
59. Salamon, A.; Di Manna, P. Empirical constraints on magnitude-distance relationships for seismically induced submarine tsunamigenic landslides. *Earth-Sci. Rev.* **2019**, *191*, 66–92. [CrossRef]
60. Synolakis, C.E.; Bardet, J.-P.; Borrero, J.C.; Davies, H.L.; Okal, E.A.; Silver, E.A.; Sweet, S.; Tappin, D. The slump origin of the 1998 Papua New Guinea tsunami. *Proc. R. Soc. Lond. Ser. A* **2002**, *458*, 763–789. [CrossRef]
61. Frucht, E.; Salamon, A.; Gal, E.; Ginat, H.; Grigorovitch, M.; Shem Tov, R.; Ward, S. A Fresh View of the Tsunami Generated by the Dead Sea Transform, 1995 Mw 7.2 Nuweiba Earthquake, along the Gulf of Elat-Aqaba. *Seism. Res. Lett.* **2019**, *90*, 1483–1493. [CrossRef]
62. Goda, K.; Mori, N.; Yasuda, T.; Prasetyo, A.; Muhammad, A.; Tsujio, D. Cascading geological hazards and risks of the 2018 Sulawesi Indonesia earthquake and sensitivity analysis of tsunami inundation simulations. *Front. Earth Sci.* **2019**, *7*, 261. [CrossRef]
63. Salamon, A.; Frucht, E.; Ward, S.N.; Gal, E.; Grigorovitch, M.; Shem-Tov, R.; Calvo, R.; Ginat, H. Tsunami Hazard Evaluation for the Head of the Gulf of Elat–Aqaba, Northeastern Red Sea. *Front. Earth Sci.* **2021**, *8*, 602462. [CrossRef]
64. Alter, T.R.; Haghighi, M.H.; Schneider, F.M.; Coppola, D.; Motagh, M.; Saul, J.; Babeyko, A.; Dahm, T.; Troll, V.R.; Tilmann, F.; et al. Complex hazard cascade culminating in the Anak Krakatau sector collapse. *Nat. Commun.* **2019**, *10*, 4339. [CrossRef] [PubMed]
65. ISC (International Seismological Centre). Available online: http://www.isc.ac.uk/iscbulletin/search/ (accessed on 24 December 2021).
66. Google Maps (2022) Map data@2022 Mapa GISrael, Scale:1:200. Available online: https://www.google.com/maps/@34.6433073,29.3987385,5.83z (accessed on 25 January 2022).

67. Salamon, A. Potential tsunamigenic sources in the eastern Mediterranean and a decision matrix for a tsunami early warning system. In *Marine Geo-Hazards in the Mediterranean*; Briand, F., Ed.; CIESM Workshop Monograph; CIESM: Monaco, 2011; Volume 42, pp. 35–42. Available online: http://www.ciesm.org/online/monographs/Nicosia.html (accessed on 25 January 2022).
68. Get Tsunami Ready. Available online: https://www.civildefence.govt.nz/get-ready/get-tsunami-ready/ (accessed on 25 January 2022).

encyclopedia MDPI

Entry

Substance Release from Polyelectrolyte Microcapsules

Egor V. Musin *, Aleksandr L. Kim and Sergey A. Tikhonenko *

Institute of Theoretical and Experimental Biophysics, Russian Academy of Science, Institutskaya St. 3, 142290 Puschino, Russia; kimerzent@gmail.com
* Correspondence: eglork@gmail.com (E.V.M.); tikhonenkosa@gmail.com (S.A.T.)

Definition: Controlled release of substance from polyelectrolyte microcapsules is a triggered degradation of the microcapsule membrane that is extensive enough to release the contained substances out into the environment. Membrane degradation can be a result of enzymatic digestion, ultrasound or light exposure, heating, application of a magnetic field, pH or ionic strength changes in the solution or bacteria-mediated processes. This technology can be used for the targeted release of drugs, and for the development of self-healing materials and new generation pesticides.

Keywords: polyelectrolyte microcapsules; decapsulation; controlled release

1. Introduction

Compared to other types of encapsulations, polyelectrolyte microcapsules have one of the main advantages—a variety of methods for the controlled release of the encapsulated substance. Due to the variability in the composition of a PMC shell, there are many ways to achieve the controlled release of the contained macromolecules. A variety of methods for the controlled release of substances from a PMC will allow the delivery of drugs to the target organ and release it locally, to create self-healing materials, pesticides (gradual release of pesticides) and genomic editing tools. Thus, the systematization of the results obtained over the past few decades on the controlled release of substances from polyelectrolyte microcapsules, will clarify the perception of the researcher in this field of study.

2. History

In the early 1990s, Decher and coauthors [1,2] were the first ones to demonstrate that it was possible to modify a surface with a layer-by-layer (LbL) technique, which later became widely used. A step-by-step buildup of a multilayer film is mediated by the formation of ionic interactions between oppositely charged areas of polyelectrolytes. Starting in 1998 [3], the LbL self-assembly technique was used as a novel tool for nano- and micro-encapsulation. In this method, multilayer polyelectrolyte films were built around different particles by the consecutive adsorption of oppositely charged polyelectrolytes via their electrostatic attraction, together with other association-mediating interactions, such as hydrogen bond formation, hydrophobic effects, charge transfer reactions, etc. Efficient microencapsulation of biologically or chemically active substances—drugs, proteins, vitamins, flavors, gas bubbles and even whole living cells—is getting more and more important for numerous purposes in the fields of biochemistry, pharmaceutics, cosmetology and catalysis [4].

The control over the microcapsule shell composition allows the creation of polyelectrolyte microcapsules (PMCs) that are able to release the encapsulated drug in response to some trigger. Decapsulation is necessary for targeted therapy development, which allows the delivery and release of a drug to the organs or tissues of interest. Such an approach minimizes the side effects of the drug and reduces the patient's rehabilitation duration. Moreover, gradual release of small amounts of the drug would allow its effect in the organism to be prolonged, for example, gradual release of insulin into the circulation. This

Citation: Musin, E.V.; Kim, A.L.; Tikhonenko, S.A. Substance Release from Polyelectrolyte Microcapsules. *Encyclopedia* **2022**, *2*, 428–440. https://doi.org/10.3390/encyclopedia2010026

Academic Editors: Raffaele Barretta, Ramesh Agarwal, Mark J. Jackson, Krzysztof Kamil Żur and Giuseppe Ruta

Received: 30 November 2021
Accepted: 1 February 2022
Published: 4 February 2022

technology is also used for the development of self-healing materials, pesticides and for cell culture cultivation.

The first triggered decomposition of PMCs was shown in the work of C. Schüler and F. Caruso (2001) [5]. They created polyelectrolyte capsules using DNA/spermidine. It is known that DNA/spermidine interactions weaken in solutions with higher ionic strength. Thus, after DNA/spermidine capsules had been placed into the 5M salt solution, the multilayers dissolved, leading to the degradation of the capsule.

Microcapsule shell degradation, in response to the pH change of the surrounding solution, appeared to be the newly developed way of decapsulation in the history of PMCs. In this case, microcapsules must consist of strong and weak polyelectrolytes, and when the pH of the solution becomes higher (in the case of polybase) or lower (in the case of polyacid) than the pKa of the weak polyelectrolyte, polyelectrolytes lose their charge, which results in capsule degradation (2002) [6].

Later on, numerous ways of decapsulation were discovered: heating (2002) [7], light-sensitive degradation (2004) [8], magnetic field application (2005) [9], microwave radiation (2006) [10], ultrasound application (2006) [11] and even bacterial spore germination (2020) [12]. Today, the most widespread technique for decapsulation is enzymatic digestion. In this case, microcapsules are created with the use of biodegradable polyelectrolytes, such as polysaccharides, polypeptides or polynucleotides [13].

3. The Ways of Encapsulated Substance Release

One of the essential questions in using PMCs as microcontainers for targeted drug delivery is the release of the encapsulated substance. Since both physical and chemical media factors strongly affect the polyelectrolyte shell of a PMC, release of the drug from the capsule can be initiated by acidity, ionic strength or polarity of the solution, glucose concentration, light, ultrasound, magnetic field application, redox state of the solution, enzymatic reactions or bacterial growth (Table 1).

3.1. *Acidity of the Solution (pH)*

The proof of principle of controlled, pH-dependent decapsulation was shown in the study of Shen et al. [14]. The authors used PMCs containing BSA gel. Such microcapsules were obtained by heating the PMCs with BSA up to 80 °C; this led to the formation of polyampholitic gel, which changed its charge in response to the change in the pH of the solution. The efficacy of the method was proved by the example of doxorubicin encapsulation: at pH > 4.8 doxorubicin was charged positively, BSA was charged negatively and an electrostatic interaction between them emerged. The interaction between BSA and doxorubicin led to the loading of the latter into the capsule. When the pH of the solution was decreased, BSA changed its charge to positive, which disrupted its electrostatic interactions with and led to the decapsulation of doxorubicin. In theory, this method can be used for reciprocal situations as well: cargo load at low pH, if the cargo has a negative charge, and decapsulation of the cargo at high pH, when BSA could change its charge to negative [14].

Table 1. Methods for controlled release of substances from polyelectrolyte microcapsules.

The Initiator of Substance Release	Microcapsule's Constitution (First Layer/Second Layer)$_{\text{Number of Repeats}}$	Released Substance	Citation
Acidity of the solution (pH)	BSA/(PAH/PSS)$_5$	Doxorubicin	[14]
Ionic strength of the solution	(PSS/PAH)$_9$/PSS	Fluorescein	[15]
Glucose	(PAD/PSS)$_6$	FITC-BSA	[16]
	(Mannan/PAH-BOH)$_7$	TRITC-BSA	[17]
Light	(PSS/PDADMAC)$_5$/PSS/DDAB/SiO_2-TiO_2	Phenol red	[18]
	MF/(PSS/PAH)$_4$PSS	FITC-dextran	[19]
	(PAH-BP/PSS)$_4$	Rhodamine B	[20]
	(PSS/PAH)$_2$-Ag-(PSS/PAH)$_2$	-	[21]
	Bt-SiO_2/PEI/PSS/PEI/PSS Bt-TiO_2/PEI/PSS/ PEI/PSS	Benzotriazole	[22]
	(PDADMAC/Au/PSS)$_4$	Alexa Fluor 555 dextran	[23]
	(PAH/PMA)$_4$ /NR	Nimbin and Doxorubicin	[24]
Ultrasound	(PSS/PAH)$_2$/Ag/(PSS/PAH)$_2$ (PSS/PAH)$_2$/Ag/(PSS/PAH)$_8$	FITC-PAH	[25]
	(PAH/DS)$_3$/AgNPs/(PAH/DS)	FITC-dextran	[26]
	(PSS/PAH)$_4$ (AuNP/PAH)$_4$	FITC-dextran	[27]
	(PAH/PSS)$_2$(PAH/Fe_3O_4)$_4$(PAH/PSS)$_2$	FITC-dextran	[11]
Magnetic field	(PAH/ Fe_3O_4 NPs)$_4$	Insulin	[28]
	(PSS/PAH)$_5$/Pd	Phenol red	[29]
	(PDADMAC/PSS)$_6$	FITC-dextran	[30]
	(PSS/PAH)/(PSS/P(Am-DDA)/ NPs/ (PAH)(PSS/PAH)$_2$	Cascade Blue-labelled dextran	[31]
Redox	(PVPON/PMA$_{SH}$)$_4$	FITC-transferrin	[32]
	(PSS-Rh/PAH)$_2$/(PSS/PAH)$_3$ (PSS-Rh/PAH)$_2$/(PSS/PAH)$_3$/PSS (PSS-Rh/PAH)$_2$/(PGA/PLL)$_4$ (PSS-Rh/PAH)$_2$/(PGA/PLL)$_4$/PGA	9-amino-acid peptide KP9	[33]
Enzymatic digestion	(PSS/P(HPMA-DMAE))$_4$	FITC-dextran	[34]
	(DS/PLA)$_4$	Rhodamine isothiocyanate	[35]
	(CT/DS)$_3$	FITC-albumin	[36]
	(CHI/ALG)$_5$	Indomethacin	[37]
Osmotic pressure	(Dex-HEMA-DMAEMA)/(PSS/PAH)$_3$	FITC-dextran	[38]
	(Dex-HEMA-DMAEMA)/(PSS/PAH)$_3$	TRITC-dextran	[39]
Spore germination	(spore forms of Bacillus subtilis)/(PSS/PAH)$_3$	FITC-dextran	[12]

3.2. Ionic Strength of the Solution

In contrast to pH-controlled decapsulation, drug release in response to the ionic strength of the solution can happen for PMCs containing both weak and strong polyelectrolytes. Antipov et al. [15] showed that the mechanism of ionic strength-mediated decapsulation is quite similar to the pH-controlled one: with an increase in ionic strength of the solution, free energy of interactions within the polyelectrolyte layers decreases leading to higher permeability of the capsules and to the release of fluorescein from the PMC ($(PSS/PAH)_9/PSS$). In another study [40], it was reported that an increase in the ionic strength of the solution resulted in the generation of local defects in the capsule and, as a result, in increased shell permeability. However, lack of linear dependence between the permeability of the capsule, salt concentration in the solution and the inability to use this approach for controlled decapsulation in vivo, resulted in a significant limitation to its applications.

3.3. Glucose Content

In the study of De Geest et al. [16], the authors created polyelectrolyte microcapsules using a glucose-sensitive polymer. Polystyrene sulphonate and copolymer of dimethylaminoethyl acrylate and 3-acrylaminophenylboronic acid were used as the main components of the capsule shell in this work. The glucose-responsive component here was a derivative of the phenylboronic acid—it could form complexes with glucose, leading to the shift in chemical equilibrium towards charged acid molecules. This resulted in increased electrostatic interactions, repulsion of microcapsule layers and capsule shell degradation [16].

Levy et al. [17] created another system based on forming ester bonds between polysaccharides and phenylboronic acid derivatives. In this case, the PMC shell contained complexes of polyacrylic acid with hemisulphate of aminophenylboronic acid, which bound with mannans through ester bonds. It was observed that the increase in simple carbohydrates (fructose, glucose, mannose and galactose) led to the disruption of the microcapsule shell. It likely occurred as a result of simple carbohydrates replacing mannans in complexes with the phenylboronic acid derivatives, which led to destabilization of the capsule shell layers and capsule degradation [17].

3.4. Light Exposure

Light exposure appears to be a convenient way to control decapsulation (Figure 1). For instance, Katagiri et al. [18] created UV-sensitive PMCs with the shell consisting of polystyrenesulfonate and polydiallyldimethylammonium layers coated with a lipid bilayer and SiO_2-TiO_2 oxide system. In this study it was shown that encapsulated low molecular weight substances could be released by UV irradiation due to TiO_2 photocatalytic activity, which promoted degradation of the polyelectrolyte layers of the capsule. The rate of capsule disruption and, as a result, the rate of the cargo release were regulated by the SiO_2:TiO_2 ratio in the outer layer of the capsule [18]. Another example of UV-controlled microcapsules was demonstrated in the study of Koo et al. [19]. The authors of this work developed PMCs with photoacid generators (PAGs) in the outer layer of the capsule shell. After UV irradiation of such capsules the PAGs released protons into the surrounding solution, which led to a local decrease in pH. Thus, long UV exposure, caused by maintaining low pH values, resulted in the swelling and disruption of the PMCs [19].

The work of Park et al. [20] is another example of UV application for controlled decapsulation. Authors used photosensitive benzophenone-modified polystyrenesulphonate (PSS) and polyallylamine (PAH) for the microcapsule shell assembly. After UV exposure, the proton in the C–H bond in the polymer layer was easily abstracted by excited benzophenone, leading to the formation of a new C–C bond between the polymers of the shell. It was demonstrated that the time of UV irradiation affected the extent of cross-linking and the pore size, which allowed the fine-tuning of the rate of the drug release from the microcapsule [20].

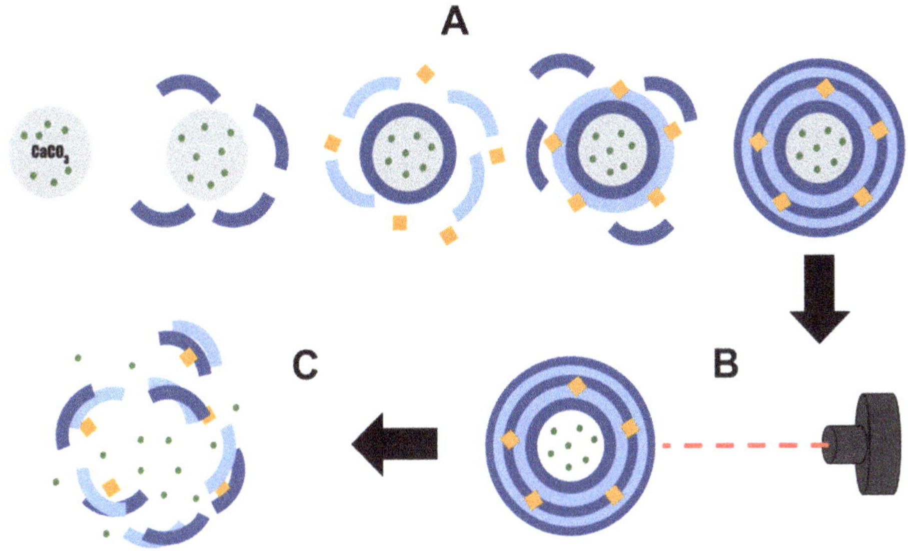

Figure 1. Diagram showing the effect of light on substance release from polyelectrolyte microcapsules. (**A**)—formation of polyelectrolyte microcapsules with metal particles; (**B**)—irradiation of microcapsules with light; (**C**)—destruction of microcapsules after light irradiation.

Other than UV light, light-controlled decapsulation can also be implemented via use of near-infrared (IR) band frequencies (700–900 nm). In contrast to UV light, IR does not have any negative effects on the living cells and can penetrate quite deeply into an organism's tissues. In the work of Skirtach et al. [21], the authors created microcapsules consisting of PAH and PSS and containing either silver nanoparticles or IR-absorbing dyes. It was reported that after exposure to laser light of IR band frequencies, multiple gaps in the PMC shell appeared and the capsule's cargo was released [21]. The same phenomenon was demonstrated by Skorb et al. [22]. In this work, they created capsules consisting of PSS and polyethylenimine (PEI), which were modified by silver nanoparticles. IR laser irradiation of such capsules resulted in the formation of pores large enough for anticorrosion cargo compound release.

In the study of Skirtach et al. [23], the authors used aggregates of gold nanoparticles to create IR-controlled PMCs. They incorporated gold nanoparticle aggregates into the microcapsule shell, which consisted of polydiallyldimethylammonium (PDADMA) and PSS. As a result, PMCs with the peak of the light absorption in the IR part of the spectrum were obtained. Gold particles absorbed the IR light and heated up, leading to the local melting of polymer layers and an increase in their permeability. At the same time, after the light exposure was stopped, the permeability of the PMC shell returned to its initial values [23].

This decapsulation method was extensively used in the application of PMCs to photothermal cancer therapy. It allowed substances to act on cancerous cells in two different ways—via a therapeutic chemical compound and via microcapsule heating. A recent study demonstrated the considerable potential of IR-absorbing microcapsules and laser-triggered decapsulation for applications in anticancer therapy [24]. Authors created PMCs containing two drugs: hydrophilic doxorubicin and hydrophobic anticancer drug, nimbin [24]. Decapsulation of these PMCs was triggered by near-infrared (NIR) light exposure and, in contrast to other studies, this work demonstrated the ability to use a low power laser (0.5 W/cm^2) for decapsulation initiation.

3.5. Ultrasound

Ultrasound-based tools are widely used for medical applications; therefore, ultrasound use for controlled drug release from PMCs appeared to be very promising since its safety and overall effect on the human organism are well described. In general, ultrasound affects the permeability of microcapsules through the cavitation effect in the liquid medium, which occurs when applied ultrasound waves have frequencies of 20 kHz or greater. Ultrasound waves cause the formation of microbubbles from the gases dissolved in the medium. Longer ultrasound exposure leads to microbubble oscillation and collapse, which is called the cavitation phenomenon. Bubble collapse, in turn, causes the redistribution of energy in the media and the generation of additional shear force between liquid layers [41]. As a result of all these processes, the PMC shell can be disturbed or destroyed. The majority of works in this field were performed based on PMCs, which consisted of PSS as a polyanion and PAH as a polycation. One of the first works in the field is the study by Skirtach et al. [25]. Using FITC-dextran as a cargo, the authors demonstrated the possibility of using ultrasound for controlled drug release and, moreover, they proved that the sensitivity of PMCs and the rate of their destruction could be increased by adding silver nanoparticles to the microcapsule shell. Similar results with the release of FITC-dextran in decapsulation were obtained in the study of Anandhakumar et al. [26], which used PAH/DS (dextran sulphate) capsules with incorporated silver particles. Another early work in the field of ultrasound-sensitive PMCs was the work by Shchukin et al. [11]. The authors used the same polycations and polyanions for PMC assembly, but they incorporated iron oxide (II-III) Fe_3O_4 into one of the middle layers. The addition of iron oxide increased the susceptibility of microcapsules to ultrasound exposure. The strongest effect on ultrasound susceptibility of PMCs was reported in the study of Kolesnikova et al. [42], where the authors used zinc oxide particles in the formulation of PAH/PSS capsules.

It is worth mentioning that the incorporation of metal into the capsule shell did not always affect its susceptibility to ultrasound in this way. For instance, De Geest et al. [27] created two types of PMCs: one consisted only from PSS and PAH, while another contained PAH and gold nanoparticles carrying carboxyl groups on their surface. When two types of microcapsules were compared using their sensitivity to ultrasound, it turned out that PMCs containing gold nanoparticles were more resistant to ultrasound exposure. Such an effect possibly occurred due to polyanionic carboxyl groups, which made the multilayered shell of the microcapsule more rigid [27].

3.6. Magnetic Field

Another example of a trigger for the controlled substance release from PMCs is the application of a magnetic field. One of the main advantages of this method is a high permeability of biological tissues to magnetic fields. In one of the pioneering works about this topic, Caruso et al. [43] created PMCs that not only contained polyelectrolytes in the shell, but also had a polystyrene core. However, the proof of concept of magnetically triggered microcapsule decapsulation was first demonstrated in 2005 by Lu et al. [9]. For this purpose, the authors used microcapsules carrying FITC-dextran as a cargo. The shell of these microcapsules consisted of PAH and PSS with incorporated cobalt nanoparticles, which were coated with gold. After the application of a magnetic field, nanoparticles began to rotate with a frequency corresponding to the frequency of oscillating magnetic field. Thus, these particles destabilized adjacent polyelectrolyte layers and increased the permeability of the PMC shell. At the same time, it was shown that the increase in permeability has its limits, which were imposed by the rate of nanoparticle rotation—if the frequency of the magnetic field became too high, nanoparticles were not able to rotate with the same corresponding rate, and the permeability increase was minimal. Later on, the possible application of capsules containing iron oxide nanoparticles was reported in the study of Zheng et al. [28]. The authors optimized conditions of insulin loading into magnetoresponsive PMCs and demonstrated magnetically controlled decapsulation, thus suggesting the use of this technology in diabetes therapy.

For targeted drug delivery implementation, it is important to create microcapsules with low permeability that can hold compounds with a low molecular weight. Such microcapsules became the main topic of the research carried out by Katagiri et al. [29]. This group developed improved magnetoresponsive PMCs containing Fe_3O_4 and coated with a lipid bilayer. Such microcapsules can be used for the transportation of substrates with different molecular weights. When a magnetic field was applied, magnetic nanoparticles Fe_3O_4 increased the permeability of the lipid bilayer of the capsule via particle heating and not through their movements [29].

At present, there have been PMCs developed containing magnetic particles both in the shell and in the core of the microcapsule. Hu et al. [30] created microcapsules containing iron pentacarbonyl $Fe(CO)_5$. Microcapsules were prepared using the layer-by-layer method. First, the magnetoresponsive core was created from iron pentacarbonyl and silicon dioxide, then polyelectrolytes were applied in layers. In the end, silicon dioxide was washed out of the capsules using sodium hydroxide solution. It was shown in this study that microcapsules constructed in such a way were distorted in response to the magnetic field application, which led to the cargo release from the capsule. In this case the fine-tuning of decapsulation was possible by changing the strength of the magnetic field: the stronger the field was, the higher the decapsulation rate was [30].

A similar technique for the fine-tuning of magnetoresponsive microcapsules was developed by Carregal-Romero et al. [31]. The authors created microcapsules containing iron oxide magnetic particles shaped as cubes, which increased their ability to heat up under the influence of the magnetic field. Another way to regulate decapsulation of magnetoresponsive capsules is based on the alteration of the magnetic particle concentration in the shell of a PMC [36]. The higher the concentration, the higher the heat capacity is, and, in turn, higher temperatures can be reached when a magnetic field is applied. Local temperature increases in the shell of the capsule led to capsule structure damage and, as a consequence, to cargo release [44].

All in all, approaches for decapsulation of magnetoresponsive microcapsules can be divided into two categories. In the first one, an alternating magnetic field is used to increase mobility and temperature of magnetic particles [9,45,46], which leads to partial or total destruction of the capsule shell. This approach caused local heating, which could be potentially harmful for healthy tissues and organs, thus limiting its use in clinical applications. The second approach is based on the use of a constant magnetic field, which could be enough for capsule deformation without them heating up [47,48]. However, this way of decapsulation usually requires fabricating microcapsules through an emulsification process, which, in turn, leads to an increase in the size of the resulting microcapsules, up to dozens or even several hundreds of micrometers. Such a large capsule size significantly limits opportunities to use this approach in targeted drug delivery.

3.7. Redox

Redox potential is varied across cellular organelles, and this feature can be used to develop PMCs that are sensitive to this kind of stimuli. One of the first works in this field was dedicated to the development of PMCs, which, in addition to polyelectrolyte compounds, contained cysteine residues in the shell [49]. When put in a solution containing DMSO, disulfide bonds were formed between cysteines, which increased the stability of the capsule shell under low pH values. However, if such capsules were placed in the reducing environment with TCEP, disulfide bonds were disrupted, and the microcapsule shell became susceptible to the action of the acidic pH. Such a system allowed a double-triggered release, which was responsive to both redox potential and the pH of the environment [49].

In another study PMCs were created that released their cargo only in response to the reducing potential of the surrounding solution. Zelikin et al. [32] created microcapsules based on polyvinylpyrrolidone and polymethacrylic acid, modified by thiol groups. Microcapsules were then placed in a strong oxidizing environment to form disulfide bonds between thiol groups. The authors demonstrated that when a reducing agent, such as dithiothreitol

(DTT), was applied to these microcapsules, the shell became gradually degraded, releasing the cargo from the capsule [32].

One outstanding example of the use of such PMCs is their possible application as peptide vaccine carriers. In an in vitro study by Rose et al. [33], it was reported that PMCs, loaded with polypeptides, could persistently present in the circulatory system of a human and be engulfed by antigen-presenting cells (APCs). The environment within the cells was reducing, which allowed the release of the cargo inside the cells in response to the change in redox potential. APCs, in turn, became able to present cargo peptides within the main histocompatibility complexes (MHC) on their surface [33].

3.8. Enzymatic Digestion

One of the alternative methods of decapsulation is enzymatic digestion of the shell of the PMC. This may be a convenient way to deliver a drug if it should act inside the cell (as it is in the case of oligonucleotides and proteins). The first example of enzymatically triggered capsules was published by De Geest et al. [34]. The authors developed capsules with two different types of shell: the first consisted of polyarginine and dextran sulphate, and the second consisted of poly(hydroxypropylmethacrylamide dimethylaminoethyl) and PSS. These capsules were then added to the African green monkey kidney cell culture. It was observed that added PMCs were actively engulfed by the cells, and the shell of the capsules was enzymatically degraded, releasing the loaded substance [34]. Such capsules were later investigated using in vivo experiments with mice [35].

One more example of enzymatically driven decapsulation of PMCs was developed in the study of Itoh et al. [36]. The authors obtained microcapsules based on chitosan and sulphate dextran, and used chitosanase as a decapsulation triggering agent. The shell degradation and controlled cargo release was demonstrated after addition of the enzyme to the capsules. Based on these results, the authors decided to go further and try to optimize the system in such a way that the drug release happened slower (so called 'sustained release' of the drug). To do this, they added an additional outer layer of positively charged chitosan. In the buffer (pH 5.6) chitosanase (pI 9.3) was also positively charged, thus, the interaction between the enzyme and the modified capsules was weakened, due to the static repulsion [36].

Most of the works in the field of enzymatically controlled capsule degradation include the use of two lytic enzymes, which are present in the human organism: pepsin (in the stomach) and hyaluronidase (in salivary and seminal glands). For pepsin-controlled capsule development, chitosan and poly(2-acrylamido-2-methylpropanesulphonic) acid were used. It was reported that pepsin action led to the degradation of the PMC shell and to the release of the encapsulated indomethacin [37]. Hyaluronidase-responsive PMCs contain the residues of hyaluronic acid [50,51] or chondroitin sulphate [52] in their shells.

As it was shown by Cardoso et al. [53], enzymatic digestion of capsules could be driven not only by action of lytic enzymes from the outside solution, but also by the action of encapsulated enzymes from within the capsule itself. One of the examples of such capsules is the case of the capsules created from chitosan and hyaluronic acid, which are described above. Together with the drug, hyaluronidase was placed into the hollow of the capsule for the triggered release of the capsule shell [53]. One more example of a similar system is the type of capsules described in the work of Borodina et al. [54]. In this work, investigators created a capsule shell consisting of poly-*L*-arginine and poly-*L*-aspartic acid. As a decapsulation agent deposited inside the capsule authors used a mix of lytic enzymes from *Streptomyces griseus* called Pronase. It was also shown that the fine-tuning of the capsule degradation rate could be achieved through the alteration of the Pronase content inside the capsule.

3.9. Osmotic Pressure

All the methods that use osmotic pressure as a decapsulation force are based on the same main principle of placing hydrogel from biodegradable material (such as dextran and its derivatives) into the capsule. In water, under the influence of osmotic forces, water molecules start to enter the capsule. This leads to an increase in osmotic pressure on the shell of the capsule, which, in the very end, results in shell damage facilitated by the simultaneous gradual hydrolysis of dextran. The control of decapsulation rate could be performed by changing the extent of crosslinking in the polyelectrolyte layers, the number of polyelectrolyte layers and the amount of the hydrogel inside the capsule [38,39].

3.10. Bacterially Driven Decapsulation

This type of decapsulation is based on the germination of encapsulated bacterial spores. The efficiency of this method was demonstrated in the work of Musin et al. [12]. The authors prepared microcapsules consisting of PSS/PAA that contained *Bacillus subtilis* spores together with FITC-dextran (Figure 2).

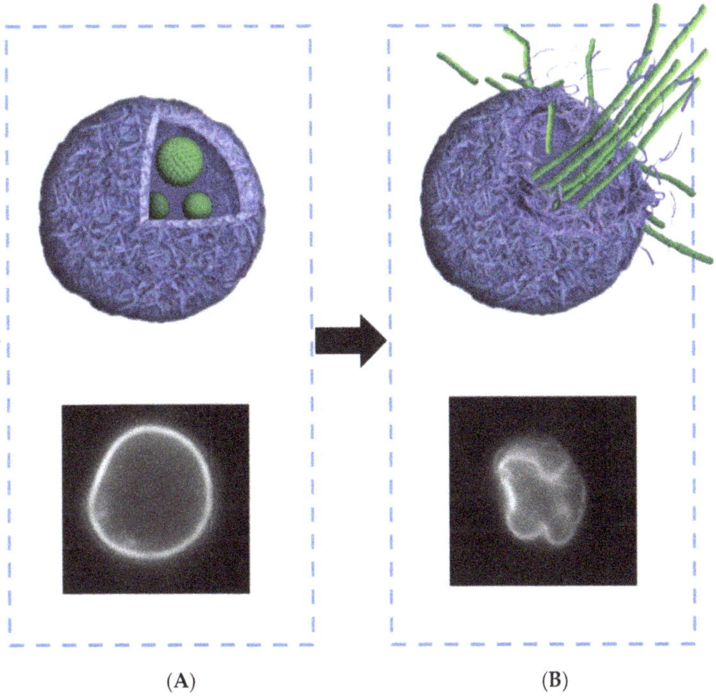

Figure 2. Scheme of bacterially driven decapsulation. (**A**)—PMC with encapsulated bacterial spores of *Bacillus subtilis*; (**B**)—release of FITC-dextran and destruction of PMC after spore germination.

When favorable conditions for bacterial growth were created, spores within the capsules became activated, since the created nanoscale shell was semi-permeable and allowed for nutrients to get into the capsule. When encapsulated bacterial spores germinated inside the microcapsule, they disturbed and destroyed the shell of the PMC, which resulted in decapsulation and release of the FITC-dextran. The main advantage of this decapsulation method was that it did not require any specialized, highly technological equipment, such as ultrasound or laser generators.

4. Practical Relevance and Applications

One of the main applications of microencapsulated substances is their triggered controlled release. The diversity of polyelectrolytes and nanoparticles used to create PMCs allows the development of a whole set of microcapsule types with various release triggers. Polyelectrolyte microcapsule decapsulation can be used in medicine, in genome editing, in the development of targeted delivery techniques, in agricultural and paint manufacturing industries and in 'smart'/functional surface development.

In medicine polyelectrolyte microcapsule are mostly used for targeted drug delivery techniques [55]. The main advantage of polyelectrolyte microcapsules as drug carriers here is that the capsules protect the drug from degradation along the way to the targeted region (cell, inflammation area, disease 'hot' spot) and they also protect the organism from the possible side effects of the drug. One important feature for successful targeted drug delivery is its controlled release from the microcapsule without any loss of therapeutic effect [56]. Versatility of LbL technology allows the production of polymeric capsules consisting of biodegradable materials with different sizes and unique surface properties, which are required for the development of successful decapsulation strategies. Another feature of PMCs, which makes them convenient to use in targeted drug delivery, is their active engulfment by several cell types, including breast cancer cells, hepatocytes, fibroblasts, colon carcinoma cells and the model kidney cell line 'Vero', as well as phagocytic, dendritic cells and macrophages. In addition to this, polyelectrolyte microcapsules can be used for genetic construct delivery in genome editing techniques [57].

In the paint manufacturing industry, PMCs are used for the development of self-healing coatings. Mikhail L. Zheludkevich and coauthors created anticorrosion coatings with a self-healing effect, which consisted of hybrid sol–gel films doped with special nanocontainers. These nanocontainers were covered with the polyelectrolyte and corrosion inhibitor (benzotriazole) layers, which were released in response to pH changes caused by the corrosion process [58]. During the anticorrosion system development, pH-sensitive components, which allowed the decapsulation corrosion inhibitors, were usually used, such as ZnO [59], SiO_2 nanoparticles [60], halloysite nanotubes [61,62], TiO_2 nanotubes [63] and others.

Another possible application of polyelectrolyte microcapsules in the agricultural industry is the controlled release of pesticides. Such a technology would allow a decrease in the rate of active compound release and prolong its action. Moreover, it may become a solution to the problem of pesticide pollution of the environment. Xiaojing Wang and Jing showed in their work the ability to encapsulate the herbicide picloram and to perform its gradual controlled decapsulation [64].

Author Contributions: Conceptualization, S.A.T. and E.V.M.; validation, S.A.T. and A.L.K.; formal analysis, S.A.T.; data curation, S.A.T. and A.L.K.; writing—original draft preparation, E.V.M. and A.L.K.; writing—review and editing, S.A.T., E.V.M. and A.L.K.; visualization, E.V.M. and A.L.K.; supervision, S.A.T.; project administration, S.A.T.; All authors have read and agreed to the published version of the manuscript.

Funding: This research was funded by the State assignment of Russian Federation: 075-01027-22-00.

Institutional Review Board Statement: Not applicable.

Informed Consent Statement: Not applicable.

Data Availability Statement: Not applicable.

Conflicts of Interest: The authors declare no conflict of interest.

Entry Link on the Encyclopedia Platform: https://encyclopedia.pub/20438.

Abbreviations

PAD	Poly 3-acrylaminophenylboronic acid: dimethyl aminoethyl acrylate
PAH-BOH	Amide bonds between the carboxylic groups of acrilic acid and aminophenylboronic acid hemisulfate
PSS	Poly(sodium 4-styrenesulfonate)
DDAB	Dioctadecyldimethylammonium bromide
MF	Melamine formaldehyde
PAH-BP	Benzophenone-modified PAH

Bt	Benzotriazole
PDADMAC	Poly(diallyldimethylammonium chloride)
PMA	Poly(methacrylic acid)
NRs	gold nanorods
DS	Dextran sulfate
PVPON	Poly(vinylpyrrolidone)
PMA_{SH}	Poly(methacrylic acid) (PMA) conjugated with cysteamine (NH2-(CH2)2)-SH)
PLL	Poly(L-lysine)
PGA	Poly(L-glutamic acid)
PSS-Rh	Rhodamine B-labelled PSS
p(HPMA-DMAE	Poly(hydroxypropylmethacrylamide dimethylaminoethyl)
CT	Chitosan
ALG	Alginate
(dex-HEMA-DMAEMA)	Microgel dextran-hydroxyethylmethacrylate dimethylaminoethyl methacrylate

References

1. Decher, G.D.; Hong, J.-D. Buildup of ultrathin multilayer films by a self-assembly process, 1 consecutive adsorption of anionic and cationic bipolar amphiphiles on charged surfaces. In *Makromolekulare Chemie. Macromolecular Symposia*; Hüthig & Wepf Verlag: Basel, Switzerland, 1991; pp. 321–327.
2. Decher, G.; Schmitt, J. Fine-Tuning of the film thickness of ultrathin multilayer films composed of consecutively alternating layers of anionic and cationic polyelectrolytes. In *Progress in Colloid & Polymer Science*; Springer Science and Business Media LLC: Berlin/Heidelberg, Germany, 2007; pp. 160–164.
3. Donath, E.; Sukhorukov, G.B.; Caruso, F.; Davis, S.A.; Möhwald, H. Novel Hollow Polymer Shells by Colloid-Templated Assembly of Polyelectrolytes. *Angew. Chemie Int.* **1998**, *37*, 2201–2205. [CrossRef]
4. Wang, C.; Ye, S.; Sun, Q.; He, C.; Ye, W.; Liu, X.; Tong, Z. Microcapsules for controlled release fabricated via layer-by-layer self-assembly of polyelectrolytes. *J. Exp. Nanosci.* **2008**, *3*, 133–145. [CrossRef]
5. Schüler, C.; Caruso, F. Decomposable Hollow Biopolymer-Based Capsules. *Biomacromolecules* **2001**, *2*, 921–926. [CrossRef] [PubMed]
6. Antipov, A.A.; Sukhorukov, G.B.; Leporatti, S.; Radtchenko, I.L.; Donath, E.; Möhwald, H. Polyelectrolyte multilayer capsule permeability control. *Colloids Surf. A Physicochem. Eng. Asp.* **2002**, *198–200*, 535–541. [CrossRef]
7. Ibarz, G.; Dähne, L.; Donath, E.; Möhwald, H. Controlled Permeability of Polyelectrolyte Capsules via Defined Annealing. *Chem. Mater.* **2002**, *14*, 4059–4062. [CrossRef]
8. Tao, X.; Li, J.; Möhwald, H. Self-Assembly, Optical Behavior, and Permeability of a Novel Capsule Based on an Azo Dye and Polyelectrolytes. *Chem. A Eur. J.* **2004**, *10*, 3397–3403. [CrossRef]
9. Lu, Z.; Prouty, M.D.; Guo, Z.; Golub, V.; Kumar, C.S.S.R.; Lvov, Y.M. Magnetic Switch of Permeability for Polyelectrolyte Microcapsules Embedded with Co@Au Nanoparticles. *Langmuir* **2005**, *21*, 2042–2050. [CrossRef]
10. Gorin, D.A.; Shchukin, D.G.; Mikhailov, A.I.; Köhler, K.; Sergeev, S.A.; Portnov, S.A.; Taranov, I.; Kislov, V.V.; Sukhorukov, G.B. Effect of microwave radiation on polymer microcapsules containing inorganic nanoparticles. *Tech. Phys. Lett.* **2006**, *32*, 70–72. [CrossRef]
11. Shchukin, D.G.; Gorin, D.A.; Möhwald, H. Ultrasonically Induced Opening of Polyelectrolyte Microcontainers. *Langmuir* **2006**, *22*, 7400–7404. [CrossRef]
12. Musin, E.V.; Kim, A.L.; Dubrovskii, A.V.; Kudryashova, E.B.; Tikhonenko, S.A. Decapsulation of Dextran by Destruction of Polyelectrolyte Microcapsule Nanoscale Shell by Bacillus subtilis Bacteria. *Nanomaterials* **2019**, *10*, 12. [CrossRef]
13. Picart, C.; Schneider, A.; Etienne, O.; Mutterer, J.; Schaaf, P.; Egles, C.; Jessel, N.; Voegel, J.-C. Controlled Degradability of Polysaccharide Multilayer Films In Vitro and In Vivo. *Adv. Funct. Mater.* **2005**, *15*, 1771–1780. [CrossRef]
14. Shen, H.-J.; Shi, H.; Ma, K.; Xie, M.; Tang, L.-L.; Shen, S.; Li, B.; Wang, X.-S.; Jin, Y. Polyelectrolyte capsules packaging BSA gels for pH-controlled drug loading and release and their antitumor activity. *Acta Biomater.* **2013**, *9*, 6123–6133. [CrossRef]
15. Antipov, A.A.; Sukhorukov, G.B.; Möhwald, H. Influence of the Ionic Strength on the Polyelectrolyte Multilayers' Permeability. *Langmuir* **2003**, *19*, 2444–2448. [CrossRef]
16. De Geest, B.G.; Jonas, A.M.; Demeester, J.; De Smedt, S.C. Glucose-Responsive Polyelectrolyte Capsules. *Langmuir* **2006**, *22*, 5070–5074. [CrossRef] [PubMed]
17. Levy, T.; Déjugnat, C.; Sukhorukov, G.B. Polymer Microcapsules with Carbohydrate-Sensitive Properties. *Adv. Funct. Mater.* **2008**, *18*, 1586–1594. [CrossRef]
18. Katagiri, K.; Koumoto, K.; Iseya, S.; Sakai, M.; Matsuda, A.; Caruso, F. Tunable UV-Responsive Organic–Inorganic Hybrid Capsules. *Chem. Mater.* **2009**, *21*, 195–197. [CrossRef]
19. Koo, H.Y.; Lee, H.-J.; Kim, J.K.; Choi, W.S. UV-triggered encapsulation and release from polyelectrolyte microcapsules decorated with photoacid generators. *J. Mater. Chem.* **2010**, *20*, 3932–3937. [CrossRef]

20. Park, M.-K.; Deng, S.; Advincula, R.C. Sustained Release Control via Photo-Cross-Linking of Polyelectrolyte Layer-by-Layer Hollow Capsules. *Langmuir* **2005**, *21*, 5272–5277. [CrossRef]
21. Skirtach, A.G.; Antipov, A.A.; Shchukin, D.G.; Sukhorukov, G.B. Remote Activation of Capsules Containing Ag Nanoparticles and IR Dye by Laser Light. *Langmuir* **2004**, *20*, 6988–6992. [CrossRef]
22. Skorb, E.V.; Sviridov, D.V.; Shchukin, D.G.; Borisenko, V.E.; Gaponenko, S.V.; Gurin, V.S. Light-controllable coatings for corrosion protection. *Phys. Chem. Appl. Nanostructures* **2009**, 515–518. [CrossRef]
23. Skirtach, A.; Karageorgiev, P.; Bédard, M.F.; Sukhorukov, G.B.; Möhwald, H. Reversibly Permeable Nanomembranes of Polymeric Microcapsules. *J. Am. Chem. Soc.* **2008**, *130*, 11572–11573. [CrossRef] [PubMed]
24. Sharma, V.; Vijay, J.; Ganesh, M.; Sundaramurthy, A. Multilayer capsules encapsulating nimbin and doxorubicin for cancer chemo-photothermal therapy. *Int. J. Pharm.* **2020**, *582*, 119350. [CrossRef] [PubMed]
25. Klitzing, R.; Möhwald, H. A realistic diffusion model for ultrathin polyelectrolyte films. *Macromolecules* **1996**, *29*, 6901–6906. [CrossRef]
26. Skirtach, A.G.; De Geest, B.G.; Mamedov, A.; Antipov, A.A.; Kotov, N.A.; Sukhorukov, G.B. Ultrasound stimulated release and catalysis using polyelectrolyte multilayer capsules. *J. Mater. Chem.* **2007**, *17*, 1050–1054. [CrossRef]
27. Anandhakumar, S.; Mahalakshmi, V.; Raichur, A.M. Silver nanoparticles modified nanocapsules for ultrasonically activated drug delivery. *Mater. Sci. Eng. C* **2012**, *32*, 2349–2355. [CrossRef]
28. Zheng, C.; Ding, Y.; Liu, X.; Wu, Y.; Ge, L. Highly magneto-responsive multilayer microcapsules for controlled release of insulin. *Int. J. Pharm.* **2014**, *475*, 17–24. [CrossRef]
29. Katagiri, K.; Nakamura, M.; Koumoto, K. Magnetoresponsive Smart Capsules Formed with Polyelectrolytes, Lipid Bilayers and Magnetic Nanoparticles. *ACS Appl. Mater. Interfaces* **2010**, *2*, 768–773. [CrossRef]
30. Hu, Y.; Liu, C.; Li, D.; Long, Y.; Song, K.; Tung, C.-H. Magnetic Compression of Polyelectrolyte Microcapsules for Controlled Release. *Langmuir* **2015**, *31*, 11195–11199. [CrossRef]
31. Carregal-Romero, S.; Guardia, P.; Yu, X.; Hartmann, R.; Pellegrino, T.; Parak, W.J. Magnetically triggered release of molecular cargo from iron oxide nanoparticle loaded microcapsules. *Nanoscale* **2015**, *7*, 570–576. [CrossRef]
32. Zelikin, A.N.; Quinn, J.F.; Caruso, F. Disulfide Cross-Linked Polymer Capsules: En Route to Biodeconstructible Systems. *Biomacromolecules* **2006**, *7*, 27–30. [CrossRef]
33. De Rose, R.; Zelikin, A.N.; Johnston, A.P.R.; Sexton, A.; Chong, S.-F.; Cortez, C.; Mulholland, W.; Caruso, F.; Kent, S.J. Binding, Internalization, and Antigen Presentation of Vaccine-Loaded Nanoengineered Capsules in Blood. *Adv. Mater.* **2008**, *20*, 4698–4703. [CrossRef]
34. De Geest, B.G.; Vandenbroucke, R.E.; Guenther, A.M.; Sukhorukov, G.B.; Hennink, W.E.; Sanders, N.N.; Demeester, J.; De Smedt, S.C. Intracellularly Degradable Polyelectrolyte Microcapsules. *Adv. Mater.* **2006**, *18*, 1005–1009. [CrossRef]
35. De Koker, S.; De Geest, B.G.; Cuvelier, C.; Ferdinande, L.; Deckers, W.; Hennink, W.E.; De Smedt, S.C.; Mertens, N. In vivo Cellular Uptake, Degradation, and Biocompatibility of Polyelectrolyte Microcapsules. *Adv. Funct. Mater.* **2007**, *17*, 3754–3763. [CrossRef]
36. Itoh, Y.; Matsusaki, M.; Kida, T.; Akashi, M. Enzyme-Responsive Release of Encapsulated Proteins from Biodegradable Hollow Capsules. *Biomacromolecules* **2006**, *7*, 2715–2718. [CrossRef] [PubMed]
37. Wang, C.; Ye, S.; Dai, L.; Liu, X.; Tong, Z. Enzymatic desorption of layer-by-layer assembled multilayer films and effects on the release of encapsulated indomethacin microcrystals. *Carbohydr. Res.* **2007**, *342*, 2237–2243. [CrossRef] [PubMed]
38. De Geest, B.G.; Déjugnat, C.; Sukhorukov, G.B.; Braeckmans, K.; De Smedt, S.C.; Demeester, J. Self-Rupturing Microcapsules. *Adv. Mater.* **2005**, *17*, 2357–2361. [CrossRef]
39. De Geest, B.; Dejugnat, C.; Verhoeven, E.; Sukhorukov, G.; Jonas, A.; Plain, J.; Demeester, J.; De Smedt, S. Layer-by-layer coating of degradable microgels for pulsed drug delivery. *J. Control. Release* **2006**, *116*, 159–169. [CrossRef]
40. Rae, J.; Ashokkumar, M.; Eulaerts, O.; von Sonntag, C.; Reisse, J.; Grieser, F. Estimation of ultrasound induced cavitation bubble temperatures in aqueous solutions. *Ultrason. Sonochemistry* **2005**, *12*, 325–329. [CrossRef]
41. Kolesnikova, T.A.; Gorin, D.A.; Fernandes, P.; Kessel, S.; Khomutov, G.B.; Fery, A.; Shchukin, D.G.; Möhwald, H. Nanocomposite Microcontainers with High Ultrasound Sensitivity. *Adv. Funct. Mater.* **2010**, *20*, 1189–1195. [CrossRef]
42. De Geest, B.G.; Skirtach, A.G.; Mamedov, A.A.; Antipov, A.A.; Kotov, N.A.; De Smedt, S.C.; Sukhorukov, G.B. Ultrasound-Triggered Release from Multilayered Capsules. *Small* **2007**, *3*, 804–808. [CrossRef]
43. Caruso, F.; Susha, A.S.; Giersig, M.; Möhwald, H. Magnetic Core-Shell Particles: Preparation of Magnetite Multilayers on Polymer Latex Microspheres. *Adv. Mater.* **1999**, *11*, 950–953. [CrossRef]
44. Katagiri, K.; Imai, Y.; Koumoto, K. Variable on-demand release function of magnetoresponsive hybrid capsules. *J. Colloid Interface Sci.* **2011**, *361*, 109–114. [CrossRef]
45. Hu, S.-H.; Tsai, C.-H.; Liao, C.-F.; Liu, D.-M.; Chen, S.-Y. Controlled Rupture of Magnetic Polyelectrolyte Microcapsules for Drug Delivery. *Langmuir* **2008**, *24*, 11811–11818. [CrossRef]
46. Degen, P.; Peschel, S.; Rehage, H. Stimulated aggregation, rotation, and deformation of magnetite-filled microcapsules in external magnetic fields. *Colloid Polym. Sci.* **2008**, *286*, 865–871. [CrossRef]
47. Long, Y.; Liu, C.; Zhao, B.; Song, K.; Yang, G.; Tung, C.-H. Bio-inspired controlled release through compression–relaxation cycles of microcapsules. *NPG Asia Mater.* **2015**, *7*, e148. [CrossRef]
48. Patel, R.; Upadhyay, R.V.; Mehta, R. Microscopic observation of magnetodeformational effects in magnetic nanocomposite micelles. *J. Phys. Condens. Matter* **2008**, *20*, 204116. [CrossRef] [PubMed]

49. Haynie, D.T.; Palath, N.; Liu, Y.; Li, A.B.; Pargaonkar, N. Biomimetic Nanostructured Materials: Inherent Reversible Stabilization of Polypeptide Microcapsules. *Langmuir* **2005**, *21*, 1136–1138. [CrossRef] [PubMed]
50. Lee, H.; Jeong, Y.; Park, T.G. Shell Cross-Linked Hyaluronic Acid/Polylysine Layer-by-Layer Polyelectrolyte Microcapsules Prepared by Removal of Reducible Hyaluronic Acid Microgel Cores. *Biomacromolecules* **2007**, *8*, 3705–3711. [CrossRef] [PubMed]
51. Szarpak, A.; Cui, D.; Dubreuil, F.; De Geest, B.G.; De Cock, L.J.; Picart, C.; Auzély-Velty, R. Designing Hyaluronic Acid-Based Layer-by-Layer Capsules as a Carrier for Intracellular Drug Delivery. *Biomacromolecules* **2010**, *11*, 713–720. [CrossRef] [PubMed]
52. Radhakrishnan, K.; Tripathy, J.; Raichur, A.M. Dual enzyme responsive microcapsules simulating an "OR" logic gate for biologically triggered drug delivery applications. *Chem. Commun.* **2013**, *49*, 5390–5392. [CrossRef]
53. Cardoso, M.J.; Caridade, S.G.; Costa, R.R.; Mano, J.F. Enzymatic Degradation of Polysaccharide-Based Layer-by-Layer Structures. *Biomacromolecules* **2016**, *17*, 1347–1357. [CrossRef]
54. Borodina, T.; Markvicheva, E.; Kunizhev, S.; Möhwald, H.; Sukhorukov, G.B.; Kreft, O. Controlled Release of DNA from Self-Degrading Microcapsules. *Macromol. Rapid Commun.* **2007**, *28*, 1894–1899. [CrossRef]
55. Kim, B.-S.; Choi, J.-W. Polyelectrolyte multilayer microcapsules: Self-assembly and toward biomedical applications. *Biotechnol. Bioprocess Eng.* **2007**, *12*, 323–332. [CrossRef]
56. Johnston, A.; Cortez, C.; Angelatos, A.S.; Caruso, F. Layer-by-layer engineered capsules and their applications. *Curr. Opin. Colloid Interface Sci.* **2006**, *11*, 203–209. [CrossRef]
57. Santos, J.L.; Nouri, A.; Fernandes, T.; Rodrigues, J.; Tomás, H. Gene delivery using biodegradable polyelectrolyte microcapsules prepared through the layer-by-layer technique. *Biotechnol. Prog.* **2012**, *28*, 1088–1094. [CrossRef] [PubMed]
58. Zheludkevich, M.L.; Shchukin, D.G.; Yasakau, K.A.; Möhwald, H.; Ferreira, M.G.S. Anticorrosion Coatings with Self-Healing Effect Based on Nanocontainers Impregnated with Corrosion Inhibitor. *Chem. Mater.* **2007**, *19*, 402–411. [CrossRef]
59. Kartsonakis, I.A.; Danilidis, I.L.; Pappas, G.S.; Kordas, G.C. Encapsulation and release of corrosion inhibitors into titania nanocontainers. *J. Nanosci. Nanotechnol.* **2010**, *10*, 5912–5920. [CrossRef]
60. Feng, Y.; Cheng, Y.F. An intelligent coating doped with inhibitor-encapsulated nanocontainers for corrosion protection of pipeline steel. *Chem. Eng. J.* **2017**, *315*, 537–551. [CrossRef]
61. Shchukin, D.G.; Lamaka, S.V.; Yasakau, K.A.; Zheludkevich, M.L.; Ferreira, M.G.S.; Möhwald, H. Active Anticorrosion Coatings with Halloysite Nanocontainers. *J. Phys. Chem. C* **2008**, *112*, 958–964. [CrossRef]
62. Lvov, Y.M.; Shchukin, D.; Moehwald, H.; Price, R.R. Halloysite Clay Nanotubes for Controlled Release of Protective Agents. *ACS Nano* **2008**, *2*, 814–820. [CrossRef]
63. Balaskas, A.; Kartsonakis, I.; Tziveleka, L.A.; Kordas, G. Improvement of anti-corrosive properties of epoxy-coated AA 2024-T3 with TiO2 nanocontainers loaded with 8-hydroxyquinoline. *Prog. Org. Coat.* **2012**, *74*, 418–426. [CrossRef]
64. Wang, X.; Zhao, J. Encapsulation of the Herbicide Picloram by Using Polyelectrolyte Biopolymers as Layer-by-Layer Materials. *J. Agric. Food Chem.* **2013**, *61*, 3789–3796. [CrossRef] [PubMed]

encyclopedia

MDPI

Entry

Reinforced Concrete Infilled Frames

Matteo Bagnoli [1,*], Ernesto Grande [1] and Gabriele Milani [2]

[1] Department of Sciences Engineering, University Guglielmo Marconi, Via Plinio 44, 00193 Roma, Italy; e.grande@unimarconi.it
[2] Department of Architecture, Built Environment and Construction Engineering, Politecnico di Milano, Piazza Leonardo da Vinci 32, 20133 Milano, Italy; gabriele.milani@polimi.it
* Correspondence: lxrmcj1bi@studenti.unimarconi.it

Definition: Masonry-Infilled Reinforced Concrete Frames are a very widespread structural typology all over the world for civil, strategic or productive uses. The damages due to these masonry panels can be life threatening to humans and can severely impact economic losses, as shown during past earthquakes. In fact, during a seismic event, most victims are caused by the collapse of buildings or due to nonstructural elements. The damage caused by an earthquake on nonstructural elements, i.e., those not belonging to the actual structural body of the building, is important for the purposes of a more general description of the effects and, of course, for economic estimates. In fact, after an earthquake, albeit of a low entity, it is very frequent to find even widespread damages of nonstructural elements causing major inconveniences even if the primary structure has reported minor damages. In recent years, many territories have been hit worldwide by strong seismic sequences, which caused widespread damages to the nonstructural elements and in particular to the masonry internal partitions and the masonry infill panels of the buildings in reinforced concrete, with damage to the floor and out-of-plane expulsions/collapses of single layers. Unfortunately, these critical issues have arisen not only in historic, but also in recent buildings with reinforced concrete, in many cases exhibiting inadequate seismic behavior, only partly attributable to the intrinsic vulnerability of the masonry panels against seismic actions. Such problems are due to the following aspects: lack of attention to construction details in the realization of the construction, use of poor-quality materials, and above all lack of design tools for the infill masonry walls. In 2018, regarding the design of nonstructural elements, the formulation of floor spectra has been recently introduced in Italy. This entry article wants to focus on all these aspects, describing the state of the art, the literature studies and the design problems to be solved.

Keywords: infilled RC frames; nonstructural elements; earthquake damages; macro-models; seismic behavior

Citation: Bagnoli, M.; Grande, E.; Milani, G. Reinforced Concrete Infilled Frames. *Encyclopedia* **2022**, 2, 473–485. https://doi.org/10.3390/encyclopedia2010030

Academic Editors: Giuseppe Ruta, Raffaele Barretta, Krzysztof Kamil Żur and Ramesh Agarwal

Received: 30 November 2021
Accepted: 2 February 2022
Published: 9 February 2022

Publisher's Note: MDPI stays neutral with regard to jurisdictional claims in published maps and institutional affiliations.

1. Introduction

In the complex of a building, are distinguished load-bearing and non-load-bearing elements. The former constitute the structure of the building and they are entrusted with the task of transmitting the vertical and horizontal actions acting on the building. The latter, on the other hand, are entrusted with other tasks:

- delimitation of the building envelope (infill);
- separation between internal environments (partitions);
- environmental comfort (acoustic insulation, thermal insulation, visual comfort, fire resistance).

In particular, the internal partitions fall within the family of vertical closures, i.e., they are those factory elements, of any shape, which constitute the vertical envelope of the built space and also represent the elements of separation between the external and internal microclimate.

Furthermore, in each structure, vertical elements and horizontal elements are also distinguished, and with them the load path is defined, which normally starts from the horizontal ones, continues in the vertical ones and leads them up to the foundation. In the case of structures in masonry, the vertical elements are made up of load-bearing walls, while in the reinforced concrete structures the vertical elements are the pillars. In the first case, therefore, load-bearing and nonbearing walls coexist, while in the second case, the structure is divided into compartments and delimited from the external environment by partitions that theoretically have no static function.

In reality, however, these non-load-bearing walls (infill walls) made within the meshes of the frames—although not considered, in the calculation phase of the structure, to be resistant to any force—influence the behavior of the structure subject to seismic events in terms of increased stiffness and resistance to lateral loads, as well as a significant increase in dissipative capacity.

In fact, the bare frames are designed with reference to bending regimes with the formation of plastic hinges at the nodes under the effect of lateral loads, while in the buffered frames, a mechanism is established in the panel that creates a bracing and traction effect in the pillars. Therefore, a different distribution of acting forces arises which gives rise to stresses not foreseen in the calculation phase, in the absence of partitions and infill panels.

This increase in stiffness, in some cases, allows the structure to adequately resist intense and prolonged seismic actions only thanks to the presence of infill walls, which allow a greater dissipation of the quantity of energy entering the system. Their contribution is demonstrated by the classic X-shaped cracks which indicate a shear failure under cyclic loads (Figure 1).

(**a**) (**b**)

Figure 1. (**a**). Classic X-shaped infill lesion with intact supporting structure; (**b**). Typical expulsion of infill masonry wall (Photos taken by the author during the L'Aquila earthquake of 6 April 2009).

However, the presence of infills does not always create a beneficial effect on the structure subject to an earthquake. In fact, due to the high stiffness, the infill panels can originate irregular plan configurations, jeopardizing a correct upstream structural configuration. These situations can also occur starting from a situation of regularity following a collapse of some wall panels which causes an imbalance in the stresses acting on the structural elements. This occurs because of the high fragility of the materials and because the breakage is a result of the loss of balance outside the plane due to the ineffectiveness of the connection with the structure and due to instability phenomena linked to the reduced thickness of the wall panels compared to the others.

In fact, the seismic behavior of buildings is significantly influenced by presence of masonry infill panels, as is now recognized by numerous studies, as well as reconnaissance for the relief of post-earthquake damage. Although in some cases the presence of rear-end collisions has positive effects on structural behavior, the walls are often susceptible to high levels of damage, even to the expulsion of the panel, particularly dangerous due to falling debris. Despite the critical importance of the infill walls to safeguard life, the presence of the panels is often neglected in the design and numerical modeling phases, as the same they are formally nonstructural elements.

Reinforced Concrete Frames with Masonry

The structures can be assimilated to real organisms consisting of a series of structural elements that collaborate with each other in a synergistic way and allow the loads to discharge to the ground. The structure, however, is not only made up of these structural elements, whose main function is precisely that of allowing the structure to exist and not to collapse under the effect of loads, but it is also made up of many elements which, although not performing a structural function, are however very important, if not fundamental, as they make the structure usable.

These elements, defined precisely as nonstructural elements, can be of an architectural nature (balconies, false ceilings, plasters, partitions, etc.), plant engineering (electrical, ventilation, heating, gas, etc.), may have a function associated with the safety of utilities (escape routes) or can simply be linked to the furniture (shelving, bookcases). However, being classified as nonstructural elements does not make them any less important. In fact, with reference to the seismic events that have struck Italy in recent years, the damage, or in some cases the actual collapse, of these nonstructural elements has caused even irreparable damage to the structural organism to which they belonged (making it unusable and therefore subject to demolition), as well as damage to people. To support the thesis according to which these elements, although not having a purely structural function, play an absolutely not marginal role in terms of safety, reference is made to the Technical Construction Standards 2018 (NTC18 [1]) which in § 7.2.3 define the nonstructural elements as follows: "By nonstructural construction elements we mean those with stiffness, strength and mass such as to significantly influence the structural response and those which, while not affecting the structural response, they are nonetheless significant for the safety and/or safety of people".

As specified by the legislation, the nonstructural elements do not have a load-bearing function, i.e., their absence does not affect the structural stability; however, they significantly influence the behavior of the structural organism to which they are associated, but above all they can cause damage to people if they are not well designed. Non-structural elements require careful design which is often not carried out in design practice, causing, as already described, serious accidents during earthquakes.

One of the most important nonstructural elements, which has attracted the attention of many researchers in the sector in recent years, is represented by infill masonry walls. The reason for this is straightforward: the most common construction type in many countries of the world, but especially in Europe (including Italy), is represented by reinforced concrete frame structures. The supporting elements of these structures, by their very nature, do not allow one subdivision between internal and external space; therefore, it is necessary to introduce elements of a nonstructural nature that are able to perform this function. In this regard, the best element for economy, ease of installation, versatility and thermal and acoustic performance is represented by the masonry infill. The masonry infill is nothing more than the set of different brick elements (bricks and/or blocks, perforated or not) connected to each other usually with mortar, but it is not uncommon to see other types of binders used (curtain walls represent an example with glue joints). They are superimposed on each other to fill the various perimeter spans of the frame structure to create a separation between the internal and external environments. Obviously, some of these infill walls will have openings which will then be filled with fixtures of various kinds. Often, to improve performance in terms of thermal and acoustic insulation, the masonry infills are made by coupling two layers of brick separated from each other by a filled cavity with insulating material or simply left empty.

In the field of residential constructions, infill panels are usually made of nonreinforced brick. On the contrary, in the field of structures for industrial use, given the larger dimensions and different needs, the type of masonry infill increasingly used is reinforced, characterized simply by the presence of reinforcement pylons placed in correspondence with the sick beds. However, it should be noted that below, with the term plugging, we will always and only refer to plugging in nonreinforced masonry.

On the basis of this description, it is easy to understand how, by its nature, the infill wall is not conceived as an element capable of fulfilling a load-bearing function, which is why it is often neglected from a structural point of view: the less attentive designers, in fact, deal with the design of the load-bearing elements (beams, pillars, curbs, floors, etc.), completely forgetting that these structural elements, during the useful life of the work, will have to collaborate with other nonstructural elements, such as curtain walls, and that the presence of these will necessarily influence it.

This way of designing is justified: the infill walls are made only after the frame structure has been completed, thus releasing them from any load-bearing function against vertical loads. In addition, the utilities and/or the intended use of the structure may change, and with them the arrangement of the openings may also change, which makes the infills susceptible to mutations and makes modifications over time difficult to predict in the design phase.

These might seem to be sufficiently valid reasons to justify the negligence of rear-end collisions in the design phase, but they are not as we will see better below.

First of all, to avoid those tensional states associated with the action of external loads that may arise inside the infills (thus giving these elements a load-bearing structural function when these, as they were conceived, do not have), it is good to realize the infills starting from the top floor of the building and then gradually descend towards the lower floors. This is a good rule of thumb that avoids loading the infill panels by creating first those of the higher floors and then cascading those of the lower floors; the vertical load associated with the elastic deformation suffered by the elements of the frame is not severe on the rear-end collisions themselves. Therefore, operating in this way, it will not be possible to consider the contribution of rear-end collisions to withstand vertical loads representing a valid and justified approximation. Another thing is the negligence of rear-end collisions with respect to horizontal actions, in particular from the stresses deriving from seismic events.

2. Numerical Modeling and Structural Response of Masonry Infills

Since the early 1960s, with the aim of adequately describing the behavior of framed structures with masonry infill, two main modeling approaches have been distinguished: macromodels and micromodels (Crisafulli, Carr and Park [2]). However, there are also other approaches (Lourenço [3]), which possess homologous and intermediate characteristics between the two mentioned above. The foremost variances between them are the accuracy and the calculation burden. The micro-modeling takes into attention all the components, while the macro-modeling considers the whole wall panel as a homogeneous unit. The latter is generally used in the case of a global response, and its influence on the behavior of reinforced concrete infilled frames is considered.

2.1. *Types of Numerical Models*

The micromodels (P. G. Asteris [4]) provide for the nonlinear finite element modeling of the reinforced concrete frame, of the infill panels and therefore of the interference and the connection between the frame and therefore the wall. Although this type of approach is more accurate, there are several drawbacks in their use, mainly related to their complexity and the associated computational burden required for numerical analyses. These characteristics make them more suitable for research uses, rather than for the design and verification of the various walls generally present in real structures. This is why, in fact, macro-models are basic logic models, which simulate the general force-deformation behavior of the wall with respect to experimental results (Lam [5]).

The macro-models that describe the behavior of the masonry panels require less calculation costs related to numerical analyses and are therefore more suitable for representing the overall behavior of the structures. Fardis and Panagiotakos [6] and P. Asteris [7] propose a series of analytical macro-models for the analysis of infilled frame structures, mainly dividing them into single strut models of two orders (Figure 2a) and multiple strut models (Figure 2b). In the first case, each panel is represented with the same diagonal strut. The more traditional form is constituted by a diagonal interlocking strut made with a similar material and having a replacement thickness as an infill panel. Lately, macro-models have also been proposed that describe the three-dimensional behavior of the infill masonry and are absolutely less expensive from the point of view of calculation than micro-models, even if these proposals have rarely been applied in recent studies.

In particular, many studies have recently addressed the problem of describing even more accurately the behavior of infill frames in steel or reinforced concrete. Among these, it is worth mentioning those of Polyakov [8], Holmes [9], Smith [10], Kligner [11], whose main purpose is to comprehend the behavior of infill frames by carrying out experimental and analytical tests. Polyakov [8] wants to specify that the connection between the infill panel and the elements of the frame is substantially found in the compression stress zones. Holmes [9] instead proposes to model the infill panel as an equivalent diagonal strut with a width equal to 1/3 of the length of the diagonal strut itself (Figure 2a).

Moreover, Mainstone [12], using an empirical formula, estimates the width of the equivalent strut: the same formula is subsequently taken up and repeated in FEMA codes 274 [13], 306 [14] and 356 [15]. The study conducted by Kadir [16] evaluates the influence of the dimensions of beams and pillars on the width of the equivalent rafter, also suggesting a well-defined formula.

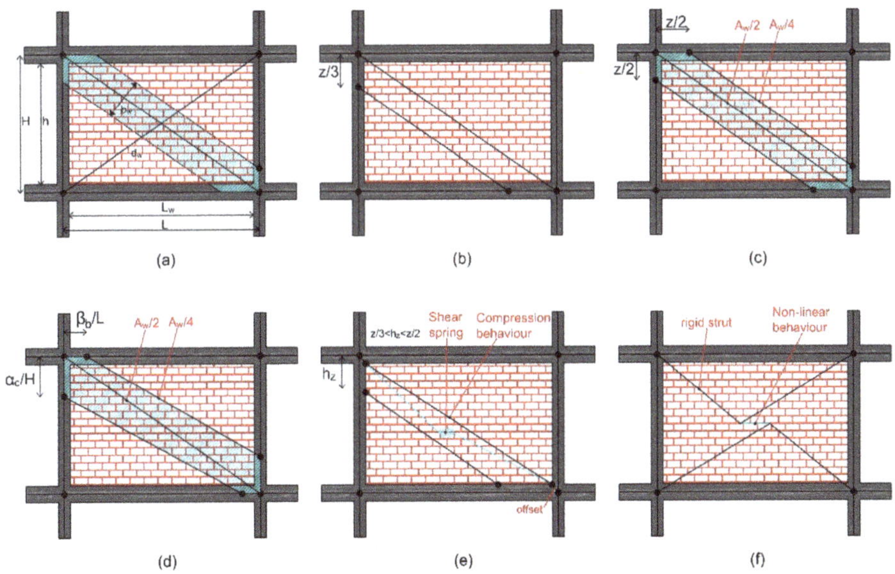

Figure 2. Marco-modeling approaches: (**a**) Holmes [9] single strut, (**b**) [10] two-strut approach, (**c**) Chrisostomou et al. [17] three-strut approach, (**d**) El-Dakhakhni et al. [18] three-strut approach, (**e**) Crisafulli [2] macro-model, (**f**) Furtado et al. [19] macro-model.

Furthermore, Chrysostomou et al. [17] (Figure 2c) and El-Dakhakhni et al. [18] (Figure 2d) suggested a six-strut approach, providing three struts per direction, to precisely reproduce the local effects due to the frame-infill interaction. As mentioned in the introduction, there are also studies concerning the use of multiple equivalent diagonal struts; among these, Crisafulli [2] has studied and verified the actual functionality and adequacy of these models and proposed a four-node panel element in which the compressive and shear behavior were accounted for separately by using a double truss mechanism and a shear spring in each direction (Figure 2e). In the end, another simplified macro-model was introduced by Furtado et al. [19]; it consists of four rigid strut elements, which are connected to the beam-column joints and a central element wherein the nonlinear behavior is lumped (Figure 2f).

Presently, research is almost totally focused on the use of single or multiple equivalent diagonal strut models, used for the description of the seismic behavior of the reinforced concrete frames, but also on the derivation of constitutive laws that simulate in the most accurate way possible the main phases of damage that characterize the behavior of the infill wall panels. Among these, Noh [20] and Huang [21] have proposed formulas to derive the main parameters that characterize the response of the monotonous and cyclic behavior of the infill panels, in both cases reinforcing the use of the equivalent strut. In these two cases, a multilinear constitutive law based on the pinching material model proposed by Lowes [22] is taken into account to schematize the degradation of strength and stiffness due to progressive damage. Although this constitutive model requires a significant number of parameters, in particular to describe the cyclic response, it allows us to simulate in a satisfactory way the response of the infill panels.

2.2. Suggested Parameters for Macro-Models

As previously indicated, Polyakov [7] uses macro-models and proposes an equivalent diagonal strut system in order to consider the effect of masonry infill panels. However, it is necessary to estimate the equivalent strut width within this numerical analysis framework. Directly related to Polyakov's proposal, Holmes [9] provides the first expression (Equation (1)) which determines the equivalent diagonal width of the strut:

$$b_w/d_w = 0.33, \tag{1}$$

where b_w and d_w are the strut width and length, respectively. Moreover, Stafford Smith [9] reinforces and specifies that the strut width must be between 0.10 and 0.25 times the diagonal length of the panel and also suggests an equation (Equation (2)) to calculate the related panel-to-frame-stiffness parameter:

$$\lambda = \sqrt[4]{((E_{w\vartheta} \cdot t_w \cdot \sin(2\vartheta))/(4 \cdot E_c \cdot I_c \cdot h_w))}, \tag{2}$$

where $E_{w\vartheta}$ is the elastic modulus of the masonry panel in the diagonal direction, t_w is the thickness of the infill panel, E_cI_c is the bending stiffness of the columns of the surrounding frame, h_w is the height of the panel and ϑ is the angle related with the aspect ratio of the panel (h_w/L_w) defined according to Equation (3):

$$\vartheta = \tan^{-1}(h_w/L_w), \tag{3}$$

where L_w is the length of the infill panel. High values of ϑ indicate that the surrounding frame is less rigid than the infill panel. Subsequently, other studies suggest equations to estimate the equivalent strut width. Mainstone [12] indicates two empirical expressions, Equations (4) and (5), based on analytical and experimental tests on various steel frames infilled with a masonry panel. As mentioned, Equation (4) has also been proposed in FEMA-306 [13].

$$b_w/d_w = 0.16 \cdot (\lambda h)^{-0.3}, \tag{4}$$

$$b_w/d_w = 0.175 \cdot (\lambda h)^{-0.4}, \tag{5}$$

Furthermore, Liauw and Kwan [23] in turn suggest a semi-empirical expression (Equation (6)), assuming $25° \leq \vartheta \leq 50°$, based on tests of infilled steel frames, used to calculate the equivalent width strut:

$$b_w/d_w = 0.95 \cdot \sin(2\vartheta)/2 \cdot \sqrt{\lambda h}, \tag{6}$$

In Bertoldi [24], the width strut (Equation (7)) is defined by two coefficients (K_1 and K_2), which are defined as a function of λh.

$$b_w/d_w = K_1/\lambda h + K_2; \{(K_1 = 1.300, K_2 = -0.178; \text{if } \lambda h < 3.14; K_1 = 0.707, K_2 = -0.010; \text{if } 3.14 < \lambda h < 7.85; K_1 = 0.470, K_2 = -0.040; \text{if } \lambda h > 7.85\}, \quad (7)$$

Decanini and Fantin [25] suggest two families of equations (Equation (8)) to be used on the basis of the state of damage of the infill panels examined. The formulas were obtained through experimental tests on reinforced concrete masonry frames under lateral load.

$$\begin{aligned} &\text{Uncracked: } \{(0.085 + 0.748/\lambda)\cdot d_w; \text{ if } \lambda h < 7.85), (0.130 + 0.393/\lambda)\cdot d_w; \text{ if } \lambda h < 7.85)\} \\ &\text{Cracked: } \{(0.010 + 0.707/\lambda)\cdot d_w; \text{ if } \lambda h < 7.85), (0.040 + 0.470/\lambda)\cdot d_w; \text{ if } \lambda h < 7.85)\}, \end{aligned} \quad (8)$$

Another equation (Equation (9)) was proposed by Paulay and Priestley [22].

$$b_w/d_w = 0.25, \quad (9)$$

It is important to note that these values are in perfect line with what was asserted by Mainstone and Holmes; in fact, they are included between the lower limit value proposed by Mainstone [12] and the upper limit value presented by Holmes [9]. Furthermore, the value proposed by Paulay and Priestley [26] corresponds to the upper limit value of the range proposed by Stafford Smith [10]. Papia [27] proposed Equation (10) to evaluate the width of the strut on the basis of the Poisson's ratio (c, β), the vertical load (k) and the geometric parameter (z):

$$b_w/d_w = kc/(z\cdot(\lambda^*)^{\beta}, \quad (10)$$

where, in particular, λ^* is dimensionless, given by Equation (11), to estimate the relative panel-to-frame-stiffness parameter.

$$\lambda^* = (E_{w\vartheta}\cdot t_w\cdot h)/(E_c\cdot A_c)\cdot(h^2/L^2 + (A_c\cdot L)/(4\cdot A_b\cdot h)), \quad (11)$$

2.3. *Constitutive Law for the Infill Masonry Wall*

For elaborate nonlinear analyses of reinforced concrete frames filled with masonry, it is essential to use an adequate constitutive law that describes the behavior in the plane of the equivalent strut. The constitutive law for an equivalent strut is generally described using a multilinear relationship obtained and derived from experimental data as a backbone to simulate both the monotone and cyclic response of the panel. Four constitutive laws present in the literature are set out and described below, two of which concern very recent studies: Bertoldi [24], De Risi [28], Panagiotakos and Fardis [5] and Huang [21].

Bertoldi's model [24] uses a four-ramification dorsal curve for the lateral force-displacement (F–Δ) relationship. The primary ascending branch represents the noncracked behavior up to the primary crack, followed by the second branch corresponding to the post-cracking behavior up to the event of maximum resistance, called peak strength (F_{peak}). The third descending ramification of the dorsal curve defines the deterioration of post-peak resistance down to residual resistance (F_{res}). The fourth branch is horizontal and identifies the residual resistance of the infill panel. The parameters necessary to derive the proposed model are the equivalent strut width (b_w), the secant stiffness at the entire cracking phase (K_{sec}), and therefore the peak resistance of the infill panel (F_{peak}). All parameters are often defined as a function of the geometric and mechanical characteristics of the infill panel and therefore of the reinforced concrete frame that houses them.

The strong point of the model suggested by Bertoldi [24] is to estimate, predict and report clearly the most probable failure mode that the infill panel undergoes and highlights. In fact, the maximum resistance of the panel depends on the expected failure mode defined, again on the basis of the mechanical properties of the infill, and corresponds to the minimum of four failure modes. In particular, the peak force (F_{peak}) is calculated considering four possible failure modes and therefore the corresponding failure stresses (Figure 3a).

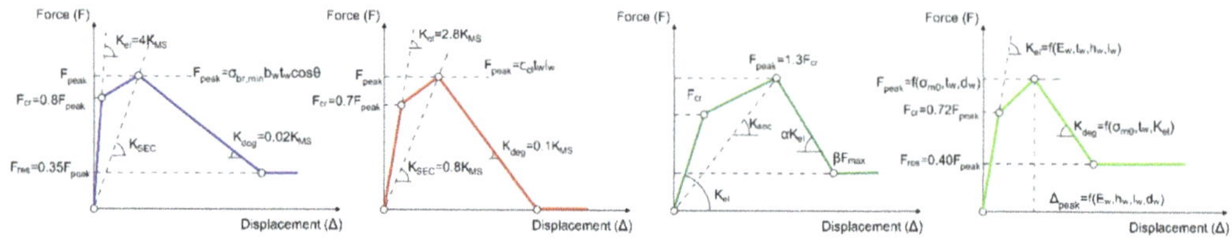

(a) (b) (c) (d)

Figure 3. Multilinear relationship models for the infill masonry panel proposed by (**a**) Bertoldi et al. [24]; (**b**) De Risi et al. [28]; (**c**) Panagiotokas et al. [6]; (**d**) Huang et al. [21].

Another constitutive model that is often employed by researchers is the modified version of the constitutive law originally proposed by Panagiotakos and Fardis [6]. In fact, within the original model, the dorsal curve is represented by four branches. In this case, the structure is composed of four branches and takes into account various states of stress: (a) initial behavior of the noncracked panel; (b) post-cracked linear response, characterized by a reduction in lateral stiffness thanks to the detachment of the infill from the containment frame; (c) post-peak softening response; (d) achievement of the residual axial resistance at a given displacement value. The model of Panagiotakos and Fardis [6] is widespread and widely used for engineering applications. It was initially obtained following the execution of 10 experimental tests performed on reinforced concrete frames filled with perforated masonry bricks that showed exactly a diagonal type break (Figure 3c).

From these starting data, supported by the analysis of an oversized database of masonry infill walls made with hollow bricks collected to be representative of the Mediterranean building heritage, recently De Risi [28] used and then modified the values of the lateral response in order to resize the dispersion with respect to the assembled database. In particular, the crack resistance (F_{cr}) and the maximum force (F_{max}), the initial noncracked stiffness (K_0), the stiffness ratio (K_0/K_{sec}) and therefore the softening-peak stiffness ratios (K_{deg}/K_{sec}) are changed.

The modification proposed by De Risi [28] significantly reduces the CoV values for the tests performed on hollow bricks with reference to the first formulation of Panagiotakos and Fardis [6]. This aspect is very important and demonstrates the accuracy of the proposed model, i.e., the average relative error of the backbone curve is 3% lower for all required parameters. Again, regarding the model suggested by De Risi [28], the primary branch corresponds to the elastic behavior up to cracking, and it is characterized by the initial noncracked stiffness (K_0), assumed to be adequate to 2.8 times the Mainstone stiffness (K_{MS}) (Figure 3b).

An interesting and very recent study is the one proposed by Huang [21] in suggesting the use of a dorsal curve calibrated also in this case on experimental results. In fact, as in the case of De Risi [28], the authors have collected a database of 264 tests performed on masonry infilled frames. Database analysis supported the proposing of an empirical model to estimate the parameters of the backbone curve for the equivalent diagonal strut. The major difference between the De Risi [28] and Huang [21] model is in the way the peak curve is obtained and differentiated; in particular, De Risi [28] started from the proposal of Panagiotakos and Fardis [5] and modified the existing semi-empirical formulations in order to reduce the dispersion with respect to their database. Instead, Huang [21] proposes empirical equations that, for the first time, correlate the different parameters of the load-bearing structure with the geometric and material properties of the buffered frame using multivariate multivariate analysis. During this study, only the median values of the backbone parameters were adopted (Figure 3d).

Unlike the work of Bertoldi [24] and De Risi [28], which provide the force-displacement curve for the equivalent strut, Huang [22] proposes formulations in terms of axial force deformation (F, Δ) for the infill strut [29,30].

3. Evaluation of the Capacity of Nonstructural Elements

Often the design of nonstructural elements is carried out by manufacturers considering only their main function, the loads deriving from their use and gravitational loads, neglecting the effects of exceptional actions such as earthquakes.

Even when product standards are available, there are few nonstructural elements for which the seismic problem is resolved at a regulatory level. As a consequence of this insufficiency of information, numerous past seismic events, even of low intensity, have brought to light performance deficiencies, highlighting how the collapse of nonstructural elements can be a source of risks for the occupants, of important service interruptions of buildings and of huge economic losses. Recognizing the importance and value of nonstructural elements, the regulatory bodies have for some years intensified the production of documents dedicated to classes of nonstructural elements or specific elements, the result of experimental and numerical research conducted.

3.1. Types of Assessment of Bearing Capacity of Nonstructural Elements

Regarding the Italian territory, the main regulatory reference is the Ministerial Decree of 17 January 2018—"Technical Standards for Construction" (NTC18 [1]) which defines, also for nonstructural elements and systems, the minimum performance targets. In §7.2.3, the NTC18 define nonstructural elements as "(...) those with stiffness, resistance and mass such as to significantly influence the structural response and those that, while not influencing the structural response, are equally significant for the safety and/or safety of people (...)".

In the Italian and European context, the standard, normative references (cogent) regarding nonstructural elements are in fact often lacking or absent. However, extending the bibliographic research effort, one can find various international standards, both general and dedicated to specific types of nonstructural elements, thanks to which it is possible to evaluate the seismic capacity of these components.

Depending on the type of nonstructural element, the rules allow us to proceed with different, alternative or complementary approaches to each other. In general, the permitted evaluation methods can be based on numerical analyses, such as classifications through experimental tests or verifications based on past experience (in the event that similar nonstructural elements have been subjected to seismic actions during past events).

3.2. The Floor Spectra

The demand for acceleration on non-load-bearing elements (masonry infills) installed at different building heights is assessed using simplified methods in the actual regulations, although in these cases, the use of simplified methods is recommended. Recent seismic events have shown how current calculation methods can lead to completely incorrect assessments of damage to these components, so it is considered necessary to review these methods in order to take a further step in the definition design of structure. In current design practice, ordinary reinforced concrete frame buildings exposed to seismic action are usually calculated using a linear structural model and equivalent static or dynamic multimodal analysis with response spectrum. Concrete buildings are generally achieved by neglecting the stiffness and resistance of nonstructural elements such as infills and partitions, which are only taken into account because of their weight and mass contribution.

In current years, the work of many researchers has been directed to the definition of simplified methodologies which can be capable of faithfully reproducing the seismic demand on nonstructural elements through comparing the floor spectra.

A study conducted by Sullivan [31] highlighted how many instances, the formulations proposed through the regulatory codes are not appropriate for the right assessment of plan accelerations. Sullivan [31] argues that the spectral amplification strongly relies upon at the damping of nonstructural components; in fact, this issue is not always taken under consideration in any of the primary actual regulatory codes. Another very important aspect concerns the influence of the superior modes in the dynamic response; the Eurocode provides a formulation dependent only on the first natural mode of vibration of the structure; however, in many structures, such as those characterized by a significant number of floors, the influence of the superior modes cannot be neglected. The equal issues had been highlighted in a study performed by Medina [32], wherein the floor spectra for frames of various heights had been evaluated considering the elastic and nonlinear behavior of the structure; this research is concluded through demonstrating the unreliability of the actual formulations contained in the regulatory codes and proposes suggestions for the assessment of plan accelerations.

At a regulatory level, the formulation of floor spectra for the design of nonstructural elements was recently introduced with the explanatory circular of NTC2018 [1].

These spectra can be determined, starting from the accelerating seismic response of the structure at the considered altitude, in the simplified hypothesis that the structure can be assumed as a harmonic forcing for the nonstructural element, taking into account the amplifications due to the dynamic effects on the single nonstructural element, related to its period of oscillation and its damping coefficient as well as to the corresponding characteristics of the structure.

In particular, for Reinforced Concrete infilled frames, the same NTC2018 [1] present a simplified formulation, in the hypothesis of trend of structural accelerations linearly increasing with height. In this case, the maximum acceleration S_a (T_a) can be determined as

$$S_a(T_a) = \alpha{\cdot}S{\cdot}(1 + z/H){\cdot}[a_p/1{+}(a_p - 1){\cdot}(1 - T_a/a{\cdot}T_1)^{\hat{}2}], \geq \alpha{\cdot}S, \text{ for } T_a < a{\cdot}T_1 \tag{12}$$

$$S_a(T_a) = \alpha{\cdot}S{\cdot}(1 + z/H){\cdot}a_p, \text{ for } a{\cdot}T_1 < T_a < b{\cdot}T_1 \tag{13}$$

$$S_a(T_a) = \alpha{\cdot}S{\cdot}(1 + z/H){\cdot}[a_p/1 + (a_p - 1){\cdot}(1 - T_a/b{\cdot}T_1)^{\hat{}2}], \geq\alpha{\cdot}S, \text{ for } T_a \geq b{\cdot}T_1 \tag{14}$$

where

α = ratio between the maximum acceleration of the ground on the subsoil type A to be considered in the limit state in question and the acceleration due to gravity;

T_a = fundamental period of vibration of the nonstructural element;

T_1 = fundamental period of vibration of the building in the considered direction;

z = height of the center of gravity of the nonstructural element measured from the foundation plane; equal to zero in structures with seismic isolation at the base;

H = the height of the construction measured from the foundation plane

S = coefficient that takes into account the category of subsoil and topographical conditions as reported in §3.2.3.2.1 [1]:

$$S = S_S{\cdot}S_T \tag{15}$$

S_S = stratigraphic amplification coefficient

S_T = topographical amplification coefficient

Please refer to chapters §3.2.3.2.1 [1] and §3.2.3.2.1 [1] of the standard for explanatory tables of S_T and S_S or parameters a, b, p defined in accordance with the fundamental vibration period.

The floor spectra described in this way (Figure 4) are generally conservative for a wide range of periods, with particular regard to nonstructural elements having a period close to the fundamental period of construction.

Because the nonstructural elements subject to evaluation are generally not positioned on the ground, but rather in any other point of the building, the motion to refer to is that of the installation point, or motion on the ground. The spectra that can be calculated and referred to in the evaluation procedures are therefore the acceleration spectra at the plane.

Today's structural design regulations, including NTC18 [1] and EC8 [33], use precisely the floor acceleration spectrum to estimate the response of an installed nonstructural element to a generic share.

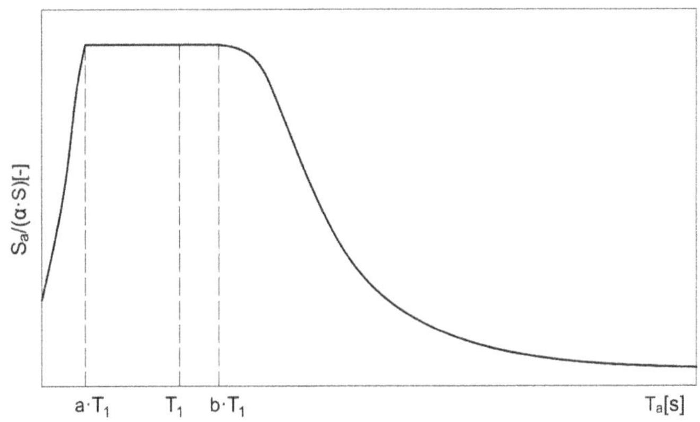

Figure 4. Proposed floor spectra built according to regulations.

In general, the floor spectra can do the following:

- consider the characteristics of dynamic amplification of the main structure, typically present around its own periods of vibration and variables along the height of the building;
- consider the viscous and hysteretic damping, possibly present when referring to the ultimate limit states, of the main structure hosting the nonstructural elements;
- consider the resonance effects between the main structure and nonstructural elements;
- consider the viscous damping of nonstructural elements;
- consider the hysteretic damping of the nonstructural elements linked to any plasticization caused by the earthquake.

It should be noted that the reduction in the ordinates of the plane spectrum, generated by the plastic hysteretic behavior of the nonstructural elements, it should be considered only in some applications. For example, this reduction must be considered in the calculation of the maximum stresses with which to dimension the connection system of the nonstructural component to the structure.

4. Conclusions

The fundamental role played by nonstructural elements in terms of safety and maintenance of functionality of buildings is now unanimously recognized. Against this, national and international standards have been developed as useful tools for designers, manufacturers and installers to ensure adequate seismic-behavior of nonstructural elements. In Italy, the main reference for construction is the Ministerial Decree 17 January 2018—"Technical Standards for Construction" (NTC18) [1] which defines, even for nonstructural elements, the minimum performance targets and the roles and responsibilities of the actors involved. Clearly, the fulfillment of the requirements required cannot be separated from the correct evaluation both of the seismic input and of the performance of nonstructural elements. For most of the nonstructural elements, the seismic input must be specified through suitable floor spectra. As discussed, these can be rigorously calculated through validated simplified methods. It should be recognized that the proposed methods greatly reduce the computational cost, if compared with methods that exploit nonlinear analyses. However, it is still necessary to have structural models sufficiently accurate to be able to grasp the dynamic behavior of the main structure. On the other hand, it is difficult to figure out a method that can solve the problem of floor spectra regardless of the modal parameters of the structure.

All this considered, there is no doubt that the infill panels influence the behavior of the structural complex, but this influence can have a double nature: on one side, there are beneficial effects linked to the increase in the resistance and dissipative capacity of the system; on the other side, we can have fragile collapses and greater stresses both locally and globally. It should also be remembered that an increase in resistance does not always correspond to an improvement in structural behavior. To improve the seismic response of a structure, either its resistance or ductility are increased. In general, it is better to increase ductility because it guarantees greater safety margins. However, it is always essential to verify whether the increase in stiffness caused by the infill panels do or do not have a positive effect on the seismic behavior of the structure. Predicting these effects is not easy because they are associated with variables characterized by a high degree of uncertainty, including the following:

- the mechanical and behavioral properties of the masonry;
- the constraint conditions that exist between the confinement frame and the infill;
- the changes to which these elements are subject during the useful life of the work;
- the ways in which they were created, therefore from the competence of the workers;
- the degree of damage connected to the seismic event;

For these reasons, acting in favor of safety, it is necessary to consider only the negative effects linked to the presence of infill masonry walls. This must be done through simple and easily manageable calculations and structural models, able to reproduce in the environment the structural behavior linked to the interaction between the load-bearing structure (the reinforced concrete frame) and the secondary elements (the infill panels) as faithfully as possible.

In conclusion, it is evident that the influence of the infills on the seismic behavior of the structure in which they are inserted is absolutely not negligible. Both the Italian and the European legislation do not give particular support to the designers in the sector who are aware that the effects that infill masonry panels have on the planned structure must be taken into account. Nevertheless, the designers are not sufficiently supported by the rules for an objective evaluation of these effects. In essence, the regulations highlight the aspects that must be treated with particular attention, but they do not specify the methods of analysis and verification that must be adopted.

This regulatory gap is linked to the difficulty of modeling, within current calculation software, infill wall, the behavior of which is highly uncertain because it is influenced by numerous factors that are difficult to evaluate experimentally. To be effective, these models should not require excessively long data processing times. To achieve this goal, the use of macromodels such as those here described becomes indispensable because once they are properly calibrated, they are able to reproduce in a certainly more faithful way the seismic behavior of the secondary masonry elements towards the structure in which they are inserted.

Author Contributions: Conceptualization, M.B.; methodology, M.B.; data curation, M.B., E.G. and G.M.; writing—original draft preparation, M.B., E.G. and G.M.; writing—review and editing, M.B., E.G. and G.M.; visualization, M.B., E.G. and G.M.; supervision, M.B. All authors have read and agreed to the published version of the manuscript.

Funding: This research received no external funding.

Institutional Review Board Statement: Not applicable.

Informed Consent Statement: Not applicable.

Data Availability Statement: The data considered in this article were derived from the cited literature.

Conflicts of Interest: The authors declare no conflict of interest.

Entry Link on the Encyclopedia Platform: https://encyclopedia.pub/20561.

References

1. Update of the *"Norme Tecniche per le Costruzioni"*, Ministerial Decree January 17, Ministry of Infrastructure and Transport, *Ordinary supplement to the "Gazzetta Ufficiale" n. 42 of 20 February 2018—General Series*. **2018**, *42-288*.
2. Crisafulli, F.J. Seismic Behaviour of Reinforced Concrete Structures with Masonry Infills. Ph.D. Thesis, Department of Civil Engineering, University of Canterbury, Christchurch, New Zealand, 1997.
3. Paulo, B.L. *Computational Strategy for Masonry Structures*; Faculdade de Engenharia da Universidade do Porto: Porto, Portugal, 1996.

4. Asteris, P.G.; Cotsovos, D.M.; Chrysostomou, C.Z.; Mohebkhah, A.; Al-Chaar, G.K. Mathematical Micromodeling of Infilled Frames: State of the Art. *Eng. Struct.* **2013**, *56*, 1905–1921. [CrossRef]
5. Lam, N.T.K.; Griffith, M.; Wilson, J.; Doherty, K. Time-history Analysis of URM Walls in out-of-Plane Flexure. *Eng. Struct.* **2003**, *25*, 743–754. [CrossRef]
6. Michael, N.F.; Panagiotakos, T.B. Seismic Design and Response of Bare and Masonry-Infilled Reinforced Concrete Buildings. Part II: Infilled Structures. *J. Earthq. Eng.* **1997**, *1*, 475–503. [CrossRef]
7. Asteris, P.; Antoniou, S.; Sophianopoulos, D.; Chrysostomou, C.Z. Mathematical Macromodeling of Infilled Frames: State of the Art. *J. Struct. Eng.* **2011**, *137*, 1508–1517. [CrossRef]
8. Polyakov, S.V. On the interaction between masonry filler walls and enclosing frame when loaded in the plane of the wall. *Transl. Earthq. Eng.* **1960**, *2*, 36–42.
9. Holmes, M. Steel frames with brickwork and concrete infilling. *Proc. Inst. Civ. Eng.* **1961**, *19*, 473–478. [CrossRef]
10. Stafford-Smith, B.; Carter, C. A method of analysis for infilled frames. *Proc. Inst. Civ. Eng.* **1969**, *44*, 31–48. [CrossRef]
11. Klingner, R.E.; Bertero, V.V. Earthquake resistance of infilled frames. *J. Struct. Div.* **1978**, *104*, 973–989. [CrossRef]
12. Mainstone, R.J. On the stiffnesses and strengths of infilled frames. *Proc. Inst. Civ. Eng.* **1971**, *49* (Suppl. IV), 57–90.
13. FEMA. *FEMA-274 Nehrp Commentary on the Guidelines for the Seismic Rehabilitation of Buildings*; Federal Emergency Management Agency: Washington, DC, USA, 1997.
14. FEMA. *FEMA-306 Evaluation of Earthquake Damaged Concrete and Masonry Wall Buildings, Basic Procedures Manual*; Federal Emergency Management Agency: Washington, DC, USA, 1999.
15. FEMA. *FEMA-356 Prestandard and Commentary for the Seismic Rehabilitation of Buildings*; Federal Emergency Management Agency: Washington, DC, USA, 2000.
16. Abdul-Kadir, M.R. The Structural Behaviour of Masonry Infill Panels in Framed Structures. Ph.D. Thesis, Department of Civil Engineering and Building Science, University of Edinburgh, Edinburgh, UK, 1974.
17. Chrysostomou, C.Z.; Gergely, P.; Abel, J.F. A six-strut model for nonlinear dynamic analysis of steel infilled frames. *Int. J. Struct. Stab. Dyn.* **2002**, *2*, 335–353. [CrossRef]
18. El-Dakhakhni, W.W.; Elgaaly, M.; Hamid, A.A. Three-strut model for concrete masonry infilled steel frames. *J. Struct. Eng.* **2003**, *129*, 177–185. [CrossRef]
19. Furtado, A.; Rodrigues, H.; Arêde, A.; Varum, H. Simplified macro-model for infill masonry walls considering the out-of-plane behaviour. *Earth. Eng. Struct. Dyn.* **2015**, *45*, 507–524. [CrossRef]
20. Noh, N.M.; Liberatore, F.M.; Tesfamariam, S. Modelling of masonry infilled RC frames subjected to cyclic loads: State of the art review and modelling with OpenSees. *Eng. Struct.* **2017**, *150*, 599–621. [CrossRef]
21. Huang, H.; Burton, H.V.; Sattar, S. Development and Utilization of a Database of Infilled Frame Experiments. *Am. Soc. Civ. Eng. Numer. Model.* **2020**, *146*. [CrossRef]
22. Lowes, L.; Mitra, N.; Altoontash, A. *A Beam-Column Joint Model for Simulating the Earthquake Response of Reinforced Concrete Frames*; PEER Report No. 2003/10; Pacific Earthquake Engineering Research Center: Berkeley, CA, USA, 2004.
23. Liauw, T.C.; Kwan, K. Nonlinear behaviour of non-integral infilled frames. *Comput. Struct.* **1984**, *18*, 551–560.
24. Bertoldi, S.H.; Decanini, L.D.; Gavarini, C. Telai tamponati soggetti ad azioni sismiche, un modello semplificato: Confronto sperimentale e numerico. In Proceedings of the 6 Convegno Nazionale L'ingegneria sismica in Italia, Perugia, Italy, 13–15 October 1993.
25. Decanini, L.D.; Fantin, G.E. Modelos Simplificados de la Mamposteria Incluida en Porticos. Caractreisticas de Rigidez y Resistencia Lateral en Estrado Limite. *J. Argent. Ing. Estruct.* **1987**, *2*, 817–836. (In Spanish)
26. Paulay, T.; Priestley, M.J. *Seismic Design of Reinforced Concrete and Masonry Buildings*; John Wiley & Sons: New York, NY, USA, 1992.
27. Papia, M.; Cavaleri, L.; Fossetti, M. Infilled frames: Developments in the evaluation of the stiffening effect of infills. *Struct. Eng. Mech.* **2003**, *16*, 675–693. [CrossRef]
28. De Risi, M.T.; Del Gaudio, C.; Ricci, P.; Verderame, G.M. In-plane behaviour and damage assessment of masonry infills with hollow clay bricks in RC frames. *Eng. Struct.* **2018**, *168*, 257–275. [CrossRef]
29. Mucedero, G.; Perrone, D.; Brunesi, E.; Monteiro, R. Numerical Modelling and Validation of the Response of Masonry Infilled RC Frames Using Experimental Testing Results. *Buildings* **2020**, *10*, 182. [CrossRef]
30. Bagnoli, M.; Grande, E.; Milani, G. Numerical Study of the In-Plane Seismic Response of RC Infilled Frames. *Constr. Mater.* **2021**, *1*, 82–94. [CrossRef]
31. Sullivan, T.; Calvi, P.M.; Bolognini, D. Evaluation of floor spectra for the seismic design of non-structural elements. *Progett. Sismica* **2015**, *6*. [CrossRef]
32. Medina, R.A.; Sankaranarayanan, R.; Kingston, K.M. Floor response spectra for light components mounted on regular moment-resisting frame structures. *Eng. Struct.* **2006**, *28*, 1927–1940. [CrossRef]
33. CEN Eurocode 8 EN 1998-1. *Design of Structures for Earthquake Resistance, Part 1. General Rules, Seismic Actions and Rules for Buildings*; European Committee for Standardisation: Brussels, Belgium, 2004.

encyclopedia

MDPI

Entry

Two-Lane Highways: Indispensable Rural Mobility

Ahmed Al-Kaisy

Civil Engineering Department, Montana State University, Bozeman, MT 59717, USA; alkaisy@montana.edu; Tel.: +1-406-994-6116

Definition: Two-lane highways refer to roadways consisting of two lanes in the cross section, one for each direction of travel. Occasionally, passing lanes may be added to one or two sides of the roadway extending the cross section to three or four lanes at those locations. In this entry, two-lane highways strictly refer to roads in rural areas meeting the previous definition and do not include urban and suburban streets.

Keywords: two-lane highways; rural; passing; platooning; access; mobility; low-volume roads

1. Introduction

Two-lane highways constitute the vast majority of roadways by length, particularly in rural areas. This is true here in the USA and in most other countries around the world. It is these highways that first brought motor vehicles to remote towns and villages over the past century and played a critical role in the growth and economies of rural communities. The dominance of two-lane highways in the current roadway networks in developed countries is driven by economics and the low level of vehicular traffic common in rural areas.

Two-lane highways serve various highway functions, from local roads serving very low volumes of local traffic to principal arteries connecting towns and small cities, and everything in between. Consequently, these highways vary in their standards, from unpaved highways in very remote areas to high-type pavement and wider cross sections for intercity routes and major arteries.

Two-lane highways, as we know them today, were mainly introduced with the introduction of the automobile. With the increase in motor vehicle traffic and the use of larger vehicles including buses and trucks, the majority of two-lane highways outside remote and frontier rural areas were paved to sustain traffic loads.

2. The Role of Two-Lane Highways in Today's Transportation System

Two-lane highways have played a critical role in the development of modern societies around the world, both in developed and developing countries. They provide essential access to rural areas with far-reaching social, economic, and lifestyle impacts.

Rural residents outside cities and urban areas need to be able to have access to nearby cities and major economic centers to satisfy their needs for food, goods, and other commodities. Further, rural residents also need access to hospitals, educational institutions, and other services that are not usually available in remote rural areas. Two-lane highways have always been used to provide this important access and the much-needed mobility to rural populations. Therefore, they take much of the credit for the continuous growth and development in rural communities and for making rural areas more livable places.

Two-lane highways also play an indispensable role in the growth of local economies in rural areas. Agriculture is an important sector of the economy in most countries and almost all agricultural activities take place in rural areas. Farmers, ranchers, and cattlemen, among others, need to be able to move their products from field to market, and much of that mobility takes place on two-lane highways. Similarly, farming-related industries, such

Citation: Al-Kaisy, A. Two-Lane Highways: Indispensable Rural Mobility. *Encyclopedia* **2022**, *2*, 625–631. https://doi.org/10.3390/encyclopedia2010042

Academic Editors: Giuseppe Ruta, Raffaele Barretta, Ramesh Agarwal and Krzysztof Kamil Żur

Received: 23 January 2022
Accepted: 1 March 2022
Published: 15 March 2022

as food processing and packaging plants, dairy products plants, etc., are all located outside cities and major urban communities, and both access and mobility are vital to the industry.

Further, two-lane highways provide critical access and mobility needed by the tourism industry. Touristic attractions and recreational activities usually exist outside cities and urban centers and many are located in remote rural areas, yet they receive a large number of visitors throughout the year or during peak seasons. Examples of those attractions are ski resorts, parks, fishing, and camping at rivers and lakes, etc. Here in the USA, national and state parks and forests receive millions of visitors every year who access these locations using two-lane highways. Additionally, two-lane highways are the only type of roads used to provide mobility within national parks and forests that extend over extensive geographic areas.

Aside from the critical roles described above, two-lane highways have also been used as major routes connecting small cities and towns. These routes usually serve as rural arteries providing essential mobility to motor vehicle traffic between small towns and urban centers.

Two-Lane Highways: Main Users

In practice, two-lane roads have primarily been designed for motor vehicle traffic, which includes passenger cars, buses, and trucks. The needs of non-motorized modes of traffic such as bicyclists and pedestrians are usually not addressed in the design of these roads. However, two-lane roads have increasingly been used by bicyclists (and occasionally by pedestrians) in touristic and recreational areas in many countries around the world. This has serious implications for roadway safety and operations. Bicyclists in rural areas tend to use shoulders, when present, and the edge of the travel lane in areas where no shoulder exists. This places bicyclists close to vehicular traffic and raises concerns about bicyclists' safety [1]. Rubie et al. [2] conducted a review of factors influencing the lateral passing distance when motor vehicles overtake bicycles. A recent study by Moll et al. [3] investigated the effect of sport cyclists on narrow two-lane rural roads in Spain and found that the presence of cyclists decreased the motorized vehicle average speed and increased followers and delays. Many other studies in the literature examined the behavior of bicyclists and drivers as they share the use of rural two-lane roads and the impacts on safety and operations [1,4–7].

3. Two-Lane Highways: Operational Characteristics

The characteristic that separates two-lane highways from other highway types is that passing occurs in the opposing lane of traffic. Specifically, passing maneuvers are restricted on these highways and are typically performed using the opposing lane when sight distance and gaps in the opposing traffic stream permit [8]. This has serious implications for traffic operation and safety. From a traffic operation perspective, the limited passing opportunities results in a higher impact of slow-moving vehicles (mainly trucks, buses, and agricultural equipment) on mobility and performance. This impact generally increases with the increase in traffic level in the two directions of travel and the proportion of slow-moving vehicles in the traffic stream.

Two-lane highways are known for a higher level of interaction between vehicles moving in the same as well as in opposing directions. This is because traffic level in one direction is a major determinant of passing opportunities and, thus, operational performance for the opposing traffic stream. Lack of passing opportunities typically results in the formation of platoons with trailing vehicles subject to additional delay. Consequently, platooning or "bunching" is an important phenomenon that is specific to two-lane highways and has serious implications for operations and safety. In the United States, the operational performance on two-lane highways, which is directly related to the platooning phenomenon, is currently estimated using two surrogate measures; the percent-time-spent-following (PTSF) and average travel speed [9]. The PTSF is defined as "the average percent of total travel time that vehicles must travel in platoons behind slower vehicles due to inability to pass on a two-lane highway" [9]. Platooning is also important on two-lane highways from a safety perspective. Drivers who are constrained by slow-moving vehicles and lack of passing opportunities may become frustrated and, therefore, tend to accept smaller gaps in the opposing traffic to perform risky passing maneuvers [10]. These risky passing maneuvers and driver distraction represent two main reasons for head-on collisions on two-lane highways. The use of passing lanes is known to alleviate this unique operational characteristic on two-lane highways with higher traffic levels. Passing lanes allow vehicles traveling at faster speeds to overtake slower-moving vehicles, thus breaking up platoons and reducing delays due to inadequate passing opportunities.

4. Two-Lane Highways: Unique Challenges

Despite the critical role two-lane highways play in providing access and mobility to rural areas, these highways pose unique challenges for highway agencies in charge of operating and maintaining the roadway network. These challenges are generally related to the following three aspects: safety, mobility, and infrastructure.

4.1. Safety Management on Two-Lane Highways

A large proportion of two-lane highways are characterized by low traffic exposure and are usually referred to as low-volume roads. These roads often provide access to remote rural areas, have lower functional class, and are built to lower standards (e.g., narrow lanes and shoulders, non-forgiving roadsides, etc.). Unlike two-lane highways along major routes, low-volume roads pose challenges for highway agencies in relation to safety management on the highway network.

In most developed countries, safety is managed on the highway system using ongoing highway safety improvement programs. These programs are funded by governments and aim to reduce the number and severity of crashes on the highway network using data-driven strategic approaches. Highway agencies have been increasingly facing tighter budgets, including funds dedicated to their ongoing safety improvement programs. One of the important steps in these programs is network screening, which is the process of analyzing the network to identify candidate safety improvement sites for further analysis and investigation. As it is not possible to conduct a detailed investigation across the entire network, network screening helps to pare down the list of sites [11]. Sites identified through network screening become candidates for safety improvement projects. Conventional screening methods using historical crash data (crash frequency, severity, and/or rates) at individual sites are the most widely used methods. While this conventional approach may work well for roads with high traffic exposure, it may not prove effective nor reasonable for low-volume two-lane highways. On one hand, low volume on these roads often results in a few sporadic crashes, and as such sites along these roads are unlikely to rank high on the list should crash frequency be used as the sole ranking criterion. On the other hand, low volumes are expected to result in higher crash rates even with only a few crashes taking place on these roads. Therefore, when using crash rates, those sites may rank high on the list, even though the few crashes occurring along these roads may be related to factors other than roadway characteristics (e.g., driver distraction, drinking and driving, etc.) [12].

Moreover, two-lane low-volume roads often have geometric and roadside features that are built to lower standards which constitute an added risk to road users, yet above-normal crash frequencies may not be observed at those locations due to low traffic volumes and the random nature of the crash occurrence. It is also important to remember that many crashes on remote low-volume roads, particularly those with lower severities, may go unreported. All these factors on low-volume roads make it very difficult to rely solely on crash history in determining sites that are good candidates for safety improvement projects. Unlike freeways, expressways, other major thoroughfares, and urban streets, two-lane highways are usually owned and operated by different levels of governments (e.g., state and local governments such as counties, townships, etc.), which adds to the complexity of safety management on these roads.

Given the challenges described above for managing safety on two-lane highways in general, and low-volume roads in particular, the following strategies are recommended to address these challenges [13]:

- Adopting a systemic approach to safety: A systemic approach to safety involves improvements that are widely implemented based on high-risk roadway features correlated with particular severe crash types [14]. This approach considers multiple locations with similar risk characteristics, selecting the preferred countermeasure(s) appropriate and affordable for widespread implementation; it is best suited for low-cost safety countermeasures. While applying the systemic approach has merits regardless of traffic exposure, it is particularly important for low-volume rural two-lane highways where crash history alone may not be adequate to identify sites for safety improvements. It is important to note that the systemic approach does not replace the conventional hot-spot analysis approach. It is a complementary technique intended to supplement site analysis and provide a more comprehensive and proactive approach to safety management efforts [14].
- Applying network screening methods sensitive to risk factors: Low traffic exposure often makes it difficult to isolate high-crash locations for safety improvements. Therefore, any network screening method that relies solely on crash history may fail to perform satisfactorily on low-volume roads. As such, it becomes imperative for network screening on low-volume roads to incorporate risk factors in identifying candidate sites for safety improvements, i.e., factors that contribute to an increased level of risk or hazard throughout the network. This approach is rational given the fact that most low-volume rural roads are built to lower standards compared with other major roads (e.g., narrow lanes, no/narrow shoulders, fixed objects at roadsides, etc.). These methods mostly utilize risk factors in conjunction with crash and traffic data.

4.2. Maintaining Acceptable Operational Performance

The unique operational characteristic of two-lane highways is that passing maneuvers occur in the opposing lane of traffic. These maneuvers can only be performed if there is adequate sight distance (thus, passing is legally allowed) and large enough gaps in the opposing traffic stream. If sight distance is restricted due to terrain or geometric alignment or if the traffic level in the opposing direction is high, then passing opportunities would become limited. This explains the high level of interaction between traffic streams in the two directions of travel, which is unique to two-lane highways. If passing opportunities are limited due to the aforementioned reasons, platoons form when faster vehicles catch up with slow-moving vehicles resulting in lower travel speeds and inferior quality of service (in practice six levels of service from A to F are used to describe the quality of service in operational analyses).

To maintain acceptable operations (i.e., an acceptable level of service) on two-lane highways, highway agencies apply certain treatments to improve passing opportunities on two-lane highways such as adding turnouts, passing lanes, or shoulder use sections at regular intervals [15]. The most popular treatment among these is the use of passing lanes where a lane is added to one or both directions of travel at regular intervals to improve traffic operations and level of service by breaking up platoons and allowing faster vehicles to overtake slower-moving vehicles. A passing lane layout in one direction of travel is shown in Figure 1.

Turnouts are also used to increase passing opportunities at locations where adding a passing lane may not be a viable option. Examples of these situations are two-lane highways in difficult terrain or in situations where adding passing lanes may not be a cost-effective solution. The use of shoulder sections is not as common in practice as the previous two treatments, because wide shoulders do not exist on many two-lane highways. When used, the shoulder-use section functions as an extended turnout. This approach enables a highway agency to promote shoulder use only where the shoulder is adequate to handle anticipated traffic loads and the need for more frequent passing opportunities has been established by the large amount of vehicle platooning [15].

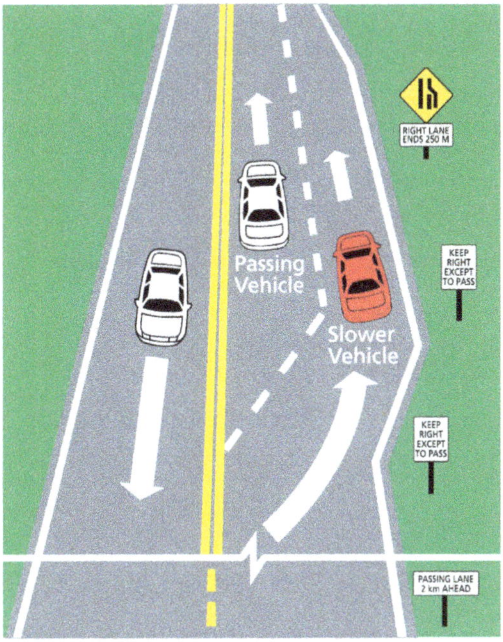

Figure 1. A drawing showing a passing lane layout [16].

4.3. Infrastructure Preservation and Maintenance

Though critical to rural access and mobility, a large proportion of rural two-lane highways only carries a small amount of traffic. The low traffic level causes these roads to receive little attention and funding for maintaining a good state of repair. While the discussion in this section applies to two-lane highways, in general, it is particularly concerned with low-volume two-lane highways in rural areas.

A recent report titled "Investment Prioritization Methods for Low-Volume Roads" by the U.S. National Academy of Sciences states that "low-volume roads are at a disadvantage relative to other roads within traditional investment prioritization processes that focus on volume-based metrics of benefit and impact" [17]. In general, the fewer the road users, the less funding is available for road maintenance and restoration, much less engineering. Consequently, low-volume two-lane highways around the world typically need reconstruction and improvement. The highest-volume highest-rate-of-return proposals receive priority for limited research funding [18]. This is consistent with other studies that confirm how difficult it is to advocate for investments in roads with very low volume on a traditional economic basis, because of the relatively small user or impact group for any given low-volume road [19]. Nevertheless, agencies see value in maintaining existing low-volume two-lane roads because of their role in providing access to rural or isolated areas and supporting economic activity (e.g., farming, logging, mining, or other industry), as well as for their network coverage role within the broader transportation system. These roads also play a role in providing access to other remotely located facilities or destinations, such as international border crossings, military facilities, and national and state parks. Further, they provide public access to essential health, education, civic, and outdoor recreational facilities. The link these roads provide between raw materials and markets is critical to economies locally and nationally in all countries around the world.

The World Bank argued that the evaluation of low-volume two-lane highway improvements must be different from that of other roadway projects because their goal is to both reduce travel costs and to support economic development and social objectives [20]. Asset management and investment prioritization processes should go beyond the traditional process, which aims to minimize lifecycle agency and user costs and consider additional factors such as the vital accessibility to remote areas and the social and economic role these highways play in addition to the mobility function. This and other related discussions in the literature all point to a common agreement that low-volume two-lane roads can be important in ways that go beyond the level of traffic they carry, and that their significance merits special consideration within the planning and resource allocation process.

5. Concluding Remarks

Two-lane highways claim the largest proportion of the highway network by length in most countries and help to extend the transportation system coverage over extensive geographic areas. These highways have unique operational characteristics as there is only a single lane for the traffic stream in each direction of travel. Passing on these highways occurs in the opposing traffic lane and is only possible when traffic conditions and sight distance permit. Therefore, platooning is a common phenomenon on two-lane highways and is a major determinant of operational performance.

Two-lane highways play a critical role in providing accessibility and mobility to rural areas and are considered indispensable for rural populations and industries given the social and economic role these highways play in all countries around the world. This role is at the core of the United Nations sustainable development goals with strategies that improve health and education, reduce inequality, and spur economic growth [21]. As such, there is a common agreement that two-lane roads can be important in ways that go beyond the level of traffic they carry. Nonetheless, two-lane highways are also unique in the challenges they pose to highway agencies owning and operating them. Three challenges were discussed, namely, safety management, maintaining acceptable operational performance, and infrastructure preservation and maintenance. An overview of these challenges and the strategies used in practice to address them were briefly presented in this entry.

Funding: This research received no external funding.

Conflicts of Interest: The author declares no conflict of interest.

Entry Link on the Encyclopedia Platform: https://encyclopedia.pub/21966.

References

1. Relph, D.; Al-Kaisy, A.; Gleason, R. Drivers' Behavior When Passing Bicyclists along Rural Recreational Routes. *Nat. Acad. Sci. Eng. Med.* **2020**, *52*, 117–130.
2. Rubie, E.; Haworth, N.; Twisk, D.; Yamamoto, N. Influences on Lateral Passing Distance When Motor Vehicles Overtake Bicycles: A Systematic Literature Review. *Transp. Rev.* **2020**, *40*, 754–773. [CrossRef]
3. Moll, S.; Lopez, G.; Garcia, A. Analysis of the Influence of Sport Cyclists on Narrow Two-Lane Rural Roads Using Instrumented Bicycles and Microsimulation. *Sustainability* **2021**, *13*, 1235. [CrossRef]
4. Chapman, J.; Noyce, D. Influence of Roadway Geometric Elements on Driver Behavior When Overtaking Bicycles on Rural Roads. *J. Traff. Transp. Eng.* **2014**, *1*, 28–38. [CrossRef]
5. Kay, J.; Savolainen, P.; Gates, T.; Datta, T. Driver Behavior during Bicycle Passing Maneuvers in Response to A Share The Road Sign Treatment. *Accid. Anal. Prev.* **2014**, *70*, 92–99. [CrossRef] [PubMed]
6. Walker, I. Drivers Overtaking Bicyclists: Objective Data on The Effects of Riding Position, Helmet Use, Vehicle Type and Apparent Gender. *Accid. Anal. Prev.* **2007**, *39*, 417–425. [CrossRef] [PubMed]
7. Trotter, M.; Kortegast, P.; Beetham, J.; Bowie, C.; Burton, J.; Thomas, J. *Improving Safety for People Who Cycle on Rural Roads*; NZ Transport Agency Research Report 589; New Zealand Transport Agency: Wellington, New Zealand, 2016.
8. Al-Kaisy, A.; Karjala, S. Indicators of Performance on Two-Lane Rural Highways: An Empirical Investigation. *Transp. Res. Rec. TRR* **2008**, *2071*, 87–97. [CrossRef]
9. Transportation Research Board. *Highway Capacity Manual*, 6th ed.; TRB, National Research Council: Washington, DC, USA, 2016.

10. Al-Kaisy, A.; Durbin, C. Platooning on Two-Lane Two-Way Highways: An Empirical Investigation. *J. Adv. Transp.* **2009**, *43*, 71–88. [CrossRef]
11. Gross, F.; Harmon, T.; Albee, M.; Himes, S.; Srinivasan, R.; Carter, D.; Dugas, M. *Evaluation of Four Network Screening Performance Measures*; Report # FHWA-SA-16-103; Federal Highway Administration, US Department of Transportation: Washington, DC, USA, 2016.
12. Al-Kaisy, A.; Huda, K. Network Screening on Low-Volume Roads: A New Proposed Method. In Proceedings of the 100th Transportation Research Board Annual Meeting, Washington, DC, USA, 21–29 January 2020.
13. Al-Kaisy, A.; Huda, K. Identifying Sites for Safety Improvements on Low-Volume Roads: Challenges and Potential Solutions. In Proceedings of the ASCE International Conference on Transportation and Development (Virtual Conference), Virtual, 8–10 June 2021.
14. Federal Highway Administration (FHWA). *A Systemic Approach to Safety: Using Risk to Drive Action*; Publication # FHWA-SA-15-054; US Department of Transportation: Washington, DC, USA, 2015. Available online: https://safety.fhwa.dot.gov/systemic/pdf/SystemicOverviewFlyer.pdf (accessed on 5 January 2022).
15. American Association of State Highway and Transportation Officials. *A Policy on Geometric Design of Highways and Streets*, 7th ed.; AASHTO: Washington, DC, USA, 2018.
16. Passing Lanes, Government of Alberta, Edmonton, Alberta, Canada. Available online: https://www.alberta.ca/passing.aspx#jumplinks-1 (accessed on 5 January 2022).
17. Stein, N.; Weisbrod, G.; Sieber, M. *NCHRP Synthesis 521: Investment Prioritization Methods for Low-Volume Roads*; The National Academy of Sciences: Washington, DC, USA, 2018; ISBN 978-0-309-39028-6. [CrossRef]
18. Coghlan, G. *Opportunities for Low-Volume Roads*; Transportation Research Board, Committee on Safety Data, Analysis, and Evaluation: Washington, DC, USA, 2015. Available online: http://onlinepubs.trb.org/onlinepubs/millennium/00068.pdf (accessed on 5 January 2022).
19. Edvardsson, K. Low Traffic Volume Roads—A Literature Review Concerning Benefit, Standard, Condition, Operation, and Maintenance, Report 775, VTI. 2013. Available online: http://www.diva-portal.org/smash/get/diva2:670643/FULLTEXT01.pdf (accessed on 14 March 2022).
20. Lebo, J.; Schelling, D. Design and Appraisal of Rural Transport Infrastructure: Ensuring Basic Access for Rural Communities. World Bank Technical Paper 496. 2001. Available online: https://documents1.worldbank.org/curated/en/227731468184131693/pdf/multi0page.pdf (accessed on 8 January 2022).
21. The United Nations. Sustainable Development: The 17 Goals. *Department of Economic and Social Affairs*. Available online: https://sdgs.un.org/goals (accessed on 20 February 2022).

encyclopedia

Entry

Numerical Solution of Desiccation Cracks in Clayey Soils

Hector U. Levatti

Division of Civil & Building Services Engineering (CBSE), London South Bank University (LSBU), London SE1 0AA, UK; levattih@lsbu.ac.uk

Definition: This entry presents the theoretical fundamentals, the mathematical formulation, and the numerical solution for the problem of desiccation cracks in clayey soils. The formulation uses two stress state variables (total stress and suction) and results in a non-symmetric and nonlinear system of transient partial differential equations. A release node algorithm technique is proposed to simulate cracking, and the strategy to implement it in the hydromechanical framework is explained in detail. This general framework was validated with experimental results, and several numerical examples were published at international conferences and in journal papers.

Keywords: desiccation cracking; hydromechanical coupling; unsaturated soil mechanics; release node technique

Citation: Levatti, H.U. Numerical Solution of Desiccation Cracks in Clayey Soils. *Encyclopedia* **2022**, *2*, 1036–1058. https://doi.org/10.3390/encyclopedia2020068

Academic Editors: Raffaele Barretta, Ramesh Agarwal, Krzysztof Kamil Żur and Giuseppe Ruta

Received: 22 April 2022
Accepted: 22 May 2022
Published: 24 May 2022

1. Introduction

Crack produced by desiccation in clayey soils is a problem that has implications in a wide range of ground-related fields. Geotechnical engineering, agriculture, mining, radioactive waste storage, tailings reservoirs, gravity dams, and public buildings are all affected by cracks due to desiccation [1–5].

The crack patterns that appear during desiccation are random, unique, and their development depends on multiple factors. Experimental approaches and numerical contributions have been made in the last 50 years [2,6–23]. Comprehensive state of the art was published in the last decade [12,24].

The process of cracking due to desiccation in clayey soils is difficult to treat numerically because the features involved are not well understood and they are dependent on one another. The process of initiation and evolution of the cracks includes a desiccation process first (a hydraulic process) and a process of shrinkage after (a mechanical process). The hydraulic and the mechanical components of the problem are both nonlinear processes. Most of the soil's properties that affect the cracking process depend on the suction (see definition of suction in Equation (3)) or moisture content values. The process is hydromechanical and coupled. Adding to the complexity is the not-yet-well-understood soil–atmosphere interaction. The mineral composition of the clay as well as the salts dissolved in the water could play a role in the process, but this influence is beyond the scope of this hydromechanical formulation.

In the past, several approaches have been presented for dealing with this problem, with models available for each phase. There are numerical approaches in the literature based on the finite element method (FEM), the discrete element method (DEM), the distinct element method (DiEM), the mesh fragmentation technique (MFT), the lattice spring model (DLSM) [2,11,25–31], etc.

There is a model [32] that presents a relatively simple application of linear elastic fracture mechanics (LEFM) to simulate tension cracks in soils. In that work, however, the soil is not drying, and the fracture toughness can be determined in a relatively easy way at the laboratory for the convenient water content of the specimen. Commercial codes, such as FLAC® [33], based on the finite difference method have been applied for simulating the curling of a high-plasticity clay subjected to desiccation with no restrictions and with no cracks [11]. Rodríguez et al. [2] developed a one-dimensional hydromechanical approach to study the initiation of a crack during a desiccation process applied to mine wastes. The model proposed by Trabelsi et al. [25] takes into account the heterogeneity of the soil and uses as the main variable the porosity to study the initiation and propagation of cracks in thin samples. The distinct element method proposed by Amarasiri et al. [27] (Amarasiri et al., 2011) is an adaptation of the commercial code UDEC® [34] for analysing desiccation cracks in soils. This approach uses an interface element to simulate the cracks but does not calculate the flow in the porous media (the variation of water content with time needs to be known). The discrete element method (DEM) has been used for the simulation of crack desiccation in thin cylindrical specimens [26] using the commercial code PFC3D® [35] in which the variation of the stiffness and the tensile strength was taken into account and drying shrinkage was imposed on each aggregate at the micro-scale. The mesh fragmentation technique [28] uses high-aspect-ratio elements between the standard elements of a mesh making the model easily adapted to standard finite element programs. The lattice spring model DLSM is a two-phase bond model [29]. Finally, there have been several attempts to use simple models of particles connected by springs or viscoelastic Maxwell elements [36].

The classical theories of continuum mechanics, unsaturated soil mechanics, and the strength of materials and concepts in fracture mechanics are the preferable framework when studying problems such as this. The mathematical problem that emerges from the classical theories in this approach is resolved by the finite element method using material parameters obtained from laboratory tests in a controlled environment. Then, the approach is theoretical, experimental, and numerical (Figure 1).

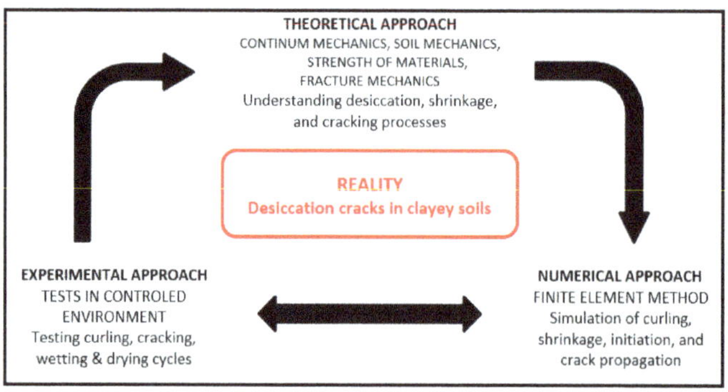

Figure 1. A theoretical, numerical, and experimental approach to the desiccation crack problem in clayey soils.

In this entry, the flow in deformable porous media is resolved by the finite element method with a $u-p$ formulation [37]. A release-node technique is used to simulate the cracking process. The capabilities of the approach have been published recently [24,38]. In this approach, desiccation, shrinkage, and cracking are solved without imposing distribution or variation of water content. The system evolves from its initial conditions in suction and displacements first. After a while, the first crack appears when the tensile strength is reached at a point in the soil's matrix. The release-node algorithm deals with the propagation of the cracks allowing changes in the suction at the boundaries at the new surfaces created by the cracks. Any other technique discussed above to tackle the cracking process can be implemented in the context of the general finite element hydromechanical model proposed here. The more complex the fracture mechanics approach is, the more parameters determined at the laboratory will be necessary and the more numerical instabilities will have to be resolved when simulating the process.

2. Mathematical Formulation and Numerical Solution for Desiccation Cracks in Clayey Soils

Although the three main physical processes involved in the formation of drying cracks (desiccation, shrinkage, and cracking) can be studied separately, to understand and model the process, it is necessary also to consider the interactions and couplings between them. Physically, the problem is coupled because the evaporation of water from the soil matrix produces matrix suction, and this matrix suction produces shrinkage and cracking. Shrinkage and cracking contribute to the desiccation process because deformation forces move the water out of the soil matrix and cracks produce new boundaries that water can use to evaporate.

Desiccation occurs because of a thermodynamical imbalance between the saturated soil mass and the environment (involving mainly temperature and relative humidity), leading to evaporation and water flow through the porous medium. When the process starts, the environment induces suction at the soil–air boundary because the relative humidity of the air is less than the relative humidity of the soil mass. For this reason, the suction is a boundary condition of the hydromechanical problem and the suction gradient in the soil specimen produces a flow of water through the porous medium towards the external surface, resulting eventually in evaporation, while in the soil's pores an evaporation–condensation process occurs constantly. Desiccation is a non-adiabatic process because energy is needed for the evaporation. This energy comes from external pressures and high temperatures that, together with the soil's shrinkage, produce changes of water pressure and of suction. This hydromechanical coupling shrinkage-flow is governed by several well-known physics laws: Darcy's law relates the relative velocity between water and soil with the hydraulic gradient and describes the movement of air when the pores are interconnected; the air dissolution in water is governed by Henry's law, and the diffusion of air in water is governed by Fick's law. The soil temperature decreases during desiccation when the energy is taken from the soil. Under non-isothermal conditions, there are other effects, such as Soret's thermal diffusion of water vapours in the air, because of pressure gradients produced by temperature gradients [39–41], vapor effusion, and Stefan's flow [42].

When suction increases, capillary effects produce shrinkage, reducing the volume of the soil specimen and reduction of the void ratio. This may happen in saturated or unsaturated conditions. At the same time, the increasing suction gives consistency to the solid matrix and increases the stiffness and the tensile strength [15,43,44]. If the deformation is restricted, then, once the tensile strength is reached, cracking develops [9]. There are three types of restrictions to shrinkage: (1) boundary conditions in displacements or stresses (e.g., friction between different portions of soil or adherence with a soil container); (2) concentration of stresses in the soil mass; or (3) heterogeneity or impurities in the soil structure [45].

Desiccation cracks can appear in saturated and/or unsaturated conditions [12]. In fact, at the beginning of the process, the behaviour of the soils is more similar to a liquid with no tensile strength at all. Tensile strength develops because of the increment in suction when the soil acquires consistency. The water loss produces increment in suction that induces unidimensional vertical shrinkage, first, and a three-dimensional shrinkage when the soil becomes stiffer. The changes in the soil properties during the process induce a change in the boundary conditions at the contact between the soil and a container (e.g., loss of adherence or crack formation). Physically, cracking is a boundary condition problem, and, for this reason, a consistent approach must deal with this boundary condition more than facing the problem from the constitutive point of view.

The model presented here is developed in terms of two separated stress variables, suction and net stress, to have the main variables directly related to the experimental results. The consequence of this choice is that coupling of the hydraulic and mechanical problems is made through the material model, and the global system of equations obtained from the finite element method is non-symmetric. Although this is only one of different approaches that can be chosen regarding this topic [46], the numerical results obtained show the present election's effectiveness.

In order to simplify the hydromechanical analysis, a mechanical constitutive model with a nonlinear elastic relation based on the stress state surface [47,48] is chosen. With this choice, the necessity of a complex stress point algorithm (plasticity model) can be avoided, and an explicit integration scheme can be used. Of course, other plastic models, such as Cam Clay, can be implemented within this general framework.

The hydraulic problem is relatively complex because of the highly nonlinear dependence of the properties and parameters on suction [7,12,49]. Although the hydromechanical analysis is at the core of the crack desiccation problem, it is crucial to have a solver for the simulation of crack initiation and propagation. Cracks are macroscopic discontinuities in the soil matrix that need to be implemented in the numerical formulation. In the present model, a release-node technique is used to simulate the initiation and propagation of a crack. This technique is relatively simple to implement and is effective to simulate the cracks induced by desiccation as is demonstrated in [38]. With this method, the tensile strength is the only parameter necessary to be determined in the laboratory. Although linear elastic fracture mechanics (LEFM) has been proposed as an alternative approach [2,32,50], the fracture parameters for soils are difficult to obtain in comparison, which is the reason why the strength of material approach is preferable.

The model presented here is based on classical theories in the context of geotechnical engineering and is consistent and relatively simple. This is preferable instead of adding additional numerical items to solve this already complex problem. However, complexity can be added gradually to include all the variables and factors that affect the physical process and the numerical results (non-isothermal processes, intrinsic and extrinsic factors, etc.).

2.1. Mathematical Formulation

In the formulation presented, the soil is assumed to be unsaturated during cracking, with negative (tensile) porewater pressure at the initial saturated or nearly saturated stage. This is convenient because conventional unsaturated soil mechanics can be used. The unsaturated soil framework allows setting up a desiccation simulation fixing hydraulic (suction) and mechanical (displacements) boundary conditions. Significant changes in hydraulic and mechanical properties that should be treated as material nonlinearities for consistent modelling are produced by changes in the degree of saturation [51]. This is particularity important in the desiccation problem because changes in volume (shrinkage) and stresses induced by the water loss in the soil mass are produced. There are three fundamental issues during the process of desiccation in soils: (1) volume change associated with water loss; (2) tensile and shear strength properties that depend on the saturation degree changes; and (3) hydraulic behaviour depending on suction or saturation degree [46].

To simplify the numerical analysis, at least regarding the mechanical constitutive relation, a nonlinear elastic model is chosen as an example in this entry. A set of parameters for this material model can be found in Appendix A and results of the model in 2D in Appendix B for Barcelona silty clay desiccated under controlled laboratory conditions [38]. This model is appropriate when the deformations are mainly volumetric, and the shear deformation does not have much relevance (suction is a spherical tensor that induces isotropic deformation). The hypotheses of the model include isothermal processes (the phases are at the same temperature and constant in time), small and slow deformations (infinitesimal quasi-static model), the medium is unsaturated and there are two fluid phases: liquid (water) and gas (dry air and water vapour), the pressure of air is constant and equal to zero (the air flows without resistance), and the state variables of the hydromechanical problem are the displacement of the soil matrix and the suction (tensile porewater pressure).

2.1.1. Mechanical and Hydraulic Constitutive Relations

To resolve the problem of desiccation cracks in clayey soils it is necessary to consider the next definitions and equations:

- Stress state variables.
- Stress–strain–suction relations.
- Stress–strain relation.
- Generalized Darcy's law for unsaturated soils and permeability tensor.
- Water retention curve.

The concept of state surfaces [47,48] is chosen for the mechanical component of this model, allowing a simple implementation of a nonlinear elastic constitutive law. For the hydraulic component, the relation between suction and degree of saturation is modelled using the van Genuchten's closed form expression, and the generalized Darcy's law [52].

Stress State Variables

Historically, for unsaturated soils, an effective stress tensor ($\boldsymbol{\sigma}'$) similar to the effective stress tensor for saturated soils was proposed [53], as shown in Equation (1):

$$\boldsymbol{\sigma}' = \boldsymbol{\sigma} - u_a \mathbf{1} + \chi(u_a - u_w)\mathbf{1} \tag{1}$$

In Equation (1), σ is the total stress tensor. The air and water pressure are, respectively, u_a and u_w , χ is a parameter that depends on the degree of saturation, the stress history, and the soil's fabric, and $1 \equiv \delta_{ij}$ is the identity tensor. In this formulation, the matrix suction and the net mean stress define the effective stress tensor $\boldsymbol{\sigma}'$ through Equation (1). Other authors argue that it is better to work with two separate variables, the net stress and the suction that can be determined at the laboratory [54–58]. This latter approach is followed in this work.

The net stress $\boldsymbol{\sigma}^{net}$ (stress over the air pressure) and the suction s are

$$\boldsymbol{\sigma}^{net} = \boldsymbol{\sigma} - u_a \mathbf{1} \tag{2}$$

$$s = u_a - u_w \tag{3}$$

Assuming small deformations, the Cauchy strain tensor $\boldsymbol{\varepsilon}$ is expressed as

$$\boldsymbol{\varepsilon} = \nabla^s \mathbf{u} = \frac{1}{2}(\mathbf{u} \otimes \nabla + \nabla \otimes \mathbf{u}) \equiv \mathbf{L}\mathbf{u} \tag{4}$$

$$\mathbf{L} = \begin{pmatrix} \frac{\partial}{\partial x} & 0 & 0 & \frac{1}{2}\frac{\partial}{\partial y} & 0 & \frac{1}{2}\frac{\partial}{\partial z} \\ 0 & \frac{\partial}{\partial y} & 0 & \frac{1}{2}\frac{\partial}{\partial x} & \frac{1}{2}\frac{\partial}{\partial z} & 0 \\ 0 & 0 & \frac{\partial}{\partial z} & 0 & \frac{1}{2}\frac{\partial}{\partial y} & \frac{1}{2}\frac{\partial}{\partial x} \end{pmatrix}^T \tag{5}$$

where ∇^s is the symmetric gradient operator and $\mathbf{u}$ is the soil matrix displacement vector. $\mathbf{L}$ is the symmetric gradient operator in matrix form when using Voigt's notation.

Stress–Strain–Suction Relations

For oedometric and triaxial deformation and considering only a half-cycle of desiccation, which is the case of the desiccation process, a nonlinear elastic constitutive approach is enough to define the changes in volume and the development of stresses in the soil matrix. The state surfaces [47,48] are experimentally obtained in the laboratory (see Figure 2) and are an interpolation in the $\{\sigma_m^{net}, s, e\}$ space. In this case, Equation (6) is used:

$$\Delta e = a_1 \Delta \ln(\sigma_m^{net} + a_4) + a_2 \Delta \ln\left(\frac{s + p_{ref}}{p_{ref}}\right) + a_3 \Delta\left[\ln(\sigma_m^{net} + a_4) \ln\left(\frac{s + p_{ref}}{p_{ref}}\right)\right] \tag{6}$$

where Δe is the void ratio increment, $\{a_1, a_2, a_3, a_4\}$ are state surface parameters calibrated from laboratory tests, and p_{ref} is a reference pressure to avoid logarithm indeterminacy.

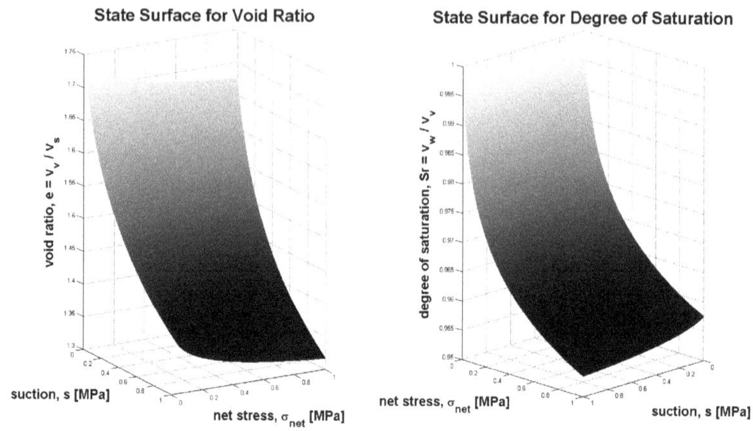

Figure 2. State surfaces.

Increments in net stresses and/or suction increase the void ratio, which is related to the volumetric deformation by

$$\varepsilon_v = -\frac{\Delta e}{1 + e_0} \tag{7}$$

where ε_v is the volumetric deformation and e_0 the initial void ratio. In the desiccation problem, deformations occur because of the decrease of void ratio triggered by increments of suction (capillary effect).

Stress–Strain Constitutive Law

The general strain–stress relation must be written in differential form because of the nonlinearity of the material behaviour; Equation (8):

$$d\boldsymbol{\sigma} = \mathbf{D} d\boldsymbol{\varepsilon} \tag{8}$$

where $\mathbf{D}$ is the fourth-order tangent stiffness tensor. The deformations are calculated by addition of a component due to the net stress plus a component due to the suction. Equation (9) considers, then, the additive deformation hypothesis:

$$d\boldsymbol{\varepsilon} = d\boldsymbol{\varepsilon}^{net} + d\boldsymbol{\varepsilon}^{s} = \mathbf{C}(\mathbb{K}, \mathbb{G}) d\boldsymbol{\sigma}^{net} + \mathbf{h}(\mathbb{K}^s) ds \tag{9}$$

In Equation (9), the parameter $\mathbb{K}$ is the volumetric modulus and $\mathbb{G}$ is the shear modulus of the soil matrix; the parameter $\mathbb{K}^s$ is the volumetric modulus due to changes in suction; the factor $\mathbf{C}$ is a fourth-order compliance tensor, and $\mathbf{h}$ is a second-order tensor.

The net stress increments can be obtained from (9):

$$d\boldsymbol{\sigma}^{net} = \mathbf{C}^{-1}(d\boldsymbol{\varepsilon} - \mathbf{h}(\mathbb{K}^s) ds) = \mathbf{D}(d\boldsymbol{\varepsilon} - \mathbf{h}(\mathbb{K}^s) ds) \tag{10}$$

In Equation (10), $\mathbf{D} = \mathbf{C}^{-1}$, is the tangent stiffness tensor.

The compliance and stiffness tensors depend on the volumetric and shear modulus $\mathbb{K}$ and $\mathbb{G}$. The suction tensor $\mathbf{h}$ depends on the volumetric suction modulus $\mathbb{K}^s$. During the desiccation process, $(\mathbb{K}, \mathbb{G}, \mathbb{K}^s)$ are variables since the volumetric strain depends on the state surface of the soil. Combining Equations (6) and (7):

$$\begin{aligned} \varepsilon_v = -\tfrac{\Delta e}{1+e_0} = -\tfrac{1}{1+e_0}\Big\{ & a_1, \Delta, \ln, (\sigma_m^{net} + a_4), +, a_2, \Delta, \ln, \left(\tfrac{s+p_{ref}}{p_{ref}}\right), + \\ & +, a_3, \Delta, \left[\ln(\sigma_m^{net} + a_4) \ln\left(\tfrac{s+p_{ref}}{p_{ref}}\right)\right]\Big\} \end{aligned} \tag{11}$$

The total strain increment and then the volumetric strain can be decomposed into net strain and suction strain increments, ε_v^{net} and ε_v^s. Incrementally, this is

$$d\varepsilon_v = d\varepsilon_v^{net} + d\varepsilon_v^s; \quad d\varepsilon_v^{net} = \mathbb{K}_t(\sigma_m^{net}, s)d\sigma_m^{net}; \quad d\varepsilon_v^s = \frac{ds}{\mathbb{K}_t^s(\sigma_m^{net}, s)} \tag{12}$$

In Equation (12), $\mathbb{K}_t(\sigma_m^{net}, s)$ and $\mathbb{K}_t^s(\sigma_m^{net}, s)$ are the tangent moduli that depend on the mean net stress and on suction. Therefore

$$d\varepsilon_v = \mathbb{K}_t(\sigma_m^{net}, s)d\sigma_m^{net} + \frac{ds}{\mathbb{K}_t^s(\sigma_m^{net}, s)} \tag{13}$$

Moreover, $d\varepsilon_v$ is

$$d\varepsilon_v(\sigma_m^{net}, s) = \frac{\partial\varepsilon_v}{\partial\sigma_m^{net}}d\sigma_m^{net} + \frac{\partial\varepsilon_v}{\partial s}ds \tag{14}$$

Studying Equations (13) and (14) allow to see the relation between the tangent moduli and the volumetric strain:

$$\begin{aligned} \frac{\partial\varepsilon_v}{\partial\sigma_m^{net}} &= \mathbb{K}_t(\sigma_m^{net}, s) \\ \frac{\partial\varepsilon_v}{\partial s} &= \frac{1}{\mathbb{K}_t^s(\sigma_m^{net}, s)} \end{aligned} \tag{15}$$

The increment of the volumetric deformation, $\Delta\varepsilon_v$, during a time increment $\Delta t = t - t_0$, can be calculated from Equation (11):

$$\begin{aligned} \Delta\varepsilon_v = -\frac{1}{1+e_0}\Big\{ & -a_1 \ln(\sigma_{m,0}^{net} + a_4). \\ & -a_2 \ln(\frac{s_0 + p_{ref}}{p_{ref}}) - a_3 \ln(\sigma_{m,0}^{net} + a_4) \ln(\frac{s_0 + p_{ref}}{p_{ref}}) \\ & +a_1 \ln(\sigma_m^{net} + a_4) + a_2 \ln(\frac{s + p_{ref}}{p_{ref}}) \\ & +a_3 \ln(\sigma_m^{net} + a_4) \ln(\frac{s + p_{ref}}{p_{ref}})\Big\} \end{aligned} \tag{16}$$

Differentiating with respect to the mean net stress σ_m^{net} and suction s:

$$\begin{aligned} \frac{\partial\varepsilon_v}{\partial\sigma_m^{net}} &= -\frac{1}{1+e_0}\left[\frac{a_1}{\sigma_m^{net}+a_4} + \frac{a_3}{\sigma_m^{net}+a_4}\ln\left(\frac{s+p_{ref}}{p_{ref}}\right)\right] \\ \frac{\partial\varepsilon_v}{\partial s} &= -\frac{1}{1+e_0}\left[\frac{a_2}{s+p_{ref}} + \frac{a_3}{s+p_{ref}}\ln\left(\sigma_m^{net} + a_4\right)\right] \end{aligned} \tag{17}$$

The tangent elastic moduli can then be obtained from Equations (15) and (17) in terms of the state surface parameters a_1, a_2, a_3, a_4, and the stress variables σ_m^{net} and s:

$$\begin{aligned} \mathbb{K}_t(\sigma_m^{net}, s) &= \frac{-a_1 - a_3 \ln\left(\frac{s+p_{ref}}{p_{ref}}\right)}{(1+e_0)\left(\sigma_m^{net}+a_4\right)} \\ \mathbb{K}_t^s(\sigma_m^{net}, s) &= \frac{(1+e_0)\left(s+p_{ref}\right)}{-a_2 - a_3 \ln\left(\sigma_m^{net}+a_4\right)} \end{aligned} \tag{18}$$

If the air pressure is assumed to be constant and equal to zero, $u_a = 0$, the elastic moduli can be calculated as in the formulas below:

$$\begin{aligned} \mathbb{K}_t(p', u_w) &= \frac{-a_1 - a_3 \ln\left(\frac{-u_w+p_{ref}}{p_{ref}}\right)}{(1+e_0)(p'+a_4)} \\ \mathbb{K}_t^s(p', u_w) &= \frac{(1+e_0)\left(-u_w+p_{ref}\right)}{-a_2 - a_3 \ln(p'+a_4)} \end{aligned} \tag{19}$$

For simplicity and because the deformation produced by the increment of suction is mainly volumetric, the Poisson's ratio can be assumed constant. Additionally, a linear elastic relation between the shear and Young's moduli can be adopted:

$$\mathbb{G}_t = \frac{3K_t(1-2\nu)}{2(1+\nu)} = \frac{3(1-2\nu)(1+e_0)(p'+a_4)}{2(1+\nu)\left[-a_1 - a_3 \ln\left(\frac{-u_w+p_{ref}}{p_{ref}}\right)\right]} \tag{20}$$

The volumetric deformation produced by changes of suction is then

$$d\varepsilon_v^s = -\frac{1}{\mathbb{K}_t^s(p', u_w)} du_w \tag{21}$$

and the stress–strain relation becomes

$$d\sigma_{ij} = D_{ijkl}\left(d\varepsilon_{kl} - \frac{d\varepsilon_v^s}{3}\delta_{kl}\right) = D_{ijkl}\left(d\varepsilon_{kl} + \frac{du_w}{3K_t^s}\delta_{kl}\right) \tag{22}$$

Finally, in matrix form, the stress–strain relation is

$$d\boldsymbol{\sigma} = \mathbf{D}(d\boldsymbol{\varepsilon} - d\boldsymbol{\varepsilon}^s) = \mathbf{D}\left(d\boldsymbol{\varepsilon} + \mathbf{m}\frac{du_w}{3K_t^s}\right) \tag{23}$$

where $\boldsymbol{\varepsilon}^s$ is the tensor of deformations due to suction and it is spherical ($\varepsilon_{ij}^s = u_w\delta_{ij}/K_t^s$) and $\mathbf{m} = \begin{bmatrix} 1 & 1 & 1 & 0 & 0 & 0 \end{bmatrix}^T$ is the identity tensor. An isotropic nonlinear elastic material is defined by the stiffness tensor $\mathbf{D}$. In this constitutive law, volumetric and shear deformations are uncoupled.

Generalized Darcy's Law for Unsaturated Soils and Permeability Tensor

The generalized Darcy's law for unsaturated soils is

$$\mathbf{q} = -\mathbf{K}(S_r)\cdot(\nabla u_w - \mathbf{g}\rho^w) \tag{24}$$

In Equation (24), $\mathbf{q}$ is the velocity of Darcy vector; ∇u_w is the gradient of the porewater pressure; $\mathbf{K}(S_r)$ is a permeability tensor that changes with water saturation degree (S_r); $\mathbf{g}$ is the gravity vector; and ρ^w is the water density.

The permeability tensor in terms of the intrinsic permeability is written as

$$\mathbf{K}(S_r) = \mathbf{k}(n)\frac{k^{rl}(S_r)}{\mu^l} \tag{25}$$

In Equation (25), k^{rl} is the relative permeability and it is nondimensional. Its values are in the range 0 to 1. This parameter changes with the degree of saturation (here, $k^{rl} = (S_r)^r$, where r is a constant). The parameter μ^l is the dynamic viscosity of water and it is temperature-dependent; $\mathbf{k}(n) = \left(\mu^l/\gamma_w\right)K1$ is the intrinsic permeability tensor, which is a function of the porosity and of the viscosity and temperature of the fluid. Finally, γ_w is the unit weight of water, and K is the hydraulic conductivity of the soil.

For cases where the hydromechanical coupling cannot be neglected, it is necessary to relate the saturated hydraulic conductivity to changes of porosity. An expression that can be used for that purpose is

$$k_{sat} = k_0 \exp[b(n-n_0)] \tag{26}$$

where k_0 is the saturated hydraulic conductivity of reference at $n = n_0$, and k_{sat} is the saturated hydraulic conductivity for porosity n. In the last equation, b is a material parameter. For saturated soils, $K = k_{sat}$.

Water Retention Curve

The van Genuchten function [59] is adopted in this formulation in order to relate changes between the degree of saturation and the suction when constant air pressure is considered:

$$S_r = \left[1 + \left(\frac{s}{P_0 \cdot f_n}\right)^{\frac{1}{1-\lambda}}\right]^{-\lambda} \quad f_n = \exp[-\eta(n - n_0)] \tag{27}$$

where λ is a material parameter and P_0 is the air-entry value for the initial porosity n_0, adopted as the reference value. Function f_n considers the changes of porosity during desiccation and its effect in the water retention curve by means of a parameter η. For non-deformable soils $f_n = 1$, because porosity is constant.

2.1.2. Governing Equations

The theoretical framework is defined using a multiphase approach where every phase (soil matrix, water, and air) fills up the entire domain and is known at the macroscopic level [60]. The process of desiccation is studied in unsaturated conditions considering the water to be in a tensile state of stress once the suction boundary condition has been defined for the hydromechanical problem. The global variables of the hydromechanical problem are the displacement vector of the soil matrix $\mathbf{u}$ and the suction s in the pores. The finite element method has been chosen for the discretisation in space and the finite differences method for the time discretisation of the problem.

The governing equations of this problem are

1. Equilibrium equation of the solid particles.
2. Balance equation of water.

Equilibrium Equations

In an unsaturated porous medium, the equilibrium equation in terms of total stresses is

$$\nabla \cdot (\boldsymbol{\sigma} - u_a \mathbf{1}) + \nabla u_a + \rho \mathbf{g} = 0 \tag{28}$$

Equation (28) is an elliptic partial differential equation where $\boldsymbol{\sigma}(\mathbf{x}, t)$ is the total stress tensor, $(\boldsymbol{\sigma} - u_a \mathbf{1})$ is the net stress tensor, u_a is the air pore pressure, ρ is the average density of the multiphase medium (soil, water, and air), and $\mathbf{g}$ is the gravity vector.

Assuming the air pressure is constant and equal to zero, $u_a = 0$, the stress state variables are $\sigma_{ij}^{net} = \sigma_{ij}$ and $s = -u_w$, and the equilibrium equation becomes

$$\nabla \cdot \boldsymbol{\sigma} + \rho \mathbf{g} = 0 \tag{29}$$

where the density of the multiphase medium is

$$\rho = (1 - n)\rho^s + n S_r \rho^w \tag{30}$$

where ρ^s is the density of the solid particles, ρ^w is the density of water, and S_r is the degree of saturation.

Balance Equations

In an unsaturated porous medium, the fluid mass balance equation (or conservation of water mass) is written as follows:

$$\nabla \cdot (\rho^w \mathbf{q}) + \frac{\partial}{\partial t}(\rho^w n S_r) = 0 \tag{31}$$

which is a parabolic equation. Assuming water to be incompressible, the water density is constant:

$$\nabla \cdot \mathbf{q} + \frac{\partial}{\partial t}(n S_r) = 0 \tag{32}$$

Combining Darcy's law, Equation (21), with the balance equation, the general flow equation is obtained:

$$\nabla\cdot[-\mathbf{K}(S_r)\cdot\nabla u_w] + \nabla\cdot[\rho^w\mathbf{K}(S_r)\cdot\mathbf{g}] + S_r\frac{\partial n}{\partial t} + n\frac{\partial S_r}{\partial t} = 0 \tag{33}$$

Adding an additional term including water compressibility, which is usual in this type of problem, a diffusive term which stabilizes the numerical solution and permits dealing with quasi-incompressible and compressible problems is obtained:

$$\nabla\cdot[-\mathbf{K}(S_r)\cdot\nabla u_w] + \nabla\cdot[\rho^w\mathbf{K}(S_r)\cdot\mathbf{g}] + S_r\frac{\partial n}{\partial t} + n\frac{\partial S_r}{\partial t} + \frac{nS_r}{K^w}\frac{\partial u_w}{\partial t} = 0 \tag{34}$$

where K^w is the water compressibility coefficient.

Assuming that the increment of porosity is equal to the volumetric deformation ($\Delta n = \varepsilon_v$), and considering that in general $S_r = S_r(\boldsymbol{\sigma}, u_w)$ and using the chain rule to evaluate $\partial S_r/\partial t$, the following equation is obtained:

$$\nabla\cdot[-\mathbf{K}(S_r)\cdot\nabla u_w] + \nabla\cdot[\rho^w\mathbf{K}(S_r)\cdot\mathbf{g}] + S_r\frac{\partial n}{\partial t} + n\frac{\partial S_r}{\partial \boldsymbol{\sigma}}\frac{\partial \boldsymbol{\sigma}}{\partial t} + n\frac{\partial S_r}{\partial u_w}\frac{\partial u_w}{\partial t} + \frac{nS_r}{K^w}\frac{\partial u_w}{\partial t} = 0 \tag{35}$$

For desiccation problems, the saturation degree depends only on the porewater pressure by means of the water retention curve, $S_r = S_r(u_w)$. Then $\partial S_r/\partial \boldsymbol{\sigma} = 0$, and Equation (35) is simplified to

$$\nabla\cdot[-\mathbf{K}(S_r)\cdot\nabla u_w] + \nabla\cdot[\rho^w\mathbf{K}(S_r)\cdot\mathbf{g}] + S_r\frac{\partial n}{\partial t} + n\frac{\partial S_r}{\partial u_w}\frac{\partial u_w}{\partial t} + \frac{nS_r}{K^w}\frac{\partial u_w}{\partial t} = 0 \tag{36}$$

This problem is highly nonlinear because all variables depend on the porewater pressure which is the main unknown variable in this strong form of the hydraulic problem: $\mathbf{K} = \mathbf{K}(u_w)$, $S_r = S_r(u_w)$, $n = n(u_w)$. Also of interest are its gradient and temporal derivative, ∇u_w and $\partial u_w/\partial t$, respectively, because they are needed to evaluate the increment of porosity, which is equivalent to the volumetric deformation. Finally, it is also necessary to evaluate the temporal derivative of the porosity, $\partial n/\partial t$.

The initial and boundary conditions for the transient hydromechanical problem are specified in terms of displacements and of porewater pressure. The initial conditions are

$$\begin{array}{c}\mathbf{u} = \mathbf{u}_0 \\ u_w = p_0\end{array} \quad \text{in } \Omega \text{ and } \Gamma \tag{37}$$

where Ω is the domain and Γ its boundary; $\mathbf{u}(\mathbf{x},\ t)$ is the displacement field of the soil matrix; $\mathbf{u}_0 = \mathbf{u}(\mathbf{x},\ 0)$ is the displacement field at the initial time $t_0 = 0$; and p_0 is the initial porewater pressure.

The Dirichlet boundary conditions for the hydromechanical problem are

$$\begin{array}{c}\mathbf{u} = \hat{u} \\ u_w = \hat{p}\end{array} \quad \text{in } \Gamma \tag{38}$$

where $\hat{\mathbf{u}}$ and $\hat{p}$ are displacements and pore water pressures at the boundary Γ of the domain.

The Newman's stress boundary conditions for the hydromechanical problem are

$$\begin{array}{c}\boldsymbol{\sigma}\cdot\mathbf{n} = \bar{t} \\ \mathbf{I}^{\mathrm{T}}\boldsymbol{\sigma} - \bar{t} = 0\end{array} \quad \text{in } \Gamma \tag{39}$$

where $\mathbf{n}$ is the vector normal to the boundary; $\bar{t}$ is the traction vector on the boundary; and $\mathbf{I}^{\mathrm{T}}$ is the following operator:

$$\mathbf{I}^{T} = \begin{pmatrix} n_x & 0 & 0 & n_y & 0 & n_z \\ 0 & n_y & 0 & n_x & n_z & 0 \\ 0 & 0 & n_z & 0 & n_y & n_x \end{pmatrix} \tag{40}$$

where n_i are the components of the unit vector $\mathbf{n}$ normal to the boundary.

Finally, the Newman's flow boundary condition for the hydromechanical problem is

$$\mathbf{K}(S_r)(-\nabla u_w + \mathbf{g}\rho^w)\cdot\mathbf{n} = q^w \quad \text{in } \Gamma \tag{41}$$

where q^w is the flow perpendicular to the boundary.

2.2. Numerical Approach

This formulation is a one-phase flow in a deforming unsaturated porous media problem known as Richard's problem [61].

2.2.1. Finite Element Approximation

The main variables and the ones that are unknown in the desiccation problem are the porewater pressure, $u_w(\mathbf{x},\ t)$, and the displacements in the soil matrix, $\mathbf{u}(\mathbf{x},\ t)$. In the framework of the finite element method, the porewater pressure can be approximated by using the shape functions, $\mathbf{N}_p$, and the pressure, $\bar{\mathbf{p}}$, in the nodes of an element. For a two-dimensional problem, the porewater is calculated as

$$u_w(x,y,t) \approx \hat{u}_w(x,y,t) = \sum_{i=1}^{n} N_i(x,y)\bar{p}_i(t) = \mathbf{N}_p\bar{\mathbf{p}} \tag{42}$$

where $\hat{u}_w(\mathbf{x},\ t)$ is the interpolated value of the porewater pressure. The variation with time of the porewater pressure is given by

$$\frac{\partial u_w}{\partial t} \approx \frac{\partial \hat{u}_w}{\partial t} = \sum_{i=1}^{n} N_i(x,y)\frac{\partial \bar{p}_i(t)}{\partial t} = \mathbf{N}_p\frac{\partial \bar{\mathbf{p}}}{\partial t} \tag{43}$$

Similarly, the displacement vector can be approximate at the nodes in a two-dimensional analysis by using the shape functions, $\tilde{\mathbf{N}}_p$, and the nodal displacements, $\bar{\mathbf{u}}$:

$$\mathbf{u}(x,y,t) \approx \hat{\mathbf{u}}(x,y,t) = \sum_{i=1}^{n} \tilde{N}_i(x,y)\bar{u}_i(t) = \tilde{\mathbf{N}}_p\bar{\mathbf{u}} \tag{44}$$

The variation with time is

$$\frac{\partial \mathbf{u}}{\partial t} \approx \frac{\partial \hat{\mathbf{u}}}{\partial t} = \sum_{i=1}^{n} \tilde{N}_i(x,y)\frac{\partial \bar{u}_i}{\partial t} = \tilde{\mathbf{N}}_p\frac{\partial \bar{\mathbf{u}}}{\partial t} \tag{45}$$

The incremental form with respect to time of the constitutive Equation (23) is expressed as

$$\frac{\partial \sigma}{\partial t} = \mathbf{D}\left(\frac{\partial \varepsilon}{\partial t} + \mathbf{m}\frac{1}{3K_t^s}\frac{\partial u_w}{\partial t}\right) \tag{46}$$

The displacement is related to the deformation by Equation [38], and after substituting (44), the following expression of the strain tensor is obtained:

$$\varepsilon = \mathbf{L}\mathbf{u} \approx \mathbf{L}\tilde{\mathbf{N}}\bar{\mathbf{u}} = \mathbf{B}\bar{\mathbf{u}} \tag{47}$$

Replacing Equations (43) and (47) in Equation (46):

$$\frac{\partial \boldsymbol{\sigma}}{\partial t} = \mathbf{DB}\frac{\partial \overline{u}}{\partial t} + \mathbf{Dm}\frac{1}{3K_t^s}\mathbf{N}_p\frac{\partial \overline{\mathbf{p}}}{\partial t} \tag{48}$$

The integral form of the equilibrium equation is

$$\int_{\Omega} \mathbf{B}^{\mathrm{T}}\frac{\partial \boldsymbol{\sigma}}{\partial t}d\Omega = \frac{\partial \mathbf{f}}{\partial t} \tag{49}$$

Replacing Equation (48) into Equation (49):

$$\int_{\Omega} \mathbf{B}^{\mathrm{T}}\left(\mathbf{DB}\frac{\partial \overline{u}}{\partial t} + \mathbf{Dm}\frac{1}{3K_t^s}\mathbf{N}_p\frac{\partial \overline{\mathbf{p}}}{\partial t}\right)d\Omega = \frac{\partial \mathbf{f}}{\partial t} \tag{50}$$

and considering that

$$\frac{\partial n}{\partial t} = \frac{\partial \varepsilon_v}{\partial t} = \mathbf{m}^T\mathbf{B}\frac{\partial \overline{\mathbf{u}}}{\partial t} \tag{51}$$

and

$$\frac{\partial S_r}{\partial t} = \frac{\partial S_r}{\partial u_w}\frac{\partial u_w}{\partial t} = \frac{\partial S_r}{\partial u_w}\mathbf{N}_p\frac{\partial \overline{\mathbf{p}}}{\partial t} \tag{52}$$

The integral form of the hydraulic problem is

$$\begin{aligned} &\int_{\Omega} \mathbf{w}^T S_r \tfrac{\partial n}{\partial t} d\Omega + \int_{\Omega} \mathbf{w}^T n \tfrac{\partial S_r}{\partial u_w}\tfrac{\partial u_w}{\partial t} d\Omega + \int_{\Omega} \mathbf{w}^T \tfrac{nS_r}{K^w}\tfrac{\partial u_w}{\partial t} d\Omega + \\ &+ \int_{\Omega} (\nabla \mathbf{w})^T \mathbf{K}(S_r)\nabla u_w d\Omega = \int_{\Omega} \rho^w (\nabla \mathbf{w})^T \mathbf{K}(S_r)\mathbf{g} d\Omega - \int_{\Gamma} \mathbf{w}^T q^w d\Gamma \end{aligned} \tag{53}$$

where $\mathbf{w}$ are the weight functions. If the weight functions are such that $\mathbf{w} = \mathbf{N}_p$, then

$$\begin{aligned} &\int_{\Omega} (\mathbf{N}_p)^T S_r \mathbf{m}^T\mathbf{B}\frac{\partial \overline{\mathbf{u}}}{\partial t} d\Omega + \int_{\Omega} (\mathbf{N}_p)^T n \frac{\partial S_r}{\partial u_w}\mathbf{N}_p\frac{\partial \overline{\mathbf{p}}}{\partial t} d\Omega + \int_{\Omega} (\mathbf{N}_p)^T \frac{nS_r}{K^w}\mathbf{N}_p\frac{\partial \overline{\mathbf{p}}}{\partial t} d\Omega \\ &\quad + \int_{\Omega} (\nabla \mathbf{N}_p)^T \mathbf{K}(S_r)\nabla \mathbf{N}_p \overline{\mathbf{p}} d\Omega \\ &\quad = \int_{\Omega} \rho^w (\nabla \mathbf{N}_p)^T \mathbf{K}(S_r)\mathbf{g} d\Omega - \int_{\Gamma} (\mathbf{N}_p)^T q^w d\Gamma \end{aligned} \tag{54}$$

The u–p Formulation

The well-known u–p formulation [37] is adopted for solving the hydromechanical problem, where $\mathbf{u}$ are the displacements and $\mathbf{p}$ is the negative porewater pressure at the gauss points of the finite elements. Equations (50) and (54) are expressed as a non-symmetric system of partial differential equations:

$$\begin{cases} \mathbf{K}_{\mathrm{T}}\frac{\partial \overline{\mathbf{u}}}{\partial t} + \mathbf{Q}_{\mathrm{T}}\frac{\partial \overline{\mathbf{p}}}{\partial t} = \frac{\partial \mathbf{f}^u}{\partial t} \\ \mathbf{P}\frac{\partial \overline{\mathbf{u}}}{\partial t} + \mathbf{S}\frac{\partial \overline{\mathbf{p}}}{\partial t} + \mathbf{H}\overline{\mathbf{p}} = \mathbf{f}^p \end{cases} \tag{55}$$

where the tangent matrices of the stress-strain problem are

Stiffness matrix:

$$\mathbf{K}_{\mathrm{T}} = \int_{\Omega} \mathbf{B}^{\mathrm{T}}\mathbf{DB}d\Omega \tag{56}$$

Coupling matrix:

$$\mathbf{Q}_{\mathrm{T}} = \int_{\Omega} \frac{1}{3\mathbb{K}_t^s} \mathbf{B}^{\mathrm{T}} \mathbf{D} \mathbf{m} \mathbf{N}_{\mathrm{p}} d\Omega \tag{57}$$

Vector of nodal forces:

$$\frac{\partial \mathbf{f}^u}{\partial t} = \int_{\Omega} \mathbf{N}_u \rho \frac{\partial \mathbf{g}}{\partial t} d\Omega + \int_{\Omega} \mathbf{N}_u \frac{\partial \bar{\mathbf{t}}}{\partial t} d\Omega \tag{58}$$

and the matrices of the flow problem are

Coupling matrix:

$$\mathbf{P} = \int_{\Omega} (\mathbf{N}_{\mathrm{p}})^{\mathrm{T}} S_r \mathbf{m}^{\mathrm{T}} \mathbf{B} d\Omega \tag{59}$$

Compressibility matrix:

$$\mathbf{S} = \int_{\Omega} (\mathbf{N}_{\mathrm{p}})^{\mathrm{T}} n \frac{\partial S_r}{\partial u_w} \mathbf{N}_{\mathrm{p}} d\Omega + \int_{\Omega} (\mathbf{N}_{\mathrm{p}})^{\mathrm{T}} \frac{n S_r}{K^w} \mathbf{N}_{\mathrm{p}} d\Omega \tag{60}$$

Permeability matrix:

$$\mathbf{H} = \int_{\Omega} (\nabla \mathbf{N}_{\mathrm{p}})^{\mathrm{T}} \mathbf{K}(S_r) \nabla \mathbf{N}_{\mathrm{p}} d\Omega \tag{61}$$

Vector of nodal flow:

$$\mathbf{f}^p = \int_{\Omega} \rho^w (\nabla \mathbf{N}_{\mathrm{p}})^{\mathrm{T}} \mathbf{K}(S_r) \mathbf{g} d\Omega - \int_{\Gamma} (\mathbf{N}_{\mathrm{p}})^{\mathrm{T}} q^w d\Gamma \tag{62}$$

If matrix notation is used, the system of partial differential Equation (55) is

$$\begin{bmatrix} 0 & 0 \\ 0 & \mathbf{H} \end{bmatrix} \begin{bmatrix} \bar{\mathbf{u}} \\ \bar{\mathbf{p}} \end{bmatrix} + \begin{bmatrix} \mathbf{K}_{\mathrm{T}} & \mathbf{Q}_{\mathrm{T}} \\ \mathbf{P} & \mathbf{S} \end{bmatrix} \begin{bmatrix} \frac{d\bar{\mathbf{u}}}{dt} \\ \frac{d\bar{\mathbf{p}}}{dt} \end{bmatrix} = \begin{bmatrix} \frac{d\mathbf{f}^u}{dt} \\ \mathbf{f}^p \end{bmatrix} \tag{63}$$

This form of hydromechanical problems in unsaturated soils is typical and can be found in the literature [62].

What is different in the derivation using separated variables is that the system is a non-symmetric system of equations because $\mathbf{Q}_{\mathrm{T}} \neq \mathbf{P}$. The constitutive Equation (46) is the one that links the hydraulic variable (pore water pressure) with the mechanical variable (total stress).

Defining the next non-symmetric matrices:

$$\mathbf{C} = \begin{bmatrix} 0 & 0 \\ 0 & \mathbf{H} \end{bmatrix}, \mathbf{D} = \begin{bmatrix} \mathbf{K}_{\mathrm{T}} & \mathbf{Q}_{\mathrm{T}} \\ \mathbf{P} & \mathbf{S} \end{bmatrix} \tag{64}$$

and defining also

$$\mathbf{X} = \begin{bmatrix} \bar{\mathbf{u}} \\ \bar{\mathbf{p}} \end{bmatrix}, \frac{d\mathbf{X}}{dt} = \begin{bmatrix} \frac{d\bar{\mathbf{u}}}{dt} \\ \frac{d\bar{\mathbf{p}}}{dt} \end{bmatrix}, \mathbf{F} = \begin{bmatrix} \frac{d\mathbf{f}^u}{dt} \\ \mathbf{f}^p \end{bmatrix} \tag{65}$$

the compact form of the hydromechanical problem becomes

$$\mathbf{C} \cdot \mathbf{X} + \mathbf{D} \cdot \frac{d\mathbf{X}}{dt} = \mathbf{F} \tag{66}$$

This is the differential equation that needs to be solved to simulate the desiccation shrinkage of unsaturated soils.

Time Integration of the Coupled Problem

In Equation (66), the unknown is $\mathbf{X}$, which is a vector that contains the unknown displacements and porewater pressures ($\bar{\mathbf{u}}$ and $\bar{\mathbf{p}}$, respectively) at the finite element mesh nodes. The time derivative can be expressed by the generalized trapezoidal method (generalized midpoint rule) as

$$\left(\frac{d\mathbf{X}}{dt}\right)_{n+\theta} = \frac{\mathbf{X}_{n+1} - \mathbf{X}_n}{\Delta t} \tag{67}$$

and the value of the unknown at point $n + \theta$ is

$$\mathbf{X}_{n+\theta} = (1-\theta)\cdot\mathbf{X}_n + \theta\cdot\mathbf{X}_{n+1} \tag{68}$$

Replacing Equations (67) and (68) in Equation (66) and multiplying both members by Δt,

$$\mathbf{C}\cdot[(1-\theta)\cdot\mathbf{X}_n + \theta\cdot\mathbf{X}_{n+1}]_{n+\theta}\cdot\Delta t + \mathbf{D}\cdot\left[\frac{\mathbf{X}_{n+1} - \mathbf{X}_n}{\Delta t}\right]_{n+\theta}\cdot\Delta t = \mathbf{F}_{n+\theta}\cdot\Delta t \tag{69}$$

Rearranging:

$$[\mathbf{C}\cdot(1-\theta)]_{n+\theta}\cdot\Delta t\cdot\mathbf{X}_n + [\mathbf{C}\cdot\theta]_{n+\theta}\cdot\Delta t\cdot\mathbf{X}_{n+1} + [\mathbf{D}]_{n+\theta}\cdot\mathbf{X}_{n+1} - [\mathbf{D}]_{n+\theta}\cdot\mathbf{X}_n = \mathbf{F}_{n+\theta}\cdot\Delta t \tag{70}$$

Placing the unknowns on the left side of the equation:

$$[\mathbf{C}\cdot\theta\Delta t + \mathbf{D}]_{n+\theta}\cdot\mathbf{X}_{n+1} = [\mathbf{D} - \mathbf{C}(1-\theta)\Delta t]_{n+\theta}\cdot\mathbf{X}_n + \mathbf{F}_{n+\theta}\cdot\Delta t \tag{71}$$

Equation (71) is a system of equations from which $\mathbf{X}_{n+1}$ can be calculated from the values of $\mathbf{X}_n$ and $\mathbf{F}_{n+\theta}$. This is an implicit method, and it is necessary to solve a system of equations in each time step. Δt is the time interval between $\mathbf{X}_n$ and $\mathbf{X}_{n+1}$, and the parameter θ varies between 0 and 1. Expanding Equation (71):

$$\begin{bmatrix} \mathbf{K}_\mathrm{T} & \mathbf{Q}_\mathrm{T} \\ \mathbf{P} & \mathbf{H}\cdot\theta\Delta\mathrm{t} + \mathbf{S} \end{bmatrix}_{n+\theta} \begin{bmatrix} \bar{\mathrm{u}} \\ \bar{\mathrm{p}} \end{bmatrix}_{n+1} = \begin{bmatrix} \mathbf{K}_\mathrm{T} & \mathbf{Q}_\mathrm{T} \\ \mathbf{P} & \mathbf{S} - \mathbf{H}(1-\theta)\Delta t \end{bmatrix}_{n+\theta} \begin{bmatrix} \bar{\mathrm{u}} \\ \bar{\mathrm{p}} \end{bmatrix}_{n} + \Delta t \begin{bmatrix} \mathbf{f}^u \\ \mathbf{f}^p \end{bmatrix}_{n+\theta} \tag{72}$$

This is a non-symmetric system of nonlinear equations that can be solved using a coupled, staggered, or uncoupled scheme.

2.2.2. Release Node Algorithm

During a process of desiccation in clayey soil, cracking occurs at some point [2,7,12]. However, there are cases where desiccation produces shrinkage but not cracking, such as, for instance, in the curling produced in unrestricted tests [63]. For this problem, the formulation presented until here is directly applicable without any addition [24].

To simulate crack formation and propagation, there are three main variables that need to be established: the value of the stress at crack initiation (σ_c), the direction that the crack follows during propagation (θ_c), and the length of the crack when it is produced (Δa)

In this approach, the tensile strength of the soil defines the stress value when cracking starts [59]. In the laboratory, the process of desiccation starts with the soil being a slurry with practically no tensile strength. After a while, the soil acquires consistency because of suction increments. During the process, the tensile strength increases first and after a while start decreasing (Figure 3a). This fact can be taken into account adopting an experimental expression to calculate the tensile strength as a function of the current suction, as proposed by [23] and shown in Equation (73).

$$\sigma_t = -0.0191w^2 + 0.6874w - 2.88 \tag{73}$$

where σ_t is the tensile stress in kPa and w is the humidity content in %.

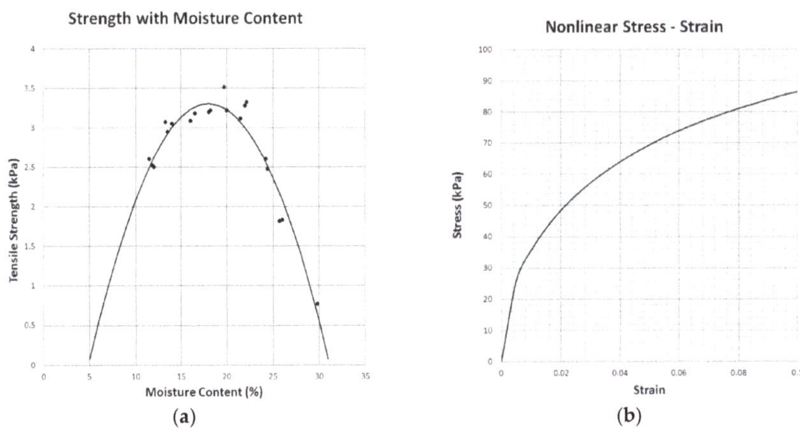

Figure 3. Crack initiation criteria: (**a**) tensile strength, moisture-content-dependent [12]; (**b**) traditional strength of material failure criteria [59].

The equation and the parameters that relate the changes in the tensile strength with humidity will be different for different soils. Equation (73) is valid only for a clay from Barcelona, but the principle is general.

The direction of crack propagation can be taken as perpendicular to the maximum principal tensile stress. Since the soil matrix is discretized by a finite element mesh, the amount of propagation of a crack can be assumed to be equal to the length of the finite element. Cracks represent a change of boundaries because new crack surfaces (new boundaries) appear at each increment, and an appropriate algorithm needs to be implemented in the finite element code. Using this method, crack propagation is modelled as a sequence of temporal boundary-value problems. The discrete growing crack is assumed as a discontinuity. In the real soil, crack propagation occurs independently of a mesh. Failure of the material at the crack process zone is a continuous process. This is a limitation, but the simulation is good enough for certain purposes. In the model and in reality, when a new crack develops, new boundaries appear, and it becomes necessary to change the displacement and suction boundary conditions in the model. Then the geometry of the problem changes and it is necessary to solve the new problem to satisfy the equilibrium and balance conditions when the crack is growing.

At any point in the soil mass where the tensile stress reaches the tensile strength, a crack starts, and it propagates if the tensile strength is reached at the crack tip. Figure 3b shows the classical Griffith fracture criterion for a nonlinear elastic material [59]. This criterion is valid for the macroscopic scale, and when the tensile strength is reached, cracking starts instantaneously. There is no damage nor plastic zone at the crack tip. Despite the limitations of this technique in terms of what happens at the cracking area, it is easily implemented in the context of the finite element method and allows a first approach to cracking. The crack-opening mechanism must be established when the crack initiates and propagates. To simplify the analysis, only Fracture Mode I is taken into consideration in in this approach.

Release Node Algorithm

A simple technique to simulate the propagation of a crack in a finite element mesh consists of releasing a node that initially had a displacement boundary condition (Figure 4a) in the case of cracks at the soil–structure interface (e.g., soil in contact with a dam or wall). By releasing the node, after the condition is reached, a new geometric boundary is free to deform, and an elemental crack geometry is added to the model (Figure 4b). In Mode I, a crack propagates following a symmetry line perpendicular to the maximum principal tensile stress. In the case of a crack at the soil–structure interface, only the displacement boundary condition needs to be released. In addition, the new surface in contact with the environment will be subjected to the suction boundary condition and this needs to be added to the model (Figure 4d). In a more general case, when the crack propagation condition is reached on the surface of the soil, the node at the crack tip needs to be split into two nodes (Figure 4f,g) generating two new surfaces and propagating the crack a length equal to the element side Δa. The new node generates a change in the node number list, node coordinates, and element connectivity, and therefore the mesh needs to be updated (this is an elemental re-meshing technique). The forces F_i between the elements (or reactions, R_i, at the boundary) that share the split node become zero after the opening and the system needs to be equilibrated. Some numerical problems can be expected because of the abrupt cancelling of these forces.

To avoid numerical instabilities, the unloading process is made gradually in several steps (Figure 4c,g).

The node release algorithm can be explained as follows:

1. Fracture criterion is reached at a point in the soil matrix.
2. Keep the external loads constant (suction profile for desiccation problems).
3. Calculate reactions at the boundary, R_i, or calculate equivalent forces that keep the split nodes together, F_i.
4. Delete the nodal restriction in the contour or split the nodes in the surface of the soil where the tensile strength was reached.
5. Add reaction R at the released node (reaction in the contour) or forces F in the node that was split.
6. Reduce the force R or F step by step to zero.
7. Add suction on the new surface created by the crack and exposed to the environment.
8. Once the model is in equilibrium, check whether the fracture criterion is reached at any other point of the geometry.
9. If yes, go to step 3 to initiate again the release node algorithm.
10. If no, continue the hydromechanical analysis.

When the crack propagation path is known, this technique is very useful and simple to implement. When desiccation is studied in the laboratory, the first cracks that appear are in the interphase between the soil and the containers, then, the first cracks directions and locations are known. For more arbitrary cracks, this technique is mesh-dependent but works relatively well if the mesh is dense enough. This method effectively calculates the stress state in the soil matrix during desiccation. This stress state is crucial for defining the initiation of the cracks and their propagation using any other alternative method to deal with the cracks.

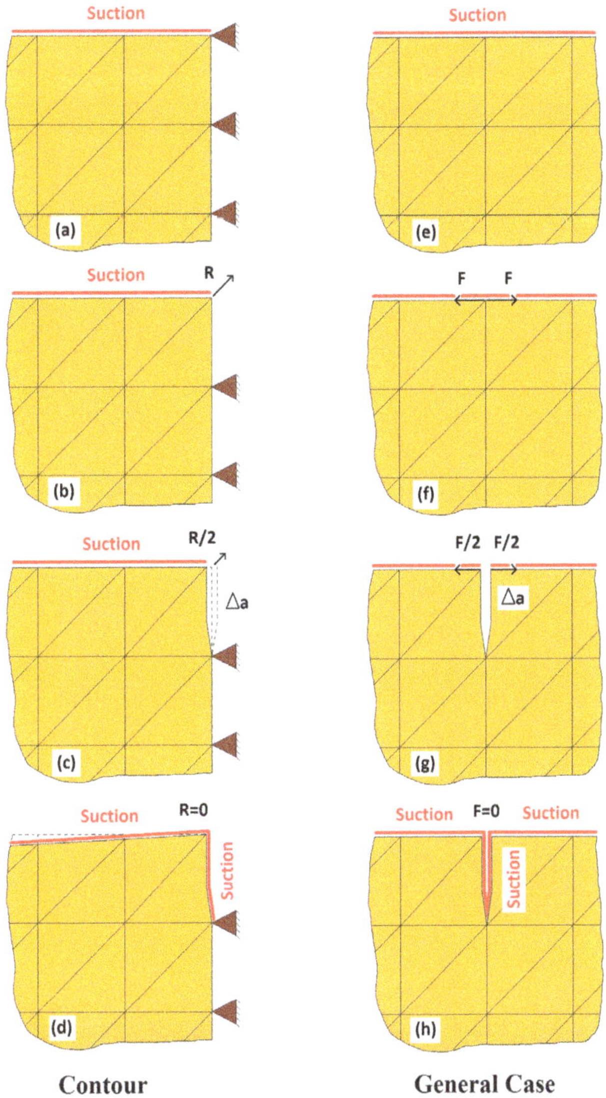

Figure 4. Release node algorithm for a contour crack and a crack starting in the middle of the soil surface. **Contour crack case:** (**a**) starting scheme in a contour, (**b**) equivalent starting scheme after reaching the cracking criteria, (**c**) reaction reduction and application of new suction condition, (**d**) new contour condition scheme with crack propagated. **General crack case**: (**e**) initial conditions, (**f**) nodes are split and equivalent forces applied, (**g**) reduction of equivalent forces, (**h**) new contour condition scheme with crack propagated.

Limitations of the Model

Several intrinsic factors define the direction of crack propagation: the tensile strength, the initial water content, anisotropy, imperfections, heterogeneity, impurities, the fracture toughness, the initial particle size, and the fabric in the field, just to mention the most important ones. If these intrinsic factors need to be added to the analysis, a more sophisticated fracture approach must be implemented in the context of the finite element hydromechanical model proposed here. Linear elastic fracture mechanics and a heterogeneous, anisotropic plastic material model should be implemented. The imperfections and impurities should be added to the geometry of the problem.

It is important to mention that the validity of the Darcy's law, symmetry of the strain tensor, and invariability of the Poisson's ratio are limitations of this model.

In this complex process, even the water density changes with temperature, salt content, permeability of the medium, etc. [64].

Aside from the geotechnical context of this entry, there are chemical, sedimentological, mineralogical, and petrological considerations that can be made to improve the model, adding the relevant parameters and physical laws. The necessary modifications and improvements of the model will depend on the area of interest and the practical application.

However, the formulation presented in this entry is general, so all these changes will be relatively simple to implement from the proposed framework. The challenge will be the determination of the material parameters and dealing with the numerical instabilities of the nonlinear material problem that every extra refinement will introduce to the formulation.

3. Conclusions and Prospects

In this entry, a mathematical formulation and a numerical solution for the analysis of desiccation cracking in clayey soils is presented. This model is hydromechanical and includes all the main variables and features that control the physical process of desiccation in clayey soils. The beauty of this approach is that it is based on fundamental principles. The unsaturated soils mechanics and strength of materials principles are used, avoiding the need for data obtained at the laboratory to force the system to behave as in reality. Once the initial conditions in suction and displacement are set, the system evolves until reaching a new state of equilibrium with the environment. The release node algorithm added to the hydromechanical core allows the study of at least a crack initiation and propagation as is shown in Appendix B.

The model offers a good balance between complexity and relatively simple implementation tools for the analysis of desiccation cracking in clayey soils. All the parameters that control the physical process can be easily determined from conventional experimental tests. The formulation is general and adaptable to more complex problems. The resulting approach helps to understand the process of desiccation cracking in soils.

This hydromechanical model without the cracking algorithm can be used for studying the curling process that soils show in the laboratory without cracks.

This model can be improved to take into consideration boundary effects such as the soil–atmosphere [65]. One way of achieving that is to simulate the fluid of the atmosphere in contact with the soil and its interaction.

The hydromechanical model can be modified to introduce more complex fracture mechanics approaches to have more control on the fracture process. Methods such as the mesh fragmentation technique can improve the treatment of multiple cracks and it can be added to the hydromechanical model presented here.

Heterogeneity and anisotropy can be added via the constitutive laws at a cost of having to determine more elastic parameters.

Imperfections can be added in the geometry or in the mesh.

If there is a need to study the process inside the pores, more sophisticated models can tackle the air dissolution in water governed by Henry's law and the diffusion of air in water governed by Fick's law. All these processes can be added at constitutive level using the same hydromechanical core and numerical solution.

If the temperature and chemical processes are relevant variables for a study, a thermal–chemical–hydromechanical formulation will be necessary. Adding the thermal component is relatively simple. The soil temperature decreases during desiccation when the energy is taken from the soil. Under non-isothermal conditions, there are other effects, such as Soret's thermal diffusion of water vapours in the air, because of pressure gradients produced by temperature gradients, vapor effusion, and Stefan's flow. All of these can be added to the formulation if necessary.

The hydromechanical core is essential to develop any other model to study the crack desiccation in clayey soils. For this reason, the mathematical formulation and the numerical solution is presented here in every single detail to allow researchers to implement it when necessary. To help with the implementation of the hydromechanical core, a set of parameters are included in Appendix A. Of course, these parameters need to be calibrated for every soil. Numerical examples can be found in previous publications of this author and a sequence of the simulation for a particular laboratory sample is presented in Appendix B. Since experimental validation is crucial for any model, results of experiments and comparison with the numerical approach are presented in Appendix C.

Additionally, the interaction between the soil and the atmosphere is crucial for the complete understanding of the process and for the moment is a topic being researched separately [65].

This model needs to be validated in three-dimensional studies [66] considering several types of soils to be properly validated for a wider range of soils and situations. This is indeed a live line of research of the author.

Funding: This research received no external funding.

Institutional Review Board Statement: Not applicable.

Informed Consent Statement: Not applicable.

Acknowledgments: Support from the Centre for Civil and Building Services Engineering (CCiBSE) at the School of the Built Environment and Architecture, London South Bank University is gratefully acknowledged.

Conflicts of Interest: The author declares no conflict of interest.

Appendix A. Set of Parameters for a Numerical Model for Barcelona Silty Clay

Table A1. Water retention curve parameters for Barcelona silty clay.

Void Ratio (e)	Porosity (n)	Dried Unit Weight γ_d (g/cm^3)	$f_n = \exp[-\eta(n - n_0)]$	λ	$1/(1 - \lambda)$
0.87	0.47	1.45	1.04	0.27	1.37
0.75	0.43	1.55	1.07	0.25	1.33
0.64	0.39	1.65	1.12	0.23	1.30
0.55	0.35	1.75	1.16	0.20	1.25

Table A2. Parameters used in the numerical analysis of Barcelona silty clay.

Mechanical Parameters						
State surface parameters				P_{ref} (MPa)	Poisson Ratio	Tensile strength
a_1 (−)	a_2 (−)	a_3 (−)	a_4 (MPa)		ν (−)	σ_c (MPa)
−0.02	−0.0025	−0.000039	0.023	0.1	0.4	0.0035
Hydraulic Parameters						
Initial permeability k_0 (m/s)			Material parameter b		Initial porosity n_0	Material parameter r
9.27×10^{-10}			25		0.6	3

Appendix B. Numerical Results of the Model for Barcelona Silty Clay

Simulation of 120 days of desiccation on a cylindrical soil sample 80 cm in diameter and 20 cm height desiccated under controlled conditions. The figure shows the evolution of the suction in the radial section of the cylindrical sample and the propagation of a crack in contact with the container of the soil in the laboratory. (a) Suction and crack after 7 days of drying; (b) suction and crack after 12 days of drying; (c) suction and crack after 37 days of drying; (d) suction and crack after 62 days of drying; (e) suction and crack after 72 days of drying; (f) suction and crack after 120 days of drying. The results show the effectivity of the release node algorithm to simulate a cracking process by desiccation.

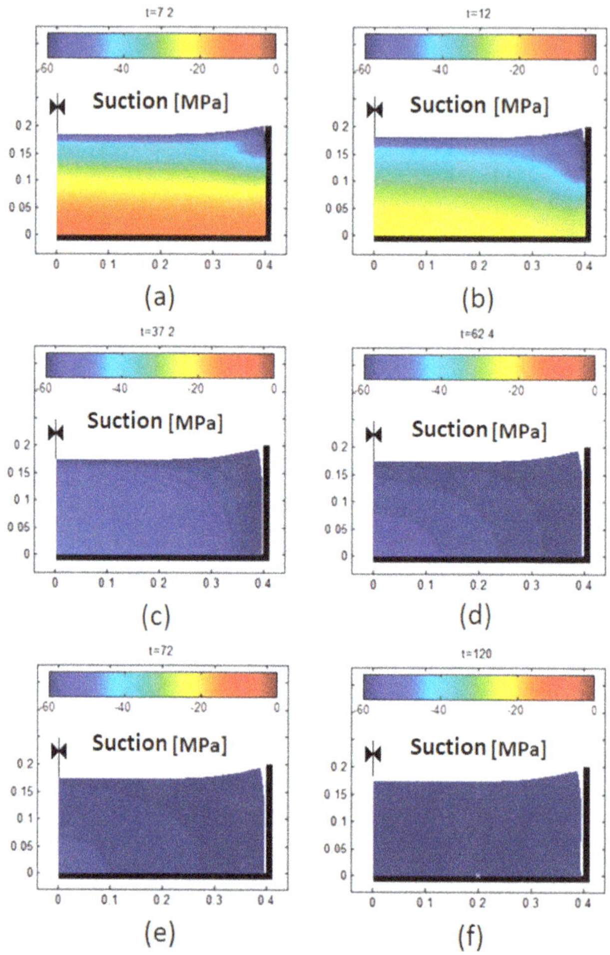

Appendix C. Comparison of Experimental and Numerical Results of the Model

Evolution of suction for different samples of 80/40 cm in diameter and 20/10 cm height in laboratory and environmental chamber conditions. The numerical simulation of radial cross-section 40 cm wide and 20 cm high is included to validate the model with the experiment.

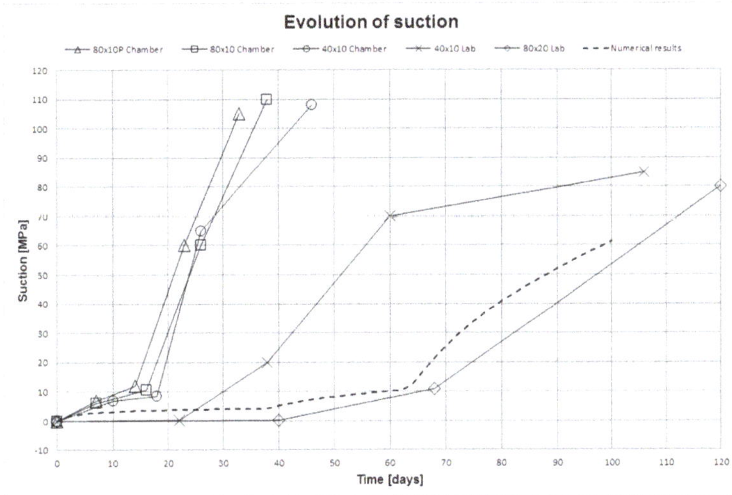

References

1. Rodríguez, R.L. Estudio Experimental de Flujo y Transporte de Cromo, Níquel Y Manganeso en Residuos de la Zona Minera de Moa (Cuba): Influencia del Comportamiento Hidromecánico. Ph.D. Thesis, UPC-BarcelonaTech, Barcelona, Spain, 2002.
2. Rodríguez, R.L.; Sánchez, M.J.; Ledesma, A.; Lloret, A. Experimental and numerical analysis of a mining waste desiccation. *Can. Geotech. J.* **2007**, *44*, 644–658. [CrossRef]
3. Tang, C.-S.; Shi, B.; Liu, C.; Zhao, L.; Wang, B. Influencing factors of geometrical structure of surface shrinkage cracks in clayey soils. *Eng. Geol.* **2008**, *101*, 204–217. [CrossRef]
4. Justo, J.L.; Vázquez, N.J.; Justo, E. Subsidence in saturated-unsaturated soils: Application to Murcia (Spain). In *Third International Conference on Unsaturated Soils*; Jucá, J.F.T., de Campos, T.M.P., Marinho, F.A.M., Eds.; Balkema: Rotterdam, The Netherlands, 2002; pp. 845–850.
5. Bronswijk, J.J.B. Modeling of water balance, cracking and subsidence in clay soils. *J. Hydrol.* **1988**, *97*, 199–212. [CrossRef]
6. Kodikara, J.; Barbour, S.L.; Fredlund, D.G. Desiccation cracking of soil layers. In *Unsaturated Soils for Asia*; Rahardjo, H., Toll, D.G., Leong, E.C., Eds.; Balkema: Rotterdam, The Netherlands, 2000; pp. 693–698.
7. Ávila, G. Estudio de la Retracción Y el Agrietamiento de Arcillas. Aplicación a la Arcilla de Bogotá. Ph.D. Thesis, UPC-BarcelonaTech, Barcelona, Spain, 2004.
8. Chertkov, V.Y. Modelling cracking stages of saturated soils as they dry and shrink. *Eur. J. Soil Sci.* **2002**, *53*, 105–118. [CrossRef]
9. Corte, A.; Higashi, A. *Experimental Research on Desiccation Cracks in Soil*; U.S. Army Materiel Command Cold Regions Research & Engineering Laboratory: Wilmette, IL, USA, 1960.
10. Hu, L.B.; Hueckel, T.; Péron, H.; Laloui, L. Modeling Evaporation, Shrinkage and Cracking of Desiccating Soils. In Proceedings of the International Association for Computer Methods and Advances in Geomechanics (IACMAG), Goa, India, 1–6 October 2008; pp. 1083–1090.
11. Kodikara, J.K.; Nahlawi, H.; Bouazza, A. Modelling of curling in desiccation clay. *Can. Geotech. J.* **2004**, *41*, 560–566. [CrossRef]
12. Lakshmikantha, M.R. Experimental and Theoretical Analysis of Cracking in Drying Soils. Ph.D. Thesis, UPC-BarcelonaTech, Barcelona, Spain, 2009.
13. Lakshmikantha, M.R.; Prat, P.C.; Ledesma, A. Characterization of crack networks in desiccating soils using image analysis techniques. In *Numerical Models in Geomechanics X*; Pande, G.N., Pietruszczak, S., Eds.; Balkema: Rotterdam, The Netherlands, 2007; pp. 167–176.
14. Lakshmikantha, M.R.; Prat, P.C.; Ledesma, A. Image analysis for the quantification of a developing crack network on a drying soil. *Geotech. Test. J.* **2009**, *32*, 505–515.
15. Lakshmikantha, M.R.; Prat, P.C.; Ledesma, A. Experimental evidences of size-effect in soil cracking. *Can. Geotech. J.* **2012**, *49*, 264–284. [CrossRef]
16. Lakshmikantha, M.R.; Prat, P.C.; Ledesma, A. Evidences of hierarchy in cracking of drying soils. *ASCE Geotech. Spec. Publ.* **2013**, *231*, 782–789.
17. Lau, J.T.K. Desiccation Cracking of Clay Soils. M.S. Thesis, University of Saskatchewan, Saskatoon, SK, Canada, 1987.
18. Morris, P.H.; Graham, J.; Williams, D.J. Cracking in drying soils. *Can. Geotech. J.* **1992**, *29*, 263–277. [CrossRef]
19. Nahlawi, H.; Kodikara, J. Laboratory experiments on desiccation cracking of thin soil layers. *Geotech. Geol. Eng.* **2006**, *24*, 1641–1664. [CrossRef]

20. Péron, H.; Hueckel, T.; Laloui, L.; Hu, L.B. Fundamentals of desiccation cracking of fine-grained soils: Experimental characterisation and mechanisms identification. *Can. Geotech. J.* **2009**, *46*, 1177–1201. [CrossRef]
21. Tang, C.-S.; Shi, B.; Liu, C.; Gao, L.; Inyang, H. Experimental Investigation of the Desiccation Cracking Behavior of Soil Layers during Drying. *J. Mater. Civ. Eng.* **2011**, *23*, 873–878. [CrossRef]
22. Tang, C.-S.; Shi, B.; Liu, C.; Suo, W.-B.; Gao, L. Experimental characterization of shrinkage and desiccation cracking in thin clay layer. *Appl. Clay Sci.* **2011**, *52*, 69–77. [CrossRef]
23. Vogel, H.J.; Hoffmann, H.; Roth, K. Studies of crack dynamics in clay soil. I: Experimental methods, results, and morphological quantification. *Geoderma* **2005**, *125*, 203–211. [CrossRef]
24. Levatti, H.U. Estudio Experimental Y Análisis Numérico de la Desecación en Suelos Arcillosos. Ph.D. Thesis, UPC-BarcelonaTech, Barcelona, Spain, 2015. Available online: https://www.tdx.cat/handle/10803/299202#page=1 (accessed on 12 January 2022).
25. Trabelsi, H.; Jamei, M.; Zenzri, H.; Olivella, S. Crack patterns in clayey soils: Experiments and modelling. *Int. J. Numer. Anal. Methods Geomech.* **2012**, *36*, 1410–1433. [CrossRef]
26. Sima, J.; Jiang, M.; Zhou, C. Modeling desiccation cracking in thin clay layer using three-dimensional discrete element method. *Proc. Am. Inst. Phys.* **2013**, *1542*, 245–248.
27. Amarasiri, A.L.; Kodikara, J.; Costa, S. Numerical modelling of desiccation cracking. *Int. J. Numer. Anal. Methods Geomech.* **2011**, *35*, 82–96. [CrossRef]
28. Sánchez, M.; Manzoli, O.L.; Guimarães, L.N. Modeling 3-D desiccation soil crack networks using a mesh fragmentation technique. *Comput. Geotech.* **2014**, *62*, 27–39. [CrossRef]
29. Gui, Y.; Zhao, G.F. Modelling of laboratory soil desiccation cracking using DLSM with a two-phase bond model. *Comput. Geotech.* **2015**, *69*, 578–587. [CrossRef]
30. Yan, C.; Luo, Z.; Zheng, Y.; Ke, W. A 2D discrete moisture diffusion model for simulating desiccation fracturing of soil. *Eng. Anal. Bound. Elem.* **2022**, *138*, 42–64. [CrossRef]
31. Nikolic, M. Discrete element model for the failure analysis of partially saturated porous media with propagating cracks represented with embedded strong discontinuities. *Comput. Methods Appl. Mech. Eng.* **2022**, *390*, 27. [CrossRef]
32. Lee, F.H.; Lo, K.W.; Lee, S.L. Tension crack development in soils. *J. Geotech. Eng. ASCE* **1988**, *114*, 915–929. [CrossRef]
33. ITASCA. *FLAC3D: Fast Lagrangian Analysis of Continua in Three Dimensions*; Version 2.0 Manual; Itasca Consulting Group: Minneapolis, MN, USA, 1997.
34. ITASCA. *UDEC: Distinct-Element Modeling of Jointed and Blocky Material in 2D*; Itasca Consulting Group: Minneapolis, MN, USA, 2017.
35. ITASCA. *PFC3D: Particle Flow Code in Three Dimensions*; Version 3.0 Manual; Itasca Consulting Group: Minneapolis, MN, USA, 2003.
36. Musielak, G.; Sliwa, T. Modeling and Numerical Simulation of Clays Cracking During Drying. *Dry. Technol.* **2015**, *33*, 1758–1767. [CrossRef]
37. Zienkiewicz, O.C.; Chan, A.H.C.; Pastor, M.; Paul, D.K.; Shiomi, T. Static and Dynamic Behaviour of Soils: A Rational Approach to Quantitative Solutions. I. Fully Saturated Problems. Proceedings of the Royal Society of London Series A. *Math. Phys. Sci.* **1990**, *429*, 285–309.
38. Levatti, H.U.; Prat, P.C.; Ledesma, A. Numerical and experimental study of initiation and propagation of desiccation cracks in clayey soils. *Comput. Geotech.* **2019**, *105*, 155–167. Available online: https://openresearch.lsbu.ac.uk/item/865wv (accessed on 12 January 2022). [CrossRef]
39. Platten, J.K. The Soret Effect: A Review of Recent Experimental Results. *J. Appl. Mech.* **2006**, *73*, 5–15. [CrossRef]
40. Ludwig, C. Diffusion zwischen ungleich erwwa"rmten orten gleich zusammengestzter lo"sungen. *Sitz Ber. Akad. Wiss. Wien. Math.-Naturw. Kl.* **1856**, *20*, 539.
41. Soret, C. Sur l'état d'équilibre que prend au point de vue de sa concentration une dissolution saline primitivement homohéne dont deux parties sont portées à des températures différentes. *Ann. Chim. Phys.* **1881**, *22*, 293–297.
42. Mitrovic, J. Josef Stefan and his evaporation–diffusion tube—the Stefan diffusion problem. *Chem. Eng. Sci.* **2012**, *75*, 279–281. [CrossRef]
43. Varsei, M.; Miller, G.A.; Hassanikhah, A. Novel approach to measuring tensile strength of compacted clayey soil during desiccation. *Int. J. Geomech.* **2016**, *16*, D4016011. [CrossRef]
44. Lakshmikantha, M.R.; Prat, P.C.; Tapia, J.; Ledesma, A. Effect of moisture content on tensile strength and fracture toughness of a silty soil. In *Unsaturated Soils: Advances in Geo-Engineering*; Toll, D.G., Augarde, C.E., Gallipoli, D., Wheeler, S.J., Eds.; CRC Press/Balkema: Leiden, The Netherlands, 2008; pp. 405–409.
45. Hueckel, T. On effective stress concept and deformation in clays subjected to environmental loads: Discussion. *Can. Geotech. J.* **1992**, *29*, 1120–1125. [CrossRef]
46. Sheng, D. Review of fundamental principles in modelling unsaturated soil behaviour. *Comput. Geotech.* **2011**, *38*, 757–776. [CrossRef]
47. Matyas, E.L.; Radhakrishna, H.S. Volume change characteristics of partially saturated soils. *Géotechnique* **1968**, *18*, 432–448. [CrossRef]
48. Lloret, A.; Alonso, E.E. State surfaces for partially saturated soils. In Proceedings of the XI International Conference on Soil Mechanics and Foundation Engineering, San Francisco, CA, USA, 12–16 August 1985; pp. 557–562.

49. Barrera, M. Estudio Experimental del Comportamiento Hidro-Mecánico de Suelos Colapsables. Ph.D. Thesis, UPC-BarcelonaTech, Barcelona, Spain, 2002.
50. Konrad, J.-M.; Ayad, R. An idealized framework for the analysis of cohesive soils undergoing desiccation. *Can. Geotech. J.* **1997**, *34*, 477–488. [CrossRef]
51. Gens, A.; Sánchez, M.; Sheng, D. On constitutive modelling of unsaturated soils. *Acta Geotech.* **2006**, *1*, 137–147. [CrossRef]
52. Van Genuchten, M.T. Closed-form equation for predicting the hydraulic conductivity of unsaturated soils. *Soil Sci. Soc. Am. J.* **1980**, *44*, 892–898. [CrossRef]
53. Bishop, A.W. The principles of effective stress. *Tek. Ukebl.* **1959**, *106*, 859–863.
54. Burland, J.B. Some aspects of the mechanical behaviour of partly saturated soil. In *Moisture Equilibrium and Moisture Changes in Soils Beneath Covered Areas*; Butterworths: Sydney, Australia, 1965; pp. 270–278.
55. Bishop, A.W.; Blight, G.E. Some aspects of effective stress in saturated and partly saturated soils. *Géotechnique* **1963**, *13*, 177–197. [CrossRef]
56. Blight, G.E. A study of effective stresses for volume change. In *Moisture Equilibrium and Moisture Changes in Soils Beneath Covered Areas*; Butterworths: Sydney, Australia, 1965; pp. 259–269.
57. Aitchison, G.D. Soil properties. In Proceedings of the 6th International Conference on Soil Mechanics and Foundation Engineering, SMFE, Montreal, QC, Canada, 8–15 September 1965; pp. 319–321.
58. Jennings, J.; Burland, J.B. Limitations to the use of effective stress in partly saturated soils. *Géotechnique* **1962**, *12*, 125–144. [CrossRef]
59. Griffith, A.A. The phenomenon of rupture and flow in solids. *Philos. Trans. Roy. Soc. Lond.* **1921**, *221*, 163–198.
60. Lewis, R.W.; Schrefler, B.A. *The Finite Element Method in the Static and Dynamic Deformation and Consolidation of Porous Media*, 2nd ed.; Wiley: Hoboken, NJ, USA, 1998.
61. Richards, L.A. Capillary conduction of liquids through porous mediums. *Physics* **1931**, *1*, 318–333. [CrossRef]
62. Schrefler, B.A.; Simoni, L.; Majorana, C.E. A general model for the mechanics of saturated-unsaturated porous materials. *Mater. Struct.* **1989**, *22*, 323–334. [CrossRef]
63. Nahlawi, H.; Kodikara, J. Experimental observations on curling of desiccating clay. In *Third International Conference on Unsaturated Soils*; Jucá, J.F.T., de Campos, T.M.P., Marinho, F.A.M., Eds.; Balkema: Rotterdam, The Netherlands, 2004; pp. 553–556.
64. Ledesma, A. Cracking in desiccating soils. In Proceedings of the 3rd European Conference on Unsaturated Soils, E-UNSAT, Paris, France, 12–14 September 2016.
65. Cui, Y. Soil-atmosphere interaction in earth structures. *J. Rock Mech. Geotech. Eng.* **2022**, *14*, 35–49. [CrossRef]
66. Zhuang, Z.; Zhu, C.; Tang, C.S.; Xu, H.; Shi, X.; Mark, V. 3D characterization of desiccation cracking in clayey soils using a structured light scanner. *Eng. Geol.* **2022**, *299*, 106566.

encyclopedia

Review

Nonlocal Elasticity for Nanostructures: A Review of Recent Achievements

Raffaele Barretta *, **Francesco Marotti de Sciarra and Marzia Sara Vaccaro**

Department of Structures for Engineering and Architecture, University of Naples Federico II, via Claudio 21, 80125 Naples, Italy
* Correspondence: rabarret@unina.it

Abstract: Recent developments in modeling and analysis of nanostructures are illustrated and discussed in this paper. Starting with the early theories of nonlocal elastic continua, a thorough investigation of continuum nano-mechanics is provided. Two-phase local/nonlocal models are shown as possible theories to recover consistency of the strain-driven purely integral theory, provided that the mixture parameter is not vanishing. Ground-breaking nonlocal methodologies based on the well-posed stress-driven formulation are shown and commented upon as effective strategies to capture scale-dependent mechanical behaviors. Static and dynamic problems of nanostructures are investigated, ranging from higher-order and curved nanobeams to nanoplates. Geometrically nonlinear problems of small-scale inflected structures undergoing large configuration changes are addressed in the framework of integral elasticity. Nonlocal methodologies for modeling and analysis of structural assemblages as well as of nanobeams laying on nanofoundations are illustrated along with benchmark applicative examples.

Keywords: nonlocal continuum mechanics; nanostructures; integral elasticity

Citation: Barretta, R.; Marotti de Sciarra, F.; Vaccaro, M.S. Nonlocal Elasticity for Nanostructures: A Review of Recent Achievements. *Encyclopedia* **2023**, *3*, 279–310. https://doi.org/10.3390/encyclopedia3010018

Academic Editor: Stefano Falcinelli

Received: 1 February 2023
Revised: 16 February 2023
Accepted: 22 February 2023
Published: 27 February 2023

1. Introduction

According to traditional ideas in continuum mechanics, constitutive equations are those intrinsic relations providing response variables at a material point of a continuum as functions of variables assessed at the same point. Thus, classical constitutive laws support the axiom of local action, stating that response variables at a material point are not affected by the state of the continuum at distant material points. However, in determining the application field of local continuum mechanics, notion of length scale plays a crucial role. Indeed, if a continuum's external characteristic length (i.e., its structural dimension or wavelength) is significantly greater than its internal characteristic length (its interatomic distance or the size of its heterogeneities), then classical constitutive laws can accurately predict the outcome. In contrast, local theories are unable to capture the effective mechanical behavior if the external and internal characteristic lengths are comparable, and nonlocality thus becomes necessary to account for long-range interaction forces. According to nonlocal continuum field theories, constitutive response at a material point of a continuum depends on the state of all points and is thus characterized by response functionals [1,2].

Nowadays, modeling and optimization of smaller and smaller smart devices represent one of the most promising fields of application of nonlocal continuum mechanics due to the growing interest in nanoscience and nanotechnology. The development of mathematical tools able to capture size effects in small-scale structures has been pushed by the increasing attention to miniaturized electromechanical devices, with several potential applications in engineering science. In this regard, the main purpose consists of conceiving effective and computationally efficient methodologies to model size-dependent behavior and design small-scale structures exploiting unconventional tools provided by nonlocal continuum mechanics, rather than time-consuming atomistic approaches [3–5].

From the mathematical point of view, nonlocal theories provide enriched constitutive laws that are not pointwise and in which long distance interactions are described by internal characteristic lengths. In [6,7], Eringen developed one of the first theories of nonlocal integral elasticity, according to which stress is the convolution integral between the elastic strain field and a proper averaging kernel governed by an internal characteristic length. Such an integral theory is thus referred to as a strain-driven nonlocal theory, as proposed in [8]. Eringen's constitutive law has been efficiently adopted to solve screw dislocation and surface wave problems, but it turned out to be inconsistent when applied to structural problems due to an incompatibility between the constitutive law and equilibrium condition. Application of the strain-driven nonlocal model to structural mechanics led to alleged paradoxical results, as detected in [9,10] and definitely clarified by [11].

In order to address the issues related to the strain-driven nonlocal theory, several formulations have been conceived in recent years. Among these improved elasticity formulations, two-phase (local and nonlocal) mixture models stand out as a useful tool to overcome the ill-posedness of Eringen's theory and effectively capture scale-dependent mechanical behaviors. A two-phase model based on a convex combination of local and strain-driven integral responses was first proposed by Eringen in [12,13]. The mixture theory of elasticity was then restored in [14,15] to formulate well-posed structural problems, assuming that the local fraction of the two-phase law was not vanishing [11]. An alternative mixture theory to bypass difficulties of the strain-driven purely nonlocal law was proposed in [16], namely the strain difference-based nonlocal model of elasticity.

To account for scale effects in nanostructures, other possible theories assume that constitutive responses depend on both elastic strain fields and higher-order gradient strain fields. Eringen's differential law and the strain gradient model of elasticity were interestingly combined by Aifantis in [17–19]. Lim et al. coupled Eringen's nonlocal law with the strain gradient elasticity [20] to address wave propagation problems in unbounded domains, leading to a higher-order differential constitutive equation. In the framework of structural mechanics, the necessity of constitutive boundary conditions associated with the differential formulation of nonlocal gradient elasticity was inferred by Barretta and Marotti de Sciarra in [21,22], where the relevant differential constitutive problem was definitely established. The theory of elastic material surfaces conceived by Gurtin and Murdoch in [23] is another important tool for the modeling of nano-mechanical behaviors. In this framework, a combination of nonlocal integral elasticity and surface elasticity was recently provided in [24] to assess the size-dependent mechanics of nanostructures. Nonlocal mathematical models have also been proposed to capture non-conventional phenomena, such as electric polarization in ferroelectric materials [25], and to address diffusion problems in heterogeneous structures [26].

A total remedy to the issues related to the strain-driven nonlocal elasticity was definitely overcome by the stress-driven integral formulation conceived by Romano and Barretta in [27]. According to this theory, size-dependent mechanical behaviors can be modeled by a new nonlocal elastic law based on a stress-driven formulation that provides a consistent approach inside the integral elasticity framework. The nonlocal elastic strain field at a point of a continuum is given as convolution integral between the local elastic strain and a scalar averaging kernel. The relevant continuum problem is well posed, and size effects due to long range interactions can be effectively captured [28–31]. In [32], the stress-driven nonlocal elasticity was generalized to a two-phase (local/nonlocal) model that is able to capture both softening and stiffening elastic responses. Moreover, the mixture methodology based on the stress-driven approach is well posed for any local fraction. This theory has been successfully applied in recent contributions, such as in [33–35].

Nowadays, growing attention is being paid to modeling and design of ultra-small structural systems exploiting consistent methodologies of nonlocal continuum mechanics. A review on the topic was contributed in [36], in which nano-mechanical behavior is investigated by means of stress-driven-based formulations of nonlocal elasticity. A brief overview can be found in [37], where a collection of works concerning applications of nonlocal theories is provided with a main reference to stress-driven formulations. In the present treatment, a comprehensive overview of new developments and outcomes in the framework of nonlocal continuum mechanics applied to nanostructures is provided. Starting from early formulations of nonlocal mechanics, recent theories of integral elasticity are illustrated and exploited to solve challenging problems of current nanotechnological interest. Innovative nonlocal methodologies to solve complex nanosystems are illustrated. A consistent approach to model nanobeams on nonlocal foundations is finally examined.

2. Eringen's Theory of Integral Elasticity

Pioneering works on nonlocal continuum mechanics were first contributed during the 20th century by Rogula [1,38], Kröner [2], Krumhansl [39] and Kunin [40]. Then, Eringen conceived an integral model of elasticity which was effectively applied to solve Rayleigh wave propagation and screw dislocation problems [6,7]. With reference to a three-dimensional continuum, Eringen's formulation is based on the idea that stress $\boldsymbol{\sigma}$ at a point $\boldsymbol{x}$ is output of a convolution integral between the elastic response to the local strain field $\boldsymbol{\varepsilon}^{el}$ and a proper attenuation function ϕ_λ described by a characteristic parameter $\lambda > 0$. Id est,

$$\boldsymbol{\sigma}(\boldsymbol{x}) = \int_\Omega \phi_\lambda(\boldsymbol{x} - \bar{\boldsymbol{x}}) \boldsymbol{E}(\bar{\boldsymbol{x}}) \boldsymbol{\varepsilon}^{el}(\bar{\boldsymbol{x}}) \, d\Omega_{\bar{\boldsymbol{x}}} \, , \tag{1}$$

where $\boldsymbol{E}$ is the local elastic stiffness tensor field and $\boldsymbol{x}$ and $\bar{\boldsymbol{x}}$ are position vectors. The constitutive law in Equation (1) is referred to as strain-driven model since $\boldsymbol{\varepsilon}^{el}$ is the source field, and specifically, it represents a Fredholm integral equation of the first kind in the unknown elastic strain field. The symbol Ω adopted in Equation (1) denotes the actual body placement, and $d\Omega_{\bar{\boldsymbol{x}}}$ denotes that integration over Ω is performed with respect to the variable $\bar{\boldsymbol{x}}$.

In order to formulate the relevant nonlocal elastic problem of equilibrium, the stress field $\boldsymbol{\sigma}$ must satisfy the equilibrium differential equation, and the total strain field $\boldsymbol{\varepsilon}$, which is the sum of the elastic $\boldsymbol{\varepsilon}^{el}$ and non-elastic $\boldsymbol{\varepsilon}^{nel}$ strain fields, must fulfill the kinematic compatibility requirement. Hence, the integro-differential formulation based on Equation (1) is obtained as follows:

$$\begin{cases} div\, \boldsymbol{\sigma}(\boldsymbol{x}) = -\mathbf{b}(\boldsymbol{x}) \, , \\ \boldsymbol{\sigma}(\boldsymbol{x}) = \int_\Omega \phi_\lambda(\boldsymbol{x} - \bar{\boldsymbol{x}}) \boldsymbol{E}(\bar{\boldsymbol{x}}) \boldsymbol{\varepsilon}^{el}(\bar{\boldsymbol{x}}) \, d\Omega_{\bar{\boldsymbol{x}}} \, , \\ \boldsymbol{\varepsilon}(\boldsymbol{x}) = sym\, \nabla \mathbf{u}(\boldsymbol{x}) \, , \end{cases} \tag{2}$$

where $\boldsymbol{x} \in \Omega$, $\mathbf{u}$ is the displacement field and $\mathbf{b}$ represents the body forces. It is worth noting that when formulating structural problems of nonlocal continua, equilibrium and kinematics do not depend on the scale under consideration. Indeed, it only affects the constitutive law that must properly account for the size effects. The averaging kernel ϕ_λ appearing in the integral constitutive law in Equation (1) is a scalar attenuation function described by a non-dimensional parameter $\lambda > 0$. Its physical dimension is $[L^{-3}]$, and the nonlocal parameter is defined as the ratio between the internal and external characteristic lengths. Moreover, the classical constitutive law of local elasticity can be recovered from Equation (1) for $\lambda \to 0^+$ at the internal points of Ω. A comprehensive analysis of the limit behaviors can be found in [41].

Let us start by applying Eringen's integral law to a one-dimensional model. Specifically, let us consider a Bernoulli–Euler beam of length L and take the abscissa x along the beam axis. By denoting with M the bending interaction field and with χ^{el} the elastic flexural curvature field, application of Eringen's theory leads to the following nonlocal elastic constitutive relation [42], id est

$$M(x) = \int_0^L \phi_\lambda(x - \bar{x})\,(k_f\,\chi^{el})(\bar{x})d\bar{x}\,, \tag{3}$$

where $k_f := I_E$ is the local elastic bending stiffness, (i.e., the second moment of the Euler–Young modulus field E on the beam cross-section). It is worth noting that the unknown elastic curvature is implicitly defined by Equation (3), since it is input of Eringen's convolution integral. As will be discussed in the following, solution of the integral Equation (3) in terms of elastic flexural curvature χ^{el} may not exist for an assigned output, provided by the bending interaction field fulfilling differential and boundary conditions of equilibrium.

2.1. Averaging Kernel and Green's Function

The attenuation kernel ϕ_λ can be chosen among Gaussian, exponential or power law function types. As suggested in [7], proper averaging kernels are represented by the Helmholtz bi-exponential function:

$$\phi_\lambda(x) = \frac{1}{2c}\exp\left(-\frac{|x|}{c}\right), \tag{4}$$

and by the normal distribution:

$$\phi_\lambda^{err}(x) = \frac{1}{c\sqrt{\pi}}\exp\left(-\frac{x^2}{c^2}\right), \tag{5}$$

where λ is defined as the ratio $\frac{c}{L}$, being c the internal characteristic length. Since the choice of different kernels leads to technically coincidental results [43], adoption of the bi-exponential function in Equation (4), as proposed by Eringen in [7], does not affect the generality but is more convenient from a theoretical and computational point of view. Indeed, the averaging function in Equation (4) is also called special kernel since it satisfies peculiar properties [11] that enable inversion of the constitutive integral equation.

Notably, the Helmholtz kernel $\phi_\lambda : \Re \to [0, +\infty[$ satisfies the following properties on the real axis:

- Symmetry and positivity

$$\phi_\lambda(x - \bar{x}) = \phi_\lambda(\bar{x} - x) \geq 0 \tag{6}$$

- Normalization

$$\int_{-\infty}^{+\infty} \phi_\lambda(x)\,dx = 1 \tag{7}$$

- Limit impulsivity

$$\lim_{\lambda \to 0^+} \int_{-\infty}^{+\infty} \phi_\lambda(x - \bar{x})\,f(\bar{x})d\bar{x} = f(x) \tag{8}$$

for any continuous function $f : \Re \to \Re$.

As proven in [27], the following proposition holds:

Proposition 1. *The Helmholtz kernel ϕ_λ is Green's function associated with the differential operator $1 - c^2\,\partial_x^2$ and satisfying the boundary conditions*

$$\begin{cases} \partial_x\phi_\lambda(0) = \frac{1}{c}\phi_\lambda(0)\,, \\ \partial_x\phi_\lambda(L) = -\frac{1}{c}\phi_\lambda(L)\,, \end{cases} \tag{9}$$

where symbol ∂_x in Equation (9) denotes derivative along the x axis. For decreasing values of λ, the kernel's support reduces, and the peak of the function increases while, for increasing λ values, more and more abscissae of the domain are involved in the convolution integral, thus increasing the propagation of long-range interactions. It is worth noting that the limit impulsivity property in Equation (8) is defined on the real axis. For bounded domains (e.g., the interval $[a,b]$ with $a,b \in \Re$), the property can be reformulated as shown below. First, let us introduce the following function:

$$\Theta(x) = \begin{cases} 1, & x \in]a,b[\,, \\ 1/2, & x = \{a,b\}\,. \end{cases} \tag{10}$$

Then, the impulsivity property in a bounded domain can be rewritten as

$$\lim_{\lambda \to 0^+} \int_a^b \phi_\lambda(x-\bar{x})\, f(\bar{x}) d\bar{x} = \Theta(x)\, f(x) \tag{11}$$

for any continuous real scalar function f. Equation (11) states that the Helmholtz function for $\lambda \to 0^+$ is the Dirac delta or the halved Dirac delta if the abscissa of evaluation is internal or external to the domain [11]; that is, for $x \in]a,b[$, in the limit, the source field $f(x)$ is found as the output of the convolution integral, while for $x = \{a,b\}$, we find the halved response $f(x)/2$. This boundary effect is caused by the fact that the kernel's support exceeds the structural domain at the exterior boundary. Therefore, in the limit for $\lambda \to 0^+$, the local constitutive law of elasticity is recovered at the interior abscissae.

As stated before, the Helmholtz function ϕ_λ satisfies the peculiar properties that enable inversion of the integral law to obtain an equivalent differential problem of Eringen's model in Equation (3), which is expressed by the following proposition:

Proposition 2. *For any $\lambda > 0$, the integral constitutive law*

$$M(x) = \int_a^b \phi_\lambda(x-\bar{x})\, (k_f\, \chi^{el})(\bar{x}) d\bar{x} \tag{12}$$

equipped with the bi-exponential kernel in Equation (4) admits either a unique solution or no solution at all, depending on whether or not the following constitutive boundary conditions are satisfied by the equilibrated bending interaction field

$$\begin{cases} \partial_x M(a) = \dfrac{1}{c} M(a)\,, \\ \partial_x M(b) = -\dfrac{1}{c} M(b)\,. \end{cases} \tag{13}$$

If Equation (13) is fulfilled by the equilibrated bending interaction field, then the unique solution χ^{el} is obtained from the second order differential equation

$$M(x) - c^2\, \partial_x^2 M(x) = k_f(x)\, \chi^{el}(x)\,. \tag{14}$$

Therefore, an equivalent differential problem is provided to explicitly find the unknown field χ^{el}. Moreover, the differential formulation represents a direct tool to detect if the Fredholm integral equation of the first kind represented by Equation (12) admits solution for an assigned output field M fulfilling the equilibrium requirements. Indeed, a unique elastic curvature χ^{el} can be found if and only if the constitutive boundary conditions in Equation (13) are fulfilled by the equilibrated bending interaction field. Otherwise, no solution exists.

Several scientific contributions have been proposed by adopting the constitutive differential law in Equation (14) without prescription of constitutive boundary conditions in Equation (13), whose necessity was first inferred in [42,44]. Paradoxical cases have been detected by applying Eringen's differential law to structural problems. The question has been definitely clarified in [45], where inconsistency of the strain-driven nonlocal formulation was shown. The constitutive boundary conditions in Equation (13) indeed relate the bending and shearing interaction fields at the boundary for any value of the characteristic length, and thus they are, in general, in contrast with the natural boundary conditions. Thus, there is a lack of solution existence due to the fact that the integral constitutive law is incompatible with equilibrium requirements.

2.2. The Alleged Paradox of the Nonlocal Elastic Cantilever

Eringen's differential law in Equation (14) has been widely applied to structural problems, disregarding the prescription of boundary conditions in Equation (13). Accordingly, application of Eringen's differential model to small-scale cantilevers under concentrated loading at the free end led to an alleged paradoxical case, as first detected by Peddieson in [9] and then by Challamel in [10]. After the discussions provided in [42,44], explanation of the paradoxical results was finally proposed in [45]. The alleged paradoxical case is discussed here. For the structural problem under consideration, the bending interaction field is univocally determined by differential and boundary conditions of equilibrium; that is, $M(x) = \mathcal{F}(L-x)$. By applying Eringen's differential law in Equation (14), we obtain

$$M(x) = k_f(x)\,\chi^{el}(x)\,, \tag{15}$$

that is, the nonlocal constitutive model collapses into the local law, and accordingly, no size effects arise. Moreover, accounting for Equation (15) in the strain-driven convolution integral in Equation (12) leads to

$$M(x) = \int_0^L \phi_\lambda(x-\bar{x})\,\mathcal{F}(L-\bar{x})d\bar{x}\,. \tag{16}$$

The output of Equation (16) provides the following bending interaction that is a nonlinear field:

$$M(x) = \frac{1}{2}\mathcal{F}L\left(\lambda\exp\left(\frac{x-L}{c}\right) - (1+\lambda)\exp\left(-\frac{x}{c}\right) + 2\left(1-\frac{x}{L}\right)\right) \tag{17}$$

which does not fulfill the differential and boundary conditions of equilibrium. Such a result has been considered a paradoxical case, but the proper interpretation is that the strain-driven integral law does not have solution. Indeed, according to Proposition 2, the equilibrated bending interaction field cannot fulfill the constitutive boundary conditions shown below:

$$\begin{cases} -\mathcal{F} = \dfrac{1}{c}\mathcal{F}L\,, \\ -\mathcal{F} = 0\,. \end{cases} \tag{18}$$

As emerged from Equation (18), the boundary conditions can be satisfied only for a vanishing applied concentrated loading that is of no applicative interest. Therefore, the relevant nonlocal problem is ill-posed due to the incompatibility between the constitutive law and equilibrium conditions.

Figures 1–3 represent the bending and shearing interactions fields and the emerging loading fields, respectively, revealing that the equilibrium requirements are clearly violated. This pathological behavior concerns all nonlocal structural problems of applicative interest, as shown in [45].

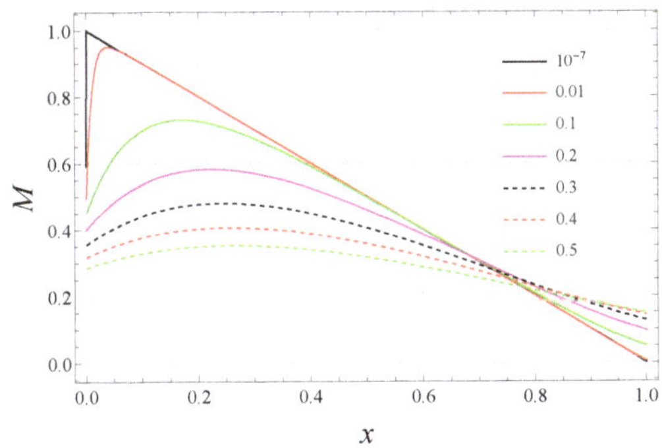

Figure 1. Bending interaction field as function of λ.

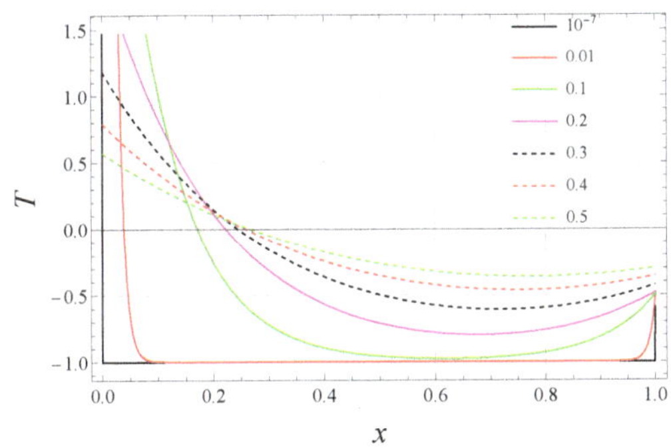

Figure 2. Shearing interaction field as function of λ.

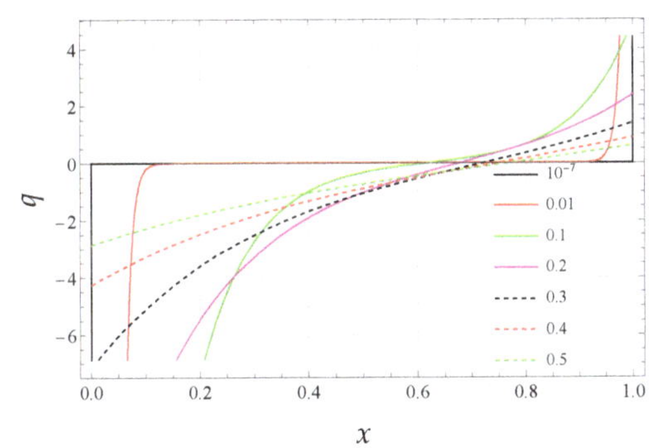

Figure 3. Emerging loading field as function of λ.

3. Strain-Driven Nonlocal Methodologies

To overcome the issues emerging from the application of the strain-driven model, a mixture model of elasticity was proposed by Eringen in [12] and then restored in [14,15]. Eringen's two-phase model was proposed in [46] for buckling analysis of slender beams, free vibration analyses of Euler–Bernoulli curved beams have been performed in [47], post-buckling of viscoelastic nanobeams has been analysed in [48], bending problem of two-phase elastic structures has been addressed in [49], mixture nonlocal integral theories have been adopted in [50] for functionally graded Timoshenko beams.

With reference to an inflected beam, Eringen's two-phase theory expresses the bending interaction field as a convex combination of the local and nonlocal responses by means of a mixture parameter. Since the mixture model is a two-parameter theory, it provides a generalization of the purely strain-driven nonlocal model and thus is able to capture the size-dependent behavior of a wide class of applications. According to the two-phase strain-driven theory, the bending interaction field M is a convex combination of the local response $s := k_f\,\chi^{el}$ and the convolution integral between the local source field s and the attenuation function ϕ_λ, id est

$$M(x) = \alpha\, s(x) + (1-\alpha) \int_0^L \phi_\lambda(x-\xi)\, s(\xi)\, d\xi\,, \tag{19}$$

where α is the mixture parameter $0 \le \alpha \le 1$ and $\lambda > 0$ is the nonlocal parameter describing the averaging kernel. For $\alpha = 0$, the purely nonlocal response is got $M(x) = \int_0^L \phi_\lambda(x-\xi)\,(k_f\,\chi^{el})(\xi)\,d\xi$ while for $\alpha = 1$, the local case $M = k_f\,\chi^{el}$ is recovered. It is worth noting that Equation (19) represents a Fredholm integral equation of the second kind in the unknown source field s (see [51,52]).

If the special kernel is adopted, then an equivalent differential problem can be derived [11]:

$$\frac{M(x)}{c^2} - \partial_x^2 M(x) = \frac{(k_f\,\chi^{el})(x)}{c^2} - \alpha\,\partial_x^2(k_f\,\chi^{el})(x)\,, \tag{20}$$

$$\begin{cases} \partial_x M(0) - \dfrac{1}{c}\,M(0) = \alpha\left(\partial_x(k_f\,\chi^{el})(0) - \dfrac{(k_f\,\chi^{el})(0)}{c}\right), \\ \partial_x M(L) + \dfrac{1}{c}\,M(L) = \alpha\left(\partial_x(k_f\,\chi^{el})(L) + \dfrac{(k_f\,\chi^{el})(L)}{c}\right). \end{cases} \tag{21}$$

The strain-driven two-phase model provides a well-posed structural problem only for a strictly positive mixture parameter $\alpha > 0$, while for $\alpha = 0$, the purely integral Eringen's law is recovered.

An alternative theory to bypass the ill-posedness of the strain-driven fully integral formulation was proposed in [16], namely the strain difference-based nonlocal elasticity model. This constitutive theory was conceived on the basis of the so-called "locality recovery condition"; that is, the nonlocal stress response is required to be uniform if the source local strain is uniform. A critical analysis of locality recovery can be found in [53]. Fulfillment of the above-mentioned requirement leads to the following constitutive local/nonlocal integral law:

$$M(x) = \Big(1-\gamma(x)\Big)\,(k_f\,\chi^{el})(x) + \int_0^L \phi_\lambda(x-\xi)\,(k_f\,\chi^{el})(\xi)\,d\xi\,, \tag{22}$$

where the function γ is defined as the integral of the attenuation kernel over the domain (i.e., $\gamma(x) := \int_0^L \phi_\lambda(x-\xi)\,d\xi$) such that Equation (22) can also be rewritten as follows:

$$M(x) = (k_f\,\chi^{el})(x) + \int_0^L \phi_\lambda(x-\xi)\,\Big((k_f\,\chi^{el})(\xi) - (k_f\,\chi^{el})(x)\Big)\,d\xi\,. \tag{23}$$

The latter form clarifies the strain difference appellation given to this elasticity theory. It is worth noting that for decreasing λ values, the function $1-\gamma(x)$ is vanishing in a non-empty core domain such that the model tends to collapse into the purely nonlocal strain-driven theory. A correction to this drawback was proposed in [54] by modifying Equation (22) as follows:

$$M(x) = \beta(x)\,(k_f\,\chi^{el})(x) + \int_0^L \phi_\lambda(x-\xi)\,(k_f\,\chi^{el})(\xi)\,d\xi\,, \tag{24}$$

where $\beta(x) := e + 1 - \gamma(x)$ being $e \ll 1$ a constant corrective factor. Equation (24) still represents a Fredholm integral equation of the second kind in the unknown field of the elastic curvature.

Among the strain-driven nonlocal methodologies, a recent contribution was provided in [55], in which the differential law in Equation (14) was reconsidered to predict the size effects in small-scale beams. It is important to underline that the differential law, without prescription of the constitutive boundary conditions, is not equivalent to Eringen's integral nonlocal theory. Thus, the differential model provided by Equation (14) represents a local constitutive law of elasticity that differs from the classical one for the term $-c^2\,\partial_x^2 M(x)$, which accounts for size effects and can be rewritten as $-c^2\,q(x)$ by exploiting the differential equilibrium condition of the Bernoulli–Euler beam theory. Equation (14) can be rewritten as $M(x) = k_f(\chi(x) - \chi^{nel}(x))$, where the term taking into account the size effects $\chi^{nel}(x) := -\frac{c^2}{k_f}\,q(x)$ is interpreted as a fictitious non-elastic curvature. This latter term was modified in [55] by including both reactive and active distributed and concentrated loadings, id est

$$\chi^{nel}(x) := -\frac{c^2}{k_f}\left(q(x) + \sum_{i=1}^{n} \mathcal{F}_i\,\delta(x - x_i) + R_0\,\delta(x) + R_L\,\delta(x - L)\right) \tag{25}$$

where $\mathcal{F}_i$ is the ith concentrated force, R_0, R_L represents the reactive forces and n is the number of concentrated loadings. The proposed differential model may not be able to capture long-range interactions in general structural problems. An exemplar case is provided by a simple cantilever under a concentrated couple, for which the fictitious non-elastic curvature is zero and thus no size effects arise.

4. The Stress-Driven Nonlocal Model

To completely overcome the ill-posedness of strain-driven formulations, a consistent theory was proposed in [27], obtained by swapping the roles of stress and elastic strain in Equation (1) to find a stress-driven integral law which is not the inverse of the previous one. The stress-driven methodology has been effectively applied to study the scale effects in axisymmetric nanoplates [56], to investigate size-dependent bending problems [57], to examine the buckling of nanobeams [58], to address vibration problems in nanorods [59], to perform nonlinear dynamic analyses of functionally graded porous nanobeams [60], to model carbon nanotubes conveying magnetic nanoflow [61] and to analyze the size-dependent buckling problems of cracked micro- and nano-cantilevers [62].

With reference to a three-dimensional continuum, the stress-driven nonlocal model is written as follows:

$$\boldsymbol{\varepsilon}^{el}(\boldsymbol{x}) = \int_{\Omega} \phi_\lambda(\boldsymbol{x} - \bar{\boldsymbol{x}})\boldsymbol{C}(\bar{\boldsymbol{x}})\boldsymbol{\sigma}(\bar{\boldsymbol{x}})\,d\Omega_{\bar{\boldsymbol{x}}}\,, \tag{26}$$

where $\boldsymbol{C} := \boldsymbol{E}^{-1}$ is the elastic compliance tensor field. The constitutive integral Equation (26) explicitly expresses the elastic strain $\boldsymbol{\varepsilon}^{el}$ as a convolution integral between the averaging kernel and the stress field $\boldsymbol{\sigma}$. The stress field $\boldsymbol{\sigma}$ must satisfy the equilibrium requirements, while the geometric strain field $\boldsymbol{\varepsilon}$, sum of the elastic and non-elastic strain fields, must fulfill the kinematic compatibility requirement.

Applied to a slender beam of length $L := b - a$, the integral law in Equation (26) is written as follows:

$$\chi^{el}(x) = \int_a^b \phi_\lambda(x - \bar{x})\,(k_f^{-1}\,M)(\bar{x})d\bar{x}\,, \tag{27}$$

where k_f^{-1} is the elastic flexural compliance. It is useful to note that according to the strain-driven model in Equation (12), the bending interaction field is expressed as

$$M(x) = \left(\phi_\lambda * (k_f\,\chi^{el})\right)(x)\,, \tag{28}$$

where $\phi_\lambda *$ is the linear convolution operator. Thus, the strain-driven law in Equation (28) provides an implicit definition of the unknown elastic curvature field χ^{el}, and it represents a Fredholm integral equation of the first kind, which is known to generally lead to no solution at all or to multiple solutions. Conversely, in the stress-driven formulation, the operator $\phi_\lambda *$ is directly applied to the equilibrated bending interaction field to explicitly find the nonlocal elastic curvature:

$$\chi^{el}(x) = \left(\phi_\lambda * (k_f^{-1} M)\right)(x) . \tag{29}$$

By virtue of the properties of the special kernel in Equation (4), the following equivalence can be stated. The nonlocal elastic curvature found by the stress-driven convolution integral

$$\chi^{el}(x) = \int_a^b \phi_\lambda(x-\bar{x})\,(k_f^{-1} M)(\bar{x})d\bar{x} \tag{30}$$

provides the unique solution of the differential problem made of the second-order differential equation

$$\chi^{el}(x) - c^2 \partial_x^2 \chi^{el}(x) = (k_f^{-1} M)(x) , \tag{31}$$

equipped with the constitutive boundary conditions

$$\begin{cases} \partial_x \chi^{el}(a) = \dfrac{1}{c}\chi^{el}(a) , \\ \partial_x \chi^{el}(b) = -\dfrac{1}{c}\chi^{el}(b) . \end{cases} \tag{32}$$

The solution in terms of the elastic curvature is unique, since the corresponding homogeneous differential problem admits only the trivial solution. According to the stress-driven nonlocal model, the relevant structural problem is well-posed since no conflict arises with equilibrium conditions. Moreover, there is no incompatibility with the kinematic boundary conditions, since the constitutive boundary conditions in Equation (32) involve only the elastic curvature field and its first derivative. This is a fundamental requirement for a constitutive law, since it must be independent of the prescribed kinematic constraints.

A basic example is provided below concerning a doubly clamped beam under uniformly distributed loading whose size-dependent behavior is modeled by the stress-driven nonlocal approach. Figure 4 shows the maximum transverse displacement adimensionalized with respect to the maximum local response versus the nonlocal parameter. The structural response exhibits a stiffening effect for increasing λ values, in agreement with the outcomes collected in [63].

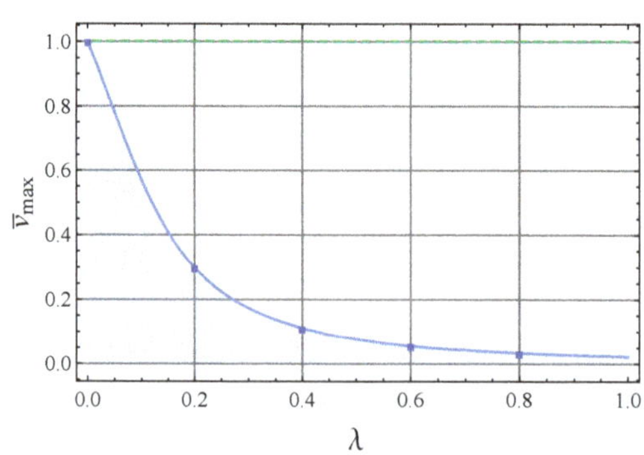

Figure 4. Clamped beam's maximum non-dimensional displacement versus nonlocal parameter.

This peculiar trend is due to the properties of the special kernel since the peak of the attenuation function reduces for increasing nonlocal parameter values, causing a decrease in elastic flexural compliance, which has a predominant effect on the extension of the kernel's support. This feature is known as the "smaller-is-stiffer" phenomenon [64], characterizing a huge class of problems in nanoengineering.

Nonlinear Mechanics of Nonlocal Elastic Beams

Recent achievements in the field of nonlocal elasticity concern the modeling of nanostructures undergoing large configuration changes. Geometrically nonlinear analysis of Timoshenko nanobeams based on the strain-driven approach was carried out in [65], the nonlinear coupled mechanics of composite nanostructures were investigated in [66], the nonlinear dynamics of carbon nanotube-based mass sensors were studied in [67], the nonlinear vibration response of nanobeams was detected in [68] by exploiting Eringen's model, nonlinear dynamic analyses of graphene sheets were carried out in [69], the geometrically nonlinear bending problem of nanobeams was explored in [70], the nonlinear behaviors in nonlocal elastic shells were modeled in [71], the post-buckling of nanotubes was addressed in [72], the nonlinear vibrations of microbeams were examined in [73], nonlinear post-buckling analysis of functionally graded Timoshenko nonlocal beams was carried out in [74], large deflection analysis of a nano-cantilever on a Winkler–Pasternak foundation was performed in [75], and thermal buckling of piezomagnetic small-scale beams was studied in [76].

In [77], the stress-driven nonlocal elasticity was efficiently applied to capture size effects in small-scale beams undergoing large configuration changes. Based on the outcomes contributed in [77], a consistent stress-driven methodology of integral elasticity is shown in this section to address applicative problems of small-scale beams characterized by geometrically nonlinear behaviors. For this purpose, let us consider a nonlocal elastic cantilever of length L whose initial configuration is characterized by a null geometric flexural curvature. The beam is subjected to a concentrated loading $\mathcal{F}$ at the free end [77], and it is parameterized as $\mathbf{r} = \{x(s), y(s)\} = \{x, y(x)\}$, where $\mathbf{r}$ is the position vector and $s \in [0, L]$ is the curvilinear abscissa. The field of the tangent unit vectors is obtained as $\mathbf{t} := \partial_s \mathbf{r}$, and its derivative $\partial_s \mathbf{t}$ is the curvature vector, whose modulus provides the exact geometric curvature, id est

$$\chi(x) = \frac{\partial_x^2 y(x)}{\left[1 + \left(\partial_x y(x)\right)^2\right]^{3/2}}, \tag{33}$$

which is coincident with the elastic one, $\chi = \chi^{el}$, since inelastic effects are not considered. Exploiting the stress-driven theory, the flexural curvature can be got as

$$\chi(x) = \int_0^l \phi_\lambda(x - \xi) \left(\frac{M}{k_f}\right)(\xi)\, d\xi, \tag{34}$$

where l is the orthogonal projection of the end point along the x axis. In Equation (34), $M = \mathcal{F}(l - x)$ is the linear bending interaction field satisfying the equilibrium requirements. By making the change of variable $z := \partial_x y$ and integrating Equation (33), we obtain

$$G(x) = \int_0^x \chi(\eta) d\eta, \tag{35}$$

where the position $G := \dfrac{z}{\sqrt{1+z^2}}$ has been done and the boundary condition $z(0) = 0$ has been prescribed. Taking into account $G := \partial_s y$ and $|\mathbf{t}| = \sqrt{(\partial_s x)^2 + G^2} = 1$, we finally obtain the differential equation in the unknown function $s(x)$, that is

$$\partial_x s(x) = \frac{1}{\sqrt{1 - G^2(x)}}, \text{ with } s(0) = 0. \tag{36}$$

Finally, the unknown length l is obtained by prescribing $s(l) = L$, and the current beam configuration $y(x)$ is got from

$$\partial_x y(x) = \frac{G(x)}{\sqrt{1 - G^2(x)}}\,, \text{ with } y(0) = 0\,. \tag{37}$$

The nonlinear problem of nonlocal elastic equilibrium described above is solved by performing an iterative solution procedure, which provides a generalization of the contribution [78] to the framework of integral elasticity. Starting from a trial projection length, l is updated at each step until the condition $s(l) = L$ is satisfied. The beam length is assumed to be $L = 100\,[\mu m]$ with a rectangular cross-section of height $5\,[\mu m]$ and base $3.5\,[\mu m]$. The Euler–Young modulus is $E = 80\,[GPa]$. Parametric analyses are performed to investigate the influence of nonlocal parameter λ and applied force $\mathcal{F}$. The results are represented in Figure 5, showing that structural responses exhibited a stiffening behavior for increasing λ values.

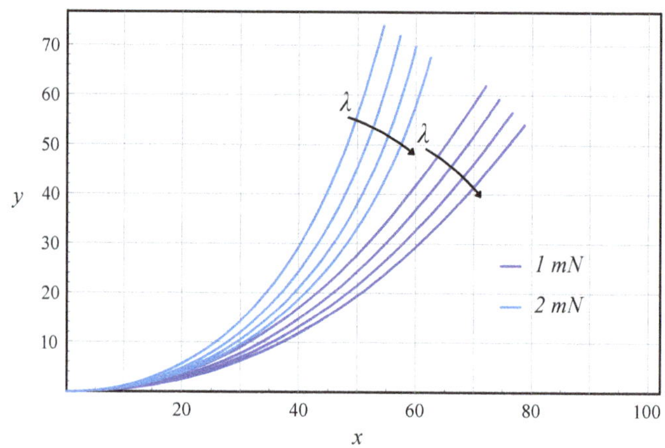

Figure 5. Current configuration $y\,[\mu m]$ of beam axis versus $x\,[\mu m]$ for $\lambda = \{0.05, 0.10, 0.15, 0.20\}$.

5. A Nonlocal Methodology for Shear Deformation Beam Theories

The stress-driven integral elasticity can be exploited to capture size-dependent behaviors of nonlocal beams whose kinematics is modeled by adopting higher-order kinematic theories [79–82]. By making reference to a three-dimensional Cauchy continuum shaped as a right prism of length L with cross-section Ω, the following vector field describes the kinematics according to the third-order theory [83,84]

$$\mathbf{u}(x,y,z) = \Big[- y\,\varphi(x) - a\,y^3 \Big(w'(x) - \varphi(x) \Big) \Big]\mathbf{i} + w(x)\mathbf{j} \tag{38}$$

where $\mathbf{i}$ is the unit vector directed along the x axis (the locus of the cross-sectional geometric centroids), while the y and z axes define the cross-sectional plane and are identified by the unit vectors $\mathbf{j}$ and $\mathbf{k} := \mathbf{i} \times \mathbf{j}$. In Equation (38), $w(x)$ is the transverse displacement field, and $\varphi(x)$ is defined such that $\varphi := -\left.\frac{\partial u_x}{\partial y}\right|_{y=0}$, while the coefficient a is $4/(3h^2)$, where h is the maximum cross-sectional dimension along the y axis. Starting from Equation (38), the following tangent deformation fields can be derived for the third-order beam: $\{\bar{\varepsilon}, \bar{\gamma}, \bar{\bar{\varepsilon}}, \bar{\bar{\gamma}}\} = \{\varphi', w' - \varphi, a\bar{\gamma}', -b\bar{\gamma}\}$, where $b = 3a$. Then, prescription of the variational equilibrium condition [85] leads to the differential system

$$\begin{cases} b\,R'(x) - Q'(x) + a\,P''(x) = q(x)\,, \\ a\,P'(x) - M'(x) + b\,R(x) - Q(x) = 0\,, \end{cases} \tag{39}$$

equipped with the boundary conditions

$$\begin{cases} (Q - b\,R - a\,P')(x_i)\,\delta w(x_i) = (-1)^i\,\mathcal{F}_i\,\delta w(x_i)\,, \\ P(x_i)\,\delta w'(x_i) = (-1)^i\,\mathcal{P}_i\,\delta w'(x_i)\,, \\ (M - a\,P)(x_i)\,\delta\varphi(x_i) = (-1)^i\,\mathcal{M}_i\,\delta\varphi(x_i)\,, \end{cases} \tag{40}$$

where $i = \{1,2\}$, $x_1 := 0$ and $x_2 := L$. In Equation (40), δw, $\delta\varphi$, $\delta w'$ are virtual kinematic fields fulfilling the homogeneous kinematic boundary conditions, $\mathcal{F}_i$, $\mathcal{M}_i$ are concentrated forces and couples, and $\mathcal{P}_i$ represents the higher-order concentrated couples, while in Equation (39), q is the distributed transverse loading. Moreover, M and Q are the bending and shearing interaction fields, respectively, while P and R are the higher-order stress resultants. By following the approach proposed in [27], it can be proven that exploiting the stress-driven integral theory of elasticity, a constitutive system can be derived that is made of two convolution integral laws

$$\begin{cases} \bar{\varepsilon}(x) = \displaystyle\int_0^L \phi_\lambda(x-\xi)\left(\frac{M(\xi)}{I_E} - \frac{a\,I_E^{(4)}\,Q'(\xi)}{\bar{A}_G I_E}\right)d\xi\,, \\ \bar{\gamma}(x) = \displaystyle\int_0^L \phi_\lambda(x-\xi)\,\frac{Q(\xi)}{\bar{A}_G}\,d\xi\,, \end{cases} \tag{41}$$

and two elastic relations involving only stress fields:

$$\begin{cases} P(x) = \dfrac{I_E^{(4)}}{I_E}\left(M(x) - a\,\dfrac{I_E^{(4)}}{\bar{A}_G}Q'(x)\right) + a\,\dfrac{I_E^{(6)}}{\bar{A}_G}Q'(x)\,, \\ R(x) = \dfrac{I_G^{(2)}}{\bar{A}_G}Q(x)\,, \end{cases} \tag{42}$$

where all functional dependencies among the tangent deformation fields have been taken into account. The equivalent differential formulation, which is useful for theoretical and computational purposes, can be found in [43], where an analytical procedure is proposed to obtain well-posed structural problems. A simply supported nanobeam under uniformly distributed loading $q = 5\,[nN/nm]$ is analyzed here. A silicon carbide beam with a Euler–Young modulus $E = 380\,[GPa]$ and Poisson's ratio $\nu = 0.3$ is examined [86]. The beam length is $L = 100\,[nm]$, and a rectangular cross-section is assumed with a height of $0.5\,L$ and base of $30\,[nm]$. From the boundary conditions in Equation (40), the pinned ends require $w(x_i) = P(x_i) = M(x_i) = 0$ for $i = \{1,2\}$.

Parametric solutions in terms of shape functions w and φ are represented in Figures 6 and 7, showing stiffening responses for increasing non-dimensional nonlocal parameter λ. Solutions to the corresponding local structural problem are recovered for $\lambda \to 0^+$. Starting from the general third-order theory, mathematical formulation of the nonlocal structural problem based on the Timoshenko beam model can be recovered as a particular case by setting $a = 0$ in Equation (38). Solutions to the relative elastostatic problem can be found in [87]. It is worth noting that the outcomes contributed in [43] amend the erroneous statements in [88] regarding the ill-posed nature of the structural problem of third-order beams based on the stress-driven nonlocal model.

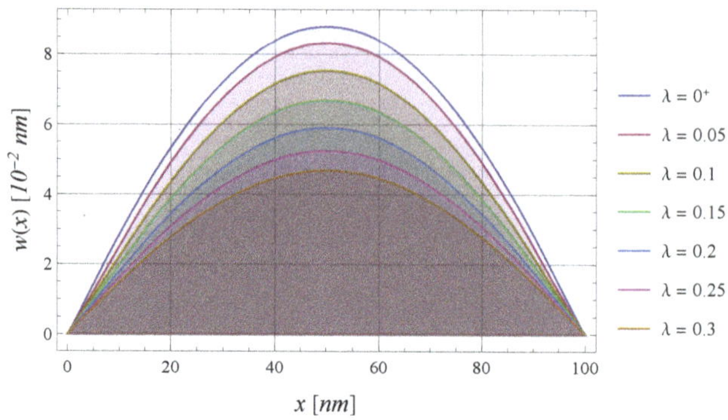

Figure 6. Simply supported beam under uniformly distributed loading: shape function w versus x.

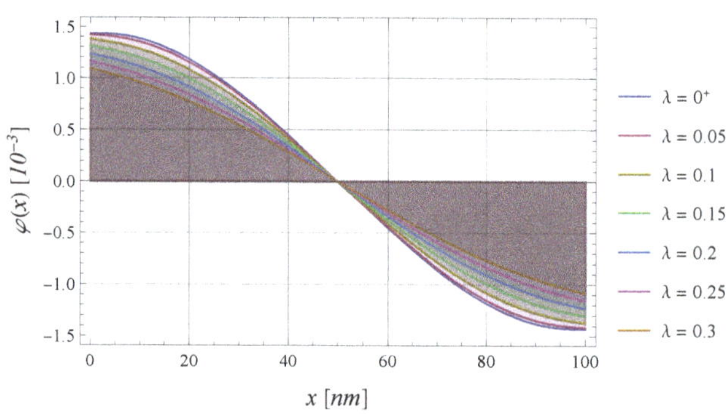

Figure 7. Simply supported beam under uniformly distributed loading: shape function φ versus x.

6. Stress-Driven Two-Phase Elasticity

Two-phase models of elasticity can be formulated by following the approach proposed in [11], providing a generalization of the stress-driven purely integral theory. Such a model leads to well-posed structural continuum problems for any value of the mixture parameter and has been effectively applied to capture the size-dependent behaviors of nanostructures, such as bending of nanobeams [32], fracture analysis of nonlocal continua [89], composite nanostructures in vibration [90], elastostatics of stubby curved nanobeams [33] and size-dependent mechanics of Bernoulli–Euler cracked nanobeams [91]. According to the stress-driven two-phase elasticity, the nonlocal response f is convex combination by means of the mixture parameter α of the local source field s and of the purely stress-driven convolution integral $\phi_\lambda * s$.

6.1. Two-Phase Elasticity for Stubby Curved Beams

The stress-driven two-phase elasticity is illustrated here with reference to curved stubby beams [33], whose kinematics is modeled by the Timoshenko theory. The treatment proposed in [33] provides a generalization of the outcomes contributed in [28] in which nonlocal mechanics of curved slender nanobeams is investigated by exploiting the stress-driven integral model and adopting a coordinate-free formulation. Moreover, particular solutions for vanishing mixture parameter can be found in [34], where Timoshenko curved beams are modeled by exploiting the purely stress-driven integral elasticity. Further contributions on the topic can be found in [92], in which nonlinear mechanics of curved nanobeams is addressed, in [93] where the time-dependent behavior of porous curved nanobeams is modeled, and in [94] where free vibration analysis of curved zigzag nanobeams is conducted.

In order to investigate the mechanics of curved nonlocal continua, the beam axis is assumed to be a regular planar curve $\mathbf{\Gamma}$ parameterized by the curvilinear abscissa $s \in [0, L]$ such that the tangent unit vector field is defined as $\mathbf{t} := \partial_s \mathbf{\Gamma}$. A local coordinate system can thus be introduced as $\{\mathbf{t}, \mathbf{t}_\perp, \mathbf{k}\}$, where $\mathbf{k} := \mathbf{t} \times \mathbf{t}_\perp$ is a uniform unit vector field and $\mathbf{t}_\perp := \mathbf{R}\mathbf{t}$ is the transversal unit vector field obtained by means of the orthogonal tensor $\mathbf{R}$ (performing the rotation with $\pi/2$ counterclockwise in the plane). According to the linearised Timoshenko beam theory, the tangent deformation field is given by $\{\varepsilon, \gamma, \chi\} : [0, L] \mapsto \Re$, which are the axial strain, shear strain and flexural curvature scalar fields, respectively. The kinematic compatibility condition requires that $\{\varepsilon = \partial_s \mathbf{v} \cdot \mathbf{t}, \gamma = \partial_s \mathbf{v} \cdot \mathbf{t}_\perp - \varphi, \chi = d_s \varphi\}$ where $\mathbf{v}$ is the displacement field of the beam axis and φ is the rotation field of the cross-sections. By duality, the stress is given by $\{N, T, M\} : [0, L] \mapsto \Re$ (i.e., the axial force, shear force and bending moment scalar fields, respectively), satisfying the following differential equations of equilibrium:

$$
\begin{cases}
\partial_s N - T\,\mathbf{t}_\perp \cdot \partial_s \mathbf{t} = -\mathbf{p} \cdot \mathbf{t}\,, \\
\partial_s T - N\,\mathbf{t} \cdot \partial_s \mathbf{t}_\perp = -\mathbf{p} \cdot \mathbf{t}_\perp\,, \\
T + \partial_s M = -\mathrm{m}\,,
\end{cases}
\tag{43}
$$

which are equipped with the boundary conditions at $\partial\mathbf{\Gamma}$:

$$
\begin{cases}
-(N\mathbf{t} + T\mathbf{t}_\perp)(0) \cdot \delta\mathbf{v}(0) = \mathbf{F}_0 \cdot \delta\mathbf{v}(0)\,, \\
(N\mathbf{t} + T\mathbf{t}_\perp)(L) \cdot \delta\mathbf{v}(L) = \mathbf{F}_L \cdot \delta\mathbf{v}(L)\,, \\
-M(0)\, \delta\varphi(0) = \mathcal{M}_0\, \delta\varphi(0)\,, \\
M(L)\, \delta\varphi(L) = \mathcal{M}_L\, \delta\varphi(L)\,,
\end{cases}
\tag{44}
$$

where $\{\delta\mathbf{v}, \delta\varphi\}$ represents any virtual displacement and rotation field fulfilling the homogeneous kinematic boundary conditions. The external force system in Equations (43) and (44) is made of distributed vector loading $\mathbf{p} : [0, L] \to V$ and bending couples $\mathrm{m} : [0, L] \to \Re$, boundary concentrated forces $\mathbf{F}_0 \in V$ and $\mathbf{F}_L \in V$ and bending couples $\mathcal{M}_0 \in \Re$ and $\mathcal{M}_L \in \Re$.

The stress-driven mixture elasticity can be formulated by introducing the vectors $\mathbf{i}$ and $\mathbf{f}$ collecting the source and output fields, respectively (i.e., $\mathbf{i} = \{\varepsilon_l^{el}, \chi_l^{el}, \gamma_l^{el}\}$; $\mathbf{f} = \{\varepsilon^{el}, \chi^{el}, \gamma^{el}\}$), where the source fields are the local elastic strains, given as

$$
\begin{cases}
\varepsilon_l^{el}(s) = \dfrac{1}{EA}\left[N + \dfrac{M}{r\,\mathbf{n} \cdot \mathbf{t}_\perp}\right](s)\,, \\
\chi_l^{el}(s) = \dfrac{M}{EJ_r}(s) + (\mathbf{n} \cdot \mathbf{t}_\perp)\dfrac{1}{r\,EA}\left[N + \dfrac{M}{r\,\mathbf{n} \cdot \mathbf{t}_\perp}\right](s)\,, \\
\gamma_l^{el}(s) = \left[\dfrac{T}{GK_r}\right](s)\,,
\end{cases}
\tag{45}
$$

where $\mathbf{n}$ is the normal unit vector, $r^{-1} := |\partial_s \mathbf{t}|$ is the scalar geometric curvature of the beam axis and J_r is the inertia moment along the bending axis η identified by the transversal unit vector $\mathbf{t}_\perp$ (i.e., $J_r = \int_\Omega \eta^2 \dfrac{r}{r - \eta\,(\mathbf{n} \cdot \mathbf{t}_\perp)}\, dA$), while K_r is the shear stiffness for curved beams, defined in [33]. It is worth noting that in Equation (45), vanishing distributed bending couples have been assumed. Thus, the stress-driven two-phase elastic law is written as follows:

$$
\mathbf{f}(s) = \alpha\, \mathbf{i}(s) + (1 - \alpha) \int_0^L \phi_\lambda(s - \xi)\, \mathbf{i}(\xi)\, d\xi\,.
\tag{46}
$$

The mixture equivalence property extended to Timoshenko curved beams provides the following equivalent differential equation:

$$\frac{\mathbf{f}(s)}{c^2} - \partial_s^2 \mathbf{f}(s) = \frac{\mathbf{i}(s)}{c^2} - \alpha\, \partial_s^2 \mathbf{i}(s) \tag{47}$$

equipped with the constitutive boundary conditions

$$\begin{cases} \partial_s \mathbf{f}(0) = \dfrac{1}{c}\, \mathbf{f}(0) + \alpha \left(\partial_s \mathbf{i}(0) - \dfrac{\mathbf{i}(0)}{c} \right), \\ \partial_s \mathbf{f}(L) = -\dfrac{1}{c}\, \mathbf{f}(L) + \alpha \left(\partial_s \mathbf{i}(L) + \dfrac{\mathbf{i}(L)}{c} \right). \end{cases} \tag{48}$$

Structural problems based on the two-phase elasticity are addressed below with reference to a silicon carbide circular nanobeam of radius $r = 20\,[nm]$. Tables 1 and 2 show the numerical results of the transverse displacements $v_{t_\perp} := \mathbf{v} \cdot \mathbf{t}_\perp$, axial displacements $v_t := \mathbf{v} \cdot \mathbf{t}$ and bending rotations φ of the following structural schemes: cantilever nanobeam under concentrated force $F = 5\,[nN]$ at the free end and slider- and roller-supported nanobeam under uniformly distributed vertical loading $q = 2\,[nN/nm]$ (along the horizontal direction) and directed upward.

Table 1. Cantilever thick nanobeam: numerical outcomes.

λ	$v_\perp(L)$ $[10^{-2}nm]$		$v_t(L)$ $[10^{-2}nm]$		$\varphi(L)$ $[10^{-3}]$		$v_{max}^{tot}(L)$ $[10^{-2}nm]$	
	$\alpha = 0.2$	$\alpha = 0.5$	$\alpha = 0.2$	$\alpha = 0.5$	$\alpha = 0.2$	$\alpha = 0.5$	$\alpha = 0.2$	$\alpha = 0.5$
0.1	2.039	2.118	−1.300	−1.349	0.9429	0.9699	2.418	2.511
0.2	1.818	1.980	−1.170	−1.268	0.8629	0.9199	2.162	2.351
0.3	1.629	1.862	−1.058	−1.198	0.7876	0.8728	1.943	2.214
0.4	1.478	1.768	−0.9655	−1.140	0.7227	0.8323	1.766	2.104
0.5	1.358	1.692	−0.8905	−1.094	0.6686	0.7985	1.624	2.015
0.6	1.261	1.632	−0.8293	−1.055	0.6235	0.7703	1.510	1.944
0.7	1.183	1.583	−0.7787	−1.024	0.5858	0.7468	1.416	1.885
0.8	1.117	1.542	−0.7364	−0.9973	0.554	0.7269	1.338	1.836
0.9	1.062	1.508	−0.7006	−0.9749	0.5269	0.7099	1.273	1.795

Table 2. Slider- and roller-supported thick nanobeam: numerical outcomes.

λ	$v_\perp(0)$ $[10^{-2}nm]$		$v_\perp(L)$ $[10^{-2}nm]$		$\varphi(L)$ $[10^{-3}]$		$v_{max}^{tot}(0)$ $[10^{-2}nm]$	
	$\alpha = 0.2$	$\alpha = 0.5$	$\alpha = 0.2$	$\alpha = 0.5$	$\alpha = 0.2$	$\alpha = 0.5$	$\alpha = 0.2$	$\alpha = 0.5$
0.1	7.233	7.522	−6.970	−7.221	−3.903	−4.03	7.233	7.522
0.2	6.447	7.031	−6.214	−6.748	−3.568	−3.821	6.447	7.031
0.3	5.783	6.616	−5.563	−6.341	−3.261	−3.629	5.783	6.616
0.4	5.252	6.284	−5.042	−6.016	−2.996	−3.464	5.252	6.284
0.5	4.830	6.020	−4.628	−5.757	−2.775	−3.325	4.830	6.020
0.6	4.489	5.807	−4.296	−5.550	−2.591	−3.210	4.489	5.807
0.7	4.211	5.633	−4.026	−5.380	−2.436	−3.113	4.211	5.633
0.8	3.980	5.489	−3.802	−5.241	−2.305	−3.032	3.980	5.489
0.9	3.786	5.367	−3.614	−5.123	−2.193	−2.962	3.786	5.367

The numerical results show a stiffening mechanical behavior for increasing nonlocal parameter λ and a softening response for increasing mixture parameter α. Such a methodology provides a generalization of the stress-driven two-phase theory to the framework of curved stubby beams. Being based on two parameters, the mixture model can be effectively exploited for the modeling and design of small-scale devices based on curved beams.

6.2. Two-Phase Elasticity for Plates

Modeling of two-dimensional nonlocal continua is a topic of current interest in the scientific literature, with a wide range of applications concerning smart ultra-small devices. A stress-driven nonlocal methodology is conceived in [56] to capture scale effects in nanoplates and then generalized in [35] on the basis of a two-phase elasticity theory. The nonlocal mechanics of two-dimensional continua is studied in [31], vibration and buckling analysis of composite nanoplates are carried out in [95], static and dynamic behaviors of nonlocal elastic plates are examined in [96], modeling of circular nanoplate actuators is addressed in [97], chemical sensing systems are proposed in [98], vibration of resonant nanoplate mass sensors is analyzed in [99], nonlinear dynamics of nanoplates is investigated in [100], magneto-electromechanical nanosensors are modeled in [101], thermoelastic damping models for rectangular micro- and nanoplate resonators are proposed in [102], free vibration of functionally graded porous nanoplates is addressed in [103], nonlinear mechanical behavior of porous sandwich nanoplates is characterized in [104], and dynamics of nanoplates is investigated in [99,105].

Stress-driven two-phase elasticity has been recently applied in [35] to capture size-dependent behaviors of two-dimensional continua modeled by the Kirchhoff plate theory. Notably, with reference to an axisymmetric annular plate of internal radius R_i and external radius R_e, a polar coordinate system r, θ, z is conveniently introduced. Moreover, in the following, ∇ will denote the gradient operator and $\otimes$ the tensor product. According to the linearized Kirchhoff theory, the flexural curvature tensor is $\chi := \nabla\nabla u$, denoted by $u : [R_i, R_e] \to \Re$ the transverse displacement field. The kinematic compatibility condition can be explicitly written as $\chi = \partial_r^2 u \, \mathbf{e}_r \otimes \mathbf{e}_r + \frac{\partial_r u}{r} \mathbf{e}_\theta \otimes \mathbf{e}_\theta$, where the eigenvalues $\partial_r^2 u$ and $\frac{\partial_r u}{r}$ are the radial χ_r and circumferential χ_θ curvatures, respectively. The equilibrated stress is given by the radial M_r and circumferential M_θ bending interaction fields, satisfying the following differential equation:

$$\frac{1}{r}\Big(\partial_r^2(M_r(r)\cdot r) - \partial_r M_\theta(r)\Big) = q(r)\,, \quad r \in \Omega\,, \tag{49}$$

equipped with the boundary condition

$$\begin{cases} M_r(r)\,\partial_r\delta u(r) = -\,\bar{M}_i\,\partial_r\delta u(r)\,, & r \in \partial\Omega_i \\ M_r(r)\,\partial_r\delta u(r) = \bar{M}_e\,\partial_r\delta u(r)\,, & r \in \partial\Omega_e \\ \Big(M_\theta(r) - \partial_r(M_r(r)\cdot r)\Big)\,\delta u(r) = -\bar{Q}_i\, r\,\delta u(r)\,, & r \in \partial\Omega_i \\ \Big(M_\theta(r) - \partial_r(M_r(r)\cdot r)\Big)\,\delta u(r) = \bar{Q}_e\, r\,\delta u(r)\,, & r \in \partial\Omega_e \end{cases} \tag{50}$$

with q transverse distributed loading, $\{\bar{M}_i, \bar{M}_e\}$ edge distributed bending couples and $\{\bar{Q}_i, \bar{Q}_e\}$ edge distributed transverse forces.

The mixture model of elasticity expresses the elastic radial curvature χ_r^{el} as

$$\chi_r^{el}(r) = \alpha\,\frac{M_r(r) - \nu M_\theta(r)}{D\,(1-\nu^2)} + (1-\alpha)\int_{R_i}^{R_e} \phi_\lambda(r-\xi)\,\frac{M_r(\xi) - \nu M_\theta(\xi)}{D\,(1-\nu^2)}\,d\xi \tag{51}$$

where ν is Poisson's ratio and D is the plate flexural stiffness.

In Equation (51), $\lambda := \dfrac{c}{(R_e - R_i)}$ is the nonlocal parameter, while α is the mixture parameter providing the purely nonlocal and local responses for $\alpha = 0$ and $\alpha = 1$, respectively. By adopting the bi-exponential special kernel, an equivalent differential formulation can be provided. Specifically, the convex combination in Equation (51) is equivalent to the following differential equation:

$$\frac{\chi_r^{el}(r)}{c^2} - \partial_r^2 \chi_r^{el}(r) = \frac{M_r(r) - \nu M_\theta(r)}{c^2 D (1-\nu^2)} - \alpha \frac{\partial_r^2 (M_r(r) - \nu M_\theta(r))}{D (1-\nu^2)}, \tag{52}$$

equipped with the constitutive boundary conditions

$$\begin{cases} \partial_r \chi_r^{el}(r)\Big|_{r=R_i} - \dfrac{1}{c}\chi_r^{el}(R_i) = \alpha \left(\dfrac{\partial_r (M_r(r) - \nu M_\theta(r))}{D(1-\nu^2)} - \dfrac{M_r(r) - \nu M_\theta(r)}{c\, D(1-\nu^2)} \right)\Bigg|_{r=R_i}, \\ \partial_r \chi_r^{el}(r)\Big|_{r=R_e} + \dfrac{1}{c}\chi_r^{el}(R_e) = \alpha \left(\dfrac{\partial_r (M_r(r) - \nu M_\theta(r))}{D(1-\nu^2)} + \dfrac{M_r(r) - \nu M_\theta(r)}{c\, D(1-\nu^2)} \right)\Bigg|_{r=R_e}. \end{cases} \tag{53}$$

The mixture elasticity is adopted to solve the structural problem of a graphene nanoplate with a Euler–Young modulus $E = 1\,[TPa]$ and Poisson's ratio $\nu = 0.25$ and external and internal radii $R_e = 30\,[nm]$ and $R_i = 3\,[nm]$, respectively. The nanoplate has clamped edges and is subjected to a uniformly distributed transverse loading $q = -10^{-3}\,[nN/nm^2]$. Parametric studies are carried out to simulate size effects and as depicted in Figure 8, a stiffening effect is obtained for increasing λ values for a fixed mixture parameter. Since it is based on two parameters, the stress-driven two-phase elasticity is able to simulate a wide class of nanotechnological applications involving miniaturized devices based on nanoplates.

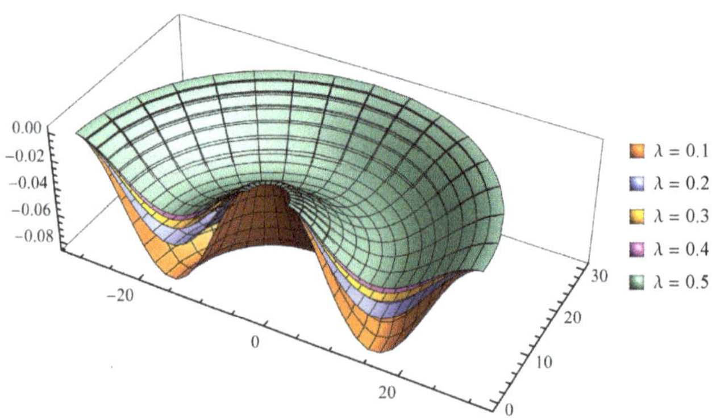

Figure 8. Nanoplate with clamped edges under uniformly distributed loading: transverse displacement fields $u\,[nm]$ for $\alpha = 0.2$.

7. Dynamics of Nanobeams

The dynamic problem of a slender nanobeam is formulated by exploiting the general two-phase local/nonlocal methodology. The beam is assumed to lay on a bed of dashpots with a viscosity η, providing a damping effect. The differential equation of the d'Alembert dynamic equilibrium for a slender beam is $\partial_x^2 M = q - \eta\,\dot{v} - m\,\ddot{v}$ where q is a transverse distributed loading and m denotes the mass per unit length. The symbol dot will be applied in the following to denote the time derivative. According to the stress-driven mixture theory of elasticity applied to slender beams, the nonlocal response $f := \chi^{el}$ is convex combination of the local source field $s := k_f^{-1} M$ and the purely stress-driven convolution integral $\phi_\lambda * s$.

By adopting the equivalent differential formulation and differentiating twice, the following equation is found:

$$\frac{1}{c^2}\,\partial_x^2\chi^{el}(x,\,t)-\partial_x^4\chi^{el}(x,\,t)=\frac{1}{c^2}\,\frac{\partial_x^2 M(x,\,t)}{k_f}-\alpha\frac{\partial_x^4 M(x,\,t)}{k_f}\,, \tag{54}$$

which can be further manipulated by prescribing the compatibility condition and the equilibrium requirements in order to find the differential equation governing the bending vibrations:

$$\frac{1}{c^2}\,\partial_x^4 v(x,\,t)-\partial_x^6 v(x,\,t)=-\frac{1}{c^2}\,\frac{m\ddot{v}(x,\,t)}{k_f}+\alpha\, m\frac{\partial_x^2\ddot{v}(x,\,t)}{k_f}\,. \tag{55}$$

With the aim of first investigating the free vibration problem of an undamped beam, vanishing loading and viscosity have been assumed in deriving Equation (55). Moreover, synchronous motions $v(x,t)$ will be investigated, which are mathematically represented by the assumption that the solution $v(x,t)$ to Equation (55) is separable in spatial and time variables (i.e., $v(x,t)=\psi(x)\,y(t)$). The following differential equations can thus be provided:

$$\begin{cases}\ddot{y}(t)+\omega^2\,y(t)=0\,,\\ \partial_x^6\psi(x)-\dfrac{1}{c^2}\partial_x^4\psi(x)+\dfrac{m\,\omega^2\,\psi(x)}{c^2\,k_f}=0\,,\end{cases} \tag{56}$$

where $\alpha=0$ is assumed. The first in Equation (56) is the differential equation governing harmonic motion, whose evaluation requires prescription of suitable initial conditions. The second in Equation (56) is a sixth-order differential equation in the unknown ψ that must be equipped with four standard boundary conditions and the following two constitutive boundary conditions:

$$\begin{cases}\partial_x^3\psi(0)-\dfrac{1}{c}\,\partial_x^2\psi(0)=0\,,\\ \partial_x^3\psi(L)+\dfrac{1}{c}\,\partial_x^2\psi(L)=0\,.\end{cases} \tag{57}$$

providing the differential problem of eigenvalues ω and eigenfunctions $\phi(x)$ which admits infinite solutions. Now, let us analyze the forced vibration problem of a damped beam governed by the following equation:

$$\frac{1}{c^2}\,\partial_x^4 v(x,\,t)-\partial_x^6 v(x,\,t)+\frac{1}{c^2}\,\frac{\eta\dot{v}(x,\,t)}{k_f}+\frac{1}{c^2}\,\frac{m\ddot{v}(x,\,t)}{k_f}=\frac{1}{c^2}\,\frac{q(x,\,t)}{k_f}\,. \tag{58}$$

The solution $v(x,t)=\sum_{j=1}^{\infty}\psi_j(z)y_j(t)$ can be inserted into Equation (58). Then, multiplying both sides by the ith eigenfunction ψ_i and integrating over the domain yield

$$\ddot{y}_i(t)+\frac{\eta}{m}\dot{y}_i(t)+\frac{k_{\lambda,i}}{m}\,y_i(t)=\frac{\int_0^L\psi_i(x)\,q(x,t)\,dx}{m}\,, \tag{59}$$

where the orthonormality property of the eigenfunctions has been taken into account. The symbol $k_{\lambda,i}$ in Equation (59) denotes the nonlocal stiffness defined in [106].

Random vibrations are now analyzed to take into account the stochastic nature of the external loadings and simulate the conditions of micro- and nanodevices subjected to environmental noise. Notably, the loading is assumed to be $q(x,t):=g(x)\,F(t)$ where g is a deterministic function and F is a stochastic process. Specifically, F is assumed to be a stationary Gaussian process with a zero mean $\mu_F:=\mathbb{E}[F(t)]$ and correlation function $R_F(\tau):=\mathbb{E}[F(t)F(t+\tau)]$. A frequency domain approach can be followed to characterize the steady state response of the output process in terms of beam displacements. The differential Equation (59) is written as

$$\ddot{Y}_i(t) + \frac{\eta}{m}\dot{Y}_i(t) + \frac{k_{\lambda,i}}{m}Y_i(t) = \frac{a_i}{m}F(t)\,, \tag{60}$$

where $a_i := \int_0^L \psi_i(x)\, g(x)dx$. The truncated Fourier transform of Equation (60) is then performed to find the response in the frequency domain:

$$\hat{Y}_i(\omega, \mathrm{T}) = \frac{1}{-\omega^2 + \frac{\eta}{m}\mathrm{i}\omega + \frac{k_{\lambda,i}}{m}} \frac{a_i}{m}\hat{F}(\omega, \mathrm{T})\,, \tag{61}$$

where the symbol ^ stands for the truncated Fourier transform in the time interval $[0, T]$ and i is the imaginary unit. By exploiting Equation (61), the analytical form of the power spectral density function can be derived:

$$S_v(x, \omega) = \sum_{j=1}^{\infty}\sum_{i=1}^{\infty} \psi_j(x)\psi_i(x) \lim_{\mathrm{T}\to\infty} \frac{1}{2\pi \mathrm{T}} \mathbb{E}\left[\hat{Y}_j^*(\omega, \mathrm{T})\hat{Y}_i(\omega, \mathrm{T})\right], \tag{62}$$

where the apex $*$ denotes the complex conjugate. Finally, the stationary variance of the beam displacements is computed by integration of the power spectral density function over the frequency domain $\sigma_v^2(x) = \int_{-\infty}^{\infty} S_v(x, \omega)\, d\omega$.

The free vibration response is investigated with reference to a slender, nonlocal elastic cantilever of a unit length. Figure 9 represents the first five nonlocal eigenfunctions for a nonlocal parameter $\lambda = 0.15$, obtained by solving Equation $(56)_2$ equipped with the standard boundary conditions and with the constitutive boundary conditions in Equation (57).

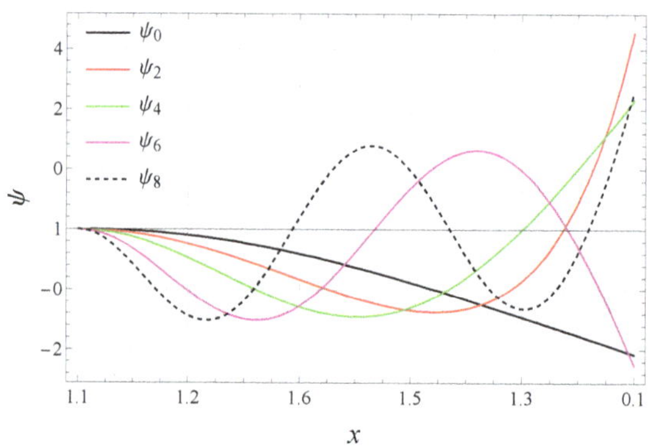

Figure 9. First five eigenfunctions of nonlocal elastic cantilever for nonlocal parameter $\lambda = 0.15$.

Random vibrations of small-scale beams are investigated in [106], where both frequency and time domain responses are evaluated. Notably, the non-stationary variance of displacements is provided by performing a Monte Carlo simulation, assuming that the beam is loaded by a stationary Gaussian process. A proper number of realizations of the stochastic input process is generated by the formula proposed by Shinozuka and Deodatis in [107]. The output process is then obtained by applying the Duhamel superposition integral, and finally the displacement samples are processed to evaluate the non-stationary variance. Further contributions exploring dynamics of nonlocal continua are provided in [108], where nanobeam-based resonators are investigated, in [109], concerning forced vibrations of dielectric elastomer-based microcantilevers, in [110], where buckling of graphene platelet-reinforced nanostructures is examined, in [111], which concerns dynamics of a piezoelectric semiconductor nanoplate, in [112], in which dynamics of nonlocal rods is examined, in [113], where transverse vibrations of nanobeams with multiple cracks are evaluated, and in [114], in which mechanical static and dynamic behaviors of microsystems are investigated providing fundamental concepts of modeling and design.

Transient response of functionally graded nanobeams is investigated in [115], buckling problem of nonhomogeneous microbeams is addressed in [116], linear and nonlinear dynamic responses of microsystems are evaluated in [117], thermo-mechanical vibration analysis of nanobeams is performed in [118], free vibration of embedded carbon and silica carbide nanotubes is analyzed in [119], dynamic analysis of nanostructures exploiting the Chebyshev–Ritz method is provided in [120], and nonlinear dynamics of nanobeams connected with fullerenes is investigated in [121].

8. Nonlocal Elasticity for Structural Assemblages

In this section, a nonlocal methodology is illustrated to account for size effects in structural assemblages. Notably, convolution integrals involving piecewise regular source fields are investigated to model beam problems involving discontinuous and concentrated loadings, non-smooth elastic and geometric properties and internal kinematic constraints, which are the most general cases when dealing with structural problems. In this context, the stress-driven integral theory of elasticity has been recently developed to capture size effects in assemblages of slender beams [122]. For this purpose, the next proposition plays a fundamental role in the development of nonlocal strategies for complex structural systems:

Proposition 3. *The nonlocal elastic curvature obtained with Equation (27) is a continuously differentiable field* $\chi^{el} \in C^1([0,L];\Re)$ *for any piecewise smooth local source field.*

Let us suppose that we have a beam partitioned into two domains of regularity. The integral constitutive law in Equation (27), equipped with the special kernel, can be explicitly written as

$$\chi^{el}(x) = \begin{cases} \dfrac{1}{2c}\displaystyle\int_0^x exp\left(\frac{\xi - x}{c}\right)\frac{M_1}{I_{E1}}(\xi)d\xi + \frac{1}{2c}\int_x^{x_d} exp\left(\frac{x-\xi}{c}\right)\frac{M_1}{I_{E1}}(\xi)d\xi \\ +\dfrac{1}{2c}\displaystyle\int_{x_d}^L exp\left(\frac{x-\xi}{c}\right)\frac{M_2}{I_{E2}}(\xi)d\xi, \qquad x \in [0, x_d], \\ \dfrac{1}{2c}\displaystyle\int_0^{x_d} exp\left(\frac{\xi - x}{c}\right)\frac{M_1}{I_{E1}}(\xi)d\xi + \frac{1}{2c}\int_{x_d}^{x} exp\left(\frac{\xi - x}{c}\right)\frac{M_2}{I_{E2}}(\xi)d\xi \\ +\dfrac{1}{2c}\displaystyle\int_{x}^L exp\left(\frac{x-\xi}{c}\right)\frac{M_2}{I_{E2}}(\xi)d\xi, \qquad x \in [x_d, L], \end{cases} \tag{63}$$

with $M_1 : [0,x_d] \mapsto \Re$ and $M_2 : [x_d, L] \mapsto \Re$ regular bending interaction fields and $I_{E1} : [0,x_d] \mapsto \Re$ and , $I_{E2} : [x_d, L] \mapsto \Re$ regular bending stiffnesses, which represents the second moment of the Euler–Young modulus field on the beam cross-sections. It can be immediately proven that the nonlocal curvature generated by the convolution integral in Equation (63) is a continuously differentiable field in $[0,L]$. Indeed, its first derivative is given by

$$\partial_x \chi^{el}(x) = \begin{cases} \frac{1}{2c^2}\left(-\int_0^x exp\left(\frac{\xi - x}{c}\right)\frac{M_1}{I_{E1}}(\xi)d\xi + \int_x^{x_d} exp\left(\frac{x-\xi}{c}\right)\frac{M_1}{I_{E1}}(\xi)d\xi \right. \\ \left. + \int_{x_d}^L exp\left(\frac{x-\xi}{c}\right)\frac{M_2}{I_{E2}}(\xi)d\xi\right), \qquad x \in [0, x_d], \\ \frac{1}{2c^2}\left(-\int_0^{x_d} exp\left(\frac{\xi - x}{c}\right)\frac{M_1}{I_{E1}}(\xi)d\xi - \int_{x_d}^x exp\left(\frac{\xi - x}{c}\right)\frac{M_2}{I_{E2}}(\xi)d\xi \right. \\ \left. + \int_x^L exp\left(\frac{x-\xi}{c}\right)\frac{M_2}{I_{E2}}(\xi)d\xi\right), \qquad x \in [x_d, L], \end{cases} \tag{64}$$

which is a continuous field in $[0, L]$. Proposition 3 plays a key role in proving the equivalent constitutive differential formulation. Indeed, the constitutive differential problem is made of the following set of differential equations, each one referring to a subdomain of regularity:

$$\begin{cases} \frac{1}{c^2}\chi_1^{el}(x) - \partial_x^2\chi_1^{el}(x) = \frac{1}{c^2}\frac{M_1}{I_{E1}}(x), & x \in [0, x_d], \\ \frac{1}{c^2}\chi_2^{el}(x) - \partial_x^2\chi_2^{el}(x) = \frac{1}{c^2}\frac{M_2}{I_{E2}}(x), & x \in [x_d, L], \end{cases} \tag{65}$$

equipped with constitutive boundary condition at $\partial[0, L]$

$$\begin{cases} \partial_x\chi_1^{el}(0) = \frac{1}{c}\chi_1^{el}(0), \\ \partial_x\chi_2^{el}(L) = -\frac{1}{c}\chi_2^{el}(L), \end{cases} \tag{66}$$

and interface boundary conditions at the internal abscissa x_d:

$$\begin{cases} \chi_1^{el}(x_d) = \chi_2^{el}(x_d), \\ \partial_x\chi_1^{el}(x_d) = \partial_x\chi_2^{el}(x_d). \end{cases} \tag{67}$$

The interface continuity conditions in Equation (67) can be prescribed by virtue of Proposition 3. These constitutive interface conditions play a fundamental role in modeling the assemblages of nonlocal elastic structures, and it is worth noting that they do not involve any convolution integrals and thus can be conveniently adopted for formulating and solving the structural problems of complex nanosystems. The constitutive conditions in Equation (67) are equivalent to those established in [123], which involve convolution integrals.

The theoretical outcomes illustrated in this Section are confirmed in the sequel by investigating a structural scheme involving a piecewise regular local field of elastic curvature induced by a concentrated loading. For this purpose, a nonlocal elastic beam with clamped and simply supported ends under a concentrated couple $\mathcal{M}$ at mid-span is analyzed, and the parametric responses will be provided for increasing the length scale parameter $\lambda = c/L$. Figure 10 shows the solutions in terms of the non-dimensional elastic curvature $\bar{\chi}^{el}(\bar{x}) = \chi^{el}(x)\frac{I_E}{\mathcal{M}}$ versus the non-dimensional abscissa $\bar{x} = x/L \in [0, 1]$. As theoretically predicted, the nonlocal fields $\bar{\chi}^{el}$ are continuously differentiable functions for $\lambda > 0$ and become more uniform as λ increases. It is interesting to analyze the asymptotic behavior for $\lambda \to 0^+$, showing that the local elastic curvature is recovered for $\bar{x} \in]0, 1] - \{\bar{x}_d\}$. At the external boundary abscissa $\bar{x} = 0$, one half of the local elastic curvature is got, and at the interior point $\bar{x} = 1/2$, the limiting response is equal to the average of the local elastic curvatures. These peculiar behaviors are due to the asymptotic behavior of the kernel. The non-dimensional fields of the transverse displacements $\bar{v}(\bar{x}) = v(x)\frac{I_E}{\mathcal{M}L^2}$ are shown in Figure 11. It is worth noting that the limiting solution is coincident with the local displacement, since the asymptotic peculiar behaviors only affect the null measure sets. The presented differential formulation, involving the prescription of constitutive boundary and interface conditions, plays a key role in conceiving a finite-element non-local methodology, as shown in [124]. Further contributions to this topic can be found in [125], where nanobeams with internal discontinuities are investigated by exploiting a mixture approach, and in [126], where strain- and stress-driven differential formulations of Timoshenko nanobeams with loading discontinuities are provided.

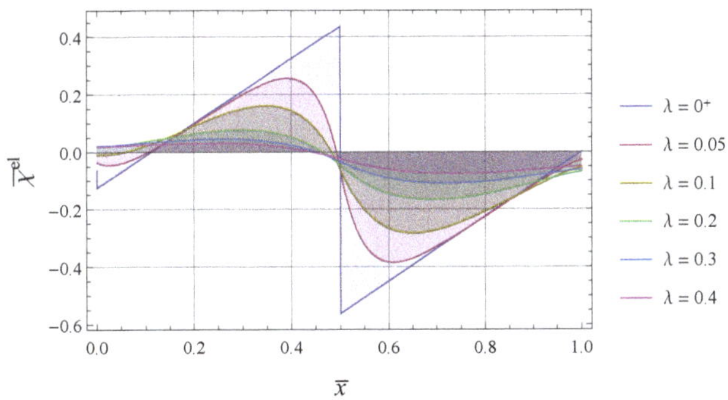

Figure 10. Beam with clamped and simply supported ends under concentrated couple $\mathcal{M}$ at mid-span $\bar{x} = 1/2$: elastic curvature $\bar{\chi}^{el}$ versus $\bar{x}$ for increasing nonlocal parameter λ.

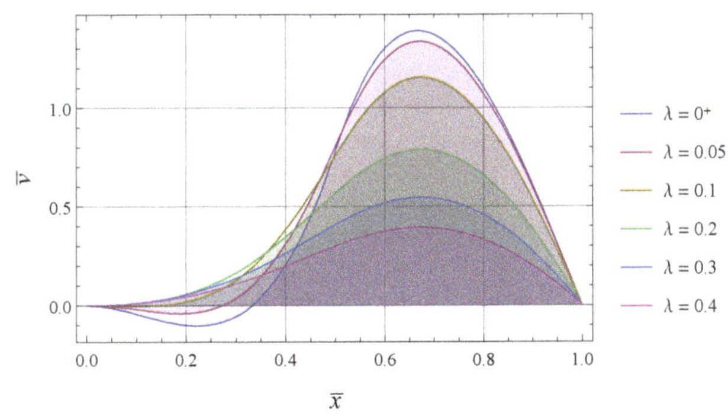

Figure 11. Beam with clamped and simply supported ends under concentrated couple $\mathcal{M}$ at mid-span $\bar{x} = 1/2$: transverse displacement $\bar{v} \cdot 10^{-2}$ versus $\bar{x}$ for increasing nonlocal parameter λ.

9. Nanostructures on Nonlocal Foundations

Many contributions in the scientific literature deal with the modeling of nanobeams resting on elastic soils, since they are involved in several biomechanical and biomedical applications of current interest. Dynamics of nanobeams on Pasternak soil is addressed in [127,128]. Buckling of small-scale structures embedded in two parameter foundations is investigated in [129,130]. Elastostatics of nanostructures on Winkler soil is examined in [131]. Free vibration of nanobeams on Pasternak soil is addressed in [132]. Nanobeams embedded in cell cytoplasm are analyzed in [133]. Nonlinear vibration analysis of nanobeams interacting with an elastic medium is carried out in [134]. The interaction between nanoshells and elastic foundations is studied in [135]. Magnetically embedded composite nanobeams are analyzed in [136]. Stability of nonlocal beams exposed to a hygro-thermo-magnetic environment and lying on elastic foundations is addressed in [137].

Vibration of functionally graded beams on a viscoelastic Winkler–Pasternak foundation is investigated in [90]. Buckling of Bernoulli–Euler nanobeams resting on a Pasternak elastic foundation is examined in [138]. Recent achievements regarding this topic are provided in [139], where mechanics of nanobeams on nano-foundations is addressed on the basis of the nonlocal model of elastic medium proposed in [140].

In order to assure that the relevant structural foundation problem is well-posed, it is necessary that the integral constitutive laws of internal and external elasticity are compatible with both the equilibrium and compatibility requirements. In the framework of nonlocal internal elasticity, the strain-driven purely integral approach proved itself to be incompatible with the equilibrium requirements, and as shown in Section 2, the equivalence property provided an effective tool to check the consistency. Indeed, according to Proposition 2, the constitutive integral law in Equation (12) admits a unique solution if and only if the equilibrated bending interaction field satisfies the constitutive boundary conditions in Equation (13). These constitutive boundary conditions are, in general, in contrast with the natural ones, and thus no solution exists to Equation (12), since the integral constitutive law is incompatible with the equilibrium requirements. In Section 4, a new integral model is shown, the stress-driven theory, which is able to overcome the intrinsic issues emerged from the strain-driven approach. Thus, source fields in the convolution integral play a fundamental role in providing a consistent constitutive theory. In the context of external integral elasticity, the Wieghardt model [141] provides a refinement of the Winkler local theory of elastic foundation by assuming that the beam deflection v is a convolution integral between a proper averaging kernel ϕ and the soil reaction field r. Concerning the Wieghardt theory, also referred to as the reaction-driven model, the following equivalence property can be stated:

Proposition 4. *For any characteristic length $c_f > 0$, the integral constitutive law*

$$v(x) = \int_0^L \phi(x - \xi, c_f) \frac{r(\xi)}{k} d\xi \tag{68}$$

equipped with the bi-exponential kernel in Equation (4) admits either a unique solution or no solution at all, depending on whether or not the interface displacement field satisfies the following constitutive boundary conditions:

$$\begin{cases} \partial_x v(0) = \dfrac{1}{c_f} v(0)\,, \\ \partial_x v(L) = -\dfrac{1}{c_f} v(L)\,. \end{cases} \tag{69}$$

If Equation (69) is fulfilled by the compatible displacement field, then the unique solution v is obtained from the second-order differential equation

$$v(x) - c_f^2\, \partial_x^2 v(x) = \frac{r(x)}{k}\,, \tag{70}$$

where k is the Winkler local stiffness of the foundation. It is apparent that an incompatibility arises between the kinematic and constitutive (Equation (69)) boundary conditions, since the latter relates to the displacement and rotation at the beam ends for any value of the characteristic parameter. These requirements are not satisfied by the kinematic boundary conditions generally involved in technical applications. A mathematical trick to overcome this issue was provided in [142], but it requires the introduction of fictitious reactive forces that have no physical meaning. An effective strategy has been then proposed in [140] to overcome the ill-posed nature of the structural problem of beams on a Wieghardt foundation. This new integral theory of foundation is based on a displacement-driven approach [140] requiring that the foundation reaction is expressed as a convolution integral driven by the interface transverse displacement:

$$r(x) = \int_0^L \phi(x-\xi,\, c_f) k\, v(\xi) d\xi\,. \tag{71}$$

As proven in [140], the constitutive integral law in Equation (71) is a consistent theory providing well-posed structural problems. Such a new model of nonlocal elastic foundation can be effectively exploited to model the size-dependent behavior of nanobeams lying on nano-foundations. Notably, the stress-driven theory of elasticity can be adopted to capture size effects in small-scale beams, while the surrounding medium can be modeled by the displacement-driven nonlocal approach. The relevant structural problem of a small-scale beam of length L and local elastic inertia $k_f := I_E$ laying on a nonlocal elastic foundation with a local stiffness k can thus be expressed as follows:

$$\begin{cases} \partial_x^2 M(x) = q(x) - r(x)\,, \\ r(x) = \displaystyle\int_0^L \phi\Big(x-\xi,\, c_f\Big) k\, v(\xi) dt\,, \\ \chi^{el}(x) = \displaystyle\int_0^L \phi(x-\xi,\, c_b) \frac{M(\xi)}{I_E} d\xi\,, \\ \partial_x^2 v(x) = \chi(x)\,, \end{cases} \tag{72}$$

where c_b is the beam's characteristic length. System in Equation (72) consists of the beam differential equation of equilibrium, the displacement-driven law of external elasticity, the stress-driven law of internal elasticity and the beam kinematic compatibility condition. It is worth noting that non-elastic effects are not taken into account in the following, so $\chi = \chi^{el}$. The structural foundation problem in Equation (72) is equipped with the standard boundary conditions involving $\{v,\, \varphi,\, M,\, T\}$, where $\varphi := \partial_x v$ and $T := -\partial_x M$. By virtue of the equivalence property proven in [139], the integro-differential problem (Equation (72)) can be reverted to a simpler differential formulation:

$$\begin{cases} c_b^2 c_f^2\, \partial_x^8 r(x) - \Big(c_b^2 + c_f^2\Big)\, \partial_x^6 r(x) + \partial_x^4 r(x) + \dfrac{k}{I_E} r(x) = \dfrac{k}{I_E} q_y(x)\,, \\ \partial_x^3 r(0) - c_f^2\, \partial_x^5 r(0) = \dfrac{1}{c_b}\Big(\partial_x^2 r(0) - c_f^2\, \partial_x^4 r(0)\Big)\,, \\ \partial_x^3 r(L) - c_f^2\, \partial_x^5 r(L) = -\dfrac{1}{c_b}\Big(\partial_x^2 r(L) - c_f^2\, \partial_x^4 r(L)\Big)\,, \\ \partial_x r(0) = \dfrac{1}{c_f} r(0)\,, \\ \partial_x r(L) = -\dfrac{1}{c_f} r(L)\,, \end{cases} \tag{73}$$

equipped with essential and natural boundary conditions involving the following fields:

$$
\begin{cases}
v(x) = \frac{1}{k} r(x) - \frac{c_f^2}{k} \partial_x^2 r(x)\,, \\
\varphi(x) = \frac{1}{k} \partial_x r(x) - \frac{c_f^2}{k} \partial_x^3 r(x)\,, \\
M(x) = \frac{I_E}{k} \partial_x^2 r(x) - \frac{I_E}{k} c_f^2 \partial_x^4 r(x) - \frac{I_E}{k} c_b^2 \partial_x^4 r(x) + \frac{I_E}{k} c_b^2 c_f^2 \partial_x^6 r(x)\,, \\
T(x) = -\frac{I_E}{k} \partial_x^3 r(x) + \frac{I_E}{k} c_f^2 \partial_x^5 r(x) + \frac{I_E}{k} c_b^2 \partial_x^5 r(x) - \frac{I_E}{k} c_b^2 c_f^2 \partial_x^7 r(x)\,.
\end{cases}
\tag{74}
$$

Effectiveness of the proposed nonlocal methodology is proven by the numerical outcomes of technical interest illustrated in [139]. As a benchmark case study, let us consider a simply supported beam on a nonlocal foundation under non-dimensional transverse loading $q^* := \frac{q\,L^3}{I_E} = 1$. Solution of the relevant structural problem in Equation (73) equipped with standard (essential and natural) boundary conditions provides the non-dimensional displacement field $v^* := \frac{v}{L}$ represented in Figure 12, for a fixed non-dimensional stiffness $k^* := \frac{k\,L^4}{I_E} = 100$, showing a softening behavior for increasing foundation parameter $\lambda_f := \frac{c_f}{L}$. Non-dimensional bending interaction fields $M^* := \frac{M\,L}{I_E}$ are represented in Figure 13 for increasing values of λ_f.

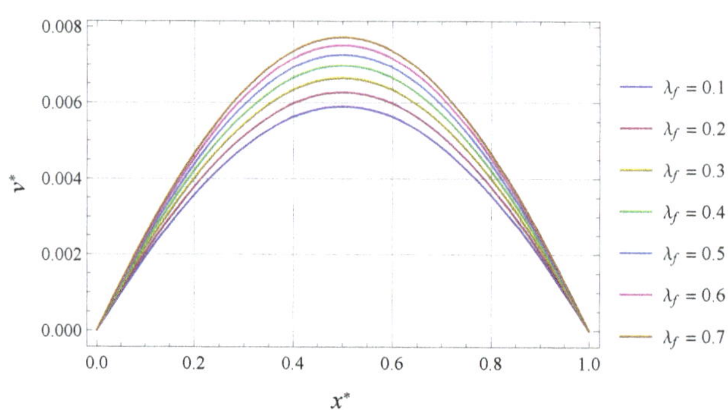

Figure 12. Non-dimensional transverse displacement fields for $k^* = 100$ and $\lambda_b = 0.2$.

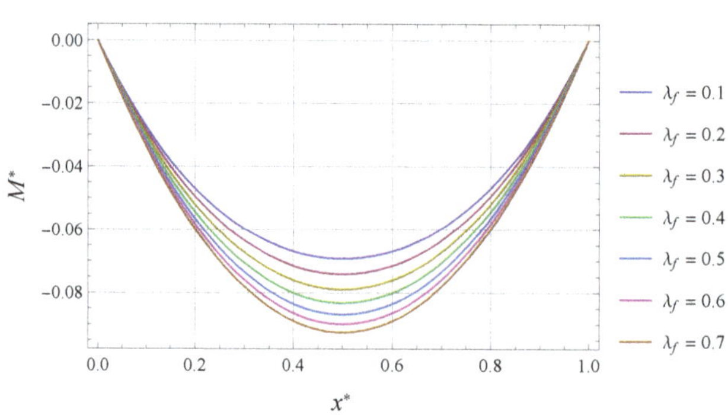

Figure 13. Non-dimensional bending interaction fields for $k^* = 100$ and $\lambda_b = 0.2$.

10. Conclusions

In the paper, recent achievements in the framework of modeling and analysis of nanostructures have been illustrated and critically discussed. A comprehensive overview on nonlocal continuum mechanics has been provided, starting from the early concepts contributed by Eringen. Alternative theories of nonlocal elasticity in the strain-driven formulation have been analyzed and discussed, such as mixture local/Eringen nonlocal model and strain difference-based theories. Stress-driven nonlocal methodologies recently proposed in literature to capture size-dependent static and dynamic behaviors of nanostructures have been collected and examined. In this framework, the stress-driven two-phase (local/nonlocal) model has been shown to be an effective tool to capture a wide class of ultra-small devices. Achievements in the field of integral elasticity applied to geometrically nonlinear mechanics of inflected nanostructures undergoing large configuration changes have been then elucidated and commented upon. A challenging issue in nonlocal mechanics has been then addressed concerning reproducibility of continuum structural problems. Nonlocal methodologies for structural assemblages have been thus inspected according to the stress-driven approach and benchmark case studies have been provided. Recent original contributions regarding mechanics of nanobeams on nonlocal foundations have been finally analyzed to address challenging applications of current interest.

Author Contributions: All the authors contributed equally to this work. All authors have read and agreed to the published version of the manuscript.

Funding: Financial support from the Italian Ministry of Education, University and Research (MIUR) in the framework of the Project PRIN—code 2017J4EAYB, Multiscale Innovative Materials and Structures (MIMS)—and from the research program ReLUIS 2020–2021 are gratefully acknowledged.

Institutional Review Board Statement: Not applicable.

Informed Consent Statement: Not applicable.

Data Availability Statement: The data presented in this study are available within this article. Further inquiries may be directed to the authors.

Conflicts of Interest: The authors declare no conflict of interest.

References

1. Rogula, D. Influence of spatial acoustic dispersion on dynamical properties of dislocations. *Bull. Pol. Acad. Sci. Tech. Sci.* **1965**, *13*, 337–385.
2. Kröner, E. Elasticity theory of materials with long range cohesive forces. *Int. J. Solids Struct.* **1967**, *3*, 731–742. [CrossRef]
3. Meo, M.; Rossi, M. Prediction of Young's modulus of single wall carbon nanotubes by molecular-mechanics based finite element modeling. *Compos. Sci. Technol.* **2006**, *66*, 1597–1605. [CrossRef]
4. Malagù, M.; Benvenuti, E.; Simone, A. One-dimensional nonlocal elasticity for tensile single-walled carbon nanotubes: A molecular structural mechanics characterization. *Eur. J. Mech.—A/Solids* **2015**, *54*, 160–170. [CrossRef]
5. Duan, K.; Li, L.; Hu, Y.; Wang, X. Enhanced interfacial strength of carbon nanotube/copper nanocomposites via Ni-coating: Molecular-dynamics insights. *Phys. E Low-Dimens. Syst. Nanostructures* **2017**, *88*, 259–264. [CrossRef]
6. Eringen, A.C. Linear theory of nonlocal elasticity and dispersion of plane waves. *Int. J. Eng. Sci.* **1972**, *10*, 425–435. [CrossRef]
7. Eringen, A.C. On differential equations of nonlocal elasticity and solutions of screw dislocation and surface waves. *J. Appl. Phys.* **1983**, *54*, 4703–4710. [CrossRef]
8. Romano, G.; Barretta, R. Stress-driven versus strain-driven nonlocal integral model for elastic nano-beams. *Compos. B. Eng.* **2017**, *114*, 184–188. [CrossRef]
9. Peddieson, J.; Buchanan, G.R.; McNitt, R.P. Application of nonlocal continuum models to nanotechnology. *Int. J. Eng. Sci.* **2003**, *41*, 305–312. [CrossRef]
10. Challamel, N.; Wang, C.M. The small length scale effect for a non-local cantilever beam: A paradox solved. *Nanotechnology* **2008**, *19*, 345703. [CrossRef]
11. Romano, G.; Barretta, R.; Diaco, M. On nonlocal integral models for elastic nano-beams. *Int. J. Mech. Sci.* **2017**, *131-132*, 490–499. [CrossRef]
12. Eringen, A.C. Theory of nonlocal elasticity and some applications. *Res. Mech.* **1987**, *21*, 313–342.
13. Eringen, A.C. *Nonlocal Continuum Field Theories*; Springer: New York, NY, USA, 2002.
14. Khodabakhshi, P.; Reddy, J. A unified integro-differential nonlocal model. *Int. J. Eng. Sci.* **2015**, *95*, 60–75. [CrossRef]

15. Pisano, A.A.; Fuschi, P. Closed form solution for a nonlocal elastic bar in tension. *Int. J. Solids Struct.* **2003**, *40*, 13–23. [CrossRef]
16. Fuschi, P.; Pisano, A.; Polizzotto, C. Size effects of small-scale beams in bending addressed with a strain-difference based nonlocal elasticity theory. *Int. J. Mech. Sci.* **2019**, *151*, 661–671. [CrossRef]
17. Aifantis, E.C. Update on a class of gradient theories. *Mech. Mater.* **2003**, *35*, 259–280. [CrossRef]
18. Aifantis, E.C. Exploring the applicability of gradient elasticity to certain micro/nano reliability problems. *Microsyst. Technol.* **2009**, *15*, 109–115. [CrossRef]
19. Aifantis, E.C. On the gradient approach–relation to Eringen's nonlocal theory. *Int. J. Eng. Sci.* **2011**, *49*, 1367–1377. [CrossRef]
20. Lim, C.; Zhang, G.; Reddy, J. A higher-order nonlocal elasticity and strain gradient theory and its Applications in wave propagation. *J. Mech. Phys. Solids* **2015**, *78*, 298–313. [CrossRef]
21. Barretta, R.; Marotti de Sciarra, F. Constitutive boundary conditions for nonlocal strain gradient elastic nano-beams. *Int. J. Eng. Sci.* **2018**, *130*, 187–198. [CrossRef]
22. Barretta, R.; Faghidian, S.; Marotti de Sciarra, F.; Vaccaro, M. Nonlocal strain gradient torsion of elastic beams: Variational formulation and constitutive boundary conditions. *Arch. Appl. Mech.* **2020**, *90*, 691–706. [CrossRef]
23. Gurtin, M.; Murdoch, A. A Continuum Theory of Elastic Material Surfaces. *Arch. Ration. Mech. Anal.* **1975**, *57*, 291–323. [CrossRef]
24. Li, L.; Lin, R.; Ng, T.Y. Contribution of nonlocality to surface elasticity. *Int. J. Eng. Sci.* **2020**, *152*, 103311. [CrossRef]
25. Carbone, L.; Gaudiello, A.; Hernández-Llanos, P. T-junction of ferroelectric wires. *ESAIM Math. Model. Numer. Anal.* **2020**, *54*, 1429–1463. [CrossRef]
26. Gaudiello, A.; Sili, A. Limit models for thin heterogeneous structures with high contrast. *J. Differ. Equ.* **2021**, *302*, 37–63. [CrossRef]
27. Romano, G.; Barretta, R. Nonlocal elasticity in nanobeams: The stress-driven integral model. *Int. J. Eng. Sci.* **2017**, *115*, 14–27. [CrossRef]
28. Barretta, R.; Marotti de Sciarra, F.; Vaccaro, M.S. On nonlocal mechanics of curved elastic beams. *Int. J. Eng. Sci.* **2019**, *144*, 103140. [CrossRef]
29. Romano, G.; Barretta, R.; Diaco, M. Iterative methods for nonlocal elasticity problems. *Contin. Mech. Thermodyn.* **2019**, *31*, 669–689. [CrossRef]
30. Sedighi, H.M.; Malikan, M. Stress-driven nonlocal elasticity for nonlinear vibration characteristics of carbon/boron-nitride hetero-nanotube subject to magneto-thermal environment. *Phys. Scr.* **2020**, *95*, 055218. [CrossRef]
31. Farajpour, A.; Howard, C.Q.; Robertson, W.S.P. On size-dependent mechanics of nanoplates. *Int. J. Eng. Sci.* **2020**, *156*, 103368. [CrossRef]
32. Barretta, R.; Fabbrocino, F.; Luciano, R.; Marotti de Sciarra, F. Closed-form solutions in stress-driven two-phase integral elasticity for bending of functionally graded nano-beams. *Phys. E Low-Dimens. Syst. Nanostructures* **2018**, *97*, 13–30. [CrossRef]
33. Vaccaro, M.S.; Pinnola, F.P.; Marotti de Sciarra, F.; Čanađija, M.; Barretta, R. Stress-driven two-phase integral elasticity for Timoshenko curved beams. *Proc. Inst. Mech. Eng. Part N J. Nanomater. Nanoeng. Nanosyst.* **2021**, *235*, 52–63. [CrossRef]
34. Zhang, P.; Qing, H.; Gao, C.F. Exact solutions for bending of Timoshenko curved nanobeams made of functionally graded materials based on stress-driven nonlocal integral model. *Compos. Struct.* **2020**, *245*, 112362. [CrossRef]
35. Vaccaro, M.S.; Sedighi, H.M. Two-phase elastic axisymmetric nanoplates. *Eng. Comput.* **2022**, *in press*. [CrossRef]
36. Farajpour, A.; Ghayesh, M.H.; Farokhi, H. A review on the mechanics of nanostructures. *Int. J. Eng. Sci.* **2018**, *133*, 231–263. [CrossRef]
37. Shariati, M.; Shishesaz, M.; Sahbafar, H.; Pourabdy, M.; Hosseini, M. A review on stress-driven nonlocal elasticity theory. *J. Comput. Appl. Mech.* **2021**, *52*, 535–552.
38. Rogula, D. Introduction to nonlocal theory of material media. In *Nonlocal Theory of Material Media*; Rogula, D., Ed.; Springer: Vienna, Austria, 1982; pp. 123–222.
39. Krumhansl, J.A. Some considerations of the relation between solid state physics and generalized continuum mechanics. In *Proceedings of the Mechanics of Generalized Continua*; Kröner, E., Ed.; Springer: Berlin/Heidelberg, Germany, 1968; pp. 298–311.
40. Kunin, I.A. The theory of elastic media with microstructure and the theory of dislocations. In *Proceedings of the Mechanics of Generalized Continua*; Kröner, E., Ed.; Springer: Berlin/Heidelberg, Germany, 1968; pp. 321–329.
41. Romano, G.; Luciano, R.; Barretta, R.; Diaco, M. Nonlocal integral elasticity in nanostructures, mixtures, boundary effects and limit behaviours. *Contin. Mech. Thermodyn.* **2018**, *30*, 641–655. [CrossRef]
42. Benvenuti, E.; Simone, A. One-dimensional nonlocal and gradient elasticity: Closed-form solution and size effect. *Mech. Res. Commun.* **2013**, *48*, 46–51. [CrossRef]
43. Vaccaro, M.S.; Barretta, R.; Marotti de Sciarra, F.; Reddy, J.N. Nonlocal integral elasticity for third-order small-scale beams. *Acta Mech.* **2022**, *233*, 2393–2403. [CrossRef]
44. Fernández-Sáez, J.; Zaera, R.; Loya, J.; Reddy, J. Bending of Euler–Bernoulli beams using Eringen's integral formulation: A paradox resolved. *Int. J. Eng. Sci.* **2016**, *99*, 107–116. [CrossRef]
45. Romano, G.; Barretta, R.; Diaco, M.; Marotti de Sciarra, F. Constitutive boundary conditions and paradoxes in nonlocal elastic nanobeams. *Int. J. Mech. Sci.* **2017**, *121*, 151–156. [CrossRef]
46. Zhu, X.; Wang, Y.; Dai, H.H. Buckling analysis of Euler–Bernoulli beams using Eringen's two-phase nonlocal model. *Int. J. Eng. Sci.* **2017**, *116*, 130–140. [CrossRef]
47. Zhang, P.; Qing, H. Free vibration analysis of Euler–Bernoulli curved beams using two-phase nonlocal integral models. *J. Vib. Control.* **2021**, *28*, 2861–2878. [CrossRef]

48. Naghinejad, M.; Ovesy, H.R. Nonlinear post-buckling analysis of viscoelastic nano-scaled beams by nonlocal integral finite element method. *ZAMM J. Appl. Math. Mech./Z. für Angew. Math. Und Mech.* **2022**, *102*, e202100148. [CrossRef]
49. Providas, E. Closed-Form Solution of the Bending Two-Phase Integral Model of Euler-Bernoulli Nanobeams. *Algorithms* **2022**, *15*, 151. [CrossRef]
50. Zhang, P.; Qing, H. Closed-form solution in bi-Helmholtz kernel based two-phase nonlocal integral models for functionally graded Timoshenko beams. *Compos. Struct.* **2021**, *265*, 113770. [CrossRef]
51. Tricomi, F.G. *Integral Equations*; Interscience: New York, NY, USA, 1957.
52. Polyanin, P.; Manzhirov, A. *Handbook of Integral Equations*, 2nd ed.; Chapman and Hall/CRC: Boca Raton, FL, USA, 2008.
53. Romano, G.; Diaco, M. On formulation of nonlocal elasticity problems. *Meccanica* **2020**, *56*, 1303–1328. [CrossRef]
54. Pisano, A.A.; Fuschi, P.; Polizzotto, C. Integral and differential approaches to Eringen's nonlocal elasticity models accounting for boundary effects with applications to beams in bending. *ZAMM J. Appl. Math. Mech./Z. für Angew. Math. Und Mech.* **2021**, *101*, e202000152. [CrossRef]
55. Pisano, A.A.; Fuschi, P.; Polizzotto, C. Euler–Bernoulli elastic beam models of Eringen's differential nonlocal type revisited within a C^0-continuous displacement framework. *Meccanica* **2021**, *56*, 2323–2337. [CrossRef]
56. Barretta, R.; Faghidian, S.A.; Marotti de Sciarra, F. Stress-driven nonlocal integral elasticity for axisymmetric nano-plates. *Int. J. Eng. Sci.* **2019**, *136*, 38–52. [CrossRef]
57. Oskouie, M.F.; Ansari, R.; Rouhi, H. Bending of Euler-Bernoulli nanobeams based on the strain- and stress-driven nonlocal integral models: A numerical approach. *Acta Mech. Sin.* **2018**, *34*, 871–882. [CrossRef]
58. Oskouie, M.F.; Ansari, R.; Rouhi, H. A numerical study on the buckling and vibration of nanobeams based on the strain- and stress-driven nonlocal integral models. *Int. J. Comput. Mater. Sci. Eng.* **2018**, *7*, 1850016.
59. Barretta, R.; Faghidian, S.A.; Luciano, R. Longitudinal vibrations of nano-rods by stress-driven integral elasticity. *Mech. Adv. Mater. Struct.* **2019**, *26*, 1307–1315. [CrossRef]
60. Qing, H.; Wei, L. Linear and nonlinear free vibration analysis of functionally graded porous nanobeam using stress-driven nonlocal integral model. *Commun. Nonlinear Sci. Numer. Simul.* **2022**, *109*, 106300. [CrossRef]
61. Mahmoudpour, E.; Esmaeili, M. Nonlinear free and forced vibration of carbon nanotubes conveying magnetic nanoflow and subjected to a longitudinal magnetic field using stress-driven nonlocal integral model. *Thin-Walled Struct.* **2021**, *166*, 108134. [CrossRef]
62. Darban, H.; Luciano, R.; Darban, R. Buckling of cracked micro- and nanocantilevers. *Acta Mech.* **2022**, *234*, 693–704. [CrossRef]
63. Abazari, A.M.; Safavi, S.M.; Rezazadeh, G.; Villanueva, L.G. Modelling the Size Effects on the Mechanical Properties of Micro/Nano Structures. *Sensors* **2015**, *15*, 28543–28562. [CrossRef]
64. Lam, D.C.C.; Yang, F.; Chong, A.C.M.; Wang, J.; Tong, P. Experiments and theory in strain gradient elasticity. *J. Mech. Phys. Solids* **2003**, *51*, 1477–1508. [CrossRef]
65. Zarepour, M.; Hosseini, S.; Akbarzadeh, A. Geometrically nonlinear analysis of Timoshenko piezoelectric nanobeams with flexoelectricity effect based on Eringen's differential model. *Appl. Math. Model.* **2019**, *69*, 563–582. [CrossRef]
66. Gholipour, A.; Ghayesh, M.H. Nonlinear coupled mechanics of functionally graded nanobeams. *Int. J. Eng. Sci.* **2020**, *150*, 103221. [CrossRef]
67. Ghaffari, S.S.; Ceballes, S.; Abdelkefi, A. Nonlinear dynamical responses of forced carbon nanotube-based mass sensors under the influence of thermal loadings. *Nonlinear Dyn.* **2020**, *100*, 1013–1035. [CrossRef]
68. Gholami, M.; Zare, E.; Alibazi, A. Applying Eringen's nonlocal elasticity theory for analyzing the nonlinear free vibration of bidirectional functionally graded Euler–Bernoulli nanobeams. *Arch. Appl. Mech.* **2021**, *91*, 2957–2971. [CrossRef]
69. Saadatmand, M.; Shahabodini, A.; Ahmadi, B.; Chegini, S. Nonlinear forced vibrations of initially curved rectangular single layer graphene sheets: An analytical approach. *Phys. E Low-Dimens. Syst. Nanostructures* **2021**, *127*, 114568. [CrossRef]
70. Lanzoni, L.; Tarantino, A.M. Bending of nanobeams in finite elasticity. *Int. J. Mech. Sci.* **2021**, *202-203*, 106500. [CrossRef]
71. Karimipour, I.; Beni, Y.T. Nonlinear dynamic analysis of nonlocal composite laminated toroidal shell segments subjected to mechanical shock. *Commun. Nonlinear Sci. Numer. Simul.* **2022**, *106*, 106105. [CrossRef]
72. Jin, Q.; Ren, Y. Nonlinear size-dependent bending and forced vibration of internal flow-inducing pre- and post-buckled FG nanotubes. *Commun. Nonlinear Sci. Numer. Simul.* **2022**, *104*, 106044. [CrossRef]
73. Alibakhshi, A.; Dastjerdi, S.; Fantuzzi, N.; Rahmanian, S. Nonlinear free and forced vibrations of a fiber-reinforced dielectric elastomer-based microbeam. *Int. J. -Non-Linear Mech.* **2022**, *144*. [CrossRef]
74. Tang, Y.; Qing, H. Size-dependent nonlinear post-buckling analysis of functionally graded porous Timoshenko microbeam with nonlocal integral models. *Commun. Nonlinear Sci. Numer. Simul.* **2023**, *116*, 106808. [CrossRef]
75. Mamandi, A. Nonlocal large deflection analysis of a cantilever nanobeam on a nonlinear Winkler-Pasternak elastic foundation and under uniformly distributed lateral load. *J. Mech. Sci. Technol.* **2023**, *37*, 813–824. [CrossRef]
76. Malikan, M.; Wiczenbach, T.; Eremeyev, V. Thermal buckling of functionally graded piezomagnetic micro- and nanobeams presenting the flexomagnetic effect. *Contin. Mech. Thermodyn.* **2022**, *34*, 1051–1066. [CrossRef]
77. Vaccaro, M.S. On geometrically nonlinear mechanics of nanocomposite beams. *Int. J. Eng. Sci.* **2022**, *173*, 103653. [CrossRef]
78. Chen, L. An integral approach for large deflection cantilever beams. *Int. J. -Non-Linear Mech.* **2010**, *45*, 301–305. [CrossRef]
79. Reddy, J.N. A simple higher-order theory for laminated composite plates. *J. Appl. Mech.* **1984**, *51*, 745–752. [CrossRef]

80. Reddy, J.N.; Wang, C.M.; Lee, K.H. Relationships between bending solutions of classical and shear deformation beam theories. *Int. J. Solids Struct.* **1997**, *34*, 3373–3384. [CrossRef]
81. Heyliger, P.R.; Reddy, J.N. A higher-order beam finite element for bending and vibration problems. *J. Sound Vib.* **1988**, *126*, 309–326. [CrossRef]
82. Reddy, J.N. *Theories and Analyses of Beams and Axisymmetric Circular Plates*; Taylor & Francis (CRC Press): Philadelphia, PA, USA, 2022.
83. Levinson, M. A new rectangular beam theory. *J. Sound Vib.* **1981**, *74*, 81–87. [CrossRef]
84. Bickford, W.B. A consistent higher order beam theory. *Dev. Theor. Appl. Mech.* **1982**, *11*, 137–150.
85. Reddy, J.N. *Energy Principles and Variational Methods in Applied Mechanics*, 2nd ed.; John Wiley & Sons: New York, NY, USA, 2002.
86. Jankowski, P.; Żur, K.K.; Kim, J.; Reddy, J. On the bifurcation buckling and vibration of porous nanobeams. *Compos. Struct.* **2020**, *250*, 112632. [CrossRef]
87. Barretta, R.; Luciano, R.; Marotti de Sciarra, F.; Ruta, G. Stress-driven nonlocal integral model for Timoshenko elastic nano-beams. *Eur. J. Mech. A/Solids* **2018**, *72*, 275–286. [CrossRef]
88. Zhang, P.; Qing, H. On well-posedness of two-phase nonlocal integral models for higher-order refined shear deformation beams. *Appl. Math. Mech.* **2021**, *42*, 931–950. [CrossRef]
89. Vantadori, S.; Luciano, R.; Scorza, D.; Darban, H. Fracture analysis of nanobeams based on the stress-driven non-local theory of elasticity. *Mech. Adv. Mater. Struct.* **2022**, *29*, 1967–1976. [CrossRef]
90. Zhang, P.; Schiavone, P.; Qing, H. Unified two-phase nonlocal formulation for vibration of functionally graded beams resting on nonlocal viscoelastic Winkler-Pasternak foundation. *Appl. Math. Mech.-Engl.* **2022**, *44*, 89–108. [CrossRef]
91. Scorza, D.; Luciano, R.; Vantadori, S. Fracture behaviour of nanobeams through Two-Phase Local/Nonlocal Stress-Driven model. *Compos. Struct.* **2022**, *280*, 114957. [CrossRef]
92. Rezaiee-Pajand, M.; Rajabzadeh-Safaei, N. Stress-driven nonlinear behavior of curved nanobeams. *Int. J. Eng. Sci.* **2022**, *178*, 103724. [CrossRef]
93. Xu, X.; Karami, B.; Shahsavari, D. Time-dependent behavior of porous curved nanobeam. *Int. J. Eng. Sci.* **2021**, *160*, 103455. [CrossRef]
94. Karamanli, A.; Vo, T.P. Finite element model for free vibration analysis of curved zigzag nanobeams. *Compos. Struct.* **2022**, *282*, 115097. [CrossRef]
95. Bacciocchi, M.; Tarantino, A.M. Analytical solutions for vibrations and buckling analysis of laminated composite nanoplates based on third-order theory and strain gradient approach. *Compos. Struct.* **2021**, *272*, 114083. [CrossRef]
96. Bacciocchi, M.; Fantuzzi, N.; Neves, A.; Ferreira, A. Vibrations and bending of thin laminated square plates with holes in gradient elasticity: A finite element solution. *Mech. Res. Commun.* **2023**, *128*. [CrossRef]
97. Lin, M.X.; Chen, C. Investigation of pull-in behavior of circular nanoplate actuator based on the modified couple stress theory. *Eng. Comput.* **2020**, *38*, 2648–2665. [CrossRef]
98. Furletov, A.; Apyari, V.; Garshev, A.; Dmitrienko, S.; Zolotov, Y. Fast and Sensitive Determination of Bioflavonoids Using a New Analytical System Based on Label-Free Silver Triangular Nanoplates. *Sensors* **2022**, *22*, 843. [CrossRef]
99. Fernández-Sáez, J.; Morassi, A.; Rubio, L.; Zaera, R. Transverse free vibration of resonant nanoplate mass sensors: Identification of an attached point mass. *Int. J. Mech. Sci.* **2019**, *150*, 217–225. [CrossRef]
100. Farajpour, A.; Żur, K.K.; Kim, J.; Reddy, J. Nonlinear frequency behaviour of magneto-electromechanical mass nanosensors using vibrating MEE nanoplates with multiple nanoparticles. *Compos. Struct.* **2021**, *260*, 113458. [CrossRef]
101. Chenghui, X.; Qu, J.; Rong, D.; Zhou, Z.; Leung, A. Theory and modeling of a novel class of nanoplate-based mass sensors with corner point supports. *Thin-Walled Struct.* **2021**, *159*, 107306.
102. Zhou, H.; Shao, D.; Song, X.; Li, P. Three-dimensional thermoelastic damping models for rectangular micro/nanoplate resonators with nonlocal-single-phase-lagging effect of heat conduction. *Int. J. Heat Mass Transf.* **2022**, *196*, 123271. [CrossRef]
103. Pham, Q.H.; Tran, T.T.; Nguyen, P.C. Nonlocal free vibration of functionally graded porous nanoplates using higher-order isogeometric analysis and ANN prediction. *Alex. Eng. J.* **2023**, *66*, 651–667. [CrossRef]
104. Al-Furjan, M.; Shan, L.; Shen, X.; Kolahchi, R.; Rajak, D.K. Combination of FEM-DQM for nonlinear mechanics of porous GPL-reinforced sandwich nanoplates based on various theories. *Thin-Walled Struct.* **2022**, *178*, 109495. [CrossRef]
105. Kawano, A.; Morassi, A.; Zaera, R. Inverse load identification in vibrating nanoplates. *Math. Methods Appl. Sci.* **2023**, *46*, 1045–1075. [CrossRef]
106. Pinnola, F.P.; Vaccaro, M.S.; Barretta, R.; Marotti de Sciarra, F. Random vibrations of stress-driven nonlocal beams with external damping. *Meccanica* **2021**, *56*, 1329–1344. [CrossRef]
107. Shinozuka, M.; Deodatis, G. Stochastic process models for earthquake ground motion. *Probabilistic Eng. Mech.* **1988**, *3*, 114–123. [CrossRef]
108. Dilena, M.; Fedele Dell'Oste, M.; Fernández-Sáez, J.; Morassi, A.; Zaera, R. Hearing distributed mass in nanobeam resonators. *Int. J. Solids Struct.* **2020**, *193*, 568–592. [CrossRef]
109. Alibakhshi, A.; Dastjerdi, S.; Malikan, M.; Eremeyev, V. Nonlinear free and forced vibrations of a dielectric elastomer-based microcantilever for atomic force microscopy. *Contin. Mech. Thermodyn.* **2022**, *in press*. [CrossRef]
110. Nguyen, P.C.; Pham, Q.H. A nonlocal isogeometric model for buckling and dynamic instability analyses of FG graphene platelets-reinforced nanoplates. *Mater. Today Commun.* **2023**, *34*, 105211. [CrossRef]

111. Zhang, Q.; Li, M.; Zhao, M. Dynamic analysis of a piezoelectric semiconductor nanoplate with surface effect. *Mater. Today Commun.* **2022**, *33*, 104406. [CrossRef]
112. Gusella, V.; Autuori, G.; Pucci, P.; Cluni, F. Dynamics of Nonlocal Rod by Means of Fractional Laplacian. *Symmetry* **2020**, *12*, 1933. [CrossRef]
113. Darban, H.; Luciano, R.; Basista, M. Free transverse vibrations of nanobeams with multiple cracks. *Int. J. Eng. Sci.* **2022**, *177*, 103703. [CrossRef]
114. Corigliano, A.; Ardito, R.; Comi, C.; Ghisi, A.; Mariani, S. *Mechanics of Microsystems*; John Wiley & Sons, Ltd.: Hoboken, NJ, USA, 2018. [CrossRef]
115. Jalaei, M.; Thai, H.T.; Civalek, O. On viscoelastic transient response of magnetically imperfect functionally graded nanobeams. *Int. J. Eng. Sci.* **2022**, *172*, 103629. [CrossRef]
116. Akgöz, B.; Civalek, O. Buckling Analysis of Functionally Graded Tapered Microbeams via Rayleigh–Ritz Method. *Mathematics* **2022**, *10*, 4429. [CrossRef]
117. Comi, C.; Corigliano, A.; Frangi, A.; Zega, V. Linear and nonlinear mechanics in MEMS; In *Silicon Sensors and Actuators: The Feynman Roadmap*; Springer: Cham, Switzerland, 2022; pp. 389–437. [CrossRef]
118. Numanouglu, H.M.; Ersoy, H.; Akgöz, B.; Civalek, O. A new eigenvalue problem solver for thermo-mechanical vibration of Timoshenko nanobeams by an innovative nonlocal finite element method. *Math. Methods Appl. Sci.* **2022**, *45*, 2592–2614. [CrossRef]
119. Civalek, O.; Uzun, B.; Yayli, M.; Akgöz, B. Size-dependent transverse and longitudinal vibrations of embedded carbon and silica carbide nanotubes by nonlocal finite element method. *Eur. Phys. J. Plus* **2020**, *135*, 381. [CrossRef]
120. Ebrahimi, F.; Barati, M.; Civalek, O. Application of Chebyshev–Ritz method for static stability and vibration analysis of nonlocal microstructure-dependent nanostructures. *Eng. Comput.* **2020**, *36*. [CrossRef]
121. Żur, K.K.; Farajpour, A.; Lim, C.; Jankowski, P. On the nonlinear dynamics of porous composite nanobeams connected with fullerenes. *Compos. Struct.* **2021**, *274*, 114356. [CrossRef]
122. Vaccaro, M.S.; Marotti de Sciarra, F.; Barretta, R. On the regularity of curvature fields in stress-driven nonlocal elastic beams. *Acta Mech.* **2021**, *232*, 2595–2603. [CrossRef]
123. Caporale, A.; Darban, H.; Luciano, R. Exact closed-form solutions for nonlocal beams with loading discontinuities. *Mech. Adv. Mater. Struct.* **2022**, *29*, 694–704. [CrossRef]
124. Pinnola, F.; Vaccaro, M.S.; Barretta, R.; Marotti de Sciarra, F. Finite element method for stress-driven nonlocal beams. *Eng. Anal. Bound. Elem.* **2022**, *134*, 22–34. [CrossRef]
125. Scorza, D.; Vantadori, S.; Luciano, R. Nanobeams with Internal Discontinuities: A Local/Nonlocal Approach. *Nanomaterials* **2021**, *11*, 2651. [CrossRef]
126. Caporale, A.; Darban, H.; Luciano, R. Nonlocal strain and stress gradient elasticity of Timoshenko nano-beams with loading discontinuities. *Int. J. Eng. Sci.* **2022**, *173*, 103620. [CrossRef]
127. Togun, N.; Bagdatli, S.M. Nonlinear Vibration of a Nanobeam on a Pasternak Elastic Foundation Based on Non-Local Euler-Bernoulli Beam Theory. *Math. Comput. Appl.* **2016**, *21*, 3. [CrossRef]
128. Togun, N. Nonlocal beam theory for nonlinear vibrations of a nanobeam resting on elastic foundation. *Bound. Value Probl.* **2016**, *2016*, 57. [CrossRef]
129. Ebrahimi, F.; Barati, M.R. Buckling analysis of nonlocal third-order shear deformable functionally graded piezoelectric nanobeams embedded in elastic medium. *J. Braz. Soc. Mech. Sci. Eng.* **2017**, *39*, 937–952. [CrossRef]
130. Darban, H.; Fabbrocino, F.; Feo, L.; Luciano, R. Size-dependent buckling analysis of nanobeams resting on two-parameter elastic foundation through stress-driven nonlocal elasticity model. *Mech. Adv. Mater. Struct.* **2021**, *28*, 2408–2416. [CrossRef]
131. Mahmoudpour, E.; Hosseini-Hashemi, S.H.; Faghidian, S.A. Nonlinear vibration analysis of FG nano-beams resting on elastic foundation in thermal environment using stress-driven nonlocal integral model. *Appl. Math. Model.* **2018**, *57*, 302–315. [CrossRef]
132. Uzun, B.; Civalek, O. Free vibration analysis Silicon nanowires surrounded by elastic matrix by nonlocal finite element method. *Adv. Nano Res.* **2019**, *7*, 99–108.
133. Koutsoumaris, C.; Eptaimeros, K. Nonlocal integral static problems of nanobeams resting on an elastic foundation. *Eur. J. Mech. A/Solids* **2021**, *89*, 104295. [CrossRef]
134. Adhikari, S.; Karlicic, D.; Liu, X. Dynamic stiffness of nonlocal damped nano-beams on elastic foundation. *Eur. J. Mech. / A Solids* **2021**, *86*, 4. [CrossRef]
135. Tran, T.T.; Tran, V.K.; Pham, Q.H.; Zenkour, A.M. Extended four-unknown higher-order shear deformation nonlocal theory for bending, buckling and free vibration of functionally graded porous nanoshell resting on elastic foundation. *Compos. Struct.* **2021**, *264*, 113737. [CrossRef]
136. Pham, Q.H.; Tran, V.K.; Tran, T.T.; Nguyen, P.C.; Malekzadeh, P. Dynamic instability of magnetically embedded functionally graded porous nanobeams using the strain gradient theory. *Alex. Eng. J.* **2022**, *61*, 10025–10044. [CrossRef]
137. Jena, S.; Chakraverty, S.; Mahesh, V.; Harursampath, D.; Sedighi, H.M. A Novel Numerical Approach for the Stability of Nanobeam Exposed to Hygro-Thermo-Magnetic Environment Embedded in Elastic Foundation. *ZAMM J. Appl. Math. Mech./Z. für Angew. Math. Und Mech.* **2022**, *102*, e202100380. [CrossRef]
138. Darban, H.; Luciano, R.; Caporale, A.; Basista, M. Modeling of buckling of nanobeams embedded in elastic medium by local-nonlocal stress-driven gradient elasticity theory. *Compos. Struct.* **2022**, *297*, 115907. [CrossRef]

139. Barretta, R.; Čanađija, M.; Luciano, R.; Marotti de Sciarra, F. On the mechanics of nanobeams on nano-foundations. *Int. J. Eng. Sci.* **2022**, *180*, 103747. [CrossRef]
140. Vaccaro, M.S.; Pinnola, F.P.; Marotti de Sciarra, F.; Barretta, R. Elastostatics of Bernoulli–Euler Beams Resting on Displacement-Driven Nonlocal Foundation. *Nanomaterials* **2021**, *11*, 573. [CrossRef]
141. Wieghardt, K. Über den Balken auf nachgiebiger Unterlage. *ZAMM - J. Appl. Math. Mech./ Z. für Angew. Math. Und Mech.* **1922**, *2*, 165–184. [CrossRef]
142. Sollazzo, A. Equilibrio della trave su suolo di Wieghardt. *Tec. Ital.* **1966**, *31*, 187–206.

MDPI
St. Alban-Anlage 66
4052 Basel
Switzerland
Tel. +41 61 683 77 34
Fax +41 61 302 89 18
www.mdpi.com

Encyclopedia Editorial Office
E-mail: encyclopedia@mdpi.com
www.mdpi.com/journal/encyclopedia

www.ingramcontent.com/pod-product-compliance
Lightning Source LLC
Chambersburg PA
CBHW041936260926
51850CB00018B/200
9783036570006